Grundlehren der mathematischen Wissenschaften 242

A Series of Comprehensive Studies in Mathematics

B. Huppert N. Blackburn

Finite Groups II

Springer-Verlag
Berlin Heidelberg New York 1982

Bertram Huppert
Mathematisches Institut der Universität
Saarstraße 21
D-6500 Mainz

Norman Blackburn
Department of Mathematics
The University
GB-Manchester M13 9 PL

ISBN 3-540-10632-4 Springer-Verlag Berlin Heidelberg New York
ISBN 0-387-10632-4 Springer-Verlag New York Heidelberg Berlin

Library of Congress Cataloging in Publication Data. Huppert, Bertram, 1927–.
Finite groups II. (Grundlehren der mathematischen Wissenschaften; 242).
Bibliography: p. Includes index. 1. Finite groups. I. Blackburn, N. (Norman).
II. Title. III. Series. QA 171.B 577. 512′.22. 81-2287.
ISBN 0-387-10632-4. AACR 2

Printed in Germany

Typesetting: Asco Trade Typesetting Limited, Chai Wan, Hong Kong.
Printing and bookbinding: Konrad Triltsch, Würzburg
2141/3140-543210

In Memoriam

Reinhold Baer (1902–1979)

Richard Brauer (1901–1977)

Preface

From "Lohengrin", Richard Wagner

At the time of the appearance of the first volume of this work in 1967, the tempestuous development of finite group theory had already made it virtually impossible to give a complete presentation of the subject in one treatise. The present volume and its successor have therefore the more modest aim of giving descriptions of the recent development of certain important parts of the subject, and even in these parts no attempt at completeness has been made.

Chapter VII deals with the representation theory of finite groups in arbitrary fields with particular attention to those of non-zero characteristic. That part of modular representation theory which is essentially the block theory of complex characters has not been included, as there are already monographs on this subject and others will shortly appear. Instead, we have restricted ourselves to such results as can be obtained by purely module-theoretical means.

In Chapter VIII, the linear (and bilinear) methods which have proved useful in questions involving nilpotent groups are discussed. A major part of this is devoted to the classification of Suzuki 2-groups (see §7); while a complete classification is not obtained, the result proved is strong enough for an application to the determination of the Zassenhaus groups in Chapter XI. The standard procedure involves the use of Lie rings, and rather than attempting a theory of the connection between nilpotent groups and Lie rings, we give a number of applications to such topics as the length of the conjugacy classes of p-groups (§9), fixed point free automorphisms of nilpotent groups (§10), the restricted Burnside problem (§12) and automorphisms of p-groups (§13). In many of these considerations, the finiteness of the group is a relatively unimportant condition, and the last two of these applications depend on the Magnus-Witt theory of the lower central series of free groups, which is described in §11.

The ground-breaking investigations of P. Hall and G. Higman on the theory of p-soluble groups form the basis of Chapter IX. These arose from the restricted Burnside problem and led first to a solution for exponent 6 (see 1.15). Then however there followed far-reaching theorems for composite exponents (4.10, 4.13, 4.17). Besides various estimates of the p-length of a p-soluble group in terms of the structure of its Sylow p-subgroups (§5), we deal with some theorems about fixed point free automorphisms of soluble groups (§6). Finally we discuss the derived notion of p-stability, which will be of considerable use in Chapter X.

The three chapters in this volume are thus all concerned with relations between finite groups and linear algebra, but otherwise they are rather independent of one another, apart from occasional technical references, of course.

The authors must apologize for the length of time which readers have had to wait for this volume. They promise that Volume III will be available within a matter of months.

It is a great pleasure to thank the many colleagues who have helped us in the preparation of this volume and its successor. In this respect the second author must give pride of place to Philip Hall, who first stimulated his interest in the subject more than 25 years ago by combining patient encouragement of a naturally pessimistic student with lectures of a beauty which seems to be lost to subsequent generations. With the writing of the book the greatest help was given by W. Gaschütz and his associates in Kiel, where each year since 1967 our sketches were read and exhaustively studied. The participants in these discussions in the course of the years were H. Bender, D. Blessenohl, W. Gaschütz, F. Gross, K. Johnsen, O.-U. Kramer, H. Laue, K.-U. Schaller and R. Schmidt. We are most grateful for the hospitality of the Mathematics Department in Kiel, without which this kind of work would not have been possible. Also we are indebted for financial assistance, enabling the two of us to meet reasonably often, to the National Science Foundation, the Alexander von Humboldt-Stiftung and the University of Manchester.

In the laborious proof-reading B. Hartley (Manchester), O. Manz, J. Pense and W. Willems (Mainz) all spent a great deal of time helping us, and we offer them our most sincere thanks. Also we thank the Manchester secretaries Kendal Anderson, Rosemary Horton and Patricia McMunn for the enormous amount of help they have given us with the typing and preparation of the manuscript.

Finally our thanks are due to Springer-Verlag and to the typesetters and printers for their patience with us and for the excellent quality of the production of this book.

July, 1981

Bertram Huppert, Mainz
Norman Blackburn, Manchester

Contents

Chapter VII. Elements of General Representation Theory 1

§ 1. Extension of the Ground-Field 4
§ 2. Splitting Fields 27
§ 3. The Number of Irreducible Modular Representations 32
§ 4. Induced Modules 44
§ 5. The Number of Indecomposable $K\mathfrak{G}$-Modules 63
§ 6. Indecomposable and Absolutely Indecomposable Modules . . 71
§ 7. Relative Projective and Relative Injective Modules 81
§ 8. The Dual Module 97
§ 9. Representations of Normal Subgroups 123
§10. One-Sided Decompositions of the Group-Ring 147
§11. Frobenius Algebras and Symmetric Algebras 165
§12. Two-Sided Decompositions of Algebras 174
§13. Blocks of p-Constrained Groups 184
§14. Kernels of Blocks 189
§15. p-Chief Factors of p-Soluble Groups 203
§16. Green's Indecomposability Theorem 223
Notes on Chapter VII 237

Chapter VIII. Linear Methods in Nilpotent Groups 238

§ 1. Central Series with Elementary Abelian Factors 239
§ 2. Jennings' Theorem 252
§ 3. Transitive Linear Groups 266
§ 4. Some Number-Theoretical Lemmas 270
§ 5. Lemmas on 2-Groups 275
§ 6. Commutators and Bilinear Mappings 286
§ 7. Suzuki 2-Groups 299
§ 8. Lie Algebras 316
§ 9. The Lie Ring Method and an Application 326
§10. Regular Automorphisms 349
§11. The Lower Central Series of Free Groups 366

§12. Remarks on the Burnside Problem 385
§13. Automorphisms of p-Groups 396
Notes on Chapter VIII 404

Chapter IX. Linear Methods and Soluble Groups 405

§1. Introduction 407
§2. Hall and Higman's Theorem B 419
§3. The Exceptional Case 429
§4. Reduction Theorems for Burnside's Problem 449
§5. Other Consequences of Theorem B 464
§6. Fixed Point Free Automorphism Groups 476
§7. p-Stability 492
§8. Soluble Groups with One Class of Involutions 503
Notes on Chapter IX 514

Bibliography 515

Index of Names 525

Index 527

Index of Symbols

$\mathbf{Aut}\,\mathfrak{X}$ XII
$S_p(\mathfrak{G})$ XII
$\gamma_n(\mathfrak{G})$ XII, 239
$\mathbf{O}_\pi(\mathfrak{G})$, $\mathbf{O}_{\pi_1,\ldots,\pi_n}(\mathfrak{G})$, $\mathbf{O}^\pi(\mathfrak{G})$ XIII, 407
$\mathfrak{G}^m$ XII
$\mathsf{A}\mathfrak{G}$, $\mathbb{Z}\mathfrak{G}$, $\mathsf{K}\mathfrak{G}$ XII
π' XIII
$\mathbf{H}(\mathsf{V})$ 12
$\mathbf{S}(\mathsf{V})$ 12
V_η 15
$\mathbf{Q}(\mathfrak{A})$ 33
$\mathbf{T}(\mathfrak{A})$ 35
$\mathsf{V}^{\mathfrak{G}}$ 44
${}^{\mathfrak{G}}\mathsf{V}$ 51
V^* 97
$\mathbf{R}(\mathfrak{S})$ 155
$\mathbf{L}(\mathfrak{S})$ 155
$\mathbf{S}_r(\mathfrak{A})$ 155
$\mathbf{S}_l(\mathfrak{A})$ 155
$\mathbf{An}_{\mathfrak{R}}(\mathsf{S})$, $\mathbf{An}_{\mathsf{V}}\mathbf{J}(\mathfrak{A})$ 155
$\mathsf{P}_{\mathfrak{G}}(\mathsf{V})$ 190
$\mathfrak{K}(\mathsf{V})$ 193
$\mathfrak{N}(\mathsf{V})$ 194
$\mathfrak{K}(\mathscr{B})$ 194
$\mathfrak{N}(\mathscr{B})$ 194
$\mathsf{K}_{\mathfrak{U}}^{\mathfrak{G}}$ 203
$\mathsf{A}_{\mathfrak{U}}^{\mathfrak{G}}$ 204
$\lambda_n(\mathfrak{G})$ 242
$\kappa_n(\mathfrak{G})$ 248
$A(n, \theta)$ 294
$l(\mathfrak{A})$ 318
$\mathfrak{g}^n$ 320
$\mathfrak{g}^{(n)}$ 323
$\mathrm{ad}\,x$ 338
$\mathrm{C}_{\mathfrak{g}}(x)$ 338
$(\mathfrak{A}$ on $\mathfrak{B})$ 409
$e_p(\mathfrak{G})$ 449
$\tau_n(\mathfrak{G})$ 449
$e_p^*(\mathfrak{G})$ 449
$SA(2, p)$ 494

Terminology and Notation

In this volume, the same terminology and notation as in Volume 1 will be used, with the following exceptions.

1. The identity mapping on a set X will be denoted by 1_X.

2. The identity element of the group $\mathfrak{G}$ will be denoted by $1_{\mathfrak{G}}$ or 1.

3. By a *section* of a group $\mathfrak{G}$ is meant a group of the form $\mathfrak{H}/\mathfrak{K}$, where $\mathfrak{K} \trianglelefteq \mathfrak{H} \leq \mathfrak{G}$.

4. If $\mathfrak{X}$ is any algebraic system, **Aut** $\mathfrak{X}$ denotes the group of all automorphisms of $\mathfrak{X}$. The group of inner automorphisms of the group $\mathfrak{G}$ is denoted by **Inn** $\mathfrak{G}$.

5. The set of Sylow p-subgroups of the finite group $\mathfrak{G}$ will be denoted by $S_p(\mathfrak{G})$.

6. The *lower central series* (III, 2.2) of the group $\mathfrak{G}$ will be denoted by

$$\mathfrak{G} = \gamma_1(\mathfrak{G}) \geq \gamma_2(\mathfrak{G}) \geq \cdots \geq \gamma_n(\mathfrak{G}) \geq \cdots;$$

here $\gamma_n(\mathfrak{G}) = [\gamma_{n-1}(\mathfrak{G}), \mathfrak{G}]$ for $n > 1$.

7. If $\mathfrak{G}$ is a group and m is a positive integer, $\mathfrak{G}^m = \langle x^m | x \in \mathfrak{G} \rangle$. Thus $\mathfrak{G}^{mn} \leq (\mathfrak{G}^m)^n$.

8. Let A be a commutative ring with identity and let $\mathfrak{G}$ be a group. The *group-ring* of $\mathfrak{G}$ over A (I, 16.6) will be denoted by $\mathsf{A}\mathfrak{G}$.

9. Let $\mathfrak{G}$ be a finite group. The field K is called a *splitting field* of $\mathfrak{G}$ if $\mathsf{K}\mathfrak{G}/\mathbf{J}(\mathsf{K}\mathfrak{G})$ is the direct sum of complete matrix algebras over K, where $\mathbf{J}(\mathsf{K}\mathfrak{G})$ is the Jacobson radical of $\mathsf{K}\mathfrak{G}$. Thus by V, 11.2a), K is a splitting field of $\mathfrak{G}$ if and only if K is a splitting field of $\mathsf{K}\mathfrak{G}/\mathbf{J}(\mathsf{K}\mathfrak{G})$.

This definition is not the same as that given in V, 11.2b), but the two definitions reduce to the same thing when $|\mathfrak{G}|$ is not divisible by char K.

10. A $\mathsf{K}\mathfrak{G}$-module M is called *absolutely irreducible* if (i) M is irreducible and (ii) $\mathrm{Hom}_{\mathsf{K}\mathfrak{G}}(\mathsf{M}, \mathsf{M}) = \mathsf{K}$.

This definition is equivalent to that given in V, 11.8; this is proved in VII, 2.2.

11. The unit matrix will be denoted by I.

12. If π is a set of primes, the complementary set of primes is denoted

by π'; thus

$$\pi' = \{p \mid p \text{ is a prime}, p \notin \pi\}.$$

13. If π is a set of primes, the product of all the normal π-subgroups of the finite group $\mathfrak{G}$ is denoted by $\mathbf{O}_\pi(\mathfrak{G})$. Thus $\mathbf{O}_\pi(\mathfrak{G})$ is the *maximal normal π-subgroup* of $\mathfrak{G}$. Clearly $\mathbf{O}_\pi(\mathfrak{G})$ is a characteristic subgroup of $\mathfrak{G}$ and $\mathbf{O}_\pi(\mathfrak{G}/\mathbf{O}_\pi(\mathfrak{G})) = 1$.

More generally, suppose that $\pi_1, \pi_2, \ldots$ are sets of primes. We define a characteristic subgroup $\mathbf{O}_{\pi_1,\ldots,\pi_i}(\mathfrak{G})$ of $\mathfrak{G}$ by induction on i: for $i > 1$,

$$\mathbf{O}_{\pi_1,\ldots,\pi_i}(\mathfrak{G})/\mathbf{O}_{\pi_1,\ldots,\pi_{i-1}}(\mathfrak{G}) = \mathbf{O}_{\pi_i}(\mathfrak{G}/\mathbf{O}_{\pi_1,\ldots,\pi_{i-1}}(\mathfrak{G})).$$

Examples. a) The Fitting subgroup of $\mathfrak{G}$ is $\prod_p \mathbf{O}_p(\mathfrak{G})$.

b) If p is a prime, the *upper p-series* of $\mathfrak{G}$ (VI, 6.1) is

$$1 \leq \mathbf{O}_{p'}(\mathfrak{G}) \leq \mathbf{O}_{p',p}(\mathfrak{G}) \leq \mathbf{O}_{p',p,p'}(\mathfrak{G}) \leq \cdots.$$

c) The maximal p-nilpotent normal subgroup of $\mathfrak{G}$ is $\mathbf{O}_{p',p}(\mathfrak{G})$. For if $\mathfrak{N}$ is a normal p-nilpotent subgroup of $\mathfrak{G}$, the normal p-complement $\mathfrak{K}$ of $\mathfrak{N}$ is a characteristic p'-subgroup of $\mathfrak{G}$. Hence $\mathfrak{K} \trianglelefteq \mathfrak{G}$, $\mathfrak{K} \leq \mathbf{O}_{p'}(\mathfrak{G})$, $\mathfrak{N}\mathbf{O}_{p'}(\mathfrak{G})/\mathbf{O}_{p'}(\mathfrak{G})$ is a normal p-subgroup and $\mathfrak{N} \leq \mathbf{O}_{p',p}(\mathfrak{G})$.

14. If π is a set of primes, $\mathbf{O}^\pi(\mathfrak{G})$ is defined to be the intersection of all the normal subgroups $\mathfrak{N}$ of $\mathfrak{G}$ for which $\mathfrak{G}/\mathfrak{N}$ is a π-group. Thus $\mathfrak{G}/\mathbf{O}^\pi(\mathfrak{G})$ is the maximal π-factor group of $\mathfrak{G}$, and $\mathbf{O}^\pi(\mathfrak{G})$ is a characteristic subgroup of $\mathfrak{G}$.

Example. If $\mathfrak{G}$ is a p-nilpotent group, $\mathbf{O}^p(\mathfrak{G})$ is the normal p-complement of $\mathfrak{G}$.

15. If $\mathfrak{F}$ is a free group, a *group-basis* of $\mathfrak{F}$ is a subset X of $\mathfrak{F}$ such that X generates $\mathfrak{F}$ and any mapping of X into a group is the restriction of some homomorphism of $\mathfrak{F}$. Such a set always exists, by definition of a free group (I, 19.1).

Chapter VII

Elements of General Representation Theory

In Chapter V, classical representation theory was studied. This is the theory of the group-ring $\mathsf{K}\mathfrak{G}$ and the $\mathsf{K}\mathfrak{G}$-modules, where K is an algebraically closed field of characteristic 0. (Many theorems remain valid under the hypothesis that K is algebraically closed and that char K does not divide the order of $\mathfrak{G}$). In this case, $\mathsf{K}\mathfrak{G}$ is semisimple and all $\mathsf{K}\mathfrak{G}$-modules are completely reducible. For many purposes it is therefore sufficient to handle the irreducible representations.

In this chapter we shall study the group-ring $\mathsf{K}\mathfrak{G}$ and the $\mathsf{K}\mathfrak{G}$-modules when K is an arbitrary field. Thus we are concerned above all with the case when char $\mathsf{K} = p$ and p is a prime divisor of the order of the group; for short we call this the *modular* case. In this case the Jacobson radical of $\mathsf{K}\mathfrak{G}$ is non-zero and not all $\mathsf{K}\mathfrak{G}$-modules are completely reducible. The number of isomorphism types of irreducible $\mathsf{K}\mathfrak{G}$-modules is the p'-class number of $\mathfrak{G}$, as long as K is sufficiently large (§ 3). The irreducible modules are determined by $\mathsf{K}\mathfrak{G}/\mathbf{J}(\mathsf{K}\mathfrak{G})$, and the divergence of $\mathsf{K}\mathfrak{G}$ from semisimplicity is determined by $\mathbf{J}(\mathsf{K}\mathfrak{G})$. Unfortunately there is no general procedure known for the determination of $\dim_{\mathsf{K}} \mathbf{J}(\mathsf{K}\mathfrak{G})$. But by using the technique of lifting idempotents from the theory of algebras, certain facts about the direct decompositions of $\mathsf{K}\mathfrak{G}$ into right ideals and two-sided ideals can be established (§ 10, 12). The decomposition of $\mathsf{K}\mathfrak{G}$ into two-sided ideals leads to the theory of blocks and is central for the further development of the theory; unfortunately no general method for finding the number of blocks is known. More detailed assertions are made by taking into account the fact that the group-ring possesses a certain self-duality, namely, it is a symmetric algebra (§ 11). Among the consequences of this self-duality is the fact that projective and injective $\mathsf{K}\mathfrak{G}$-modules are the same. If

$$\mathsf{K}\mathfrak{G} = \mathsf{P}_1 \oplus \cdots \oplus \mathsf{P}_n$$

is a decomposition of $\mathsf{K}\mathfrak{G}$ into indecomposable right ideals P_i, then all types of indecomposable projective $\mathsf{K}\mathfrak{G}$-modules occur among the P_i.

Further, each P_i has just one maximal submodule, namely $\mathsf{P}_i\mathbf{J}(\mathsf{K}\mathfrak{G})$, and P_i has just one minimal submodule S_i, which is isomorphic to $\mathsf{P}_i/\mathsf{P}_i\mathbf{J}(\mathsf{K}\mathfrak{G})$. Also P_i is determined to within isomorphism by S_i. Thus the top and bottom composition factors of P_i are known, but the complete composition structure of P_i can only rarely be determined. We therefore restrict ourselves to the investigation of the multiplicities of the composition factors of P_i. This yields the Cartan matrix C of $\mathsf{K}\mathfrak{G}$. The calculation of the elementary divisors of C from the centralizers of the p'-elements of $\mathfrak{G}$ is possible by deep theorems of Richard Brauer, which, however, will not be presented in this chapter.

In this way some information about the indecomposable projective $\mathsf{K}\mathfrak{G}$-modules can be obtained, but the general indecomposable $\mathsf{K}\mathfrak{G}$-module is almost unapproachable. If char $\mathsf{K} = p$, there is only a finite number of types of indecomposable $\mathsf{K}\mathfrak{G}$-modules if and only if the Sylow p-subgroups of $\mathfrak{G}$ are cyclic (§ 5). On the one hand this fact leads in the further development of the theory to the deep results of Brauer and Dade on groups with cyclic Sylow p-subgroups, but on the other hand it presents difficulties for the development of the general theory which have not yet been overcome.

In spite of these difficulties, some useful general facts about $\mathsf{K}\mathfrak{G}$-modules have been proved. Among these are the theory of the induced module and the reciprocity theorems (§ 4), the theorems of Clifford type about the relations between $\mathsf{K}\mathfrak{G}$-modules and $\mathsf{K}\mathfrak{N}$-modules for a normal subgroup $\mathfrak{N}$ of $\mathfrak{G}$ (§ 9) and the duality theory of $\mathsf{K}\mathfrak{G}$-modules (§ 8).

What are the aims of a general theory of group-rings? We mention here two lines of development.

(1) If $\mathfrak{G}$ is a p-soluble group, then the p-chief factors of $\mathfrak{G}$ yield irreducible $\mathsf{K}\mathfrak{G}$-modules in a natural way, where $\mathsf{K} = GF(p)$. (It is not very important that K need not be a splitting field for $\mathfrak{G}$, since the theory of the Schur index is trivial for finite fields.) On the one hand, one would like to know what place these representations, obtained so directly from the structure of $\mathfrak{G}$, have in the general theory (§ 15). On the other hand, abundant knowledge of irreducible $\mathsf{K}\mathfrak{G}$-modules of a given p-soluble group $\mathfrak{G}$ is often necessary for the construction of more complicated p-soluble groups.

(2) Another application is much better developed; analogously to local number theory there is a local theory of the characters of $\mathfrak{G}$ over $\mathbb{C}$. This is developed in the following way.

Let L be a field of characteristic 0 with a non-Archimedean valuation, let $\mathfrak{o}$ be the ring of integers in L, let $\mathfrak{p}$ be the maximal ideal of $\mathfrak{o}$ and let $\mathsf{K} = \mathfrak{o}/\mathfrak{p}$. We choose L large enough to be a splitting field for $\mathfrak{G}$. As in V, 12.5, each $\mathsf{L}\mathfrak{G}$-module may be regarded as obtained from an $\mathfrak{o}\mathfrak{G}$-module by extending the domain of coefficients. If M is an $\mathfrak{o}\mathfrak{G}$-module,

$M/M\mathfrak{p}$ can be made into a $K\mathfrak{G}$-module in a natural way. Thus the theory of $K\mathfrak{G}$-modules appears as a first approximation to the theory of $\mathfrak{o}\mathfrak{G}$-modules. If $\mathfrak{o}$ is supposed to be complete with respect to its valuation, then as in Hensel's lemma we can build up the $\mathfrak{o}\mathfrak{G}$-module from the $K\mathfrak{G}$-module by successive approximation. The result is a theory of characters in $\mathfrak{o}$ and thus a local representation theory for the prime p. Amongst other results this yields refinements of the classical orthogonality relations which have been drawn upon for the proofs of deep assertions about the structure of finite groups. The "local to global" step from the local theory to a theory of $D\mathfrak{G}$-modules, where D is a Dedekind ring, has been only partially successful up to now. The results thus obtained have played no part in the structure theory of finite groups.

The results of this chapter and the consequent modular representation theory are above all the work of Richard Brauer. Since 1936 he has systematically built up this theory and made it into a more and more delicate instrument for the investigation of finite groups.

We shall assume that the reader is familiar with the following simple facts about projective and injective modules. The proofs may be found in MACLANE [1]. Let $\mathfrak{R}$ be an arbitrary ring with 1.

(1) An $\mathfrak{R}$-module P is called *projective* if any diagram

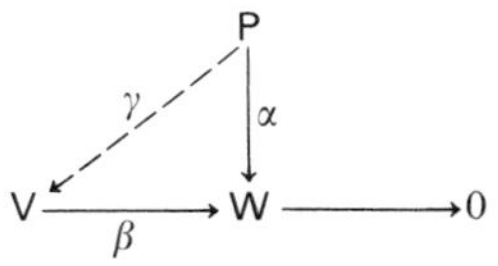

can be completed by adding γ; more precisely, if V, W are $\mathfrak{R}$-modules, $\alpha \in \mathrm{Hom}_{\mathfrak{R}}(P, W)$, $\beta \in \mathrm{Hom}_{\mathfrak{R}}(V, W)$ and β is an epimorphism, there exists $\gamma \in \mathrm{Hom}_{\mathfrak{R}}(P, V)$ such that $\alpha = \gamma\beta$ (p. 20).

(2) An $\mathfrak{R}$-module is projective if and only if it is a direct summand of a free module. Any finitely generated projective $\mathfrak{R}$-module is a direct summand of a finitely generated free $\mathfrak{R}$-module (p. 21).

(3) If P is a projective $\mathfrak{R}$-module and

$$0 \to V \xrightarrow{\alpha} W \to P \to 0$$

is an exact sequence of $\mathfrak{R}$-modules, then there exists an $\mathfrak{R}$-submodule P' of W isomorphic to P such that $W = V\alpha \oplus P'$ (p. 24).

(4) If

$$0 \to U \to V \to W \to 0$$

is an exact sequence of $\mathfrak{R}$-modules and P is a projective $\mathfrak{R}$-module, then

$$0 \to \mathrm{Hom}_{\mathfrak{R}}(\mathsf{P}, \mathsf{U}) \to \mathrm{Hom}_{\mathfrak{R}}(\mathsf{P}, \mathsf{V}) \to \mathrm{Hom}_{\mathfrak{R}}(\mathsf{P}, \mathsf{W}) \to 0$$

is an exact sequence of Abelian groups. If $\mathfrak{R}$ is a K-algebra, this is an exact sequence of vector spaces over K (p. 24).

(5) Direct summands of projective modules are projective. Direct sums of projective modules are projective.

(6) An $\mathfrak{R}$-module J is called *injective* if any diagram

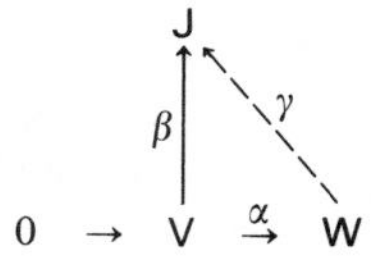

can be completed by adding γ; more precisely, if V, W are $\mathfrak{R}$-modules, $\alpha \in \mathrm{Hom}_{\mathfrak{R}}(\mathsf{V}, \mathsf{W})$, $\beta \in \mathrm{Hom}_{\mathfrak{R}}(\mathsf{V}, \mathsf{J})$ and α is a monomorphism, then there exists $\gamma \in \mathrm{Hom}_{\mathfrak{R}}(\mathsf{W}, \mathsf{J})$ such that $\beta = \alpha\gamma$ (p. 92).

(7) An $\mathfrak{R}$-module J is injective if and only if any diagram

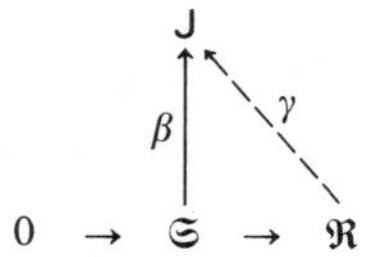

can be completed by adding γ; here $\mathfrak{S}$ is a right ideal of $\mathfrak{R}$ (p. 92).

(8) An injective submodule of an $\mathfrak{R}$-module is a direct summand (p. 92).

(9) Direct summands of injective modules are injective. Direct sums of a finite number of injective modules are injective.

§ 1. Extension of the Ground-Field

In this section we consider the behaviour of group-rings and modules under extension of the ground field.

1.1 Definition. (V, 11.1) Suppose that the field L is an extension of the field K.

a) If $\mathfrak{A}$ is a K-algebra, then $\mathfrak{A} \otimes_{\mathsf{K}} \mathsf{L}$ becomes an L-algebra, multiplication being given by

$$(a_1 \otimes \lambda_1)(a_2 \otimes \lambda_2) = a_1 a_2 \otimes \lambda_1 \lambda_2$$

for all a_1, a_2 in $\mathfrak{A}$ and λ_1, λ_2 in L. We denote this algebra by $\mathfrak{A}_{\mathsf{L}}$. If $\{a_1, \ldots, a_n\}$ is a K-basis of $\mathfrak{A}$ and

$$a_i a_j = \sum_{k=1}^{n} c_{ijk} a_k$$

with $c_{ijk} \in \mathsf{K}$, then $\{a_1 \otimes 1, \ldots, a_n \otimes 1\}$ is an L-basis of $\mathfrak{A}_{\mathsf{L}}$ and

$$(a_i \otimes 1)(a_j \otimes 1) = \sum_{k=1}^{n} c_{ijk}(a_k \otimes 1).$$

In particular $\dim_{\mathsf{L}} \mathfrak{A}_{\mathsf{L}} = \dim_{\mathsf{K}} \mathfrak{A}$.

b) If V is an $\mathfrak{A}$-module, the vector space $\mathsf{V} \otimes_{\mathsf{K}} \mathsf{L}$ becomes an $\mathfrak{A}_{\mathsf{L}}$-module V_{L} if we put

$$(v \otimes \lambda_1)(a \otimes \lambda_2) = va \otimes \lambda_1 \lambda_2$$

for $v \in \mathsf{V}$, $a \in \mathfrak{A}$ and $\lambda_1, \lambda_2 \in \mathsf{L}$. We have $\dim_{\mathsf{L}} \mathsf{V}_{\mathsf{L}} = \dim_{\mathsf{K}} \mathsf{V}$.

c) If $\mathfrak{A}$ is a K-algebra and $\mathfrak{B}$ is a K-subspace of $\mathfrak{A}$, there is an L-homomorphism ε of $\mathfrak{B} \otimes_{\mathsf{K}} \mathsf{L}$ into $\mathfrak{A} \otimes_{\mathsf{K}} \mathsf{L}$ in which $(b \otimes \lambda)\varepsilon = b \otimes \lambda$ ($b \in \mathfrak{B}$, $\lambda \in \mathsf{L}$). We write $\operatorname{im} \varepsilon = \mathfrak{B}_{\mathsf{L}}$. Note that ε is a monomorphism, for if T is a K-basis of L, $\mathfrak{B} \otimes_{\mathsf{K}} \mathsf{L} = \bigoplus_{t \in \mathsf{T}} \mathfrak{B} \otimes t$ and $\mathfrak{A} \otimes_{\mathsf{K}} \mathsf{L} = \bigoplus_{t \in \mathsf{T}} \mathfrak{A} \otimes t$. If $\mathfrak{B}$ is a subring of $\mathfrak{A}$, $\mathfrak{B}_{\mathsf{L}}$ is a subring of $\mathfrak{A}_{\mathsf{L}}$; if $\mathfrak{B}$ is an ideal of $\mathfrak{A}$, $\mathfrak{B}_{\mathsf{L}}$ is an ideal of $\mathfrak{A}_{\mathsf{L}}$.

1.2 Lemma. *Suppose that L is an extension of the field K.*

a) *If $\mathfrak{A}_1, \ldots, \mathfrak{A}_k$ are K-algebras, then*

$$(\mathfrak{A}_1 \oplus \cdots \oplus \mathfrak{A}_k)_{\mathsf{L}} \cong (\mathfrak{A}_1)_{\mathsf{L}} \oplus \cdots \oplus (\mathfrak{A}_k)_{\mathsf{L}}.$$

b) *If $\mathfrak{A}$ is a K-algebra and $(\mathfrak{A})_m$ is the complete matrix ring of degree m over $\mathfrak{A}$, then $((\mathfrak{A})_m)_{\mathsf{L}} \cong (\mathfrak{A}_{\mathsf{L}})_m$.*

c) *If $\mathfrak{A}$ is a K-algebra and $\mathfrak{J}$ is a two-sided ideal of $\mathfrak{A}$, then $(\mathfrak{A}/\mathfrak{J})_{\mathsf{L}} \cong \mathfrak{A}_{\mathsf{L}}/\mathfrak{J}_{\mathsf{L}}$.*

d) *If* $\mathfrak{A}$ *is a finite-dimensional* K*-algebra, then* $\mathbf{J}(\mathfrak{A})_L \subseteq \mathbf{J}(\mathfrak{A}_L)$.

e) *If* V *and* W *are* $\mathfrak{A}$*-modules and* $V \supseteq W$, *then* $(V/W) \otimes_K L$ *and* V_L/W_L *are isomorphic* $\mathfrak{A}_L$*-modules.*

Proof. a) It is easily checked that there is an isomorphism α of $(\mathfrak{A}_1 \oplus \cdots \oplus \mathfrak{A}_k)_L$ onto $(\mathfrak{A}_1)_L \oplus \cdots \oplus (\mathfrak{A}_k)_L$ such that

$$((a_1, \ldots, a_k) \otimes \lambda)\alpha = (a_1 \otimes \lambda, \ldots, a_k \otimes \lambda) \quad (a_i \in \mathfrak{A}_i, \lambda \in L).$$

b) There is an isomorphism β for which

$$((a_{ij}) \otimes \lambda)\beta = (a_{ij} \otimes \lambda) \quad (a_{ij} \in \mathfrak{A}, \lambda \in L).$$

c) There is an L-algebra epimorphism γ of $\mathfrak{A}_L$ onto $(\mathfrak{A}/\mathfrak{J})_L$ in which

$$(a \otimes \lambda)\gamma = (a + \mathfrak{J}) \otimes \lambda \quad (a \in \mathfrak{A}, \lambda \in L).$$

If T is a K-basis of L, $\mathfrak{A}_L = \bigoplus_{t \in T} \mathfrak{A} \otimes t$, so $\ker \gamma = \mathfrak{J}_L$.

d) By V, 2.4a), $\mathbf{J}(\mathfrak{A})$ is nilpotent. Suppose that $\mathbf{J}(\mathfrak{A})^n = 0$. Then $(\mathbf{J}(\mathfrak{A})_L)^n = 0$. Thus $\mathbf{J}(\mathfrak{A})_L$ is a nilpotent ideal of $\mathfrak{A}_L$. Hence by V, 2.4b), $\mathbf{J}(\mathfrak{A})_L \subseteq \mathbf{J}(\mathfrak{A}_L)$.

e) The proof is similar to that of c). **q.e.d.**

1.3 Examples. Suppose that L is an extension of the field K.

a) We have $(K\mathfrak{G})_L \cong L\mathfrak{G}$.

By 1.1a), $(K\mathfrak{G})_L$ has the L-basis $\{g \otimes 1 \mid g \in \mathfrak{G}\}$ and

$$(g_1 \otimes 1)(g_2 \otimes 1) = g_1 g_2 \otimes 1.$$

Hence the mapping α of $(K\mathfrak{G})_L$ into $L\mathfrak{G}$ given by

$$\left(\sum_{g \in \mathfrak{G}} g \otimes \lambda_g\right)\alpha = \sum_{g \in \mathfrak{G}} \lambda_g g \quad (\lambda_g \in L)$$

is an L-algebra isomorphism of $(K\mathfrak{G})_L$ onto $L\mathfrak{G}$.

b) By 1.2a) and b), for

$$\mathfrak{A} \cong \bigoplus_{i=1}^{k} (K)_{n_i},$$

we get immediately

$$\mathfrak{A}_{\mathsf{L}} \cong \bigoplus_{i=1}^{k} (\mathsf{L})_{n_i}.$$

In the following lemma, some elementary facts about fields are collected for later use.

1.4 Lemma. a) *Suppose that* $0 \neq f \in \mathsf{K}[t]$ *and* L *is an extension field of* K. *Let* $f = \prod_{i=1}^{r} g_i^{m_i}$ *be the decomposition of* f *in* $\mathsf{L}[t]$ *with pairwise non-associated irreducible polynomials* g_i. *Then*

$$(\mathsf{K}[t]/f\mathsf{K}[t])_{\mathsf{L}} \cong \mathsf{L}[t]/f\mathsf{L}[t] \cong \bigoplus_{i=1}^{r} \mathsf{L}[t]/g_i^{m_i}\mathsf{L}[t].$$

b) *Suppose* $\mathsf{K} = GF(q)$ *and* $\mathsf{L}_i = GF(q^{n_i})$ $(i = 1, 2)$. *Let* d *be the greatest common divisor and* k *the least common multiple of* n_1 *and* n_2. *Then*

$$\mathsf{L}_1 \otimes_{\mathsf{K}} \mathsf{L}_2 \cong GF(q^k) \oplus \cdots \oplus GF(q^k),$$

with d *direct summands on the right.*

c) *Let* L_1 *be a separable extension of* K *and* L_2 *any extension of* K. *Then*

$$\mathsf{L}_1 \otimes_{\mathsf{K}} \mathsf{L}_2 \cong \mathsf{F}_1 \oplus \cdots \oplus \mathsf{F}_r,$$

where the fields F_i *are separable extensions of* L_2.

Proof. a) We have

$$\begin{aligned}(\mathsf{K}[t]/f\mathsf{K}[t])_{\mathsf{L}} &\cong (\mathsf{L} \otimes_{\mathsf{K}} \mathsf{K}[t])/(\mathsf{L} \otimes_{\mathsf{K}} f\mathsf{K}[t]) \qquad \text{(by 1.2c))}\\ &\cong \mathsf{L}[t]/f\mathsf{L}[t],\end{aligned}$$

The mapping α of $\mathsf{L}[t]$ into $\bigoplus_{i=1}^{r} \mathsf{L}[t]/g_i^{m_i}\mathsf{L}[t]$ given by

$$h\alpha = (h + g_1^{m_1}\mathsf{L}[t], \ldots, h + g_r^{m_r}\mathsf{L}[t]) \quad (h \in \mathsf{L}[t])$$

is obviously an L-algebra homomorphism, and $h \in \ker \alpha$ if and only if $g_i^{m_i}$ divides h for all $i = 1, \ldots, r$. Thus $\ker \alpha = f\mathsf{L}[t]$. By the Chinese remainder theorem for the principal ideal ring $\mathsf{L}[t]$, the system of congruences

$$h \equiv h_i \pmod{g_i^{m_i}} \qquad (i = 1, \ldots, r)$$

has a solution h in $\mathsf{L}[t]$ for any given $h_1, \ldots, h_r$. Hence α is an epimorphism. Thus

$$\mathsf{L}[t]/f\mathsf{L}[t] = \mathsf{L}[t]/\ker\alpha \cong \bigoplus_{i=1}^{r} \mathsf{L}[t]/g_i^{m_i}\mathsf{L}[t].$$

b) We have $\mathsf{L}_1 \cong \mathsf{K}[t]/f\mathsf{K}[t]$ for some irreducible and separable polynomial f in $\mathsf{K}[t]$ of degree n_1. As f has no multiple roots in any extension field of K, we have a decomposition $f = g_1 \cdots g_r$ in $\mathsf{L}_2[t]$ with irreducible, pairwise non-associated polynomials g_i. Thus by a),

$$\mathsf{L}_1 \otimes_{\mathsf{K}} \mathsf{L}_2 \cong \bigoplus_{i=1}^{r} \mathsf{L}_2[t]/g_i\mathsf{L}_2[t].$$

If r_i is the degree of g_i, we have

$$\mathsf{L}_2[t]/g_i\mathsf{L}_2[t] \cong GF(q^{n_2 r_i}).$$

Writing $m_i = n_2 r_i$ we get

$$\mathsf{L}_1 \otimes_{\mathsf{K}} \mathsf{L}_2 \cong \bigoplus_{i=1}^{r} GF(q^{m_i})$$

with $n_2 | m_i$. As this decomposition of $\mathsf{L}_1 \otimes_{\mathsf{K}} \mathsf{L}_2$ into simple algebras is unique, we have also $n_1 | m_i$, by symmetry. Hence $k | m_i$ for all $i = 1, \ldots, r$.

In L_i the identical relation $a^{q^{n_i}} = a$ holds. Hence $a^{q^k} = a$. For any $c = \sum_j a_j \otimes b_j$ with $a_j \in \mathsf{L}_1$, $b_j \in \mathsf{L}_2$, we conclude, since $\mathsf{L}_1 \otimes_{\mathsf{K}} \mathsf{L}_2$ is commutative and char $\mathsf{K} = p$, that

$$c^{q^k} = \left(\sum_j a_j \otimes b_j\right)^{q^k} = \sum_j a_j^{q^k} \otimes b_j^{q^k} = \sum_j a_j \otimes b_j = c.$$

This identical relation in $\mathsf{L}_1 \otimes_{\mathsf{K}} \mathsf{L}_2$ is naturally inherited by all the epimorphic images $GF(q^{m_i})$ of $\mathsf{L}_1 \otimes_{\mathsf{K}} \mathsf{L}_2$. Thus the polynomial $t^{q^k} - t$ has q^{m_i} roots in $GF(q^{m_i})$, and hence $m_i \leq k$. Since $k | m_i$, this forces $m_i = k$ for all i. Comparing dimensions over K,

$$n_1 n_2 = \dim_{\mathsf{K}}(\mathsf{L}_1 \otimes_{\mathsf{K}} \mathsf{L}_2) = \dim_{\mathsf{K}}(GF(q^k) \oplus \cdots \oplus GF(q^k)) = kr,$$

and thus r is the greatest common divisor of n_1 and n_2.

c) As $L_1 : K$ is separable, we have

$$L_1 = K(a) \cong K[t]/fK[t]$$

for some irreducible, separable polynomial f in $K[t]$. Hence in $L_2[t]$ we obtain

$$f = \prod_{i=1}^{r} g_i$$

for non-associated irreducible, separable polynomials g_i. Thus 1.4 a) implies that

$$L_1 \otimes_K L_2 \cong L_2[t]/fL_2[t] \cong \bigoplus_{i=1}^{r} L_2[t]/g_i L_2[t].$$

As g_i is separable, $L_2[t]/g_i L_2[t]$ is a field F_i, which is separable over L_2. **q.e.d.**

Now we can describe the behaviour of the radical of a group-ring under extension of the ground-field. We remark that all results in this section remain true for K-algebras $\mathfrak{A}$ which can be defined over a finite subfield K_0 of K, that is, for which there exists a K_0-algebra $\mathfrak{A}_0$ such that $\mathfrak{A} \cong \mathfrak{A}_0 \otimes_{K_0} K$.

1.5 Theorem. *Suppose that* L *is an extension of the field* K.

a) *We have*

$$\mathbf{J}((K\mathfrak{G}/\mathbf{J}(K\mathfrak{G}))_L) = 0$$

and

$$\mathbf{J}(K\mathfrak{G} \otimes_K L) = \mathbf{J}(K\mathfrak{G}) \otimes_K L.$$

b) *If* $\{a_1, \ldots, a_s\}$ *is a* K*-basis of* $\mathbf{J}(K\mathfrak{G})$, *it is also an* L*-basis of* $\mathbf{J}(L\mathfrak{G})$. *In particular for all fields* L *of characteristic p we have*

$$\dim_L \mathbf{J}(L\mathfrak{G}) = \dim_{GF(p)} \mathbf{J}(GF(p)\mathfrak{G}).$$

Proof. a) If char $K = 0$, we have

$$\mathbf{J}(K\mathfrak{G}) = 0 = \mathbf{J}(K\mathfrak{G} \otimes_K L) \quad \text{(by 1.3a) and V, 2.7).}$$

Hence we assume char $\mathsf{K} = p$. We write $\mathsf{K}_0 = GF(p)$ and show first that

$$\mathbf{J}(\mathsf{K}_0\mathfrak{G}) \otimes_{\mathsf{K}_0} \mathsf{K} = \mathbf{J}(\mathsf{K}_0\mathfrak{G} \otimes_{\mathsf{K}_0} \mathsf{K}).$$

By Wedderburn's theorem,

$$\mathsf{K}_0\mathfrak{G}/\mathbf{J}(\mathsf{K}_0\mathfrak{G}) \cong \bigoplus_{i=1}^{r} (\mathsf{F}_i)_{n_i},$$

where the F_i are finite division rings. By another well-known theorem of Wedderburn, the F_i are fields. Hence $\mathsf{F}_i = GF(p^{d_i})$ for some d_i. We conclude that

$$\begin{aligned}(\mathsf{K}_0\mathfrak{G} \otimes_{\mathsf{K}_0} \mathsf{K})/(\mathbf{J}(\mathsf{K}_0\mathfrak{G}) \otimes_{\mathsf{K}_0} \mathsf{K}) &\cong (\mathsf{K}_0\mathfrak{G}/\mathbf{J}(\mathsf{K}_0\mathfrak{G}))_{\mathsf{K}} \quad \text{(by 1.2c))} \\ &\cong \bigoplus_{i=1}^{r} (\mathsf{F}_i \otimes_{\mathsf{K}_0} \mathsf{K})_{n_i} \quad \text{(by 1.2a) and b)).}\end{aligned}$$

If f_i is an irreducible polynomial of degree d_i in $\mathsf{K}_0[t]$, we have

$$\mathsf{F}_i \cong \mathsf{K}_0[t]/f_i\mathsf{K}_0[t].$$

As f_i has no multiple roots in any extension field of K_0, we have a decomposition

$$f_i = \prod_{j=1}^{s_i} g_{ij},$$

with irreducible polynomials g_{ij} from $\mathsf{K}[t]$, pairwise non-associated for fixed i. By 1.4a), it follows that

$$\mathsf{F}_i \otimes_{\mathsf{K}_0} \mathsf{K} \cong \bigoplus_{j=1}^{s_i} \mathsf{K}[t]/g_{ij}\mathsf{K}[t].$$

We denote the field $\mathsf{K}[t]/g_{ij}\mathsf{K}[t]$ by F_{ij}. Thus

$$(\mathsf{F}_i \otimes_{\mathsf{K}_0} \mathsf{K})_{n_i} \cong \left(\bigoplus_{j=1}^{s_i} \mathsf{F}_{ij}\right)_{n_i} \cong \bigoplus_{j=1}^{s_i} (\mathsf{F}_{ij})_{n_i}.$$

Hence

$$(\mathsf{K}_0\mathfrak{G}/\mathbf{J}(\mathsf{K}_0\mathfrak{G}))_{\mathsf{K}} \cong \bigoplus_{i=1}^{r}\bigoplus_{j=1}^{s_i}(\mathsf{F}_{ij})_{n_i}$$

is semisimple. By 1.2d),

$$\mathbf{J}(\mathsf{K}_0\mathfrak{G}) \otimes_{\mathsf{K}_0} \mathsf{K} \subseteq \mathbf{J}(\mathsf{K}_0\mathfrak{G} \otimes_{\mathsf{K}_0} \mathsf{K});$$

since also

$$(\mathsf{K}_0\mathfrak{G} \otimes_{\mathsf{K}_0} \mathsf{K})/(\mathbf{J}(\mathsf{K}_0\mathfrak{G}) \otimes_{\mathsf{K}_0} \mathsf{K}) \cong (\mathsf{K}_0\mathfrak{G}/\mathbf{J}(\mathsf{K}_0\mathfrak{G}))_{\mathsf{K}}$$

is semisimple,

$$\mathbf{J}(\mathsf{K}_0\mathfrak{G}) \otimes_{\mathsf{K}_0} \mathsf{K} = \mathbf{J}(\mathsf{K}_0\mathfrak{G} \otimes_{\mathsf{K}_0} \mathsf{K}).$$

Now suppose that char $\mathsf{K} = p$ and L is any extension of K. By 1.3a) and the results proved already,

$$\begin{aligned}\dim_{\mathsf{K}} \mathbf{J}(\mathsf{K}\mathfrak{G}) &= \dim_{\mathsf{K}} \mathbf{J}(\mathsf{K}_0\mathfrak{G} \otimes_{\mathsf{K}_0} \mathsf{K}) = \dim_{\mathsf{K}}(\mathbf{J}(\mathsf{K}_0\mathfrak{G}) \otimes_{\mathsf{K}_0} \mathsf{K}) \\ &= \dim_{\mathsf{K}_0} \mathbf{J}(\mathsf{K}_0\mathfrak{G}) = \dim_{\mathsf{L}} \mathbf{J}(\mathsf{L}\mathfrak{G}).\end{aligned}$$

Also by 1.2d),

$$\mathbf{J}(\mathsf{K}\mathfrak{G}) \otimes_{\mathsf{K}} \mathsf{L} \subseteq \mathbf{J}(\mathsf{K}\mathfrak{G} \otimes_{\mathsf{K}} \mathsf{L}).$$

Equality of the dimensions shows that

$$\mathbf{J}(\mathsf{K}\mathfrak{G}) \otimes_{\mathsf{K}} \mathsf{L} = \mathbf{J}(\mathsf{K}\mathfrak{G} \otimes_{\mathsf{K}} \mathsf{L}).$$

As

$$\begin{aligned}(\mathsf{K}\mathfrak{G}/\mathbf{J}(\mathsf{K}\mathfrak{G}))_{\mathsf{L}} &\cong (\mathsf{K}\mathfrak{G} \otimes_{\mathsf{K}} \mathsf{L})/(\mathbf{J}(\mathsf{K}\mathfrak{G}) \otimes_{\mathsf{K}} \mathsf{L}) \quad \text{(see 1.2c))} \\ &= (\mathsf{K}\mathfrak{G} \otimes_{\mathsf{K}} \mathsf{L})/\mathbf{J}(\mathsf{K}\mathfrak{G} \otimes_{\mathsf{K}} \mathsf{L})\end{aligned}$$

is semisimple, we see that

$$\mathbf{J}((\mathsf{K}\mathfrak{G}/\mathbf{J}(\mathsf{K}\mathfrak{G}))_{\mathsf{L}}) = 0.$$

b) In a) we saw that $\dim_K \mathbf{J}(K\mathfrak{G}) = \dim_L \mathbf{J}(L\mathfrak{G})$. Hence we have only to prove that $\mathbf{J}(K\mathfrak{G}) \subseteq \mathbf{J}(L\mathfrak{G})$. Let $\mathfrak{J}$ be the L-subspace of $L\mathfrak{G}$ spanned by $\mathbf{J}(K\mathfrak{G})$. Since $\mathbf{J}(K\mathfrak{G})$ is a nilpotent ideal of $K\mathfrak{G}$, $\mathfrak{J}$ is a nilpotent ideal of $L\mathfrak{G}$. Thus $\mathbf{J}(L\mathfrak{G}) \supseteq \mathfrak{J} \supseteq \mathbf{J}(K\mathfrak{G})$. **q.e.d.**

To show that complete reducibility is preserved under field extensions, we first prove a simple lemma, which will also be frequently used later on.

1.6 Lemma. *Let $\mathfrak{A}$ be an algebra of finite dimension over a field* K *and let* V *be an $\mathfrak{A}$-module.*

a) *$V/V\mathbf{J}(\mathfrak{A})$ is a completely reducible $\mathfrak{A}$-module. If* W *is a submodule of* V *such that* V/W *is completely reducible, then* $W \supseteq V\mathbf{J}(\mathfrak{A})$.

b) *The set*

$$\mathbf{An}_V\mathbf{J}(\mathfrak{A}) = \{v \mid v \in V,\ v\mathbf{J}(\mathfrak{A}) = 0\}$$

is a completely reducible submodule of V. *If* W *is a completely reducible submodule of* V, *then* $W \subseteq \mathbf{An}_V\mathbf{J}(\mathfrak{A})$.

Proof. a) As $\overline{\mathfrak{A}} = \mathfrak{A}/\mathbf{J}(\mathfrak{A})$ is a semisimple algebra of finite dimension over K and $V/V\mathbf{J}(\mathfrak{A})$ can be regarded as an $\overline{\mathfrak{A}}$-module, $V/V\mathbf{J}(\mathfrak{A})$ is completely reducible as an $\overline{\mathfrak{A}}$-module and also as an $\mathfrak{A}$-module. Conversely, if V/W is completely reducible, then $(V/W)\mathbf{J}(\mathfrak{A}) = 0$ and thus $V\mathbf{J}(\mathfrak{A}) \subseteq W$.

b) By similar arguments the submodule W is completely reducible if and only if $W\mathbf{J}(\mathfrak{A}) = 0$, that is, if $W \subseteq \mathbf{An}_V\mathbf{J}(\mathfrak{A})$. **q.e.d.**

1.7 Definition. Let $\mathfrak{A}$ be an algebra of finite dimension over a field K and let V be an $\mathfrak{A}$-module.

a) We call the largest completely reducible factor module $V/V\mathbf{J}(\mathfrak{A})$ of V the *head* of V and denote it by $\mathbf{H}(V)$.

b) We call the largest completely reducible submodule $\mathbf{An}_V\mathbf{J}(\mathfrak{A})$ of V the *socle* of V and denote it by $\mathbf{S}(V)$.

1.8 Theorem. *Suppose that* V *is a* $K\mathfrak{G}$*-module and* L *is an extension of* K. *Then the* $L\mathfrak{G}$*-module* V_L *is completely reducible if and only if* V *is completely reducible.*

Proof. By 1.5,

$$V_L\mathbf{J}(K\mathfrak{G} \otimes_K L) = (V \otimes_K L)(\mathbf{J}(K\mathfrak{G}) \otimes_K L) = V\mathbf{J}(K\mathfrak{G}) \otimes_K L.$$

Hence $V_L \mathbf{J}(K\mathfrak{G} \otimes_K L) = 0$ if and only if $V\mathbf{J}(K\mathfrak{G}) = 0$. But by 1.6, $V\mathbf{J}(K\mathfrak{G}) = 0$ and $V_L \mathbf{J}(K\mathfrak{G} \otimes_K L) = 0$ are equivalent to the complete reducibility of V and V_L respectively. **q.e.d.**

1.9 Remark. In 1.5 a), we showed that $(K\mathfrak{G}/\mathbf{J}(K\mathfrak{G}))_L$ is semisimple for every extension L of K. This property is called the *separability* of $K\mathfrak{G}/\mathbf{J}(K\mathfrak{G})$. By a theorem of Wedderburn (see DEURING [2], p. 23), in any algebra $\mathfrak{A}$ for which $\mathfrak{A}/\mathbf{J}(\mathfrak{A})$ is separable, there exists a semisimple subalgebra $\mathfrak{S}$ ($\cong \mathfrak{A}/\mathbf{J}(\mathfrak{A})$) such that $\mathfrak{A} = \mathfrak{S} \oplus \mathbf{J}(\mathfrak{A})$. (This is only a direct sum of K-vector spaces; in general $\mathfrak{S}$ is not an ideal of $\mathfrak{A}$.) For any two such complements $\mathfrak{S}_1$, $\mathfrak{S}_2$ of $\mathbf{J}(\mathfrak{A})$, there exists $j \in \mathbf{J}(\mathfrak{A})$ such that

$$(1 + j)^{-1}\mathfrak{S}_1(1 + j) = \mathfrak{S}_2$$

(see CURTIS and REINER [1], p. 491). The proof of this theorem is technically related to the proof of the Zassenhaus theorem about the existence and conjugacy of complements of normal Hall subgroups in finite groups. The existence of complements of $\mathbf{J}(K\mathfrak{G})$ in $K\mathfrak{G}$, however, has not found important applications in representation theory.

1.10 Lemma. *Let K be any field of prime characteristic p. Then*

$$K\mathfrak{G}/\mathbf{J}(K\mathfrak{G}) = \bigoplus_{i=1}^{k} (L_i)_{n_i},$$

where the L_i are finite, separable extensions of K.

Proof. We put $K_0 = GF(p)$. Then we have

$$K_0\mathfrak{G}/\mathbf{J}(K_0\mathfrak{G}) = \bigoplus_{i=1}^{m} (F_i)_{n_i}$$

for finite division rings F_i, which by Wedderburn's theorem are fields. This implies that

$$\begin{aligned} K\mathfrak{G}/\mathbf{J}(K\mathfrak{G}) &\cong (K_0\mathfrak{G} \otimes_{K_0} K)/\mathbf{J}(K_0\mathfrak{G}) \otimes_{K_0} K \quad &&\text{(by 1.5)} \\ &\cong (K_0\mathfrak{G}/\mathbf{J}(K_0\mathfrak{G})) \otimes_{K_0} K &&\text{(by 1.2c))} \\ &\cong \bigoplus_{i=1}^{m} (F_i \otimes_{K_0} K)_{n_i} &&\text{(by 1.2a), b)).} \end{aligned}$$

As F_i is separable over K_0, it follows from 1.4c) that

$$\mathsf{F}_i \otimes_{\mathsf{K}_0} \mathsf{K} \cong \bigoplus_{j=1}^{m_i} \mathsf{F}_{ij}$$

for separable extensions F_{ij} of K. Hence

$$\mathsf{K}\mathfrak{G}/\mathbf{J}(\mathsf{K}\mathfrak{G}) \cong \bigoplus_{i=1}^{m} \bigoplus_{j=1}^{m_i} (\mathsf{F}_{ij})_{n_i}. \qquad \textbf{q.e.d.}$$

1.11 Theorem. *Let* K *be a field of prime characteristic* p.

a) *If* ϕ *is the character of a* $\mathsf{K}\mathfrak{G}$*-module* V, *then* $\phi(g) = \phi(g_{p'})$ *for any element* g *of* $\mathfrak{G}$, *where* $g_{p'}$ *is the* p'*-part of* g.

b) *Let* $\mathfrak{X}$ *be the set of* p'*-elements of* $\mathfrak{G}$. *Let* $\mathsf{V}_1, \ldots, \mathsf{V}_k$ *be all the irreducible* $\mathsf{K}\mathfrak{G}$*-modules to within isomorphism, and let* ϕ_i *be the character of* V_i. *Then* $\phi_1, \ldots, \phi_k$ *are* K*-linearly independent on* $\mathfrak{X}$.

Proof. a) Let $p^a m$ be the order of g, where $(m, p) = 1$. Then there exist integers i, j such that $im + jp^a = 1$, and $g_{p'} = g^{jp^a}$. Hence if ξ is an eigen-value of the linear transformation $v \to vg$ of V (in some extension of K), $\xi^{im+jp^a} = \xi$ and $\xi^{mp^a} = 1$. Since char $\mathsf{K} = p$, it follows that $\xi^m = 1$ and $\xi^{jp^a} = \xi$. But $\phi(g)$ is the sum of the various eigen-values ξ and $\phi(g_{p'})$ is the corresponding sum of the ξ^{jp^a}, so $\phi(g_{p'}) = \phi(g)$.

b) By Lemma 1.10, we have

$$\mathsf{K}\mathfrak{G}/\mathbf{J}(\mathsf{K}\mathfrak{G}) \cong \bigoplus_{i=1}^{k} (\mathsf{L}_i)_{n_i},$$

where L_i is separable over K. Let δ_i be the corresponding epimorphism of $\mathsf{K}\mathfrak{G}$ onto $(\mathsf{L}_i)_{n_i}$ and let γ_i be the faithful irreducible representation over K of $(\mathsf{L}_i)_{n_i}$ on the module of row vectors with coefficients in L_i. Then $\delta_1\gamma_1, \ldots, \delta_k\gamma_k$ are the inequivalent irreducible representations of $\mathsf{K}\mathfrak{G}$ and every irreducible representation of $\mathsf{K}\mathfrak{G}$ is equivalent to one of $\delta_1\gamma_1, \ldots, \delta_k\gamma_k$. We suppose the numbering so chosen that $\delta_i\gamma_i$ is the representation of $\mathfrak{G}$ on V_i. Thus if $a \in \mathsf{K}\mathfrak{G}$,

$$\phi_i(a) = \operatorname{tr}(a\delta_i\gamma_i).$$

If $C \in (\mathsf{L}_i)_{n_i}$, it is easy to see that

$$\operatorname{tr} C\gamma_i = \operatorname{tr}_{\mathsf{L}_i:\mathsf{K}}(\operatorname{tr} C),$$

so

$$\phi_i(a) = \operatorname{tr}_{\mathsf{L}_i:\mathsf{K}}(\operatorname{tr} a\delta_i) \quad (i = 1, \ldots, k).$$

Since L_i is separable over K, there exists $b_i \in \mathsf{L}_i$ such that $\operatorname{tr}_{\mathsf{L}_i:\mathsf{K}} b_i = 1$ (see BOURBAKI [1], p. 155). Let a_i be an element of $\mathsf{K}\mathfrak{G}$ for which

$$a_i\delta_i = \begin{pmatrix} b_i & 0 \ldots 0 \\ 0 & 0 \ldots 0 \\ \vdots & \vdots \quad \vdots \\ 0 & 0 \ldots 0 \end{pmatrix}, \qquad a_i\delta_j = 0 \quad (j \neq i).$$

Then $\phi_i(a_i) = \operatorname{tr}_{\mathsf{L}_i:\mathsf{K}} b_i = 1$ and $\phi_j(a_i) = 0 \quad (j \neq i)$. Hence $\phi_1, \ldots, \phi_k$ are linearly independent. It follows from a) that their restrictions to $\mathfrak{X}$ are also linearly independent. **q.e.d.**

In particular, it follows from 1.11 that if $\phi_i(g) = \phi_j(g)$ for all $g \in \mathfrak{G}$, then $i = j$. This is not true for arbitrary representations over a field of finite characteristic, although it is true for arbitrary representations over $\mathbb{C}$.

1.12 Lemma. *Let* V, W *be* $\mathsf{K}\mathfrak{G}$*-modules and let* L *be an extension of* K. *Then*

$$\operatorname{Hom}_{\mathsf{L}\mathfrak{G}}(\mathsf{V}_\mathsf{L}, \mathsf{W}_\mathsf{L}) \cong \operatorname{Hom}_{\mathsf{K}\mathfrak{G}}(\mathsf{V}, \mathsf{W}) \otimes_\mathsf{K} \mathsf{L}.$$

If V = W, *the two sides of this are isomorphic* L*-algebras.*

Proof. The proof is identical with that of V, 11.9. **q.e.d.**

We have seen that complete reducibility is preserved under field extensions. We study now how irreducible modules behave under Galois extensions of the ground field.

1.13 Definition. Let L be a Galois extension of K with Galois group $\mathfrak{H}$. Let V by any $\mathsf{L}\mathfrak{G}$-module. For every $\eta \in \mathfrak{H}$, we define the $\mathsf{L}\mathfrak{G}$-module V_η as follows.

Let V_η be an Abelian group for which there is an additive isomorphism ι_η of V onto V_η. We define the structure of a vector space over L on V_η by the formula

$$\lambda(v\iota_\eta) = ((\lambda\eta^{-1})v)\iota_\eta \quad (v \in \mathsf{V}, \lambda \in \mathsf{L}).$$

If $\{v_1, \ldots, v_n\}$ is an L-basis of V, then $\{v_1 \iota_\eta, \ldots, v_n \iota_\eta\}$ is an L-basis of V_η. Hence $\dim_L V = \dim_L V_\eta$.

We now define the structure of an $L\mathfrak{G}$-module on V_η by putting

$$(v\iota_\eta)g = (vg)\iota_\eta \quad (v \in V, g \in \mathfrak{G}).$$

It is easy to see that V_η indeed becomes an $L\mathfrak{G}$-module. We consider the matrix representation of $\mathfrak{G}$ on V_η.

Suppose that $\{v_1, \ldots, v_n\}$ is an L-basis of V and that

$$v_i g = \sum_{j=1}^{n} a_{ij}(g) v_j \quad (i = 1, \ldots, n),$$

where $a_{ij}(g) \in L$. Then $\{v_1 \iota_\eta, \ldots, v_n \iota_\eta\}$ is an L-basis of V_η and

$$(v_i \iota_\eta)g = (v_i g)\iota_\eta = \left(\sum_{j=1}^{n} a_{ij}(g) v_j\right)\iota_\eta$$
$$= \sum_{j=1}^{n} (a_{ij}(g)\eta)(v_j \iota_\eta).$$

Hence the matrix representation of $\mathfrak{G}$ on V_η is conjugate under η to the matrix representation of $\mathfrak{G}$ on V. Thus if χ is the character of $\mathfrak{G}$ on V, the character ψ of $\mathfrak{G}$ on V_η is given by

$$\psi(g) = \sum_{i=1}^{n} a_{ii}(g)\eta = \chi(g)\eta.$$

We write $\psi = \chi_\eta$.

If W is an $L\mathfrak{G}$-submodule of V, it is obvious that $W\iota_\eta$ is an $L\mathfrak{G}$-submodule of V_η. Hence V_η is irreducible if and only if V is irreducible.

1.14 Lemma. *Let* L *be a Galois extension of* K *with Galois group* $\mathfrak{H}$. *Let* V *be a* $K\mathfrak{G}$*-module and let* W *be an irreducible* $L\mathfrak{G}$*-module. If* W *is isomorphic to a submodule of* V_L, *then for all* $\eta \in \mathfrak{H}$ *the algebraic conjugate* W_η *is also isomorphic to a submodule of* V_L.

Proof. By assumption, there exists an α such that

$$0 \neq \alpha \in \operatorname{Hom}_{L\mathfrak{G}}(W, V_L).$$

We consider the mapping $\alpha' = \iota_\eta^{-1}\alpha(1 \otimes \eta)$ of W_η into V_L. We show first that α' is L-linear.

Suppose that $w \in \mathsf{W}_\eta$, $\lambda \in \mathsf{L}$ and

$$w\iota_\eta^{-1}\alpha = \sum_i v_i \otimes \lambda_i,$$

where $v_i \in \mathsf{V}$ and $\lambda_i \in \mathsf{L}$. Then

$$\begin{aligned}(\lambda w)\alpha' &= ((\lambda\eta^{-1})(w\iota_\eta^{-1}))\alpha(1 \otimes \eta)\\ &= ((\lambda\eta^{-1})(w\iota_\eta^{-1}\alpha))(1 \otimes \eta)\\ &= \sum_i (v_i \otimes \lambda_i(\lambda\eta^{-1}))(1 \otimes \eta).\\ &= \sum_i v_i \otimes (\lambda_i\eta)\lambda\\ &= \lambda((w\iota_\eta^{-1}\alpha)(1 \otimes \eta)) = \lambda(w\alpha').\end{aligned}$$

Next we show that α' is an $\mathsf{L}\mathfrak{G}$-homomorphism. For $w \in \mathsf{W}_\eta$ and $g \in \mathfrak{G}$, we have

$$\begin{aligned}(wg)\alpha' &= (wg)\iota_\eta^{-1}\alpha(1 \otimes \eta)\\ &= ((w\iota_\eta^{-1})g)\alpha(1 \otimes \eta) && \text{(by definition of } \mathsf{W}_\eta)\\ &= ((w\iota_\eta^{-1}\alpha)g)(1 \otimes \eta) && \text{(as } \alpha \text{ is an } \mathsf{L}\mathfrak{G}\text{-homomorphism)}\\ &= \left(\sum_i v_i g \otimes \lambda_i\right)(1 \otimes \eta)\\ &= \sum_i v_i g \otimes \lambda_i\eta\\ &= \left(\sum_i v_i \otimes \lambda_i\right)(1 \otimes \eta)g\\ &= w\iota_\eta^{-1}\alpha(1 \otimes \eta)g = (w\alpha')g.\end{aligned}$$

Since $\alpha \neq 0$, $\alpha' \neq 0$. As W_η is irreducible, it follows that it is isomorphic to a submodule of V_L. **q.e.d.**

1.15 Lemma. *Let K be a field of prime characteristic, L any extension of K and V an irreducible $\mathsf{K}\mathfrak{G}$-module. Then*

$$\mathsf{V}_\mathsf{L} \cong \mathsf{W}_1 \oplus \cdots \oplus \mathsf{W}_r$$

for non-isomorphic irreducible $\mathsf{L}\mathfrak{G}$*-modules* $\mathsf{W}_1, \ldots, \mathsf{W}_r$.

Proof. The fields L_i, occuring in 1.10, are by Wedderburn's theorem antiisomorphic to the endomorphism rings of the irreducible $\mathsf{K}\mathfrak{G}$-modules. Hence, in particular, $\mathsf{F} = \mathrm{Hom}_{\mathsf{K}\mathfrak{G}}(\mathsf{V}, \mathsf{V})$ is a field.
By 1.8, V_L is completely reducible. Suppose that

$$\mathsf{V}_\mathsf{L} = \mathsf{H}_1 \oplus \cdots \oplus \mathsf{H}_r,$$

where the H_i are the homogeneous components of V_L; this means that

$$\mathsf{H}_i \cong \mathsf{W}_i \oplus \underset{s_i}{\cdots} \oplus \mathsf{W}_i$$

for irreducible $\mathsf{L}\mathfrak{G}$-modules W_i, and $\mathsf{W}_i \not\cong \mathsf{W}_j$ for $i \neq j$. We obtain

$$\mathrm{Hom}_{\mathsf{L}\mathfrak{G}}(\mathsf{V}_\mathsf{L}, \mathsf{V}_\mathsf{L}) \cong \bigoplus_{i=1}^{r} \mathrm{Hom}_{\mathsf{L}\mathfrak{G}}(\mathsf{H}_i, \mathsf{H}_i) \cong \bigoplus_{i=1}^{r} (\mathsf{F}_i)_{s_i},$$

where $\mathsf{F}_i = \mathrm{Hom}_{\mathsf{L}\mathfrak{G}}(\mathsf{W}_i, \mathsf{W}_i)$ is a division algebra. But by 1.12,

$$\mathrm{Hom}_{\mathsf{L}\mathfrak{G}}(\mathsf{V}_\mathsf{L}, \mathsf{V}_\mathsf{L}) \cong \mathrm{Hom}_{\mathsf{K}\mathfrak{G}}(\mathsf{V}, \mathsf{V}) \otimes_\mathsf{K} \mathsf{L} \cong \mathsf{F} \otimes_\mathsf{K} \mathsf{L}.$$

Hence $\mathrm{Hom}_{\mathsf{L}\mathfrak{G}}(\mathsf{V}_\mathsf{L}, \mathsf{V}_\mathsf{L})$ is commutative, and this forces $s_i = 1$ for all i.
q.e.d.

1.16 Theorem. *Let* L *be a Galois extension of* K *with Galois group* $\mathfrak{H}$, *and let* V *be an* $\mathsf{L}\mathfrak{G}$*-module. We denote* V, *regarded as a* $\mathsf{K}\mathfrak{G}$*-module, by* V_0.

a) $\mathsf{V}_0 \otimes_\mathsf{K} \mathsf{L} \cong \bigoplus_{\eta \in \mathfrak{H}} \mathsf{V}_\eta$.

b) *If* χ, χ_0 *are the characters of* V, V_0 *respectively, then*

$$\chi_0(g) = \sum_{\eta \in \mathfrak{H}} \chi(g)\eta = \mathrm{tr}_{\mathsf{L}:\mathsf{K}}\chi(g).$$

c) *If* V *is a completely reducible* $\mathsf{L}\mathfrak{G}$*-module,* V_0 *is a completely reducible* $\mathsf{K}\mathfrak{G}$*-module.*

d) *If* V *is an irreducible* $\mathsf{L}\mathfrak{G}$*-module, then*

$$\mathsf{V}_0 \cong \mathsf{W} \oplus \cdots \oplus \mathsf{W}$$

for some irreducible $\mathsf{K}\mathfrak{G}$*-module* W.

e) *Suppose that* K *is of prime characteristic and* V *is an irreducible* $\mathsf{L}\mathfrak{G}$*-module. We define the subfield* K_χ *of* L *by the formula*

$$\mathsf{K}_\chi = \mathsf{K}(\chi(g) \mid g \in \mathfrak{G}).$$

Let

$$\mathfrak{U} = \{\eta \mid \eta \in \mathfrak{H},\ a\eta = a \quad \text{for all } a \in \mathsf{K}_\chi\}$$

and let $\{\eta_1, \ldots, \eta_m\}$ *be a transversal of* $\mathfrak{U}$ *in* $\mathfrak{H}$. *Then*

$$\mathsf{V}_0 \cong \mathsf{W} \oplus \underset{s}{\cdots} \oplus \mathsf{W},$$

where W *is irreducible,* $s = (\mathsf{L} : \mathsf{K}_\chi)$ *and* $\mathsf{W}_\mathsf{L} \cong \bigoplus_{i=1}^m \mathsf{V}_{\eta_i}$. *In particular,* V_0 *is irreducible if and only if* $\mathsf{L} = \mathsf{K}_\chi$.

Proof. a) We again denote by ι_η the mapping of V onto V_η as in 1.13. We define a mapping ϱ of $\mathsf{V}_0 \otimes_\mathsf{K} \mathsf{L}$ into $\bigoplus_{\eta \in \mathfrak{H}} \mathsf{V}_\eta$ by putting

$$(v \otimes \lambda)\varrho = (\lambda(v\iota_\eta))_{\eta \in \mathfrak{H}}$$

for $v \in \mathsf{V}_0$, $\lambda \in \mathsf{L}$. Obviously ϱ is well-defined. We show that ϱ is L-linear.

For $v \in \mathsf{V}_0$ and λ_1, λ_2 in L, we have

$$\begin{aligned}(\lambda_2(v \otimes \lambda_1))\varrho &= (v \otimes \lambda_1\lambda_2)\varrho = ((\lambda_1\lambda_2)(v\iota_\eta))_{\eta \in \mathfrak{H}}\\ &= \lambda_2((\lambda_1(v\iota_\eta))_{\eta \in \mathfrak{H}}) = \lambda_2((v \otimes \lambda_1)\varrho).\end{aligned}$$

For $v \in \mathsf{V}_0$, $\lambda \in \mathsf{L}$ and $g \in \mathfrak{G}$, we obtain

$$\begin{aligned}((v \otimes \lambda)g)\varrho &= (vg \otimes \lambda)\varrho = (\lambda((vg)\iota_\eta))_{\eta \in \mathfrak{H}}\\ &= (\lambda((v\iota_\eta)g))_{\eta \in \mathfrak{H}} = ((\lambda(v\iota_\eta))g)_{\eta \in \mathfrak{H}}\\ &= ((\lambda(v\iota_\eta))_{\eta \in \mathfrak{H}})g = ((v \otimes \lambda)\varrho)g.\end{aligned}$$

Hence ϱ is an $\mathsf{L}\mathfrak{G}$-homomorphism. Now

$$\dim_\mathsf{L}(\mathsf{V}_0 \otimes_\mathsf{K} \mathsf{L}) = \dim_\mathsf{K} \mathsf{V}_0 = (\mathsf{L} : \mathsf{K}) \dim_\mathsf{L} \mathsf{V} = \dim_\mathsf{L} \bigoplus_{\eta \in \mathfrak{H}} \mathsf{V}_\eta.$$

Thus, to prove that ϱ is an isomorphism, it suffices to show that $\ker \varrho = 0$.

Let $\{v_1, \ldots, v_n\}$ be an L-basis of V and $\{a_1, \ldots, a_m\}$ a K-basis of L. Then

$$\{a_i v_j | i = 1, \ldots, m; j = 1, \ldots, n\}$$

is a K-basis of V_0. Hence every element v of $\mathsf{V}_0 \otimes_{\mathsf{K}} \mathsf{L}$ can be written in the form

$$v = \sum_{i,j} a_i v_j \otimes \lambda_{ij} \qquad (\lambda_{ij} \in \mathsf{L}).$$

If $v \in \ker \varrho$, we obtain

$$0 = \sum_{i,j} \lambda_{ij}((a_i v_j)\iota_\eta) = \left(\sum_{i,j} (\lambda_{ij}\eta^{-1}) a_i v_j\right)\iota_\eta$$

for every $\eta \in \mathfrak{H}$. As $\{v_1, \ldots, v_n\}$ is an L-basis of V, this implies that

$$0 = \left(\sum_i (\lambda_{ij}\eta^{-1}) a_i\right)\eta = \sum_i \lambda_{ij}(a_i\eta)$$

for all $\eta \in \mathfrak{H}$. But by the separability of L over K,

$$\det(a_i\eta)_{i,\eta} \neq 0$$

(see BOURBAKI [1], p. 119). Hence $\lambda_{ij} = 0$ for all i, j and $v = 0$. This shows that $\ker \rho = 0$.

b) This follows at once from a).

c) If V is completely reducible, then

$$\begin{aligned} 0 &= \mathsf{V}\mathbf{J}(\mathsf{L}\mathfrak{G}) && \text{(by 1.6)} \\ &\supseteq \mathsf{V}\mathbf{J}(\mathsf{K}\mathfrak{G}) && \text{(by 1.5b))} \\ &= \mathsf{V}_0\mathbf{J}(\mathsf{K}\mathfrak{G}). \end{aligned}$$

Hence V_0 is a completely reducible $\mathsf{K}\mathfrak{G}$-module, by 1.6.

d) Let W be an irreducible $\mathsf{K}\mathfrak{G}$-submodule of V_0. Then by 1.12 and a),

$$\begin{aligned} 0 \neq \operatorname{Hom}_{\mathsf{K}\mathfrak{G}}(\mathsf{W}, \mathsf{V}_0) \otimes_{\mathsf{K}} \mathsf{L} &\cong \operatorname{Hom}_{\mathsf{L}\mathfrak{G}}(\mathsf{W}_{\mathsf{L}}, (\mathsf{V}_0)_{\mathsf{L}}) \\ &= \operatorname{Hom}_{\mathsf{L}\mathfrak{G}}\left(\mathsf{W}_{\mathsf{L}}, \bigoplus_{\eta \in \mathfrak{H}} \mathsf{V}_\eta\right). \end{aligned}$$

Hence $\operatorname{Hom}_{\mathsf{L}\mathfrak{G}}(\mathsf{W}_{\mathsf{L}}, \mathsf{V}_\eta) \neq 0$ for some $\eta \in \mathfrak{H}$. By 1.8, W_{L} is completely reducible, so V_η is isomorphic to a submodule of W_{L}. Thus by 1.14, V

also is isomorphic to a submodule of W_L. If W_1 and W_2 are irreducible submodules of V_0, then $(W_1)_L$ and $(W_2)_L$ have direct summands isomorphic to V. Thus

$$0 \neq \mathrm{Hom}_{L\mathfrak{G}}((W_1)_L, (W_2)_L) \cong \mathrm{Hom}_{K\mathfrak{G}}(W_1, W_2) \otimes_K L.$$

This forces $W_1 \cong W_2$. Hence

$$V_0 \cong \underbrace{W \oplus \cdots \oplus W}_{s}$$

for some s.

e) Now suppose that K is of prime characteristic. Let χ be the character of V. If $\chi = \chi_\eta$, then $V \cong V_\eta$ by 1.11. Thus the isomorphism type of V_{η_i} appears among the V_η exactly $|\mathfrak{U}| = (L : K_\chi)$ times. Hence it follows from a) and d) that

$$\underbrace{W_L \oplus \cdots \oplus W_L}_{s} \cong V_0 \otimes_K L \cong \bigoplus_{\eta \in \mathfrak{H}} V_\eta \cong \bigoplus_{i=1}^{m} (\underbrace{V_{\eta_i} \oplus \cdots \oplus V_{\eta_i}}_{(L:K_\chi)}).$$

By 1.15, the multiplicity of any irreducible direct summand of W_L is 1. Hence

$$W_L \cong \bigoplus_{i=1}^{m} V_{\eta_i} \quad \text{and} \quad s = (L : K_\chi). \qquad \textbf{q.e.d.}$$

1.16 has the important consequence that in the case of finite fields the Schur index is 1.

1.17 Theorem (R. BRAUER [4]). *Let* L *be a finite field and* V *an irreducible* $L\mathfrak{G}$*-module. Let* K *be a subfield of* L *such that* $\chi(g) \in K$ *for all* $g \in \mathfrak{G}$, *where* χ *is the character of* V. *Then the representation of* $\mathfrak{G}$ *on* V *can be realized over* K; *that is, there exists an irreducible* $K\mathfrak{G}$*-module* W *such that* $V \cong W_L$.

Proof. Certainly L is a Galois extension of K. Now all V_η have the same character and are therefore isomorphic, by 1.11. Thus by 1.16e), $W_L \cong V$. **q.e.d.**

1.18 Theorem. *Suppose that* L *is a Galois extension of* K *with Galois group* $\mathfrak{H}$.

a) *Let* V *be an irreducible* $\mathsf{L}\mathfrak{G}$*-module. Then there exists exactly one irreducible* $\mathsf{K}\mathfrak{G}$*-module* W *(to within isomorphism) such that* V *is isomorphic to a direct summand of* W_{L}.

b) *(cf. V, 13.3) Suppose that* W *is an irreducible* $\mathsf{K}\mathfrak{G}$*-module. Then*

$$\mathsf{W}_{\mathsf{L}} \cong \mathsf{V}_1 \oplus \cdots \oplus \mathsf{V}_r$$

with irreducible $\mathsf{L}\mathfrak{G}$*-modules* V_i, *and all* V_i *are conjugate under automorphisms in* $\mathfrak{H}$ *in the sense of 1.13. Every isomorphism type appears among the* V_i *with the same multiplicity. Finally r divides* $(\mathsf{L} : \mathsf{K})$.

Proof. a) We have

$$\mathsf{K}\mathfrak{G}/\mathbf{J}(\mathsf{K}\mathfrak{G}) \cong \bigoplus_i \mathsf{W}_i,$$

where the W_i are irreducible $\mathsf{K}\mathfrak{G}$-modules. Thus it follows that

$$(\mathsf{K}\mathfrak{G}/\mathbf{J}(\mathsf{K}\mathfrak{G}))_{\mathsf{L}} \cong \bigoplus_i (\mathsf{W}_i)_{\mathsf{L}},$$

and by 1.8, the $(\mathsf{W}_i)_{\mathsf{L}}$ are completely reducible $\mathsf{L}\mathfrak{G}$-modules. However,

$$\begin{aligned}(\mathsf{K}\mathfrak{G}/\mathbf{J}(\mathsf{K}\mathfrak{G}))_{\mathsf{L}} &\cong (\mathsf{K}\mathfrak{G} \otimes_{\mathsf{K}} \mathsf{L})/(\mathbf{J}(\mathsf{K}\mathfrak{G}) \otimes_{\mathsf{K}} \mathsf{L}) && \text{(by 1.2e))}\\ &= (\mathsf{K}\mathfrak{G} \otimes_{\mathsf{K}} \mathsf{L})/\mathbf{J}(\mathsf{K}\mathfrak{G} \otimes_{\mathsf{K}} \mathsf{L}) && \text{(by 1.5a))}\\ &\cong \mathsf{L}\mathfrak{G}/\mathbf{J}(\mathsf{L}\mathfrak{G}) && \text{(by 1.3a)).}\end{aligned}$$

As the irreducible $\mathsf{L}\mathfrak{G}$-module V is isomorphic to a direct summand of $\mathsf{L}\mathfrak{G}/\mathbf{J}(\mathsf{L}\mathfrak{G})$ (considered as an $\mathsf{L}\mathfrak{G}$-module), V is isomorphic to a direct summand of some $(\mathsf{W}_i)_{\mathsf{L}}$.

Suppose that V is isomorphic to a direct summand of $(\mathsf{W}_i)_{\mathsf{L}}$ and $(\mathsf{W}_j)_{\mathsf{L}}$. Then by 1.12,

$$0 \neq \operatorname{Hom}_{\mathsf{L}\mathfrak{G}}((\mathsf{W}_i)_{\mathsf{L}}, (\mathsf{W}_j)_{\mathsf{L}}) \cong \operatorname{Hom}_{\mathsf{K}\mathfrak{G}}(\mathsf{W}_i, \mathsf{W}_j) \otimes_{\mathsf{K}} \mathsf{L}.$$

This shows that $\operatorname{Hom}_{\mathsf{K}\mathfrak{G}}(\mathsf{W}_i, \mathsf{W}_j) \neq 0$. Since W_i and W_j are irreducible, $\mathsf{W}_i \cong \mathsf{W}_j$.

b) Let V be isomorphic to an irreducible direct summand of W_{L}. By 1.14 every conjugate V_η of V $(\eta \in \mathfrak{H})$ is also isomorphic to a direct summand of the completely reducible module W_{L}.

We consider the $\mathsf{K}\mathfrak{G}$-module V_0. By 1.16d),

$$\mathsf{V}_0 \cong \mathsf{U} \oplus \cdots \oplus \mathsf{U}$$

for some irreducible $\mathsf{K}\mathfrak{G}$-module U. By 1.16a),

$$\bigoplus_{\eta \in \mathfrak{H}} \mathsf{V}_\eta \cong \mathsf{V}_0 \otimes_{\mathsf{K}} \mathsf{L} \cong \mathsf{U}_{\mathsf{L}} \oplus \cdots \oplus \mathsf{U}_{\mathsf{L}}.$$

Thus U_{L} and W_{L} contain a submodule isomorphic to V. Hence $\mathsf{U} \cong \mathsf{W}$, by a). From

$$\bigoplus_{\eta \in \mathfrak{H}} \mathsf{V}_\eta \cong \mathsf{V}_0 \otimes_{\mathsf{K}} \mathsf{L} \cong \mathsf{W}_{\mathsf{L}} \oplus \cdots \oplus \mathsf{W}_{\mathsf{L}}, \tag{1}$$

we conclude that

$$\mathsf{W}_{\mathsf{L}} \cong \mathsf{V}_1 \oplus \cdots \oplus \mathsf{V}_r,$$

where each V_i is isomorphic to one of the conjugates V_η of V. Further, if s is the number of summands W_{L} in (1), $rs = |\mathfrak{H}| = (\mathsf{L} : \mathsf{K})$, so r divides $(\mathsf{L} : \mathsf{K})$. As the η for which $\mathsf{V}_\eta \cong \mathsf{V}$ form a subgroup of $\mathfrak{H}$, all isomorphism types of irreducible modules appear in $\bigoplus_{\eta \in \mathfrak{H}} \mathsf{V}_\eta$ with the same multiplicity. Hence by (1), this is also true for W_{L}. **q.e.d.**

1.19 Corollary. *Suppose that* L *is a Galois extension of* K *and* W *is an irreducible* $\mathsf{K}\mathfrak{G}$*-module. If* $\dim_{\mathsf{K}} \mathsf{W}$ *is coprime to* $(\mathsf{L} : \mathsf{K})$, *then* W_{L} *is an irreducible* $\mathsf{L}\mathfrak{G}$*-module.*

Proof. By 1.18,

$$\mathsf{W}_{\mathsf{L}} \cong \mathsf{V}_1 \oplus \cdots \oplus \mathsf{V}_r,$$

where the V_i are irreducible $\mathsf{L}\mathfrak{G}$-modules all of the same dimension, and r divides $(\mathsf{L} : \mathsf{K})$. But we also have

$$r \dim_{\mathsf{L}} \mathsf{V}_1 = \dim_{\mathsf{L}} \mathsf{W}_{\mathsf{L}} = \dim_{\mathsf{K}} \mathsf{W}.$$

This forces $r = 1$, so W_{L} is irreducible. **q.e.d.**

For the sake of completeness, we show that the analogue of Theorem 1.18b) is also true for indecomposable modules.

1.20 Theorem. *Suppose that* L *is a Galois extension of* K *with Galois group* $\mathfrak{H}$. *Suppose further that* W *is an indecomposable* $\mathsf{K}\mathfrak{G}$*-module. Then*

$$\mathsf{W}_\mathsf{L} \cong \mathsf{V}_1 \oplus \cdots \oplus \mathsf{V}_r,$$

where the V_i *are indecomposable* $\mathsf{L}\mathfrak{G}$*-modules and all the* V_i *are conjugate under automorphisms in* $\mathfrak{H}$ *in the sense of 1.13. Every isomorphism type appears among the* V_i *with the same multiplicity, which is a divisor of* $|\mathfrak{H}| = (\mathsf{L}:\mathsf{K})$.

Proof. (DRESS). Suppose that

$$\mathsf{W}_\mathsf{L} \cong \bigoplus_{i=1}^{t} (\mathsf{V}_i \oplus \underset{s_i}{\cdots} \oplus \mathsf{V}_i),$$

where the V_i are indecomposable, pairwise non-isomorphic $\mathsf{L}\mathfrak{G}$-modules. This implies that

$$(\mathsf{W}_\mathsf{L})_0 \cong \bigoplus_{i=1}^{t} (\mathsf{V}_i \oplus \underset{s_i}{\cdots} \oplus \mathsf{V}_i)_0.$$

Since W is a $\mathsf{K}\mathfrak{G}$-module, $(\mathsf{W}_\mathsf{L})_0$ is the direct sum of $(\mathsf{L}:\mathsf{K})$ copies of W, so by the Krull-Schmidt theorem,

$$(\mathsf{V}_1)_0 \cong \mathsf{W} \oplus \cdots \oplus \mathsf{W}.$$

Hence

$$\mathsf{W}_\mathsf{L} \oplus \cdots \oplus \mathsf{W}_\mathsf{L} \cong (\mathsf{V}_1)_0 \otimes_\mathsf{K} \mathsf{L} \cong \bigoplus_{\eta \in \mathfrak{H}} (\mathsf{V}_1)_\eta,$$

by 1.16a). Let

$$\mathfrak{U} = \{\eta \mid \eta \in \mathfrak{H}, (\mathsf{V}_1)_\eta \cong \mathsf{V}_1\}$$

and let $\eta_1, \ldots, \eta_s$ be a transversal of $\mathfrak{U}$ in $\mathfrak{H}$. Then

$$\mathsf{W}_\mathsf{L} \oplus \cdots \oplus \mathsf{W}_\mathsf{L} \cong \bigoplus_{\eta \in \mathfrak{H}} (\mathsf{V}_1)_\eta \cong \bigoplus_{i=1}^{s} ((\mathsf{V}_1)_{\eta_i} \oplus \cdots \oplus (\mathsf{V}_1)_{\eta_i}),$$

where each direct sum of the $(\mathsf{V}_1)_{\eta_i}$ contains $|\mathfrak{U}|$ copies. By the Krull-Schmidt theorem, then,

$$\mathsf{W}_\mathsf{L} \cong \bigoplus_{i=1}^{s} ((\mathsf{V}_1)_{\eta_i} \oplus \cdots \oplus (\mathsf{V}_1)_{\eta_i}),$$

where the number k of copies of $(V_1)_{\eta_i}$ is independent of i and divides $|\mathfrak{H}|$. **q.e.d.**

1.21 Theorem. *Suppose that* L *is an extension of* K *and that* V, W *are* $K\mathfrak{G}$*-modules such that* $W \otimes_K L$ *is isomorphic to a direct summand of* $V \otimes_K L$. *Then* W *is isomorphic to a direct summand of* V.

Proof. a) First suppose that W is indecomposable. It follows from the hypothesis that there exist mappings

$$\alpha \in \mathrm{Hom}_{L\mathfrak{G}}(W \otimes_K L, V \otimes_K L), \quad \beta \in \mathrm{Hom}_{L\mathfrak{G}}(V \otimes_K L, W \otimes_K L)$$

such that $\alpha\beta = 1$. By 1.12,
$\mathrm{Hom}_{L\mathfrak{G}}(W \otimes_K L, V \otimes_K L) \cong \mathrm{Hom}_{K\mathfrak{G}}(W, V) \otimes_K L$, so it follows that

$$\left(\sum_i \alpha_i \otimes \lambda_i\right)\left(\sum_j \beta_j \otimes \lambda_j'\right) = 1$$

for certain $\alpha_i \in \mathrm{Hom}_{K\mathfrak{G}}(W, V)$, $\beta_j \in \mathrm{Hom}_{K\mathfrak{G}}(V, W)$. Thus $\alpha_i\beta_j$ lies in the K-algebra $\mathfrak{A} = \mathrm{Hom}_{K\mathfrak{G}}(W, W)$. Suppose that $\alpha_i\beta_j \in \mathbf{J}(\mathfrak{A})$ for all i, j. Then $\alpha_i\beta_j \otimes \lambda_i\lambda_j' \in \mathbf{J}(\mathfrak{A}) \otimes_K L \subseteq \mathbf{J}(\mathfrak{A}_L)$ by 1.2d), so

$$1 = \sum_{i,j} \alpha_i\beta_j \otimes \lambda_i\lambda_j' \in \mathbf{J}(\mathfrak{A}_L),$$

a contradiction. Hence there exist i, j for which $\alpha_i\beta_j \notin \mathbf{J}(\mathfrak{A})$. It follows from V, 2.4b) that the right ideal $\mathfrak{J}$ of $\mathfrak{A}$ generated by $\alpha_i\beta_j$ is not a nil ideal. Thus $\mathfrak{J}$ contains an element $\alpha_i\beta_j\gamma$ which is not nilpotent. Since W is indecomposable, it follows from Fitting's lemma (I, 10.7) that $\alpha_i\beta_j\gamma$ is an automorphism of W. Thus $\alpha_i\beta_j$ is a non-singular linear transformation of W, and it follows easily that $W\alpha_i \cong W$ and

$$V = W\alpha_i \oplus (\ker \beta_j).$$

b) To deal with the general case, write

$$W = W_1 \oplus \cdots \oplus W_n,$$

where $W_1, \ldots, W_n$ are indecomposable $K\mathfrak{G}$-modules. We use induction on n. By a), $V \cong W_1 \oplus U$ for some $K\mathfrak{G}$-module U. But also $V_L \cong W_L \oplus X$ for some $L\mathfrak{G}$-module X, so

$$(W_1)_L \oplus U_L \cong V_L \cong W_L \oplus X \cong (W_1)_L \oplus \cdots \oplus (W_n)_L \oplus X.$$

By the Krull-Schmidt theorem,

$$\begin{aligned} U_L &\cong (W_2)_L \oplus \cdots \oplus (W_n)_L \oplus X \\ &\cong (W_2 \oplus \cdots \oplus W_n)_L \oplus X. \end{aligned}$$

By the inductive hypothesis, $W_2 \oplus \cdots \oplus W_n$ is a direct summand of U, so W is a direct summand of V. **q.e.d.**

1.22 Theorem (DEURING [1], E. NOETHER). *Let* V, W *be* $K\mathfrak{G}$*-modules. If there exists an extension* L *of* K *such that* V_L, W_L *are isomorphic* $L\mathfrak{G}$*-modules, then* V, W *are isomorphic* $K\mathfrak{G}$*-modules.*

Proof. This follows at once from 1.21. **q.e.d.**

Exercises

1) Let L be a Galois extension of K with Galois group $\mathfrak{H}$. Let ρ be an absolutely irreducible representation of $\mathfrak{G}$ in L for which $\operatorname{tr}\rho(g) \in K$ for all $g \in \mathfrak{G}$. Prove the following.

a) Since ρ and ρ^η ($\eta \in \mathfrak{H}$) have the same trace, they are equivalent, and there exist matrices $A(\eta)$ with coefficients in L such that

$$\rho^\eta(g) = A(\eta)\rho(g)A(\eta)^{-1}$$

for all $\eta \in \mathfrak{H}$ and all $g \in \mathfrak{G}$.

b) For all η_1, η_2 in $\mathfrak{H}$, we have

$$A(\eta_1\eta_2) = c(\eta_1, \eta_2)(A(\eta_1)\eta_2)A(\eta_2)$$

for some $c(\eta_1, \eta_2) \in L^\times$.

c) From $A(\eta_1(\eta_2\eta_3)) = A((\eta_1\eta_2)\eta_3)$, derive

$$(c(\eta_1, \eta_2)\eta_3)c(\eta_1\eta_2, \eta_3) = c(\eta_1, \eta_2\eta_3)c(\eta_2, \eta_3).$$

Hence c is a 2-cocycle on $\mathfrak{H}$ with values in $L^\times$ (in the sense of I, § 17).

d) Suppose that c is a coboundary. Then

$$c(\eta_1, \eta_2) = (b(\eta_1)\eta_2)b(\eta_2)b(\eta_1\eta_2)^{-1}$$

for some $b(\eta) \in L^\times$. If we write $A'(\eta) = b(\eta)A(\eta)$, then $A'(\eta_1\eta_2) = (A'(\eta_1)\eta_2)A'(\eta_2)$. Hence A' is a cocycle on $\mathfrak{H}$ with values in the group

$GL(n, \mathsf{L})$ (where n is the degree of ρ). An extension of Hilbert's Theorem 90 says that $\mathbf{H}^1(\mathfrak{H}, GL(n, \mathsf{L})) = 1$ (see SERRE [1], p. 159). This means that $A'(\eta) = (B\eta)B^{-1}$ for some $B \in GL(n, \mathsf{L})$. Then

$$(B^{-1}\rho(g)B)\eta = B^{-1}\rho(g)B$$

for all $\eta \in \mathfrak{H}$ and $g \in \mathfrak{G}$.

e) Show that ρ can be realized in K if and only if c is a coboundary.

2) Now prove Theorem 1.17 by using the fact that $\mathbf{H}^2(\mathfrak{H}, \mathsf{L}^\times) = 1$ if K is finite and L is a Galois extension of K with Galois group $\mathfrak{H}$.

3) Give an example of a reducible representation that cannot be realized in the smallest field containing all the traces of the matrices in the representation.

§ 2. Splitting Fields

We now turn to the discussion of splitting fields. Most of the results of this section are true for algebras $\mathfrak{A}$ for which $\mathfrak{A}/\mathbf{J}(\mathfrak{A})$ is separable (cf. 1.9), but for the sake of simplicity we shall often restrict ourselves to group-rings.

2.1 Definition. Let $\mathfrak{A}$ be an algebra over K of finite dimension.

a) An irreducible $\mathfrak{A}$-module V is called *absolutely irreducible* if $\mathrm{Hom}_{\mathfrak{A}}(\mathsf{V}, \mathsf{V}) = \mathsf{K}$.
(Observe that in general, neither $\mathrm{Hom}_{\mathfrak{A}}(\mathsf{V}, \mathsf{V}) = \mathsf{K}$ nor the irreducibility of V implies the other).

b) K is called a *splitting field* of $\mathfrak{A}$ if

$$\mathfrak{A}/\mathbf{J}(\mathfrak{A}) \cong \bigoplus_{i=1}^{k} (\mathsf{K})_{n_i}$$

for some n_i.

c) If $\mathfrak{G}$ is a group, a splitting field of $\mathsf{K}\mathfrak{G}$ is called a splitting field for $\mathfrak{G}$.

2.2 Lemma. *Let V be an irreducible $\mathsf{K}\mathfrak{G}$-module. Then the following statements are equivalent.*

a) *V is absolutely irreducible.*

b) *For any extension L of K, the $\mathsf{L}\mathfrak{G}$-module V_{L} is irreducible.*

c) *If $\hat{\mathsf{K}}$ is the algebraic closure of K, then $\mathsf{V}_{\hat{\mathsf{K}}}$ is irreducible.*

Proof. a) ⇒ b): By 1.8, V_L is completely reducible and by 1.12,

$$\mathrm{Hom}_{L\mathfrak{G}}(V_L, V_L) \cong \mathrm{Hom}_{K\mathfrak{G}}(V, V) \otimes_K L = K \otimes_K L \cong L.$$

Hence 0 and 1 are the only idempotents in $\mathrm{Hom}_{L\mathfrak{G}}(V_L, V_L)$, and the completely reducible $L\mathfrak{G}$-module V_L is therefore irreducible.

b) ⇒ c): This is trivial.

c) ⇒ a): $V_{\hat{K}}$ is irreducible by assumption, so by Schur's lemma and 1.12,

$$\hat{K} = \mathrm{Hom}_{\hat{K}\mathfrak{G}}(V_{\hat{K}}, V_{\hat{K}}) \cong \mathrm{Hom}_{K\mathfrak{G}}(V, V) \otimes_K \hat{K}.$$

This shows that

$$\dim_K \mathrm{Hom}_{K\mathfrak{G}}(V, V) = \dim_{\hat{K}} \mathrm{Hom}_{\hat{K}\mathfrak{G}}(V_{\hat{K}}, V_{\hat{K}}) = 1.$$

Hence $\mathrm{Hom}_{K\mathfrak{G}}(V, V) = K$ and V is absolutely irreducible. **q.e.d.**

The two concepts introduced in 2.1 are closely related.

2.3 Theorem. *The following assertions are equivalent.*

a) *Every irreducible* $K\mathfrak{G}$*-module is absolutely irreducible.*

b) K *is a splitting field for* $K\mathfrak{G}$.

Proof. Let $V_1, \ldots, V_k$ be all the irreducible $K\mathfrak{G}$-modules (to within isomorphism). As $V_i\mathbf{J}(K\mathfrak{G}) = 0$, the V_i are all the irreducible $K\mathfrak{G}/\mathbf{J}(K\mathfrak{G})$-modules. By Wedderburn's theorem,

$$K\mathfrak{G}/\mathbf{J}(K\mathfrak{G}) \cong \bigoplus_{i=1}^{k} (\mathfrak{C}_i)_{n_i},$$

where the division algebra $\mathfrak{C}_i$ is antiisomorphic to

$$\mathrm{Hom}_{K\mathfrak{G}}(V_i, V_i) = \mathrm{Hom}_{K\mathfrak{G}/\mathbf{J}(K\mathfrak{G})}(V_i, V_i).$$

By 2.1, the absolute irreducibility of V_i means that $\mathrm{Hom}_{K\mathfrak{G}}(V_i, V_i) = K$. As the Wedderburn decomposition is unique, the assertion follows. **q.e.d.**

2.4 Theorem. a) *Let* K *be a splitting field for* $K\mathfrak{G}$ *and* $V_1, \ldots, V_k$ *all the irreducible* $K\mathfrak{G}$*-modules (to within isomorphism). Then for any extension*

L *of* K, *the* $\mathsf{L}\mathfrak{G}$-*modules* $(\mathsf{V}_i)_\mathsf{L}$ $(i = 1, \ldots, k)$ *are all the irreducible* $\mathsf{L}\mathfrak{G}$-*modules (to within isomorphism).*

b) *Let* L *be an algebraically closed extension of* K *and* $\mathsf{V}_1', \ldots, \mathsf{V}_k'$ *all the irreducible* $\mathsf{L}\mathfrak{G}$-*modules (to within isomorphism). Suppose that for every i there exists a* $\mathsf{K}\mathfrak{G}$-*module* V_i *(necessarily irreducible) such that* $\mathsf{V}_i' \cong (\mathsf{V}_i)_\mathsf{L}$. *Then* K *is a splitting field for* $\mathsf{K}\mathfrak{G}$ *and* $\mathsf{V}_1, \ldots, \mathsf{V}_k$ *are all the irreducible* $\mathsf{K}\mathfrak{G}$-*modules (to within isomorphism).*

Proof. a) As K is a splitting field for $\mathsf{K}\mathfrak{G}$, we have

$$\mathsf{K}\mathfrak{G}/\mathbf{J}(\mathsf{K}\mathfrak{G}) \cong \bigoplus_{i=1}^{k} (\mathsf{K})_{n_i}$$

for some n_i. By 2.3, V_i is absolutely irreducible, hence $(\mathsf{V}_i)_\mathsf{L}$ is irreducible by 2.2. For $i \neq j$, by 1.12,

$$\mathrm{Hom}_{\mathsf{L}\mathfrak{G}}((\mathsf{V}_i)_\mathsf{L}, (\mathsf{V}_j)_\mathsf{L}) \cong \mathrm{Hom}_{\mathsf{K}\mathfrak{G}}(\mathsf{V}_i, \mathsf{V}_j) \otimes_\mathsf{K} \mathsf{L} = 0.$$

This shows that $(\mathsf{V}_i)_\mathsf{L} \ncong (\mathsf{V}_j)_\mathsf{L}$ for $i \neq j$. We also have

$$\begin{aligned} \mathsf{L}\mathfrak{G}/\mathbf{J}(\mathsf{L}\mathfrak{G}) &\cong (\mathsf{K}\mathfrak{G} \otimes_\mathsf{K} \mathsf{L})/\mathbf{J}(\mathsf{K}\mathfrak{G} \otimes_\mathsf{K} \mathsf{L}) \\ &= (\mathsf{K}\mathfrak{G} \otimes_\mathsf{K} \mathsf{L})/(\mathbf{J}(\mathsf{K}\mathfrak{G}) \otimes_\mathsf{K} \mathsf{L}) \quad \text{(by 1.5a))} \\ &\cong \mathsf{K}\mathfrak{G}/\mathbf{J}(\mathsf{K}\mathfrak{G}) \otimes_\mathsf{K} \mathsf{L} \quad \text{(by 1.2c))} \\ &\cong \bigoplus_{i=1}^{k} ((\mathsf{K})_{n_i} \otimes_\mathsf{K} \mathsf{L}) \cong \bigoplus_{i=1}^{k} (\mathsf{L})_{n_i} \quad \text{(by 1.2a), b)).} \end{aligned}$$

By Wedderburn's theorem, $\mathsf{L}\mathfrak{G}/\mathbf{J}(\mathsf{L}\mathfrak{G})$ and $\mathsf{L}\mathfrak{G}$ have exactly k irreducible modules (to within isomorphism), and these must be $(\mathsf{V}_1)_\mathsf{L}, \ldots, (\mathsf{V}_k)_\mathsf{L}$.

b) Let W be any irreducible $\mathsf{K}\mathfrak{G}$-module. Then W_L contains an irreducible submodule, and this is isomorphic to some V_i'. Hence by 1.12,

$$0 \neq \mathrm{Hom}_{\mathsf{L}\mathfrak{G}}(\mathsf{V}_i', \mathsf{W}_\mathsf{L}) \cong \mathrm{Hom}_{\mathsf{K}\mathfrak{G}}(\mathsf{V}_i, \mathsf{W}) \otimes_\mathsf{K} \mathsf{L}.$$

As V_i and W are irreducible $\mathsf{K}\mathfrak{G}$-modules, we deduce from $\mathrm{Hom}_{\mathsf{K}\mathfrak{G}}(\mathsf{V}_i, \mathsf{W}) \neq 0$ that $\mathsf{W} \cong \mathsf{V}_i$.

We show that all the V_i are absolutely irreducible. Since L is algebraically closed, L contains a subfield L_0 isomorphic to the algebraic closure $\hat{\mathsf{K}}$ of K. From

$$V_i' \cong V_i \otimes_K L \cong (V_i \otimes_K L_0) \otimes_{L_0} L,$$

we see that $V_i \otimes_K L_0$ is irreducible. Hence V_i is absolutely irreducible by 2.2, and by 2.3, K is a splitting field for $K\mathfrak{G}$. **q.e.d.**

2.5 Lemma. *Let K be any field, L an algebraically closed extension of K and $\mathfrak{G}$ a finite group. Then there exists a splitting field F for $\mathfrak{G}$ such that $K \subseteq F \subseteq L$ and $F : K$ is a Galois extension.*

Proof. Let L_0 be the algebraic closure of the prime field P of L in L. As L_0 is algebraically closed, L_0 is a splitting field for $\mathfrak{G}$. Let $V_1', \ldots, V_k'$ be all the irreducible $L_0\mathfrak{G}$-modules to within isomorphism. Let $\{v_{i1}, \ldots, v_{i,n_i}\}$ be an L_0-basis for V_i' and

$$v_{ij}g = \sum_{l=1}^{n_i} a_{jl}^i(g)v_{il} \quad (g \in \mathfrak{G},\ a_{jl}^i(g) \in L_0).$$

Let M be the subfield of L generated over the prime field P of L by all the finitely many $a_{jl}^i(g)$ and their algebraic conjugates over P. As every $a_{jl}^i(g)$ is algebraic over P, we have $(M : P) < \infty$.

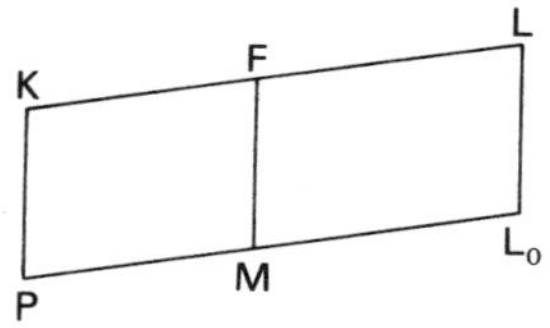

If char $K = 0$, then $M : P$ is trivially a separable extension. If char $K > 0$, then P and M are finite, so again $M : P$ is separable. Thus $M : P$ is a Galois extension. We put $F = K(M)$. From Galois theory we know that $F : K$ is also a Galois extension and $K \subseteq F \subseteq L$.

We still have to show that F is a splitting field for $\mathfrak{G}$. It suffices to show that M is a splitting field for $\mathfrak{G}$. We define an $M\mathfrak{G}$-module V_i by

$$V_i = \bigoplus_{j=1}^{n_i} Mv_{ij}$$

and

$$v_{ij}g = \sum_{l=1}^{n_i} a_{jl}^i(g)v_{il} \quad (g \in \mathfrak{G}).$$

Then $(V_i)_{L_0} \cong V_i'$. Thus by 2.4b), M is a splitting field for $\mathfrak{G}$. **q.e.d.**

The determination of splitting fields for $\mathfrak{G}$ is a very delicate problem in characteristic 0. In V, 19.11 we proved that if n is the exponent of $\mathfrak{G}$, the field of n-th roots of unity over the rational field $\mathbb{Q}$ is always a splitting field for $\mathfrak{G}$. But also smaller fields may suffice, and in general there is no uniquely determined smallest splitting field.

We now show that for characteristic p the answer to this question is much simpler. This is due basically to the fact that every finite division ring is commutative.

2.6 Theorem (R. BRAUER). a) *Let* L *be an algebraically closed field of characteristic p and $\phi_1, \ldots, \phi_k$ the characters of all the irreducible* L$\mathfrak{G}$-*modules. Then*

$$K = GF(p)(\phi_i(g) | g \in \mathfrak{G}, i = 1, \ldots, k)$$

is the unique smallest splitting field for $\mathfrak{G}$ of characteristic p (to within isomorphism).

b) *Suppose that $\mathfrak{G}$ is of exponent $p^b m$, where $(p, m) = 1$. Let f be the smallest positive integer such that $p^f \equiv 1(m)$. Then $GF(p^f)$ is a splitting field for $\mathfrak{G}$ and all subgroups of $\mathfrak{G}$.*

Proof. a) By 2.5, there exists a splitting field F for $\mathfrak{G}$ such that $F : GF(p)$ is a Galois extension and $F \subseteq L$. Let V_i $(i = 1, \ldots, k)$ be the types of irreducible F$\mathfrak{G}$-modules. Then by 2.4a), the $(V_i)_L$ are the types of irreducible L$\mathfrak{G}$-modules. Hence the character of V_i is also the character of $(V_i)_L$, so by assumption has values in K. Thus by 1.17, there exist irreducible K$\mathfrak{G}$-modules V_i' such that $V_i \cong (V_i')_F$. Then

$$(V_i)_L \cong (V_i' \otimes_K F) \otimes_F L \cong V_i' \otimes_K L,$$

and so by 2.4b) K is a splitting field for $\mathfrak{G}$.

It is obvious that K is the smallest splitting field, for the values $\phi_i(g)$ of the irreducible characters must belong to every splitting field, by 2.4a).

b) Choose L as in a). The field $GF(p^f)$ contains all m-th roots of unity in L. Let ϱ_i be a representation of $\mathfrak{G}$ with character ϕ_i, and let $\varepsilon_{i1}(g), \ldots, \varepsilon_{in_i}(g)$ be the eigen-values of $\varrho_i(g)$ (in L). As

$$\varrho_i(g)^{p^b m} = \varrho_i(g^{p^b m}) = I$$

and char $L = p$,

$$0 = \varepsilon_{ij}(g)^{p^b m} - 1 = (\varepsilon_{ij}(g)^m - 1)^{p^b}.$$

Hence $\varepsilon_{ij}(g)^m = 1$, so $\varepsilon_{ij}(g) \in GF(p^f)$ and $\phi_i(g) \in GF(p^f)$. Now we apply a). **q.e.d.**

§ 3. The Number of Irreducible Modular Representations

Suppose first that K is an algebraically closed field the characteristic of which does not divide the order of the group $\mathfrak{G}$. The following statements about the irreducible representations of the group-ring $K\mathfrak{G}$ hold.

(1) If $h(\mathfrak{G})$ is the class-number of $\mathfrak{G}$, there are exactly $h(\mathfrak{G})$ irreducible representations of $K\mathfrak{G}$ (to within isomorphism) (V, 5.1).

(2) If the degrees of the irreducible representations of $K\mathfrak{G}$ are n_i $(i = 1, \ldots, h(\mathfrak{G}))$, then

$$|\mathfrak{G}| = \sum_{i=1}^{h(g)} n_i^2.$$

This relation follows from the assertion that the semisimple algebra $K\mathfrak{G}$ is the direct sum of complete matrix algebras of degree n_i over K (V, 5.1).

(3) In addition we have the assertion that the n_i are divisors of $|\mathfrak{G}|$ (V, 5.12).

What appears in place of these assertions when K is an algebraically closed field of arbitrary characteristic p? In place of (2) we obviously have

$$(2') \qquad \dim_K K\mathfrak{G}/\mathbf{J}(K\mathfrak{G}) = \sum_i n_i^2,$$

where the n_i are the K-dimensions of the irreducible $K\mathfrak{G}$-modules and $\mathbf{J}(K\mathfrak{G})$ is the Jacobson radical of $K\mathfrak{G}$. If p is a divisor of $|\mathfrak{G}|$, then by V, 2.7, $K\mathfrak{G}$ is not semisimple, but we know little about the dimension of $\mathbf{J}(K\mathfrak{G})$. Thus the formula (2′) is of little use.

Assertion (3) still holds if $\mathfrak{G}$ is p-soluble (see SWAN [1] and 9.21), but is false in general (see Example 3.10). It is conjectured that in all cases the highest power of p dividing n_i is a divisor of $|\mathfrak{G}|$.[1]

However a general form of (1) is known, which will be the subject of this section.

---

1 This conjecture is wrong. S. Norton showed recently that McLaughlin's simple group of order $2^7 \cdot 3^6 \cdot 5^3 \cdot 7 \cdot 11$ has in characteristic 2 an absolutely irreducible module of dimension $2^9 \cdot 7$.

3.1 Definition. If $\mathfrak{A}$ is an algebra over the field K, we denote by $\mathbf{Q}(\mathfrak{A})$ the K-subspace of $\mathfrak{A}$ spanned by all elements $ab - ba$ with $a, b \in \mathfrak{A}$.

3.2 Lemma. a) *If $\mathfrak{A}$ is a complete matrix algebra over K, then*

$$\mathbf{Q}(\mathfrak{A}) = \{A \mid A \in \mathfrak{A}, \operatorname{tr} A = 0\},$$

and $\dim_{\mathsf{K}} \mathfrak{A}/\mathbf{Q}(\mathfrak{A}) = 1$.

b) *If $\mathfrak{G}$ is a finite group, $\mathbf{Q}(\mathsf{K}\mathfrak{G})$ is spanned over K by all elements of the form $x - y$, where x, y are conjugate elements of $\mathfrak{G}$. If $\{g_1, \ldots, g_h\}$ is a complete set of representatives of the conjugacy classes of $\mathfrak{G}$, then $\{g_i + \mathbf{Q}(\mathsf{K}\mathfrak{G}) \mid i = 1, \ldots, h\}$ is a K-basis of $\mathsf{K}\mathfrak{G}/\mathbf{Q}(\mathsf{K}\mathfrak{G})$.*

c) *If $\mathfrak{A}_1, \ldots, \mathfrak{A}_m$ are ideals of $\mathfrak{A}$ and $\mathfrak{A} = \bigoplus_{i=1}^m \mathfrak{A}_i$, then*

$$\mathbf{Q}(\mathfrak{A}) = \bigoplus_{i=1}^{m} \mathbf{Q}(\mathfrak{A}_i).$$

Proof. a) Suppose that $\mathfrak{A}$ is the algebra of all $n \times n$ matrices with coefficients in K. We denote by X the K-subspace which consists of all such matrices of trace 0. Since $\operatorname{tr}(AB - BA) = 0$ for all A, B in $\mathfrak{A}$, we have $\mathbf{Q}(\mathfrak{A}) \subseteq \mathsf{X}$. Let $\{E_{ij} \mid i, j = 1, \ldots, n\}$ be the usual basis of $\mathfrak{A}$ for which $E_{ij}E_{kl} = \delta_{jk}E_{il}$. Then

$$E_{ij} = E_{i1}E_{1j} - E_{1j}E_{i1} \in \mathbf{Q}(\mathfrak{A})$$

for $i \neq j$ and

$$E_{ii} - E_{11} = E_{i1}E_{1i} - E_{1i}E_{i1} \in \mathbf{Q}(\mathfrak{A})$$

for $i = 1, \ldots, n$. Thus we obtain

$$n^2 - 1 \leq \dim_{\mathsf{K}} \mathbf{Q}(\mathfrak{A}) \leq \dim_{\mathsf{K}} \mathsf{X} = n^2 - 1,$$

and so $\mathbf{Q}(\mathfrak{A}) = \mathsf{X}$. Hence $\dim_{\mathsf{K}} \mathfrak{A}/\mathbf{Q}(\mathfrak{A}) = \dim_{\mathsf{K}} \mathfrak{A}/\mathsf{X} = 1$.

b) If x, y are conjugate elements of $\mathfrak{G}$, $x = g^{-1}yg$ for some $g \in \mathfrak{G}$ and

$$x - y = g^{-1}(yg) - (yg)g^{-1} \in \mathbf{Q}(\mathsf{K}\mathfrak{A}).$$

It is clear that $\mathbf{Q}(\mathsf{K}\mathfrak{G})$ is spanned over K by all $uv - vu$ with u, v in $\mathfrak{G}$; since however $vu = u^{-1}(uv)u$, it follows that $\mathbf{Q}(\mathsf{K}\mathfrak{G})$ is spanned by all $x - y$, where x, y are conjugate elements of $\mathfrak{G}$. Now the g_i and all $g - g_i$

for which g is conjugate to but distinct from g_i form a K-basis of $\mathsf{K}\mathfrak{G}$. Since $\mathbf{Q}(\mathsf{K}\mathfrak{G})$ is spanned by the $g - g_i$ for which g is conjugate to g_i, it follows that

$$\{g_i + \mathbf{Q}(\mathsf{K}\mathfrak{G}) | i = 1, \ldots, h\}$$

is a K-basis of $\mathsf{K}\mathfrak{G}/\mathbf{Q}(\mathsf{K}\mathfrak{G})$.

c) This is obvious. **q.e.d.**

It should be observed that statement (1) on page 32 is an easy consequence of Lemma 3.2. In the modular case, however, we need further preparations.

3.3 Lemma. *Suppose that* char $\mathsf{K} = p$ *and* $\mathfrak{A}$ *is an algebra over* K.

a) *If* $a \in \mathbf{Q}(\mathfrak{A})$, *then* $a^p \in \mathbf{Q}(\mathfrak{A})$.

b) *If* a, b *are elements of* $\mathfrak{A}$, *then*

$$(a + b)^{p^m} - a^{p^m} - b^{p^m} \in \mathbf{Q}(\mathfrak{A})$$

for all natural numbers m.

Proof. We first prove b) for $m = 1$. Clearly, $(a + b)^p - a^p - b^p$ is the sum of the $2^p - 2$ products $c_1 \cdots c_p$, where $c_i = a$ or $c_i = b$, but not all c_i are equal to a and not all c_i are equal to b. Now if $2 \leq i \leq p$,

$$c_i \cdots c_p c_1 \cdots c_{i-1} - c_1 \cdots c_p \in \mathbf{Q}(\mathfrak{A}).$$

Thus

$$\begin{aligned} c_1 \cdots c_p + c_2 \cdots c_p c_1 + \cdots + c_p c_1 \cdots c_{p-1} \\ \equiv p c_1 \cdots c_p \equiv 0 \mod \mathbf{Q}(\mathfrak{A}). \end{aligned}$$

Hence $(a + b)^p - a^p - b^p$ lies in $\mathbf{Q}(\mathfrak{A})$.

To prove a), we first observe that

$$(ab)^p - (ba)^p = a((ba)^{p-1}b) - ((ba)^{p-1}b)a \in \mathbf{Q}(\mathfrak{A}).$$

But by b) for $m = 1$,

$$(ab - ba)^p - (ab)^p + (ba)^p \in \mathbf{Q}(\mathfrak{A}).$$

Hence $(ab - ba)^p \in \mathbf{Q}(\mathfrak{A})$. It now follows from b) for $m = 1$ that $c^p \in \mathbf{Q}(\mathfrak{A})$ for all $c \in \mathbf{Q}(\mathfrak{A})$.

We prove b) by induction on m. The case $m = 1$ has already been settled. If $m > 1$, write

$$(a + b)^{p^{m-1}} = a^{p^{m-1}} + b^{p^{m-1}} + c;$$

thus $c \in \mathbf{Q}(\mathfrak{A})$ by the inductive hypothesis. By the case $m = 1$,

$$(a + b)^{p^m} \equiv a^{p^m} + b^{p^m} + c^p \mod \mathbf{Q}(\mathfrak{A}).$$

But $c^p \in \mathbf{Q}(\mathfrak{A})$ by a), so the assertion is clear. **q.e.d.**

3.4 Definition. Suppose that char $\mathsf{K} = p$ and that $\mathfrak{A}$ is an algebra over K. We define

$$\mathbf{T}(\mathfrak{A}) = \{a \mid a \in \mathfrak{A},\ a^{p^m} \in \mathbf{Q}(\mathfrak{A}) \quad \text{for all sufficiently large } m\}.$$

3.5 Lemma. $\mathbf{T}(\mathfrak{A})$ *is a* K*-subspace of* $\mathfrak{A}$ *containing* $\mathbf{Q}(\mathfrak{A})$ *and all the nilpotent elements of* $\mathfrak{A}$.

Proof. By 3.3a) and Definition 3.4, $\mathbf{Q}(\mathfrak{A}) \subseteq \mathbf{T}(\mathfrak{A})$. If $t \in \mathbf{T}(\mathfrak{A})$ and $c \in \mathsf{K}$, then for all sufficiently large m,

$$(ct)^{p^m} = c^{p^m} t^{p^m} \in \mathbf{Q}(\mathfrak{A}),$$

since $\mathbf{Q}(\mathfrak{A})$ is a K-subspace. If t_1, t_2 are in $\mathbf{T}(\mathfrak{A})$, then it follows from 3.3b) that for all sufficiently large m,

$$(t_1 + t_2)^{p^m} \equiv t_1^{p^m} + t_2^{p^m} \equiv 0 \mod \mathbf{Q}(\mathfrak{A}).$$

Thus $\mathbf{T}(\mathfrak{A})$ is a K-subspace of $\mathfrak{A}$.

If a is a nilpotent element of $\mathfrak{A}$, then $a^{p^m} = 0 \in \mathbf{Q}(\mathfrak{A})$ for some m, so $a \in \mathbf{T}(\mathfrak{A})$. **q.e.d.**

3.6 Lemma. a) *If* $\mathfrak{A}$ *is a complete matrix algebra over* K, *then*

$$\mathbf{T}(\mathfrak{A}) = \mathbf{Q}(\mathfrak{A}) = \{A \mid A \in \mathfrak{A},\ \operatorname{tr} A = 0\},$$

and $\dim_{\mathsf{K}} \mathfrak{A}/\mathbf{T}(\mathfrak{A}) = 1$.

b) *Suppose that* char $\mathsf{K} = p$, $\mathfrak{G}$ *is a finite group and that* $\{g_1, \ldots, g_k\}$ *is a complete set of representatives of those conjugacy classes of* $\mathfrak{G}$ *which consist of elements of order prime to p (the p′-elements). Then* $\{g_i + \mathbf{T}(\mathsf{K}\mathfrak{G}) \mid i = 1, \ldots, k\}$ *is a* K*-basis of* $\mathsf{K}\mathfrak{G}/\mathbf{T}(\mathsf{K}\mathfrak{G})$.

c) *If* $\mathfrak{A}_1, \ldots, \mathfrak{A}_m$ *are ideals in* $\mathfrak{A}$ *and* $\mathfrak{A} = \bigoplus_{i=1}^m \mathfrak{A}_i$, *then* $\mathbf{T}(\mathfrak{A}) = \bigoplus_{i=1}^m \mathbf{T}(\mathfrak{A}_i)$.

Proof. a) $\mathfrak{A}$ contains an element E_{11} for which $E_{11}^2 = E_{11}$ and $\operatorname{tr} E_{11} = 1$. By 3.2a), $E_{11} \notin \mathbf{Q}(\mathfrak{A})$. For all m we have

$$E_{11}^{p^m} = E_{11} \notin \mathbf{Q}(\mathfrak{A}),$$

so $E_{11} \notin \mathbf{T}(\mathfrak{A})$. Since $\mathbf{T}(\mathfrak{A}) \supseteq \mathbf{Q}(\mathfrak{A})$ and $\dim_{\mathsf{K}} \mathfrak{A}/\mathbf{Q}(\mathfrak{A}) = 1$, it follows that $\mathbf{T}(\mathfrak{A}) = \mathbf{Q}(\mathfrak{A})$.

b) Suppose that p^a is the highest power of p which divides $|\mathfrak{G}|$. For $g \in \mathfrak{G}$, write $g = g_p g_{p'} = g_{p'} g_p$, where g_p is a p-element and $g_{p'}$ a p'-element. Since g and $g_{p'}$ commute, we see that for all $n \geq 0$,

$$\begin{aligned}(g - g_{p'})^{p^{a+n}} &= (g_p g_{p'})^{p^{a+n}} - (g_{p'})^{p^{a+n}} \\ &= (g_{p'})^{p^{a+n}} - (g_{p'})^{p^{a+n}} = 0.\end{aligned}$$

Thus $g - g_{p'} \in \mathbf{T}(\mathsf{K}\mathfrak{G})$. Now $g_{p'}$ is conjugate to g_i for some i, and $g_{p'} - g_i \in \mathbf{Q}(\mathsf{K}\mathfrak{G})$ by 3.2b). Thus $g - g_i \in \mathbf{T}(\mathsf{K}\mathfrak{G})$. Hence the $g_i + \mathbf{T}(\mathsf{K}\mathfrak{G})$ span $\mathsf{K}\mathfrak{G}/\mathbf{T}(\mathsf{K}\mathfrak{G})$. To show that they are linearly independent, suppose that

$$r = \sum_{i=1}^{k} c_i g_i \in \mathbf{T}(\mathsf{K}\mathfrak{G})$$

with $c_i \in \mathsf{K}$. If $|\mathfrak{G}| = p^a q$, choose m so that $p^m \equiv 1(q)$ and

$$r^{p^m} \in \mathbf{Q}(\mathsf{K}\mathfrak{G}).$$

By 3.3b),

$$0 \equiv \left(\sum_{i=1}^{k} c_i g_i\right)^{p^m} \equiv \sum_{i=1}^{k} c_i^{p^m} g_i^{p^m} \equiv \sum_{i=1}^{k} c_i^{p^m} g_i \mod \mathbf{Q}(\mathsf{K}\mathfrak{G}).$$

By 3.2b), $c_i = 0$ $(i = 1, \ldots, k)$.

c) This follows at once from Definition 3.4 and 3.2c). **q.e.d.**

3.7 Lemma. *If* char $\mathsf{K} = p$ *and* $\mathfrak{A}$ *is an algebra of finite dimension over* K, *then*

$$\mathbf{T}(\mathfrak{A}/\mathbf{J}(\mathfrak{A})) = \mathbf{T}(\mathfrak{A})/\mathbf{J}(\mathfrak{A}).$$

Proof. By V, 2.4, $\mathbf{J}(\mathfrak{A})$ is nilpotent. Hence $\mathbf{J}(\mathfrak{A}) \subseteq \mathbf{T}(\mathfrak{A})$ by 3.5. Trivially,

$$\mathbf{Q}(\mathfrak{A}/\mathbf{J}(\mathfrak{A})) = (\mathbf{Q}(\mathfrak{A}) + \mathbf{J}(\mathfrak{A}))/\mathbf{J}(\mathfrak{A}).$$

It follows at once that

$$\mathbf{T}(\mathfrak{A})/\mathbf{J}(\mathfrak{A}) \subseteq \mathbf{T}(\mathfrak{A}/\mathbf{J}(\mathfrak{A})).$$

Suppose that $r + \mathbf{J}(\mathfrak{A}) \in \mathbf{T}(\mathfrak{A}/\mathbf{J}(\mathfrak{A}))$. Then for all sufficiently large m,

$$r^{p^m} + \mathbf{J}(\mathfrak{A}) = (r + \mathbf{J}(\mathfrak{A}))^{p^m} \in (\mathbf{Q}(\mathfrak{A}) + \mathbf{J}(\mathfrak{A}))/\mathbf{J}(\mathfrak{A}).$$

Since $\mathbf{Q}(\mathfrak{A}) \subseteq \mathbf{T}(\mathfrak{A})$ and $\mathbf{J}(\mathfrak{A}) \subseteq \mathbf{T}(\mathfrak{A})$, it follows that $r^{p^m} \in \mathbf{T}(\mathfrak{A})$. From the definition of $\mathbf{T}(\mathfrak{A})$, it then follows that $r^{p^n} \in \mathbf{Q}(\mathfrak{A})$ for all sufficiently large n, and so $r \in \mathbf{T}(\mathfrak{A})$. Hence $\mathbf{T}(\mathfrak{A}/\mathbf{J}(\mathfrak{A})) = \mathbf{T}(\mathfrak{A})/\mathbf{J}(\mathfrak{A})$. **q.e.d.**

3.8 Theorem (R. BRAUER [6]). *Let* K *be a field of characteristic* p *and* $\mathfrak{A}$ *a finite dimensional* K*-algebra. Suppose that* $\mathfrak{A}/\mathbf{J}(\mathfrak{A})$ *is a direct sum of complete matrix algebras over* K*. Then* $\dim_{\mathsf{K}} \mathfrak{A}/\mathbf{T}(\mathfrak{A})$ *is the number of isomorphism types of irreducible* $\mathfrak{A}$*-modules.*

Proof. Suppose that

$$\mathfrak{A}/\mathbf{J}(\mathfrak{A}) \cong \mathfrak{A}_1 \oplus \cdots \oplus \mathfrak{A}_m$$

where $\mathfrak{A}_1, \ldots, \mathfrak{A}_m$ are complete matrix algebras over K. Thus m is the number of isomorphism types of irreducible $\mathfrak{A}$-modules. By 3.7 and 3.6c),

$$\mathbf{T}(\mathfrak{A})/\mathbf{J}(\mathfrak{A}) = \mathbf{T}(\mathfrak{A}/\mathbf{J}(\mathfrak{A})) \cong \mathbf{T}(\mathfrak{A}_1) \oplus \cdots \oplus \mathbf{T}(\mathfrak{A}_m).$$

Hence

$$\begin{aligned}\dim_{\mathsf{K}} \mathfrak{A}/\mathbf{T}(\mathfrak{A}) &= \dim_{\mathsf{K}}(\mathfrak{A}/\mathbf{J}(\mathfrak{A}))/(\mathbf{T}(\mathfrak{A})/\mathbf{J}(\mathfrak{A})) \\ &= \sum_{i=1}^{m} \dim_{\mathsf{K}} \mathfrak{A}_i/\mathbf{T}(\mathfrak{A}_i).\end{aligned}$$

By 3.6a), $\dim_{\mathsf{K}} \mathfrak{A}_i/\mathbf{T}(\mathfrak{A}_i) = 1$. Thus $\dim_{\mathsf{K}} \mathfrak{A}/\mathbf{T}(\mathfrak{A}) = m$. **q.e.d.**

3.9 Theorem (R. BRAUER [2]). *Let* K *be a splitting field of characteristic* p *for* $\mathfrak{G}$*. Then the number of isomorphism types of irreducible* $\mathsf{K}\mathfrak{G}$*-modules is equal to the number of conjugacy classes of* $\mathfrak{G}$ *which consist of* p'*-elements.*

Proof. This follows at once from 3.8 and 3.6b). **q.e.d.**

3.10 Example (R. BRAUER, NESBITT [1]). We determine the irreducible representations of $\mathfrak{G} = SL(2, p)$ over an algebraically closed field K of characteristic p.

a) First we prove that $SL(2, p)$ has exactly p conjugacy classes which consist of p'-elements. Since conjugate elements have the same trace, this may be done by showing that p'-elements of $SL(2, p)$ with the same trace are conjugate and that every element of $GF(p)$ is the trace of some p'-element in $SL(2, p)$.

Certainly there exist p'-elements in $SL(2, p)$ with trace 2 and -2, namely I and $-I$. If $s \in GF(p)$ and $s \neq \pm 2$, put

$$A(s) = \begin{pmatrix} 0 & 1 \\ -1 & s \end{pmatrix}.$$

Thus $A(s) \in SL(2, p)$ and $\operatorname{tr} A(s) = s$. We show that $A(s)$ is a p'-element. Otherwise the p-component $A(s)_p$ of $A(s)$ is conjugate to an element

$$C = \begin{pmatrix} 1 & a \\ 0 & 1 \end{pmatrix}$$

for some $a \neq 0$, and $A(s)$ is conjugate to an element in the centralizer of C. But the centralizer of C in $SL(2, p)$ consists of elements of the form

$$\pm \begin{pmatrix} 1 & c \\ 0 & 1 \end{pmatrix},$$

and all these elements have trace ± 2. Thus $A(s)$ is a p'-element.

Let A be a p'-element of $SL(2, p)$ of trace s. Then

$$f(t) = t^2 - st + 1$$

is the characteristic polynomial of A. If $s = \pm 2, f(t) = (t \pm 1)^2$ and A thus has the eigen-values 1, 1 or $-1, -1$. Hence A is conjugate in $GL(2, p)$ to a matrix of the form $\pm C$ (C as above), and $A^{2p} = I$. Since A is a p'-element, it follows easily that $A = \pm I$. If $s \neq \pm 2$, then A is not a scalar multiple of I, so there is a vector v_1 such that v_1 and $v_2 = v_1 A$ are linearly independent. By the Cayley-Hamilton theorem,

$$0 = f(A) = A^2 - sA + I.$$

Thus

$$v_2 A = v_1 A^2 = -v_1 + s v_2.$$

Hence A is conjugate in $GL(2, p)$ to $A(s)$. We have to show that A is conjugate even in $SL(2, p)$ to $A(s)$. Suppose that

$$X^{-1} A X = A(s)$$

with $X \in GL(2, p)$. If we can find $Y \in \mathbf{C}_{GL(2,p)}(A(s))$ such that $\det XY = 1$, then

$$(XY)^{-1} A(XY) = Y^{-1} A(s) Y = A(s)$$

and $XY \in SL(2, p)$. It thus suffices to show that the determinants of the elements of $\mathbf{C}_{GL(2,p)}(A(s))$ take all values in $GF(p)^{\times}$. A simple calculation shows that

$$\mathbf{C}_{GL(2,p)}(A(s)) = \left\{ \begin{pmatrix} x & y \\ -y & x + sy \end{pmatrix} \middle| \, x, y \in GF(p), x^2 + sxy + y^2 \neq 0 \right\}.$$

Thus we have to show that the quadratic form $x^2 + sxy + y^2$ represents all values in $GF(p)$. This is trivial for $p = 2$. If p is odd, then

$$x^2 + sxy + y^2 = (x + \tfrac{1}{2} sy)^2 + (1 - \tfrac{1}{4} s^2) y^2.$$

If $1 - \frac{1}{4}s^2$ is not a square, it is trivial that this form represents all values in $GF(p)$. Otherwise, $1 - \frac{1}{4}s^2$ is a non-zero square since $s \neq \pm 2$; thus the form is equivalent to $x^2 + y^2$ and represents all values, by II, 10.6.

b) We now construct irreducible $\mathsf{K}\mathfrak{G}$-modules V_m $(0 \leq m \leq p - 1)$ for which $\dim_{\mathsf{K}} \mathsf{V}_m = m + 1$ (cf. V, 5.13). It then follows from a) and 3.9 that these V_m are all the irreducible $\mathsf{K}\mathfrak{G}$-modules to within isomorphism.

Let V_m be the K-vector space of homogeneous polynomials of degree m in the independent variables x and y. If $A = (a_{ij}) \in SL(2, p)$, we put

$$(x^i y^{m-i}) A = (a_{11} x + a_{12} y)^i (a_{21} x + a_{22} y)^{m-i}.$$

Then V_m obviously becomes a $\mathsf{K}\mathfrak{G}$-module of K-dimension $m + 1$. We now show that V_m is irreducible for $0 \leq m \leq p - 1$. Let W be a non-zero submodule of V_m and suppose that

$$0 \neq f = \sum_{j=0}^{n} a_j x^j y^{m-j} \in \mathsf{W},$$

where $a_n \neq 0$ and $n \leq m$. For $t \in GF(p)$, we put

$$S(t) = \begin{pmatrix} 1 & t \\ 0 & 1 \end{pmatrix} \quad \text{and} \quad T(t) = \begin{pmatrix} 1 & 0 \\ t & 1 \end{pmatrix}.$$

Then

$$fS(t) = \sum_{j=0}^{n} a_j(x + ty)^j y^{m-j} = \sum_{j=0}^{n} f_j(x, y)t^j \in \mathsf{W}$$

for suitable polynomials f_j, where $f_0 = f$ and $f_n = a_n y^m$. But

$$\sum_{t=1}^{p-1} t^{-i}(fS(t)) \in \mathsf{W}.$$

Since in $GF(p)$

$$\sum_{t=1}^{p-1} t^i = \begin{cases} -1 & \text{if } p-1 \text{ divides } i, \\ 0 & \text{otherwise,} \end{cases}$$

we obtain

$$\sum_{t=1}^{p-1} t^{-i}(fS(t)) = \sum_{t=1}^{p-1} t^{-i} \sum_{j=0}^{n} f_j t^j = \sum_{j=0}^{n} f_j \sum_{t=1}^{p-1} t^{j-i} = -\sum_{p-1 \mid j-i} f_j$$

$$= \begin{cases} -f_i & \text{for } 1 \leq i \leq p-2, 1 \leq i \leq n, \\ -f_0 - f_{p-1} & \text{for } i = 0, p-1 = n = m. \end{cases}$$

Since $f_0 = f \in \mathsf{W}$, it follows that $f_i \in \mathsf{W}$ for all $i = 1, \ldots, n$. In particular, since $f_n = a_n y^m$, we have $y^m \in \mathsf{W}$. A further application of the same method shows that W also contains the element

$$\sum_{t=1}^{p-1} t^{-j}(y^m T(t)) = \sum_{t=1}^{p-1} t^{-j}(tx + y)^m$$

$$= \begin{cases} \binom{m}{j} x^j y^{m-j} & \text{for } 0 \leq j \leq m < p-1 \\ & \text{and } 0 < j < p-1 = m, \\ x^{p-1} + y^{p-1} & \text{for } m = p-1, j = 0. \end{cases}$$

As $y^{p-1} \in \mathsf{W}$ has already been shown in the case $m = p - 1$, it follows that $x^j y^{m-j} \in \mathsf{W}$ for all j such that $0 \leq j \leq m$. Thus $\mathsf{W} = \mathsf{V}_m$ and V_m is irreducible for $0 \leq m \leq p - 1$.

c) Now (but not earlier), we can also determine $\dim_{\mathsf{K}} \mathbf{J}(\mathsf{K}\mathfrak{G})$. From what we have proved it follows that

$$\dim_{\mathsf{K}} \mathbf{J}(\mathsf{K}\mathfrak{G}) = \dim_{\mathsf{K}} \mathsf{K}\mathfrak{G} - \sum_{i=1}^{p} i^2 = p(p^2 - 1) - \frac{p}{6}(p + 1)(2p + 1)$$

$$= \frac{p(p + 1)(4p - 7)}{6}.$$

We thus see that $\dim_{\mathsf{K}} \mathbf{J}(\mathsf{K}\mathfrak{G})$ accounts for a quite substantial part of the dimension of $\mathsf{K}\mathfrak{G}$.

d) It is easy to find cases where the dimension of an irreducible module does not divide the order of the group. Namely $SL(2, 7)$ has an irreducible module of dimension 5, but 5 does not divide the order of $SL(2, 7)$.

3.11 Theorem (BERMAN [1]). *Let* K *be any field. Suppose that* $\mathfrak{G}$ *is a finite group and define* $\mathfrak{X}$ *to be (i) the whole of* $\mathfrak{G}$ *if* char $\mathsf{K} = 0$, *(ii) the set of* p'*-elements of* $\mathfrak{G}$ *if* char K *is a prime p. Let m be (i)* $|\mathfrak{G}|$ *if* char $\mathsf{K} = 0$, *(ii) the greatest* p'*-divisor of* $|\mathfrak{G}|$ *if* char $\mathsf{K} = p$. *Let* $\mathsf{L} = \mathsf{K}(\xi)$ *for some primitive m-th root of unity* ξ. *Let A be the set of integers a for which there exists a* K*-automorphism* α *of* L *such that* $\xi\alpha = \xi^a$. *Then there is an equivalence relation* $\sim$ *on* $\mathfrak{X}$ *in which* $x \sim y$ *if and only if y is conjugate to* x^a *for some* $a \in A$, *and the number of inequivalent irreducible representations of* $\mathfrak{G}$ *in* K *is the number of equivalence classes under* $\sim$.

Proof. If $a \in A$, then $(a, m) = 1$, so the mapping $x \to x^a$ is a bijective mapping of $\mathfrak{X}$ onto $\mathfrak{X}$. It follows easily that $\sim$ is an equivalence relation. Let s denote the number of equivalence classes. Thus s is the dimension of the K-space M of mappings f of $\mathfrak{X}$ into K having the property that $f(x) = f(y)$ whenever y is conjugate to x^a for some $a \in A$. Now if $\phi_1, \ldots, \phi_r$ are the characters of all the inequivalent irreducible representations of $\mathfrak{G}$ in K, then $\phi_1, \ldots, \phi_r$ are K-linearly independent on $\mathfrak{X}$; this follows from 1.11 if char $\mathsf{K} = p$ and from V, 5.8 if char $\mathsf{K} = 0$. It is to be shown that $r = s$, and this will be accomplished if we show that $\phi_1, \ldots, \phi_r$ span M.

Let F be a splitting field for $\mathfrak{G}$ such that $\mathsf{F} : \mathsf{K}$ is a Galois extension (cf. 2.5) and $\mathsf{F} \supseteq \mathsf{L}$.

First, suppose that

$$\phi_i = \psi_1 + \psi_2 + \cdots$$

is the decomposition of ϕ_i into absolutely irreducible constituents in F. Thus if $x \in \mathfrak{X}$, each $\psi_j(x)$ is a sum of powers of ξ and $\psi_j(x^a)$ is the sum of the corresponding powers of ξ^a. Hence if α is a K-automorphism of F and $\xi\alpha = \xi^a$,

$$\psi_j(x^a) = \psi_j(x)\alpha.$$

Thus $\phi_i(x)\alpha = \phi_i(x^a)$. Since $\phi_i(x) \in \mathsf{K}$, it follows that $\phi_i(x) = \phi_i(x^a)$. Hence $\phi_i \in \mathsf{M}$.

Next suppose that χ is the character of an absolutely irreducible representation of $\mathfrak{G}$ in F. By 1.18, χ is a component of some ϕ_i and

$$\phi_i = r_i(\chi + \chi' + \cdots),$$

where $\chi, \chi', \ldots$ are conjugate under K-automorphisms of F. Since $\phi_1, \ldots, \phi_r$ are linearly independent, $\phi_i \neq 0$ and r_i is not divisible by char K. Hence $\sum_\alpha \chi\alpha$, where $\chi\alpha$ runs through all distinct algebraic conjugates of χ, is a K-multiple of ϕ_i.

It is to be proved that any element σ of M is a K-linear combination of $\phi_1, \ldots, \phi_r$. Then σ is a class-function on $\mathfrak{X}$ into F. By 1.11 or V, 5.8, the absolutely irreducible characters of $\mathfrak{G}$ are F-linearly independent on $\mathfrak{X}$, and by 3.9, their number is the number of classes of $\mathfrak{G}$ contained in $\mathfrak{X}$. Thus the space of class-functions on $\mathfrak{X}$ is spanned by the absolutely irreducible characters, and we can write

$$\sigma = \sum_i c_i\chi_i,$$

where $c_i \in \mathsf{F}$ and $\chi_1, \chi_2, \ldots$ are distinct absolutely irreducible characters of $\mathfrak{G}$.

Let α be an automorphism of F over K. Then $\xi\alpha = \xi^a$ for some $a \in A$. Hence for every $x \in \mathfrak{X}$, we obtain

$$\sum_i c_i(\chi_i\alpha)(x) = \sum_i c_i\chi_i(x^a) = \sigma(x^a) = \sigma(x) = \sum_i c_i\chi_i(x).$$

So $c_i = c_j$ whenever $\chi_i\alpha = \chi_j$. Hence σ is an F-linear combination of the $\sum_\alpha \chi_i\alpha$ (summed over the different algebraic conjugates of χ_i), and from above it follows that σ is an F-linear combination of the ϕ_i, say

$$\sigma = \sum_{i=1}^r d_i\phi_i$$

with $d_i \in \mathsf{F}$. Since, however, σ and the ϕ_i all have values in K, we obtain for every automorphism α of F over K

$$\sigma = \sum_{i=1}^{r} (d_i\alpha)\phi_i.$$

As the ϕ_i $(i = 1, \ldots, r)$ are linearly independent over K, there exist $x_j \in \mathfrak{X}$ $(j = 1, \ldots, r)$ such that the matrix $(\phi_i(x_j))$ $(i, j = 1, \ldots, r)$ is non-singular. Hence the d_i are uniquely determined by

$$\sigma(x_j) = \sum_{i=1}^{r} d_i\phi_i(x_j) \quad (j = 1, \ldots, r).$$

It follows that $d_i = d_i\alpha \in \mathsf{K}$. So σ is a K-linear combination of the ϕ_i, as required. **q.e.d.**

3.12 Theorem. a) *The number of isomorphism types of irreducible $\mathbb{Q}\mathfrak{G}$-modules is equal to the number of conjugacy classes of cyclic subgroups of $\mathfrak{G}$.*

b) *If r is the number of real conjugacy classes of $\mathfrak{G}$ and $2s$ is the number of non-real ones, then $r + s$ is the number of isomorphism types of irreducible $\mathbb{R}\mathfrak{G}$-modules.*

Proof. a) In the notation of 3.11, A consists of all integers prime to $|\mathfrak{G}|$, so $x \sim y$ if and only if $\langle x \rangle$ is conjugate to $\langle y \rangle$.

b) In this case $a \in A$ if and only if $a \equiv \pm 1$ $(|\mathfrak{G}|)$, so $x \sim y$ if and only if x is conjugate to $y^{\pm 1}$. The assertion follows since the class C is real if and only if $C = C^{-1}$. **q.e.d.**

Exercises

4) Let p be an odd prime, K an algebraically closed field of characteristic p and $\mathfrak{G} = GL(2, p)$.

a) Show as in 3.10a) that $\mathfrak{G}$ has exactly $p(p - 1)$ conjugacy classes of p'-elements.

b) Let $\mathfrak{G}$ operate on $\mathsf{V}_m = \bigoplus_{i=0}^{m} \mathsf{K}x^i y^{m-i}$ as in 3.10b) and let ρ_m denote the representation of $\mathfrak{G}$ on V_m. Define $\rho_{m,n}$ by

$$\rho_{m,n}(g) = (\det g)^n \rho_m(g)$$

for $g \in \mathfrak{G}$. Show that the $\rho_{m,n}$ with $0 \leq m \leq p-1$ and $0 \leq n \leq p-2$ are representatives of all the irreducible representations of $\mathfrak{G}$ over K. (To show that the representations $\rho_{m,n}$ are pairwise inequivalent, consider

$$g = \begin{pmatrix} 0 & 1 \\ -t & s \end{pmatrix},$$

where $\langle t \rangle = GF(p)^{\times}$, and show that tr $\rho_m(g) \neq 0$ for some choice of s.)

§ 4. Induced Modules

The principal tool for the construction of modules over the group-ring $\mathsf{K}\mathfrak{G}$ is the operation of forming the induced module, and we shall now discuss the formal properties of this. We shall prove the reciprocity theorems of Nakayama, which are generalizations of the Frobenius reciprocity theorem. We shall obtain two dual assertions which both reduce to the Frobenius reciprocity theorem in the case of completely reducible modules. For the sake of logical clarity, we shall temporarily use not only the induced modules but also the coinduced modules. With applications to group-rings over rings of p-adic integers in mind, we consider group-rings $\mathsf{A}\mathfrak{G}$ of finite groups $\mathfrak{G}$ over an arbitrary commutative ring A with identity.

4.1 Definition. Let $\mathfrak{U}$ be a subgroup of $\mathfrak{G}$.

a) If V is an $\mathsf{A}\mathfrak{U}$-module, we put

$$\mathsf{V}^{\mathfrak{G}} = \mathsf{V} \otimes_{\mathsf{A}\mathfrak{U}} \mathsf{A}\mathfrak{G}.$$

By defining $(v \otimes a)g = v \otimes ag$ for all $v \in \mathsf{V}$, $a \in \mathsf{A}\mathfrak{G}$ and $g \in \mathfrak{G}$, $\mathsf{V}^{\mathfrak{G}}$ becomes an $\mathsf{A}\mathfrak{G}$-module, the module *induced* from V.

b) If V_1, V_2 are $\mathsf{A}\mathfrak{U}$-modules and $\alpha \in \operatorname{Hom}_{\mathsf{A}\mathfrak{U}}(\mathsf{V}_1, \mathsf{V}_2)$, we put $\alpha^{\mathfrak{G}} = \alpha \otimes 1$. Thus $\alpha^{\mathfrak{G}}$ is a mapping of $\mathsf{V}_1^{\mathfrak{G}}$ into $\mathsf{V}_2^{\mathfrak{G}}$. In fact, $\alpha^{\mathfrak{G}}$ is an $\mathsf{A}\mathfrak{G}$-homomorphism of $\mathsf{V}_1^{\mathfrak{G}}$ into $\mathsf{V}_2^{\mathfrak{G}}$, for if $g \in \mathfrak{G}$,

$$\begin{aligned} ((v \otimes a)g)\alpha^{\mathfrak{G}} &= (v \otimes ag)(\alpha \otimes 1) = v\alpha \otimes ag = (v\alpha \otimes a)g \\ &= ((v \otimes a)\alpha^{\mathfrak{G}})g \end{aligned}$$

for all $v \in \mathsf{V}_1$, $a \in \mathsf{A}\mathfrak{G}$.

c) If W is an $\mathsf{A}\mathfrak{G}$-module, we denote by $\mathsf{W}_{\mathfrak{U}}$ the $\mathsf{A}\mathfrak{U}$-module obtained from W by restricting the domain of operators to $\mathsf{A}\mathfrak{U}$.

4.2 Theorem. *Suppose that $\mathfrak{U} \leq \mathfrak{G}$. Then $\cdot^{\mathfrak{G}}$ is a covariant exact functor from the category of $\mathsf{A}\mathfrak{U}$-modules into the category of $\mathsf{A}\mathfrak{G}$-modules. This means the following.*

a) *If $\mathsf{V}_1, \mathsf{V}_2, \mathsf{V}_3$ are $\mathsf{A}\mathfrak{U}$-modules and $\alpha \in \mathrm{Hom}_{\mathsf{A}\mathfrak{U}}(\mathsf{V}_1, \mathsf{V}_2)$, $\beta \in \mathrm{Hom}_{\mathsf{A}\mathfrak{U}}(\mathsf{V}_2, \mathsf{V}_3)$, then $(\alpha\beta)^{\mathfrak{G}} = \alpha^{\mathfrak{G}}\beta^{\mathfrak{G}}$.*

b) *If*

$$0 \to \mathsf{V}_1 \xrightarrow{\alpha} \mathsf{V}_2 \xrightarrow{\beta} \mathsf{V}_3 \to 0$$

is an exact sequence of $\mathsf{A}\mathfrak{U}$-modules, then

$$0 \to \mathsf{V}_1^{\mathfrak{G}} \xrightarrow{\alpha^{\mathfrak{G}}} \mathsf{V}_2^{\mathfrak{G}} \xrightarrow{\beta^{\mathfrak{G}}} \mathsf{V}_3^{\mathfrak{G}} \to 0$$

is an exact sequence of $\mathsf{A}\mathfrak{G}$-modules. If α is the inclusion mapping, this implies that

$$\mathsf{V}_2^{\mathfrak{G}}/\mathsf{V}_1^{\mathfrak{G}} \cong \mathsf{V}_3^{\mathfrak{G}} \cong (\mathsf{V}_2/\mathsf{V}_1)^{\mathfrak{G}}.$$

Proof. a) For $v_1 \in \mathsf{V}_1$ and $a \in \mathsf{A}\mathfrak{G}$, we have

$$\begin{aligned}(v_1 \otimes a)\alpha^{\mathfrak{G}}\beta^{\mathfrak{G}} &= ((v_1 \otimes a)(\alpha \otimes 1))(\beta \otimes 1) = v_1\alpha\beta \otimes a \\ &= (v_1 \otimes a)(\alpha\beta)^{\mathfrak{G}}.\end{aligned}$$

b) We have to show that

(1) $\alpha^{\mathfrak{G}}$ is a monomorphism;

(2) $\mathrm{im}\, \alpha^{\mathfrak{G}} = \ker \beta^{\mathfrak{G}}$;

(3) $\beta^{\mathfrak{G}}$ is an epimorphism.

Obviously (3) holds. To prove (1) and (2), we consider $\ker \gamma^{\mathfrak{G}}$ and $\mathrm{im}\, \gamma^{\mathfrak{G}}$ for $\gamma \in \mathrm{Hom}_{\mathsf{A}\mathfrak{U}}(\mathsf{W}_1, \mathsf{W}_2)$, where $\mathsf{W}_1, \mathsf{W}_2$ are any $\mathsf{A}\mathfrak{U}$-modules. Let T be a transversal of $\mathfrak{U}$ in $\mathfrak{G}$. Thus

$$\mathsf{A}\mathfrak{G} = \bigoplus_{t \in \mathsf{T}} (\mathsf{A}\mathfrak{U})t$$

and

$$\mathsf{W}_i^{\mathfrak{G}} = \bigoplus_{t \in \mathsf{T}} \mathsf{W}_i \otimes t.$$

Now if

$$\sum_{t \in \mathsf{T}} w_t \otimes t \in \ker \gamma^{\mathfrak{G}},$$

then

$$\sum_{t \in \mathsf{T}} w_t\gamma \otimes t = 0,$$

hence $w_t\gamma = 0$ and $w_t \in \ker \gamma$ for all $t \in \mathsf{T}$. Thus

$$\ker \gamma^{\mathfrak{G}} = \ker \gamma \otimes_{\mathsf{A}\mathfrak{U}} \mathsf{A}\mathfrak{G}.$$

Similarly,

$$\operatorname{im} \gamma^{\mathfrak{G}} = \operatorname{im} \gamma \otimes_{\mathsf{A}\mathfrak{U}} \mathsf{A}\mathfrak{G}.$$

Now we conclude that

$$\ker \alpha^{\mathfrak{G}} = \ker \alpha \otimes_{\mathsf{A}\mathfrak{U}} \mathsf{A}\mathfrak{G} = 0,$$

hence (1) holds; and

$$\operatorname{im} \alpha^{\mathfrak{G}} = \operatorname{im} \alpha \otimes_{\mathsf{A}\mathfrak{U}} \mathsf{A}\mathfrak{G} = \ker \beta \otimes_{\mathsf{A}\mathfrak{U}} \mathsf{A}\mathfrak{G} = \ker \beta^{\mathfrak{G}},$$

hence (2) holds. **q.e.d.**

The following theorem contains the most important formal properties of induced modules.

4.3 Theorem (D. G. HIGMAN [2]). *Suppose that $\mathfrak{U} \leq \mathfrak{G}$ and that T is a transversal of $\mathfrak{U}$ in $\mathfrak{G}$ (that is, $|\mathfrak{U}x \cap \mathsf{T}| = 1$ for all $x \in \mathfrak{G}$).*

a) *Suppose that V is an $\mathsf{A}\mathfrak{U}$-module and η is the mapping of V into $(\mathsf{V}^{\mathfrak{G}})_{\mathfrak{U}}$ defined by $v\eta = v \otimes 1$ (for $v \in \mathsf{V}$). Then η is an $\mathsf{A}\mathfrak{U}$-monomorphism of V onto a direct summand of $(\mathsf{V}^{\mathfrak{G}})_{\mathfrak{U}}$.*

b) *If W is an $\mathsf{A}\mathfrak{G}$-module and ε is the mapping of $(\mathsf{W}_{\mathfrak{U}})^{\mathfrak{G}}$ into W defined by*

$$\left(\sum_{t \in \mathsf{T}} w_t \otimes t\right)\varepsilon = \sum_{t \in \mathsf{T}} w_t t \quad (w_t \in \mathsf{W}),$$

then ε *is an* $A\mathfrak{G}$*-epimorphism of* $(W_{\mathfrak{U}})^{\mathfrak{G}}$ *onto* W *and*

$$((W_{\mathfrak{U}})^{\mathfrak{G}})_{\mathfrak{U}} = (\ker \varepsilon)_{\mathfrak{U}} \oplus W_{\mathfrak{U}}\eta.$$

c) *If* W *is an* $A\mathfrak{G}$*-module and* μ *is the mapping of* W *into* $(W_{\mathfrak{U}})^{\mathfrak{G}}$ *defined by*

$$w\mu = \sum_{t \in T} wt^{-1} \otimes t \quad (w \in W),$$

then μ *is an* $A\mathfrak{G}$*-monomorphism of* W *into* $(W_{\mathfrak{U}})^{\mathfrak{G}}$; μ *is independent of the choice of the transversal* T *and* $(W\mu)_{\mathfrak{U}}$ *is a direct summand of* $((W_{\mathfrak{U}})^{\mathfrak{G}})_{\mathfrak{U}}$. *Further*

$$w\mu\varepsilon = |\mathfrak{G} : \mathfrak{U}|w$$

for all $w \in W$. *If* $|\mathfrak{G} : \mathfrak{U}|$ *has an inverse in the ring* A, *then*

$$(W_{\mathfrak{U}})^{\mathfrak{G}} = W\mu \oplus \ker \varepsilon.$$

Proof. a) Obviously, η is an $A\mathfrak{U}$-monomorphism. We have

$$V^{\mathfrak{G}} = \bigoplus_{t \in T} V \otimes t = (V \otimes 1) \oplus V',$$

where

$$V' = \bigoplus_{t \in T, \notin \mathfrak{U}} V \otimes t.$$

It remains to show that V' is an $A\mathfrak{U}$-module. But if $t \notin \mathfrak{U}$ and $u \in \mathfrak{U}$ then $tu = u't'$, where $u' \in \mathfrak{U}$, $t' \in T$ and $t' \notin \mathfrak{U}$; thus

$$(v \otimes t)u = vu' \otimes t' \in V'$$

for all $v \in V$.

b) Suppose that $g \in \mathfrak{G}$. For each $t \in T$, write $tg = (tgt'^{-1})t'$, where $t' \in T$ and $tgt'^{-1} \in \mathfrak{U}$. As t runs through T, so does t'. Hence

$$\left(\sum_{t \in T} w_t \otimes t\right)g = \sum_{t' \in T} w_t(tgt'^{-1}) \otimes t'.$$

Applying ε, we obtain

$$\left(\left(\sum_{t\in\mathsf{T}} w_t \otimes t\right)g\right)\varepsilon = \sum_{t'\in\mathsf{T}} w_t t g t'^{-1} t' = \sum_{t\in\mathsf{T}} w_t t g = \left(\left(\sum_{t\in\mathsf{T}} w_t \otimes t\right)\varepsilon\right)g.$$

It follows that ε is an $A\mathfrak{G}$-epimorphism of $(\mathsf{W}_{\mathfrak{U}})^{\mathfrak{G}}$ onto W. Obviously

$$(\mathsf{W}_{\mathfrak{U}})\eta \cap \ker\varepsilon = 0.$$

Since

$$\sum_{t\in\mathsf{T}} w_t \otimes t - \left(\sum_{t\in\mathsf{T}} w_t t\right) \otimes 1 \in \ker\varepsilon,$$

it follows that

$$((\mathsf{W}_{\mathfrak{U}})^{\mathfrak{G}})_{\mathfrak{U}} = (\ker\varepsilon)_{\mathfrak{U}} \oplus \mathsf{W}_{\mathfrak{U}}\eta.$$

c) Suppose that T' is another transversal of $\mathfrak{U}$ in $\mathfrak{G}$. Then given $t \in \mathsf{T}$, we can write $t = u_t t'$ with $u_t \in \mathfrak{U}$ and $t' \in \mathsf{T}'$. Then

$$\sum_{t\in\mathsf{T}} wt^{-1} \otimes t = \sum_{t'\in\mathsf{T}'} wt'^{-1}u_t^{-1} \otimes u_t t' = \sum_{t'\in\mathsf{T}'} wt'^{-1} \otimes t'.$$

Thus μ is independent of the choice of the transversal.
It follows that for all $g \in \mathfrak{G}$,

$$(wg)\mu = \sum_{t\in\mathsf{T}} wgt^{-1} \otimes t = \sum_{t\in\mathsf{T}} wgt^{-1} \otimes (tg^{-1})g$$

$$= \left(\sum_{t\in\mathsf{T}} w(tg^{-1})^{-1} \otimes tg^{-1}\right)g = (w\mu)g.$$

Hence μ is an $A\mathfrak{G}$-homomorphism. Since

$$(\mathsf{W}_{\mathfrak{U}})^{\mathfrak{G}} = \bigoplus_{t\in\mathsf{T}} \mathsf{W}_{\mathfrak{U}} \otimes t,$$

μ is a monomorphism. If $\mathsf{W}' = \bigoplus_{t\in\mathsf{T},\, t\notin\mathfrak{U}} \mathsf{W}_{\mathfrak{U}} \otimes t$, then as in a), W' is an $A\mathfrak{U}$-submodule. Now W' is a complement of $(\mathsf{W}\mu)_{\mathfrak{U}}$. For on the one hand, if

$$w\mu = \sum_{t\in\mathsf{T}} wt^{-1} \otimes t \in \mathsf{W}\mu \cap \mathsf{W}',$$

then $w = 0$. And on the other hand, if $\mathsf{T} \cap \mathfrak{U} = \{t_1\}$, we have

$$\sum_{t \in \mathsf{T}} w_t \otimes t - \sum_{t \in \mathsf{T}} (w_{t_1} t_1) t^{-1} \otimes t \in \mathsf{W}'.$$

This shows that

$$\mathsf{W}' \oplus (\mathsf{W}\mu)_{\mathfrak{U}} = ((\mathsf{W}_{\mathfrak{U}})^{\mathfrak{G}})_{\mathfrak{U}}.$$

Finally

$$w\mu\varepsilon = \sum_{t \in \mathsf{T}} wt^{-1}t = |\mathfrak{G} : \mathfrak{U}| w.$$

If $|\mathfrak{G} : \mathfrak{U}|$ is a unit in A, write $\varepsilon' = |\mathfrak{G} : \mathfrak{U}|^{-1}\varepsilon$. Then $\mu\varepsilon' = 1$ and $(y - y\varepsilon'\mu)\varepsilon' = 0$ for all $y \in (\mathsf{W}_{\mathfrak{U}})^{\mathfrak{G}}$. Thus

$$(\mathsf{W}_{\mathfrak{U}})^{\mathfrak{G}} = \operatorname{im} \mu \oplus \ker \varepsilon' = \mathsf{W}\mu \oplus \ker \varepsilon.$$ **q.e.d.**

We shall now characterize $\mathsf{V}^{\mathfrak{G}}$ by a universal property.

4.4 Theorem. *Suppose that* $\mathfrak{U} \leq \mathfrak{G}$.

a) *Suppose that* V *is an* $\mathsf{A}\mathfrak{U}$*-module,* η *is the mapping in 4.3a) and* W *is an* $\mathsf{A}\mathfrak{G}$*-module. For each* $\alpha \in \operatorname{Hom}_{\mathsf{A}\mathfrak{U}}(\mathsf{V}, \mathsf{W}_{\mathfrak{U}})$, *there exists exactly one* $\alpha' \in \operatorname{Hom}_{\mathsf{A}\mathfrak{G}}(\mathsf{V}^{\mathfrak{G}}, \mathsf{W})$ *such that* $\alpha = \eta\alpha'$.

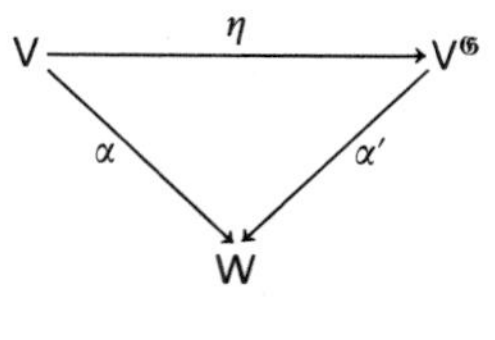

b) *Suppose that* V *is an* $\mathsf{A}\mathfrak{U}$*-module,* V' *is an* $\mathsf{A}\mathfrak{G}$*-module and* $\varrho \in \operatorname{Hom}_{\mathsf{A}\mathfrak{U}}(\mathsf{V}, \mathsf{V}'_{\mathfrak{U}})$. *Suppose that, given an* $\mathsf{A}\mathfrak{G}$*-module* W *and given* $\alpha \in \operatorname{Hom}_{\mathsf{A}\mathfrak{U}}(\mathsf{V}, \mathsf{W}_{\mathfrak{U}})$, *there always exists exactly one* $\alpha' \in \operatorname{Hom}_{\mathsf{A}\mathfrak{G}}(\mathsf{V}', \mathsf{W})$ *such that* $\varrho\alpha' = \alpha$. *Then* V' *and* $\mathsf{V}^{\mathfrak{G}}$ *are isomorphic* $\mathsf{A}\mathfrak{G}$*-modules.*

Proof. a) There exists an A-homomorphism α' of $\mathsf{V}^{\mathfrak{G}} = \mathsf{V} \otimes_{\mathsf{A}\mathfrak{U}} \mathsf{A}\mathfrak{G}$ into W such that $(v \otimes a)\alpha' = (v\alpha)a$ for all $v \in \mathsf{V}$ and $a \in \mathsf{A}\mathfrak{G}$, since α is an $\mathsf{A}\mathfrak{U}$-homomorphism. Since

$$((v \otimes a)g)\alpha' = (v \otimes ag)\alpha' = (v\alpha)(ag) = ((v\alpha)a)g = ((v \otimes a)\alpha')g$$

for all $g \in \mathfrak{G}$, α' is an $\mathsf{A}\mathfrak{G}$-homomorphism. Further

$$v\eta\alpha' = (v \otimes 1)\alpha' = (v\alpha)1 = v\alpha$$

for all $v \in \mathsf{V}$, so $\eta\alpha' = \alpha$. To establish the uniqueness of α', it suffices to show that if $\beta \in \mathrm{Hom}_{\mathsf{A}\mathfrak{G}}(\mathsf{V}^{\mathfrak{G}}, \mathsf{W})$ and $\eta\beta = 0$, then $\beta = 0$. This is so because

$$0 = (v\eta\beta)g = ((v \otimes 1)\beta)g = (v \otimes g)\beta$$

and the $v \otimes g$ (with $v \in \mathsf{V}$, $g \in \mathfrak{G}$) generate $\mathsf{V}^{\mathfrak{G}}$ as an A-module.

b) We have the following commutative diagrams.

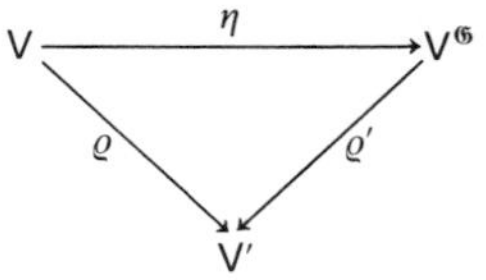

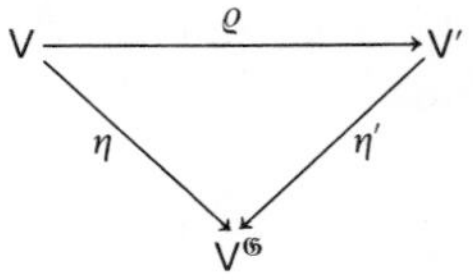

where $\varrho = \eta\varrho'$ and $\eta = \varrho\eta'$. It follows that $\eta = \eta(\varrho'\eta')$ and $\varrho = \varrho(\eta'\varrho')$. We also have the following commutative diagrams.

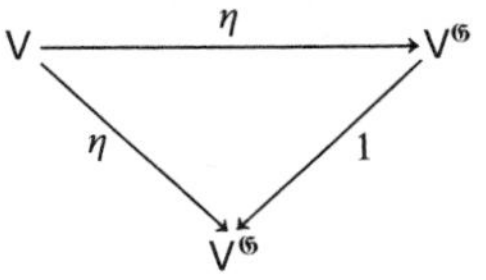

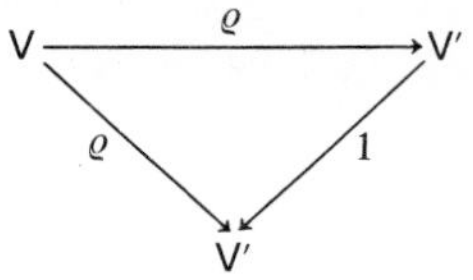

On account of the uniqueness assertions, it follows from these that $\varrho'\eta'$ is the identity mapping on $\mathsf{V}^{\mathfrak{G}}$ and $\eta'\varrho'$ is the identity mapping on V'. Thus ϱ' is an $\mathsf{A}\mathfrak{G}$-isomorphism of $\mathsf{V}^{\mathfrak{G}}$ onto V'. **q.e.d.**

A conceptual proof of the first reciprocity theorem of Nakayama follows easily from 4.4 (cf. V, 16.6).

4.5 Theorem (NAKAYAMA). *Suppose that* $\mathfrak{U} \leq \mathfrak{G}$, *that* V *is an* $\mathsf{A}\mathfrak{U}$-*module and that* W *is an* $\mathsf{A}\mathfrak{G}$-*module. Then*

$$\mathrm{Hom}_{\mathsf{A}\mathfrak{G}}(\mathsf{V}^{\mathfrak{G}}, \mathsf{W}) \cong \mathrm{Hom}_{\mathsf{A}\mathfrak{U}}(\mathsf{V}, \mathsf{W}_{\mathfrak{U}}).$$

(This is an isomorphism of A-*modules, of course).*

Proof. Let η be the $A\mathfrak{U}$-monomorphism of V into $(V^{\mathfrak{G}})_{\mathfrak{U}}$ defined in 4.3a). If $\beta \in \mathrm{Hom}_{A\mathfrak{G}}(V^{\mathfrak{G}}, W)$, then $\eta\beta \in \mathrm{Hom}_{A\mathfrak{U}}(V, W_{\mathfrak{U}})$. We put $\beta\vartheta = \eta\beta$; thus ϑ is an A-monomorphism of $\mathrm{Hom}_{A\mathfrak{G}}(V^{\mathfrak{G}}, W)$ into $\mathrm{Hom}_{A\mathfrak{U}}(V, W_{\mathfrak{U}})$. Given $\alpha \in \mathrm{Hom}_{A\mathfrak{U}}(V, W_{\mathfrak{U}})$, by 4.4 there exists a unique α' such that $\alpha = \eta\alpha' = \alpha'\vartheta$. Thus ϑ is an isomorphism. **q.e.d.**

Before we evaluate the assertion of 4.5, we prove a dual assertion. Our proofs will be most transparent if we temporarily introduce a functor ${}^{\mathfrak{G}}\cdot$ dual to $\cdot^{\mathfrak{G}}$.

4.6 Definition. If $\mathfrak{G}$ is a finite group and V is an A-module, $H = \mathrm{Hom}_A(A\mathfrak{G}, V)$ is, of course, just the direct sum of $|\mathfrak{G}|$ copies of V. H becomes an $A\mathfrak{G}$-module if we define ϕg (for $\phi \in H$, $g \in \mathfrak{G}$) by

$$a(\phi g) = (ga)\,\phi \quad (a \in A\mathfrak{G}).$$

For

$$a((\phi g)g') = (g'a)(\phi g) = (gg'a)\phi = a(\phi(gg')).$$

If now $\mathfrak{U} \leq \mathfrak{G}$ and V is an $A\mathfrak{U}$-module, $\mathrm{Hom}_{A\mathfrak{U}}(A\mathfrak{G}, V)$ is an $A\mathfrak{G}$-submodule of H, for if $\phi \in \mathrm{Hom}_{A\mathfrak{U}}(A\mathfrak{G}, V)$ and $g \in \mathfrak{G}$, then

$$(au)(\phi g) = (gau)\phi = ((ga)\phi)u = (a(\phi g))u$$

for all $a \in A\mathfrak{G}$, $u \in \mathfrak{U}$. The $A\mathfrak{G}$-module $\mathrm{Hom}_{A\mathfrak{U}}(A\mathfrak{G}, V)$ is called the $A\mathfrak{G}$-module *coinduced* by V and will be denoted by ${}^{\mathfrak{G}}V$.

If V_1, V_2 are $A\mathfrak{U}$-modules and $\alpha \in \mathrm{Hom}_{A\mathfrak{U}}(V_1, V_2)$, then for any $\phi \in {}^{\mathfrak{G}}V_1 = \mathrm{Hom}_{A\mathfrak{U}}(A\mathfrak{G}, V_1)$ we put $\phi({}^{\mathfrak{G}}\alpha) = \phi\alpha$. Thus ${}^{\mathfrak{G}}\alpha$ is a mapping of ${}^{\mathfrak{G}}V_1$ into ${}^{\mathfrak{G}}V_2$. In fact, ${}^{\mathfrak{G}}\alpha$ is an $A\mathfrak{G}$-homomorphism of ${}^{\mathfrak{G}}V_1$ into ${}^{\mathfrak{G}}V_2$, since if $\phi \in {}^{\mathfrak{G}}V_1$ and $g \in \mathfrak{G}$, then

$$a((\phi g)({}^{\mathfrak{G}}\alpha)) = (a(\phi g))\alpha = ((ga)\phi)\alpha = a((\phi({}^{\mathfrak{G}}\alpha))g)$$

for all $a \in A\mathfrak{G}$.

We now prove for ${}^{\mathfrak{G}}V$ the assertions dual to 4.3–4.5.

4.7 Theorem. *Suppose that* $\mathfrak{U} \leq \mathfrak{G}$. *Then* ${}^{\mathfrak{G}}\cdot$ *is a covariant exact functor from the category of* $A\mathfrak{U}$*-modules into the category of* $A\mathfrak{G}$*-modules.*

Proof. Suppose that V_1, V_2, V_3 are $\mathsf{A}\mathfrak{U}$-modules and that $\alpha \in \mathrm{Hom}_{\mathsf{A}\mathfrak{U}}(\mathsf{V}_1, \mathsf{V}_2)$, $\beta \in \mathrm{Hom}_{\mathsf{A}\mathfrak{U}}(\mathsf{V}_2, \mathsf{V}_3)$. Then obviously ${}^{\mathfrak{G}}(\alpha\beta) = ({}^{\mathfrak{G}}\alpha)({}^{\mathfrak{G}}\beta)$. Thus ${}^{\mathfrak{G}}\cdot$ is a covariant functor.

Let T be a transversal of $\mathfrak{U}$ in $\mathfrak{G}$; thus $\mathfrak{G} = \bigcup_{t \in \mathsf{T}} \mathfrak{U}t$.

For $\phi \in {}^{\mathfrak{G}}\mathsf{V}_1$, $\phi \in \ker {}^{\mathfrak{G}}\alpha$ is equivalent to

$$0 = a(\phi\, {}^{\mathfrak{G}}\alpha) = (a\phi)\alpha \quad \text{for all } a \in \mathsf{A}\mathfrak{G}.$$

This means that $a\phi \in \ker \alpha$. Hence

$$\ker {}^{\mathfrak{G}}\alpha = \mathrm{Hom}_{\mathsf{A}\mathfrak{U}}(\mathsf{A}\mathfrak{G}, \ker \alpha) = {}^{\mathfrak{G}}(\ker \alpha).$$

Obviously, $\mathrm{im}\, {}^{\mathfrak{G}}\alpha \subseteq \mathrm{Hom}_{\mathsf{A}\mathfrak{U}}(\mathsf{A}\mathfrak{G}, \mathrm{im}\, \alpha)$. Suppose that $\beta \in \mathrm{Hom}_{\mathsf{A}\mathfrak{U}}(\mathsf{A}\mathfrak{G}, \mathrm{im}\, \alpha)$. Given $t \in \mathsf{T}$, $t^{-1}\beta = v_t\alpha$ for some $v_t \in \mathsf{V}_1$. Now $\mathsf{A}\mathfrak{G} = \bigoplus_{t \in \mathsf{T}} t^{-1}\, \mathsf{A}\mathfrak{U}$ is a free $\mathsf{A}\mathfrak{U}$-mudule with basis $\{t^{-1} | t \in \mathsf{T}\}$. Thus there exists

$$\phi \in \mathrm{Hom}_{\mathsf{A}\mathfrak{U}}(\mathsf{A}\mathfrak{G}, \mathsf{V}_1) = {}^{\mathfrak{G}}\mathsf{V}_1$$

such that $t^{-1}\phi = v_t$ for all $t \in \mathsf{T}$. Then

$$t^{-1}(\phi({}^{\mathfrak{G}}\alpha)) = t^{-1}\phi\alpha = v_t\alpha = t^{-1}\beta.$$

This shows that $\phi({}^{\mathfrak{G}}\alpha) = \beta$. Thus

$$\mathrm{im}\, {}^{\mathfrak{G}}\alpha = \mathrm{Hom}_{\mathsf{A}\mathfrak{U}}(\mathsf{A}\mathfrak{G}, \mathrm{im}\, \alpha) = {}^{\mathfrak{G}}(\mathrm{im}\, \alpha).$$

It now follows easily that ${}^{\mathfrak{G}}\cdot$ is an exact functor. **q.e.d.**

4.8 Theorem. *Suppose that $\mathfrak{U} \le \mathfrak{G}$ and that T is a transversal of $\mathfrak{U}$ in $\mathfrak{G}$.*

a) *Suppose that V is an $\mathsf{A}\mathfrak{U}$-module. The mapping ϱ defined by*

$$\phi\varrho = 1\phi \quad (\textit{for } \phi \in {}^{\mathfrak{G}}\mathsf{V})$$

is an $\mathsf{A}\mathfrak{U}$-epimorphism of $({}^{\mathfrak{G}}\mathsf{V})_{\mathfrak{U}}$ onto V, and $\ker \varrho$ is a direct summand of $({}^{\mathfrak{G}}\mathsf{V})_{\mathfrak{U}}$.

b) *If W is an $\mathsf{A}\mathfrak{G}$-module, the mapping σ defined by*

$$a(w\sigma) = wa \quad (w \in \mathsf{W}, a \in \mathsf{A}\mathfrak{G})$$

is an $\mathsf{A}\mathfrak{G}$-monomorphism of W into ${}^{\mathfrak{G}}(\mathsf{W}_{\mathfrak{U}})$, and $(\mathsf{W}\sigma)_{\mathfrak{U}}$ is a direct summand of $({}^{\mathfrak{G}}(\mathsf{W}_{\mathfrak{U}}))_{\mathfrak{U}}$.

c) *If W is an $\mathsf{A}\mathfrak{G}$-module, the mapping τ defined by*

$$\phi\tau = \sum_{t \in \mathsf{T}} (t^{-1}\phi)t \quad (\phi \in {}^{\mathfrak{G}}(\mathsf{W}_{\mathfrak{U}}))$$

is an $\mathsf{A}\mathfrak{G}$*-epimorphism of* ${}^{\mathfrak{G}}(\mathsf{W}_{\mathfrak{U}})$ *onto* W. τ *is independent of the choice of the transversal* T, *and* $(\ker \tau)_{\mathfrak{U}}$ *is a direct summand of* $({}^{\mathfrak{G}}(\mathsf{W}_{\mathfrak{U}}))_{\mathfrak{U}}$. *Further*

$$w\sigma\tau = |\mathfrak{G} : \mathfrak{U}|w$$

for all $w \in \mathsf{W}$. *If* $|\mathfrak{G} : \mathfrak{U}|$ *has an inverse in the ring* A, *then*

$${}^{\mathfrak{G}}(\mathsf{W}_{\mathfrak{U}}) = \mathsf{W}\sigma \oplus \ker \tau.$$

Proof. a) Since

$$(\phi u)\varrho = 1(\phi u) = u\phi = (1\phi)u = (\phi\varrho)u$$

for $\phi \in {}^{\mathfrak{G}}\mathsf{V}$ and $u \in \mathfrak{U}$, ϱ is an $\mathsf{A}\mathfrak{U}$-homomorphism. If $v \in \mathsf{V}$, define $\phi \in \mathrm{Hom}_{\mathsf{A}}(\mathsf{A}\mathfrak{G}, \mathsf{V})$ by putting $u\phi = vu$ for $u \in \mathfrak{U}$ and $g\phi = 0$ for $g \in \mathfrak{G} - \mathfrak{U}$. Then $\phi \in {}^{\mathfrak{G}}\mathsf{V}$ and $\phi\varrho = v$. Thus ϱ is an epimorphism. Now

$$\ker \varrho = \{\phi | \phi \in {}^{\mathfrak{G}}\mathsf{V},\ 1\phi = 0\} = \{\phi | \phi \in {}^{\mathfrak{G}}\mathsf{V},\ u\phi = 0 \quad \text{for all } u \in \mathfrak{U}\}.$$

If

$$\mathsf{V}' = \{\phi | \phi \in {}^{\mathfrak{G}}\mathsf{V},\ g\phi = 0 \quad \text{for all } g \in \mathfrak{G} - \mathfrak{U}\},$$

then V' is an $\mathsf{A}\mathfrak{U}$-submodule and

$$({}^{\mathfrak{G}}\mathsf{V})_{\mathfrak{U}} = \mathsf{V}' \oplus \ker \varrho.$$

b) If $w \in \mathsf{W}$, then $w\sigma \in {}^{\mathfrak{G}}(\mathsf{W}_{\mathfrak{U}})$ since

$$(au)(w\sigma) = w(au) = (wa)u = (a(w\sigma))u$$

for all $a \in \mathsf{A}\mathfrak{G}$, $u \in \mathfrak{U}$. σ is an $\mathsf{A}\mathfrak{G}$-homomorphism since if $w \in \mathsf{W}$ and $g \in \mathfrak{G}$,

$$a((wg)\sigma) = (wg)a = (ga)(w\sigma) = a((w\sigma)g).$$

If $w\sigma = 0$, then

$$0 = 1(w\sigma) = w1 = w;$$

thus σ is an $\mathsf{A}\mathfrak{G}$-monomorphism of W into ${}^{\mathfrak{G}}(\mathsf{W}_{\mathfrak{U}})$. If

$$\mathsf{X} = \{\phi \mid \phi \in {}^{\mathfrak{G}}(\mathsf{W}_{\mathfrak{U}}), (\mathsf{A}\mathfrak{U})\phi = 0\},$$

X is an $\mathsf{A}\mathfrak{U}$-submodule of $({}^{\mathfrak{G}}(\mathsf{W}_{\mathfrak{U}}))_{\mathfrak{U}}$. If $w\sigma \in \mathsf{W}\sigma \cap \mathsf{X}$, then

$$0 = 1(w\sigma) = w1 = w,$$

so $\mathsf{W}\sigma \cap \mathsf{X} = 0$. Further, for all $\phi \in {}^{\mathfrak{G}}(\mathsf{W}_{\mathfrak{U}})$,

$$1(\phi - (1\phi)\sigma) = 1\phi - 1\phi = 0.$$

As ϕ and $(1\phi)\sigma$ are $\mathsf{A}\mathfrak{U}$-homomorphisms, it follows that $\phi - (1\phi)\sigma \in \mathsf{X}$ and thus $({}^{\mathfrak{G}}(\mathsf{W}_{\mathfrak{U}}))_{\mathfrak{U}} = (\mathsf{W}\sigma)_{\mathfrak{U}} \oplus \mathsf{X}$. Hence

$$({}^{\mathfrak{G}}(\mathsf{W}_{\mathfrak{U}}))_{\mathfrak{U}} = (\mathsf{W}\sigma)_{\mathfrak{U}} \oplus \mathsf{X}.$$

c) Suppose that T' is another transversal of $\mathfrak{U}$ in $\mathfrak{G}$. Then, given $t \in \mathsf{T}$, we can write $t = u_t t'$ with $u_t \in \mathfrak{U}$ and $t' \in \mathsf{T}'$. Then

$$\sum_{t \in \mathsf{T}} (t^{-1}\phi)t = \sum_{t' \in \mathsf{T}'} ((t'^{-1}u_t^{-1})\phi)u_t t' = \sum_{t' \in \mathsf{T}'} (t'^{-1}\phi)t'.$$

Thus τ is independent of the choice of the transversal T. It follows that for all $g \in \mathfrak{G}$,

$$\begin{aligned}(\phi g)\tau &= \sum_{t \in \mathsf{T}} (t^{-1}(\phi g))t = \sum_{t \in \mathsf{T}} ((gt^{-1})\phi)t \\ &= \sum_{t \in \mathsf{T}} ((tg^{-1})^{-1}\phi)(tg^{-1})g = (\phi\tau)g.\end{aligned}$$

Thus $\tau \in \operatorname{Hom}_{\mathsf{A}\mathfrak{G}}({}^{\mathfrak{G}}(\mathsf{W}_{\mathfrak{U}}), \mathsf{W})$.

Suppose that $\mathsf{T} \cap \mathfrak{U} = \{t_1\}$. Since $\mathsf{A}\mathfrak{G}$ is a free $\mathsf{A}\mathfrak{U}$-module with basis $\{t^{-1} \mid t \in \mathsf{T}\}$, it follows that given $w \in \mathsf{W}$, there exists $\phi \in \operatorname{Hom}_{\mathsf{A}\mathfrak{U}}(\mathsf{A}\mathfrak{G}, \mathsf{W}_{\mathfrak{U}})$ such that $t_1^{-1}\phi = wt_1^{-1}$ and $t^{-1}\phi = 0$ for $t \in \mathsf{T}$, $t \neq t_1$. Thus $\phi\tau = w$. Hence τ is an $\mathsf{A}\mathfrak{G}$-epimorphism of ${}^{\mathfrak{G}}(\mathsf{W}_{\mathfrak{U}})$ onto W. Put

$$\mathsf{Y} = \{\phi \mid \phi \in {}^{\mathfrak{G}}(\mathsf{W}_{\mathfrak{U}}), g\phi = 0 \quad \text{for all } g \in \mathfrak{G} - \mathfrak{U}\}.$$

Then Y is an $\mathsf{A}\mathfrak{U}$-submodule of $({}^{\mathfrak{G}}(\mathsf{W}_{\mathfrak{U}}))_{\mathfrak{U}}$. If $\phi \in \ker \tau \cap \mathsf{Y}$, then

$$0 = \phi\tau = \sum_{t \in \mathsf{T}} (t^{-1}\phi)t = (t_1^{-1}\phi)t_1.$$

This forces $t_1^{-1}\phi = 0$, and since $\phi \in \mathsf{Y}$, $\phi = 0$. Thus $\ker\tau \cap \mathsf{Y} = 0$. If $\phi \in {}^{\mathfrak{G}}(\mathsf{W}_{\mathfrak{U}})$, we define $\psi \in {}^{\mathfrak{G}}(\mathsf{W}_{\mathfrak{U}})$ by

$$t^{-1}\psi = t^{-1}\phi \quad \text{for } t_1 \neq t \in \mathsf{T}$$

and

$$t_1^{-1}\psi = -\sum_{t_1 \neq t \in \mathsf{T}} (t^{-1}\phi)tt_1^{-1}.$$

Then $\psi \in \ker\tau$ and $\phi - \psi \in \mathsf{Y}$. Thus

$${}^{\mathfrak{G}}(\mathsf{W}_{\mathfrak{U}}) = \ker\tau \oplus \mathsf{Y}.$$

Finally, if $w \in \mathsf{W}$,

$$w\sigma\tau = \sum_{t\in\mathsf{T}} (t^{-1}(w\sigma))t = \sum_{t\in\mathsf{T}} wt^{-1}t = |\mathfrak{G}:\mathfrak{U}|w.$$

Suppose that $|\mathfrak{G}:\mathfrak{U}|$ has an inverse in A. If we write

$$\tau' = |\mathfrak{G}:\mathfrak{U}|^{-1}\tau,$$

then $\sigma\tau' = 1$, so $(\phi\tau'\sigma - \phi)\tau' = 0$ for all $\phi \in {}^{\mathfrak{G}}(\mathsf{W}_{\mathfrak{U}})$. Thus

$${}^{\mathfrak{G}}(\mathsf{W}_{\mathfrak{U}}) = \operatorname{im}\sigma \oplus \ker\tau' = \mathsf{W}\sigma \oplus \ker\tau. \qquad \textbf{q.e.d.}$$

We also characterize ${}^{\mathfrak{G}}\mathsf{V}$ by a universal property.

4.9 Theorem. *Suppose that* $\mathfrak{U} \leq \mathfrak{G}$.

a) *Suppose that* V *is an* $\mathsf{A}\mathfrak{U}$*-module,* W *is an* $\mathsf{A}\mathfrak{G}$*-module and* $\alpha \in \operatorname{Hom}_{\mathsf{A}\mathfrak{U}}(\mathsf{W}_{\mathfrak{U}}, \mathsf{V})$. *Then there exists precisely one* $\alpha' \in \operatorname{Hom}_{\mathsf{A}\mathfrak{G}}(\mathsf{W}, {}^{\mathfrak{G}}\mathsf{V})$ *such that* $\alpha = \alpha'\varrho$, *where* ϱ *is the mapping defined in 4.8a).*

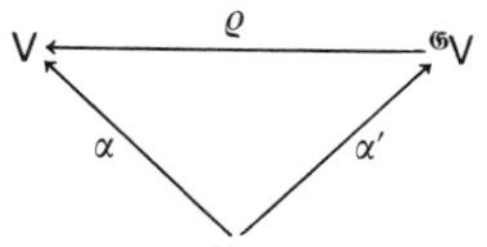

b) *Suppose that* V *is an* $\mathsf{A}\mathfrak{U}$*-module,* V' *is an* $\mathsf{A}\mathfrak{G}$*-module and* $\pi \in \operatorname{Hom}_{\mathsf{A}\mathfrak{U}}(\mathsf{V}'_{\mathfrak{U}}, \mathsf{V})$. *Suppose that, given an* $\mathsf{A}\mathfrak{G}$*-module* W *and*

$\alpha \in \mathrm{Hom}_{\mathsf{A}\mathfrak{U}}(\mathsf{W}_{\mathfrak{U}}, \mathsf{V})$, *there always exists exactly one* $\alpha' \in \mathrm{Hom}_{\mathsf{A}\mathfrak{G}}(\mathsf{W}, \mathsf{V}')$ *such that* $\alpha = \alpha'\pi$. *Then* V' *and* ${}^{\mathfrak{G}}\mathsf{V}$ *are isomorphic* $\mathsf{A}\mathfrak{G}$*-modules.*

Proof. a) Define the mapping $w\alpha'$ $(w \in \mathsf{W})$ of $\mathsf{A}\mathfrak{G}$ into V by

$$a(w\alpha') = (wa)\alpha \quad (a \in \mathsf{A}\mathfrak{G}).$$

Then $w\alpha' \in {}^{\mathfrak{G}}\mathsf{V}$, and α' is an A-homomorphism such that

$$(w\alpha')\varrho = 1(w\alpha') = w\alpha$$

for all $w \in \mathsf{W}$. Thus $\alpha = \alpha'\varrho$. Further, α' is an $\mathsf{A}\mathfrak{G}$-homomorphism, since if $a \in \mathsf{A}\mathfrak{G}$, $w \in \mathsf{W}$ and $g \in \mathfrak{G}$, then

$$a((w\alpha')g) = (ga)(w\alpha') = (wga)\alpha = a((wg)\alpha').$$

To verify the uniqueness of α', it obviously suffices to show that if $\beta \in \mathrm{Hom}_{\mathsf{A}\mathfrak{G}}(\mathsf{W}, {}^{\mathfrak{G}}\mathsf{V})$ and $\beta\varrho = 0$, then $\beta = 0$. Thus we must show that $g(w\beta) = 0$ for all $g \in \mathfrak{G}$, $w \in \mathsf{W}$. But

$$g(w\beta) = 1((w\beta)g) = 1((wg)\beta) = (wg)(\beta\varrho) = 0.$$

b) The assertion of b) follows similarly to that of 4.4b). **q.e.d.**

The second reciprocity theorem of Nakayama follows immediately from 4.9.

4.10 Theorem (NAKAYAMA). *Suppose that* $\mathfrak{U} \leq \mathfrak{G}$, *that* V *is an* $\mathsf{A}\mathfrak{U}$*-module and that* W *is an* $\mathsf{A}\mathfrak{G}$*-module. Then*

$$\mathrm{Hom}_{\mathsf{A}\mathfrak{G}}(\mathsf{W}, {}^{\mathfrak{G}}\mathsf{V}) \cong \mathrm{Hom}_{\mathsf{A}\mathfrak{U}}(\mathsf{W}_{\mathfrak{U}}, \mathsf{V}).$$

Proof. The mapping $\beta \to \beta\varrho$ yields an A-homomorphism, and as in 4.5, this is an isomorphism of $\mathrm{Hom}_{\mathsf{A}\mathfrak{G}}(\mathsf{W}, {}^{\mathfrak{G}}\mathsf{V})$ onto $\mathrm{Hom}_{\mathsf{A}\mathfrak{U}}(\mathsf{W}_{\mathfrak{U}}, \mathsf{V})$ by 4.9a). **q.e.d.**

So far we have only rarely used the finiteness of $\mathfrak{G}$. (In 4.3c) and 4.8c) the finiteness of $|\mathfrak{G}:\mathfrak{U}|$ was used.) The essential point is that for finite $|\mathfrak{G}:\mathfrak{U}|$, $\cdot^{\mathfrak{G}}$ and ${}^{\mathfrak{G}}\cdot$ are the same.

4.11 Theorem. *Suppose that* $\mathfrak{U} \leq \mathfrak{G}$ *and that* $|\mathfrak{G}:\mathfrak{U}|$ *is finite. For each* $\mathsf{A}\mathfrak{U}$*-module* V *there is an* $\mathsf{A}\mathfrak{G}$*-isomorphism* μ_{V} *of* ${}^{\mathfrak{G}}\mathsf{V}$ *onto* $\mathsf{V}^{\mathfrak{G}}$ *such that if* $\alpha \in \mathrm{Hom}_{\mathsf{A}\mathfrak{U}}(\mathsf{V}_1, \mathsf{V}_2)$, *the diagram*

$$\begin{array}{ccc} {}^{\mathfrak{G}}\mathsf{V}_1 & \xrightarrow{\ {}^{\mathfrak{G}}\alpha\ } & {}^{\mathfrak{G}}\mathsf{V}_2 \\ \Big\downarrow{\scriptstyle \mu_{\mathsf{V}_1}} & & \Big\downarrow{\scriptstyle \mu_{\mathsf{V}_2}} \\ \mathsf{V}_1^{\mathfrak{G}} & \xrightarrow[\ \alpha^{\mathfrak{G}}\ ]{} & \mathsf{V}_2^{\mathfrak{G}} \end{array}$$

is always commutative. (Thus the μ_{V} define a natural equivalence between the functors ${}^{\mathfrak{G}}\cdot$ and $\cdot^{\mathfrak{G}}$; see MACLANE [1], *p. 29.) In particular,* $\mathsf{V}^{\mathfrak{G}} \cong {}^{\mathfrak{G}}\mathsf{V}$.

Proof. Let T be a transversal of $\mathfrak{U}$ in $\mathfrak{G}$. For each $\mathsf{A}\mathfrak{U}$-module V, we define a mapping μ_{V} of ${}^{\mathfrak{G}}\mathsf{V} = \mathrm{Hom}_{\mathsf{A}\mathfrak{U}}(\mathsf{A}\mathfrak{G}, \mathsf{V})$ into $\mathsf{V}^{\mathfrak{G}} = \mathsf{V} \otimes_{\mathsf{A}\mathfrak{U}} \mathsf{A}\mathfrak{G}$ by putting

$$\phi\mu_{\mathsf{V}} = \sum_{t \in \mathsf{T}} t^{-1}\phi \otimes t \quad (\phi \in {}^{\mathfrak{G}}\mathsf{V}).$$

Obviously μ_{V} is an A-monomorphism. Since $\mathsf{A}\mathfrak{G}$ is a free $\mathsf{A}\mathfrak{U}$-module with basis $\{t^{-1} \mid t \in \mathsf{T}\}$, there exists $\phi \in \mathrm{Hom}_{\mathsf{A}\mathfrak{U}}(\mathsf{A}\mathfrak{G}, \mathsf{V})$ such that the $t^{-1}\phi$ take preassigned values in V. Thus μ_{V} is surjective, and so μ_{V} is an A-isomorphism. It is an $\mathsf{A}\mathfrak{G}$-isomorphism, since if $g \in \mathfrak{G}$ and $tg = u_t t'$ with $u_t \in \mathfrak{U}$ and $t' \in \mathsf{T}$, then

$$\begin{aligned}(\phi\mu_{\mathsf{V}})g &= \sum_{t \in \mathsf{T}} t^{-1}\phi \otimes tg = \sum_{t \in \mathsf{T}} t^{-1}\phi \otimes u_t t' = \sum_{t \in \mathsf{T}} (t^{-1}u_t)\phi \otimes t' \\ &= \sum_{t' \in \mathsf{T}} (g t'^{-1})\phi \otimes t' = \sum_{t' \in \mathsf{T}} t'^{-1}(\phi g) \otimes t' = (\phi g)\mu_{\mathsf{V}}.\end{aligned}$$

To verify that the μ_{V} define a natural equivalence between $\cdot^{\mathfrak{G}}$ and ${}^{\mathfrak{G}}\cdot$, we have to establish the commutativity of the diagram displayed in the theorem. For $\phi \in {}^{\mathfrak{G}}\mathsf{V}_1$, we have

$$\begin{aligned}\phi({}^{\mathfrak{G}}\alpha)\mu_{\mathsf{V}_2} &= \sum_{t \in \mathsf{T}} t^{-1}(\phi({}^{\mathfrak{G}}\alpha)) \otimes t = \sum_{t \in \mathsf{T}} t^{-1}\phi\alpha \otimes t \\ &= \sum_{t \in \mathsf{T}} (t^{-1}\phi \otimes t)(\alpha \otimes 1) = \phi\mu_{\mathsf{V}_2}(\alpha^{\mathfrak{G}}).\end{aligned}$$

q.e.d.

We shall now show that Theorems 4.5 and 4.10 can be used to prove a generalization of the Frobenius reciprocity theorem. We replace the ring A by a field K. We shall need a simple lemma.

4.12 Lemma. *Let K be an arbitrary field. Let W be an irreducible $\mathsf{K}\mathfrak{G}$-module and let X be an arbitrary $\mathsf{K}\mathfrak{G}$-module.*

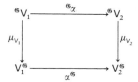

a) *There exists a uniquely determined smallest submodule* X' *of* X *such that* X/X' *is a completely reducible* $K\mathfrak{G}$*-module and all composition factors of* X/X' *are isomorphic to* W. *If* X/X' *is the direct sum of k submodules isomorphic to* W, *then*

$$k \dim_K \mathrm{Hom}_{K\mathfrak{G}}(W, W) = \dim_K \mathrm{Hom}_{K\mathfrak{G}}(X, W).$$

In particular, if K *is a splitting field for* $\mathfrak{G}$, *we have*

$$k = \dim_K \mathrm{Hom}_{K\mathfrak{G}}(X, W).$$

b) *There exists a uniquely determined largest submodule* X'' *of* X *such that* X'' *is a completely reducible* $K\mathfrak{G}$*-module and all composition factors of* X'' *are isomorphic to* W. *If* X'' *is the direct sum of l submodules isomorphic to* W, *then*

$$l \dim_K \mathrm{Hom}_{K\mathfrak{G}}(W, W) = \dim_K \mathrm{Hom}_{K\mathfrak{G}}(W, X).$$

In particular, if K *is a splitting field for* $\mathfrak{G}$, *then*

$$l = \dim_K \mathrm{Hom}_{K\mathfrak{G}}(W, X).$$

Proof. a) Suppose that X/X_1 and X/X_2 are completely reducible factor modules of X. By 1.6, $X/(X_1 \cap X_2)$ is also completely reducible. If all the composition factors of X/X_1 and X/X_2 are isomorphic to W, so are those of $X/(X_1 \cap X_2)$, since

$$X_1/(X_1 \cap X_2) \cong (X_1 + X_2)/X_2.$$

Suppose that $0 \neq \alpha \in \mathrm{Hom}_{K\mathfrak{G}}(X, W)$. As W is irreducible, $X/\ker\alpha \cong W$. It follows from the above that $\ker\alpha \supseteq X'$. Thus if

$$X/X' = W_1 \oplus \cdots \oplus W_k,$$

where $W_i \cong W$, then

$$\mathrm{Hom}_{K\mathfrak{G}}(X, W) \cong \mathrm{Hom}_{K\mathfrak{G}}(X/X', W) \cong \bigoplus_{i=1}^{k} \mathrm{Hom}_{K\mathfrak{G}}(W_i, W).$$

It follows that

$$\dim_K \mathrm{Hom}_{K\mathfrak{G}}(X, W) = k \dim_K \mathrm{Hom}_{K\mathfrak{G}}(W, W).$$

If K is a splitting field for $\mathfrak{G}$, then $\mathrm{Hom}_{\mathsf{K}\mathfrak{G}}(\mathsf{W}, \mathsf{W}) = \mathsf{K}$, whence

$$\dim_{\mathsf{K}} \mathrm{Hom}_{\mathsf{K}\mathfrak{G}}(\mathsf{X}, \mathsf{W}) = k.$$

b) The existence of X'' is proved similarly, and so as in a),

$$\dim_{\mathsf{K}} \mathrm{Hom}_{\mathsf{K}\mathfrak{G}}(\mathsf{W}, \mathsf{X}) = \dim_{\mathsf{K}} \mathrm{Hom}_{\mathsf{K}\mathfrak{G}}(\mathsf{W}, \mathsf{X}'') = l \dim_{\mathsf{K}} \mathrm{Hom}_{\mathsf{K}\mathfrak{G}}(\mathsf{W}, \mathsf{W}).$$

q.e.d.

4.13 Theorem. *Let* K *be an arbitrary field and suppose that* $\mathfrak{U} \leq \mathfrak{G}$. *Suppose further that* V *is an irreducible* $\mathsf{K}\mathfrak{U}$*-module and that* W *is an irreducible* $\mathsf{K}\mathfrak{G}$*-module.*

a) *Let k be the multiplicity of* W *as a composition factor in the head of* $\mathsf{V}^{\mathfrak{G}}$, *and let l be the multiplicity of* V *as a composition factor in the socle of* $\mathsf{W}_{\mathfrak{U}}$. *Then*

$$k \dim_{\mathsf{K}} \mathrm{Hom}_{\mathsf{K}\mathfrak{G}}(\mathsf{W}, \mathsf{W}) = l \dim_{\mathsf{K}} \mathrm{Hom}_{\mathsf{K}\mathfrak{U}}(\mathsf{V}, \mathsf{V}).$$

In particular, if K *is a splitting field for* $\mathfrak{U}$ *and* $\mathfrak{G}$, *then* $k = l$.

b) *Let m be the multiplicity of* W *as a composition factor in the socle of* $\mathsf{V}^{\mathfrak{G}}$, *and let n be the multiplicity of* V *as a composition factor in the head of* $\mathsf{W}_{\mathfrak{U}}$. *Then*

$$m \dim_{\mathsf{K}} \mathrm{Hom}_{\mathsf{K}\mathfrak{G}}(\mathsf{W}, \mathsf{W}) = n \dim_{\mathsf{K}} \mathrm{Hom}_{\mathsf{K}\mathfrak{U}}(\mathsf{V}, \mathsf{V}).$$

If K *is a splitting field for* $\mathfrak{U}$ *and* $\mathfrak{G}$, *then* $m = n$.

Proof. a) We have

$$\begin{aligned} k \dim_{\mathsf{K}} \mathrm{Hom}_{\mathsf{K}\mathfrak{G}}(\mathsf{W}, \mathsf{W}) &= \dim_{\mathsf{K}} \mathrm{Hom}_{\mathsf{K}\mathfrak{G}}(\mathsf{V}^{\mathfrak{G}}, \mathsf{W}) && \text{(by 4.12a))} \\ &= \dim_{\mathsf{K}} \mathrm{Hom}_{\mathsf{K}\mathfrak{U}}(\mathsf{V}, \mathsf{W}_{\mathfrak{U}}) && \text{(by 4.5)} \\ &= l \dim_{\mathsf{K}} \mathrm{Hom}_{\mathsf{K}\mathfrak{U}}(\mathsf{V}, \mathsf{V}) && \text{(by 4.12b)).} \end{aligned}$$

b) Similarly

$$\begin{aligned} m \dim_{\mathsf{K}} \mathrm{Hom}_{\mathsf{K}\mathfrak{G}}(\mathsf{W}, \mathsf{W}) &= \dim_{\mathsf{K}} \mathrm{Hom}_{\mathsf{K}\mathfrak{G}}(\mathsf{W}, \mathsf{V}^{\mathfrak{G}}) && \text{(by 4.12b))} \\ &= \dim_{\mathsf{K}} \mathrm{Hom}_{\mathsf{K}\mathfrak{G}}(\mathsf{W}, {}^{\mathfrak{G}}\mathsf{V}) && \text{(by 4.11)} \\ &= \dim_{\mathsf{K}} \mathrm{Hom}_{\mathsf{K}\mathfrak{U}}(\mathsf{W}_{\mathfrak{U}}, \mathsf{V}) && \text{(by 4.10)} \\ &= n \dim_{\mathsf{K}} \mathrm{Hom}_{\mathsf{K}\mathfrak{U}}(\mathsf{V}, \mathsf{V}) && \text{(by 4.12a)).} \end{aligned}$$

q.e.d.

4.14 Example. Let K be a splitting field for $\mathfrak{G}$. We apply 4.13 with $\mathfrak{U} = 1$ and $\mathsf{V} = \mathsf{K}$. Then $\mathsf{V}^{\mathfrak{G}} = \mathsf{K}\mathfrak{G}$. By 4.13a), the multiplicity of an irreducible $\mathsf{K}\mathfrak{G}$-module W in the head $\mathsf{K}\mathfrak{G}/\mathbf{J}(\mathsf{K}\mathfrak{G})$ of $\mathsf{K}\mathfrak{G}$ is equal to the multiplicity of K in the socle of $\mathsf{W}_{\mathfrak{U}}$, and this is $\dim_{\mathsf{K}}\mathsf{W}$. This assertion is, of course, a consequence of the Wedderburn theorem long known to us. But 4.13b) yields a new result, namely, the multiplicity of W in the socle of $\mathsf{V}^{\mathfrak{G}} = \mathsf{K}\mathfrak{G}$ is equal to the multiplicity of K in $\mathsf{W}_{\mathfrak{U}}/(\mathsf{W}_{\mathfrak{U}})\mathbf{J}(\mathsf{K}\mathfrak{U})$, which is also $\dim_{\mathsf{K}}\mathsf{W}$. This result will also be a consequence of our considerations in § 11.

For later use, we add the following lemma.

4.15 Lemma. a) *Suppose that* $\mathfrak{U} \leq \mathfrak{G}$. *Let* V *be a* $\mathsf{K}\mathfrak{U}$*-module and let* W *be a* $\mathsf{K}\mathfrak{G}$*-module. Then*

$$(\mathsf{V} \otimes_{\mathsf{K}} \mathsf{W}_{\mathfrak{U}})^{\mathfrak{G}} \cong \mathsf{V}^{\mathfrak{G}} \otimes_{\mathsf{K}} \mathsf{W}.$$

b) *If* $\mathfrak{N} \trianglelefteq \mathfrak{G}$ *and* W *is a* $\mathsf{K}\mathfrak{G}$*-module, then*

$$(\mathsf{W}_{\mathfrak{N}})^{\mathfrak{G}} \cong \mathsf{W} \otimes_{\mathsf{K}} \mathsf{K}(\mathfrak{G}/\mathfrak{N}),$$

where the $\mathsf{K}\mathfrak{G}$*-module structure of* $\mathsf{K}(\mathfrak{G}/\mathfrak{N})$ *is given by* $(\mathfrak{N}g_1)g_2 = \mathfrak{N}g_1g_2$

Proof. a) Let T be a transversal of $\mathfrak{U}$ in $\mathfrak{G}$. We have

$$(\mathsf{V} \otimes_{\mathsf{K}} \mathsf{W}_{\mathfrak{U}})^{\mathfrak{G}} = \bigoplus_{t \in \mathsf{T}} (\mathsf{V} \otimes \mathsf{W}) \otimes t$$

and

$$\mathsf{V}^{\mathfrak{G}} \otimes_{\mathsf{K}} \mathsf{W} = \bigoplus_{t \in \mathsf{T}} (\mathsf{V} \otimes t) \otimes \mathsf{W}.$$

We define a K-linear mapping α from $(\mathsf{V} \otimes_{\mathsf{K}} \mathsf{W}_{\mathfrak{U}})^{\mathfrak{G}}$ into $\mathsf{V}^{\mathfrak{G}} \otimes_{\mathsf{K}} \mathsf{W}$ by

$$((v \otimes w) \otimes t)\alpha = (v \otimes t) \otimes wt \quad (v \in \mathsf{V}, w \in \mathsf{W}, t \in \mathsf{T}).$$

It is easily checked that α is bijective. For $g \in \mathfrak{G}$ and $tg = ut'$ with $u \in \mathfrak{U}$ and $t' \in \mathsf{T}$, we have

$$\begin{aligned}(((v \otimes w) \otimes t)g)\alpha &= ((v \otimes w) \otimes tg)\alpha = ((v \otimes w) \otimes ut')\alpha \\ &= ((vu \otimes wu) \otimes t')\alpha = (vu \otimes t') \otimes wut'\end{aligned}$$

and

$$(((v \otimes w) \otimes t)\alpha)g = ((v \otimes t) \otimes wt)g = (v \otimes tg) \otimes wtg$$
$$= (vu \otimes t') \otimes wut'.$$

Hence α is a $\mathsf{K}\mathfrak{G}$-isomorphism.

b) Let K be the trivial $\mathsf{K}\mathfrak{N}$-module of dimension 1. By a),

$$(\mathsf{W}_{\mathfrak{N}} \otimes_{\mathsf{K}} \mathsf{K})^{\mathfrak{G}} \cong \mathsf{W} \otimes_{\mathsf{K}} \mathsf{K}^{\mathfrak{G}}.$$

The assertion follows at once since $\mathsf{W}_{\mathfrak{N}} \otimes_{\mathsf{K}} \mathsf{K} \cong \mathsf{W}_{\mathfrak{N}}$ and $\mathsf{K}^{\mathfrak{G}} \cong \mathsf{K}(\mathfrak{G}/\mathfrak{N})$.
q.e.d.

We mention finally that the first cohomology group can be defined by using coinduced modules.

4.16 Theorem. *Let* V *be an* $\mathsf{A}\mathfrak{G}$*-module,* $\mathsf{H} = \mathrm{Hom}_{\mathsf{A}}(\mathsf{A}\mathfrak{G}, \mathsf{V})$ *and* $\mathsf{J} = \mathrm{Hom}_{\mathsf{A}\mathfrak{G}}(\mathsf{A}\mathfrak{G}, \mathsf{V})$, *considered as* $\mathsf{A}\mathfrak{G}$*-modules as in 4.6. If*

$$\mathsf{C} = \mathbf{C}_{\mathsf{H}}(\mathfrak{G}) \quad \textit{and} \quad \mathsf{Z}/\mathsf{J} = \mathbf{C}_{\mathsf{H}/\mathsf{J}}(\mathfrak{G}),$$

then $\mathbf{H}^1(\mathfrak{G}, \mathsf{V}) \cong \mathsf{Z}/(\mathsf{C} + \mathsf{J})$.

Proof. If $\phi \in \mathsf{H}$ and the mapping f of $\mathfrak{G}$ into V is defined by

$$f(g) = g\phi - 1\phi \quad (g \in \mathfrak{G}),$$

then $\phi \in \mathsf{Z}$ if and only if $f \in \mathbf{Z}^1(\mathfrak{G}, \mathsf{V})$. For $\phi \in \mathsf{Z}$ is equivalent to $\phi g_1 - \phi \in \mathsf{J}$ for all $g_1 \in \mathfrak{G}$, or

$$g_2(\phi g_1 - \phi) = (1(\phi g_1 - \phi))g_2$$

for all $g_1, g_2 \in \mathfrak{G}$. But this is equivalent to

$$(g_1 g_2)\phi - g_2\phi = (g_1\phi)g_2 - (1\phi)g_2,$$

or

$$f(g_1 g_2) - f(g_2) = f(g_1)g_2.$$

Hence if $\phi \in \mathsf{Z}$ and we put $f(g) = g\phi - 1\phi$ for all $g \in \mathfrak{G}$, the mapping $\phi \to f + \mathbf{B}^1(\mathfrak{G}, \mathsf{V})$ is a homomorphism of Z into $\mathbf{H}^1(\mathfrak{G}, \mathsf{V})$. Given $f \in \mathbf{Z}^1(\mathfrak{G}, \mathsf{V})$, we define $\phi \in \mathsf{H}$ by

$$\left(\sum_{g\in\mathfrak{G}} a_g g\right)\phi = \sum_{g\in\mathfrak{G}} a_g f(g).$$

Since $f(1) = 0$, $f(g) = g\phi - 1\phi$, so $\phi \in \mathsf{Z}$. Hence the above mapping is an epimorphism. We must prove that the kernel M of it is $\mathsf{C} + \mathsf{J}$. Now

$$\mathsf{M} = \{\phi|\phi \in \mathsf{Z}, \text{ there exists } v \in \mathsf{V} \text{ such that } g\phi - 1\phi = v(g-1) \text{ for all } g \in \mathfrak{G}\},$$

$$\mathsf{C} = \{\phi|\phi \in \mathsf{H}, \phi g = \phi \quad \text{for all } g \in \mathfrak{G}\}$$
$$= \{\phi|\phi \in \mathsf{H}, g\phi = 1\phi \quad \text{for all } g \in \mathfrak{G}\}$$

and

$$\mathsf{J} = \{\phi|\phi \in \mathsf{H}, (xg)\phi = (x\phi)g \quad \text{for all } x, g \in \mathfrak{G}\}$$
$$= \{\phi|\phi \in \mathsf{H}, g\phi = (1\phi)g \quad \text{for all } g \in \mathfrak{G}\}.$$

Hence $\mathsf{C} + \mathsf{J} \subseteq \mathsf{M}$. Conversely, suppose $\phi \in \mathsf{Z}$ and $g\phi - 1\phi = v(g-1)$. If $\psi \in \mathsf{J}$ is defined by $g\psi = vg$, then $\phi - \psi \in \mathsf{C}$. Hence $\mathsf{C} + \mathsf{J} = \mathsf{M}$.

q.e.d.

Exercises

5) Prove the following by applying the universal properties 4.4 and 4.9.

a) If V is a projective $\mathsf{A}\mathfrak{U}$-module, then $\mathsf{V}^{\mathfrak{G}}$ is a projective $\mathsf{A}\mathfrak{G}$-module.

b) If V is an injective $\mathsf{A}\mathfrak{U}$-module, then ${}^{\mathfrak{G}}\mathsf{V}$ is an injective $\mathsf{A}\mathfrak{G}$-module.

6) Deduce from the universal properties that if $\mathfrak{U}_1 \leq \mathfrak{U}_2 \leq \mathfrak{G}$ and V is a $\mathsf{K}\mathfrak{U}_1$-module, then

$$(\mathsf{V}^{\mathfrak{U}_2})^{\mathfrak{G}} \cong \mathsf{V}^{\mathfrak{G}} \quad \text{and} \quad {}^{\mathfrak{G}}({}^{\mathfrak{U}_2}\mathsf{V}) \cong {}^{\mathfrak{G}}\mathsf{V}.$$

7) (WILLEMS [1]). Suppose that $\mathfrak{U} \leq \mathfrak{G}$ and that K is a field. Let V be a $\mathsf{K}\mathfrak{U}$-module. If $\mathsf{V}^{\mathfrak{G}}$ is a completely reducible $\mathsf{K}\mathfrak{G}$-module, then V is a completely reducible $\mathsf{K}\mathfrak{U}$-module.

(Hint: Let W be a $\mathsf{K}\mathfrak{U}$-submodule of V. Take a projection $\pi \in \mathrm{Hom}_{\mathsf{K}\mathfrak{G}}(\mathsf{V}^{\mathfrak{G}}, \mathsf{V}^{\mathfrak{G}})$ with $\mathrm{im}\,\pi = \mathsf{W}^{\mathfrak{G}}$. Write

$$(v \otimes t)\pi = \sum_{t'\in\mathsf{T}} v\pi_{t,t'} \otimes t',$$

where T is a transversal of 𝔘 in 𝔊. Show that π_{t_1,t_1} is a K𝔘-projection of V onto W, where $\{t_1\} = \mathsf{T} \cap \mathfrak{U}$.)

§ 5. The Number of Indecomposable K𝔊-Modules

If char K does not divide $|\mathfrak{G}|$, any K𝔊-module is completely reducible, and knowledge of all irreducible K𝔊-modules is to a large extent sufficient for the description of all K𝔊-modules. If, however, K𝔊 is not semisimple, a description of all indecomposable K𝔊-modules is required. Unfortunately, this can only be carried out in a few cases, because if char K $= p$, the number of indecomposable K𝔊-modules is finite if and only if the Sylow p-subgroups of 𝔊 are cyclic. This will be shown in this section.

It is easy to see that in case of an infinite field K of characteristic p there are infinitely many non-isomorphic indecomposable K𝔊-modules of dimension 2 for the elementary Abelian group $\mathfrak{G} = \langle g_1 \rangle \times \langle g_2 \rangle$ of type (p, p). Namely, for any $a \in \mathsf{K}$, consider the K𝔊-module with K-basis $\{v_1, v_2\}$ in which

$$v_1 g_1 = v_1 \qquad v_1 g_2 = v_1$$

$$v_2 g_1 = v_1 + v_2 \qquad v_2 g_2 = a v_1 + v_2.$$

But in Theorem 5.1 the emphasis is on the fact that indecomposable modules of arbitrarily large dimensions exist.

We begin by considering p-groups.

5.1 Theorem (D. G. HIGMAN [3]). *Suppose that* K *is a field of characteristic* p *and that* 𝔊 *is a non-cyclic* p*-group. Then there exist indecomposable* K𝔊*-modules of* K*-dimension 1, 3, 5, ...*

Proof (CURTIS, REINER [1]). It follows from III, 7.1 that any non-cyclic p-group possesses an elementary Abelian factor group of type (p, p). Thus, to prove the theorem, we may assume that 𝔊 is elementary Abelian of type (p, p). Suppose then, that $\mathfrak{G} = \langle g_1 \rangle \times \langle g_2 \rangle$ and $g_i^p = 1$.

Let n be a natural number and let V be a vector space over K of dimension $2n + 1$ with K-basis $\{v_0, v_1, \dots, v_n, w_1, \dots, w_n\}$. Since char K $= p$, V becomes a K𝔊-module if we put

$$v_i g_1 = v_i \qquad v_i g_2 = v_i \qquad (0 \le i \le n),$$

$$w_i g_1 = w_i + v_i \qquad w_i g_2 = w_i + v_{i-1} \qquad (1 \leq i \leq n).$$

We show that V is an indecomposable $\mathsf{K}\mathfrak{G}$-module. Let π be the projection of V onto the K-subspace $\mathsf{W} = \langle w_1, \ldots, w_n \rangle$ with $\ker \pi = \langle v_0, v_1, \ldots, v_n \rangle$. Obviously $g_1 - 1$ induces a K-isomorphism of W onto $\langle v_1, \ldots, v_n \rangle$ and $g_2 - 1$ induces a K-isomorphism of W onto $\langle v_0, v_1, \ldots, v_{n-1} \rangle$.

Suppose that $\mathsf{V} = \mathsf{V}_1 \oplus \mathsf{V}_2$ is a non-trivial decomposition of V as a $\mathsf{K}\mathfrak{G}$-module. Let $n_i = \dim_{\mathsf{K}} \mathsf{V}_i\pi$ $(i = 1, 2)$. Then

$$\mathsf{W} = \mathsf{V}\pi = \mathsf{V}_1\pi + \mathsf{V}_2\pi,$$

so

$$n = \dim_{\mathsf{K}} \mathsf{W} \leq \dim_{\mathsf{K}} \mathsf{V}_1\pi + \dim_{\mathsf{K}} \mathsf{V}_2\pi = n_1 + n_2.$$

Hence

$$\dim_{\mathsf{K}} \mathsf{V}_1 + \dim_{\mathsf{K}} \mathsf{V}_2 = \dim_{\mathsf{K}} \mathsf{V} = 2n + 1 < (2n_1 + 1) + (2n_2 + 1).$$

Thus $\dim_{\mathsf{K}} \mathsf{V}_i < 2n_i + 1$ for either $i = 1$ or $i = 2$. Hence for this i, $n_i > 0$. Since

$$\mathsf{V}_i/(\mathsf{V}_i \cap \ker \pi) \cong \mathsf{V}_i\pi$$

and $\dim_{\mathsf{K}} \mathsf{V}_i\pi = n_i$, it follows that

$$\dim_{\mathsf{K}}(\mathsf{V}_i \cap \ker \pi) < n_i + 1.$$

But

$$\mathsf{V}_i(g_1 - 1) + \mathsf{V}_i(g_2 - 1) \subseteq \mathsf{V}_i \cap \ker \pi.$$

So

$$\dim_{\mathsf{K}}(\mathsf{V}_i(g_1 - 1) + \mathsf{V}_i(g_2 - 1)) \leq n_i.$$

Note that for all $v \in \mathsf{V}$, we have $v - v\pi \in \langle v_0, \ldots, v_n \rangle$, so

$$(v - v\pi)(g - 1) = 0$$

for all $g \in \mathfrak{G}$, and $v(g - 1) = v\pi(g - 1)$. Thus

$$V_i(g - 1) = V_i\pi(g - 1)$$

and

$$\dim_K(V_i\pi(g_1 - 1) + V_i\pi(g_2 - 1)) \leq n_i.$$

But $g_1 - 1$ operates monomorphically on im π, so

$$\dim_K(V_i\pi(g_1 - 1)) = \dim_K V_i\pi = n_i.$$

Hence

$$V_i\pi(g_2 - 1) \subseteq V_i\pi(g_1 - 1).$$

Since $n_i > 0$, there exists a greatest integer k such that $k \leq n$ and

$$V_i\pi \subseteq \langle w_k, \ldots, w_n \rangle.$$

Hence there exists $w \in V_i\pi$ such that

$$w = c_k w_k + \cdots + c_n w_n$$

and $c_k \neq 0$. Thus

$$w(g_2 - 1) = c_k v_{k-1} + \cdots + c_n v_{n-1} \notin \langle v_k, \ldots, v_n \rangle.$$

But $V_i\pi(g_1 - 1) \subseteq \langle v_k, \ldots, v_n \rangle$, which gives a contradiction. **q.e.d.**

For cyclic p-groups, however, the situation is quite different. Before we study the cyclic case, we have to prove a general result about group-rings of p-groups.

5.2 Theorem. *Let K be a field of characteristic p and $\mathfrak{G}$ a p-group.*

a) $K\mathfrak{G}$ *has precisely one irreducible representation, namely, the unit representation.*

b) $\mathbf{J}(K\mathfrak{G})$ *is the augmentation ideal;*

$$\mathbf{J}(K\mathfrak{G}) = \left\{ \sum_{g \in \mathfrak{G}} a_g g \,\middle|\, \sum_{g \in \mathfrak{G}} a_g = 0 \right\}.$$

$\mathbf{J}(K\mathfrak{G})$ *is the only maximal right ideal and the only maximal left ideal of* $K\mathfrak{G}$.

c) *If* $\mathfrak{R}$ *is a proper right ideal of* $\mathsf{K}\mathfrak{G}$, *then* $\mathsf{K}\mathfrak{G}/\mathfrak{R}$ *is an indecomposable* $\mathsf{K}\mathfrak{G}$*-module. In particular, the free* $\mathsf{K}\mathfrak{G}$*-module with one generator is indecomposable.*

d) $\mathsf{K}\mathfrak{G}$ *has precisely one minimal right ideal, namely* $\mathsf{K}\sum_{g\in\mathfrak{G}} g$.

e) *Every non-zero submodule of* $\mathsf{K}\mathfrak{G}$ *is indecomposable.*

Proof. a) This is V, 5.16.

b) By V, 2.1, $\mathbf{J}(\mathsf{K}\mathfrak{G})$ is the intersection of the kernels of all the irreducible representations of $\mathsf{K}\mathfrak{G}$. It follows from a) that $\mathbf{J}(\mathsf{K}\mathfrak{G})$ is the kernel of the unit representation; but this is the augmentation ideal. By V, 2.2, $\mathbf{J}(\mathsf{K}\mathfrak{G})$ is the intersection of all maximal right ideals of $\mathsf{K}\mathfrak{G}$. Since $\dim_{\mathsf{K}} \mathsf{K}\mathfrak{G}/\mathbf{J}(\mathsf{K}\mathfrak{G}) = 1$, it follows that $\mathbf{J}(\mathsf{K}\mathfrak{G})$ is the only maximal right ideal of $\mathsf{K}\mathfrak{G}$. By symmetry, $\mathbf{J}(\mathsf{K}\mathfrak{G})$ is also the only maximal left ideal of $\mathsf{K}\mathfrak{G}$.

c) $\mathsf{K}\mathfrak{G}/\mathfrak{R}$ has precisely one maximal submodule, namely $\mathbf{J}(\mathsf{K}\mathfrak{G})/\mathfrak{R}$. Hence $\mathsf{K}\mathfrak{G}/\mathfrak{R}$ is indecomposable.

d) If $\mathfrak{J}$ is a minimal right ideal of $\mathsf{K}\mathfrak{G}$, the representation of $\mathfrak{G}$ on $\mathfrak{J}$ is the unit representation, by a). Thus $\mathfrak{J} = \mathsf{K}a$ for some $a \in \mathsf{K}\mathfrak{G}$ and $ag = a$ for all $g \in \mathfrak{G}$. It is easy to see that a then is a scalar multiple of $\sum_{g\in\mathfrak{G}} g$.

e) This follows immediately from d). **q.e.d.**

More detailed results about the group-ring of a p-group over a field of characteristic p will be proved in the next chapter (see VIII, § 2). We shall prove later (in 10.3, 10.14 and 11.6) that assertions similar to Theorem 5.2 hold for any projective indecomposable $\mathsf{K}\mathfrak{G}$-module P; namely, the socle and head of P are irreducible and isomorphic.

5.3 Theorem. *Suppose that* $\mathfrak{G} = \langle g \rangle$ *is a cyclic group and* K *is a field.*

a) *Any indecomposable* $\mathsf{K}\mathfrak{G}$*-module is an epimorphic image of* $\mathsf{K}\mathfrak{G}$.

b) *Suppose that* $|\mathfrak{G}| = p^n$ *and* $\operatorname{char} \mathsf{K} = p$.

(1) *For* $1 \leq i \leq p^n$, $\mathsf{K}\mathfrak{G}$ *has precisely one* $\mathsf{K}\mathfrak{G}$*-submodule of codimension i, namely* $\mathbf{J}(\mathsf{K}\mathfrak{G})^i$. $\mathbf{J}(\mathsf{K}\mathfrak{G})^i$ *is the* $\mathsf{K}\mathfrak{G}$*-submodule generated by* $(g-1)^i$.

(2) *For* $1 \leq i \leq p^n$, *put* $\mathsf{V}_i = \mathsf{K}\mathfrak{G}/\mathbf{J}(\mathsf{K}\mathfrak{G})^i$. *Then* V_i *is an indecomposable* $\mathsf{K}\mathfrak{G}$*-module of dimension i, and* V_i *has a* K*-basis* $\{v_1, \ldots, v_i\}$ *such that*

$$v_1 g = v_1, \; v_j g = v_{j-1} + v_j \quad (j = 2, \ldots, i).$$

(3) *Every indecomposable* $\mathsf{K}\mathfrak{G}$*-module is isomorphic to* V_i *for some i.*

Proof. a) Let V be any indecomposable $\mathsf{K}\mathfrak{G}$-module, and let σ be the linear transformation $v \to vg$ of V. Let t be an indeterminate over K. Then V becomes a finitely generated $\mathsf{K}[t]$-module if we put

$$vf(t) = vf(\sigma) \quad (v \in \mathsf{V}, f \in \mathsf{K}[t]).$$

Since $\mathsf{K}[t]$ is a principal ideal domain, V is the direct sum of cyclic $\mathsf{K}[t]$-submodules (I, 13.10). But a $\mathsf{K}[t]$-submodule is the same as a $\mathsf{K}\mathfrak{G}$-submodule. Since V is indecomposable, it follows that V is a cyclic $\mathsf{K}\mathfrak{G}$-module. Thus there is an epimorphism of $\mathsf{K}\mathfrak{G}$ onto V.

b) There is an K-algebra epimorphism α of $\mathsf{K}[t]$ onto $\mathsf{K}\mathfrak{G}$ such that $t\alpha = g$, and

$$\ker \alpha = (t^{p^n} - 1)\mathsf{K}[t] = (t - 1)^{p^n}\mathsf{K}[t].$$

Hence

$$\mathsf{K}\mathfrak{G} \cong \mathsf{K}[t]/(t - 1)^{p^n}\mathsf{K}[t].$$

The submodules of $\mathsf{K}\mathfrak{G}$ correspond to the ideals $\mathfrak{J}$ of $\mathsf{K}[t]$ for which

$$\mathfrak{J} \supseteq (t - 1)^{p^n}\mathsf{K}[t].$$

As $\mathsf{K}[t]$ is a principal ideal domain, $\mathfrak{J} = f\mathsf{K}[t]$, where $f \mid (t - 1)^{p^n}$. Hence $\mathfrak{J} = (t - 1)^i\mathsf{K}[t]$ for some i $(0 \leq i \leq p^n)$ and

$$\mathfrak{J}\alpha = (g - 1)^i\mathsf{K}\mathfrak{G}.$$

By 5.2, the radical $\mathbf{J}(\mathsf{K}\mathfrak{G})$ of $\mathsf{K}\mathfrak{G}$ is spanned by the elements

$$g^i - 1 = (g - 1)(1 + g + \cdots + g^{i-1}).$$

Hence

$$\mathbf{J}(\mathsf{K}\mathfrak{G}) = (g - 1)\mathsf{K}\mathfrak{G}.$$

Thus the submodules of $\mathsf{K}\mathfrak{G}$ are the

$$\mathbf{J}(\mathsf{K}\mathfrak{G})^i = (g - 1)^i\mathsf{K}\mathfrak{G} \quad (0 \leq i \leq p^n),$$

and

$$\dim_{\mathsf{K}} \mathsf{K}\mathfrak{G}/\mathbf{J}(\mathsf{K}\mathfrak{G})^i = \dim_{\mathsf{K}} \mathsf{K}[t]/(t - 1)^i\mathsf{K}[t] = \text{degree of } (t - 1)^i = i.$$

This proves (1).

By 5.2c), $\mathsf{V}_i = \mathsf{K}\mathfrak{G}/\mathbf{J}(\mathsf{K}\mathfrak{G})^i$ is indecomposable.

If we put

$$v_j = (g - 1)^{i-j} + \mathbf{J}(\mathsf{K}\mathfrak{G})^i \quad (j = 1, \ldots, i-1), \quad v_i = 1 + \mathbf{J}(\mathsf{K}\mathfrak{G})^i,$$

then $\{v_1, \ldots, v_i\}$ is a K-basis of V_i and

$$v_1 g = v_1, v_j g = v_{j-1} + v_j \quad (j = 2, \ldots, i).$$

Thus (2) is proved.

Suppose that V is any indecomposable $\mathsf{K}\mathfrak{G}$-module. By a), $\mathsf{V} \cong \mathsf{K}\mathfrak{G}/\mathsf{W}$ for some $\mathsf{K}\mathfrak{G}$-submodule W. By (1), $\mathsf{W} = (g-1)^i\mathsf{K}\mathfrak{G}$ for some i, hence

$$\mathsf{V} \cong \mathsf{K}\mathfrak{G}/(g-1)^i\mathsf{K}\mathfrak{G} \cong \mathsf{V}_i. \qquad \textbf{q.e.d.}$$

5.4 Theorem (D. G. HIGMAN [3], KASCH, KNESER, KUPISCH [1]). *Suppose that K is a field of characteristic p and that $\mathfrak{P}$ is a Sylow p-subgroup of $\mathfrak{G}$.*

a) *If $\mathfrak{P}$ is not cyclic, there exist indecomposable $\mathsf{K}\mathfrak{G}$-modules of arbitrarily large K-dimension. In particular, there exist infinitely many isomorphism types of indecomposable $\mathsf{K}\mathfrak{G}$-modules.*

b) *If $\mathfrak{P}$ is cyclic, there exist at most $|\mathfrak{G}|$ non-isomorphic indecomposable $\mathsf{K}\mathfrak{G}$-modules. Any indecomposable $\mathsf{K}\mathfrak{G}$-module is an epimorphic image of $\mathsf{K}\mathfrak{G}$.*

Proof. a) Since $\mathfrak{P}$ is not cyclic, it follows from 5.1 that for each natural number k there exists an indecomposable $\mathsf{K}\mathfrak{P}$-module V_k of dimension $2k+1$. By 4.3a), V_k is isomorphic to a direct summand of $(\mathsf{V}_k^{\mathfrak{G}})_{\mathfrak{P}}$. Thus, by the Krull-Schmidt theorem, V_k is isomorphic to a direct summand of $\mathsf{W}_{\mathfrak{P}}$ for some indecomposable direct summand W of $\mathsf{V}_k^{\mathfrak{G}}$. Hence

$$\dim_{\mathsf{K}} \mathsf{W} \geq \dim_{\mathsf{K}} \mathsf{V}_k = 2k + 1.$$

b) Now suppose that $\mathfrak{P}$ is cyclic. By 5.3, there are $|\mathfrak{P}|$ indecomposable $\mathsf{K}\mathfrak{P}$-modules $\mathsf{V}_1, \mathsf{V}_2, \ldots,$ where $\dim_{\mathsf{K}} \mathsf{V}_i = i$, and every indecomposable $\mathsf{K}\mathfrak{P}$-module is isomorphic to one of them. We show that any indecomposable $\mathsf{K}\mathfrak{G}$-module V is isomorphic to a direct summand of $\mathsf{V}_j^{\mathfrak{G}}$ for some j satisfying $j \leq \dim_{\mathsf{K}} \mathsf{V}$. Indeed, $|\mathfrak{G} : \mathfrak{P}|$ has an inverse in K. Hence by 4.3c), V is isomorphic to a direct summand of $(\mathsf{V}_{\mathfrak{P}})^{\mathfrak{G}}$. Now $\mathsf{V}_{\mathfrak{P}} = \bigoplus_i \mathsf{U}_i$ for certain indecomposable $\mathsf{K}\mathfrak{P}$-modules U_i, so by the Krull-Schmidt theorem, V is isomorphic to a direct summand of $\mathsf{U}_i^{\mathfrak{G}}$ for some i and $\dim_{\mathsf{K}} \mathsf{V} \geq \dim_{\mathsf{K}} \mathsf{U}_i$. Thus V is isomorphic to a direct summand of some $\mathsf{V}_j^{\mathfrak{G}}$, where $j \leq \dim_{\mathsf{K}} \mathsf{V}$. The number of non-isomorphic indecomposable direct summands of $\mathsf{V}_j^{\mathfrak{G}}$ of dimension at least j is at most $|\mathfrak{G} : \mathfrak{P}|$, since $\dim_{\mathsf{K}} \mathsf{V}_j^{\mathfrak{G}} = |\mathfrak{G} : \mathfrak{P}| j$. Hence there are at most $|\mathfrak{G}|$ non-isomorphic indecomposable $\mathsf{K}\mathfrak{G}$-modules.

By 5.3, V_j is an epimorphic image of K𝔓. By 4.2, $\cdot^{\mathfrak{G}}$ is an exact functor. Thus $V_j^{\mathfrak{G}}$ is an epimorphic image of $(\mathsf{K}\mathfrak{P})^{\mathfrak{G}} \cong \mathsf{K}\mathfrak{G}$. Thus any direct summand of $V_j^{\mathfrak{G}}$ is an epimorphic image of K𝔊. **q.e.d.**

Theorem 5.4 suggests that the theory of representations over a field of characteristic p is particularly simple if the Sylow p-subgroups of 𝔊 are cyclic. For the case in which p^2 does not divide $|\mathfrak{G}|$, Brauer developed a deep and rich theory in 1941 (R. BRAUER [3]). This theory was extended only much later to the case of cyclic Sylow p-subgroups (DADE [2]).

We look more closely into a special case which we shall need later.

5.5 Theorem (SRINIVASAN [1]). *Let 𝔊 be a group with an Abelian p-complement 𝔎 and let 𝔓 be a Sylow p-subgroup of 𝔊. Let* K *be an algebraically closed field of characteristic p.*

a) *If* $|\mathfrak{K}| = k$ *and*

$$\mathsf{K}\mathfrak{K} = \bigoplus_{i=1}^{k} \mathsf{K}e_i,$$

where $e_i, \ldots, e_k$ *are mutually orthogonal idempotents, then*

$$\mathsf{K}\mathfrak{G} = \bigoplus_{i=1}^{k} \mathsf{P}_i,$$

where $\mathsf{P}_i = (\mathsf{K}e_i)^{\mathfrak{G}}$. *Also* P_i *is an indecomposable* K𝔊*-module and* $(\mathsf{P}_i)_{\mathfrak{P}}$ *is isomorphic to* K𝔓.

b) *Suppose that also* $\mathfrak{P} \trianglelefteq \mathfrak{G}$. *Then each* P_i *has exactly one maximal submodule* M_i, *and* $\dim_{\mathsf{K}} \mathsf{P}_i/\mathsf{M}_i = 1$. *If*

$$e_i x = \alpha_i(x) e_i \quad \text{for } x \in \mathfrak{K},$$

then $\mathsf{P}_i/\mathsf{M}_i$ *is a* K𝔎*-module with character* α_i. *In particular,* $\mathsf{P}_i \not\cong \mathsf{P}_j$ *for* $i \neq j$.

c) *If 𝔓 is cyclic and normal in 𝔊, then for* $1 \leq i \leq |\mathfrak{P}|$, *𝔊 has exactly* $|\mathfrak{G}/\mathfrak{P}|$ *non-isomorphic indecomposable* K𝔊*-modules of* K*-dimension i. These are already indecomposable as* K𝔓*-modules, and they are all the indecomposable* K𝔊*-modules to within isomorphism.*

Proof. a) Since the order of 𝔎 is not divisible by p and K is algebraically closed, there exist mutually orthogonal idempotents e_i ($i = 1, \ldots, k$) in K𝔎 such that

$$1 = e_1 + \cdots + e_k \quad \text{and} \quad \mathsf{K}\mathfrak{K} = \bigoplus_{i=1}^{k} \mathsf{K}e_i.$$

Hence

$$\mathsf{K}\mathfrak{G} \cong (\mathsf{K}\mathfrak{K})^{\mathfrak{G}} = \bigoplus_{i=1}^{k} \mathsf{P}_i,$$

where $\mathsf{P}_i = (\mathsf{K}e_i)^{\mathfrak{G}}$. As $\mathfrak{P}$ is a transversal of $\mathfrak{K}$ in $\mathfrak{G}$, we have

$$(\mathsf{K}e_i)^{\mathfrak{G}} = \bigoplus_{y \in \mathfrak{P}} \mathsf{K}e_i \otimes y = e_i \otimes \mathsf{K}\mathfrak{P}.$$

This shows that $(\mathsf{P}_i)_{\mathfrak{P}} \cong \mathsf{K}\mathfrak{P}$. Hence $(\mathsf{P}_i)_{\mathfrak{P}}$ is indecomposable, by 5.2c). Thus P_i is indecomposable.

b) By 5.2b),

$$\mathbf{J}(\mathsf{K}\mathfrak{P}) = \left\{ \sum_{y \in \mathfrak{P}} a_y y \,\middle|\, \sum_{y \in \mathfrak{P}} a_y = 0 \right\}.$$

Hence for every $x \in \mathfrak{K}$ and $z \in \mathbf{J}(\mathsf{K}\mathfrak{P})$,

$$(e_i \otimes z)x = e_i x \otimes x^{-1}zx = \alpha_i(x)e_i \otimes x^{-1}zx \in e_i \otimes \mathbf{J}(\mathsf{K}\mathfrak{P}).$$

Thus $\mathsf{M}_i = e_i \otimes \mathbf{J}(\mathsf{K}\mathfrak{P})$ is a $\mathsf{K}\mathfrak{G}$-submodule of P_i of codimension 1. In particular, for $x \in \mathfrak{K}$,

$$(e_i \otimes 1)x = \alpha_i(x)(e_i \otimes 1)$$

and

$$\mathsf{P}_i = \mathsf{M}_i \oplus \mathsf{K}(e_i \otimes 1).$$

By a) and 5.2b), M_i is the only maximal submodule of P_i.

c) For $1 \leq j \leq |\mathfrak{P}|$ we consider the $\mathsf{K}\mathfrak{G}$-modules

$$\mathsf{V}_{ij} = \mathsf{P}_i/(e_i \otimes \mathbf{J}(\mathsf{K}\mathfrak{P})^j) = (e_i \otimes \mathsf{K}\mathfrak{P})/(e_i \otimes \mathbf{J}(\mathsf{K}\mathfrak{P})^j).$$

Then $\dim_{\mathsf{K}} \mathsf{V}_{ij} = j$ by 5.3b). Since $(\mathsf{V}_{ij})_{\mathfrak{P}}$ is indecomposable, certainly all V_{ij} are indecomposable $\mathsf{K}\mathfrak{G}$-modules. The representation of $\mathfrak{K}$ on the head $(e_i \otimes \mathsf{K}\mathfrak{P})/(e_i \otimes \mathbf{J}(\mathsf{K}\mathfrak{P}))$ of V_{ij} belongs to the character α_i. Hence

$V_{ij} \not\cong V_{i'j'}$ if $(i, j) \neq (i', j')$. Thus the V_{ij} $(i = 1, \ldots, |\mathfrak{K}|; j = 1, \ldots, |\mathfrak{P}|)$ are $|\mathfrak{K}|\ |\mathfrak{P}| = |\mathfrak{G}|$ indecomposable non-isomorphic $K\mathfrak{G}$-modules. By 5.4b) there are no other isomorphism types of indecomposable $K\mathfrak{G}$-modules. **q.e.d.**

5.6 Remarks. a) By a theorem of ROITER [1], Theorem 5.1 can be strengthened as follows. Let K be an infinite field of characteristic p and $\mathfrak{G}$ a non-cyclic p-group. Then for infinitely many natural numbers n there are infinitely many types of indecomposable $K\mathfrak{G}$-modules of dimension n.

b) For the elementary Abelian group of order 4, all indecomposable modules over a field of characteristic 2 have been determined (BAŠEV [1]; HELLER, REINER [1]). The same has been done for dihedral 2-groups (RINGEL [1]).

c) In contrast to b), the class of indecomposable $K\mathfrak{G}$-modules for an elementary Abelian group $\mathfrak{G}$ of type (p, p) for $p > 2$ is extremely large. BRENNER [1] has proved the following result.

Let $\mathfrak{H}$ be any local algebra of finite dimension (cf. 6.1). There exists an indecomposable $K\mathfrak{G}$-module V such that $\mathfrak{S} = \mathrm{Hom}_{K\mathfrak{G}}(V, V)$ has a nilpotent 2-sided ideal $\mathfrak{J}$ such that $\mathfrak{S}/\mathfrak{J} \cong \mathfrak{H}$.

Exercises

8) Let K be a field of characteristic p and $\mathfrak{G}$ an elementary Abelian group of type (p, p). Show that the $K\mathfrak{G}$-modules mentioned at the beginning of § 5 are indecomposable and pairwise non-isomorphic.

§ 6. Indecomposable and Absolutely Indecomposable Modules

6.1 Definition. a) Let $\mathfrak{R}$ be a ring (with identity element). The units of $\mathfrak{R}$ are those elements in $\mathfrak{R}$ that have left and right inverses in $\mathfrak{R}$. (On account of the associativity of the multiplication in $\mathfrak{R}$, both inverses coincide.) If the non-units of $\mathfrak{R}$ form a two-sided ideal of $\mathfrak{R}$, then $\mathfrak{R}$ is called a *local ring*.

b) For any ring $\mathfrak{R}$ (with identity element) we define the *Jacobson radical* $\mathbf{J}(\mathfrak{R})$ of $\mathfrak{R}$ as the intersection of the kernels of all irreducible $\mathfrak{R}$-modules. As in V, 2.2, $\mathbf{J}(\mathfrak{R})$ is the intersection of all maximal right ideals of $\mathfrak{R}$. (We remark that $\mathbf{J}(\mathfrak{R})$ in general need not be nilpotent.)

6.2 Lemma. a) *If $\mathfrak{R}$ is a local ring, the Jacobson radical* $\mathbf{J}(\mathfrak{R})$ *of* $\mathfrak{R}$ *is the set of all non-units of* $\mathfrak{R}$ *and* $\mathfrak{R}/\mathbf{J}(\mathfrak{R})$ *is a division ring.*

b) *If* $\mathfrak{R}/\mathbf{J}(\mathfrak{R})$ *is a division ring, then* $\mathfrak{R}$ *is a local ring.*

Proof. a) We denote by $\mathfrak{J}$ the two-sided ideal of non-units of the local ring $\mathfrak{R}$. If $\mathfrak{M}$ is any maximal right ideal of $\mathfrak{R}$, then $\mathfrak{M}$ contains only non-units of $\mathfrak{R}$. Hence $\mathfrak{M} \subseteq \mathfrak{J}$, so $\mathfrak{M} = \mathfrak{J}$. Thus $\mathfrak{J}$ is the only maximal right ideal of $\mathfrak{R}$ and $\mathfrak{J} = \mathbf{J}(\mathfrak{R})$.

If $a \in \mathfrak{R} - \mathbf{J}(\mathfrak{R})$, then a is a unit of $\mathfrak{R}$ and thus has a right inverse in $\mathfrak{R}$. Then certainly $a + \mathbf{J}(\mathfrak{R})$ has a right inverse in $\mathfrak{R}/\mathbf{J}(\mathfrak{R})$. Thus $\mathfrak{R}/\mathbf{J}(\mathfrak{R})$ is a division ring.

b) Suppose now that $\mathfrak{R}/\mathbf{J}(\mathfrak{R})$ is a division ring. We show first that if a has no right inverse in $\mathfrak{R}$, then $a \in \mathbf{J}(\mathfrak{R})$.

$a\mathfrak{R} \neq \mathfrak{R}$, so $a\mathfrak{R}$ lies in a maximal right ideal $\mathfrak{M}$ of $\mathfrak{R}$. But also $\mathbf{J}(\mathfrak{R}) \subseteq \mathfrak{M}$, so $a\mathfrak{R} + \mathbf{J}(\mathfrak{R}) \subseteq \mathfrak{M}$. Hence $(a\mathfrak{R} + \mathbf{J}(\mathfrak{R}))/\mathbf{J}(\mathfrak{R})$ is a proper right ideal of the division ring $\mathfrak{R}/\mathbf{J}(\mathfrak{R})$. This forces $(a\mathfrak{R} + \mathbf{J}(\mathfrak{R}))/\mathbf{J}(\mathfrak{R}) = 0$, hence $a \in \mathbf{J}(\mathfrak{R})$.

Now let b be a non-unit in $\mathfrak{R}$ and suppose b has a right inverse a. If a has a right inverse c, then $b = c$. But this is impossible as b is not a unit. Hence a has no right inverse and so $a \in \mathbf{J}(\mathfrak{R})$. But then

$$1 = ba \in \mathbf{J}(\mathfrak{R}),$$

a contradiction. Hence b has no right inverse and thus lies in $\mathbf{J}(\mathfrak{R})$. Thus $\mathbf{J}(\mathfrak{R})$ is the set of non-units of $\mathfrak{R}$, and $\mathfrak{R}$ is a local ring. **q.e.d.**

6.3 Lemma. *If $\mathfrak{R}$ is a local ring, 0 and 1 are the only idempotents in $\mathfrak{R}$.*

Proof. Let $\mathfrak{R}$ be a local ring and e an idempotent of $\mathfrak{R}$. If e is a unit, then $e = 1$, for if $ea = 1$, then

$$e = e(ea) = e^2a = ea = 1.$$

Similarly, if the idempotent $1 - e$ is a unit, then $e = 0$. Thus if $e \neq 0$ and $e \neq 1$, then e and $1 - e$ are non-units. Thus by 6.2a), there follows the contradiction

$$1 = e + (1 - e) \in \mathbf{J}(\mathfrak{R}). \qquad \textbf{q.e.d.}$$

6.4 Theorem (FITTING). *Suppose that $\mathfrak{R}$ is a ring and* V *an $\mathfrak{R}$-module with maximal and minimal condition for submodules. The following assertions are equivalent.*

a) V *is an indecomposable* $\mathfrak{R}$*-module.*
b) $\mathfrak{S} = \mathrm{Hom}_{\mathfrak{R}}(\mathsf{V}, \mathsf{V})$ *is a local ring.*

Proof. a) $\Rightarrow$ b): Suppose $\alpha \in \mathfrak{S}$. By I, 10.7 either α is an automorphism of V or α is nilpotent. We show that the non-units of $\mathfrak{S}$ form a two-sided ideal in $\mathfrak{S}$.

If α and β are non-units of $\mathfrak{S}$, then $\alpha + \beta$ is a non-unit by I, 10.10. If α is a non-unit of $\mathfrak{S}$, then α is nilpotent, so $\ker \alpha \supset 0$ and $\operatorname{im} \alpha \subset \mathsf{V}$. Hence for any $\gamma \in \mathfrak{S}$,

$$\ker(\alpha\gamma) \supseteq \ker \alpha \supset 0$$

and

$$\operatorname{im}(\gamma\alpha) \subseteq \operatorname{im} \alpha \subset \mathsf{V}.$$

Thus $\alpha\gamma$ and $\gamma\alpha$ are non-units of $\mathfrak{S}$, and $\mathfrak{S}$ is a local ring.

b) $\Rightarrow$ a): If $\mathfrak{S}$ is a local ring, then by 6.3, 0 and 1 are the only idempotents in $\mathfrak{S}$. But as every direct decomposition $\mathsf{V} = \mathsf{V}_1 \oplus \mathsf{V}_2$ of V gives rise to an idempotent π in $\mathfrak{S}$ such that $\operatorname{im} \pi = \mathsf{V}_1$ and $\ker \pi = \mathsf{V}_2$, V has to be indecomposable. **q.e.d.**

The only idempotents of the ring $\mathbb{Z}$ of rational integers are 0 and 1, but $\mathbb{Z}$ is not a local ring. We show that this cannot happen for rings with minimum condition for right ideals.

6.5 Theorem. *Let* $\mathfrak{R}$ *be a ring satisfying the minimum condition for right ideals. If 0 and 1 are the only idempotents in* $\mathfrak{R}$*, then* $\mathfrak{R}$ *is a local ring.*

Proof. Let $\mathfrak{S} = \mathrm{Hom}_{\mathfrak{R}}(\mathfrak{R}, \mathfrak{R})$. Since the elements of $\mathfrak{S}$ are precisely the left multiplications by elements of $\mathfrak{R}$, $\mathfrak{S}$ is antiisomorphic to $\mathfrak{R}$. Thus by hypothesis, 0 and 1 are the only idempotents in $\mathfrak{S}$. Hence $\mathfrak{R}$ is an indecomposable $\mathfrak{R}$-module. But by a well-known theorem of Hopkins (see KAPLANSKY [1], p. 134), $\mathfrak{R}$ also satisfies the maximum condition for right ideals. Hence by 6.4, $\mathfrak{S}$ is a local ring. Thus $\mathfrak{R}$ is a local ring. **q.e.d.**

Next we introduce the notion of absolutely indecomposable modules. The definition is similar to 2.1.

6.6 Definition. A $\mathsf{K}\mathfrak{G}$-module V is called *absolutely indecomposable* if

$$\dim_{\mathsf{K}} \mathrm{Hom}_{\mathsf{K}\mathfrak{G}}(\mathsf{V}, \mathsf{V})/\mathbf{J}(\mathrm{Hom}_{\mathsf{K}\mathfrak{G}}(\mathsf{V}, \mathsf{V})) = 1,$$

which means that the K-algebra $\mathrm{Hom}_{K\mathfrak{G}}(V, V)/\mathbf{J}(\mathrm{Hom}_{K\mathfrak{G}}(V, V))$ is isomorphic to K.

6.7 Lemma. *Let* V *be a* $K\mathfrak{G}$*-module.*

a) *If* V *is absolutely indecomposable, then for any extension* L *of* K, V_L *is an absolutely indecomposable* $L\mathfrak{G}$*-module.*

b) *If* K *is algebraically closed and* V *is an indecomposable* $K\mathfrak{G}$*-module, then* V *is absolutely indecomposable.*

Proof. We write $\mathfrak{S} = \mathrm{Hom}_{K\mathfrak{G}}(V, V)$. By 1.12, we have

$$\mathrm{Hom}_{L\mathfrak{G}}(V_L, V_L) \cong \mathfrak{S} \otimes_K L.$$

By hypothesis, $\mathfrak{S}/\mathbf{J}(\mathfrak{S}) \cong K$. Hence by 1.2c),

$$L \cong K \otimes_K L \cong (\mathfrak{S}/\mathbf{J}(\mathfrak{S})) \otimes_K L \cong \mathfrak{S}_L/\mathbf{J}(\mathfrak{S})_L.$$

By 1.2d), $\mathbf{J}(\mathfrak{S})_L \subseteq \mathbf{J}(\mathfrak{S}_L)$. As $\mathfrak{S}_L$ is a ring with identity element, it follows that $\mathbf{J}(\mathfrak{S})_L = \mathbf{J}(\mathfrak{S}_L)$ and $\mathfrak{S}_L/\mathbf{J}(\mathfrak{S}_L) \cong L$. Hence $\mathfrak{S}_L$ is a local ring by 6.2b), and V_L is indecomposable by 6.4.

b) As V is indecomposable, $\mathrm{Hom}_{K\mathfrak{G}}(V, V)$ is a local ring, by 6.4. Then by 6.2a),

$$\mathrm{Hom}_{K\mathfrak{G}}(V, V)/\mathbf{J}(\mathrm{Hom}_{K\mathfrak{G}}(V, V))$$

is a division algebra of finite dimension over K. As K is algebraically closed, it follows that

$$\mathrm{Hom}_{K\mathfrak{G}}(V, V)/\mathbf{J}(\mathrm{Hom}_{K\mathfrak{G}}(V, V)) \cong K,$$

and V is absolutely indecomposable. **q.e.d.**

It is natural to conjecture by analogy with 2.2 that the indecomposability of V_L for every extension L of K implies the absolute indecomposability of V. But this is not true in general, as is shown by the following example.

6.8 Example (GREEN). Let $\mathfrak{G} = \langle g_1 \rangle \times \langle g_2 \rangle$ be an elementary Abelian group of order p^2.

a) Let K be a field of characteristic p and let A be a non-singular $n \times n$ matrix such that the characteristic polynomial f of A is irreducible. Let V be a $K\mathfrak{G}$-module for the matrix representation

$$g_1 \to \begin{pmatrix} I & 0 \\ I & I \end{pmatrix}, \qquad g_2 \to \begin{pmatrix} I & 0 \\ A & I \end{pmatrix}$$

of $\mathfrak{G}$ of degree $2n$. Then $\mathrm{Hom}_{\mathsf{K}\mathfrak{G}}(\mathsf{V}, \mathsf{V})/\mathbf{J}(\mathrm{Hom}_{\mathsf{K}\mathfrak{G}}(\mathsf{V}, \mathsf{V}))$ is isomorphic to the field $\mathsf{K}[t]/f\mathsf{K}[t]$.

To prove this, we first check that every matrix which commutes with the images of g_1 and g_2 is of the form

$$\begin{pmatrix} X & 0 \\ Y & X \end{pmatrix},$$

where $XA = AX$. Thus there is an isomorphism α of $\mathrm{Hom}_{\mathsf{K}\mathfrak{G}}(\mathsf{V}, \mathsf{V})$ onto the algebra $\mathfrak{A}$ of all such matrices. Let

$$\mathfrak{B} = \{X \mid X \in (\mathsf{K})_n, \quad XA = AX\},$$

and let β be the epimorphism of $\mathfrak{A}$ onto $\mathfrak{B}$ given by

$$\begin{pmatrix} X & 0 \\ Y & X \end{pmatrix}\beta = X.$$

If $\mathfrak{J} = \ker \alpha\beta$, then $\mathfrak{J}$ is nilpotent, since

$$\mathfrak{J}\alpha = \left\{\begin{pmatrix} 0 & 0 \\ Y & 0 \end{pmatrix} \middle| \, Y \in (\mathsf{K})_n\right\}.$$

Also $\mathrm{Hom}_{\mathsf{K}\mathfrak{G}}(\mathsf{V}, \mathsf{V})/\mathfrak{J} \cong \mathfrak{B}$. Since the characteristic polynomial f of A is irreducible, f is also the minimum polynomial of A.

Let W be the space of dimension n on which A operates and $0 \neq w \in \mathsf{W}$. Then $\langle w, wA, \ldots, wA^{n-1}\rangle$ is an A-invariant subspace of W. As the characteristic polynomial f of A is irreducible,

$$\mathsf{W} = \langle w, wA, \ldots, wA^{n-1}\rangle.$$

Suppose $X \in \mathfrak{B}$ and

$$wX = \sum_{j=0}^{n-1} \lambda_j w A^j \quad (\lambda_j \in \mathsf{K}).$$

If we define the polynomial h by $h = \sum_{j=0}^{n-1} \lambda_j t^j$, then

$$wX = wh(A).$$

Then

$$wA^jX = wXA^j = wh(A)A^j = wA^jh(A),$$

so $X = h(A)$. Hence

$$\mathfrak{B} = \left\{\sum_{j=0}^{n-1} \lambda_j A^j \middle| \lambda_j \in \mathsf{K}\right\}$$

and $\mathfrak{B} \cong \mathsf{K}[t]/f\mathsf{K}[t]$. This is a field, since f is irreducible. Hence $\mathrm{Hom}_{\mathsf{K}\mathfrak{G}}(\mathsf{V}, \mathsf{V})/\mathfrak{J}$ is a field and $\mathbf{J}(\mathrm{Hom}_{\mathsf{K}\mathfrak{G}}(\mathsf{V}, \mathsf{V})) = \mathfrak{J}$. Therefore

$$\mathrm{Hom}_{\mathsf{K}\mathfrak{G}}(\mathsf{V}, \mathsf{V})/\mathbf{J}(\mathrm{Hom}_{\mathsf{K}\mathfrak{G}}(\mathsf{V}, \mathsf{V}))$$

is isomorphic to the field $\mathsf{K}[t]/f\mathsf{K}[t]$.

b) Now let $\mathsf{K} = GF(p)(x)$ be the field of all rational functions in an indeterminate x over $GF(p)$ and let A be the $p \times p$ matrix

$$\begin{pmatrix} 0 & 1 & 0 & \dots & 0 \\ 0 & 0 & 1 & \dots & 0 \\ \vdots & \vdots & \vdots & & \vdots \\ 0 & 0 & 0 & & 1 \\ x & 0 & 0 & \dots & 0 \end{pmatrix}.$$

The characteristic polynomial $t^p - x$ of A is irreducible over K (see VAN DER WAERDEN [2], vol. I, p. 184). Thus if V is the $\mathsf{K}\mathfrak{G}$-module constructed in a), $\mathrm{Hom}_{\mathsf{K}\mathfrak{G}}(\mathsf{V}, \mathsf{V})/\mathbf{J}(\mathrm{Hom}_{\mathsf{K}\mathfrak{G}}(\mathsf{V}, \mathsf{V}))$ is not isomorphic to K and V is not absolutely indecomposable.

Nevertheless, for every extension L of K,

$$\mathrm{Hom}_{\mathsf{L}\mathfrak{G}}(\mathsf{V}_\mathsf{L}, \mathsf{V}_\mathsf{L}) \cong \mathrm{Hom}_{\mathsf{K}\mathfrak{G}}(\mathsf{V}, \mathsf{V}) \otimes_\mathsf{K} \mathsf{L}$$

is a local ring and hence V_L is indecomposable, by 6.4. To show this, we write $\mathfrak{S} = \mathrm{Hom}_{\mathsf{K}\mathfrak{G}}(\mathsf{V}, \mathsf{V})$. Then by 1.2c) and 1.4a),

$$\begin{aligned} \mathfrak{S}_\mathsf{L}/\mathbf{J}(\mathfrak{S})_\mathsf{L} \cong (\mathfrak{S}/\mathbf{J}(\mathfrak{S})) \otimes_\mathsf{K} \mathsf{L} &\cong \mathsf{K}[t]/(t^p - x)\mathsf{K}[t] \otimes_\mathsf{K} \mathsf{L} \\ &\cong \mathsf{L}[t]/(t^p - x)\mathsf{L}[t]. \end{aligned}$$

By 1.2d), $\mathbf{J}(\mathfrak{S})_{\mathsf{L}} \subseteq \mathbf{J}(\mathfrak{S}_{\mathsf{L}})$. If $t^p - x$ is irreducible in $\mathsf{L}[t]$, then $\mathfrak{S}_{\mathsf{L}}/\mathbf{J}(\mathfrak{S})_{\mathsf{L}}$ is a field, hence $\mathbf{J}(\mathfrak{S})_{\mathsf{L}} = \mathbf{J}(\mathfrak{S}_{\mathsf{L}})$, and $\mathfrak{S}_{\mathsf{L}}$ is a local ring by 6.2b).

If $t^p - x$ is reducible in $\mathsf{L}[t]$, then $t^p - x = (t - z)^p$ for some $z \in \mathsf{L}$ (see VAN DER WAERDEN, as above). Then

$$\mathfrak{i} = (t - z)\mathsf{L}[t]/(t^p - x)\mathsf{L}[t]$$

is an ideal in $\mathsf{L}[t]/(t^p - x)\mathsf{L}[t]$ for which $(\mathsf{L}[t]/(t^p - x)\mathsf{L}[t])/\mathfrak{i}$ is isomorphic to L and $\mathfrak{i}^p = 0$. If $\mathfrak{J}/\mathbf{J}(\mathfrak{S})_{\mathsf{L}}$ is the corresponding ideal of $\mathfrak{S}_{\mathsf{L}}/\mathbf{J}(\mathfrak{S})_{\mathsf{L}}$, then $\mathfrak{J}$ is a two-sided ideal for which $\mathfrak{S}_{\mathsf{L}}/\mathfrak{J} \cong \mathsf{L}$ and $\mathfrak{J}^p \subseteq \mathbf{J}(\mathfrak{S})_{\mathsf{L}}$. As $\mathbf{J}(\mathfrak{S})_{\mathsf{L}}$ is nilpotent, $\mathfrak{J}$ is nilpotent. This shows that $\mathfrak{J} = \mathbf{J}(\mathfrak{S}_{\mathsf{L}})$ and $\mathfrak{S}_{\mathsf{L}}/\mathbf{J}(\mathfrak{S}_{\mathsf{L}}) \cong \mathsf{L}$. So $\mathfrak{S}_{\mathsf{L}}$ is a local ring in this case also, and V_{L} is indecomposable.

The fact that the full analogue of 2.2 does not hold is not very serious for our purpose, since it does hold for perfect fields.

6.9 Theorem. *Let* K *be a perfect field and* V *a* $\mathsf{K}\mathfrak{G}$*-module. Then the following are equivalent.*

a) V *is absolutely indecomposable.*

b) *For every extension* L *of* K, V_{L} *is indecomposable.*

c) *If* $\hat{\mathsf{K}}$ *is the algebraic closure of* K, $\mathsf{V}_{\hat{\mathsf{K}}}$ *is indecomposable.*

Proof. By 6.7a), a) implies b)., Trivially, b) implies c). To prove that c) implies a), suppose that $\mathsf{V}_{\hat{\mathsf{K}}}$ is indecomposable and write $\mathfrak{S} = \mathrm{Hom}_{\mathsf{K}\mathfrak{G}}(\mathsf{V}, \mathsf{V})$. Thus $\mathfrak{S}_{\hat{\mathsf{K}}} \cong \mathrm{Hom}_{\hat{\mathsf{K}}\mathfrak{G}}(\mathsf{V}_{\hat{\mathsf{K}}}, \mathsf{V}_{\hat{\mathsf{K}}})$. As $\mathsf{V}_{\hat{\mathsf{K}}}$ is indecomposable, certainly V itself is indecomposable, so $\mathfrak{S}$ and $\mathfrak{S}_{\hat{\mathsf{K}}}$ are local rings. Thus $\mathfrak{S}_{\hat{\mathsf{K}}}/\mathbf{J}(\mathfrak{S}_{\hat{\mathsf{K}}})$ is a division algebra of finite dimension over the algebraically closed field $\hat{\mathsf{K}}$ and is thus isomorphic to $\hat{\mathsf{K}}$. Write $\mathfrak{D} = \mathfrak{S}/\mathbf{J}(\mathfrak{S})$. It is to be shown that the division ring $\mathfrak{D}$ is isomorphic to K. We have

$$\mathfrak{D} \otimes_{\mathsf{K}} \hat{\mathsf{K}} \cong \mathfrak{S}_{\hat{\mathsf{K}}}/\mathbf{J}(\mathfrak{S})_{\hat{\mathsf{K}}}.$$

Let a be an element of $\mathbf{J}(\mathfrak{D} \otimes_{\mathsf{K}} \hat{\mathsf{K}})$. Then a is nilpotent and lies in $\mathfrak{D} \otimes_{\mathsf{K}} \mathsf{L}$ for some finite extension L of K; thus a generates a nilpotent ideal in $\mathfrak{D} \otimes_{\mathsf{K}} \mathsf{L}$. But since K is perfect, $\mathbf{J}(\mathfrak{D} \otimes_{\mathsf{K}} \mathsf{L}) = 0$ (see VAN DER WAERDEN [3], vol. II, p. 77). Hence $a = 0$, and $\mathbf{J}(\mathfrak{D} \otimes_{\mathsf{K}} \hat{\mathsf{K}}) = 0$. Thus $\mathfrak{S}_{\hat{\mathsf{K}}}/\mathbf{J}(\mathfrak{S})_{\hat{\mathsf{K}}}$ is a semisimple algebra and $\mathbf{J}(\mathfrak{S}_{\hat{\mathsf{K}}}) = \mathbf{J}(\mathfrak{S})_{\hat{\mathsf{K}}}$. Thus $\mathfrak{D} \otimes_{\mathsf{K}} \hat{\mathsf{K}}$ is isomorphic to $\mathfrak{S}_{\hat{\mathsf{K}}}/\mathbf{J}(\mathfrak{S}_{\hat{\mathsf{K}}})$ and hence to $\hat{\mathsf{K}}$. Thus

$$\dim_{\mathsf{K}} \mathfrak{D} = \dim_{\hat{\mathsf{K}}}(\mathfrak{D} \otimes \hat{\mathsf{K}}) = 1$$

and $\mathfrak{D}$ is isomorphic to K. Hence V is absolutely indecomposable.

q.e.d.

For a fixed $\mathsf{K}\mathfrak{G}$-module V, there exists a decomposition of V_L into absolutely indecomposable $\mathsf{L}\mathfrak{G}$-modules in a finite extension L of K.

6.10 Theorem. *Let* V *be a* $\mathsf{K}\mathfrak{G}$*-module. Then there exists a finite extension* L *of* K *such that*

$$\mathsf{V}_\mathsf{L} = \mathsf{V}_1 \oplus \cdots \oplus \mathsf{V}_s$$

for absolutely indecomposable $\mathsf{L}\mathfrak{G}$*-modules* V_i.

Proof. For any extension L of K, the number of summands in a direct decomposition of V_L is bounded by $\dim_\mathsf{K} \mathsf{V}$. Hence we may choose a finite extension L of K such that

$$\mathsf{V}_\mathsf{L} = \mathsf{V}_1 \oplus \cdots \oplus \mathsf{V}_s,$$

where the V_i are indecomposable and s is as large as possible. Then for every finite extension L' of L, $(\mathsf{V}_i)_{\mathsf{L}'}$ is indecomposable. Write $\mathfrak{S}_i = \operatorname{Hom}_{\mathsf{L}\mathfrak{G}}(\mathsf{V}_i, \mathsf{V}_i)$. By extending L if necessary, we may suppose that for every $i = 1, \ldots, s$, $\dim_\mathsf{L} \mathbf{J}(\mathfrak{S}_i)$ is as large as possible; this means that

$$\mathbf{J}(\mathfrak{S}_i) \otimes_\mathsf{L} \mathsf{L}' = \mathbf{J}(\mathfrak{S}_i \otimes_\mathsf{L} \mathsf{L}')$$

for every finite extension L' of L. We have to show that $\dim_\mathsf{L} \mathfrak{S}_i/\mathbf{J}(\mathfrak{S}_i) = 1$ for all i.

Since V_i is indecomposable, $\mathfrak{S}_i/\mathbf{J}(\mathfrak{S}_i)$ is a division algebra $\mathfrak{D}_i$. Suppose that $\dim_\mathsf{L} \mathfrak{D}_i \neq 1$. Denote by $\hat{\mathsf{L}}$ the algebraic closure of L. Now $\mathfrak{D}_i \otimes_\mathsf{L} \hat{\mathsf{L}}$ is not a division ring, so it has zero divisors. Suppose that $xy = 0$, where

$$0 \neq x = \sum_{j=1}^{m} a_j \otimes \lambda_j, \quad 0 \neq y = \sum_{k=1}^{n} b_k \otimes \lambda'_k,$$

and a_j, b_k are in $\mathfrak{D}_i$, λ_j, λ'_k are in $\hat{\mathsf{L}}$. Then λ_j, λ'_k are algebraic over L, so they all lie in a finite extension L' of L. Thus x, y are zero divisors in $\mathfrak{D}_i \otimes_\mathsf{L} \mathsf{L}'$. As $(\mathsf{V}_i)_{\mathsf{L}'}$ is indecomposable,

$$\mathfrak{S}_i \otimes_\mathsf{L} \mathsf{L}' \cong \operatorname{Hom}_{\mathsf{L}'\mathfrak{G}}((\mathsf{V}_i)_{\mathsf{L}'}, (\mathsf{V}_i)_{\mathsf{L}'})$$

is a local ring. But we have

$$\mathbf{J}(\mathfrak{S}_i) \otimes_{\mathsf{L}} \mathsf{L}' = \mathbf{J}(\mathfrak{S}_i \otimes_{\mathsf{L}} \mathsf{L}'),$$

and

$$(\mathfrak{S}_i \otimes_{\mathsf{L}} \mathsf{L}')/(\mathbf{J}(\mathfrak{S}_i) \otimes_{\mathsf{L}} \mathsf{L}') \cong \mathfrak{D}_i \otimes_{\mathsf{L}} \mathsf{L}'$$

is not a division algebra. This contradiction shows that $\dim_{\mathsf{L}} \mathfrak{D}_i = 1$, so V_i is absolutely indecomposable. **q.e.d.**

6.11 Example. Suppose that $\mathfrak{G} = \langle g_1 \rangle \times \langle g_2 \rangle$ is elementary Abelian of order p^2 and that L is a field of characteristic p. For every $a \in \mathsf{L}$ we define a representation ϱ_a of $\mathfrak{G}$ by

$$\varrho_a(g_1) = \begin{pmatrix} 1 & 0 \\ 1 & 1 \end{pmatrix}, \quad \varrho_a(g_2) = \begin{pmatrix} 1 & 0 \\ a & 1 \end{pmatrix}.$$

ϱ_a is indecomposable, for even its restriction to $\langle g_1 \rangle$ in indecomposable.

In which subfield K of L can ϱ_a be realized? ϱ_a can be realized in K if and only if there exists a non-singular 2×2 matrix $C = (c_{ij})$ with coefficients c_{ij} in some extension of L such that we have

$$C\varrho_a(g_1)C^{-1} = \frac{1}{d}\begin{pmatrix} * & -c_{12}^2 \\ c_{22}^2 & * \end{pmatrix} \in (\mathsf{K})_2$$

and

$$C\varrho_a(g_2)C^{-1} = \frac{1}{d}\begin{pmatrix} * & -ac_{12}^2 \\ ac_{22}^2 & * \end{pmatrix} \in (\mathsf{K})_2,$$

where $d = \det C$. As $c_{12} \neq 0$ or $c_{22} \neq 0$, this forces $a \in \mathsf{K}$. Hence $GF(p)(a)$ is the smallest field in which ϱ_a can be realized.

This observation leads to several remarks.

a) If we choose $\mathsf{K} = GF(p)$ and a transcendental over K, then ρ_a cannot be realized in the algebraic closure of K.

b) Choose $\mathsf{K} = GF(p)$ and a_i such that $(\mathsf{K}(a_i) : \mathsf{K}) = i$ $(i = 1, 2, \ldots)$. Then all ϱ_{a_i} can be realized in the algebraic closure of K, but not in any finite extension L of K.

c) Suppose that a is algebraic over $\mathsf{K} = GF(p)$ and $\mathsf{L} = \mathsf{K}(a)$. Let $\mathsf{V}_a = \mathsf{L}v_1 \oplus \mathsf{L}v_2$ be an $\mathsf{L}\mathfrak{G}$-module for ϱ_a, with

$$v_1 g_1 = v_1 \qquad v_1 g_2 = v_1$$

$$v_2 g_1 = v_1 + v_2 \qquad v_2 g_2 = a v_1 + v_2.$$

We denote V_a, regarded as a $\mathsf{K}\mathfrak{G}$-module, by V_a^0. Thus $\dim_{\mathsf{K}} \mathsf{V}_a^0 = 2(\mathsf{L}:\mathsf{K})$. Suppose that $\{l_1, \ldots, l_s\}$ is a K-basis of L and

$$l_i a = \sum_{j=1}^{s} a_{ij} l_j \quad (a_{ij} \in \mathsf{K}).$$

Thus $A = (a_{ij})$ is the matrix of the K-linear transformation $l \to la$ of L. Hence the minimum polynomial of A is the minimum polynomial f of a over K and is therefore irreducible. Also, since the degree of f is $s = (\mathsf{L}:\mathsf{K})$, f is also the characteristic polynomial of A. Now if ϱ_a^0 is the matrix representation of $\mathfrak{G}$ relative to the K-basis

$$\{l_1 v_1, \ldots, l_s v_1, l_1 v_2, \ldots, l_s v_2\}$$

of V_a^0, then

$$\varrho_a^0(g_1) = \begin{pmatrix} I & 0 \\ I & I \end{pmatrix}, \quad \varrho_a^0(g_2) = \begin{pmatrix} I & 0 \\ A & I \end{pmatrix}.$$

Hence by 6.8a),

$$\mathrm{Hom}_{\mathsf{K}\mathfrak{G}}(\mathsf{V}_a^0, \mathsf{V}_a^0)/\mathbf{J}(\mathrm{Hom}_{\mathsf{K}\mathfrak{G}}(\mathsf{V}_a^0, \mathsf{V}_a^0)) \cong \mathsf{K}[t]/f\mathsf{K}[t] \cong \mathsf{K}(a).$$

Thus by 6.4, V_a^0 is indecomposable.

The characteristic roots of the matrix A are the s distinct conjugates $a = a_1, \ldots, a_s$ of a over K. Hence there exists a matrix $Z \in (\mathsf{L})_s$ such that $Z^{-1}AZ = B$, where B is the diagonal matrix $(\delta_{ij} a_i)$. If we write

$$W = \begin{pmatrix} Z & 0 \\ 0 & Z \end{pmatrix},$$

then

$$W^{-1}\varrho_a^0(g_1)W = \begin{pmatrix} I & 0 \\ I & I \end{pmatrix} \quad \text{and} \quad W^{-1}\varrho_a^0(g_2)W = \begin{pmatrix} I & 0 \\ B & I \end{pmatrix}.$$

Thus $\mathsf{V}_a^0 \otimes_{\mathsf{K}} \mathsf{L}$ has an L-basis $\{w_1, \ldots, w_s, w_1', \ldots, w_s'\}$ such that

$$w_i g_1 = w_i \qquad w_i g_2 = w_i$$

$$w_i' g_1 = w_i + w_i' \qquad w_i' g_2 = a_i w_i + w_i'.$$

This shows that

$$\mathsf{V}_a^0 \otimes_{\mathsf{K}} \mathsf{L} \cong \langle w_1, w_1' \rangle \oplus \cdots \oplus \langle w_s, w_s' \rangle,$$

where $\langle w_i, w_i' \rangle \cong \mathsf{V}_{a_i}$ is an absolutely indecomposable $\mathsf{L}\mathfrak{G}$-module. By the remark, it is clear that for no proper subfield L_1 of L the module $\mathsf{V}_a^0 \otimes_{\mathsf{K}} \mathsf{L}_1$ is a sum of absolutely indecomposable $\mathsf{L}_1\,\mathfrak{G}$-modules. Hence in the situation of Theorem 6.10, no universal choice of L for all $\mathsf{K}\mathfrak{G}$-modules is possible.

Exercises

9) Let $\mathfrak{R}$ be an Artinian ring. Show that the following statements are equivalent.

a) $\mathfrak{R}$ is a local ring.

b) Every element of $\mathfrak{R}$ is either nilpotent or invertible.

10) Let V be a $\mathsf{K}\mathfrak{G}$-module and $\hat{\mathsf{K}}$ the algebraic closure of K. Then the following statements are equivalent.

a) V_{L} is indecomposable for every extension L of K.

b) $\mathsf{V}_{\hat{\mathsf{K}}}$ is indecomposable.

c) $\mathsf{V}_{\hat{\mathsf{K}}}$ is absolutely indecomposable.

11) Let $\mathfrak{G}$ be an elementary Abelian group of order p^2. Prove that the $\mathsf{K}\mathfrak{G}$-module V of 5.1 is absolutely indecomposable.

§ 7. Relative Projective and Relative Injective Modules

As in § 4, A denotes a commutative ring with identity.

7.1 Definition. Suppose $\mathfrak{U} \leq \mathfrak{G}$.

a) An $\mathsf{A}\mathfrak{G}$-module V is called $(\mathfrak{G}, \mathfrak{U})$-*projective*, if whenever

$$0 \to \mathsf{T} \xrightarrow{\alpha} \mathsf{U} \to \mathsf{V} \to 0$$

is an exact sequence of $A\mathfrak{G}$-modules and $(T\alpha)_{\mathfrak{U}}$ is a direct summand of $U_{\mathfrak{U}}$, then $T\alpha$ is a direct summand of U.

b) An $A\mathfrak{G}$-module V is called $(\mathfrak{G}, \mathfrak{U})$-*injective*, if whenever

$$0 \to V \xrightarrow{\alpha} U \to T \to 0$$

is an exact sequence of $A\mathfrak{G}$-modules and $(V\alpha)_{\mathfrak{U}}$ is a direct summand of $U_{\mathfrak{U}}$, then $V\alpha$ is a direct summand of U.

If A is a field, the $(\mathfrak{G}, 1)$-projective modules are just the projective $A\mathfrak{G}$-modules, and the $(\mathfrak{G}, 1)$-injective modules are the injective $A\mathfrak{G}$-modules. More generally, an $A\mathfrak{G}$-module V is projective (injective) if and only if it is $(\mathfrak{G}, 1)$-projective ($(\mathfrak{G}, 1)$-injective) and V is a projective (injective) A-module (see Exercise 14).

To study the $(\mathfrak{G}, \mathfrak{U})$-projective and $(\mathfrak{G}, \mathfrak{U})$-injective modules, we introduce Gaschütz operators.

7.2 Definition. Suppose that T is a transversal of $\mathfrak{U}$ in $\mathfrak{G}$. If V is an $A\mathfrak{G}$-module, we call γ a $(\mathfrak{G}, \mathfrak{U})$-*Gaschütz operator* on V if $\gamma \in \mathrm{Hom}_{A\mathfrak{U}}(V_{\mathfrak{U}}, V_{\mathfrak{U}})$ and

$$v \sum_{t \in T} t^{-1} \gamma t = v$$

for all $v \in V$. Obviously, this condition is independent of the choice of the transversal T.

7.3 Lemma. *Suppose that* $\mathfrak{U} \leq \mathfrak{G}$.

a) *If the* $A\mathfrak{G}$*-modules* V_i $(i \in I)$ *all have* $(\mathfrak{G}, \mathfrak{U})$*-Gaschütz operators, so does* $\bigoplus_{i \in I} V_i$.

b) *If* V *has a* $(\mathfrak{G}, \mathfrak{U})$*-Gaschütz operator, so does any direct summand of* V.

c) *If* $|\mathfrak{G} : \mathfrak{U}|$ *has an inverse in* A, *any* $A\mathfrak{G}$*-module has a* $(\mathfrak{G}, \mathfrak{U})$*-Gaschütz operator.*

d) *If* W *is an* $A\mathfrak{U}$*-module, then* $W^{\mathfrak{G}} = W \otimes_{A\mathfrak{U}} A\mathfrak{G}$ *has a* $(\mathfrak{G}, \mathfrak{U})$*-Gaschütz operator.*

e) *The regular* $A\mathfrak{G}$*-module* $A\mathfrak{G}$ *has a* $(\mathfrak{G}, \mathfrak{U})$*-Gaschütz operator.*

Proof. a) If γ_i is a $(\mathfrak{G}, \mathfrak{U})$-Gaschütz operator on V_i, we define γ on $\bigoplus_{i \in I} V_i$ such that γ_i is the restriction of γ to V_i $(i \in I)$. Then γ is obviously a $(\mathfrak{G}, \mathfrak{U})$-Gaschütz operator on $\bigoplus_{i \in I} V_i$.

b) Suppose that $V = V_1 \oplus V_2$, where V_1 and V_2 are $A\mathfrak{G}$-modules, and γ is a $(\mathfrak{G}, \mathfrak{U})$-Gaschütz operator on V. Let π_i ($i = 1, 2$) be the projection of V onto V_i defined by the decomposition $\mathsf{V} = \mathsf{V}_1 \oplus \mathsf{V}_2$. Then $\gamma\pi_i$ induces an $\mathsf{A}\mathfrak{U}$-homomorphism of V_i into V_i, and for all $v_i \in \mathsf{V}_i$ we have

$$v_i = \left(v_i \sum_{t \in \mathsf{T}} t^{-1}\gamma t\right)\pi_i = v_i \sum_{t \in \mathsf{T}} t^{-1}\gamma\pi_i t,$$

since π_i is an $\mathsf{A}\mathfrak{G}$-homomorphism. Thus $\gamma\pi_i$ is a $(\mathfrak{G}, \mathfrak{U})$-Gaschütz operator on V_i.

c) If V is an $\mathsf{A}\mathfrak{G}$-module and γ is defined by

$$v\gamma = |\mathfrak{G} : \mathfrak{U}|^{-1} v$$

for all $v \in \mathsf{V}$, then γ is a $(\mathfrak{G}, \mathfrak{U})$-Gaschütz operator on V.

d) Suppose that T is a transversal of $\mathfrak{U}$ in $\mathfrak{G}$, so chosen that $1 \in \mathsf{T}$. Then

$$\mathsf{W}^{\mathfrak{G}} = \bigoplus_{t \in \mathsf{T}} \mathsf{W} \otimes t.$$

We define γ by

$$\left(\sum_{t \in \mathsf{T}} w_t \otimes t\right)\gamma = w_1 \otimes 1 \quad (w_t \in \mathsf{W}).$$

It is easily checked that γ is an $\mathsf{A}\mathfrak{U}$-homomorphism. For all $w \in \mathsf{W}$ we have

$$(w \otimes t') \sum_{t \in \mathsf{T}} t^{-1}\gamma t = \sum_{t \in \mathsf{T}} (w \otimes t' t^{-1})\gamma t = w \otimes t'.$$

Thus γ is a $(\mathfrak{G}, \mathfrak{U})$-Gaschütz operator on $\mathsf{W}^{\mathfrak{G}}$.

e) We apply d) to

$$\mathsf{A}\mathfrak{G} \cong \mathsf{A}\mathfrak{U} \otimes_{\mathsf{A}\mathfrak{U}} \mathsf{A}\mathfrak{G} = (\mathsf{A}\mathfrak{U})^{\mathfrak{G}}.$$ **q.e.d.**

We shall make several applications of the following simple lemma.

7.4 Lemma. *Suppose that* V *and* W *are* $\mathsf{A}\mathfrak{G}$*-modules, that* $\mathfrak{U} \leq \mathfrak{G}$ *and that* T *is a transversal of* $\mathfrak{U}$ *in* $\mathfrak{G}$. *Suppose that* $\alpha \in \mathrm{Hom}_{\mathsf{A}\mathfrak{U}}(\mathsf{V}_{\mathfrak{U}}, \mathsf{W}_{\mathfrak{U}})$. *If* α' *is the mapping defined by*

$$v\alpha' = \sum_{t \in \mathsf{T}} vt^{-1}\alpha t \quad (v \in \mathsf{V}),$$

then α' *is independent of the choice of* T *and* $\alpha' \in \mathrm{Hom}_{\mathsf{A}\mathfrak{G}}(\mathsf{V}, \mathsf{W})$.

Proof. Suppose that $t' = u_t t$ for each $t \in \mathsf{T}$ with $u_t \in \mathfrak{U}$. Then

$$\sum_{t \in \mathsf{T}} vt'^{-1}\alpha t' = \sum_{t \in \mathsf{T}} vt^{-1}u_t^{-1}\alpha u_t t = \sum_{t \in \mathsf{T}} vt^{-1}\alpha u_t^{-1}u_t t = v\alpha'.$$

Thus α' is independent of the choice of T. It follows that for all $g \in \mathfrak{G}$,

$$(vg)\alpha' = \sum_{t \in \mathsf{T}} vgt^{-1}\alpha t = \left(\sum_{t \in \mathsf{T}} v(tg^{-1})^{-1}\alpha(tg^{-1})\right)g = (v\alpha')g.$$

This shows that $\alpha' \in \mathrm{Hom}_{\mathsf{A}\mathfrak{G}}(\mathsf{V}, \mathsf{W})$. **q.e.d.**

7.5 Theorem (D. G. HIGMAN [4]). *Suppose that* $\mathfrak{U} \leq \mathfrak{G}$ *and that* V *is an* $\mathsf{A}\mathfrak{G}$*-module. The following assertions are equivalent.*

a) V *is* $(\mathfrak{G}, \mathfrak{U})$*-projective.*

b) V *is isomorphic to a direct summand of* $(\mathsf{V}_{\mathfrak{U}})^{\mathfrak{G}}$.

c) *There exists an* $\mathsf{A}\mathfrak{U}$*-module* W *such that* V *is isomorphic to a direct summand of* $\mathsf{W}^{\mathfrak{G}}$.

d) V *has a* $(\mathfrak{G}, \mathfrak{U})$*-Gaschütz operator.*

Proof. a) ⇒ b): The mapping ε of 4.3b) is an $\mathsf{A}\mathfrak{G}$-epimorphism of $(\mathsf{V}_{\mathfrak{U}})^{\mathfrak{G}}$ onto V and $(\ker \varepsilon)_{\mathfrak{U}}$ is a direct summand of $((\mathsf{V}_{\mathfrak{U}})^{\mathfrak{G}})_{\mathfrak{U}}$. By hypothesis, V is $(\mathfrak{G}, \mathfrak{U})$-projective, so $\ker \varepsilon$ is a direct summand of $(\mathsf{V}_{\mathfrak{U}})^{\mathfrak{G}}$. Thus

$$(\mathsf{V}_{\mathfrak{U}})^{\mathfrak{G}} = \ker \varepsilon \oplus \mathsf{V}',$$

where $\mathsf{V}' \cong \mathsf{V}$.

b) ⇒ c): Take $\mathsf{W} = \mathsf{V}_{\mathfrak{U}}$.

c) ⇒ d): This follows at once from 7.3d) and 7.3b).

d) ⇒ a): Suppose that γ is a $(\mathfrak{G}, \mathfrak{U})$-Gaschütz operator on V and that α is an $\mathsf{A}\mathfrak{G}$-epimorphism of an $\mathsf{A}\mathfrak{G}$-module U onto V for which $(\ker \alpha)_{\mathfrak{U}}$ is a direct summand of $\mathsf{U}_{\mathfrak{U}}$; say

$$\mathsf{U}_{\mathfrak{U}} = (\ker \alpha)_{\mathfrak{U}} \oplus \mathsf{W},$$

where W is an $\mathsf{A}\mathfrak{U}$-module. The restriction α' of α to W is an $\mathsf{A}\mathfrak{U}$-isomorphism of W onto $\mathsf{V}_{\mathfrak{U}}$. Let $\beta = \alpha'^{-1}$. Thus $\beta\alpha = 1_{\mathsf{V}}$. Also

$$\alpha\gamma\beta \in \mathrm{Hom}_{A\mathfrak{U}}(U_{\mathfrak{U}}, U_{\mathfrak{U}}).$$

Let T be a transversal of $\mathfrak{U}$ in $\mathfrak{G}$ and define δ by

$$u\delta = \sum_{t \in T} ut^{-1}\alpha\gamma\beta t \quad (u \in U).$$

Then $\delta \in \mathrm{Hom}_{A\mathfrak{G}}(U, U)$, by Lemma 7.4, and

$$u\delta\alpha = \sum_{t \in T} ut^{-1}\alpha\gamma\beta t\alpha = \sum_{t \in T} u\alpha t^{-1}\gamma\beta\alpha t = \sum_{t \in T} u\alpha t^{-1}\gamma t = u\alpha.$$

This shows that $U = U\delta + \ker \alpha$. If $u\delta \in U\delta \cap \ker \alpha$, then

$$u\delta = \sum_{t \in T} ut^{-1}\alpha\gamma\beta t = \sum_{t \in T} u\alpha t^{-1}\gamma\beta t = 0.$$

Thus $U = U\delta \oplus \ker \alpha$. Hence V is $(\mathfrak{G}, \mathfrak{U})$-projective. **q.e.d.**

The dual to 7.5 is the following.

7.6 Theorem (D. G. HIGMAN [4]). *Suppose that $\mathfrak{U} \leq \mathfrak{G}$ and that V is an $A\mathfrak{G}$-module. The following assertions are equivalent.*

a) *V is $(\mathfrak{G}, \mathfrak{U})$-injective.*

b) *V is isomorphic to a direct summand of $(V_{\mathfrak{U}})^{\mathfrak{G}}$.*

c) *There exists an $A\mathfrak{U}$-module W such that V is isomorphic to a direct summand of $W^{\mathfrak{G}}$.*

d) *V has a $(\mathfrak{G}, \mathfrak{U})$-Gaschütz operator.*

Proof. a) $\Rightarrow$ b): The mapping μ of 4.3c) is an $A\mathfrak{G}$-monomorphism of V into $(V_{\mathfrak{U}})^{\mathfrak{G}}$, and $(V\mu)_{\mathfrak{U}}$ is a direct summand of $((V_{\mathfrak{U}})^{\mathfrak{G}})_{\mathfrak{U}}$. By hypothesis, V is $(\mathfrak{G}, \mathfrak{U})$-injective, so $V\mu$ is a direct summand of $(V_{\mathfrak{U}})^{\mathfrak{G}}$.

b) $\Rightarrow$ c) $\Rightarrow$ d): As in 7.5.

d) $\Rightarrow$ a): Suppose that γ is a $(\mathfrak{G}, \mathfrak{U})$-Gaschütz operator on V, that V is an $A\mathfrak{G}$-submodule of the $A\mathfrak{G}$-module U and that $U_{\mathfrak{U}} = V_{\mathfrak{U}} \oplus W$ for some $A\mathfrak{U}$-submodule W. Let π be the projection of $U_{\mathfrak{U}}$ onto $V_{\mathfrak{U}}$ with kernel W, and define $\delta \in \mathrm{Hom}_A(U, V)$ by

$$u\delta = \sum_{t \in T} ut^{-1}\pi\gamma t,$$

where T is a transversal of $\mathfrak{U}$ in $\mathfrak{G}$. If $v \in V$, then $v\delta = v$. Thus δ is a projection of U onto V, and $U = V \oplus \ker \delta$. By 7.4, δ is an $A\mathfrak{G}$-homomorphism, so $\ker \delta$ is an $\mathsf{A}\mathfrak{G}$-submodule. Hence V is a $(\mathfrak{G}, \mathfrak{U})$-injective $\mathsf{A}\mathfrak{G}$-module. **q.e.d.**

7.7 Theorem (D. G. HIGMAN [4]). *Suppose that* $\mathfrak{U} \leq \mathfrak{G}$.

a) *The* $(\mathfrak{G}, \mathfrak{U})$*-projective* $\mathsf{A}\mathfrak{G}$*-modules are precisely the same as the* $(\mathfrak{G}, \mathfrak{U})$*-injective* $\mathsf{A}\mathfrak{G}$*-modules.*

b) *If* $|\mathfrak{G} : \mathfrak{U}|$ *has an inverse in* A, *any* $\mathsf{A}\mathfrak{G}$*-module is* $(\mathfrak{G}, \mathfrak{U})$*-projective and* $(\mathfrak{G}, \mathfrak{U})$*-injective.*

c) *Suppose that* $\mathfrak{P}$ *is a Sylow p-subgroup of* $\mathfrak{G}$ *and that* $\mathfrak{P} \leq \mathfrak{U} \leq \mathfrak{G}$. *Suppose further that* A *is either a field of characteristic p or a commutative local ring for which the residue class field is of characteristic p. Then any* $\mathsf{A}\mathfrak{G}$*-module is* $(\mathfrak{G}, \mathfrak{U})$*-projective and* $(\mathfrak{G}, \mathfrak{U})$*-injective.*

Proof. a) This assertion follows from 7.5 and 7.6.

b) This follows from 7.3c) and 7.5 or 7.6.

c) This is a consequence of b). **q.e.d.**

Now we consider the important case where the ring A is a field, denoted as usual by K.

7.8 Theorem (GASCHÜTZ [1]). *Let* V *be a* $\mathsf{K}\mathfrak{G}$*-module. Then the following assertions are equivalent.*

a) V *is projective.*

b) V *is injective.*

c) V *has a* $(\mathfrak{G}, 1)$*-Gaschütz operator.*

Proof. This follows immediately from 7.5 and 7.6, as $(\mathfrak{G}, 1)$-projectivity (injectivity) for $\mathsf{K}\mathfrak{G}$-modules means the same as projectivity (injectivity). **q.e.d.**

7.9 Remarks. Those algebras $\mathfrak{R}$ of finite dimension over K, for which $\mathfrak{R}$ is an injective right $\mathfrak{R}$-module, are called quasi-Frobenius algebras. (Such an algebra $\mathfrak{R}$ is also an injective left $\mathfrak{R}$-module.) For quasi-Frobenius algebras the notions of projective and injective modules coincide. Conversely, it can be proved that if $\mathfrak{R}$ is a ring for which every projective (injective) module is injective (projective), then $\mathfrak{R}$ satisfies the minimum conditions for left and right ideals and is an injective $\mathfrak{R}$-module.

Quasi-Frobenius algebras can also be characterized by the fact that the formation of annihilators induces a duality of the lattice of right ideals onto the lattice of left ideals. In § 11 we shall derive this important property for group-rings over fields from the fact that group-rings are symmetric algebras.

7.10 Lemma. *Suppose that* $\mathfrak{U} \leq \mathfrak{V} \leq \mathfrak{G}$.

a) *Suppose that* V_i $(i \in \mathsf{I})$ *is an* $\mathsf{A}\mathfrak{G}$*-module. Then* $\bigoplus_{i \in \mathsf{I}} \mathsf{V}_i$ *is* $(\mathfrak{G}, \mathfrak{V})$*-projective if and only if all the* V_i *are* $(\mathfrak{G}, \mathfrak{V})$*-projective.*

b) *If* V *is* $(\mathfrak{G}, \mathfrak{V})$*-projective, then* V *is also* $(\mathfrak{G}, \mathfrak{V}^g)$*-projective for all* $g \in \mathfrak{G}$.

c) *If* V *is* $(\mathfrak{G}, \mathfrak{U})$*-projective, then* V *is also* $(\mathfrak{G}, \mathfrak{V})$*-projective.*

d) *If* V *is* $(\mathfrak{G}, \mathfrak{V})$*-projective and* $\mathsf{V}_{\mathfrak{V}}$ *is* $(\mathfrak{V}, \mathfrak{U})$*-projective, then* V *is also* $(\mathfrak{G}, \mathfrak{U})$*-projective.*

e) *If* W *is a* $(\mathfrak{V}, \mathfrak{U})$*-projective* $\mathsf{A}\mathfrak{V}$*-module, then any direct summand of* $\mathsf{W}^{\mathfrak{G}}$ *is* $(\mathfrak{G}, \mathfrak{U})$*-projective.*

Proof. a) This follows from 7.5 by using 7.3a), b).

b) Suppose that T is a transversal of $\mathfrak{V}$ in $\mathfrak{G}$. By 7.5, there exists a $(\mathfrak{G}, \mathfrak{V})$-Gaschütz operator γ on V. Then T^g is a transversal of $\mathfrak{V}^g$ in $\mathfrak{G}$, and $g^{-1}\gamma g$ is a $(\mathfrak{G}, \mathfrak{V}^g)$-Gaschütz operator on V. By 7.5, V is $(\mathfrak{G}, \mathfrak{V}^g)$-projective.

c) By 7.5, there exists an $\mathsf{A}\mathfrak{U}$-module W such that V is a direct summand of $\mathsf{W}^{\mathfrak{G}}$. If we put $\mathsf{W}' = \mathsf{W}^{\mathfrak{V}}$, then $\mathsf{W}^{\mathfrak{G}} \cong \mathsf{W}'^{\mathfrak{G}}$. Thus V is a direct summand of $\mathsf{W}'^{\mathfrak{G}}$, and V is $(\mathfrak{G}, \mathfrak{V})$-projective by 7.5.

d) Suppose that

$$0 \to \mathsf{T} \xrightarrow{\alpha} \mathsf{U} \to \mathsf{V} \to 0$$

is an exact sequence of $\mathsf{A}\mathfrak{G}$-modules, and suppose that $(\mathsf{T}\alpha)_{\mathfrak{U}}$ is a direct summand of $\mathsf{U}_{\mathfrak{U}}$. As $\mathsf{V}_{\mathfrak{V}}$ is $(\mathfrak{V}, \mathfrak{U})$-projective, $(\mathsf{T}\alpha)_{\mathfrak{V}}$ is a direct summand of $\mathsf{U}_{\mathfrak{V}}$. As V is also $(\mathfrak{G}, \mathfrak{V})$-projective, it follows that $\mathsf{T}\alpha$ is a direct summand of U. Thus V is $(\mathfrak{G}, \mathfrak{U})$-projective.

e) By 7.5, there exists an $\mathsf{A}\mathfrak{U}$-module V such that $\mathsf{V}^{\mathfrak{V}} \cong \mathsf{W} \oplus \mathsf{W}'$ for some $\mathsf{A}\mathfrak{V}$-module W'. Thus

$$\mathsf{V}^{\mathfrak{G}} \cong \mathsf{W}^{\mathfrak{G}} \oplus \mathsf{W}'^{\mathfrak{G}}.$$

Hence $\mathsf{W}^{\mathfrak{G}}$ is $(\mathfrak{G}, \mathfrak{U})$-projective. By a), so is any direct summand of $\mathsf{W}^{\mathfrak{G}}$. **q.e.d.**

7.11 Theorem. a) *If* V *is a projective* $\mathsf{A}\mathfrak{G}$*-module and* $\mathfrak{U} \leq \mathfrak{G}$, *then* $\mathsf{V}_{\mathfrak{U}}$ *is a projective* $\mathsf{A}\mathfrak{U}$*-module.*

b) *Suppose that* V *is an* $\mathsf{A}\mathfrak{G}$*-module and that* $\mathfrak{U}$ *is a subgroup of* $\mathfrak{G}$ *for which* $|\mathfrak{G} : \mathfrak{U}|$ *has an inverse in* A. *If* $\mathsf{V}_{\mathfrak{U}}$ *is a* $(\mathfrak{U}, 1)$*-projective* $\mathsf{A}\mathfrak{U}$*-module, then* V *is a* $(\mathfrak{G}, 1)$*-projective* $\mathsf{A}\mathfrak{G}$*-module.*

Proof. a) Since $(\mathsf{A}\mathfrak{G})_{\mathfrak{U}}$ is a free $\mathsf{A}\mathfrak{U}$-module, the restriction of any free $\mathsf{A}\mathfrak{G}$-module to $\mathfrak{U}$ is a free $\mathsf{A}\mathfrak{U}$-module. Thus the restrictions of projective $\mathsf{A}\mathfrak{G}$-modules are projective $\mathsf{A}\mathfrak{U}$-modules.

b) By 7.7b), V is $(\mathfrak{G}, \mathfrak{U})$-projective. By hypothesis $\mathsf{V}_{\mathfrak{U}}$ is $(\mathfrak{U}, 1)$-projective. Hence by 7.10d), V is also $(\mathfrak{G}, 1)$-projective. **q.e.d.**

Before we consider the consequences of 7.11, we remark that the relative statement corresponding to 7.11a) is in general not true.

7.12 Example. Let $\mathfrak{G} = \langle a, b, c\rangle$ be the non-Abelian group of order p^3 and exponent p for $p > 2$, where

$$a^p = b^p = c^p = 1, \quad [b, a] = c.$$

Let K be any field of characteristic p. We take $\mathfrak{U} = \langle a\rangle$, $\mathfrak{V} = \langle a, c\rangle$. We regard K as a module for the trivial representation of $\mathfrak{U}$. Then by 7.3d), $\mathsf{K}^{\mathfrak{G}}$ is $(\mathfrak{G}, \mathfrak{U})$-projective. We have

$$\mathsf{K}^{\mathfrak{G}} = \bigoplus_{i,j=0}^{p-1} \mathsf{K}x_{ij},$$

where $x_{ij} = 1 \otimes b^i c^j$. Thus the module-structure of $(\mathsf{K}^{\mathfrak{G}})_{\mathfrak{V}}$ is described by

(1) $$x_{ij}c = x_{i,j+1} \quad (j \text{ modulo } p)$$

and

(2) $$\begin{aligned} x_{ij}a &= 1 \otimes b^i c^j a = 1 \otimes ab^i c^{i+j} = 1a \otimes b^i c^{i+j} \\ &= 1 \otimes b^i c^{i+j} = x_{i,i+j}. \end{aligned}$$

If we write

$$\mathsf{V}_i = \bigoplus_{j=0}^{p-1} \mathsf{K}x_{ij},$$

we obtain the decomposition

$$(\mathsf{K}^{\mathfrak{G}})_{\mathfrak{V}} = \mathsf{V}_0 \oplus \cdots \oplus \mathsf{V}_{p-1}.$$

as a direct sum of $\mathsf{K}\mathfrak{V}$-modules. To show that $(\mathsf{K}^{\mathfrak{G}})_{\mathfrak{V}}$ is not $(\mathfrak{V}, \mathfrak{U})$-projective, it suffices to show that V_1 is not $(\mathfrak{V}, \mathfrak{U})$-projective, by 7.3. Suppose that γ is a $(\mathfrak{V}, \mathfrak{U})$-Gaschütz operator on V_1. Then

(3) $$x_{1j} = \sum_{l=0}^{p-1} x_{1j}c^{-l}\gamma c^l$$

for all $j = 0, \ldots, p-1$. Suppose that

(4) $$x_{1j}\gamma = \sum_{k=0}^{p-1} a_{jk}x_{1k}$$

with $a_{jk} \in \mathsf{K}$. As γ is a $\mathsf{K}\mathfrak{U}$-homomorphism, we have

$$(x_{1j}a)\gamma = x_{1,j+1}\gamma = \sum_{k=0}^{p-1} a_{j+1,k}x_{1k}$$
$$= (x_{1j}\gamma)a = \sum_{k=0}^{p-1} a_{jk}x_{1,k+1} = \sum_{k=0}^{p-1} a_{j,k-1}x_{1k}.$$

This shows that

(5) $$a_{j+1,k} = a_{j,k-1}.$$

Now (3), (1) and (4) imply that

$$x_{1j} = \sum_{l=0}^{p-1} x_{1,j-l}\gamma c^l = \sum_{k,l=0}^{p-1} a_{j-l,k}x_{1k}c^l$$
$$= \sum_{k,l=0}^{p-1} a_{j-l,k}x_{1,k+l} = \sum_{k,l=0}^{p-1} a_{j-l,k-l}x_{1k}.$$

This shows that

$$\delta_{jk} = \sum_{l=0}^{p-1} a_{j-l,k-l}.$$

But now (5) gives the contradiction

$$1 = \sum_{l=0}^{p-1} a_{j-l,j-l} = pa_{00} = 0.$$

Hence $\mathsf{K}^{\mathfrak{G}}$ is $(\mathfrak{G}, \mathfrak{U})$-projective, but $(\mathsf{K}^{\mathfrak{G}})_{\mathfrak{V}}$ is not $(\mathfrak{V}, \mathfrak{U})$-projective.

7.13 Lemma. *Suppose that* $\mathfrak{U} \leq \mathfrak{V} \leq \mathfrak{G}$ *and* $\mathfrak{U} \trianglelefteqq \mathfrak{G}$. *If* V *is a* $(\mathfrak{G}, \mathfrak{U})$-*projective* $\mathsf{A}\mathfrak{G}$-*module, then* $\mathsf{V}_{\mathfrak{V}}$ *is* $(\mathfrak{V}, \mathfrak{U})$-*projective.*

Proof. As V is $(\mathfrak{G}, \mathfrak{U})$-projective, V is a direct summand of $\mathsf{W}^{\mathfrak{G}}$ for some $\mathsf{A}\mathfrak{U}$-module W, by 7.5. By Mackey's lemma, $\mathsf{V}_{\mathfrak{V}}$ is a direct summand of

$$(\mathsf{W}^{\mathfrak{G}})_{\mathfrak{V}} = \bigoplus_{i=1}^{k} ((\mathsf{W} \otimes g_i)_{\mathfrak{U}^{g_i} \cap \mathfrak{V}})^{\mathfrak{V}} = \bigoplus_{i=1}^{k} ((\mathsf{W} \otimes g_i)_{\mathfrak{U}})^{\mathfrak{V}}$$

for some $g_i \in \mathfrak{G}$. Thus $\mathsf{V}_{\mathfrak{V}}$ is $(\mathfrak{V}, \mathfrak{U})$-projective, by 7.3d) and b). **q.e.d.**

7.14 Theorem. *Suppose that* char $\mathsf{K} = p$ *and that* $\mathfrak{P}$ *is a Sylow p-subgroup of* $\mathfrak{G}$. *Let* V *be a* $\mathsf{K}\mathfrak{G}$-*module. Then* V *is projective if and only if* $\mathsf{V}_{\mathfrak{P}}$ *is a projective* $\mathsf{K}\mathfrak{P}$-*module.*

Proof. This follows trivially from 7.11. **q.e.d.**

We see from 7.14 that the projective $\mathsf{K}\mathfrak{P}$-modules, where $\mathfrak{P}$ is a p-group and K is a field of characteristic p, are particularly important. Fortunately these are easy to describe.

7.15 Theorem. *Let* $\mathfrak{P}$ *be a p-group and* K *any field of characteristic p. Then any finitely generated projective* $\mathsf{K}\mathfrak{P}$-*module is free.*

Proof. Let P be a finitely generated projective $\mathsf{K}\mathfrak{P}$-module. Then there exists a finitely generated free $\mathsf{K}\mathfrak{P}$-module F such that $\mathsf{F} = \mathsf{P} \oplus \mathsf{P}'$. Since P and P' are finite-dimensional vector spaces over K, there exist decompositions

$$\mathsf{P} = \bigoplus_{i} \mathsf{P}_i \quad \text{and} \quad \mathsf{P}' = \bigoplus_{j} \mathsf{P}'_j,$$

where P_i and P'_j are indecomposable $\mathsf{K}\mathfrak{P}$-modules. But by 5.2, $\mathsf{K}\mathfrak{P}$ is an indecomposable $\mathsf{K}\mathfrak{P}$-module. By the Krull-Schmidt theorem (I, 12.4), each P_i must be a free $\mathsf{K}\mathfrak{P}$-module with one generator. Thus P is a free $\mathsf{K}\mathfrak{P}$-module. **q.e.d.**

The assertion of Theorem 7.15 is also valid for not necessarily finitely generated projective modules. Indeed, Kaplansky has shown that any projective module over a local ring is free, and under the assumptions of Theorem 7.15, $\mathsf{K}\mathfrak{P}$ is a local ring by 5.2.

We shall often use the following consequence of 7.15.

7.16 Corollary (DICKSON). *Suppose that* char $\mathsf{K} = p$ *and that* P *is a finitely generated projective* $\mathsf{K}\mathfrak{G}$*-module. If* $\mathfrak{P}$ *is a Sylow p-subgroup of* $\mathfrak{G}$*, then* $|\mathfrak{P}|$ *divides* $\dim_{\mathsf{K}} \mathsf{P}$.

Proof. By 7.11a), $\mathsf{P}_{\mathfrak{P}}$ is a projective $\mathsf{K}\mathfrak{P}$-module. By 7.15, $\mathsf{P}_{\mathfrak{P}}$ is free. Hence $|\mathfrak{P}|$ divides $\dim_{\mathsf{K}} \mathsf{P}$. **q.e.d.**

How does the process of inducing influence projectivity of modules?

7.17 Theorem. *Suppose that* $\mathfrak{U} \leq \mathfrak{G}$ *and that* V *is an* $\mathsf{A}\mathfrak{U}$*-module. Then the induced module* $\mathsf{V}^{\mathfrak{G}}$ *is a* $(\mathfrak{G}, 1)$*-projective* $\mathsf{A}\mathfrak{G}$*-module if and only if* V *is a* $(\mathfrak{U}, 1)$*-projective* $\mathsf{A}\mathfrak{U}$*-module.*

Proof. a) Suppose first that $\mathsf{V}^{\mathfrak{G}}$ is a $(\mathfrak{G}, 1)$-projective $\mathsf{A}\mathfrak{G}$-module. By 4.3a), $(\mathsf{V}^{\mathfrak{G}})_{\mathfrak{U}} = \mathsf{V}\eta \oplus \mathsf{V}'$, where V' is an $\mathsf{A}\mathfrak{U}$-module and η is an $\mathsf{A}\mathfrak{U}$-monomorphism. But $(\mathsf{V}^{\mathfrak{G}})_{\mathfrak{U}}$ is a $(\mathfrak{U}, 1)$-projective $\mathsf{A}\mathfrak{U}$-module by 7.13, hence $\mathsf{V}\eta$ is also $(\mathfrak{U}, 1)$-projective. As V is isomorphic to $\mathsf{V}\eta$, V is a $(\mathfrak{U}, 1)$-projective $\mathsf{A}\mathfrak{U}$-module.

b) Suppose now that V is a $(\mathfrak{U}, 1)$-projective $\mathsf{A}\mathfrak{U}$-module. By 7.10e), $\mathsf{V}^{\mathfrak{G}}$ is $(\mathfrak{G}, 1)$-projective. **q.e.d.**

Besides the induction process we occasionally consider the "inflation process". We prove the following as a further application of Gaschütz operators.

7.18 Theorem. *Suppose that* $\mathfrak{N} \trianglelefteq \mathfrak{G}$ *and that* V *is a non-zero* $\mathsf{K}(\mathfrak{G}/\mathfrak{N})$*-module. Thus* V *becomes a* $\mathsf{K}\mathfrak{G}$*-module, if* vg $(v \in \mathsf{V}, g \in \mathfrak{G})$ *is defined to be* $v(\mathfrak{N}g)$*. Then the following assertions are equivalent.*

a) V *is a projective* $\mathsf{K}\mathfrak{G}$*-module.*

b) *The characteristic of* K *does not divide* $|\mathfrak{N}|$ *and* V *is a projective* $\mathsf{K}(\mathfrak{G}/\mathfrak{N})$*-module.*

Proof. If β is any K-linear mapping of V into V, then

$$v \sum_{g \in \mathfrak{G}} g^{-1}\beta g = v \sum_{g \in \mathfrak{G}} (\mathfrak{N}g)^{-1}\beta(\mathfrak{N}g) = |\mathfrak{N}| v \sum_{h \in \mathfrak{G}/\mathfrak{N}} h^{-1}\beta h$$

for all $v \in \mathsf{V}$.

a) $\Rightarrow$ b): Suppose first that V is a projective $\mathsf{K}\mathfrak{G}$-module. By 7.8, there exists a $(\mathfrak{G}, 1)$-Gaschütz operator γ on V. Thus

$$v = v \sum_{g \in \mathfrak{G}} g^{-1}\gamma g = |\mathfrak{N}| v \sum_{h \in \mathfrak{G}/\mathfrak{N}} h^{-1}\gamma h$$

for all $v \in \mathsf{V}$. Since $\mathsf{V} \neq 0$, it follows that char K does not divide $|\mathfrak{N}|$ and that $\gamma' = |\mathfrak{N}|\gamma$ is a $(\mathfrak{G}/\mathfrak{N}, 1)$-Gaschütz operator on the $\mathsf{K}(\mathfrak{G}/\mathfrak{N})$-module V.

b) ⇒ a): If char K does not divide $|\mathfrak{N}|$ and γ' is a $(\mathfrak{G}/\mathfrak{N}, 1)$-Gaschütz operator on the $\mathsf{K}(\mathfrak{G}/\mathfrak{N})$-module V, then from

$$v = v \sum_{h \in \mathfrak{G}/\mathfrak{N}} h^{-1}\gamma' h = |\mathfrak{N}|^{-1} v \sum_{g \in \mathfrak{G}} g^{-1}\gamma' g,$$

it follows that $|\mathfrak{N}|^{-1}\gamma'$ is a $(\mathfrak{G}, 1)$-Gaschütz operator on the $\mathsf{K}\mathfrak{G}$-module V.

q.e.d.

7.19 Theorem. *Let* V, W *be* $\mathsf{K}\mathfrak{G}$*-modules.*

a) *If* F *is a free* $\mathsf{K}\mathfrak{G}$*-module of rank* r, *then* $\mathsf{V} \otimes_{\mathsf{K}} \mathsf{F}$ *is a free* $\mathsf{K}\mathfrak{G}$*-module of rank* $r(\dim_{\mathsf{K}} \mathsf{V})$.

b) *Suppose that* $\mathfrak{U} \leq \mathfrak{B} \leq \mathfrak{G}$. *If* V *is* $(\mathfrak{G}, \mathfrak{B})$*-projective and* $\mathsf{W}_{\mathfrak{B}}$ *is* $(\mathfrak{B}, \mathfrak{U})$*-projective, then* $\mathsf{V} \otimes_{\mathsf{K}} \mathsf{W}$ *is* $(\mathfrak{G}, \mathfrak{U})$*-projective.*

c) *If* W *is projective, then* $\mathsf{V} \otimes_{\mathsf{K}} \mathsf{W}$ *is also projective.*

Proof. a) We have $\mathsf{F} \cong \mathsf{F}_0^{\mathfrak{G}}$, where F_0 is the K-module of dimension r. Hence by 4.15a),

$$\mathsf{V} \otimes_{\mathsf{K}} \mathsf{F} \cong \mathsf{V} \otimes_{\mathsf{K}} \mathsf{F}_0^{\mathfrak{G}} \cong (\mathsf{V} \otimes_{\mathsf{K}} \mathsf{F}_0)^{\mathfrak{G}}.$$

The assertion follows at once, since $\dim_{\mathsf{K}}(\mathsf{V} \otimes_{\mathsf{K}} \mathsf{F}_0) = r(\dim_{\mathsf{K}} \mathsf{V})$.

b) Let S be a transversal of $\mathfrak{B}$ in $\mathfrak{G}$ and let T be a transversal of $\mathfrak{U}$ in $\mathfrak{B}$. Then TS is a transversal of $\mathfrak{U}$ in $\mathfrak{G}$. Let

$$\gamma \in \mathrm{Hom}_{\mathsf{K}\mathfrak{B}}(\mathsf{V}_{\mathfrak{B}}, \mathsf{V}_{\mathfrak{B}})$$

be a $(\mathfrak{G}, \mathfrak{B})$-Gaschütz operator on V and let

$$\delta \in \mathrm{Hom}_{\mathsf{K}\mathfrak{U}}(\mathsf{W}_{\mathfrak{U}}, \mathsf{W}_{\mathfrak{U}})$$

be a $(\mathfrak{B}, \mathfrak{U})$-Gaschütz operator on $\mathsf{W}_{\mathfrak{B}}$. Then

$$\gamma \otimes \delta \in \mathrm{Hom}_{\mathsf{K}\mathfrak{U}}((\mathsf{V} \otimes_{\mathsf{K}} \mathsf{W})_{\mathfrak{U}}, (\mathsf{V} \otimes_{\mathsf{K}} \mathsf{W})_{\mathfrak{U}})$$

is a $(\mathfrak{G}, \mathfrak{U})$-Gaschütz operator on $\mathsf{V} \otimes_{\mathsf{K}} \mathsf{W}$; for if $v \in \mathsf{V}$, $w \in \mathsf{W}$,

$$(v \otimes w) \sum_{s \in \mathsf{S}, t \in \mathsf{T}} s^{-1}t^{-1}(\gamma \otimes \delta)ts$$

$$= \sum_{s \in \mathsf{S}, t \in \mathsf{T}} vs^{-1}t^{-1}\,\gamma ts \otimes ws^{-1}t^{-1}\delta ts$$

$$= \sum_{s \in \mathsf{S}, t \in \mathsf{T}} vs^{-1}\gamma s \otimes ws^{-1}t^{-1}\delta ts,$$

as $\gamma \in \mathrm{Hom}_{\mathsf{K}\mathfrak{V}}(\mathsf{V}_{\mathfrak{V}}, \mathsf{V}_{\mathfrak{V}})$ and $\mathsf{T} \subseteq \mathfrak{V}$. Thus

$$(v \otimes w) \sum_{s \in \mathsf{S}, t \in \mathsf{T}} s^{-1}t^{-1}(\gamma \otimes \delta)ts$$

$$= \sum_{s \in \mathsf{S}} \left(vs^{-1}\gamma s \otimes \left(\sum_{t \in \mathsf{T}} (ws^{-1})t^{-1}\delta t \right) s \right)$$

$$= \sum_{s \in \mathsf{S}} (vs^{-1}\gamma s \otimes (ws^{-1})s)$$

$$= \left(\sum_{s \in \mathsf{S}} vs^{-1}\gamma s \right) \otimes w = v \otimes w.$$

c) This is an immediate consequence of b). **q.e.d.**

Before considering consequences for the radical of a group-ring, we make the following remark.

7.20 Theorem. *Let $\mathfrak{U}$ be a subgroup of $\mathfrak{G}$ and K a field the characteristic of which does not divide $|\mathfrak{G} : \mathfrak{U}|$. Let V be a $\mathsf{K}\mathfrak{G}$-module.*

a) *If W is a $\mathsf{K}\mathfrak{G}$-submodule of V such that $\mathsf{W}_{\mathfrak{U}}$ is a direct summand of $\mathsf{V}_{\mathfrak{U}}$, then W is a direct summand of V.*

b) *If $\mathsf{V}_{\mathfrak{U}}$ is a completely reducible $\mathsf{K}\mathfrak{U}$-module, then V is a completely reducible $\mathsf{K}\mathfrak{G}$-module.*

Proof. a) By 7.7b), W is $(\mathfrak{G}, \mathfrak{U})$-injective, hence by Definition 7.1b) W is a direct summand of V.

b) Now a) applies to every $\mathsf{K}\mathfrak{G}$-submodule of V. Hence V is completely reducible. **q.e.d.**

7.21 Theorem. *Suppose that $\mathfrak{N}$ is a normal subgroup of $\mathfrak{G}$.*

a) $\mathbf{J}(\mathsf{K}\mathfrak{N}) \subseteq \mathbf{J}(\mathsf{K}\mathfrak{G})$.

b) (GREEN, STONEHEWER [1], VILLAMAYOR [1]). *If* char K *does not divide* $|\mathfrak{G}/\mathfrak{N}|$, *then* $\mathbf{J}(\mathsf{K}\mathfrak{G}) = \mathbf{J}(\mathsf{K}\mathfrak{N})\mathsf{K}\mathfrak{G}$ *and*

$$\dim_{\mathsf{K}} \mathbf{J}(\mathsf{K}\mathfrak{G}) = (\dim_{\mathsf{K}} \mathbf{J}(\mathsf{K}\mathfrak{N}))|\mathfrak{G}/\mathfrak{N}|.$$

c) (WILLEMS [1]). *If* $\mathbf{J}(\mathsf{K}\mathfrak{G}) = \mathbf{J}(\mathsf{K}\mathfrak{N})\mathsf{K}\mathfrak{G}$, *then* char K *does not divide* $|\mathfrak{G}/\mathfrak{N}|$.

Proof. a) Suppose that $x \in \mathbf{J}(\mathsf{K}\mathfrak{N})$ and let V be an irreducible $\mathsf{K}\mathfrak{G}$-module. By Clifford's theorem (V, 17.3a)), $\mathsf{V}_{\mathfrak{N}}$ is a completely reducible $\mathsf{K}\mathfrak{G}$-module, say

$$\mathsf{V}_{\mathfrak{N}} = \mathsf{V}_1 \oplus \cdots \oplus \mathsf{V}_r,$$

where the V_i are irreducible $\mathsf{K}\mathfrak{N}$-modules. Since $x \in \mathbf{J}(\mathsf{K}\mathfrak{N})$, $\mathsf{V}_i x = 0$, so also $\mathsf{V}x = 0$. As this is the case for all irreducible $\mathsf{K}\mathfrak{G}$-modules V, it follows that $x \in \mathbf{J}(\mathsf{K}\mathfrak{G})$.

b) Suppose that $x \in \mathbf{J}(\mathsf{K}\mathfrak{G})$. If T is a transversal of $\mathfrak{N}$ in $\mathfrak{G}$ containing 1, then since $\mathsf{K}\mathfrak{G} = \bigoplus_{t \in \mathsf{T}} (\mathsf{K}\mathfrak{N})t$, we can write x in the form $x = \sum_{t \in \mathsf{T}} x_t t$ with uniquely determined elements x_t of $\mathsf{K}\mathfrak{N}$. We show that $x_t \in \mathbf{J}(\mathsf{K}\mathfrak{N})$.

To do this, let W be an irreducible $\mathsf{K}\mathfrak{N}$-module. We form the induced $\mathsf{K}\mathfrak{G}$-module

$$\mathsf{V} = \mathsf{W}^{\mathfrak{G}} = \bigoplus_{t \in \mathsf{T}} \mathsf{W} \otimes t.$$

If $w \in \mathsf{W}$ and $u \in \mathfrak{N}$, then

$$(w \otimes t)u = w \otimes (tut^{-1})t = wu^{t^{-1}} \otimes t.$$

Thus $\mathsf{W} \otimes t$ is an irreducible $\mathsf{K}\mathfrak{N}$-module and $\mathsf{V}_{\mathfrak{N}}$ is a completely reducible $\mathsf{K}\mathfrak{N}$-module. Since char K does not divide $|\mathfrak{G}/\mathfrak{N}|$, it follows from 7.20 that V is also a completely reducible $\mathsf{K}\mathfrak{G}$-module. Since $x \in \mathbf{J}(\mathsf{K}\mathfrak{G})$, we conclude that $\mathsf{V}x = 0$. In particular, we see that for all $w \in \mathsf{W}$,

$$0 = (w \otimes 1)x = \sum_{t \in \mathsf{T}} wx_t \otimes t.$$

Thus $wx_t = 0$, so $\mathsf{W}x_t = 0$. As this holds for all irreducible $\mathsf{K}\mathfrak{N}$-modules W, we see that $x_t \in \mathbf{J}(\mathsf{K}\mathfrak{N})$.

It follows from this together with a) that

$$\mathbf{J}(\mathsf{K}\mathfrak{G}) = \bigoplus_{t \in \mathsf{T}} \mathbf{J}(\mathsf{K}\mathfrak{N})t = \mathbf{J}(\mathsf{K}\mathfrak{N})(\mathsf{K}\mathfrak{G})$$

and

$$\dim_{\mathsf{K}} \mathbf{J}(\mathsf{K}\mathfrak{G}) = |\mathfrak{G}/\mathfrak{N}| \dim_{\mathsf{K}} \mathbf{J}(\mathsf{K}\mathfrak{N}).$$

c) We consider $\mathsf{K}(\mathfrak{G}/\mathfrak{N})$ as a $\mathsf{K}\mathfrak{G}$-module. Take $x \in \mathbf{J}(\mathsf{K}\mathfrak{N})$. As $\mathbf{J}(\mathsf{K}\mathfrak{N})$ is contained in the augmentation ideal

$$\left\{ \sum_{u \in \mathfrak{N}} a_u(u - 1) \,\middle|\, a_u \in \mathsf{K} \right\}$$

of $\mathsf{K}\mathfrak{N}$, we have $x = \sum_{u \in \mathfrak{N}} a_u(u - 1)$ for some $a_u \in \mathsf{K}$. Hence for all $g \in \mathfrak{G}$,

$$(g\mathfrak{N})x = \sum_{u \in \mathfrak{N}} a_u(g(u - 1)\mathfrak{N}) = 0.$$

Thus by our assumption

$$\mathsf{K}(\mathfrak{G}/\mathfrak{N})\mathbf{J}(\mathsf{K}\mathfrak{G}) = \mathsf{K}(\mathfrak{G}/\mathfrak{N})\mathbf{J}(\mathsf{K}\mathfrak{N})\mathsf{K}\mathfrak{G} = 0.$$

So $\mathsf{K}(\mathfrak{G}/\mathfrak{N})$ is a completely reducible $\mathsf{K}\mathfrak{G}$-module by 1.6. Hence $\mathsf{K}(\mathfrak{G}/\mathfrak{N})$ is also a completely reducible $\mathsf{K}(\mathfrak{G}/\mathfrak{N})$-module. Thus char K does not divide $|\mathfrak{G}/\mathfrak{N}|$. **q.e.d.**

7.22 Theorem. *Suppose that $\mathfrak{G}$ has a normal Sylow p-subgroup $\mathfrak{P}$ and that* char $\mathsf{K} = p$. *Then $\mathbf{J}(\mathsf{K}\mathfrak{G})$ is spanned over K by all $x - y$ with $x, y \in \mathfrak{G}$ and $xy^{-1} \in \mathfrak{P}$. Also*

$$\dim_{\mathsf{K}} \mathbf{J}(\mathsf{K}\mathfrak{G}) = |\mathfrak{G}| - |\mathfrak{G}/\mathfrak{P}|.$$

Proof. This follows immediately from 5.2 and 7.21. **q.e.d.**

Finally, we mention a fact about regular submodules.

7.23 Theorem. *If V is a $\mathsf{K}\mathfrak{G}$-module, L is an extension of K and V_{L} has an $\mathsf{L}\mathfrak{G}$-submodule isomorphic to $\mathsf{L}\mathfrak{G}$, then V has a $\mathsf{K}\mathfrak{G}$-submodule isomorphic to $\mathsf{K}\mathfrak{G}$.*

Proof. By 7.8, $\mathsf{L}\mathfrak{G}$ is an injective $\mathsf{L}\mathfrak{G}$-module, so V_{L} has a direct summand $\mathsf{L}\mathfrak{G}$-isomorphic to $\mathsf{L}\mathfrak{G} = (\mathsf{K}\mathfrak{G})_{\mathsf{L}}$. By 1.21, V has a direct summand isomorphic to $\mathsf{K}\mathfrak{G}$. **q.e.d.**

Exercises

12) Suppose that $\mathfrak{U}_1, \dots, \mathfrak{U}_m$ are subgroups of $\mathfrak{G}$ and that the greatest common divisor of the orders $|\mathfrak{U}_i|$ is 1. If V is an $\mathsf{A}\mathfrak{G}$-module which is $(\mathfrak{G}, \mathfrak{U}_i)$-projective for all i, then V is also $(\mathfrak{G}, 1)$-projective.

13) Suppose that D is a Dedekind domain. Show that a finitely generated D$\mathfrak{G}$-module V is projective if and only if for every prime ideal $\mathfrak{p}$ of D the localization $V_\mathfrak{p}$ is a projective $D_\mathfrak{g}\mathfrak{G}$-module.
(Hint: a) Show that if $V_\mathfrak{p}$ is $(\mathfrak{G}, 1)$-projective for all $\mathfrak{p}$, then V is $(\mathfrak{G}, 1)$-projective.

b) Apply the following theorem: If V is a finitely generated D-module, then V is a projective D-module if and only if for every prime ideal $\mathfrak{p}$ of D, $V_\mathfrak{p}$ is a projective $D_\mathfrak{p}$-module. (See BOURBAKI [2], p. 138)).

14) Suppose that A is a commutative ring and that V is an A$\mathfrak{G}$-module. Then V is an injective A$\mathfrak{G}$-module if and only if V is $(\mathfrak{G}, 1)$-injective and V is an injective A-module. And V is a projective A$\mathfrak{G}$-module if and only if V is $(\mathfrak{G}, 1)$-projective and V is a projective A-module.

15) a) Prove 7.10c), d), e) by constructing Gaschütz operators.

b) Prove 7.11a) by constructing a $(\mathfrak{U}, 1)$-Gaschütz operator for $V_\mathfrak{U}$. Why does a similar construction not work in the relative case?

16) Let $V = V(n, q)$ be an n-dimensional vector space over the field $K = GF(q)$ and let $\mathfrak{G} = GL(n, q)$ act naturally on V. For which pairs (n, q) is V a projective K$\mathfrak{G}$-module?

17) Suppose that $\mathfrak{U} \leq \mathfrak{G}$ and that V is a K$\mathfrak{U}$-module.

a) If $\dim_K V = 1$, then $V^\mathfrak{G}$ is a projective K$\mathfrak{G}$-module if and only if char K does not divide $|\mathfrak{U}|$.

b) If char K does not divide $|\mathfrak{U}|$ and V is an irreducible K$\mathfrak{U}$-module, then $V^\mathfrak{G}$ is isomorphic to a direct summand of K$\mathfrak{G}$.

18) Suppose that $\mathfrak{N} \trianglelefteq \mathfrak{G}$ and that K is an arbitrary field.

a) If V is a projective K$\mathfrak{G}$-module, then

$$V_0 = \{v | v \in V, vg = v \quad \text{for all } g \in \mathfrak{N}\}$$

is a projective $K(\mathfrak{G}/\mathfrak{N})$-module, where the module structure of V_0 is defined by $v(\mathfrak{N}g) = vg$ for $v \in V_0$ and $g \in \mathfrak{G}$.

b) If V is a projective K$\mathfrak{G}$-module and we put

$$V_1 = \langle vg - v | v \in V, g \in \mathfrak{N} \rangle,$$

then V/V_1 is a projective $K(\mathfrak{G}/\mathfrak{N})$-module, where the module structure of V/V_1 is defined by $(v + V_1)(\mathfrak{N}g) = vg + V_1$.

19) Suppose that P is a projective K$\mathfrak{G}$-module.

a) If L is an extension of K, then P_L is a projective L$\mathfrak{G}$-module.

b) If K_0 is a subfield of K such that $(K : K_0) < \infty$, then the $K_0\mathfrak{G}$-module P_0, obtained as in 1.16 from P, is projective.

c) If in addition to b), K is a Galois extension of K_0, then for any automorphism η of K over K_0, the algebraic conjugate P_η of P is projective.

§ 8. The Dual Module

In this section we collect some elementary results about the dual module, many of which will be used later on.

8.1 Definition. Suppose that K is an arbitrary field and V a K$\mathfrak{G}$-module. The K-vector space $V^* = \mathrm{Hom}_K(V, K)$ dual to V becomes a K$\mathfrak{G}$-module if we put

$$v(fg) = (vg^{-1})f$$

for all $v \in V$, $f \in V^*$ and $g \in \mathfrak{G}$. For obviously $fg \in V^*$ and $f1 = f$; also if $v \in V$, $f \in V^*$ and g_1, g_2 are in $\mathfrak{G}$, then

$$v(f(g_1g_2)) = (vg_2^{-1}g_1^{-1})f = (vg_2^{-1})(fg_1) = v((fg_1)g_2).$$

We give a more explicit description of the dual module.

8.2 Lemma. *Let V be a K$\mathfrak{G}$-module with K-basis $\{v_1, \ldots, v_n\}$. Let $\{f_1, \ldots, f_n\}$ be the dual basis of V^*; thus $v_if_j = \delta_{ij}$. If*

$$v_ig = \sum_{j=1}^{n} a_{ij}(g)v_j \quad (g \in \mathfrak{G},\ a_{ij}(g) \in K),$$

then

$$f_ig = \sum_{j=1}^{n} a_{ji}(g^{-1})f_j.$$

(In terms of matrices, this shows that if ϱ is a matrix representation

corresponding to V, *then the representation* ϱ^*, *where* $\varrho^*(g) = \varrho(g^{-1})^t$, *corresponds to* V*).

Proof. By definition of V*,

$$v_i(f_jg) = (v_ig^{-1})f_j = \left(\sum_{k=1}^{n} a_{ik}(g^{-1})v_k\right)f_j = a_{ij}(g^{-1})$$
$$= v_i\left(\sum_{k=1}^{n} a_{kj}(g^{-1})f_k\right).$$

This shows that

$$f_jg = \sum_{k=1}^{n} a_{kj}(g^{-1})f_k.$$ **q.e.d.**

In this section we want to develop a duality theory for $K\mathfrak{G}$-modules. We use the corresponding statements for K-modules without proof. These statements are either standard results of the duality theory of finite-dimensional vector spaces or follow easily from these.

8.3 Lemma. *Suppose that* V *is a finitely generated* $K\mathfrak{G}$-*module.*

a) *If* W *is a* $K\mathfrak{G}$-*submodule of* V, *then*

$$W^{\perp} = \{f \mid f \in V^*, \quad wf = 0 \quad \text{for all } w \in W\}$$

is a $K\mathfrak{G}$-*submodule of* V*. *If* $W_2 \subseteq W_1 \subseteq V$, *where* W_1, W_2 *are* $K\mathfrak{G}$-*submodules*, $(W_1/W_2)^*$ *and* $W_2^{\perp}/W_1^{\perp}$ *are isomorphic* $K\mathfrak{G}$-*modules.*

b) *If* V *and* W *are finitely generated* $K\mathfrak{G}$-*modules and* $\alpha \in \mathrm{Hom}_{K\mathfrak{G}}(V, W)$, *then the mapping* α^* *of* W* *into* V* *defined by*

$$v(f\alpha^*) = (v\alpha)f \quad (v \in V, f \in W^*)$$

is a $K\mathfrak{G}$-*homomorphism.*

If $\alpha \in \mathrm{Hom}_{K\mathfrak{G}}(V, W)$ *and* $\beta \in \mathrm{Hom}_{K\mathfrak{G}}(W, X)$, *then* $(\alpha\beta)^* = \beta^*\alpha^*$. *If*

$$0 \to V \overset{\alpha}{\to} W \overset{\beta}{\to} X \to 0$$

is an exact sequence of $K\mathfrak{G}$-*modules, then also*

$$0 \to X^* \overset{\beta^*}{\to} W^* \overset{\alpha^*}{\to} V^* \to 0$$

is exact. (is a contravariant and exact functor on the category of finitely generated* $K\mathfrak{G}$*-modules.)*

c) *Let* τ *be the canonical isomorphism of* V *onto* V^{**} *defined by*

$$f(v\tau) = vf \quad (v \in V, f \in V^*).$$

Then τ *is a* $K\mathfrak{G}$*-isomorphism of* V *onto* V^{**}.

d) *If* $V = V_1 \oplus V_2$ *is a decomposition of* V *into* $K\mathfrak{G}$*-submodules* V_1, V_2, *then* $V^* = V_1^\perp \oplus V_2^\perp$ *is a decomposition of* V^* *into* $K\mathfrak{G}$*-submodules. We have* $V_1^\perp \cong V_2^*$ *and* $V_2^\perp \cong V_1^*$.

e) V *is irreducible if and only if* V^* *is irreducible.*

f) V *is indecomposable if and only if* V^* *is indecomposable.*

g) *If* U *is a* $K\mathfrak{G}$*-submodule of* V, $[U, \mathfrak{G}]^\perp/U^\perp = \mathbf{C}_{V^*/U^\perp}(\mathfrak{G})$ *and* $[U^\perp, \mathfrak{G}] = W^\perp$, *where* $W/U = \mathbf{C}_{V/U}(\mathfrak{G})$.

Proof. a) Let σ be the restriction mapping of V^* into W^*. From

$$w((fg)\sigma) = w(fg) = (wg^{-1})f = (wg^{-1})(f\sigma) = w((f\sigma)g)$$

for $w \in W, f \in V^*$ and $g \in \mathfrak{G}$, it follows that σ is a $K\mathfrak{G}$-homomorphism. Then σ is an epimorphism of V^* onto W^* with kernel $W^\perp$. Thus $W^\perp$ is a $K\mathfrak{G}$-submodule of V^*, and $W^* \cong V^*/W^\perp$.

Now $(W_1/W_2)^*$ is isomorphic to the $K\mathfrak{G}$-submodule

$$\{f \mid f \in W_1^*, \quad wf = 0 \quad \text{for all } w \in W_2\}$$

of W_1^*, and in the isomorphism $W_1^* \cong V^*/W_1^\perp$, this corresponds to $W_2^\perp/W_1^\perp$.

b) It is well-known that $\alpha^* \in \mathrm{Hom}_K(W^*, V^*)$ and that the resulting functor is contravariant and exact. Also, α^* is a $K\mathfrak{G}$-homomorphism, since

$$\begin{aligned} v((fg)\alpha^*) &= (v\alpha)(fg) = ((v\alpha)g^{-1})f = ((vg^{-1})\alpha)f \\ &= (vg^{-1})(f\alpha^*) = (v((f\alpha^*)g) \end{aligned}$$

for all $v \in V$, $f \in W^*$ and $g \in \mathfrak{G}$.

c) For $v \in V$, $f \in V^*$ and $g \in \mathfrak{G}$, we have

$$f((v\tau)g) = (fg^{-1})(v\tau) = v(fg^{-1}) = (vg)f = f((vg)\tau).$$

This shows that $\tau \in \mathrm{Hom}_{K\mathfrak{G}}(V, V^{**})$, and it is well-known that τ is an isomorphism.

d) The direct decomposition $V^* = V_1^\perp \oplus V_2^\perp$ of the K-space V^* is well-known. By a), it is a direct decomposition of V^* into $K\mathfrak{G}$-submodules.

e) If V^* is irreducible, then V is irreducible by a). If V is irreducible, then V^{**} is irreducible by c), hence V^* is irreducible by a).

f) This follows similarly from d) and c).

g) If $f \in V^*$, f lies in $[U, \mathfrak{G}]^\perp$ if and only if $(u(g-1))f = 0$ for all $u \in U, g \in \mathfrak{G}$. Since $(u(g-1))f = u(f(g^{-1}-1))$, this condition is equivalent to $f(g^{-1}-1) \in U^\perp$ for all $g \in \mathfrak{G}$. Hence $[U, \mathfrak{G}]^\perp/U^\perp = \mathbf{C}_{V^*/U^\perp}(\mathfrak{G})$.

Applying this to the subspace $U^\perp$ of V^*, we find that for $v \in V$, $f(v\tau) = 0$ for all $f \in [U^\perp, \mathfrak{G}]$ if and only if $f'((v\tau)(g-1)) = 0$ for all $f' \in U^\perp, g \in \mathfrak{G}$. Thus $vf = 0$ for all $f \in [U^\perp, \mathfrak{G}]$ if and only if $v(g-1) \in U$ for all $g \in \mathfrak{G}$. Thus if $W/U = \mathbf{C}_{V/U}(\mathfrak{G})$,

$$W = \{v \mid v \in V, \quad vf = 0 \quad \text{for all } f \in [U^\perp, \mathfrak{G}]\}.$$

Hence $W^\perp = [U^\perp, \mathfrak{G}]$. **q.e.d.**

8.4 Lemma. a) *Let* V *be a finitely generated* $K\mathfrak{G}$*-module and let* L *be an extension of* K. *Then* $(V \otimes_K L)^*$ *and* $V^* \otimes_K L$ *are isomorphic* $L\mathfrak{G}$*-modules.*

b) *If* V_1 *and* V_2 *are finitely generated* $K\mathfrak{G}$*-modules, then* $(V_1 \otimes_K V_2)^*$ *and* $V_1^* \otimes_K V_2^*$ *are isomorphic* $K\mathfrak{G}$*-modules.*

Proof. a) As is well-known, there exists an L-isomorphism α of $V^* \otimes_K L$ onto $(V \otimes_K L)^*$ such that

$$(v \otimes l_1)((f \otimes l_2)\alpha) = l_1 l_2(vf)$$

for all $v \in V$, $f \in V^*$ and l_1, l_2 in L. For $g \in \mathfrak{G}$,

$$\begin{aligned}(v \otimes l_1)((f \otimes l_2)g\alpha) &= (v \otimes l_1)((fg \otimes l_2)\alpha) = l_1 l_2(v(fg)) \\ &= l_1 l_2((vg^{-1})f) = (vg^{-1} \otimes l_1)((f \otimes l_2)\alpha) \\ &= ((v \otimes l_1)g^{-1})((f \otimes l_2)\alpha) \\ &= (v \otimes l_1)(((f \otimes l_2)\alpha)g).\end{aligned}$$

Thus α is an $L\mathfrak{G}$-isomorphism.

b) There exists a K-isomorphism β of $V_1^* \otimes_K V_2^*$ onto $(V_1 \otimes_K V_2)^*$ such that for $v_i \in V_i$ and $f_i \in V_i^*$ $(i = 1, 2)$

$$(v_1 \otimes v_2)((f_1 \otimes f_2)\beta) = (v_1 f_1)(v_2 f_2).$$

For every $g \in \mathfrak{G}$ we have

$$\begin{aligned}(v_1 \otimes v_2)((f_1 \otimes f_2)g\beta) &= (v_1 \otimes v_2)((f_1 g \otimes f_2 g)\beta) = (v_1(f_1 g))(v_2(f_2 g)) \\ &= ((v_1 g^{-1})f_1)((v_2 g^{-1})f_2) \\ &= (v_1 g^{-1} \otimes v_2 g^{-1})((f_1 \otimes f_2)\beta) \\ &= (v_1 \otimes v_2)((f_1 \otimes f_2)\beta g).\end{aligned}$$

Hence β is a $\mathsf{K}\mathfrak{G}$-isomorphism. **q.e.d.**

8.5 Lemma. *Let* V *be a finitely generated* $\mathsf{K}\mathfrak{G}$*-module.*

a) *The mapping* α *for which*

$$(v \otimes f)\alpha = vf \quad (v \in \mathsf{V}, f \in \mathsf{V}^*)$$

is a $\mathsf{K}\mathfrak{G}$*-epimorphism of* $\mathsf{V} \otimes_{\mathsf{K}} \mathsf{V}^*$ *onto* K*, where* K *is regarded as a module for the trivial representation of* $\mathfrak{G}$.

b) *Let* $\{v_1, \ldots, v_n\}$ *be a* K*-basis of* V *and let* $\{f_1, \ldots, f_n\}$ *be the dual basis of* V^**; thus* $v_i f_j = \delta_{ij}$. *Then*

$$\left(\sum_{i=1}^{n} v_i \otimes f_i\right)g = \sum_{i=1}^{n} v_i \otimes f_i$$

for all $g \in \mathfrak{G}$. *Hence* $\mathsf{V} \otimes_{\mathsf{K}} \mathsf{V}^*$ *contains a submodule* V_0 *isomorphic to* K.

c) *If* char K *does not divide* $\dim_{\mathsf{K}} \mathsf{V}$, *then* V_0 *is a direct summand of the* $\mathsf{K}\mathfrak{G}$*-module* $\mathsf{V} \otimes_{\mathsf{K}} \mathsf{V}^*$.

d) *If* char K *divides* $\dim_{\mathsf{K}} \mathsf{V}$, *then* K *appears as a composition factor in* $\mathsf{V} \otimes_{\mathsf{K}} \mathsf{V}^*$ *with multiplicity at least 2.*

Proof. a) We have for every $v \in \mathsf{V}, f \in \mathsf{V}^*$ and $g \in \mathfrak{G}$,

$$(vg \otimes fg)\alpha = (vg)(fg) = (vgg^{-1})f = vf = (vf)g.$$

Hence α is a $\mathsf{K}\mathfrak{G}$-epimorphism of $\mathsf{V} \otimes_{\mathsf{K}} \mathsf{V}^*$ onto K.

b) By 8.2,

$$v_i g = \sum_{j=1}^{n} a_{ij}(g)v_j \quad \text{and} \quad f_i g = \sum_{j=1}^{n} a_{ji}(g^{-1})f_j.$$

Hence

$$\left(\sum_{i=1}^{n} v_i \otimes f_i\right)g = \sum_{i=1}^{n} v_i g \otimes f_i g = \sum_{i,j,k=1}^{n} a_{ij}(g)a_{ki}(g^{-1})v_j \otimes f_k$$

$$= \sum_{j,k=1}^{n} \sum_{i=1}^{n} a_{ki}(g^{-1})a_{ij}(g)v_j \otimes f_k$$

$$= \sum_{j,k=1}^{n} \delta_{kj}v_j \otimes f_k = \sum_{j=1}^{n} v_j \otimes f_j.$$

As $\{v_1, \ldots, v_n\}$ is a K-basis of V and $f_i \neq 0$, we have $\sum_{i=1}^{n} v_i \otimes f_i \neq 0$. Hence

$$\mathsf{V}_0 = \mathsf{K} \sum_{i=1}^{n} v_i \otimes f_i$$

is a $\mathsf{K}\mathfrak{G}$-submodule isomorphic to K.

c) In the notation of a) and b),

$$\left(\sum_{i=1}^{n} v_i \otimes f_i\right)\alpha = \sum_{i=1}^{n} v_i f_i = n.$$

If char K does not divide n, we conclude that $\mathsf{V} \otimes_{\mathsf{K}} \mathsf{V}^* = \ker \alpha \oplus \mathsf{V}_0$.

d) If char K divides n, then $(\sum_{i=1}^{n} v_i \otimes f_i)\alpha = 0$. From

$$0 \subset \mathsf{V}_0 \subseteq \ker \alpha \subset \mathsf{V} \otimes_{\mathsf{K}} \mathsf{V}^*$$

and

$$\mathsf{V}_0 \cong (\mathsf{V} \otimes_{\mathsf{K}} \mathsf{V}^*)/\ker \alpha \cong \mathsf{K},$$

the assertion follows. **q.e.d.**

8.6 Theorem. *Suppose that* V *and* W *are irreducible* $\mathsf{K}\mathfrak{G}$*-modules. Then the following assertions, in which* K *is regarded as the module for the trivial representation of* $\mathfrak{G}$*, are equivalent.*

a) $\mathsf{V}^* \cong \mathsf{W}$.

b) $\mathsf{V} \otimes_{\mathsf{K}} \mathsf{W}$ *has a factor module isomorphic to* K.

c) $\mathsf{V} \otimes_{\mathsf{K}} \mathsf{W}$ *has a submodule isomorphic to* K.

Proof. By 8.5, a) implies b) and c).

To show that b) implies a), suppose that α is a $K\mathfrak{G}$-epimorphism of $V \otimes_K W$ onto K. We define a mapping β of W into V^* by putting

$$v(w\beta) = (v \otimes w)\alpha \quad (v \in V, w \in W).$$

Then for all $g \in \mathfrak{G}$,

$$\begin{aligned} v((w\beta)g) &= (vg^{-1})(w\beta) \quad \text{(by the module structure of } V^*) \\ &= (vg^{-1} \otimes w)\alpha = (v \otimes wg)g^{-1}\alpha = (v \otimes wg)\alpha g^{-1} \\ &= (v \otimes wg)\alpha \quad \text{(since } K \text{ is a trivial } K\mathfrak{G}\text{-module)} \\ &= v((wg)\beta). \end{aligned}$$

Thus β is a $K\mathfrak{G}$-homomorphism of W into V^*. As W and V^* are irreducible and $\beta \neq 0$, it follows that $W \cong V^*$.

To prove that c) implies a), suppose that

$$0 \to K \to V \otimes_K W$$

is an exact sequence of $K\mathfrak{G}$-modules. By 8.3b), the sequence

$$(V \otimes_K W)^* \to K^* \to 0$$

is also exact. By 8.4b), $(V \otimes_K W)^* \cong V^* \otimes_K W^*$. Also $K^* \cong K$, so since b) implies a), $V^{**} \cong W^*$. Hence $V^* \cong W$ by 8.3c). **q.e.d.**

We now give an example to show that for irreducible $K\mathfrak{G}$-modules V and W, the existence of a composition factor of $V \otimes_K W$ isomorphic to K does not imply the duality of V and W.

8.7 Example. Let $\mathfrak{G} = SL(2, 5)$. We consider the $K\mathfrak{G}$-modules, where K is a field of characteristic 5. By 3.10, there are five irreducible $K\mathfrak{G}$-modules to within isomorphism, namely

$$V_i = Kx^i \oplus Kx^{i-1}y \oplus \cdots \oplus Ky^i \quad (i = 0, 1, 2, 3, 4).$$

Thus $-I$ operates non-trivially on V_1 and V_3, but trivially on V_0, V_2 and V_4. Hence every composition factor of $V_2 \otimes_K V_4$ is isomorphic to one of the modules V_0, V_2 or V_4. If n_i is the multiplicity of V_i as a composition factor of $V_2 \otimes_K V_4$, then

$$15 = \dim_K V_2 \otimes_K V_4 = n_0 + 3n_2 + 5n_4.$$

Also, if ϱ_i denotes the representation of $\mathfrak{G}$ on V_i and

$$\chi_i(g) = \operatorname{tr} \varrho_i(g) \quad \text{for all } g \in \mathfrak{G},$$

then

$$\chi_2(g)\chi_4(g) = \operatorname{tr}(\varrho_2(g) \otimes \varrho_4(g)) = n_0\chi_0(g) + n_2\chi_2(g) + n_4\chi_4(g).$$

Applying this with

$$g = \begin{pmatrix} 2 & 0 \\ 0 & -2 \end{pmatrix} \quad \text{and} \quad g = \begin{pmatrix} 1 & 1 \\ -1 & 0 \end{pmatrix}$$

gives the congruences

$$n_0 - n_2 + n_4 \equiv -1 \quad \text{and} \quad n_0 - n_4 \equiv 0 \pmod{5}.$$

It follows that $n_0 = n_4 = 1$ and $n_2 = 3$. Thus $\mathsf{V}_2 \otimes_{\mathsf{K}} \mathsf{V}_4$ has a composition factor isomorphic to K, but V_2 and V_4 are not dual since they have different dimensions.

8.8 Lemma. *Let* V *and* W *be finitely generated* $\mathsf{K}\mathfrak{G}$*-modules.*

a) $\operatorname{Hom}_{\mathsf{K}}(\mathsf{V}, \mathsf{W})$ *becomes a* $\mathsf{K}\mathfrak{G}$*-module if we put*

$$v(\alpha g) = ((vg^{-1})\alpha)g$$

for all $v \in \mathsf{V}$, $\alpha \in \operatorname{Hom}_{\mathsf{K}}(\mathsf{V}, \mathsf{W})$ *and* $g \in \mathfrak{G}$. *Then*

$$\operatorname{Hom}_{\mathsf{K}\mathfrak{G}}(\mathsf{V}, \mathsf{W}) = \{\alpha \mid \alpha \in \operatorname{Hom}_{\mathsf{K}}(\mathsf{V}, \mathsf{W}), \quad \alpha g = \alpha \text{ for all } g \in \mathfrak{G}\}$$

is the set of all elements of $\operatorname{Hom}_{\mathsf{K}}(\mathsf{V}, \mathsf{W})$ *which are fixed under* $\mathfrak{G}$.

b) *The mapping* β *of* $\mathsf{V}^* \otimes_{\mathsf{K}} \mathsf{W}$ *into* $\operatorname{Hom}_{\mathsf{K}}(\mathsf{V}, \mathsf{W})$ *defined by putting*

$$v((f \otimes w)\beta) = (vf)w \quad (v \in \mathsf{V}, f \in \mathsf{V}^*, w \in \mathsf{W})$$

is a $\mathsf{K}\mathfrak{G}$*-isomorphism of* $\mathsf{V}^* \otimes_{\mathsf{K}} \mathsf{W}$ *onto the* $\mathsf{K}\mathfrak{G}$*-module* $\operatorname{Hom}_{\mathsf{K}}(\mathsf{V}, \mathsf{W})$.

Proof. a) It is easy to check that $\alpha g \in \operatorname{Hom}_{\mathsf{K}}(\mathsf{V}, \mathsf{W})$, $\alpha 1 = \alpha$ and $(\alpha g_1)g_2 = \alpha(g_1 g_2)$ for all $\alpha \in \operatorname{Hom}_{\mathsf{K}}(\mathsf{V}, \mathsf{W})$ and $g_i \in \mathfrak{G}$. Thus $\operatorname{Hom}_{\mathsf{K}}(\mathsf{V}, \mathsf{W})$ is a $\mathsf{K}\mathfrak{G}$-module. Also $\alpha g = \alpha$ if and only if $v\alpha = ((vg^{-1})\alpha)g$ for all $v \in \mathsf{V}$, that is, $(v\alpha)g = (vg)\alpha$ for all $v \in \mathsf{V}$. Thus $\operatorname{Hom}_{\mathsf{K}\mathfrak{G}}(\mathsf{V}, \mathsf{W})$ is the set of elements of $\operatorname{Hom}_{\mathsf{K}}(\mathsf{V}, \mathsf{W})$ left fixed by all elements of $\mathfrak{G}$.

b) It is well-known that β is a K-isomorphism of $V^* \otimes_K W$ onto $\mathrm{Hom}_K(V, W)$. For $v \in V$, $f \in V^*$, $w \in W$ and $g \in \mathfrak{G}$, we have

$$v((f \otimes w)g\beta) = v((fg \otimes wg)\beta) = (v(fg))(wg) = ((vg^{-1})f)(wg)$$

and

$$v((f \otimes w)\beta g) = ((vg^{-1})((f \otimes w)\beta))g = ((vg^{-1})f)(wg).$$

Thus β is a $K\mathfrak{G}$-isomorphism. **q.e.d.**

8.9 Lemma. *Let V_1 and V_2 be finitely generated $K\mathfrak{G}$-modules. We denote by $\mathbf{B}(V_1, V_2)$ the K-vector space of K-bilinear forms on $V_1 \times V_2$.*

a) *$\mathbf{B}(V_1, V_2)$ becomes a $K\mathfrak{G}$-module if we put*

$$(fg)(v_1, v_2) = f(v_1 g^{-1}, v_2 g^{-1})$$

for $v_i \in V_i$ $(i = 1, 2)$, $f \in \mathbf{B}(V_1, V_2)$ and $g \in \mathfrak{G}$.

b) *$V_1^* \otimes_K V_2^*$ and $\mathbf{B}(V_1, V_2)$ are isomorphic $K\mathfrak{G}$-modules; in fact, the mapping γ defined by*

$$((f_1 \otimes f_2)\gamma)(v_1, v_2) = (v_1 f_1)(v_2 f_2) \quad (v_i \in V_i,\ f_i \in V_i^*)$$

is a $K\mathfrak{G}$-isomorphism.

Proof. a) is easily checked.

b) It is well-known that γ is a K-isomorphism of $V_1^* \otimes_K V_2^*$ onto $\mathbf{B}(V_1, V_2)$. For $v_i \in V_i$, $f_i \in V_i^*$ and $g \in \mathfrak{G}$, we have

$$\begin{aligned}(((f_1 \otimes f_2)g)\gamma)(v_1, v_2) &= ((f_1 g \otimes f_2 g)\gamma)(v_1, v_2)\\ &= (v_1(f_1 g))(v_2(f_2 g))\\ &= (v_1 g^{-1})f_1 (v_2 g^{-1})f_2\\ &= ((f_1 \otimes f_2)\gamma)(v_1 g^{-1}, v_2 g^{-1})\\ &= (((f_1 \otimes f_2)\gamma)g)(v_1, v_2).\end{aligned}$$

Thus γ is a $K\mathfrak{G}$-isomorphism. **q.e.d.**

8.10 Lemma. *Let V_1, V_2 be finitely generated $K\mathfrak{G}$-modules.*

a) *Suppose that $\alpha \in \mathrm{Hom}_K(V_2, V_1^*)$, f is a K-bilinear form on $V_1 \times V_2$ and*

$$f(v_1, v_2) = v_1(v_2\alpha)$$

for all $v_i \in \mathsf{V}_i$. *Then* α *is a* $\mathsf{K}\mathfrak{G}$*-homomorphism if and only if* f *is* $\mathfrak{G}$*-invariant; (that is,* $f(v_1 g, v_2 g) = f(v_1, v_2)$ *for all* $v_i \in \mathsf{V}_i$ *and all* $g \in \mathfrak{G}$*).*

b) $\mathsf{V}_2 \cong \mathsf{V}_1^*$ *if and only if there exists a non-singular* $\mathfrak{G}$*-invariant bilinear form on* $\mathsf{V}_1 \times \mathsf{V}_2$.

Proof. a) This follows from

$$\begin{aligned} f(v_1 g, v_2 g) - f(v_1, v_2) &= (v_1 g)(v_2 g\alpha) - (v_1 g g^{-1})(v_2\alpha) \\ &= (v_1 g)(v_2 g\alpha - v_2\alpha g). \end{aligned}$$

b) It is clear that in a), α is an isomorphism if and only if f is non-singular. The assertion thus follows from a). **q.e.d.**

8.11 Theorem (GOW). *Suppose that* char $\mathsf{K} \neq 2$ *and that* V *is an indecomposable* $\mathsf{K}\mathfrak{G}$*-module for which* $\mathsf{V} \cong \mathsf{V}^*$.

a) *There exists a non-singular symmetric or symplectic* $\mathfrak{G}$*-invariant form on* V.

b) *If* V *is absolutely indecomposable, there cannot exist both symmetric and symplectic non-singular* $\mathfrak{G}$*-invariant bilinear forms on* V.

Proof. Let $\mathfrak{S} = \operatorname{Hom}_{\mathsf{K}\mathfrak{G}}(\mathsf{V}, \mathsf{V})$ and $\mathfrak{J} = \mathbf{J}(\mathfrak{S})$.

a) By hypothesis, there exists a $\mathsf{K}\mathfrak{G}$-isomorphism α of V onto V^*. Let α^* be the dual linear mapping of $\mathsf{V}^{**} = \mathsf{V}$ onto V^*, defined by

$$w(v\alpha) = v(w\alpha^*)$$

for all $v, w \in \mathsf{V}$. By 8.3b), c) α^* is a $\mathsf{K}\mathfrak{G}$-isomorphism. Put $\beta = \alpha^*\alpha^{-1}$. Then $\beta \in \mathfrak{S}$ and

$$w(v\alpha) = v(w\beta\alpha)$$

for all $v, w \in \mathsf{V}$.

For $\varepsilon = \pm 1$, a bilinear form f_ε on V is defined by putting

$$f_\varepsilon(v, w) = v(w\alpha) + \varepsilon w(v\alpha).$$

By 8.10, f_ε is $\mathfrak{G}$-invariant, f_1 is symmetric and f_{-1} is symplectic. Suppose that f_1, f_{-1} are both singular. Since

$$f_\varepsilon(v, w) = v((w + \varepsilon w\beta)\alpha),$$

it follows that $1_V + \beta$ and $1_V - \beta$ are both singular. They are therefore non-units of the local ring $\mathfrak{S}$ and thus lie in $\mathfrak{J}$. Hence $2 \cdot 1_V \in \mathfrak{J}$ and then $1_V \in \mathfrak{J}$ since char $K \neq 2$. This is a contradiction. So either f_1 or f_{-1} is non-singular.

b) Suppose that f_1, f_2 are non-singular $\mathfrak{G}$-invariant bilinear forms on V, where f_1 is symmetric and f_2 is symplectic. Then there exist K-isomorphisms α_1, α_2 of V onto V* such that

$$w(v\alpha_1) = f_1(v, w) \quad \text{and} \quad w(v\alpha_2) = f_2(v, w).$$

Thus α_1, α_2 are K$\mathfrak{G}$-isomorphisms, $\alpha_1^* = \alpha_1$ and $\alpha_2^* = -\alpha_2$. Put $\beta = \alpha_1\alpha_2^{-1}$, then $\beta \in \mathfrak{S}$. From $\alpha_1 = \beta\alpha_2$ we obtain $\alpha_1^* = \alpha_2^*\beta^*$, so $\alpha_1 = -\alpha_2\beta^*$. Thus

$$\alpha_1\beta^*\alpha_1^{-1} = \beta\alpha_2\beta^*\alpha_1^{-1} = -\beta\alpha_1\alpha_1^{-1} = -\beta.$$

Since V is absolutely indecomposable, $\mathfrak{S}/\mathfrak{J} \cong K$, so $\beta = a1_V + \gamma$ for some $a \in K$ and $\gamma \in \mathfrak{J}$. Then $\beta^* = a1_{V^*} + \gamma^*$ and by 8.3b), $\gamma^* \in \mathrm{Hom}_{K\mathfrak{G}}(V^*, V^*)$. Thus $\alpha_1\gamma^*\alpha_1^{-1} \in \mathfrak{S}$. Since γ is singular, so is γ^*, whence $\alpha_1\gamma^*\alpha_1^{-1} \in \mathfrak{J}$. Thus

$$-a1_V - \gamma = -\beta = \alpha_1\beta^*\alpha_1^{-1} = a1_V + \alpha_1\gamma^*\alpha_1^{-1}$$

and hence $2a1_V \in \mathfrak{J}$. Since char $K \neq 2$, this implies that $a = 0$. Thus $\beta \in \mathfrak{J}$, β is singular and $\alpha_1 = \beta\alpha_2$ is not an isomorphism, a contradiction.

q.e.d.

Theorem 8.11 is wrong for fields of characteristic 2. Let $\mathfrak{G} = SL(2, 5)$ and let K be an algebraically closed field of characteristic 2. Then there exists an indecomposable projective K$\mathfrak{G}$-module $P \cong P^*$ with two composition factors which are both isomorphic to an irreducible module of dimension 4; also P does not carry a $\mathfrak{G}$-invariant non-singular symmetric form, and, since char $K = 2$, this implies that P carries no non-singular symplectic form (WILLEMS [2]).

8.12 Theorem. *Suppose that* V *is an absolutely irreducible* K$\mathfrak{G}$*-module. If* $V \cong V^*$, *then to within a scalar multiple, there exists only one non-zero* $\mathfrak{G}$*-invariant bilinear form* f *on* V, *and this form is non-singular. If* char K *is different from 2,* f *is symmetric or symplectic.*

Proof. We denote by $\mathbf{B}_0(V)$ the space of all $\mathfrak{G}$-invariant bilinear forms on V; thus $f \in \mathbf{B}_0(V)$ if and only if

$$f(v_1, v_2) = f(v_1 g, v_2 g) = (fg^{-1})(v_1, v_2)$$

for all $v_i \in \mathsf{V}$ and all $g \in \mathfrak{G}$.

By 8.8b) and 8.9b),

$$\mathrm{Hom}_{\mathsf{K}}(\mathsf{V}, \mathsf{V}) \cong \mathsf{V}^* \otimes_{\mathsf{K}} \mathsf{V} \cong \mathsf{V}^* \otimes_{\mathsf{K}} \mathsf{V}^* \cong \mathbf{B}(\mathsf{V}, \mathsf{V}).$$

Hence by 8.8a), $\mathrm{Hom}_{\mathsf{K}\mathfrak{G}}(\mathsf{V}, \mathsf{V}) \cong \mathbf{B}_0(\mathsf{V})$. Since V is absolutely irreducible, it follows that

$$\dim_{\mathsf{K}} \mathbf{B}_0(\mathsf{V}) = \dim_{\mathsf{K}} \mathrm{Hom}_{\mathsf{K}\mathfrak{G}}(\mathsf{V}, \mathsf{V}) = 1.$$

The existence of a $\mathfrak{G}$-invariant non-singular form follows from 8.10, and that the form is symmetric or symplectic follows from 8.11a). **q.e.d.**

A corresponding result can be proved in the case of characteristic 2.

8.13 Theorem (FONG [2]). *Let K be a perfect field of characteristic 2 and let V be an irreducible $\mathsf{K}\mathfrak{G}$-module. If $\mathsf{V} \cong \mathsf{V}^*$ and V is not a module for the trivial representation of $\mathfrak{G}$, then there exists a non-singular $\mathfrak{G}$-invariant symplectic form on V. In particular, $\dim_{\mathsf{K}} \mathsf{V}$ is even.*

Proof. If $\dim_{\mathsf{K}} \mathsf{V} = 1$, say $\mathsf{V} = \mathsf{K}v$, then $\mathsf{V} \cong \mathsf{V}^*$ implies $vg = vg^{-1}$ for all $g \in \mathfrak{G}$. Since $vg = av$ for some $a \in \mathsf{K}^\times$, we conclude that $a^2 = 1$, hence $a = 1$ as char $\mathsf{K} = 2$. Hence $\dim_{\mathsf{K}} \mathsf{V} \geq 2$.

By 8.10, there exists a non-singular $\mathfrak{G}$-invariant bilinear form f on V. Put $q(v) = f(v, v)$. Thus

$$q(v_1 + v_2) = q(v_1) + q(v_2) + [v_1, v_2],$$

where

$$[v_1, v_2] = f(v_1, v_2) + f(v_2, v_1).$$

We distinguish two cases.

Case 1. Suppose that

$$q(v_1 + v_2) = q(v_1) + q(v_2)$$

for all $v_i \in \mathsf{V}$. Choose linearly independent elements v_1, v_2 in V, as we may since $\dim_{\mathsf{K}} \mathsf{V} \geq 2$. If $q(v_1) \neq 0$, there exists an a in the perfect field

K such that

$$q(av_1 + v_2) = a^2q(v_1) + q(v_2) = 0.$$

Thus the set

$$\mathsf{U} = \{v \mid v \in \mathsf{V}, q(v) = 0\}$$

is a non-zero $\mathfrak{G}$-invariant subspace of V. Since V is irreducible, U = V. Thus

$$0 = q(v) = f(v, v)$$

for all $v \in \mathsf{V}$, and f is symplectic.

Case 2. If the hypothesis of case 1 does not hold, $[,]$ is a non-zero $\mathfrak{G}$-invariant bilinear form on V. Since char K = 2, $[,]$ is symplectic. Let R be the radical of $[,]$, defined by

$$\mathsf{R} = \{v \mid v \in \mathsf{V}, [v, w] = 0 \quad \text{for all } w \in \mathsf{V}\}.$$

Since $[,]$ is non-zero, R $\neq$ V. But R is $\mathfrak{G}$-invariant and V is irreducible, so R = 0 and $[\,,]$ is non-singular. **q.e.d.**

8.14 Definition. Let V be a vector space of finite dimension over the field K.

a) A *quadratic form* q on V is a mapping of V into K such that

$$q(kv) = k^2q(v) \quad \text{and} \quad q(v_1 + v_2) = q(v_1) + q(v_2) + f(v_1, v_2)$$

for all v, v_1, v_2 in V and $k \in \mathsf{K}$, where f is a bilinear form on V.

b) A quadratic form q on V is called *non-singular* if the corresponding bilinear form f is non-singular.

c) Let V be a K$\mathfrak{G}$-module. A quadratic form q on V is called $\mathfrak{G}$-*invariant* if $q(vg) = q(v)$ for all $v \in \mathsf{V}$, $g \in \mathfrak{G}$.

(In this case, the corresponding bilinear form f is obviously also $\mathfrak{G}$-invariant.)

8.15 Remark. a) If char K $\neq$ 2 and q is a quadratic form on V, then the bilinear form f defined by

$$f(v_1, v_2) = q(v_1 + v_2) - q(v_1) - q(v_2)$$

is obviously symmetric and $f(v, v) = 2q(v)$. Hence we can forget about q and work only with the bilinear form f.

But if char $\mathsf{K} = 2$, then $f(v, v) = 0$ for all $v \in \mathsf{V}$, hence f is symplectic.

b) If char $\mathsf{K} \neq 2$ and V is an absolutely irreducible self-dual $\mathsf{K}\mathfrak{G}$-module, then V carries a $\mathfrak{G}$-invariant bilinear form, which is either symmetric or symplectic (8.12). If K is a perfect field of characteristic 2 and $\mathsf{V} \neq \mathsf{K}$, then by 8.13 V carries a non-singular $\mathfrak{G}$-invariant symplectic form. We can ask whether V carries in addition a $\mathfrak{G}$-invariant quadratic form.

8.16 Definition. Let K be a field of characteristic 2 and V a $\mathsf{K}\mathfrak{G}$-module.

a) The K-linear mapping σ defined by

$$(v \otimes w)\sigma = w \otimes v$$

is a $\mathsf{K}\mathfrak{G}$-isomorphism of $\mathsf{V} \otimes_{\mathsf{K}} \mathsf{V}$.

b) We put

$$\mathbf{S}(\mathsf{V}) = \{t \mid t \in \mathsf{V} \otimes_{\mathsf{K}} \mathsf{V}, t\sigma = t\}$$

and

$$\mathbf{A}(\mathsf{V}) = \{t(\sigma - 1) \mid t \in \mathsf{V} \otimes_{\mathsf{K}} \mathsf{V}\} = \operatorname{im}(\sigma - 1).$$

Then $\mathbf{S}(\mathsf{V})$ and $\mathbf{A}(\mathsf{V})$ are $\mathsf{K}\mathfrak{G}$-submodules of $\mathsf{V} \otimes_{\mathsf{K}} \mathsf{V}$. As

$$(\sigma - 1)^2 = \sigma^2 - 1 = 0,$$

we obtain $\mathbf{A}(\mathsf{V}) \subseteq \mathbf{S}(\mathsf{V})$. (If char $\mathsf{K} \neq 2$, then obviously $\mathsf{V} \otimes_{\mathsf{K}} \mathsf{V} = \mathbf{S}(\mathsf{V}) \oplus \mathbf{A}(\mathsf{V})$.)

8.17 Lemma. *Let K be a perfect field of characteristic 2 and V a finitely generated $\mathsf{K}\mathfrak{G}$-module.*

a) $(\mathsf{V} \otimes_{\mathsf{K}} \mathsf{V})/\mathbf{S}(\mathsf{V}) \cong \mathbf{A}(\mathsf{V})$.

b) $\mathbf{S}(\mathsf{V})/\mathbf{A}(\mathsf{V}) \cong \mathsf{V}^{(2)}$, *where $\mathsf{V}^{(2)}$ is the $\mathsf{K}\mathfrak{G}$-module algebraically conjugate to V under the automorphism $a \to a^2$ of the perfect field K.*

c) *If V carries a $\mathfrak{G}$-invariant symplectic form f which is not identically zero, then there exists a $\mathsf{K}\mathfrak{G}$-epimorphism δ of $\mathbf{A}(\mathsf{V})$ onto the trivial $\mathsf{K}\mathfrak{G}$-module K such that*

$$(v \otimes w + w \otimes v)\delta = f(v, w).$$

Proof. a) $\sigma - 1$ is a $\mathsf{K}\mathfrak{G}$-epimorphism of $\mathsf{V} \otimes_{\mathsf{K}} \mathsf{V}$ onto $\mathbf{A}(\mathsf{V})$, and the kernel of $\sigma - 1$ is obviously $\mathbf{S}(\mathsf{V})$.

b) Let $\{v_1, \ldots, v_n\}$ be a K-basis of V. Then

$$\mathbf{A}(\mathsf{V}) = \langle v_i \otimes v_j + v_j \otimes v_i | i, j = 1, \ldots, n; i \neq j \rangle$$

and

$$\mathbf{S}(\mathsf{V}) = \langle v_i \otimes v_i, \mathbf{A}(\mathsf{V}) | i = 1, \ldots, n \rangle,$$

where $\langle \mathsf{T} \rangle$ denotes the vector space over K spanned by T. Hence

$$\{v_i \otimes v_i + \mathbf{A}(\mathsf{V}) | i = 1, \ldots, n\}$$

is a K-basis of $\mathbf{S}(\mathsf{V})/\mathbf{A}(\mathsf{V})$. If

$$v_i g = \sum_{j=1}^{n} a_{ij}(g) v_j \quad (a_{ij}(g) \in \mathsf{K}),$$

then

$$\begin{aligned}(v_i \otimes v_i + \mathbf{A}(\mathsf{V}))g &= v_i g \otimes v_i g + \mathbf{A}(\mathsf{V}) \\ &= \sum_{j,k=1}^{n} a_{ij}(g) a_{ik}(g) v_j \otimes v_k + \mathbf{A}(\mathsf{V}) \\ &= \sum_{j=1}^{n} a_{ij}(g)^2 v_j \otimes v_j + \mathbf{A}(\mathsf{V}).\end{aligned}$$

This shows that $\mathbf{S}(\mathsf{V})/\mathbf{A}(\mathsf{V}) \cong \mathsf{V}^{(2)}$.

c) The mapping $\sigma - 1$ induces a $\mathsf{K}\mathfrak{G}$-isomorphism τ of $(\mathsf{V} \otimes_{\mathsf{K}} \mathsf{V})/\mathbf{S}(\mathsf{V})$ onto $\mathbf{A}(\mathsf{V})$ such that

$$(v \otimes w + \mathbf{S}(\mathsf{V}))\tau = v \otimes w + w \otimes v.$$

The mapping $\phi \in \operatorname{Hom}_{\mathsf{K}}(\mathsf{V} \otimes_{\mathsf{K}} \mathsf{V}, \mathsf{K})$ defined by

$$(v \otimes w)\phi = f(v, w)$$

is a $\mathsf{K}\mathfrak{G}$-homomorphism of $\mathsf{V} \otimes_{\mathsf{K}} \mathsf{V}$ into the trivial $\mathsf{K}\mathfrak{G}$-module K, as f is $\mathfrak{G}$-invariant. As f is symplectic, we obtain

$$(v \otimes w + w \otimes v)\phi = f(v, w) + f(w, v) = 0$$

and

$$(v \otimes v)\phi = f(v, v) = 0.$$

This shows $\mathbf{S}(\mathsf{V}) \subseteq \ker \phi$. Hence there exists a $\mathsf{K}\mathfrak{G}$-epimorphism $\bar{\phi}$ of $(\mathsf{V} \otimes_{\mathsf{K}} \mathsf{V})/\mathbf{S}(\mathsf{V})$ onto K such that

$$(v \otimes w + \mathbf{S}(\mathsf{V}))\bar{\phi} = f(v, w).$$

If we define δ by $\delta = \tau^{-1}\bar{\phi}$, then

$$(v \otimes w + w \otimes v)\delta = (v \otimes w + \mathbf{S}(\mathsf{V}))\bar{\phi} = f(v, w). \qquad \textbf{q.e.d.}$$

8.18 Lemma. *Let K be a field of characteristic 2 and V a $\mathsf{K}\mathfrak{G}$-module. Let $\{v_1, \ldots, v_n\}$ be a K-basis of V and q a quadratic form on V. We put $q_{ii} = q(v_i)$ and $q_{ij} = f(v_i, v_j)$ $(i \neq j)$, where f is defined by*

$$f(v, v') = q(v + v') - q(v) - q(v').$$

We define $\alpha \in \mathrm{Hom}_{\mathsf{K}}(\mathbf{S}(\mathsf{V}), \mathsf{K})$ by

$$(v_i \otimes v_i)\alpha = q_{ii}$$

$$(v_i \otimes v_j + v_j \otimes v_i)\alpha = q_{ij} \quad (i \neq j).$$

Then q is $\mathfrak{G}$-invariant if and only if $\alpha \in \mathrm{Hom}_{\mathsf{K}\mathfrak{G}}(\mathbf{S}(\mathsf{V}), \mathsf{K})$.

Proof. For $v = \sum_{i=1}^{n} x_i v_i$ $(x_i \in \mathsf{K})$, we have

$$q(v) = \sum_{i \leq j} x_i x_j q_{ij}.$$

Now q is $\mathfrak{G}$-invariant if and only if

$$(1) \qquad q(v_i g) = q(v_i) \quad (i = 1, \ldots, n), \quad \text{and}$$

$$(2) \qquad f(v_i g, v_j g) = f(v_i, v_j) \quad (i \neq j)$$

for all $g \in \mathfrak{G}$. Suppose that

$$v_i g = \sum_{j=1}^{n} a_{ij}(g) v_j.$$

Then (1) is equivalent to

$$(1') \qquad q_{ii} = q(v_i) = q(v_i g) = \sum_{j=1}^{n} a_{ij}(g)^2 q_{jj} + \sum_{j<k} a_{ij}(g) a_{ik}(g) q_{jk}.$$

Also (2) is equivalent to

$$(2') \qquad q_{ij} = f(v_i, v_j) = f(v_i g, v_j g) = \sum_{k \neq l} a_{ik}(g) a_{jl}(g) q_{kl} \quad (i \neq j).$$

On the other hand, α is in $\mathrm{Hom}_{\mathsf{K}\mathfrak{G}}(\mathbf{S}(\mathsf{V}), \mathsf{K})$ if and only if

$$\begin{aligned} q_{ii} = q_{ii} g &= (v_i \otimes v_i)\alpha g = (v_i g \otimes v_i g)\alpha \\ &= \left(\sum_{j,k} a_{ij}(g) a_{ik}(g) v_j \otimes v_k\right)\alpha \\ &= \sum_{j=1}^{n} a_{ij}(g)^2 q_{jj} + \sum_{j<k} a_{ij}(g) a_{ik}(g) q_{jk} \end{aligned}$$

and for $i \neq j$

$$\begin{aligned} q_{ij} = q_{ij} g &= (v_i \otimes v_j + v_j \otimes v_i)\alpha g \\ &= (v_i g \otimes v_j g + v_j g \otimes v_i g)\alpha \\ &= \sum_{k,l} a_{ik}(g) a_{jl}(g)(v_k \otimes v_l + v_l \otimes v_k)\alpha \\ &= \sum_{k \neq l} a_{ik}(g) a_{jl}(g) q_{kl}. \end{aligned}$$

Hence the assertion follows. **q.e.d.**

8.19 Theorem. *Let K be a perfect field of characteristic 2 and V an irreducible $\mathsf{K}\mathfrak{G}$-module not isomorphic to the trivial $\mathsf{K}\mathfrak{G}$-module. Then the following assertions are equivalent.*

a) *There exists a non-singular $\mathfrak{G}$-invariant quadratic form on V.*

b) *There exists a $\mathsf{K}\mathfrak{G}$-epimorphism of $\mathbf{S}(\mathsf{V})$ onto the trivial $\mathsf{K}\mathfrak{G}$-module K.*

Proof. Let $\{v_1, \ldots, v_n\}$ be a K-basis of V.

a) ⇒ b): Let q be a non-singular $\mathfrak{G}$-invariant quadratic form on V and f the corresponding symplectic form. We put

$$q_{ii} = q(v_i) \quad \text{and} \quad q_{ij} = f(v_i, v_j) \quad (i \neq j).$$

If we define $\alpha \in \mathrm{Hom}_{\mathsf{K}}(\mathbf{S}(\mathsf{V}), \mathsf{K})$ by

$$(v_i \otimes v_i)\alpha = q_{ii} \quad \text{and} \quad (v_i \otimes v_j + v_j \otimes v_i)\alpha = q_{ij} \quad (i \neq j),$$

then $\alpha \in \mathrm{Hom}_{\mathsf{K}\mathfrak{G}}(\mathbf{S}(\mathsf{V}), \mathsf{K})$ by 8.18, and obviously $\alpha \neq 0$.

b) ⇒ a): Suppose that $0 \neq \alpha \in \mathrm{Hom}_{\mathsf{K}\mathfrak{G}}(\mathbf{S}(\mathsf{V}), \mathsf{K})$. We put

$$q_{ii} = (v_i \otimes v_i)\alpha \quad \text{and} \quad q_{ij} = (v_i \otimes v_j + v_j \otimes v_i)\alpha \quad \text{for } i \neq j.$$

By 8.18, this defines a $\mathfrak{G}$-invariant quadratic form q on V. We have to show that q is non-singular.

By 8.17b),

$$\mathbf{S}(\mathsf{V})/\mathbf{A}(\mathsf{V}) \cong \mathsf{V}^{(2)} \ncong \mathsf{K}.$$

As V is irreducible, so is $\mathsf{V}^{(2)}$. Thus $\mathbf{A}(\mathsf{V}) \subsetneq \ker \alpha$ and hence

$$f(v_i, v_j) = q_{ij} = (v_i \otimes v_j + v_j \otimes v_i)\alpha \neq 0$$

for some $i \neq j$. The $\mathfrak{G}$-invariant symplectic form f on V is thus not the zero form. Therefore

$$\mathsf{W} = \{w \mid w \in \mathsf{V}, f(v, w) = 0 \quad \text{for all } v \in \mathsf{V}\}$$

is a proper $\mathsf{K}\mathfrak{G}$-submodule of V. As V is irreducible, we obtain $\mathsf{W} = 0$. Hence f is non-singular, and q is a non-singular $\mathfrak{G}$-invariant quadratic form on V.

q.e.d.

8.20 Theorem (WILLEMS [2]). *Let K be a perfect field of characteristic 2 and let V be an irreducible self-dual $\mathsf{K}\mathfrak{G}$-module which is not isomorphic to the trivial module.*

a) *If V does not carry a non-singular $\mathfrak{G}$-invariant quadratic form, then there exists an indecomposable $\mathsf{K}\mathfrak{G}$-module M and a submodule N such that M/N is the trivial module and $\mathsf{N} \cong \mathsf{V}^{(2)}$. (By 10.5 this means that $\mathsf{V}^{(2)}$ is a composition factor of $\mathsf{P}_1\mathbf{J}(\mathsf{K}\mathfrak{G})/\mathsf{P}_1\mathbf{J}(\mathsf{K}\mathfrak{G})^2$, where P_1 denotes the indecomposable projective $\mathsf{K}\mathfrak{G}$-module for which $\mathsf{P}_1/\mathsf{P}_1\mathbf{J}(\mathsf{K}\mathfrak{G}) \cong \mathsf{K}$.)*

b) *If $\mathfrak{G}$ is soluble, then V carries a non-singular $\mathfrak{G}$-invariant quadratic form.*

Proof. a) As $\mathsf{V} \cong \mathsf{V}^* \not\cong \mathsf{K}$, by 8.13 there exists a non-singular symplectic $\mathfrak{G}$-invariant form on V. Thus by 8.17c), there exists a $\mathsf{K}\mathfrak{G}$-epimorphism δ of $\mathbf{A}(\mathsf{V})$ onto K. We put $\mathsf{W} = \ker \delta$ and $\mathsf{U} = \mathbf{S}(\mathsf{V})/\mathsf{W}$.
Then

$$\mathsf{U}/(\mathbf{A}(\mathsf{V})/\mathsf{W}) \cong \mathbf{S}(\mathsf{V})/\mathbf{A}(\mathsf{V}) \cong \mathsf{V}^{(2)} \quad \text{(by 8.17b))}$$

and

$$\mathbf{A}(\mathsf{V})/\mathsf{W} \cong \mathsf{K}.$$

By hypothesis, V does not carry a $\mathfrak{G}$-invariant non-singular quadratic form, so the trivial module K is not an epimorphic image of $\mathbf{S}(\mathsf{V})$, by 8.19. Thus U is indecomposable. Then U^* is also indecomposable and has a composition series

$$0 \subset \mathsf{X} \subset \mathsf{U}^*$$

such that $\mathsf{U}^*/\mathsf{X} \cong \mathsf{K}$ and

$$\mathsf{X} \cong (\mathsf{V}^{(2)})^* \cong (\mathsf{V}^*)^{(2)} \cong \mathsf{V}^{(2)}.$$

b) Obviously, we can assume that V is a faithful $\mathsf{K}\mathfrak{G}$-module. We put

$$f = |\mathbf{O}_{2'}(\mathfrak{G})|^{-1} \sum_{g \in \mathbf{O}_{2'}(\mathfrak{G})} g.$$

As $\mathbf{O}_{2'}(\mathfrak{G}) \trianglelefteq \mathfrak{G}$, f is a central idempotent of $\mathsf{K}\mathfrak{G}$.

Suppose now that V does not carry a non-singular $\mathfrak{G}$-invariant quadratic form. Then by a) there exists an indecomposable $\mathsf{K}\mathfrak{G}$-module U^* such that

$$\mathsf{U}^*/\mathsf{X} \cong \mathsf{K} \quad \text{and} \quad \mathsf{X} \cong \mathsf{V}^{(2)}.$$

As multiplication by f induces the identity on U^*/X, certainly $\mathsf{U}^*f \neq 0$. As U^* is indecomposable, $\mathsf{U}^*f = \mathsf{U}^*$. But then for every $u \in \mathsf{U}^*$ and every $g \in \mathbf{O}_{2'}(\mathfrak{G})$ we obtain

$$ug = (uf)g = u(fg) = uf = u.$$

So $\mathbf{O}_{2'}(\mathfrak{G})$ operates trivially on U^*, hence also on $\mathsf{X} \cong \mathsf{V}^{(2)}$. As X is irreducible, the normal 2-subgroup $\mathbf{O}_{2',2}(\mathfrak{G})/\mathbf{O}_{2'}(\mathfrak{G})$ of $\mathfrak{G}/\mathbf{O}_{2'}(\mathfrak{G})$ operates trivially on X, by V, 5.17). But as $\mathfrak{G}$ is soluble, $\mathbf{O}_{2',2}(\mathfrak{G}) > 1$.

This is a contradiction, for $\mathfrak{G}$ operates faithfully on V and on X $\cong$ $V^{(2)}$. Hence V carries a non-singular $\mathfrak{G}$-invariant quadratic form. **q.e.d.**

8.21 Application. a) Suppose that $K = GF(2^f)$, $\mathfrak{G} = Sp(2m, 2^f)$ and that V is the natural K$\mathfrak{G}$-module of dimension $2m$ over K. If $(2m, 2^f) \neq (2, 2)$, then

$$|Sp(2m, 2^f)| > |O_{\pm}(2m, 2^f)|$$

for both types of the orthogonal groups $O_{\pm}$ (see II, 10.16d)). Thus there is no $\mathfrak{G}$-invariant quadratic form on V of the kind described in Theorem 8.20. Hence by 8.20a), there exists an indecomposable K$\mathfrak{G}$-module M such that M/N $\cong$ K and N is an algebraic conjugate of V. Using Exercise 34 we see that $\mathbf{H}^1(\mathfrak{G}, N) \neq 0$. Thus also $\mathbf{H}^1(\mathfrak{G}, V) \neq 0$. (In fact, $|\mathbf{H}^1(\mathfrak{G}, V)| = 2^f$ (POLLATSEK [1]).

b) A similar argument applies for the natural module $V = V(4, q)$ for the Suzuki group $Sz(q)$ (see XI, § 3). There does exist on V a non-singular $Sz(q)$-invariant symplectic form (XI, 3.8). If there were an invariant quadratic form, we would obtain a monomorphism of the simple group $Sz(q)$ into some orthogonal group $O(4, q)$ and indeed into $PSO'(4, q)$. But if the quadratic form is equivalent to $x_1x_2 + x_3x_4$, then

$$PSO'(4, q) \cong PSL(2, q) \times PSL(2, q);$$

and in the other case,

$$PSO'(4, q) \cong PSL(2, q^2)$$

(see VAN DER WAERDEN [1], p. 20 and p. 26; cf. exerc. 33). Hence $PSO'(4, q)$ has Abelian Sylow 2-subgroups, whereas $Sz(q)$ does not.

By the same argument as in a), we obtain $\mathbf{H}^1(Sz(q), V) \neq 0$.

8.22 Remark. Let K be a splitting field for $\mathfrak{G}$. It follows easily from V, 13.7 that if char K = 0, the number of irreducible self-dual K$\mathfrak{G}$-modules is equal to the number of real conjugacy classes in $\mathfrak{G}$, that is, of classes fixed under formation of inverses. It can be proved by using Brauer characters that if char K = p, the number of isomorphism types of irreducible self-dual K$\mathfrak{G}$-modules is equal to the number of real p'-classes.

8.23 Lemma. a) K$\mathfrak{G}$ $\cong$ (K$\mathfrak{G}$)*.

b) *If* P *is a finitely generated projective* K$\mathfrak{G}$*-module, then* P* *is also projective.*

Proof. a) We define a K-bilinear form f on $K\mathfrak{G} \times K\mathfrak{G}$ by the rule

$$f(g_1, g_2) = \begin{cases} 1 & \text{if} \quad g_1 = g_2 \\ 0 & \text{if} \quad g_1 \neq g_2. \end{cases}$$

Then f is non-singular and

$$f(ag, bg) = f(a, b)$$

for all $a, b \in K\mathfrak{G}$, $g \in \mathfrak{G}$. Hence by 8.10, $K\mathfrak{G} \cong (K\mathfrak{G})^*$.

b) There exists a finitely generated free $K\mathfrak{G}$-module F such that $F = P_1 \oplus P_2$, where $P_1 \cong P$. By 8.3d),

$$F^* = P_1^\perp \oplus P_2^\perp \cong P_2^* \oplus P_1^*.$$

As $F = K\mathfrak{G} \oplus \cdot_m \cdot \oplus K\mathfrak{G}$ for some m, we find by repeated application of 8.3d) and by a) that

$$F^* \cong (K\mathfrak{G})^* \oplus \cdot_m \cdot \oplus (K\mathfrak{G})^* \cong K\mathfrak{G} \oplus \cdot_m \cdot \oplus K\mathfrak{G} \cong F.$$

From $F \cong P_2^* \oplus P_1^*$ and $P_1 \cong P$, it follows that P^* is projective. (We could also easily construct a $(\mathfrak{G}, 1)$-Gaschütz operator for P^*.) **q.e.d.**

8.24 Applications. a) Let $\mathfrak{G} = SL(2, p)$ and, as in 3.10, let V_i be the $K\mathfrak{G}$-module of homogeneous polynomials in two variables of degree i over $K = GF(p)$. By 3.10, every absolutely irreducible $K\mathfrak{G}$-module is isomorphic to one of the V_i for $0 \leq i \leq p - 1$. By 8.3e) and 8.4a), V_2^* is absolutely irreducible of dimension 3, so $V_2^* \cong V_2$. It follows from 8.12 that if p is odd, there exists a non-singular $\mathfrak{G}$-invariant symmetric or symplectic form f on V_2. As $\dim_K V_2 = 3$ is odd, f has to be symmetric. Hence we obtain a homomorphism of $SL(2, p)$ into the orthogonal group $O(3, p)$ on V_2 belonging to the form f. The kernel of the representation of $SL(2, p)$ on V_2 is obviously $\langle -I \rangle$. So $PSL(2, p)$ is isomorphic to a subgroup of $O(3, p)$. Comparing orders, we easily derive the isomorphism $PSL(2, p) \cong O(3, p)'$ of II, 10.11.

The same method works if p is replaced by p^f (p odd). The proof that $V_2 \cong V_2^*$ requires additional arguments in this case, as $SL(2, p^f)$ has more than one absolutely irreducible module of dimension 3, namely the modules conjugate to V_2 under automorphisms of the field $GF(p^f)$.

b) Let $K = GF(q)$, where q is odd. Let V be a vector space of dimension 4 over K and let $\mathfrak{G}$ be the group of all linear transformations of V onto itself of determinant 1; thus $\mathfrak{G} = SL(4, q)$. Then the components

$\mathsf{V}^{(i)}$ $(i = 0, \ldots, 4)$ of the exterior algebra of V are $\mathsf{K}\mathfrak{G}$-modules; in particular, $\mathsf{V}^{(4)}$ is one-dimensional, and if $\mathsf{V}^{(4)} = \mathsf{K}w$, then $wg = (\det g)w = w$ for all $g \in \mathfrak{G}$. For $w_1, w_2 \in \mathsf{V}^{(2)}$ we have $w_1 \wedge w_2 = w_2 \wedge w_1$. If we put

$$w_1 \wedge w_2 = f(w_1, w_2)w,$$

then f is a symmetric bilinear form on $\mathsf{V}^{(2)}$. It is easy to see that f is non-singular. Then for all $g \in SL(4, q)$,

$$f(w_1g, w_2g)w = w_1g \wedge w_2g = (w_1 \wedge w_2)g = w_1 \wedge w_2 = f(w_1, w_2)w.$$

Hence f is invariant under $SL(4, q)$. (Compare also Exercise 31). As the kernel of the representation of $SL(4, q)$ on $\mathsf{V}^{(2)}$ is precisely $\langle -I \rangle$, we obtain a monomorphism of $SL(4, q)/\langle -I \rangle$ into an orthogonal group $O(6, q)$. Now

$$|SL(4, q)/\langle -I \rangle| = \tfrac{1}{2}q^6(q^4 - 1)(q^3 - 1)(q^2 - 1) \quad \text{(II, 6.2)}$$

and

$$|O(6, q)'| = \tfrac{1}{2}q^6(q^3 - \varepsilon)(q^4 - 1)(q^2 - 1) \quad \text{(II, 10.16)},$$

where $\varepsilon = 1$ or $\varepsilon = -1$. Hence $\varepsilon = 1$ and

$$SL(4, q)/\langle -I \rangle \cong O(6, q)',$$

where the orthogonal group $O(6, q)$ belongs to a space of index 3. (Of course it may be seen directly that $\mathsf{V}^{(2)}$ has index 3.) Thus

$$PO(6, q)' \cong PSL(4, q),$$

the isomorphism connected with Plücker's geometry of lines.

We conclude this section with a lemma of "Clifford type" involving the $\mathsf{K}\mathfrak{G}$-structure of $\mathrm{Hom}_{\mathsf{K}}(\mathsf{V}, \mathsf{W})$. It is related to V, 17.5.

8.25 Lemma. *Suppose that* $\mathfrak{N} \trianglelefteq \mathfrak{G}$. *Suppose that* V *and* W *are* $\mathsf{K}\mathfrak{G}$-*modules,* $\mathsf{W}_{\mathfrak{N}}$ *is absolutely irreducible and that*

$$\mathsf{V}_{\mathfrak{N}} = \mathsf{W}_1 \oplus \cdots \oplus \mathsf{W}_r,$$

where $W_i \cong W_{\mathfrak{N}}$ $(i = 1, \ldots, r)$. *Then* $\mathrm{Hom}_{K\mathfrak{N}}(W_{\mathfrak{N}}, V_{\mathfrak{N}})$ *is a* $K\mathfrak{G}$*-submodule of* $\mathrm{Hom}_K(W, V)$ *and*

$$V \cong W \otimes_K \mathrm{Hom}_{K\mathfrak{N}}(W_{\mathfrak{N}}, V_{\mathfrak{N}}).$$

Proof. $\mathrm{Hom}_{K\mathfrak{N}}(W_{\mathfrak{N}}, V_{\mathfrak{N}})$ is a $K\mathfrak{G}$-submodule of $\mathrm{Hom}_K(W, V)$, since by 8.8 it is the set of elements of $\mathrm{Hom}_K(W, V)$ left fixed by all elements of the normal subgroup $\mathfrak{N}$ of $\mathfrak{G}$.

We define a K-linear mapping δ of $W \otimes_K \mathrm{Hom}_{K\mathfrak{N}}(W_{\mathfrak{N}}, V_{\mathfrak{N}})$ into V by putting

$$(w \otimes \alpha)\delta = w\alpha \quad (w \in W, \alpha \in \mathrm{Hom}_{K\mathfrak{N}}(W_{\mathfrak{N}}, V_{\mathfrak{N}})).$$

It is easily checked that δ is well-defined. Let α_i be a $K\mathfrak{N}$-isomorphism of $W_{\mathfrak{N}}$ onto W_i. Then

$$(W_{\mathfrak{N}} \otimes \alpha_i)\delta = W_{\mathfrak{N}}\alpha_i = W_i.$$

Hence $\mathrm{im}\,\delta = V$.

From

$$\dim_K \mathrm{Hom}_{K\mathfrak{N}}(W_{\mathfrak{N}}, V_{\mathfrak{N}}) = \dim_K \bigoplus_{j=1}^{r} \mathrm{Hom}_{K\mathfrak{N}}(W_{\mathfrak{N}}, W_i) = \dim_K \bigoplus_{j=1}^{r} K = r,$$

we obtain

$$\dim_K V = r \dim_K W = \dim_K(W \otimes_K \mathrm{Hom}_{K\mathfrak{N}}(W_{\mathfrak{N}}, V_{\mathfrak{N}})).$$

Hence δ is an isomorphism. It is a $K\mathfrak{G}$-isomorphism, since if $w \in W$, $\alpha \in \mathrm{Hom}_{K\mathfrak{N}}(W_{\mathfrak{N}}, V_{\mathfrak{N}})$ and $g \in \mathfrak{G}$, then

$$\begin{aligned}((w \otimes \alpha)g)\delta &= (wg \otimes \alpha g)\delta = (wg)(\alpha g) = ((wgg^{-1})\alpha)g = (w\alpha)g \\ &= ((w \otimes \alpha)\delta)g.\end{aligned}$$

q.e.d.

Exercises

20) Suppose that V is a finitely generated $K\mathfrak{G}$-module. By 8.8, $\mathrm{Hom}_K(V, V)$ is a $K\mathfrak{G}$-module. Show that the trace mapping on $\mathrm{Hom}_K(V, V)$ is a $K\mathfrak{G}$-homomorphism of $\mathrm{Hom}_K(V, V)$ onto the trivial $K\mathfrak{G}$-module K. Compare this result with 8.5a).

21) Give a "natural" proof of 8.5b) using 8.8b).

22) Let P be a projective K$\mathfrak{G}$-module. Show that the dual module P* of P is projective by constructing a ($\mathfrak{G}$, 1)-Gaschütz operator on P*.

23) Let K be a perfect field of characteristic 2 and V an absolutely irreducible K$\mathfrak{G}$-module. Show that the K-space of $\mathfrak{G}$-invariant quadratic forms on V has dimension at most 1.
(Hint: Let q_1 and q_2 be $\mathfrak{G}$-invariant quadratic forms on V. We can assume that the corresponding symplectic forms, defined by

$$(v_1, v_2)_i = q_i(v_1 + v_2) - q_i(v_1) - q_i(v_2),$$

are the same. Show that there exists $v \in \mathsf{V}$, $v \neq 0$, such that $q_1(v) = q_2(v)$ and calculate $q_i(va)$ for $a \in \mathsf{K}\mathfrak{G}$.)

24) (WILLEMS) Let B_0 be the K-space of $\mathfrak{G}$-invariant bilinear forms on K$\mathfrak{G}$.

a) The forms f_h $(h \in \mathfrak{G})$ defined by

$$f_h(g_1, g_2) = \begin{cases} 1 & \text{if} \quad g_2 = hg_1 \\ 0 & \text{otherwise} \end{cases}$$

form a K-basis of B_0.

b) If we define a multiplication $*$ on B_0 by

$$(f_1 * f_2)(g_1, g_2) = \sum_{x \in \mathfrak{G}} f_1(g_1, x) f_2(x, g_2),$$

then B_0 becomes a K-algebra with unit element.

c) The K-linear mapping α defined by $g\alpha = f_g$ is an antiisomorphism of K$\mathfrak{G}$ onto B_0.

d) Suppose $a \in \mathsf{K}\mathfrak{G}$. Then $a\alpha$ is a non-singular bilinear form on K$\mathfrak{G}$ if and only if a is a unit of K$\mathfrak{G}$.

e) Suppose $a = \sum_{g \in \mathfrak{G}} a_g g \in \mathsf{K}\mathfrak{G}$ and char $\mathsf{K} \neq 2$. Then $a\alpha$ is symmetric (symplectic) if and only if $a_{g^{-1}} = a_g$ $(a_{g^{-1}} = -a_g)$ for all $g \in \mathfrak{G}$. What is the corresponding result if char $\mathsf{K} = 2$?

f) Now suppose that char $\mathsf{K} = p$ and that p divides $|\mathfrak{G}|$. Show that there exist "many" linearly independent non-singular $\mathfrak{G}$-invariant symmetric bilinear forms on K$\mathfrak{G}$. If $p \neq 2$, there does not exist a non-singular symplectic $\mathfrak{G}$-invariant form on K$\mathfrak{G}$.

25) Let V be a finitely generated $\mathsf{K}\mathfrak{G}$-module and ϱ a representation of $\mathfrak{G}$ on V. If $\mathsf{V} \cong \mathsf{V}^*$ and ε is an eigen-value of $\varrho(g)$ $(g \in \mathfrak{G})$ in some extension field of K, then ε^{-1} is also an eigen-value of $\varrho(g)$.

26) Consider the n-dimensional vector space $\mathsf{V} = \mathsf{V}(n, q)$ over $\mathsf{K} = GF(q)$ as a module for $\mathfrak{G} = GL(n, q)$ in the natural way. Suppose that $n > 1$.

a) $\mathsf{V} \not\cong \mathsf{V}^*$ as $\mathsf{K}\mathfrak{G}$-modules if $q > 2$ or $n > 2$.

b) Consider now V as a module for $\mathfrak{H} = SL(n, q)$. Then for any q, $\mathsf{V} \cong \mathsf{V}^*$ only if $n = 2$.

27) Take $\mathsf{K} = GF(q)$ and $\mathsf{L} = GF(q^n)$. Let the multiplicative group $\mathsf{L}^\times$ of L operate on the K-vector space L by $x \to ax$. Show that L is a self-dual $\mathsf{L}^\times$-module if and only if $q^n = 2, 3$.

28) Suppose that we have the same situation as in Exercise 27. Let $\mathfrak{U}$ be a subgroup of $\mathsf{L}^\times$ and suppose that L is an irreducible $\mathsf{K}\mathfrak{U}$-module. If L is a self-dual $\mathsf{K}\mathfrak{U}$-module, then either $n = 1$ and $|\mathfrak{U}| \leq 2$ or $n = 2m$ is even and $|\mathfrak{U}|$ divides $q^m + 1$.

29) Let $\mathsf{V} = \mathsf{V}(2m, q)$ be a non-singular symplectic space.

a) If $\mathfrak{Z}$ is a cyclic subgroup of the symplectic group $Sp(2m, q)$ operating irreducibly on V, then $|\mathfrak{Z}|$ divides $q^m + 1$.

b) $Sp(2m, q)$ has a cyclic subgroup of order $q^m + 1$ which operates irreducibly on V. (See II, 9.23.)

30) Let $\mathsf{V} = \mathsf{V}(n, q)$ be a vector space over $GF(q)$ with a non-singular symmetric bilinear form. Suppose that q is odd and $n > 1$.

a) If the orthogonal group $O(n, q)$ of isometries of V has a cyclic subgroup $\mathfrak{Z}$ which operates irreducibly on V, then $n = 2m$ is even, $|\mathfrak{Z}|$ divides $q^m + 1$ and V has index $m - 1$. (For the last statement observe that in the case of index m, V has exactly $(q^m - 1)(q^{m-1} - 1)$ vectors $v \neq 0$ for which $(v, v) = 0$, and $\mathfrak{Z}$ operates fixed point freely on these vectors.)

b) If $n = 2m$ and V has index $m - 1$, there is a cyclic subgroup $\mathfrak{Z}$ of $O(2m, q)$ of order $q^m + 1$ operating irreducibly on V. (See HUPPERT [3]).

31) Let V be a $\mathsf{K}\mathfrak{G}$-module with $\dim_{\mathsf{K}} \mathsf{V} = n$. Let ϱ be the representation of $\mathfrak{G}$ on V and assume that $\det \varrho(g) = 1$ for all $g \in \mathfrak{G}$. Consider the homogeneous components $\mathsf{V}^{(i)}$ of the exterior algebra of V as $\mathsf{K}\mathfrak{G}$-

modules and prove that $V^{(i)*} \cong V^{(n-i)}$. (Construct a non-singular $\mathfrak{G}$-invariant bilinear form on $V^{(i)} \times V^{(n-i)}$.)

32) Suppose that V is a $K\mathfrak{G}$-module and that L is an extension field of K. We denote by $B_0(V)$ the space of $\mathfrak{G}$-invariant bilinear forms on V. Interpret and prove the formula $B_0(V_L) \cong B_0(V) \otimes_K L$.

33) Prove the isomorphism $PSL(2, q^2) \cong PSO(4, q)'$, where $O(4, q)$ is the orthogonal group of a space of dimension 4 and index 1 over $GF(q)$ (q odd), by the following argument.

Write $K = GF(q)$, $L = GF(q^2)$ and let α be the only non-trivial automorphism of L over K. Let $V = V(2, q^2)$ be the vector space of dimension 2 over L, on which $\mathfrak{G} = SL(2, q^2)$ operates naturally. We denote by V_α the $L\mathfrak{G}$-module conjugate to V under α and put $W = V \otimes_K V_\alpha$.

a) As $SL(2, q^2) = Sp(2, q^2)$, there is a $\mathfrak{G}$-invariant non-singular symplectic form on V. Construct from this a non-singular $\mathfrak{G}$-invariant symmetric bilinear form on W.

b) Prove $\operatorname{Hom}_{L\mathfrak{G}}(W, W) = L$ and show that W is an absolutely irreducible $L\mathfrak{G}$-module.
(Observe first that W restricted to a Sylow p-subgroup of $SL(2, q^2)$ has only one irreducible submodule if char $K = p$.)

c) Show that there exists a $K\mathfrak{G}$-module W_0 and a non-singular symmetric $\mathfrak{G}$-invariant bilinear form f_0 on W_0 such that $W \cong W_0 \otimes_K L$.

d) Show that this gives a monomorphism of $PSL(2, q^2)$ into $PSO(4, q)'$. Show by comparing orders that this is an isomorphism and that the form f_0 on W_0 has index 1.

34) Let W be a $K\mathfrak{G}$-module and V a submodule of W such that $W/V \cong K$ is the trivial $K\mathfrak{G}$-module. Suppose $W = V \oplus Kw$ as a vector space over K.

a) If the mapping f from $\mathfrak{G}$ to V is defined by

$$wg = w + f(g),$$

then f is a 1-cocycle in $\mathbf{Z}^1(\mathfrak{G}, V)$.

b) $f \in \mathbf{B}^1(\mathfrak{G}, V)$ if and only if V is a direct summand of W.

c) Suppose in addition that V is irreducible. Then $\mathbf{H}^1(\mathfrak{G}, V) \neq 0$ if and only if V is isomorphic to a composition factor of $P_1\mathbf{J}(K\mathfrak{G})/P_1\mathbf{J}(K\mathfrak{G})^2$ where P_1 is the projective $K\mathfrak{G}$-module characterized by $P_1/P_1\mathbf{J}(K\mathfrak{G}) \cong K$.

§ 9. Representations of Normal Subgroups

Clifford's theorem (V, 17.3) states that if $\mathfrak{N} \trianglelefteq \mathfrak{G}$, K is any field and V is an irreducible K$\mathfrak{G}$-module, the restriction $V_{\mathfrak{N}}$ is a completely reducible K$\mathfrak{N}$-module and its composition factors form a class of $\mathfrak{G}$-conjugate K$\mathfrak{N}$-modules. These results are very general; in particular, there is no restriction on the characteristic of K. In this section, we shall consider the extent to which the same is true for indecomposable K$\mathfrak{G}$-modules, and we shall attempt to extend the results of V, § 17 to the modular case. We begin with a counterexample which shows that if V is an indecomposable K$\mathfrak{G}$-module, the indecomposable components of $V_{\mathfrak{N}}$ may have different dimensions.

9.1 Example. Let $\mathfrak{G} = \langle g_1 \rangle \times \langle g_2 \rangle$ be an elementary Abelian group of order p^2 and let K be a field of characteristic p. Let V be a 3-dimensional vector space over K and let $\{v_1, v_2, v_3\}$ be a K-basis of V. Then V becomes a K$\mathfrak{G}$-module if we put

$$
\begin{aligned}
v_1 g_1 &= v_1 & v_1 g_2 &= v_1 \\
v_2 g_1 &= v_2 & v_2 g_2 &= v_2 \\
v_3 g_1 &= v_1 + v_3 & v_3 g_2 &= v_2 + v_3.
\end{aligned}
$$

An easy calculation shows that

$$
\mathrm{Hom}_{K\mathfrak{G}}(V, V) \cong \left\{ \begin{pmatrix} a & 0 & 0 \\ 0 & a & 0 \\ b & c & a \end{pmatrix} \middle| \, a, b, c \in K \right\}.
$$

Thus $\mathrm{Hom}_{K\mathfrak{G}}(V, V)/\mathbf{J}(\mathrm{Hom}_{K\mathfrak{G}}(V, V))$ is isomorphic to K, hence $\mathrm{Hom}_{K\mathfrak{G}}(V, V)$ is a local ring and V is absolutely indecomposable. If we put $\mathfrak{N} = \langle g_2 \rangle$, then

$$
V_{\mathfrak{N}} = \langle v_1 \rangle \oplus \langle v_2, v_3 \rangle
$$

is a decomposition of $V_{\mathfrak{N}}$ into indecomposable K$\mathfrak{N}$-modules of different dimensions.

This example shows that there is no analogue of Clifford's theorem (V, 17.3) for indecomposable modules. But an analogue will be proved under special conditions. Before doing so, we introduce the following notation, which will be used frequently in this section.

9.2 Definition. Suppose that $\mathfrak{N} \trianglelefteq \mathfrak{G}$. If V_1 and V_2 are $\mathsf{K}\mathfrak{N}$-modules, V_1 and V_2 are said to be $\mathfrak{G}$-*conjugate* if there exists a K-isomorphism θ of V_1 onto V_2 and an element g of $\mathfrak{G}$ such that

$$(v\theta)x = (vx^g)\theta$$

for all $v \in \mathsf{V}_1$, $x \in \mathfrak{N}$. Thus if W is a $\mathsf{K}\mathfrak{G}$-module and V a $\mathsf{K}\mathfrak{N}$-submodule of $\mathsf{W}_{\mathfrak{N}}$, $\mathsf{V}g$ is a $\mathsf{K}\mathfrak{N}$-submodule of $\mathsf{W}_{\mathfrak{N}}$ $\mathfrak{G}$-conjugate to V. Also, if V is a $\mathsf{K}\mathfrak{N}$-module and $g \in \mathfrak{G}$, then $\mathsf{V} \otimes g$ is a $\mathsf{K}\mathfrak{N}$-submodule of $(\mathsf{V}^{\mathfrak{G}})_{\mathfrak{N}}$ $\mathfrak{G}$-conjugate to V, for $\mathsf{V} \otimes 1$ is a $\mathsf{K}\mathfrak{N}$-submodule of $(\mathsf{V}^{\mathfrak{G}})_{\mathfrak{N}}$ $\mathsf{K}\mathfrak{N}$-isomorphic to V. The *inertia subgroup* of V in $\mathfrak{G}$ is defined by

$$\{g \mid g \in \mathfrak{G}, \mathsf{V} \otimes g \cong \mathsf{V}\} \quad (\text{V}, 17.6).$$

9.3 Theorem (NAKAYAMA [3]). *Suppose that $\mathfrak{N} \trianglelefteq \mathfrak{G}$, that K is a field and V is an indecomposable $\mathsf{K}\mathfrak{G}$-module. If V is a $(\mathfrak{G}, \mathfrak{N})$-projective $\mathsf{K}\mathfrak{G}$-module, then $\mathsf{V}_{\mathfrak{N}}$ is a direct sum of $\mathfrak{G}$-conjugate indecomposable $\mathsf{K}\mathfrak{N}$-modules, each isomorphism type occurring with the same multiplicity in the sense of the Krull-Schmidt theorem. (By 7.7b), this theorem can be applied if* char K *does not divide* $|\mathfrak{G}/\mathfrak{N}|$.)

Proof. By 7.5, V is isomorphic to a direct summand of $(\mathsf{V}_{\mathfrak{N}})^{\mathfrak{G}}$. By the Krull-Schmidt theorem, there is an indecomposable direct summand W of $\mathsf{V}_{\mathfrak{N}}$ such that V is isomorphic to a direct summand of $\mathsf{W}^{\mathfrak{G}}$. If T is a transversal of $\mathfrak{N}$ in $\mathfrak{G}$, then

$$\mathsf{W}^{\mathfrak{G}} = \bigoplus_{t \in \mathsf{T}} \mathsf{W} \otimes t.$$

Since $\mathsf{W} \otimes t$ is a $\mathsf{K}\mathfrak{N}$-module conjugate under $\mathfrak{G}$ to W, $\mathsf{W} \otimes t$ is indecomposable. A further application of the Krull-Schmidt theorem shows that

$$\mathsf{V}_{\mathfrak{N}} = \mathsf{V}_1 \oplus \cdots \oplus \mathsf{V}_k, \tag{1}$$

where each V_i is $\mathsf{K}\mathfrak{N}$-isomorphic to $\mathsf{W} \otimes t$ for some $t \in \mathsf{T}$. Thus V_i is $\mathsf{K}\mathfrak{N}$-isomorphic to $\mathsf{V}_1 g$ for some $g \in \mathfrak{G}$. But

$$\mathsf{V}_{\mathfrak{N}} = (\mathsf{V}_{\mathfrak{N}})g = \mathsf{V}_1 g \oplus \cdots \oplus \mathsf{V}_k g \tag{2}$$

is also a decomposition of $\mathsf{V}_{\mathfrak{N}}$ into indecomposable $\mathsf{K}\mathfrak{N}$-submodules, so by the Krull-Schmidt theorem, the multiplicity of V_i in (1) is equal to that of $\mathsf{V}_1 g$ in (2) and hence to that of V_1 in (1). **q.e.d.**

The following statement, which is dual to Clifford's theorem, is sometimes useful.

9.4 Theorem. *Suppose that* $\mathfrak{N} \trianglelefteq \mathfrak{G}$, *that* K *is a field of characteristic* p *and* p *does not divide* $|\mathfrak{G}/\mathfrak{N}|$. *If* V *is a completely reducible* $\mathsf{K}\mathfrak{N}$*-module, then* $\mathsf{V}^{\mathfrak{G}}$ *is a completely reducible* $\mathsf{K}\mathfrak{G}$*-module.*

Proof. If T is a transversal of $\mathfrak{N}$ in $\mathfrak{G}$, we have

$$(\mathsf{V}^{\mathfrak{G}})_{\mathfrak{N}} = \bigoplus_{t \in \mathsf{T}} \mathsf{V} \otimes t,$$

where each $\mathsf{V} \otimes t$ is a completely reducible $\mathsf{K}\mathfrak{N}$-module. Hence $(\mathsf{V}^{\mathfrak{G}})_{\mathfrak{N}}$ is completely reducible. As char K does not divide $|\mathfrak{G}/\mathfrak{N}|$, $\mathsf{V}^{\mathfrak{G}}$ is completely reducible by 7.20b). **q.e.d.**

9.5 Remark. The conditions in 9.4 are necessary. The fact that $\mathfrak{N} \trianglelefteq \mathfrak{G}$ alone is not sufficient to imply the conclusion of 9.4 is shown by the example where $\mathfrak{G}$ is a non-identity p-group, $\mathfrak{N} = 1$, char $\mathsf{K} = p$ and V is the trivial $\mathsf{K}\mathfrak{N}$-module. Then $\mathsf{V}^{\mathfrak{G}} \cong \mathsf{K}\mathfrak{G}$ is indecomposable by 5.2, but $\mathsf{V}^{\mathfrak{G}}$ is certainly not completely reducible.

Also in the case when $\mathfrak{N} \ntrianglelefteq \mathfrak{G}$ and char K does not divide $|\mathfrak{G} : \mathfrak{N}|$, the conclusion of 9.4 need not hold. To see this, let $\mathfrak{G}$ be the group $SL(2, 3)$ of order 24, let $\mathfrak{N}$ be a Sylow 3-subgroup of $\mathfrak{G}$ and let K be an algebraically closed field of characteristic 3. Let V be the trivial $\mathsf{K}\mathfrak{N}$-module of K-dimension 1. We consider the $\mathsf{K}\mathfrak{G}$-module $\mathsf{V}^{\mathfrak{G}}$.

To within isomorphism, $\mathfrak{G}$ has exactly three irreducible modules V_i ($i = 1, 2, 3$), and $\dim_{\mathsf{K}} \mathsf{V}_i = i$, by 3.10. By Nakayama's reciprocity theorem (4.13), the multiplicity n_i of V_i in the head $\overline{\mathsf{V}} = \mathsf{V}^{\mathfrak{G}}/\mathsf{V}^{\mathfrak{G}}\mathbf{J}(\mathsf{K}\mathfrak{G})$ of $\mathsf{V}^{\mathfrak{G}}$ is equal to the multiplicity of V in the socle of $(\mathsf{V}_i)_{\mathfrak{N}}$. By 5.2, V is the only irreducible $\mathsf{K}\mathfrak{N}$-module, so $n_1 = 1$, $n_2 \geq 1$ and $n_3 \geq 1$. Now V_2 is a module for the natural representation of $SL(2, 3)$ of degree 2, so $(\mathsf{V}_2)_{\mathfrak{N}}$ is not a trivial $\mathsf{K}\mathfrak{N}$-module. Thus $n_2 = 1$. From

$$3 + 3n_3 = n_1 + 2n_2 + 3n_3 = \dim_{\mathsf{K}} \overline{\mathsf{V}} \leq \dim_{\mathsf{K}} \mathsf{V}^{\mathfrak{G}} = |\mathfrak{G} : \mathfrak{N}| = 8,$$

if follows that $n_3 = 1$. Thus $\dim_{\mathsf{K}} \overline{\mathsf{V}} = 6$, so $\mathsf{V}^{\mathfrak{G}}\mathbf{J}(\mathsf{K}\mathfrak{G}) \neq 0$. Hence $\mathsf{V}^{\mathfrak{G}}$ is not completely reducible.

Next we prove a generalization of V, 17.11.

9.6 Theorem (H. N. WARD [2], WILLEMS [1]). *Suppose that* $\mathfrak{N} \trianglelefteq \mathfrak{G}$, K *is an arbitrary field and* V *is an indecomposable* $\mathsf{K}\mathfrak{N}$*-module. Let* $\mathfrak{I}$ *be the inertia group of* V *in* $\mathfrak{G}$ *and suppose that*

$$\mathsf{V}^{\mathfrak{I}} = \mathsf{V}_1 \oplus \cdots \oplus \mathsf{V}_r,$$

where the V_i *are indecomposable* $\mathsf{K}\mathfrak{I}$*-modules.*

a) $\mathsf{V}_i^{\mathfrak{G}}$ *is an indecomposable* $\mathsf{K}\mathfrak{G}$*-module* ($i = 1, \ldots, r$).

b) *If* V_i *is irreducible for some* i, *then* $\mathsf{V}_i^{\mathfrak{G}}$ *and* V *are irreducible.*

c) $\mathsf{V}_i^{\mathfrak{G}} \cong \mathsf{V}_j^{\mathfrak{G}}$ *if and only if* $\mathsf{V}_i \cong \mathsf{V}_j$.

Proof. a) Let T be a transversal of $\mathfrak{I}$ in $\mathfrak{G}$ containing 1.

Since $\mathfrak{N} \trianglelefteq \mathfrak{I}$, $(\mathsf{V}^{\mathfrak{I}})_{\mathfrak{N}}$ is the direct sum of $\mathsf{K}\mathfrak{N}$-modules of the form $\mathsf{V} \otimes s$ ($s \in \mathfrak{I}$). Since $\mathfrak{I}$ is the inertia group of V, $\mathsf{V} \otimes s \cong \mathsf{V}$. Hence by the Krull-Schmidt theorem,

$$(\mathsf{V}_i)_{\mathfrak{N}} \cong \mathsf{V} \oplus \underset{n_i}{\cdots} \oplus \mathsf{V} \tag{3}$$

for some n_i. Thus for any $g \in \mathfrak{G}$,

$$\mathsf{V}_i \otimes g \cong \mathsf{V} \otimes g \oplus \underset{n_i}{\cdots} \oplus \mathsf{V} \otimes g.$$

This implies that

$$(\mathsf{V}_i^{\mathfrak{G}})_{\mathfrak{N}} \cong \left(\bigoplus_{t \in \mathsf{T}} \mathsf{V}_i \otimes t\right)_{\mathfrak{N}} \cong \bigoplus_{t \in \mathsf{T}} (\mathsf{V} \otimes t \oplus \underset{n_i}{\cdots} \oplus \mathsf{V} \otimes t)_{\mathfrak{N}}. \tag{4}$$

Suppose now that $\mathsf{V}_i^{\mathfrak{G}} = \mathsf{W} \oplus \mathsf{X}$ is a direct decomposition of the $\mathsf{K}\mathfrak{G}$-module $\mathsf{V}_i^{\mathfrak{G}}$. By 4.3a), V_i is isomorphic to a direct summand of $(\mathsf{V}_i^{\mathfrak{G}})_{\mathfrak{I}}$. By the Krull-Schmidt theorem, we may suppose that V_i is isomorphic to a direct summand of $\mathsf{W}_{\mathfrak{I}}$. Suppose that

$$\mathsf{W}_{\mathfrak{I}} = \mathsf{V}_i' \oplus \mathsf{W}' \quad \text{and} \quad \mathsf{V}_i \cong \mathsf{V}_i'.$$

Restricting to $\mathfrak{N}$, we see by (3) that

$$\mathsf{W}_{\mathfrak{N}} = \mathsf{W}_1 \oplus \cdots \oplus \mathsf{W}_{n_i} \oplus \mathsf{W}'_{\mathfrak{N}},$$

where $\mathsf{W}_1 \cong \cdots \cong \mathsf{W}_{n_i} \cong \mathsf{V}$. As W is a $\mathsf{K}\mathfrak{G}$-module,

$$\mathsf{W}_{\mathfrak{N}} = \mathsf{W}_{\mathfrak{N}}t = \mathsf{W}_1 t \oplus \cdots \oplus \mathsf{W}_{n_i} t \oplus \mathsf{W}'_{\mathfrak{N}}t,$$

and $\mathsf{W}_j t$, $\mathsf{V} \otimes t$ are isomorphic $\mathsf{K}\mathfrak{N}$-modules. Hence $\mathsf{V} \otimes t$ appears as a direct summand of $\mathsf{W}_{\mathfrak{N}}$ with multiplicity at least n_i. If t, t' are distinct elements of T, then $\mathsf{V} \otimes t$ and $\mathsf{V} \otimes t'$ are non-isomorphic $\mathsf{K}\mathfrak{N}$-modules. Hence we see from (4) that $(\mathsf{V}_i^{\mathfrak{G}})_{\mathfrak{N}}$ is isomorphic to a direct summand of $\mathsf{W}_{\mathfrak{N}}$. Since $\dim_{\mathsf{K}} \mathsf{V}_i^{\mathfrak{G}} \geq \dim_{\mathsf{K}} \mathsf{W}$, we obtain $\mathsf{X} = 0$. Thus $\mathsf{V}_i^{\mathfrak{G}}$ is indecomposable.

b) Suppose now that V_i is irreducible. Let W be any irreducible factor module of $\mathsf{V}_i^{\mathfrak{G}}$. Then by Nakayama's reciprocity theorem (4.13), V_i is isomorphic to a $\mathsf{K}\mathfrak{J}$-submodule of $\mathsf{W}_{\mathfrak{J}}$. Since $(\mathsf{V}^{\mathfrak{J}})_{\mathfrak{N}}$ is the direct sum of $\mathsf{K}\mathfrak{N}$-modules $\mathsf{V} \otimes s$ ($s \in \mathfrak{J}$) and $\mathfrak{J}$ is the inertia group of V, $(\mathsf{V}_i)_{\mathfrak{N}}$ is the direct sum of n_i submodules isomorphic to V for some n_i. Hence $\mathsf{W}_{\mathfrak{N}}$ contains a submodule X which is the direct sum of n_i submodules isomorphic to V. As $\mathsf{W}_{\mathfrak{N}}$ is completely reducible, the indecomposable $\mathsf{K}\mathfrak{N}$-module V is irreducible. Thus if $g \in \mathfrak{G}$, $\mathsf{W}_{\mathfrak{N}}$ contains the $\mathsf{K}\mathfrak{N}$-submodule $\mathsf{X}g$, which is the direct sum of n_i submodules isomorphic to $\mathsf{V} \otimes g$.

Again, let T be a transversal of $\mathfrak{J}$ in $\mathfrak{G}$. Then the $\mathsf{K}\mathfrak{N}$-modules $\mathsf{V} \otimes t$ ($t \in \mathsf{T}$) are irreducible and non-isomorphic, and they all appear as direct summands of $\mathsf{W}_{\mathfrak{N}}$ with multiplicity at least n_i. Thus

$$\dim_{\mathsf{K}} \mathsf{W} \geq |\mathfrak{G} : \mathfrak{J}| n_i \dim_{\mathsf{K}} \mathsf{V} = |\mathfrak{G} : \mathfrak{J}| \dim_{\mathsf{K}} \mathsf{V}_i = \dim_{\mathsf{K}} \mathsf{V}_i^{\mathfrak{G}}.$$

Hence $\mathsf{V}_i^{\mathfrak{G}} = \mathsf{W}$ is irreducible.

c) Suppose that $\mathsf{V}_i^{\mathfrak{G}} \cong \mathsf{V}_j^{\mathfrak{G}}$, but $\mathsf{V}_i \not\cong \mathsf{V}_j$. By 4.3a),

$$(\mathsf{V}_i^{\mathfrak{G}})_{\mathfrak{J}} = \mathsf{V}'_i \oplus \mathsf{V}''_i \quad \text{and} \quad (\mathsf{V}_j^{\mathfrak{G}})_{\mathfrak{J}} = \mathsf{V}'_j \oplus \mathsf{V}''_j,$$

where V'_i, V'_j, V''_i, V''_j are $\mathsf{K}\mathfrak{J}$-modules, $\mathsf{V}'_i \cong \mathsf{V}_i$ and $\mathsf{V}'_j \cong \mathsf{V}_j$. By the Krull-Schmidt theorem, V_i is isomorphic to a $\mathsf{K}\mathfrak{J}$-direct summand of V''_j. Thus $(\mathsf{V}_i)_{\mathfrak{N}}$ is isomorphic to a $\mathsf{K}\mathfrak{N}$-direct summand of $(\mathsf{V}''_j)_{\mathfrak{N}}$. Using (3) and (4), it follows that $(\mathsf{V}_i)_{\mathfrak{N}}$ is isomorphic to a $\mathsf{K}\mathfrak{N}$-direct summand of

$$\bigoplus_{1 \neq t \in \mathsf{T}} (\mathsf{V} \otimes t \oplus \underset{n_j}{\cdots} \oplus \mathsf{V} \otimes t).$$

Hence by (3), V is $\mathsf{K}\mathfrak{N}$-isomorphic to $\mathsf{V} \otimes t$ for some $t \in \mathsf{T}$, $t \neq 1$. This contradicts the definition of the inertia group $\mathfrak{J}$, since $t \notin \mathfrak{J}$. **q.e.d.**

9.7 Remark. The following is a generalization of V, 17.12.

Suppose that $\mathfrak{N} \trianglelefteq \mathfrak{G}$ and that $(|\mathfrak{N}|, |\mathfrak{G}/\mathfrak{N}|) = 1$. Let K be an algebraically closed field of any characteristic. Let W be an irreducible $K\mathfrak{N}$-module the inertia group of which is $\mathfrak{G}$. Then there exists an irreducible $K\mathfrak{G}$-module V such that $V_{\mathfrak{N}} \cong W$.

A short proof of this result, using Brauer characters, was recently given by GOW [1].[1] We shall prove a special case which will be used later. For this we need the following lemma.

9.8 Lemma. *Suppose that* $\mathfrak{N} \trianglelefteq \mathfrak{G}$ *and* $\mathfrak{G} = \mathfrak{N}\langle g\rangle$. *Let* ϱ *be a homomorphism of* $\mathfrak{N}$ *into a group* $\mathfrak{H}$. *Suppose that there exists an element* z *of* $\mathfrak{H}$ *such that*

$$x^g\varrho = z^{-1}(x\varrho)z \quad \textit{for all } x \in \mathfrak{N}$$

and $g^n\varrho = z^n$, *where* $n = |\mathfrak{G}/\mathfrak{N}|$. *Then there exists a homomorphism* σ *of* $\mathfrak{G}$ *into* $\mathfrak{H}$ *such that* ϱ *is the restriction of* σ *to* $\mathfrak{N}$ *and* $g\sigma = z$.

Proof. We define σ by putting $(g^i x)\sigma = z^i(x\varrho)$ for all $x \in \mathfrak{N}$ and $0 \leq i < n$. Then $(g^j x)\sigma = z^j(x\varrho)$ for any integer j, since if $j = nq + i$ and $0 \leq i < n$, then

$$\begin{aligned}(g^j x)\sigma = (g^i g^{nq} x)\sigma = z^i((g^{nq}x)\varrho) &= z^i(g^n\varrho)^q(x\varrho)\\ &= z^{i+nq}(x\varrho) = z^j(x\varrho).\end{aligned}$$

Thus if x, y are elements of $\mathfrak{N}$ and i, j are positive integers,

$$\begin{aligned}(g^i x g^j y)\sigma &= (g^{i+j}x^{g^j}y)\sigma = z^{i+j}(x^{g^j}y)\varrho = z^{i+j}(x^{g^j}\varrho)(y\varrho)\\ &= z^{i+j}z^{-j}(x\varrho)z^j(y\varrho) = z^i(x\varrho)z^j(y\varrho)\\ &= (g^i x)\sigma(g^j y)\sigma.\end{aligned}$$

Hence σ is a homomorphism of $\mathfrak{G}$ into $\mathfrak{H}$ with the required properties.

q.e.d.

9.9 Theorem. *Let* K *be an algebraically closed field of characteristic* p *and let* $\mathfrak{N}$ *be a normal subgroup of* $\mathfrak{G}$ *for which* $\mathfrak{G}/\mathfrak{N}$ *is cyclic. Let* V *be a* $K\mathfrak{N}$-*module the inertia group in* $\mathfrak{G}$ *of which is* $\mathfrak{G}$ *itself.*

a) (SRINIVASAN [1]) *If* V *is an irreducible* $K\mathfrak{N}$-*module, there exists a (necessarily irreducible)* $K\mathfrak{G}$-*module* W *such that* $W_{\mathfrak{N}} \cong V$.

b) (WILLEMS [1]) *If* $\mathfrak{G}/\mathfrak{N}$ *is a* p'-*group and* V *an indecomposable* $K\mathfrak{N}$-*module, there exists a* $K\mathfrak{G}$-*module* W *such that* $W_{\mathfrak{N}} \cong V$.

---

1 See also ISAACS [1].

Proof. Let $n = |\mathfrak{G}/\mathfrak{N}|$. By hypothesis, $\mathfrak{G} = \mathfrak{N}\langle h\rangle$ for some element h. We denote the representation of $\mathfrak{N}$ on V by ϱ. Since $\mathfrak{G}$ is the inertia group of ϱ, there exists a non-singular K-linear mapping y of V onto itself such that for all $x \in \mathfrak{N}$

$$\varrho(x^h) = y^{-1}\varrho(x)y. \tag{5}$$

Thus

$$y^{-n}\varrho(x)y^n = \varrho(x^{h^n}) = \varrho(h^n)^{-1}\varrho(x)\varrho(h^n).$$

This shows that

$$\varrho(h^n)y^{-n} \in \mathrm{Hom}_{\mathsf{K}\mathfrak{N}}(\mathsf{V}, \mathsf{V}). \tag{6}$$

a) Now V is irreducible. Since K is algebraically closed, it follows from (6) and Schur's lemma that

$$\varrho(h^n)y^{-n} = a1$$

for some $a \in \mathsf{K}^\times$. Choose $b \in \mathsf{K}^\times$ such that $b^n = a$ and put $z = by$. Then

$$\varrho(x^h) = z^{-1}\varrho(x)z$$

for all $x \in \mathfrak{N}$ and $\varrho(h^n) = z^n$. By 9.8, the assertion follows.

b) Since K is algebraically closed, it follows that

$$\mathrm{Hom}_{\mathsf{K}\mathfrak{N}}(\mathsf{V}, \mathsf{V}) = \mathsf{K}1 \oplus \mathbf{J}(\mathrm{Hom}_{\mathsf{K}\mathfrak{N}}(\mathsf{V}, \mathsf{V})).$$

Hence it follows from (6) that

$$\varrho(h^n)y^{-n} = a1 + \alpha, \quad \text{where} \quad a \in \mathsf{K}^\times \quad \text{and} \quad \alpha \in \mathbf{J}(\mathrm{Hom}_{\mathsf{K}\mathfrak{N}}(\mathsf{V}, \mathsf{V})). \tag{7}$$

Again we choose $b \in \mathsf{K}^\times$ such that $b^n = a$ and put $z = by$. Thus

$$\varrho(x^h) = z^{-1}\varrho(x)z$$

for all $x \in \mathfrak{N}$ and

$$\varrho(h^n)z^{-n} = 1 + \beta \tag{8}$$

where β is a nilpotent element of $\mathrm{Hom}_{\mathsf{K}\mathfrak{N}}(\mathsf{V}, \mathsf{V})$. It follows from (8) that

$$\begin{aligned} z^{-n}\varrho(h^n) &= \varrho(h^n)^{-1}\varrho(h^n)z^{-n}\varrho(h^n) \\ &= \varrho(h^n)^{-1}(1 + \beta)\varrho(h^n) = 1 + \beta, \end{aligned}$$

since $h^n \in \mathfrak{N}$ and $1 + \beta \in \mathrm{Hom}_{\mathsf{K}\mathfrak{N}}(\mathsf{V}, \mathsf{V})$. Thus

$$\varrho(h^n)z^{-n} = z^{-n}\varrho(h^n). \tag{9}$$

Since β is nilpotent, there exists an integer k such that $\beta^{p^k} = 0$. Hence it follows from (8) and (9) that

$$\varrho(h^{np^k})z^{-np^k} = (1 + \beta)^{p^k} = 1 + \beta^{p^k} = 1.$$

Write $h^{p^k} = g$. Since $|\mathfrak{G}/\mathfrak{N}|$ is not divisible by p, $\mathfrak{G} = \mathfrak{N}\langle g\rangle$. Also

$$\varrho(x^g) = \varrho(x^{h^{p^k}}) = z^{-p^k}\varrho(x)z^{p^k}$$

and $\varrho(g^n) = z^{np^k}$. Thus the assertions again follow from 9.8. **q.e.d.**

We turn now to analogues of V, 17.12b). For this we need the following lemmas.

9.10 Lemma. *Suppose that* $\mathfrak{N} \trianglelefteq \mathfrak{G}$ *and that* V *is a* $\mathsf{K}\mathfrak{G}$*-module. For* $\alpha \in \mathrm{Hom}_{\mathsf{K}\mathfrak{N}}(\mathsf{V}_{\mathfrak{N}}, \mathsf{V}_{\mathfrak{N}})$ *and* $g \in \mathfrak{G}$, *we define* α^g *by*

$$v\alpha^g = vg^{-1}\alpha g \quad (v \in \mathsf{V}).$$

Then $\alpha^g \in \mathrm{Hom}_{\mathsf{K}\mathfrak{N}}(\mathsf{V}_{\mathfrak{N}}, \mathsf{V}_{\mathfrak{N}})$, and $\alpha \to \alpha^g$ *is an algebra automorphism of* $\mathrm{Hom}_{\mathsf{K}\mathfrak{N}}(\mathsf{V}_{\mathfrak{N}}, \mathsf{V}_{\mathfrak{N}})$.

Proof. Suppose $v \in \mathsf{V}$, $g \in \mathfrak{G}$, $h \in \mathfrak{N}$ and $\alpha \in \mathrm{Hom}_{\mathsf{K}\mathfrak{N}}(\mathsf{V}_{\mathfrak{N}}, \mathsf{V}_{\mathfrak{N}})$. Then

$$\begin{aligned} (vh)\alpha^g &= (vh)g^{-1}\alpha g = ((vg^{-1})(ghg^{-1}))\alpha g = ((vg^{-1})\alpha)(ghg^{-1})g \\ &= vg^{-1}\alpha gh = (v\alpha^g)h. \end{aligned}$$

Thus $\alpha^g \in \mathrm{Hom}_{\mathsf{K}\mathfrak{N}}(\mathsf{V}_{\mathfrak{N}}, \mathsf{V}_{\mathfrak{N}})$. It is obvious that $\alpha \to \alpha^g$ is K-linear and bijective. If $\alpha, \beta \in \mathrm{Hom}_{\mathsf{K}\mathfrak{N}}(\mathsf{V}_{\mathfrak{N}}, \mathsf{V}_{\mathfrak{N}})$ and $v \in \mathsf{V}$, then

$$v(\alpha\beta)^g = vg^{-1}\alpha\beta g = (vg^{-1}\alpha g)g^{-1}\beta g = v\alpha^g\beta^g.$$ **q.e.d.**

9.11 Lemma. *Suppose $\mathfrak{N} \trianglelefteq \mathfrak{G}$. Let V be a $\mathsf{K}\mathfrak{G}$-module and let W_1, W_2 be $\mathsf{K}\mathfrak{G}$-modules of equal dimension m, on which $\mathfrak{N}$ acts trivially.*

a) *There is a* K*-monomorphism τ of*

$$\mathrm{Hom}_{\mathsf{K}\mathfrak{N}}(\mathsf{V}_{\mathfrak{N}} \otimes_{\mathsf{K}} (\mathsf{W}_1)_{\mathfrak{N}}, \mathsf{V}_{\mathfrak{N}} \otimes_{\mathsf{K}} (\mathsf{W}_2)_{\mathfrak{N}})$$

into the complete matrix ring $(\mathrm{Hom}_{\mathsf{K}\mathfrak{N}}(\mathsf{V}_{\mathfrak{N}}, \mathsf{V}_{\mathfrak{N}}))_m$ defined as follows. Let $\{w_1, \ldots, w_m\}$ be a K*-basis of W_1 and let $\{w'_1, \ldots, w'_m\}$ be a* K*-basis of W_2. If $\alpha \in \mathrm{Hom}_{\mathsf{K}\mathfrak{N}}(\mathsf{V}_{\mathfrak{N}} \otimes_{\mathsf{K}} (\mathsf{W}_1)_{\mathfrak{N}}, \mathsf{V}_{\mathfrak{N}} \otimes_{\mathsf{K}} (\mathsf{W}_2)_{\mathfrak{N}})$ and*

$$(v \otimes w_i)\alpha = \sum_{j=1}^{m} v\alpha_{ij} \otimes w'_j \quad (v\alpha_{ij} \in \mathsf{V}),$$

we put $\alpha\tau = (\alpha_{ij})$.

b) *If $\mathsf{W}_1 = \mathsf{W}_2$ and $w'_i = w_i$, then τ is an algebra monomorphism of*

$$\mathrm{Hom}_{\mathsf{K}\mathfrak{N}}(\mathsf{V}_{\mathfrak{N}} \otimes_{\mathsf{K}} (\mathsf{W}_1)_{\mathfrak{N}}, \mathsf{V}_{\mathfrak{N}} \otimes_{\mathsf{K}} (\mathsf{W}_1)_{\mathfrak{N}})$$

into the complete matrix ring $(\mathrm{Hom}_{\mathsf{K}\mathfrak{N}}(\mathsf{V}_{\mathfrak{N}}, \mathsf{V}_{\mathfrak{N}}))_m$.

c) *Suppose that*

$$w_i g = \sum_{j=1}^{m} a_{ij}(g) w_j \quad \text{and} \quad w'_i g = \sum_{j=1}^{m} b_{ij}(g) w'_j.$$

If $\alpha \in \mathrm{Hom}_{\mathsf{K}\mathfrak{G}}(\mathsf{V} \otimes_{\mathsf{K}} \mathsf{W}_1, \mathsf{V} \otimes_{\mathsf{K}} \mathsf{W}_2)$ and $\alpha\tau = (\alpha_{ij})$, then

$$\sum_{j=1}^{m} a_{ij}(g)\alpha_{jk} = \sum_{j=1}^{m} \alpha_{ij}^{g} b_{jk}(g).$$

Proof. a) For every $\alpha \in \mathrm{Hom}_{\mathsf{K}}(\mathsf{V} \otimes_{\mathsf{K}} \mathsf{W}_1, \mathsf{V} \otimes_{\mathsf{K}} \mathsf{W}_2)$, there exist uniquely defined $\alpha_{ij} \in \mathrm{Hom}_{\mathsf{K}}(\mathsf{V}, \mathsf{V})$ such that

$$(v \otimes w_i)\alpha = \sum_{j=1}^{m} v\alpha_{ij} \otimes w'_j \quad (i = 1, \ldots, m).$$

Obviously, the mapping τ defined by $\alpha\tau = (\alpha_{ij})$ is a K-monomorphism. If $x \in \mathfrak{N}$ and $\alpha \in \mathrm{Hom}_{\mathsf{K}\mathfrak{N}}(\mathsf{V}_{\mathfrak{N}} \otimes_{\mathsf{K}} (\mathsf{W}_1)_{\mathfrak{N}}, \mathsf{V}_{\mathfrak{N}} \otimes_{\mathsf{K}} (\mathsf{W}_2)_{\mathfrak{N}})$, then

$$\sum_{j=1}^{m} (vx)\alpha_{ij} \otimes w'_j = (vx \otimes w_i)\alpha = (vx \otimes w_i x)\alpha = (v \otimes w_i)x\alpha$$

$$= (v \otimes w_i)\alpha x = \left(\sum_{j=1}^{m} v\alpha_{ij} \otimes w_j'\right)x$$

$$= \sum_{j=1}^{m} (v\alpha_{ij})x \otimes w_j'.$$

Comparing coefficients of w_j', we obtain $\alpha_{ij} \in \mathrm{Hom}_{\mathsf{K}\mathfrak{N}}(\mathsf{V}_{\mathfrak{N}}, \mathsf{V}_{\mathfrak{N}})$.

b) Suppose that α, β are elements of $\mathrm{Hom}_{\mathsf{K}\mathfrak{N}}(\mathsf{V}_{\mathfrak{N}} \otimes_{\mathsf{K}} (\mathsf{W}_1)_{\mathfrak{N}}, \mathsf{V}_{\mathfrak{N}} \otimes_{\mathsf{K}} (\mathsf{W}_1)_{\mathfrak{N}})$. If $\alpha\tau = (\alpha_{ij})$ and $\beta\tau = (\beta_{ij})$, then

$$(v \otimes w_i)\alpha\beta = ((v \otimes w_i)\alpha)\beta = \left(\sum_{j=1}^{m} v\alpha_{ij} \otimes w_j\right)\beta$$

$$= \sum_{k=1}^{m}\left(\sum_{j=1}^{m} v\alpha_{ij}\beta_{jk} \otimes w_k\right).$$

Hence $(\alpha\beta)\tau = (\alpha\tau)(\beta\tau)$.

c) If $g \in \mathfrak{G}$ and $\alpha \in \mathrm{Hom}_{\mathsf{K}\mathfrak{G}}(\mathsf{V} \otimes_{\mathsf{K}} \mathsf{W}_1, \mathsf{V} \otimes_{\mathsf{K}} \mathsf{W}_2)$, then

$$((v \otimes w_i)g)\alpha = (vg \otimes w_i g)\alpha = \left(\sum_{j=1}^{m} vg \otimes a_{ij}(g)w_j\right)\alpha$$

$$= \sum_{j=1}^{m}\sum_{k=1}^{m} (vg)\alpha_{jk} \otimes a_{ij}(g)w_k'$$

and

$$((v \otimes w_i)\alpha)g = \left(\sum_{j=1}^{m} v\alpha_{ij} \otimes w_j'\right)g$$

$$= \sum_{k=1}^{m}\sum_{j=1}^{m} (v\alpha_{ij})g \otimes b_{jk}(g)w_k'.$$

This implies that

$$\sum_{j=1}^{m} (vg)\alpha_{jk}a_{ij}(g) = \sum_{j=1}^{m} (v\alpha_{ij})gb_{jk}(g).$$

Substituting v for vg, we obtain

$$v \sum_{j=1}^{m} a_{ij}(g)\alpha_{jk} = \sum_{j=1}^{m} (vg^{-1})\alpha_{ij} g b_{jk}(g)$$

$$= v \sum_{j=1}^{m} \alpha_{ij}^{g} b_{jk}(g). \qquad \textbf{q.e.d.}$$

9.12 Theorem (H. N. WARD [2]; HUPPERT, WILLEMS [1]). *Suppose that $\mathfrak{N} \trianglelefteq \mathfrak{G}$, K is an arbitrary field and V is a $\mathsf{K}\mathfrak{G}$-module.*

a) *If $\mathsf{V}_{\mathfrak{N}}$ is absolutely indecomposable and W is an indecomposable $\mathsf{K}(\mathfrak{G}/\mathfrak{N})$-module, then $\mathsf{V} \otimes_{\mathsf{K}} \mathsf{W}$ is an indecomposable $\mathsf{K}\mathfrak{G}$-module.*

b) *If $\mathsf{V}_{\mathfrak{N}}$ is absolutely irreducible and W is an irreducible $\mathsf{K}(\mathfrak{G}/\mathfrak{N})$-module, then $\mathsf{V} \otimes_{\mathsf{K}} \mathsf{W}$ is an irreducible $\mathsf{K}\mathfrak{G}$-module.*

c) *If $\mathsf{V}_{\mathfrak{N}}$ is absolutely indecomposable and W_1, W_2 are $\mathsf{K}(\mathfrak{G}/\mathfrak{N})$-modules such that $\mathsf{V} \otimes_{\mathsf{K}} \mathsf{W}_1 \cong \mathsf{V} \otimes_{\mathsf{K}} \mathsf{W}_2$, then $\mathsf{W}_1 \cong \mathsf{W}_2$.*

Proof. a) Suppose that π is an idempotent in $\mathrm{Hom}_{\mathsf{K}\mathfrak{G}}(\mathsf{V} \otimes_{\mathsf{K}} \mathsf{W}, \mathsf{V} \otimes_{\mathsf{K}} \mathsf{W})$. We want to show that $\pi = 0$ or $\pi = 1$. Let $\{w_1, \ldots, w_m\}$ be a K-basis of W and

$$w_i g = \sum_{j=1}^{m} a_{ij}(g) w_j \quad (g \in \mathfrak{G}).$$

We apply Lemma 9.11 with $\mathsf{W}_1 = \mathsf{W}_2 = \mathsf{W}$ and $w_i = w_i'$ $(i = 1, \ldots, m)$. Thus we obtain $\pi\tau = (\pi_{ij})$, where

$$\sum_{j=1}^{m} a_{ij}(g)\pi_{jk} = \sum_{j=1}^{m} \pi_{ij}^{g} a_{jk}(g). \tag{10}$$

By 9.11b), $\pi\tau = (\pi_{ij})$ is an idempotent of $(\mathrm{Hom}_{\mathsf{K}\mathfrak{N}}(\mathsf{V}_{\mathfrak{N}}, \mathsf{V}_{\mathfrak{N}}))_m$. As $\mathsf{V}_{\mathfrak{N}}$ is absolutely indecomposable, we have the decomposition

$$\mathrm{Hom}_{\mathsf{K}\mathfrak{N}}(\mathsf{V}_{\mathfrak{N}}, \mathsf{V}_{\mathfrak{N}}) = \mathsf{K}1_{\mathsf{V}} \oplus \mathbf{J}(\mathrm{Hom}_{\mathsf{K}\mathfrak{N}}(\mathsf{V}_{\mathfrak{N}}, \mathsf{V}_{\mathfrak{N}})).$$

Suppose that

$$\pi_{ij} = p_{ij} 1_{\mathsf{V}} + \varrho_{ij} \quad (i, j = 1, \ldots, m),$$

where $p_{ij} \in \mathsf{K}$ and $\varrho_{ij} \in \mathbf{J}(\mathrm{Hom}_{\mathsf{K}\mathfrak{N}}(\mathsf{V}_{\mathfrak{N}}, \mathsf{V}_{\mathfrak{N}}))$. Then

$$\pi_{ij}^{g} = p_{ij} 1_{\mathsf{V}} + \varrho_{ij}^{g},$$

where $\varrho_{ij}^g \in \mathbf{J}(\mathrm{Hom}_{K\mathfrak{N}}(V_{\mathfrak{N}}, V_{\mathfrak{N}}))$ by 9.10. Comparing the coefficients of 1_V in (10), we obtain

$$\sum_{j=1}^{m} a_{ij}(g)p_{jk} = \sum_{j=1}^{m} p_{ij}a_{jk}(g).$$

Hence the mapping π', defined by

$$w_i\pi' = \sum_{j=1}^{m} p_{ij}w_j \quad (i = 1, \ldots, m)$$

is a $K\mathfrak{G}$-homomorphism of W. As (π_{ij}) is idempotent, π' is also idempotent. Hence the indecomposability of W implies $\pi' = 0$ or $\pi' = 1$. Replacing π by $1 - \pi$ if necessary, we can assume that $\pi' = 0$, hence

$$\pi_{ij} \in \mathbf{J}(\mathrm{Hom}_{K\mathfrak{N}}(V_{\mathfrak{N}}, V_{\mathfrak{N}}))$$

for all i, j. As $\mathbf{J}(\mathrm{Hom}_{K\mathfrak{N}}(V_{\mathfrak{N}}, V_{\mathfrak{N}}))$ is a nilpotent ideal, this implies that

$$(\pi_{ij})^t = 0$$

for sufficiently large t. Hence $\pi = \pi^t = 0$, and $V \otimes_K W$ is indecomposable.

b) Let U be a non-zero $K\mathfrak{G}$-submodule of $V \otimes_K W$ and let u be a non-zero element of U. Write

$$u = \sum_{i=1}^{k} v_i \otimes w_i \quad (v_i \in V,\ w_i \in W),$$

where $w_1 \neq 0$ and $v_1, \ldots, v_k$ are linearly independent over K. Since $V_{\mathfrak{N}}$ is absolutely irreducible, $K = \mathrm{Hom}_{K\mathfrak{N}}(V_{\mathfrak{N}}, V_{\mathfrak{N}})$. Hence by Jacobson's density lemma (V, 4.2), given $v \in V$, there exists

$$a = \sum_{x \in \mathfrak{N}} a_x x \in K\mathfrak{N} \quad (a_x \in K)$$

such that $v_1 a = v$ and $v_i a = 0$ for $2 \leq i \leq k$. Since $\mathfrak{N}$ operates trivially on W, we obtain

$$(v_i \otimes w_i)a = \sum_{x \in \mathfrak{N}} a_x(v_i \otimes w_i)x = \sum_{x \in \mathfrak{N}} a_x(v_i x \otimes w_i) = v_i a \otimes w_i$$

and $ua = v \otimes w_1$. Thus $\mathsf{V} \otimes w_1 \subseteq \mathsf{U}$. Hence for all $g \in \mathfrak{G}$,

$$\mathsf{V} \otimes w_1 g = \mathsf{V}g \otimes w_1 g = (\mathsf{V} \otimes w_1)g \subseteq \mathsf{U}g = \mathsf{U}.$$

Since $w_1 \neq 0$ and W is irreducible, it follows that $\mathsf{U} = \mathsf{V} \otimes_{\mathsf{K}} \mathsf{W}$. Hence $\mathsf{V} \otimes_{\mathsf{K}} \mathsf{W}$ is irreducible.

c) Now suppose that α is a $\mathsf{K}\mathfrak{G}$-isomorphism of $\mathsf{V} \otimes_{\mathsf{K}} \mathsf{W}_1$ onto $\mathsf{V} \otimes_{\mathsf{K}} \mathsf{W}_2$. Then certainly $\dim_{\mathsf{K}} \mathsf{W}_1 = \dim_{\mathsf{K}} \mathsf{W}_2$, and we can apply 9.11. Thus in the notation of 9.11c), we have $\alpha\tau = (\alpha_{ij})$ and

$$\sum_{j=1}^{m} a_{ij}(g)\alpha_{jk} = \sum_{j=1}^{m} \alpha_{ij}^{g} b_{jk}(g). \tag{11}$$

Again we obtain

$$\alpha_{ij} = c_{ij} 1_{\mathsf{V}} + \gamma_{ij},$$

where $c_{ij} \in \mathsf{K}$ and $\gamma_{ij} \in \mathbf{J}(\mathrm{Hom}_{\mathsf{K}\mathfrak{N}}(\mathsf{V}_{\mathfrak{N}}, \mathsf{V}_{\mathfrak{N}}))$. Hence (11) implies that

$$\sum_{j=1}^{m} a_{ij}(g)c_{jk} = \sum_{j=1}^{m} c_{ij} b_{jk}(g).$$

Thus the mapping α', defined by

$$w_i \alpha' = \sum_{j=1}^{m} c_{ij} w_j' \quad (i = 1, \ldots, m)$$

is in $\mathrm{Hom}_{\mathsf{K}\mathfrak{G}}(\mathsf{W}_1, \mathsf{W}_2)$. Application of the same procedure to the inverse of α shows that α' has an inverse. Hence W_1 and W_2 are isomorphic $\mathsf{K}\mathfrak{G}$-modules. **q.e.d.**

9.13 Corollary. *Suppose that $\mathfrak{N} \trianglelefteq \mathfrak{G}$ and that V_1, V_2 are $\mathsf{K}\mathfrak{G}$-modules for which $(\mathsf{V}_1)_{\mathfrak{N}}$ and $(\mathsf{V}_2)_{\mathfrak{N}}$ are isomorphic absolutely irreducible $\mathsf{K}\mathfrak{N}$-modules. Then there exists a $\mathsf{K}(\mathfrak{G}/\mathfrak{N})$-module W of K-dimension 1 such that $\mathsf{V}_2 \cong \mathsf{V}_1 \otimes_{\mathsf{K}} \mathsf{W}$.*

Proof. By 4.15b),

$$\mathsf{V}_1 \otimes_{\mathsf{K}} \mathsf{K}(\mathfrak{G}/\mathfrak{N}) \cong ((\mathsf{V}_1)_{\mathfrak{N}})^{\mathfrak{G}} \cong ((\mathsf{V}_2)_{\mathfrak{N}})^{\mathfrak{G}} \cong \mathsf{V}_2 \otimes_{\mathsf{K}} \mathsf{K}(\mathfrak{G}/\mathfrak{N}).$$

Since $\mathsf{K}(\mathfrak{G}/\mathfrak{N})$ has the trivial module as a quotient module,

$V_2 \otimes_K K(\mathfrak{G}/\mathfrak{N})$ has a quotient module isomorphic to V_2. Hence V_2 is isomorphic to a composition factor of $V_1 \otimes_K K(\mathfrak{G}/\mathfrak{N})$. Let

$$0 = U_0 \subset U_1 \subset \cdots \subset U_m = K(\mathfrak{G}/\mathfrak{N})$$

be a composition series of the $K\mathfrak{G}$-module $K(\mathfrak{G}/\mathfrak{N})$. Then by 9.12b),

$$(V_1 \otimes_K U_i)/(V_1 \otimes_K U_{i-1}) \cong V_1 \otimes_K (U_i/U_{i-1})$$

is an irreducible $K\mathfrak{G}$-module. Thus by the Jordan-Hölder theorem, $V_2 \cong V_1 \otimes_K W$ for some irreducible $K(\mathfrak{G}/\mathfrak{N})$-module W. Since $\dim_K V_1 = \dim_K V_2$, certainly $\dim_K W = 1$. **q.e.d.**

From Theorem 9.12 we can easily deduce results about the irreducible and indecomposable modules of direct products.

9.14 Theorem. *Let K be a splitting field for $\mathfrak{G}_1$ and $\mathfrak{G}_2$. Let $V_1, \ldots, V_s$ be all the irreducible $K\mathfrak{G}_1$-modules and let $W_1, \ldots, W_t$ be all the irreducible $K\mathfrak{G}_2$-modules (to within isomorphism). Then the $V_i \otimes_K W_j$, regarded as $K(\mathfrak{G}_1 \times \mathfrak{G}_2)$-modules with*

$$(v_i \otimes w_j)(g_1 g_2) = v_i g_1 \otimes w_j g_2 \quad (v_i \in V_i,\ w_j \in W_j,\ g_k \in \mathfrak{G}_k)$$

are all the irreducible $K(\mathfrak{G}_1 \times \mathfrak{G}_2)$-modules, to within isomorphism.

Proof. We apply 9.12b) to the normal subgroup $\mathfrak{G}_1$ of $\mathfrak{G}_1 \times \mathfrak{G}_2 = \mathfrak{G}$. V_i can be regarded as a $K\mathfrak{G}$-module on which $\mathfrak{G}_2$ operates trivially. By hypothesis, the restriction of this module to $\mathfrak{G}_1$ is absolutely irreducible. Hence by 9.12b), the $V_i \otimes_K W_j$ are irreducible $K\mathfrak{G}$-modules.

Suppose that $V_i \otimes_K W_j$ and $V_k \otimes_K W_l$ are isomorphic $K\mathfrak{G}$-modules. Restriction to $\mathfrak{G}_1$ shows that

$$V_i \underbrace{\oplus \cdots \oplus}_{\dim_K W_j} V_i \cong V_k \underbrace{\oplus \cdots \oplus}_{\dim_K W_l} V_k.$$

By the Jordan-Hölder theorem, $V_i \cong V_k$, so $i = k$. Similarly $j = l$.

Finally, it must be shown that the $V_i \otimes_K W_j$ are all the irreducible $K\mathfrak{G}$-modules to within isomorphism. We have $s = l(\mathfrak{G}_1)$ and $t = l(\mathfrak{G}_2)$, where $l(\mathfrak{X})$ is the number of conjugacy classes of $\mathfrak{X}$ if char $K = 0$ and $l(\mathfrak{X})$ is the number of p'-classes of $\mathfrak{X}$ if char $K = p$, by V, 5.1 and 3.9. Thus

$$st = l(\mathfrak{G}_1)l(\mathfrak{G}_2) = l(\mathfrak{G}_1 \times \mathfrak{G}_2)$$

is the number of isomorphism types of irreducible $\mathsf{K}(\mathfrak{G}_1 \times \mathfrak{G}_2)$-modules.
q.e.d.

The statement corresponding to 9.14 for indecomposable modules is unfortunately not as complete.

9.15 Theorem. *Let* K *be an algebraically closed field.*

a) *If* V_i *is an indecomposable* $\mathsf{K}\mathfrak{G}_i$*-module* ($i = 1, 2$), *then* $\mathsf{V}_1 \otimes_{\mathsf{K}} \mathsf{V}_2$ *is an indecomposable* $\mathsf{K}(\mathfrak{G}_1 \times \mathfrak{G}_2)$*-module.*

b) *If* V_i, W_i *are indecomposable* $\mathsf{K}\mathfrak{G}_i$*-modules* ($i = 1, 2$) *and* $\mathsf{V}_1 \otimes_{\mathsf{K}} \mathsf{V}_2 \cong \mathsf{W}_1 \otimes_{\mathsf{K}} \mathsf{W}_2$, *then* $\mathsf{V}_i \cong \mathsf{W}_i$ ($i = 1, 2$).

c) *Every indecomposable* $\mathsf{K}(\mathfrak{G}_1 \times \mathfrak{G}_2)$*-module is of the type* $\mathsf{V}_1 \otimes_{\mathsf{K}} \mathsf{V}_2$ *for suitable* $\mathsf{K}\mathfrak{G}_i$*-modules* V_i *if and only if at least one of the orders* $|\mathfrak{G}_1|$, $|\mathfrak{G}_2|$ *is prime to* char K.

Proof. a), b) The proofs of these assertions are the same as those of 9.14, using 9.12a), c) instead of 9.12b) and the Krull-Schmidt theorem instead of the Jordan-Hölder theorem.

c) In the case when char $\mathsf{K} = 0$, the assertion follows from 9.14, since in this case there is no difference between irreducibility and indecomposibility. Suppose that char $\mathsf{K} = p$.

First suppose that p does not divide $|\mathfrak{G}_2|$. We show (following CONLON [1]) that any indecomposable $\mathsf{K}(\mathfrak{G}_1 \times \mathfrak{G}_2)$-module V is of the type $\mathsf{V}_1 \otimes_{\mathsf{K}} \mathsf{V}_2$. Since p does not divide $|\mathfrak{G}_1 \times \mathfrak{G}_2 : \mathfrak{G}_1|$, V is isomorphic to a direct summand of $(\mathsf{V}_{\mathfrak{G}_1})^{\mathfrak{G}_1 \times \mathfrak{G}_2}$, by 4.3c). By the Krull-Schmidt theorem, there exists an indecomposable direct summand W of $\mathsf{V}_{\mathfrak{G}_1}$ such that V is isomorphic to a direct summand of $\mathsf{W}^{\mathfrak{G}_1 \times \mathfrak{G}_2}$. Now

$$\mathsf{W}^{\mathfrak{G}_1 \times \mathfrak{G}_2} = \bigoplus_{g \in \mathfrak{G}_2} \mathsf{W} \otimes g,$$

and for $g_i \in \mathfrak{G}_i$, $w \in \mathsf{W}$, $g \in \mathfrak{G}_2$

$$(w \otimes g)g_1 g_2 = wg_1 \otimes gg_2.$$

This shows that $\mathsf{W}^{\mathfrak{G}_1 \times \mathfrak{G}_2}$ is isomorphic to $\mathsf{W} \otimes_{\mathsf{K}} \mathsf{K}\mathfrak{G}_2$. Let

$$\mathsf{K}\mathfrak{G}_2 = \bigoplus_i \mathsf{U}_i$$

be a decomposition of $\mathsf{K}\mathfrak{G}_2$ into indecomposable $\mathsf{K}\mathfrak{G}_2$-modules U_i. (Since p does not divide $|\mathfrak{G}_2|$, the U_i are in fact irreducible.) Then V is isomorphic to a direct summand of $\bigoplus_i(\mathsf{W} \otimes_{\mathsf{K}} \mathsf{U}_i)$. By a), each $\mathsf{W} \otimes_{\mathsf{K}} \mathsf{U}_i$ is an indecomposable $\mathsf{K}(\mathfrak{G}_1 \times \mathfrak{G}_2)$-module. Hence by the Krull-Schmidt theorem, $\mathsf{V} \cong \mathsf{W} \otimes_{\mathsf{K}} \mathsf{U}_i$ for some i.

Now suppose that char $\mathsf{K} = p$ divides $|\mathfrak{G}_1|$ and $|\mathfrak{G}_2|$. Let $\mathfrak{P}_i$ be a Sylow p-subgroup of $\mathfrak{G}_i$. We put $\mathfrak{G} = \mathfrak{G}_1 \times \mathfrak{G}_2$ and $\mathfrak{P} = \mathfrak{P}_1 \times \mathfrak{P}_2$. Let $\mathfrak{R}_i$ be a subgroup of $\mathfrak{P}_i$ of index p. By 5.1, there exists an indecomposable $\mathsf{K}\mathfrak{P}$-module W such that $\dim_{\mathsf{K}}\mathsf{W} > p^2$ and $\mathfrak{R}_1 \times \mathfrak{R}_2$ operates trivially on W. If W is of the type $\mathsf{W}_1 \otimes_{\mathsf{K}} \mathsf{W}_2$ for $\mathsf{K}\mathfrak{P}_i$-modules W_i, then W_i is indecomposable and $\mathfrak{R}_i$ operates trivially on W_i. Thus $\dim_{\mathsf{K}}\mathsf{W}_i \leq p$ by 5.3. This gives the contradiction $\dim_{\mathsf{K}}\mathsf{W} \leq p^2$. Hence W is not of the type $\mathsf{W}_1 \otimes_{\mathsf{K}} \mathsf{W}_2$ for $\mathsf{K}\mathfrak{P}_i$-modules W_i.

Put $\mathsf{V} = \mathsf{W}^{\mathfrak{G}}$. Suppose that

$$\mathsf{V} = \mathsf{V}_1 \oplus \cdots \oplus \mathsf{V}_s,$$

where the V_i are indecomposable $\mathsf{K}\mathfrak{G}$-modules. By 4.3, W is isomorphic to a direct summand of $(\mathsf{W}^{\mathfrak{G}})_{\mathfrak{P}}$. By the Krull-Schmidt theorem, W is isomorphic to a direct summand of $(\mathsf{V}_k)_{\mathfrak{P}}$ for some k. Suppose that $\mathsf{V}_k = \mathsf{X}_1 \otimes_{\mathsf{K}} \mathsf{X}_2$, where X_i is an indecomposable $\mathsf{K}\mathfrak{G}_i$-module ($i = 1, 2$). If

$$(\mathsf{X}_i)_{\mathfrak{P}_i} = \bigoplus_j \mathsf{X}_{ij} \quad (i = 1, 2),$$

where the X_{ij} are indecomposable $\mathsf{K}\mathfrak{P}_i$-modules, then

$$(\mathsf{V}_k)_{\mathfrak{P}} \cong (\mathsf{X}_1)_{\mathfrak{P}_1} \otimes_{\mathsf{K}} (\mathsf{X}_2)_{\mathfrak{P}_2} \cong \bigoplus_{j,l} \mathsf{X}_{1j} \otimes_{\mathsf{K}} \mathsf{X}_{2l}.$$

By a), each $\mathsf{X}_{1j} \otimes_{\mathsf{K}} \mathsf{X}_{2l}$ is an indecomposable $\mathsf{K}\mathfrak{P}$-module. Thus by the Krull-Schmidt theorem, we get the contradiction $\mathsf{W} \cong \mathsf{X}_{1j} \otimes_{\mathsf{K}} \mathsf{X}_{2l}$ for some j, l. Thus V_k is an indecomposable $\mathsf{K}\mathfrak{G}$-module not of the form $\mathsf{X}_1 \otimes_{\mathsf{K}} \mathsf{X}_2$ for indecomposable $\mathsf{K}\mathfrak{G}_i$-modules X_i. **q.e.d.**

9.16 Lemma (FEIT [3]). *Let $\mathfrak{G} = \mathfrak{G}_1 \times \mathfrak{G}_2$, let V_i be a $\mathsf{K}\mathfrak{G}_i$-module ($i = 1, 2$) and let $\mathsf{V} = \mathsf{V}_1 \otimes_{\mathsf{K}} \mathsf{V}_2$.*

a) $\mathbf{C}_{\mathsf{V}}(\mathfrak{G}) = \mathbf{C}_{\mathsf{V}_1}(\mathfrak{G}_1) \otimes_{\mathsf{K}} \mathbf{C}_{\mathsf{V}_2}(\mathfrak{G}_2)$.

b) $\mathrm{Hom}_{\mathsf{K}\mathfrak{G}}(\mathsf{V}, \mathsf{V})$ *and* $\mathrm{Hom}_{\mathsf{K}\mathfrak{G}_1}(\mathsf{V}_1, \mathsf{V}_1) \otimes_{\mathsf{K}} \mathrm{Hom}_{\mathsf{K}\mathfrak{G}_2}(\mathsf{V}_2, \mathsf{V}_2)$ *are isomorphic* K-*algebras.*

Proof. a) If $\{w_1, \ldots, w_n\}$ is a K-basis of V_2, then $\sum_{i=1}^n v_i \otimes w_i$ ($v_i \in \mathsf{V}_1$) lies in $\mathbf{C}_{\mathsf{V}}(\mathfrak{G}_1 \times 1)$ if and only if $v_i \in \mathbf{C}_{\mathsf{V}_1}(\mathfrak{G}_1)$ for all $i = 1, \ldots, n$. Thus

$\mathbf{C}_V(\mathfrak{G}) \subseteq \mathbf{C}_{V_1}(\mathfrak{G}_1) \otimes_K V_2$. Application of the same argument to $\mathbf{C}_{V_1}(\mathfrak{G}_1) \otimes_K V_2$ yields

$$\mathbf{C}_V(\mathfrak{G}) \subseteq \mathbf{C}_{V_1}(\mathfrak{G}_1) \otimes_K \mathbf{C}_{V_2}(\mathfrak{G}_2).$$

Thus

$$\mathbf{C}_V(\mathfrak{G}) = \mathbf{C}_{V_1}(\mathfrak{G}_1) \otimes_K \mathbf{C}_{V_2}(\mathfrak{G}_2).$$

b) We put $H_i = \mathrm{Hom}_K(V_i, V_i)$ and $H = \mathrm{Hom}_K(V, V)$. By V, 9.14, there exists a K-algebra isomorphism τ of $H_1 \otimes_K H_2$ onto H such that

$$(v_1 \otimes v_2)((\alpha_1 \otimes \alpha_2)\tau) = v_1\alpha_1 \otimes v_2\alpha_2$$

for all $v_i \in V_i$, $\alpha_i \in H_i$ $(i = 1, 2)$. We regard H_1, H_2, H as $K\mathfrak{G}_1$-, $K\mathfrak{G}_2$-, $K\mathfrak{G}$-modules respectively in the sense of 8.8. Then τ is a $K\mathfrak{G}$-isomorphism, for if $v_i \in V_i$, $\alpha_i \in H_i$ and $g_i \in \mathfrak{G}_i$, then

$$\begin{aligned}(v_1 \otimes v_2)(((\alpha_1 \otimes \alpha_2)g_1g_2)\tau) &= (v_1 \otimes v_2)((\alpha_1 g_1 \otimes \alpha_2 g_2)\tau)\\ &= v_1(\alpha_1 g_1) \otimes v_2(\alpha_2 g_2)\\ &= ((v_1 g_1^{-1})\alpha_1)g_1 \otimes ((v_2 g_2^{-1})\alpha_2)g_2\\ &= ((v_1 g_1^{-1} \otimes v_2 g_2^{-1})(\alpha_1 \otimes \alpha_2)\tau)g_1g_2\\ &= (v_1 \otimes v_2)(((\alpha_1 \otimes \alpha_2)\tau)g_1g_2).\end{aligned}$$

Hence τ carries $\mathbf{C}_{H_1 \otimes_K H_2}(\mathfrak{G})$ onto $\mathbf{C}_H(\mathfrak{G})$. By a),

$$\mathbf{C}_{H_1 \otimes_K H_2}(\mathfrak{G}) = \mathbf{C}_{H_1}(\mathfrak{G}_1) \otimes_K \mathbf{C}_{H_2}(\mathfrak{G}_2).$$

So $\mathbf{C}_{H_1}(\mathfrak{G}_1) \otimes_K \mathbf{C}_{H_2}(\mathfrak{G}_2)$ and $\mathbf{C}_H(\mathfrak{G})$ are isomorphic K-algebras. The assertion now follows from the observation that

$$\mathbf{C}_H(\mathfrak{G}) = \mathrm{Hom}_{K\mathfrak{G}}(V, V) \quad \text{and} \quad \mathbf{C}_{H_i}(\mathfrak{G}_i) = \mathrm{Hom}_{K\mathfrak{G}_i}(V_i, V_i).$$ **q.e.d.**

9.17 Theorem. *Let* $\mathfrak{G} = \mathfrak{G}_1 \times \mathfrak{G}_2$, *let* V_i *be a* $K\mathfrak{G}_i$*-module and let* $V = V_1 \otimes_K V_2$.

a) *If* V_1 *is indecomposable and* V_2 *is absolutely indecomposable, then* V *is an indecomposable* $K\mathfrak{G}$*-module.*

b) *If* V_1 *and* V_2 *are absolutely indecomposable, then* V *is absolutely indecomposable.*

Proof. Let $\mathfrak{S}_i = \operatorname{Hom}_{K\mathfrak{G}_i}(V_i, V_i)$ $(i = 1, 2)$. We consider the mapping α from $\mathfrak{S}_1 \otimes_K \mathfrak{S}_2$ onto $\mathfrak{S}_1/\mathbf{J}(\mathfrak{S}_1) \otimes_K \mathfrak{S}_2/\mathbf{J}(\mathfrak{S}_2)$ defined by

$$(\sigma_1 \otimes \sigma_2)\alpha = (\sigma_1 + \mathbf{J}(\mathfrak{S}_1)) \otimes (\sigma_2 + \mathbf{J}(\mathfrak{S}_2))$$

for $\sigma_i \in \mathfrak{S}_i$ $(i = 1, 2)$. Then α is an algebra epimorphism, and the kernel of α is the ideal

$$\mathfrak{I} = \mathbf{J}(\mathfrak{S}_1) \otimes_K \mathfrak{S}_2 + \mathfrak{S}_1 \otimes_K \mathbf{J}(\mathfrak{S}_2).$$

Thus

$$(\mathfrak{S}_1/\mathbf{J}(\mathfrak{S}_1)) \otimes_K (\mathfrak{S}_2/\mathbf{J}(\mathfrak{S}_2)) \cong (\mathfrak{S}_1 \otimes_K \mathfrak{S}_2)/\mathfrak{I}.$$

Now $\mathfrak{S}_1/\mathbf{J}(\mathfrak{S}_1)$ is a division algebra since V_1 is indecomposable and $\mathfrak{S}_2/\mathbf{J}(\mathfrak{S}_2) = K$ since V_2 is absolutely indecomposable. Hence

$$(\mathfrak{S}_1 \otimes_K \mathfrak{S}_2)/\mathfrak{I} \cong \mathfrak{S}_1/\mathbf{J}(\mathfrak{S}_1)$$

is a division algebra, and so $\mathfrak{I} \supseteq \mathbf{J}(\mathfrak{S}_1 \otimes_K \mathfrak{S}_2)$. But $\mathfrak{I}$ is nilpotent, so $\mathfrak{I} = \mathbf{J}(\mathfrak{S}_1 \otimes_K \mathfrak{S}_2)$ and

$$(\mathfrak{S}_1 \otimes_K \mathfrak{S}_2)/\mathbf{J}(\mathfrak{S}_1 \otimes_K \mathfrak{S}_2) \cong \mathfrak{S}_1/\mathbf{J}(\mathfrak{S}_1).$$

If we put $\mathfrak{S} = \operatorname{Hom}_{K\mathfrak{G}}(V, V)$, then by 9.16b)

$$\mathfrak{S}/\mathbf{J}(\mathfrak{S}) \cong \mathfrak{S}_1/\mathbf{J}(\mathfrak{S}_1).$$

Thus $\mathfrak{S}$ is a local ring and hence V is indecomposable. In case b), we obtain

$$\mathfrak{S}/\mathbf{J}(\mathfrak{S}) \cong \mathfrak{S}_1/\mathbf{J}(\mathfrak{S}_1) = K,$$

and V is absolutely indecomposable. **q.e.d.**

9.18 Theorem (ROTH [1], SCHWARZ [1]). *Let K be a splitting field of $\mathfrak{G}$. Suppose that $\mathfrak{N}$ is a normal subgroup of $\mathfrak{G}$ such that $\mathfrak{G}/\mathfrak{N}$ is Abelian and* char K *does not divide $|\mathfrak{G}/\mathfrak{N}|$. Let V be an irreducible $K\mathfrak{G}$-module. By Clifford's theorem, there is an irreducible $K\mathfrak{N}$-module W such that*

$$V_{\mathfrak{N}} \cong e \bigoplus_{i=1}^{m} W \otimes g_i,$$

where $\{g_1, \ldots, g_n\}$ *is a transversal of the inertia subgroup* $\mathfrak{I}$ *of* W *in* $\mathfrak{G}$. *Then*

$$\mathsf{W}^{\mathfrak{G}} \cong e \bigoplus_{j=1}^{r} \mathsf{V} \otimes_{\mathsf{K}} \mathsf{U}_j,$$

where the U_j *are irreducible* $\mathsf{K}\mathfrak{G}$*-modules of* K*-dimension 1 on which* $\mathfrak{N}$ *operates trivially. The* $\mathsf{V} \otimes_{\mathsf{K}} \mathsf{U}_j$ *are irreducible and pairwise non-isomorphic. Also*

$$e^2 rm = |\mathfrak{G}/\mathfrak{N}|.$$

Proof. It follows from the hypotheses on K and $\mathfrak{G}/\mathfrak{N}$ that

$$\mathsf{K}(\mathfrak{G}/\mathfrak{N}) = \bigoplus_{j=1}^{n} \mathsf{U}_j,$$

where $n = |\mathfrak{G}/\mathfrak{N}|$ and $\mathsf{U}_1, \ldots, \mathsf{U}_n$ are all the irreducible $\mathsf{K}(\mathfrak{G}/\mathfrak{N})$-modules to within isomorphism; further $\dim_{\mathsf{K}} \mathsf{U}_i = 1$ for all i and the U_i form a group under the tensor product operation. Now $\mathsf{W} \otimes g_i$ is a $\mathsf{K}\mathfrak{N}$-module and $(\mathsf{W} \otimes g_i)^{\mathfrak{G}} \cong \mathsf{W}^{\mathfrak{G}}$. Hence by 4.15b),

$$(\mathsf{V}_{\mathfrak{N}})^{\mathfrak{G}} \cong \mathsf{V} \otimes_{\mathsf{K}} \mathsf{K}(\mathfrak{G}/\mathfrak{N}) \cong \bigoplus_{j=1}^{n} (\mathsf{V} \otimes_{\mathsf{K}} \mathsf{U}_j),$$

so

$$em\mathsf{W}^{\mathfrak{G}} \cong (\mathsf{V}_{\mathfrak{N}})^{\mathfrak{G}} \cong \bigoplus_{j=1}^{n} \mathsf{V} \otimes_{\mathsf{K}} \mathsf{U}_j.$$

Let

$$\mathfrak{S} = \{\mathsf{U}_j | \mathsf{V} \otimes_{\mathsf{K}} \mathsf{U}_j \cong \mathsf{V}\}.$$

Then $\mathfrak{S}$ is a subgroup of the group formed by the U_i. Let $\{\mathsf{U}_1, \ldots, \mathsf{U}_r\}$ be a transversal of $\mathfrak{S}$ in this group. Then

$$em\mathsf{W}^{\mathfrak{G}} \cong |\mathfrak{S}| \bigoplus_{j=1}^{r} \mathsf{V} \otimes_{\mathsf{K}} \mathsf{U}_j.$$

Since $r|\mathfrak{S}| = |\mathfrak{G}/\mathfrak{N}|$, this gives

$$\mathsf{W}^{\mathfrak{G}} \cong \frac{|\mathfrak{G}/\mathfrak{N}|}{emr} \bigoplus_{j=1}^{r} \mathsf{V} \otimes_{\mathsf{K}} \mathsf{U}_j.$$

In particular, $\mathsf{W}^{\mathfrak{G}}$ is completely reducible and the socle of $\mathsf{W}^{\mathfrak{G}}$ is $\mathsf{W}^{\mathfrak{G}}$ itself. Hence by Nakayama's lemma (4.13b)), the multiplicity of V in $\mathsf{W}^{\mathfrak{G}}$ is equal to the multiplicity of W in the head of $\mathsf{V}_{\mathfrak{N}}$, and this is e. Hence

$$|\mathfrak{G}/\mathfrak{N}| = e^2 rm. \qquad \textbf{q.e.d.}$$

By repeated application of 9.18, we can easily prove that if $\mathfrak{G}$ is soluble and K is a splitting field of $\mathfrak{G}$ such that char K does not divide $|\mathfrak{G}|$, then the dimension of any irreducible $\mathsf{K}\mathfrak{G}$-module divides $|\mathfrak{G}|$. Of course, this is true for any group $\mathfrak{G}$, by V, 12.11. But this method can be used to obtain a result even in the case when char K does divide $|\mathfrak{G}|$. To do so, we need the following lemma, which is of independent interest.

9.19 Lemma. *Let K be a finite or algebraically closed field of characteristic p. Let $\mathfrak{N}$ be a normal subgroup of $\mathfrak{G}$ of index a power of p. Let V be an irreducible $\mathsf{K}\mathfrak{G}$-module for which all the irreducible $\mathsf{K}\mathfrak{N}$-submodules of $\mathsf{V}_{\mathfrak{N}}$ are isomorphic. Then V is an irreducible $\mathsf{K}\mathfrak{N}$-module.*

Proof. (GREEN). First suppose that K is finite. Let W be an irreducible $\mathsf{K}\mathfrak{N}$-submodule of $\mathsf{V}_{\mathfrak{N}}$ and let $q = |\mathrm{Hom}_{\mathsf{K}\mathfrak{N}}(\mathsf{W}, \mathsf{W})|$. Put

$$f(j) = 1 + q + \cdots + q^{j-1} \quad (j \geq 1).$$

We prove by induction on j that $f(j)$ is the number of irreducible $\mathsf{K}\mathfrak{N}$-submodules of any $\mathsf{K}\mathfrak{N}$-submodule U of $\mathsf{V}_{\mathfrak{N}}$ of composition length j. This is obvious if $j = 1$. For $j > 1$, we observe that by Clifford's theorem, $\mathsf{V}_{\mathfrak{N}}$ is completely reducible. Hence so is U, and we can write $\mathsf{U} = \mathsf{W}_1 \oplus \mathsf{U}_1$, where $\mathsf{W}_1 \cong \mathsf{W}$. By the inductive hypothesis, the number of irreducible $\mathsf{K}\mathfrak{N}$-submodules of U_1 is $f(j-1)$. Now if W^* is an irreducible $\mathsf{K}\mathfrak{N}$-submodule of U not contained in U_1, then $\mathsf{W}^* \cap \mathsf{U}_1 = 0$, so W^* is of the form

$$\{w + w\alpha \mid w \in \mathsf{W}_1\}$$

for some $\alpha \in \mathrm{Hom}_{\mathsf{K}\mathfrak{N}}(\mathsf{W}_1, \mathsf{U}_1)$. Thus the number of irreducible $\mathsf{K}\mathfrak{N}$-submodules of U is

$$f(j-1) + |\mathrm{Hom}_{\mathsf{K}\mathfrak{N}}(\mathsf{W}_1, \mathsf{U}_1)|.$$

Since U_1 is the direct sum of $j-1$ $\mathsf{K}\mathfrak{N}$-submodules isomorphic to W, this number is

$$f(j-1) + |\mathrm{Hom}_{\mathsf{K}\mathfrak{N}}(\mathsf{W}, \mathsf{W})|^{j-1} = f(j-1) + q^{j-1} = f(j).$$

Hence the number of irreducible $\mathsf{K}\mathfrak{N}$-submodules of $\mathsf{V}_{\mathfrak{N}}$ is $f(l)$, where l is the composition length of $\mathsf{V}_{\mathfrak{N}}$. Since q is a power of p, $f(l) \equiv 1(p)$. But a Sylow p-subgroup $\mathfrak{P}$ of $\mathfrak{G}$ permutes the irreducible $\mathsf{K}\mathfrak{N}$-submodules of $\mathsf{V}_{\mathfrak{N}}$ among themselves, so there exists an irreducible $\mathsf{K}\mathfrak{N}$-submodule W of $\mathsf{V}_{\mathfrak{N}}$ invariant under $\mathfrak{P}$. Hence W is invariant under $\mathfrak{P}\mathfrak{N} = \mathfrak{G}$. Since V is an irreducible $\mathsf{K}\mathfrak{G}$-module, $\mathsf{W} = \mathsf{V}$, and V is an irreducible $\mathsf{K}\mathfrak{N}$-module.

Now suppose that K is algebraically closed, and let k be the number of p'-conjugacy classes of $\mathfrak{G}$. By 2.6, there exists a finite splitting field L for all the subgroups of $\mathfrak{G}$, and since K is algebraically closed, we may suppose that $\mathsf{L} \subseteq \mathsf{K}$. Then by 3.9, there are k non-isomorphic irreducible $\mathsf{L}\mathfrak{G}$-modules $\mathsf{V}_1, \ldots, \mathsf{V}_k$. By 2.4, $\mathsf{V} \cong (\mathsf{V}_i)_{\mathsf{K}}$ for some i. Similarly, the irreducible $\mathsf{K}\mathfrak{N}$-submodules of $\mathsf{V}_{\mathfrak{N}}$ are isomorphic to W_{K} for some irreducible $\mathsf{L}\mathfrak{N}$-module W. Hence by 1.22 the irreducible submodules of $(\mathsf{V}_i)_{\mathfrak{N}}$ are all isomorphic to W. By the finite case, $(\mathsf{V}_i)_{\mathfrak{N}}$ is an irreducible $\mathsf{L}\mathfrak{N}$-module, and $\mathsf{V}_{\mathfrak{N}}$ is irreducible. **q.e.d.**

Combining 9.18 and 9.19, we obtain the following.

9.20 Theorem (SWAN [1]). *Suppose that $\mathfrak{G}/\mathfrak{N}$ is soluble and that K is an algebraically closed field. Let V be an irreducible $\mathsf{K}\mathfrak{G}$-module. Then the composition length of $\mathsf{V}_{\mathfrak{N}}$ divides $|\mathfrak{G}/\mathfrak{N}|$.*

Proof. This is proved by induction on $|\mathfrak{G}/\mathfrak{N}|$. Suppose that $\mathfrak{N} < \mathfrak{G}$. As $\mathfrak{G}/\mathfrak{N}$ is soluble, there exists $\mathfrak{M} \trianglelefteq \mathfrak{G}$ such that $\mathfrak{M} \geq \mathfrak{N}$ and $\mathfrak{M}/\mathfrak{N}$ is a non-identity Abelian group of prime-power order. By Clifford's theorem,

$$\mathsf{V}_{\mathfrak{M}} = \mathsf{U}_1 \oplus \cdots \oplus \mathsf{U}_s$$

for certain $\mathfrak{G}$-conjugate irreducible $\mathsf{K}\mathfrak{M}$-modules $\mathsf{U}_1, \ldots, \mathsf{U}_s$. By the inductive hypothesis, s divides $|\mathfrak{G}/\mathfrak{M}|$. We distinguish two cases.

Case 1. Suppose that char K does not divide $|\mathfrak{M}/\mathfrak{N}|$. By V, 17.3f) (Clifford's theorem),

$$(\mathsf{U}_1)_{\mathfrak{N}} \cong e \bigoplus_{i=1}^{m} \mathsf{W} \otimes g_i,$$

where $\{g_1, \ldots, g_m\}$ is a transversal of the inertia subgroup of W. By 9.18, *em* divides $|\mathfrak{M}/\mathfrak{N}|$. As U_j is $\mathfrak{G}$-conjugate to U_1, each $(\mathsf{U}_j)_{\mathfrak{N}}$ also splits into *em* irreducible parts. Hence the composition length of $\mathsf{V}_{\mathfrak{N}}$ is *sem*, and this divides $|\mathfrak{G}/\mathfrak{M}|\,|\mathfrak{M}/\mathfrak{N}| = |\mathfrak{G}/\mathfrak{N}|$.

Case 2. Suppose that char $\mathsf{K} = p$ and that $\mathfrak{M}/\mathfrak{N}$ is a p-group. By Clifford's theorem,

$$(\mathsf{U}_1)_{\mathfrak{N}} = \mathsf{X}_1 \oplus \cdots \oplus \mathsf{X}_r,$$

where the X_i are the homogeneous components of $(\mathsf{U}_1)_{\mathfrak{N}}$, and $\mathsf{X}_1, \ldots, \mathsf{X}_r$ are transitively permuted by $\mathfrak{M}$. Let $\mathfrak{T}$ be the stabiliser of X_1; thus $|\mathfrak{M} : \mathfrak{T}| = r$. By V, 17.3e), X_1 is an irreducible $\mathsf{K}\mathfrak{T}$-module. Hence by 9.19, X_1 is an irreducible $\mathsf{K}\mathfrak{N}$-module. Thus the composition length of $(\mathsf{U}_1)_{\mathfrak{N}}$ is r, as is that of each $(\mathsf{U}_j)_{\mathfrak{N}}$. Hence the composition length of $\mathsf{V}_{\mathfrak{N}}$ is sr, and sr divides

$$|\mathfrak{G}/\mathfrak{M}|\,|\mathfrak{M} : \mathfrak{T}| = |\mathfrak{G} : \mathfrak{T}|.$$

Since $\mathfrak{N} \leq \mathfrak{T}$, sr divides $|\mathfrak{G}/\mathfrak{N}|$. **q.e.d.**

9.21 Theorem. *Suppose that $\mathfrak{G}$ is soluble and that K is an algebraically closed field. Let V be an irreducible $\mathsf{K}\mathfrak{G}$-module and let $\mathfrak{A}$ be a normal Abelian subgroup of $\mathfrak{G}$. Then* $\dim_{\mathsf{K}} \mathsf{V}$ *divides* $|\mathfrak{G}/\mathfrak{A}|$ *(cf. V, 17.10).*

Proof. As $\mathfrak{A}$ is Abelian,

$$\mathsf{V}_{\mathfrak{A}} = \mathsf{W}_1 \oplus \cdots \oplus \mathsf{W}_m,$$

where $\dim_{\mathsf{K}} \mathsf{W}_i = 1$. Thus the composition length m of $\mathsf{V}_{\mathfrak{A}}$ is equal to $\dim_{\mathsf{K}} \mathsf{V}$, and by 9.20, this divides $|\mathfrak{G}/\mathfrak{A}|$. **q.e.d.**

9.22 Remark. Theorem 9.21 remains valid under the weaker hypothesis that K is algebraically closed, char $\mathsf{K} = p$ and $\mathfrak{G}$ is p-soluble (DADE [3], SWAN [1]).

9.23 Theorem. *Let K be an algebraically closed field of characteristic p, let $\mathfrak{G}$ be a soluble group and p^a the highest power of p dividing $|\mathfrak{G}|$. If V is an irreducible $\mathsf{K}\mathfrak{G}$-module and p^a divides* $\dim_{\mathsf{K}} \mathsf{V}$, *then V is a projective $\mathsf{K}\mathfrak{G}$-module.*
(This is a partial converse of Dickson's theorem 7.16.)

Proof. We proceed by induction on $|\mathfrak{G}|$. Let $\mathfrak{N}$ be a maximal normal subgroup of $\mathfrak{G}$. Then $|\mathfrak{G}/\mathfrak{N}|$ is a prime.

Case 1. Suppose that $p \neq |\mathfrak{G}/\mathfrak{N}|$. By 9.20, we obtain

$$\mathsf{V}_{\mathfrak{N}} = \mathsf{W}_1 \oplus \cdots \oplus \mathsf{W}_r,$$

where the W_j are irreducible $\mathsf{K}\mathfrak{N}$-modules of the same dimension and r divides $|\mathfrak{G}/\mathfrak{N}|$. Hence p does not divide r, and so p^a divides $|\mathfrak{N}|$ and $\dim_{\mathsf{K}} \mathsf{W}_j$. By induction, each W_j is a projective $\mathsf{K}\mathfrak{N}$-module. Hence V is a projective $\mathsf{K}\mathfrak{G}$-module by 7.11b).

Case 2. Now suppose $|\mathfrak{G}/\mathfrak{N}| = p$. By Clifford's theorem we have

$$\mathsf{V}_{\mathfrak{N}} = e \bigoplus_{j=1}^{m} \mathsf{W}_j,$$

where the W_j are irreducible, non-isomorphic $\mathfrak{G}$-conjugate $\mathsf{K}\mathfrak{N}$-modules. By 9.20, em divides $|\mathfrak{G}/\mathfrak{N}| = p$.

If $\mathsf{V}_{\mathfrak{N}}$ is irreducible, then p^a divides $\dim_{\mathsf{K}} \mathsf{V}_{\mathfrak{N}}$ but not $|\mathfrak{N}|$. This contradicts 9.21. Hence $em = p$. By 9.19, $e = p$ is impossible. Hence $e = 1$ and $m = p$. Then $\mathsf{V} \cong \mathsf{W}_1^{\mathfrak{G}}$ by V, 17.3e). Now p^{a-1} divides $\dim_{\mathsf{K}} \mathsf{W}_1$ and is the highest power of p dividing $|\mathfrak{N}|$. Hence by induction W_1 is a projective $\mathsf{K}\mathfrak{N}$-module. Then V is a projective $\mathsf{K}\mathfrak{G}$-module by 7.17.

q.e.d.

9.24 Remark. 9.23 is not true for arbitrary finite groups.

Let $\mathfrak{G}$ be the first simple group of Janko of order

$$175560 = 2^3 \cdot 3 \cdot 5 \cdot 7 \cdot 11 \cdot 19.$$

Let K be an algebraically closed field of characteristic 2. Then there are two types of irreducible $\mathsf{K}\mathfrak{G}$-modules of dimension $56 = 2^3 \cdot 7$. These are not projective; the corresponding projective modules have dimension $552 = 2^3 \cdot 3 \cdot 23$ (LANDROCK, MICHLER [1]).

Exercises

35) Suppose that $\mathfrak{N} \trianglelefteq \mathfrak{G}$ and that V is an irreducible $\mathsf{K}\mathfrak{N}$-module. Let $\mathfrak{J}$ be the inertia group of V in $\mathfrak{G}$ and suppose that

$$\mathsf{V}^{\mathfrak{J}} = \mathsf{V}_1 \oplus \cdots \oplus \mathsf{V}_r,$$

where the V_i are irreducible $\mathsf{K}\mathfrak{I}$-modules. Show that if W is an irreducible $\mathsf{K}\mathfrak{G}$-module and V is isomorphic to a direct summand of $\mathsf{W}_{\mathfrak{N}}$, then $\mathsf{W} \cong \mathsf{V}_i^{\mathfrak{G}}$ for some i.

a) Prove this by using characters in the case $\mathsf{K} = \mathbb{C}$.

b) Prove it in the general case by arguments similar to those in the proof of 9.6.

36) Show that the statement in Exercise 35 does not remain true if the word "irreducible" is replaced by "indecomposable". (Consider a group of type (p, p) and use representations as in 6.11.)

37) (WILLEMS [1]) Let K be an algebraically closed field and suppose $\mathfrak{N} \trianglelefteq \mathfrak{G}$. Let V be an irreducible $\mathsf{K}\mathfrak{N}$-module and let $\mathfrak{I}$ be the inertia group of V in $\mathfrak{G}$. Then the following statements are equivalent.

a) $\mathsf{V}^{\mathfrak{G}}$ is an irreducible $\mathsf{K}\mathfrak{G}$-module.

b) $\mathsf{V}^{\mathfrak{I}}$ is an irreducible $\mathsf{K}\mathfrak{I}$-module.

c) $\mathfrak{I} = \mathfrak{N}$.

(In the proof of b) $\Rightarrow$ c), assume that $\mathfrak{N} < \mathfrak{I}$ and take a subgroup $\mathfrak{U}$ such that $\mathfrak{N} < \mathfrak{U} \leq \mathfrak{I}$ and $\mathfrak{U}/\mathfrak{N}$ is cyclic. By 9.9, there is a $\mathsf{K}\mathfrak{U}$-module W such that $\mathsf{W}_{\mathfrak{N}} \cong \mathsf{V}$. Consider $\mathsf{V}^{\mathfrak{U}}$.)

38) a) Show that there is no analogue of Corollary 9.13 for indecomposable modules.

b) Suppose that K is an algebraically closed field and $\mathfrak{N} \trianglelefteq \mathfrak{G}$. Let $\mathsf{V}_1, \mathsf{V}_2$ be $\mathsf{K}\mathfrak{G}$-modules such that $(\mathsf{V}_1)_{\mathfrak{N}}$ and $(\mathsf{V}_2)_{\mathfrak{N}}$ are indecomposable and isomorphic. If V_2 is $(\mathfrak{G}, \mathfrak{N})$-projective, there exists a $\mathsf{K}(\mathfrak{G}/\mathfrak{N})$-module W such that $\mathsf{V}_2 \cong \mathsf{V}_1 \otimes_{\mathsf{K}} \mathsf{W}$ and $\dim_{\mathsf{K}} \mathsf{W} = 1$.

39) Let K be a field of characteristic p and $\mathfrak{G} = \langle g \rangle$ a cyclic group of order p^n. Which indecomposable representations of $\langle g^p \rangle$ can be extended to representations of $\mathfrak{G}$?

40) Suppose that K is algebraically closed and that char K does not divide $|\mathfrak{G}/\mathfrak{N}|$. Let V be a $\mathsf{K}\mathfrak{G}$-module and let $\mathsf{W}_1, \ldots, \mathsf{W}_k$ be all the irreducible $\mathsf{K}(\mathfrak{G}/\mathfrak{N})$-modules (to within isomorphism). Suppose that the $\mathsf{V} \otimes_{\mathsf{K}} \mathsf{W}_i$ are irreducible non-isomorphic $\mathsf{K}\mathfrak{G}$-modules. Show that $\mathsf{V}_{\mathfrak{N}}$ is irreducible

a) by character calculations for char $\mathsf{K} = 0$,

b) by module-theoretic considerations in the general case.

(Hint: a) Let ϱ be the regular character of $\mathfrak{G}/\mathfrak{N}$, lifted to $\mathfrak{G}$, and calculate $(\chi, \chi\varrho)_{\mathfrak{G}}$, where χ is the character of V.

b) On the one hand

$$(\mathsf{V}_{\mathfrak{N}})^{\mathfrak{G}} \cong \mathsf{V} \otimes_{\mathsf{K}} \mathsf{K}(\mathfrak{G}/\mathfrak{N}) \cong \bigoplus_{i=1}^{k} (\dim_{\mathsf{K}} \mathsf{W}_i)(\mathsf{V} \otimes_{\mathsf{K}} \mathsf{W}_i),$$

where V itself appears with multiplicity only 1. On the other hand, if $\mathsf{V}_{\mathfrak{N}} = \mathsf{V}_1 \oplus \cdots \oplus \mathsf{V}_s$ for irreducible $\mathsf{K}\mathfrak{N}$-modules V_i, then V appears in each (completely reducible) $\mathsf{V}_i^{\mathfrak{G}}$ with multiplicity at least 1.)

41) Give an example to show that Theorem 9.9b) is not true in general without the assumption that $\mathfrak{G}/\mathfrak{N}$ is a p'-group.

42) Let $\mathsf{K} = GF(2)$ and consider $\mathsf{V} = GF(4)$ as an irreducible $\mathsf{K}\mathfrak{Z}$-module, where $\mathfrak{Z}$ is the cyclic group of order 3. Show that $\mathsf{V} \otimes_{\mathsf{K}} \mathsf{V}$ is not an indecomposable $\mathsf{K}(\mathfrak{Z} \times \mathfrak{Z})$-module.

43) Suppose that $\mathsf{K}\mathfrak{G} = \bigoplus_i \mathsf{P}_i$, where the P_i are indecomposable projective $\mathsf{K}\mathfrak{G}$-modules. Show that

$$\dim_{\mathsf{K}} \bigoplus_{i,j,k,l} \mathrm{Hom}_{\mathsf{K}\mathfrak{G}}(\mathsf{P}_i \otimes_{\mathsf{K}} \mathsf{P}_j, \mathsf{P}_k \otimes_{\mathsf{K}} \mathsf{P}_l) = |\mathfrak{G}|^3$$

and

$$\dim_{\mathsf{K}} \bigoplus_{i,j,k,l} \mathrm{Hom}_{\mathsf{K}\mathfrak{G}}(\mathsf{P}_i, \mathsf{P}_k) \otimes_{\mathsf{K}} \mathrm{Hom}_{\mathsf{K}\mathfrak{G}}(\mathsf{P}_j, \mathsf{P}_l) = |\mathfrak{G}|^2.$$

Hence there exist indices i,j,k,l such that

$$\mathrm{Hom}_{\mathsf{K}\mathfrak{G}}(\mathsf{P}_i, \mathsf{P}_k) \otimes_{\mathsf{K}} \mathrm{Hom}_{\mathsf{K}\mathfrak{G}}(\mathsf{P}_j, \mathsf{P}_l) \subset \mathrm{Hom}_{\mathsf{K}\mathfrak{G}}(\mathsf{P}_i \otimes_{\mathsf{K}} \mathsf{P}_j, \mathsf{P}_k \otimes_{\mathsf{K}} \mathsf{P}_l).$$

44) Suppose $\mathfrak{N} \trianglelefteq \mathfrak{G}$. Let V be a $\mathsf{K}\mathfrak{G}$-module such that $\mathsf{V}_{\mathfrak{N}}$ is absolutely indecomposable. Then the following assertions are equivalent.

a) $\mathsf{V}_{\mathfrak{N}}$ is a projective $\mathsf{K}\mathfrak{N}$-module and char K does not divide $|\mathfrak{G}/\mathfrak{N}|$.
b) V is a projective $\mathsf{K}\mathfrak{G}$-module.

(Use 9.12.)

§ 10. One-Sided Decompositions of the Group-Ring

Let $\mathfrak{A}$ be an algebra of finite dimension over a field K. In the sequel we write $\overline{\mathfrak{A}} = \mathfrak{A}/\mathbf{J}(\mathfrak{A})$. As $\overline{\mathfrak{A}}$ is semisimple, it follows from V, 3.3 that there is a decomposition

$$\overline{1} = \overline{e}_1 + \cdots + \overline{e}_n \tag{1}$$

of the unit element $\overline{1}$ of $\overline{\mathfrak{A}}$ into mutually orthogonal idempotents $\overline{e}_i$, such that

$$\overline{\mathfrak{A}} = \overline{e}_1 \overline{\mathfrak{A}} \oplus \cdots \oplus \overline{e}_n \overline{\mathfrak{A}}$$

is a direct decomposition of $\overline{\mathfrak{A}}$ into irreducible $\overline{\mathfrak{A}}$-modules $\overline{e}_i \overline{\mathfrak{A}}$. We now wish to obtain from (1) a decomposition

$$1 = e_1 + \cdots + e_n,$$

where $e_1, \ldots, e_n$ are mutually orthogonal idempotents of $\mathfrak{A}$ such that $\overline{e}_i = e_i + \mathbf{J}(\mathfrak{A})$. This process is known as "lifting" the idempotents $\overline{e}_i$. With an application in mind in which $\mathfrak{A}$ will not be a ring with minimum condition but the group-ring of a finite group over a ring of p-adic integers, we proceed at first rather more generally than is necessary for the purpose of this section.

10.1 Lemma. *Let $\mathfrak{R}$ be a ring and let $\mathfrak{J}$ be a two-sided ideal of $\mathfrak{R}$ for which $\bigcap_{n=1}^{\infty} \mathfrak{J}^n = 0$. We put $\mathfrak{J}^0 = \mathfrak{R}$. For $a, b \in \mathfrak{R}$ we define*

$$d(a, b) = \begin{cases} 0 & \text{if } a = b \\ 2^{-k} & \text{if } a - b \in \mathfrak{J}^k, \text{ but } a - b \notin \mathfrak{J}^{k+1}. \end{cases}$$

a) *d is a metric on $\mathfrak{R}$ and the so-called ultrametric inequality*

$$d(a, b) \leq \max(d(a, c), d(c, b))$$

holds for all $a, b, c \in \mathfrak{R}$. We call the Hausdorff topology defined on $\mathfrak{R}$ by d the $\mathfrak{J}$-adic topology on $\mathfrak{R}$.

b) *With the topology defined in a), $\mathfrak{R}$ is a topological ring; that is, the mappings $(a, b) \to a + b, a - b, ab$ are continuous. We call the topological ring $\mathfrak{R}$ an $\mathfrak{J}$-adic ring.*

c) *The sequence (a_n) of elements $a_n \in \mathfrak{R}$ is a Cauchy sequence if and only if $(a_{n+1} - a_n)$ is a null sequence.*

Proof. a) It is obvious that for $a, b \in \mathfrak{R}$, we have $d(a, b) \geq 0$, $d(a, b) = d(b, a)$; and $d(a, b) = 0$ is equivalent to

$$a - b \in \bigcap_{n=1}^{\infty} \mathfrak{J}^n = 0.$$

Suppose that $d(a, c) = 2^{-m}$ and $d(c, b) = 2^{-n}$, where $n \leq m$. Then $a - c \in \mathfrak{J}^m$ and $c - b \in \mathfrak{J}^n$. Since $\mathfrak{J}^m \subseteq \mathfrak{J}^n$, it follows that

$$a - b = (a - c) + (c - b) \in \mathfrak{J}^m + \mathfrak{J}^n = \mathfrak{J}^n.$$

Hence

$$d(a, b) \leq 2^{-n} = \max(d(a, c), d(c, b)) \leq d(a, c) + d(c, b)$$

and thus d is a metric on $\mathfrak{R}$.

b) If $d(a, a_0) \leq 2^{-n}$ and $d(b, b_0) \leq 2^{-n}$, then

$$(a \pm b) - (a_0 \pm b_0) = (a - a_0) \pm (b - b_0) \in \mathfrak{J}^n$$

and

$$ab - a_0 b_0 = (a - a_0)b + a_0(b - b_0) \in \mathfrak{J}^n\mathfrak{R} + \mathfrak{R}\mathfrak{J}^n = \mathfrak{J}^n.$$

Thus addition, subtraction and multiplication are continuous operations, and $\mathfrak{R}$ is a topological ring.

c) Suppose that $(a_{n+1} - a_n)$ is a null sequence, that is,

$$a_{n+1} - a_n \in \mathfrak{J}^k \quad \text{for all } n \geq n(k).$$

For $n \geq n(k)$ and $m \geq 0$ we obtain

$$a_{n+m} - a_n = \sum_{i=n}^{m+n-1} (a_{i+1} - a_i) \in \mathfrak{J}^k,$$

so $d(a_{n+m}, a_n) \leq 2^{-k}$. Thus (a_n) is a Cauchy sequence in $\mathfrak{R}$. **q.e.d.**

We come now to the lifting of idempotents.

10.2 Theorem. *Let $\mathfrak{R}$ be an $\mathfrak{J}$-adic ring which is complete with respect to the $\mathfrak{J}$-adic topology. We put $\overline{\mathfrak{R}} = \mathfrak{R}/\mathfrak{J}$.*

a) *There exist polynomials p_n $(n = 0, 1, \ldots)$ with rational integral coefficients such that $p_n(0) = 0$, having the following properties.*

If $\bar{e} = e_0 + \mathfrak{J}$ is an idempotent in $\overline{\mathfrak{R}}$, the limit e of the sequence $(p_n(e_0))$ exists, and e is an idempotent in $\mathfrak{R}$ for which $e - e_0 \in \mathfrak{J}$. If $a \in \mathfrak{R}$ and $e_0 a = a e_0$ (or $e_0 a = 0$ or $a e_0 = 0$), then $ea = ae$ (or $ea = 0$ or $ae = 0$).

b) *Suppose that $\bar{1} = \bar{e}_1 + \cdots + \bar{e}_n$, where $\bar{e}_i \in \overline{\mathfrak{R}}$, $\bar{e}_i^2 = \bar{e}_i$ and*

$\bar{e}_i \bar{e}_j = 0$ *for* $i \neq j$. *Then there exist mutually orthogonal idempotents* e_i *in* $\mathfrak{R}$ *such that* $1 = e_1 + \cdots + e_n$ *and* $e_i + \mathfrak{I} = \bar{e}_i$.

Proof. a) We define polynomials p_n in $\mathbb{Z}[t]$ recursively by $p_0(t) = t$ and

$$p_n = 3p_{n-1}^2 - 2p_{n-1}^3$$

for $n > 0$. Then $p_n(0) = 0$ for all n. We have

$$p_0(e_0) - e_0 = 0 \in \mathfrak{I} \quad \text{and} \quad p_0(e_0)^2 - p_0(e_0) = e_0^2 - e_0 \in \mathfrak{I}.$$

We prove by induction on n that

$$p_n(e_0) - e_0 \in \mathfrak{I} \quad \text{and} \quad p_n(e_0)^2 - p_n(e_0) \in \mathfrak{I}^{2^n}.$$

This is true for $n = 0$. For $n > 0$, we have

$$\begin{aligned} p_n(e_0) - e_0 &= 3p_{n-1}(e_0)^2 - 2p_{n-1}(e_0)^3 - e_0 \\ &\equiv 3e_0^2 - 2e_0^3 - e_0 \equiv 0 \mod \mathfrak{I}, \end{aligned}$$

and

$$\begin{aligned} &p_n(e_0)^2 - p_n(e_0) \\ &= 4(p_{n-1}(e_0)^2 - p_{n-1}(e_0))^3 - 3(p_{n-1}(e_0)^2 - p_{n-1}(e_0))^2 \\ &\in (\mathfrak{I}^{2^{n-1}})^3 + (\mathfrak{I}^{2^{n-1}})^2 = \mathfrak{I}^{2^n}. \end{aligned}$$

Thus

$$p_{n+1}(e_0) - p_n(e_0) = (1 - 2p_n(e_0))(p_n(e_0)^2 - p_n(e_0)) \in \mathfrak{I}^{2^n}.$$

Hence by 10.1, $(p_n(e_0))$ is a Cauchy sequence. Since by hypothesis $\mathfrak{R}$ is complete, the limit e of this sequence exists. Since $\mathfrak{R}$ is a topological ring and

$$p_n(e_0)^2 - p_n(e_0) \in \mathfrak{I}^{2^n},$$

we obtain

$$e^2 - e = \lim_{n \to \infty} (p_n(e_0)^2 - p_n(e_0)) = 0.$$

Thus $e^2 = e$. Since the set

$$\mathfrak{I} = \{a | d(a, 0) \leq \tfrac{1}{2}\}$$

is closed, it follows from $p_n(e_0) - e_0 \in \mathfrak{I}$ that $e - e_0 \in \mathfrak{I}$.

If $e_0 a = a e_0$, then $p_n(e_0)a = ap_n(e_0)$ for all n, so

$$ea = \lim_{n\to\infty} p_n(e_0)a = \lim_{n\to\infty} ap_n(e_0) = ae.$$

Since $p_n(0) = 0$, it follows similarly from $e_0 a = 0$ that $ea = 0$ and from $ae_0 = 0$ that $ae = 0$.

b) We construct a set $e_1, \ldots, e_m$ of mutually orthogonal idempotents e_i in $\mathfrak{R}$ such that $e_i + \mathfrak{I} = \bar{e}_i$ for $1 \leq m < n$ by induction on m. For $m = 1$ this is done by applying a). For $m > 1$, there exists a set of mutually orthogonal idempotents $e_1, \ldots, e_{m-1}$ for which $e_i + \mathfrak{I} = \bar{e}_i$, by the inductive hypothesis. If $\bar{e}_m = a + \mathfrak{I}$ (with $a \in \mathfrak{R}$), put

$$\begin{aligned} b = a - a(e_1 + \cdots + e_{m-1}) - (e_1 + \cdots + e_{m-1})a \\ + (e_1 + \cdots + e_{m-1})a(e_1 + \cdots + e_{m-1}). \end{aligned}$$

Then $e_j b = be_j = 0$ for $1 \leq j \leq m - 1$ on account of the mutual orthogonality of the e_i. Since $\bar{e}_j\bar{e}_m = \bar{e}_m\bar{e}_j = 0$ for $1 \leq j \leq m - 1$, we have $e_j a \in \mathfrak{I}$ and $ae_j \in \mathfrak{I}$. Thus $b + \mathfrak{I} = a + \mathfrak{I} = \bar{e}_m$. Hence by a), there exists an indempotent e_m in $\mathfrak{R}$ for which $e_m - b \in \mathfrak{I}$ and $e_j e_m = e_m e_j = 0$ for $1 \leq j \leq m - 1$.

We apply this result with $m = n - 1$ and then put

$$e_n = 1 - (e_1 + \cdots + e_{n-1}).$$

Then $e_n e_j = e_j e_n = 0$ for $j < n$ and $e_n^2 = e_n$. Also

$$e_n + \mathfrak{I} = \bar{1} - (\bar{e}_1 + \cdots + \bar{e}_{n-1}) = \bar{e}_n. \qquad \textbf{q.e.d.}$$

From 10.2 we deduce a basic theorem about direct decompositions of not necessarily semisimple algebras.

10.3 Theorem. *Let* $\mathfrak{A}$ *be an algebra of finite dimension over a field* K. *We put* $\overline{\mathfrak{A}} = \mathfrak{A}/\mathbf{J}(\mathfrak{A})$. *Let*

$$\overline{\mathfrak{A}} = \bar{e}_1\overline{\mathfrak{A}} \oplus \cdots \oplus \bar{e}_n\overline{\mathfrak{A}}$$

be a direct decomposition of $\overline{\mathfrak{A}}$ *into minimal right ideals* $\bar{e}_i\overline{\mathfrak{A}}$, *where the* $\bar{e}_i$ *are mutually orthogonal idempotents in* $\overline{\mathfrak{A}}$ *for which* $\bar{1} = \bar{e}_1 + \cdots + \bar{e}_n$.

a) *There exist mutually orthogonal idempotents* e_i *in* $\mathfrak{A}$ *such that*

$$e_i + \mathbf{J}(\mathfrak{A}) = \bar{e}_i \quad \textit{and} \quad 1 = e_1 + \cdots + e_n.$$

b) $\mathfrak{A} = e_1\mathfrak{A} \oplus \cdots \oplus e_n\mathfrak{A}$.

c) $e_i\mathbf{J}(\mathfrak{A})$ *is the only maximal submodule of* $e_i\mathfrak{A}$, *and* $e_i\mathfrak{A}/e_i\mathbf{J}(\mathfrak{A}) \cong \bar{e}_i\overline{\mathfrak{A}}$. *Hence* $e_i\mathfrak{A}$ *is an indecomposable projective* $\mathfrak{A}$-*module the head of which is irreducible and isomorphic to* $\bar{e}_i\overline{\mathfrak{A}}$.

Proof. By V, 2.4, $\mathbf{J}(\mathfrak{A})$ is nilpotent. Thus the $\mathbf{J}(\mathfrak{A})$-adic topology of $\mathfrak{A}$ introduced in 10.1 is discrete, and $\mathfrak{A}$ is trivially complete. Hence a) follows from 10.2. The assertion b) follows at once from a).

If $a \in e_i\mathfrak{A} \cap \mathbf{J}(\mathfrak{A})$, then $a = e_ib$ for some $b \in \mathfrak{A}$. Thus

$$a = e_ib = e_i^2b = e_ia \in e_i\mathbf{J}(\mathfrak{A}).$$

Hence $e_i\mathfrak{A} \cap \mathbf{J}(\mathfrak{A}) = e_i\mathbf{J}(\mathfrak{A})$. Therefore

$$\begin{aligned} e_i\mathfrak{A}/e_i\mathbf{J}(\mathfrak{A}) &= e_i\mathfrak{A}/(e_i\mathfrak{A} \cap \mathbf{J}(\mathfrak{A})) \cong (e_i\mathfrak{A} + \mathbf{J}(\mathfrak{A}))/\mathbf{J}(\mathfrak{A}) \\ &= (e_i + \mathbf{J}(\mathfrak{A}))(\mathfrak{A}/\mathbf{J}(\mathfrak{A})) = \bar{e}_i\overline{\mathfrak{A}}. \end{aligned}$$

As $\bar{e}_i\overline{\mathfrak{A}}$ is an irreducible $\mathfrak{A}$-module, $e_i\mathbf{J}(\mathfrak{A})$ is certainly a maximal submodule of $e_i\mathfrak{A}$. If $\mathfrak{M}$ is any maximal submodule of $e_i\mathfrak{A}$, then $e_i\mathfrak{A}/\mathfrak{M}$ is irreducible and is therefore annihilated by $\mathbf{J}(\mathfrak{A})$. It follows that

$$e_i\mathbf{J}(\mathfrak{A}) = (e_i\mathfrak{A})\mathbf{J}(\mathfrak{A}) \subseteq \mathfrak{M},$$

so $e_i\mathbf{J}(\mathfrak{A})$ is the only maximal submodule of $e_i\mathfrak{A}$. This implies the indecomposability of $e_i\mathfrak{A}$, since a decomposable module possesses more than one maximal submodule. **q.e.d.**

We now need an elementary lemma about finitely generated modules.

10.4 Lemma. *Let* $\mathfrak{R}$ *be any ring with unit element and* V *a finitely generated* $\mathfrak{R}$-*module.*

a) *Every proper submodule* W *of* V *is contained in a maximal submodule of* V.

b) *(Nakayama's lemma; cf. vol. I, p. 643) If* W *is an* $\mathfrak{R}$-*submodule of* V *for which* $\mathsf{V} = \mathsf{W} + \mathsf{V}\mathbf{J}(\mathfrak{R})$, *then* $\mathsf{W} = \mathsf{V}$.

Proof. a) Suppose that $\mathsf{V} = \sum_{i=1}^{k} v_i\mathfrak{R}$. Consider the set

$$\mathscr{S} = \{\mathsf{X} \mid \mathsf{X} \text{ is a submodule of } \mathsf{V} \text{ and } \mathsf{W} \subseteq \mathsf{X} \subset \mathsf{V}\}.$$

As $\mathsf{W} \in \mathscr{S}$, $\mathscr{S}$ is not empty. We show that $\mathscr{S}$ is inductive. To do this, let $\mathscr{H}$ be a chain in $\mathscr{S}$. Then $\mathsf{Y} = \bigcup_{\mathfrak{X} \in \mathscr{H}} \mathsf{X}$ is a submodule of V and $\mathsf{W} \subseteq \mathsf{Y}$. Suppose that $\mathsf{Y} = \mathsf{V}$. Then for every $i = 1, \dots, k$, there exists $\mathsf{X}_i \in \mathscr{H}$ such that $v_i \in \mathsf{X}_i$. As $\mathscr{H}$ is a chain, some X_j contains X_i for all $i = 1, \dots, k$, so $\mathsf{V} \subseteq \mathsf{X}_i$, a contradiction. Hence $\mathscr{S}$ is inductive. Thus by Zorn's lemma, $\mathscr{S}$ has maximal elements, and these are maximal submodules of V containing W.

b) Suppose that $\mathsf{W} \subset \mathsf{V}$. By a) there exists a maximal submodule X of V such that $\mathsf{W} \subseteq \mathsf{X}$. Then V/X is an irreducible $\mathfrak{R}$-module and is therefore annihilated by $\mathbf{J}(\mathfrak{R})$. Hence $\mathsf{V}\mathbf{J}(\mathfrak{R}) \subseteq \mathsf{X}$. Thus

$$\mathsf{X} \supseteq \mathsf{W} + \mathsf{V}\mathbf{J}(\mathfrak{R}) = \mathsf{V}.$$

This is a contradiction, hence $\mathsf{W} = \mathsf{V}$. **q.e.d.**

10.5 Lemma. *Let $\mathfrak{R}$ be an arbitrary ring with unit element.*

a) *Suppose that P is a projective $\mathfrak{R}$-module and V is a finitely generated $\mathfrak{R}$-module. If there exists an $\mathfrak{R}$-epimorphism α of $\mathsf{P}/\mathsf{P}\mathbf{J}(\mathfrak{R})$ onto $\mathsf{V}/\mathsf{V}\mathbf{J}(\mathfrak{R})$, then there is an $\mathfrak{R}$-epimorphism γ of P onto V such that*

$$x\gamma + \mathsf{V}\mathbf{J}(\mathfrak{R}) = (x + \mathsf{P}\mathbf{J}(\mathfrak{R}))\alpha$$

for all $x \in \mathsf{P}$.

b) *If P is a finitely generated projective $\mathfrak{R}$-module, $\beta \in \operatorname{Hom}_{\mathfrak{R}}(\mathsf{P}, \mathsf{P})$ and $x\beta - x \in \mathsf{P}\mathbf{J}(\mathfrak{R})$ for all $x \in \mathsf{P}$, then β is an automorphism of P.*

Proof. a) Let ν_{V}, ν_{P} denote the natural epimorphisms of V onto $\mathsf{V}/\mathsf{V}\mathbf{J}(\mathfrak{R})$ and P onto $\mathsf{P}/\mathsf{P}\mathbf{J}(\mathfrak{R})$ respectively; thus

$$v\nu_{\mathsf{V}} = v + \mathsf{V}\mathbf{J}(\mathfrak{R}) \quad (v \in \mathsf{V}) \quad \text{and} \quad y\nu_{\mathsf{P}} = y + \mathsf{P}\mathbf{J}(\mathfrak{R}) \quad (y \in \mathsf{P}).$$

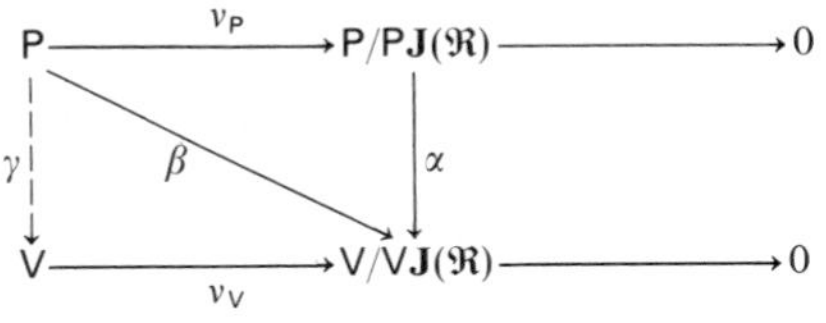

Then $\beta = \nu_{\mathsf{P}}\alpha$ is an epimorphism of P onto $\mathsf{V}/\mathsf{V}\mathbf{J}(\mathfrak{R})$. Since P is projective, there exists $\gamma \in \operatorname{Hom}_{\mathfrak{R}}(\mathsf{P}, \mathsf{V})$ such that $\gamma\nu_{\mathsf{V}} = \beta = \nu_{\mathsf{P}}\alpha$. Now β is an epimorphism; thus, given $v \in \mathsf{V}$, there exists $y \in \mathsf{P}$ such that

$$v + \mathsf{V}\mathbf{J}(\mathfrak{R}) = y\beta = y\gamma v_{\mathsf{V}} = y\gamma + \mathsf{V}\mathbf{J}(\mathfrak{R}).$$

Thus $\mathsf{V} = \mathsf{P}\gamma + \mathsf{V}\mathbf{J}(\mathfrak{R})$. Since V is finitely generated, it follows from 10.4b) that $\mathsf{V} = \mathsf{P}\gamma$. Thus γ is an epimorphism of P onto V.

b) It follows from the hypothesis that $\mathsf{P} = \mathsf{P}\beta + \mathsf{P}\mathbf{J}(\mathfrak{R})$. Since P is finitely generated, we get $\mathsf{P} = \mathsf{P}\beta$ by 10.4b). Thus β is an epimorphism. Hence $\mathsf{P}/\ker\beta \cong \mathsf{P}$ is projective and we obtain $\mathsf{P} = \ker\beta \oplus \mathsf{P}'$ for some submodule P' isomorphic to P. Thus

$$\mathsf{P}\mathbf{J}(\mathfrak{R}) = (\ker\beta)\mathbf{J}(\mathfrak{R}) \oplus \mathsf{P}'\mathbf{J}(\mathfrak{R}).$$

Since $\ker\beta \cap \mathsf{P}' = 0$, it follows by Dedekind's identity that

$$\mathsf{P}\mathbf{J}(\mathfrak{R}) \cap \ker\beta = (\ker\beta)\mathbf{J}(\mathfrak{R}).$$

Hence for $x \in \ker\beta$ we have

$$x = x - x\beta \in \mathsf{P}\mathbf{J}(\mathfrak{R}) \cap \ker\beta = (\ker\beta)\mathbf{J}(\mathfrak{R}).$$

Thus $\ker\beta = (\ker\beta)\mathbf{J}(\mathfrak{R})$. But $\ker\beta$ is finitely generated since it is a direct summand of P. Thus by 10.4b), $\ker\beta = 0$ and β is a monomorphism. Hence β is an isomorphism of P onto P. **q.e.d.**

10.6 Theorem. *Let $\mathfrak{R}$ be an arbitrary ring with unit element and suppose that P_1, P_2 are finitely generated projective $\mathfrak{R}$-modules. Then*

$$\mathsf{P}_1/\mathsf{P}_1\mathbf{J}(\mathfrak{R}) \cong \mathsf{P}_2/\mathsf{P}_2\mathbf{J}(\mathfrak{R})$$

if and only if $\mathsf{P}_1 \cong \mathsf{P}_2$.

Proof. If $\mathsf{P}_1 \cong \mathsf{P}_2$, then certainly $\mathsf{P}_1/\mathsf{P}_1\mathbf{J}(\mathfrak{R}) \cong \mathsf{P}_2/\mathsf{P}_2\mathbf{J}(\mathfrak{R})$.

Now suppose that α is an isomorphism of $\mathsf{P}_1/\mathsf{P}_1\mathbf{J}(\mathfrak{R})$ onto $\mathsf{P}_2/\mathsf{P}_2\mathbf{J}(\mathfrak{R})$. We apply 10.5a) to α and to α^{-1}. Thus there exists an epimorphism β of P_1 onto P_2 such that

$$x\beta + \mathsf{P}_2\mathbf{J}(\mathfrak{R}) = (x + \mathsf{P}_1\mathbf{J}(\mathfrak{R}))\alpha$$

for all $x \in \mathsf{P}_1$, and there exists an epimorphism γ of P_2 onto P_1 such that

$$y\gamma + \mathsf{P}_1\mathbf{J}(\mathfrak{R}) = (y + \mathsf{P}_2\mathbf{J}(\mathfrak{R}))\alpha^{-1}$$

for all $y \in \mathsf{P}_2$. Thus if $x \in \mathsf{P}_1$,

$$x\beta\gamma + \mathsf{P}_1\mathbf{J}(\mathfrak{R}) = (x\beta + \mathsf{P}_2\mathbf{J}(\mathfrak{R}))\alpha^{-1} = x + \mathsf{P}_1\mathbf{J}(\mathfrak{R}).$$

It follows from 10.5b) that $\beta\gamma$ is an automorphism of P_1. Hence β is a monomorphism and indeed an isomorphism of P_1 onto P_2. **q.e.d.**

10.7 Definition. a) Let $\mathfrak{R}$ be a ring and V an $\mathfrak{R}$-module. If S is a subset of V, we define the annihilator of S by

$$\mathbf{An}_{\mathfrak{R}}(\mathsf{S}) = \{r \mid r \in \mathfrak{R},\ \mathsf{S}r = 0\}.$$

Obviously $\mathbf{An}_{\mathfrak{R}}(\mathsf{S})$ is a right ideal in $\mathfrak{R}$.

If $\mathfrak{S}$ is a subset of $\mathfrak{R}$, we put

$$\mathbf{An}_{\mathsf{V}}(\mathfrak{S}) = \{v \mid v \in \mathsf{V},\ v\mathfrak{S} = 0\}.$$

If $\mathfrak{S}$ is a left ideal of $\mathfrak{R}$, then $\mathbf{An}_{\mathsf{V}}(\mathfrak{S})$ is an $\mathfrak{R}$-submodule of V.

b) If $\mathfrak{S}$ is a subset of the ring $\mathfrak{R}$, we put

$$\mathbf{R}(\mathfrak{S}) = \{r \mid r \in \mathfrak{R},\ \mathfrak{S}r = 0\}$$

and

$$\mathbf{L}(\mathfrak{S}) = \{r \mid r \in \mathfrak{R},\ r\mathfrak{S} = 0\}.$$

Obviously $\mathbf{R}(\mathfrak{S})$ is a right ideal and $\mathbf{L}(\mathfrak{S})$ is a left ideal in $\mathfrak{R}$. If $\mathfrak{S}$ is a right ideal of $\mathfrak{R}$, then $\mathbf{R}(\mathfrak{S})$ is a two-sided ideal of $\mathfrak{R}$; if $\mathfrak{S}$ is a left ideal of $\mathfrak{R}$, then $\mathbf{L}(\mathfrak{S})$ is a two-sided ideal of $\mathfrak{R}$. If $\mathfrak{S}_1$, $\mathfrak{S}_2$ are additive subgroups of $\mathfrak{R}$, then

$$\mathbf{R}(\mathfrak{S}_1 + \mathfrak{S}_2) = \mathbf{R}(\mathfrak{S}_1) \cap \mathbf{R}(\mathfrak{S}_2)$$

and

$$\mathbf{R}(\mathfrak{S}_1 \cap \mathfrak{S}_2) \supseteq \mathbf{R}(\mathfrak{S}_1) + \mathbf{R}(\mathfrak{S}_2);$$

analogous statements hold for the left annihilators.

c) Let $\mathfrak{A}$ be an algebra of finite dimension over a field K. We denote by $\mathbf{S}_r(\mathfrak{A})$ the socle of the right $\mathfrak{A}$-module $\mathfrak{A}$ and by $\mathbf{S}_l(\mathfrak{A})$ the socle of the left $\mathfrak{A}$-module $\mathfrak{A}$. Thus by 1.6, $\mathbf{S}_l(\mathfrak{A}) = \mathbf{R}(\mathbf{J}(\mathfrak{A}))$ and $\mathbf{S}_r(\mathfrak{A}) = \mathbf{L}(\mathbf{J}(\mathfrak{A}))$. By the remarks in b), $\mathbf{S}_r(\mathfrak{A})$ and $\mathbf{S}_l(\mathfrak{A})$ are both two-sided ideals of $\mathfrak{A}$.

10.8 Remark. If $\mathfrak{A}$ is an algebra of finite dimension, then $\mathfrak{A}/\mathbf{J}(\mathfrak{A})$ is the largest completely reducible (right or left) factor module of $\mathfrak{A}$. But it is not true in general that $\mathbf{S}_r(\mathfrak{A}) = \mathbf{S}_l(\mathfrak{A})$; this is shown by the example of the ring of all triangular matrices over a field (see Exercise 45). If however $\mathfrak{A}$ is the group-ring of a finite group over a field, then $\mathbf{S}_r(\mathfrak{A}) = \mathbf{S}_l(\mathfrak{A})$, as we shall see in 11.6b).

We now study the projective modules over an algebra.

10.9 Theorem. *Let $\mathfrak{A}$ be an algebra of finite dimension over a field* K.

a) *Any finitely generated projective $\mathfrak{A}$-module P is determined to within isomorphism by its head $\mathsf{P}/\mathsf{P}\mathbf{J}(\mathfrak{A})$.*

b) *If V is a finitely generated completely reducible $\mathfrak{A}$-module, there exists a finitely generated projective $\mathfrak{A}$-module P such that $\mathsf{V} \cong \mathsf{P}/\mathsf{P}\mathbf{J}(\mathfrak{A})$.*

c) *A finitely generated projective $\mathfrak{A}$-module P is indecomposable if and only if $\mathsf{P}/\mathsf{P}\mathbf{J}(\mathfrak{A})$ is irreducible. Any finitely generated projective indecomposable $\mathfrak{A}$-module is isomorphic to one of the right ideals $e_i\mathfrak{A}$ of 10.3.*

Proof. a) This is 10.6.

b) Since

$$(\mathsf{P}_1 \oplus \mathsf{P}_2)/(\mathsf{P}_1 \oplus \mathsf{P}_2)\mathbf{J}(\mathfrak{A}) \cong \mathsf{P}_1/\mathsf{P}_1\mathbf{J}(\mathfrak{A}) \oplus \mathsf{P}_2/\mathsf{P}_2\mathbf{J}(\mathfrak{A}),$$

we can assume that V is irreducible. Since $\mathsf{V}\mathbf{J}(\mathfrak{A}) = 0$, V may be regarded as an irreducible $\overline{\mathfrak{A}}$-module, where $\overline{\mathfrak{A}} = \mathfrak{A}/\mathbf{J}(\mathfrak{A})$. Let

$$\overline{\mathfrak{A}} = \bar{e}_1\overline{\mathfrak{A}} \oplus \cdots \oplus \bar{e}_n\overline{\mathfrak{A}}$$

be a direct decomposition of $\overline{\mathfrak{A}}$ into minimal right ideals $\bar{e}_i\overline{\mathfrak{A}}$, where the $\bar{e}_i$ are mutually orthogonal idempotents in $\overline{\mathfrak{A}}$ for which $\bar{1} = \bar{e}_1 + \cdots + \bar{e}_n$. Thus $\mathsf{V} \cong \bar{e}_i\overline{\mathfrak{A}}$ for some i. But by 10.3c), $\bar{e}_i\overline{\mathfrak{A}} \cong e_i\mathfrak{A}/e_i\mathbf{J}(\mathfrak{A})$ for the indecomposable projective $\mathfrak{A}$-module $e_i\mathfrak{A}$.

c) Suppose that $\mathsf{P}/\mathsf{P}\mathbf{J}(\mathfrak{A})$ is reducible. Since $\mathsf{P}/\mathsf{P}\mathbf{J}(\mathfrak{A})$ is completely reducible, there exist non-zero completely reducible modules V_1, V_2 such that

$$\mathsf{P}/\mathsf{P}\mathbf{J}(\mathfrak{A}) = \mathsf{V}_1 \oplus \mathsf{V}_2.$$

By b), $\mathsf{V}_i \cong \mathsf{P}_i/\mathsf{P}_i\mathbf{J}(\mathfrak{A})$ for some projective module P_i. Hence

$$\mathsf{P}/\mathsf{P}\mathbf{J}(\mathfrak{A}) \cong (\mathsf{P}_1 \oplus \mathsf{P}_2)/(\mathsf{P}_1 \oplus \mathsf{P}_2)\mathbf{J}(\mathfrak{A}).$$

By a), $\mathsf{P} \cong \mathsf{P}_1 \oplus \mathsf{P}_2$, so P is decomposable.

If $\mathsf{P}/\mathsf{P}\mathbf{J}(\mathfrak{A})$ is irreducible, then $\mathsf{P}\mathbf{J}(\mathfrak{A})$ is the only maximal submodule of P, so P is indecomposable.

If P is a projective indecomposable finitely generated $\mathfrak{A}$-module, $\mathsf{P}/\mathsf{P}\mathbf{J}(\mathfrak{A})$ is irreducible. As in b), we have

$$\mathsf{P}/\mathsf{P}\mathbf{J}(\mathfrak{A}) \cong \bar{e}_i\bar{\mathfrak{A}} \cong e_i\mathfrak{A}/e_i\mathbf{J}(\mathfrak{A})$$

for some i, so $\mathsf{P} \cong e_i\mathfrak{A}$ by a). **q.e.d.**

The iteration of the formation of head and socle of a module brings us to the Loewy series.

10.10 Definition. Suppose that $\mathfrak{A}$ is an algebra of finite dimension and V an $\mathfrak{A}$-module.

a) The *lower Loewy series*

$$\mathsf{V} = \mathsf{V}_0 \supseteq \mathsf{V}_1 \supseteq \mathsf{V}_2 \supseteq \cdots$$

of V is defined by $\mathsf{V}_n = \mathsf{V}\mathbf{J}(\mathfrak{A})^n$ $(n = 0, 1, \ldots)$.

b) Dually, the *upper Loewy series*

$$0 = \mathsf{S}_0 \subseteq \mathsf{S}_1 \subseteq \mathsf{S}_2 \subseteq \cdots$$

of V is defined by $\mathsf{S}_n/\mathsf{S}_{n-1} = \mathbf{S}(\mathsf{V}/\mathsf{S}_{n-1})$ for $n > 0$.

10.11 Theorem. *Let $\mathfrak{A}$ be an algebra of finite dimension over K and V an $\mathfrak{A}$-module.*

a) *There exists an integer k (called the Loewy length of V) such that*

$$\mathsf{V}\mathbf{J}(\mathfrak{A})^{k-1} \supset \mathsf{V}\mathbf{J}(\mathfrak{A})^k = 0$$

and

$$\mathsf{V} = \mathsf{V}_0 \supset \mathsf{V}_1 \supset \cdots \supset \mathsf{V}_k = 0.$$

b) *Also*

$$0 = \mathsf{S}_0 \subset \mathsf{S}_1 \subset \cdots \subset \mathsf{S}_k = \mathsf{V},$$

and $\mathsf{S}_n = \mathbf{An}_{\mathsf{V}}(\mathbf{J}(\mathfrak{A})^n)$.

Proof. a) The existence of k follows from the fact that $\mathbf{J}(\mathfrak{A})$ is nilpotent. If $\mathsf{V}_i = \mathsf{V}_{i+1}$ and $i < k$, then

$$\mathsf{V}\mathbf{J}(\mathfrak{A})^{k-1} = \mathsf{V}_i\mathbf{J}(\mathfrak{A})^{k-i-1} = \mathsf{V}_{i+1}\mathbf{J}(\mathfrak{A})^{k-i-1} = \mathsf{V}\mathbf{J}(\mathfrak{A})^k = 0,$$

a contradiction.

b) From $(\mathsf{S}_{n+1}/\mathsf{S}_n)\mathbf{J}(\mathfrak{A}) = 0$, it follows immediately that $\mathsf{S}_n \subseteq \mathbf{An}_\mathsf{V}(\mathbf{J}(\mathfrak{A})^n)$. We establish equality by induction on n. For $n = 0$, it is trivial. For $n > 0$, we have

$$\mathbf{An}_\mathsf{V}(\mathbf{J}(\mathfrak{A})^n)\mathbf{J}(\mathfrak{A}) \subseteq \mathbf{An}_\mathsf{V}(\mathbf{J}(\mathfrak{A})^{n-1}) = \mathsf{S}_{n-1}$$

by the inductive hypothesis. It follows that $\mathbf{An}_\mathsf{V}(\mathbf{J}(\mathfrak{A})^n) \subseteq \mathsf{S}_n$. Thus $\mathbf{An}_\mathsf{V}(\mathbf{J}(\mathfrak{A})^n) = \mathsf{S}_n$ and in particular $\mathsf{S}_k = \mathsf{V}$.

If there exists $i < k$ such that $\mathbf{An}_\mathsf{V}(\mathbf{J}(\mathfrak{A})^i) = \mathbf{An}_\mathsf{V}(\mathbf{J}(\mathfrak{A})^{i+1})$, then it follows from $(\mathsf{V}\mathbf{J}(\mathfrak{A})^{k-i-1})\mathbf{J}(\mathfrak{A})^{i+1} = \mathsf{V}\mathbf{J}(\mathfrak{A})^k = 0$ that

$$0 = (\mathsf{V}\mathbf{J}(\mathfrak{A})^{k-i-1})\mathbf{J}(\mathfrak{A})^i = \mathsf{V}\mathbf{J}(\mathfrak{A})^{k-1},$$

a contradiction. Thus

$$0 = \mathsf{S}_0 \subset \mathsf{S}_1 \subset \cdots \subset \mathsf{S}_k = \mathsf{V}.$$ **q.e.d.**

In many cases, the projective indecomposable $\mathsf{K}\mathfrak{G}$-module the head of which yields the unit representation of $\mathfrak{G}$ can be described group-theoretically.

10.12 Theorem. *Let* K *be a field of characteristic* p.

a) *Suppose that* $\mathfrak{H}$ *is a subgroup of* $\mathfrak{G}$ *of order prime to* p. *If* P *is a projective indecomposable* $\mathsf{K}\mathfrak{G}$*-module the head of which yields the unit representation of* $\mathfrak{G}$, *then* P *is isomorphic to a direct summand of* $\mathsf{V}^\mathfrak{G}$, *where* V *is a module for the unit representation of* $\mathfrak{H}$. *In particular,* $\dim_\mathsf{K}\mathsf{P} \leq |\mathfrak{G}:\mathfrak{H}|$.

b) *Suppose further that* $\mathfrak{G}$ *possesses a* p*-complement* $\mathfrak{Q}$. *If* V *is the module for the unit representation of* $\mathfrak{Q}$, *then* $\mathsf{V}^\mathfrak{G}$ *is the projective indecomposable* $\mathsf{K}\mathfrak{G}$*-module the head of which yields the unit representation of* $\mathfrak{G}$. *The element*

$$e = |\mathfrak{Q}|^{-1} \sum_{x \in \mathfrak{Q}} x$$

is an idempotent and $e\mathsf{K}\mathfrak{G} \cong \mathsf{V}^\mathfrak{G}$.

Proof. a) Since $p \nmid |\mathfrak{H}|$, V is a projective $\mathsf{K}\mathfrak{H}$-module. By 7.17, $\mathsf{V}^\mathfrak{G}$ is a projective $\mathsf{K}\mathfrak{G}$-module.

Let W be the 1-dimensional $\mathsf{K}\mathfrak{G}$-module corresponding to the unit representation of $\mathfrak{G}$. We denote by k the multiplicity of W in the head of $\mathsf{V}^{\mathfrak{G}}$ and by l the multiplicity of V in the socle of $\mathsf{W}_{\mathfrak{H}}$. Then $l = 1$, and by 4.13a),

$$k = k \dim_{\mathsf{K}} \mathrm{Hom}_{\mathsf{K}\mathfrak{G}}(\mathsf{W}, \mathsf{W}) = l \dim_{\mathsf{K}} \mathrm{Hom}_{\mathsf{K}\mathfrak{H}}(\mathsf{V}, \mathsf{V}) = 1.$$

If $\mathsf{V}^{\mathfrak{G}} = \mathsf{P}_1 \oplus \cdots \oplus \mathsf{P}_r$ with indecomposable projective $\mathsf{K}\mathfrak{G}$-modules P_i, then there exists some i such that $\mathsf{P}_i/\mathsf{P}_i\mathbf{J}(\mathsf{K}\mathfrak{G}) \cong \mathsf{W}$. Hence $\mathsf{P}_i \cong \mathsf{P}$. Finally,

$$\dim_{\mathsf{K}} \mathsf{P} \leq \dim_{\mathsf{K}} \mathsf{V}^{\mathfrak{G}} = |\mathfrak{G} : \mathfrak{H}|.$$

b) By Dickson's theorem (7.16) we know that the order $|\mathfrak{P}|$ of a Sylow p-subgroup $\mathfrak{P}$ of $\mathfrak{G}$ divides $\dim_{\mathsf{K}} \mathsf{P}$. By a),

$$\dim_{\mathsf{K}} \mathsf{P} \leq |\mathfrak{G} : \mathfrak{Q}| = |\mathfrak{P}| = \dim_{\mathsf{K}} \mathsf{V}^{\mathfrak{G}}.$$

Thus $\mathsf{V}^{\mathfrak{G}} \cong \mathsf{P}$.

Since $\operatorname{char} \mathsf{K} \nmid |\mathfrak{Q}|$, e is an idempotent in $\mathsf{K}\mathfrak{G}$. Clearly $ex = e$ for all $x \in \mathfrak{Q}$. If T is a transversal of $\mathfrak{Q}$ in $\mathfrak{G}$, then

$$e\mathsf{K}\mathfrak{G} = \bigoplus_{t \in \mathsf{T}} \mathsf{K}(et).$$

If $g \in \mathfrak{G}$ and $tg = xt'$ with $x \in \mathfrak{Q}$ and $t' \in \mathsf{T}$, then

$$(et)g = ext' = et'.$$

But if $\mathsf{V} = \mathsf{K}v$, then also $(v \otimes t)g = v \otimes t'$. Hence $e\mathsf{K}\mathfrak{G} \cong \mathsf{V}^{\mathfrak{G}}$. **q.e.d.**

10.13 Theorem. *Suppose that $\mathfrak{A}$ is an algebra of finite dimension over K and that $\mathfrak{A}$ is an injective $\mathfrak{A}$-module. (By 7.8, the group-ring of a finite group over any field has this property.) Let P be a finitely generated indecomposable projective $\mathfrak{A}$-module.*

a) *Then the socle $\mathbf{S}(\mathsf{P})$ of P is the unique minimal submodule of P and $\mathsf{P}\mathbf{S}_r(\mathfrak{A}) = \mathbf{S}(\mathsf{P})$. In particular, $\mathbf{S}(\mathsf{P})$ is irreducible.*

b) *If P_1 and P_2 are indecomposable projective $\mathfrak{A}$-modules, both finitely generated, then $\mathsf{P}_1 \cong \mathsf{P}_2$ if and only if $\mathbf{S}(\mathsf{P}_1) \cong \mathbf{S}(\mathsf{P}_2)$.*

c) *If V is an irreducible $\mathfrak{A}$-module, there exists an indecomposable projective $\mathfrak{A}$-module P such that $\mathbf{S}(\mathsf{P}) \cong \mathsf{V}$.*

Proof. a) Since $\mathfrak{A}$ is an injective $\mathfrak{A}$-module, so is any finitely generated free $\mathfrak{A}$-module and hence also any finitely generated projective $\mathfrak{A}$-module.

We consider the socle $\mathsf{S} = \mathbf{S}(\mathsf{P})$ of P. Suppose $\alpha \in \mathrm{Hom}_{\mathfrak{A}}(\mathsf{S}, \mathsf{S})$. If we denote by ι the embedding of S in P, we have the following diagram.

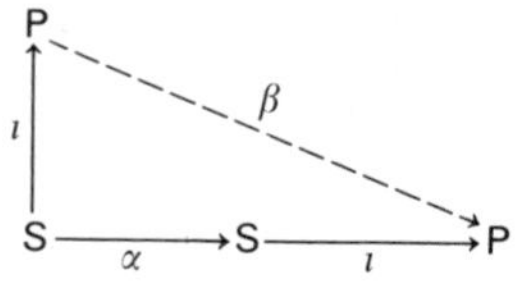

As P is injective, there exists $\beta \in \mathrm{Hom}_{\mathfrak{A}}(\mathsf{P}, \mathsf{P})$ such that for all $s \in \mathsf{S}$,

$$s\beta = s\iota\beta = s\alpha\iota = s\alpha.$$

Hence restriction to S is an algebra epimorphism of $\mathrm{Hom}_{\mathfrak{A}}(\mathsf{P}, \mathsf{P})$ onto $\mathrm{Hom}_{\mathfrak{A}}(\mathsf{S}, \mathsf{S})$. If S is reducible, it is decomposable and then $\mathrm{Hom}_{\mathfrak{A}}(\mathsf{S}, \mathsf{S})$ is not a local ring. As every epimorphic image of a local ring is again local, we conclude that $\mathrm{Hom}_{\mathfrak{A}}(\mathsf{P}, \mathsf{P})$ is not a local ring, in contradiction to the indecomposability of P. Thus S is irreducible.

As P is a direct summand of a free $\mathfrak{A}$-module F, we obtain

$$\mathsf{S} = \mathbf{S}(\mathsf{P}) = \mathbf{S}(\mathsf{F}) \cap \mathsf{P} = \mathsf{F}\mathbf{S}_r(\mathfrak{A}) \cap \mathsf{P} = \mathsf{P}\mathbf{S}_r(\mathfrak{A}).$$

b) Now let γ be an isomorphism of $\mathbf{S}(\mathsf{P}_1)$ onto $\mathbf{S}(\mathsf{P}_2)$. From the diagram

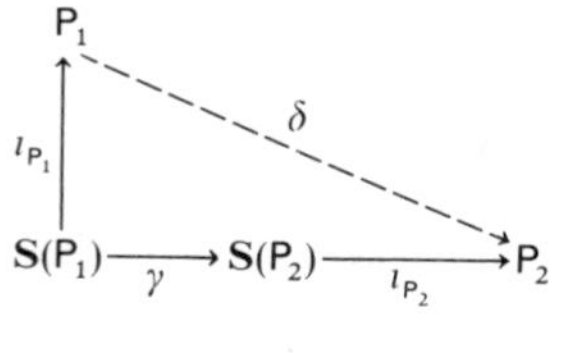

it follows that there exists a homomorphism δ of P_1 into P_2 such that γ is the restriction of δ to $\mathbf{S}(\mathsf{P}_1)$. As

$$\ker\delta \cap \mathbf{S}(\mathsf{P}_1) = \ker\gamma = 0$$

and $\mathbf{S}(\mathsf{P}_1)$ is the only minimal submodule of P_1, we conclude that $\ker\delta = 0$. Hence $\mathsf{P}_1\delta \cong \mathsf{P}_1$ is an injective submodule of P_2. Thus $\mathsf{P}_2 = \mathsf{P}_1\delta \oplus \mathsf{P}_2'$ for some P_2'. The indecomposability of P_2 forces $\mathsf{P}_2 = \mathsf{P}_1\delta \cong \mathsf{P}_1$.

c) By 10.9, the number of isomorphism types of indecomposable projective finitely generated $\mathfrak{A}$-modules is equal to the number of types of irreducible $\mathfrak{A}$-modules. Thus by b) all isomorphism types of irreducible $\mathfrak{A}$-modules must appear among the socles of indecomposable projective $\mathfrak{A}$-modules. **q.e.d.**

We remark that in a) P is the injective envelope of $\mathbf{S}(\mathsf{P})$ in the sense of the theorem of Eckmann and Schopf (see MACLANE [1], p. 103). Then b) follows from the fact that the injective envelope is uniquely determined.

Every module is an epimorphic image of a projective module. But in general an indecomposable module need not be an epimorphic image of an indecomposable projective module. We can state the following lemma.

10.14 Lemma. *Let* $\mathsf{V} \neq 0$ *be a* $\mathsf{K}\mathfrak{G}$*-module of finite dimension over* K.

a) V *is an epimorphic image of an indecomposable projective* $\mathsf{K}\mathfrak{G}$*-module* P *if and only if the head* $\mathbf{H}(\mathsf{V})$ *of* V *is irreducible. Then* $\mathbf{H}(\mathsf{V}) \cong \mathbf{H}(\mathsf{P})$.

b) V *is isomorphic to a submodule of an indecomposable projective* $\mathsf{K}\mathfrak{G}$*-module* P *if and only if the socle* $\mathbf{S}(\mathsf{V})$ *of* V *is irreducible. Then* $\mathbf{S}(\mathsf{V}) \cong \mathbf{S}(\mathsf{P})$.

Proof. a) Suppose first that $\mathsf{V} = \mathsf{P}\alpha$ for an indecomposable projective module P and $\alpha \in \mathrm{Hom}_{\mathsf{K}\mathfrak{G}}(\mathsf{P}, \mathsf{V})$. By 10.9c), $\mathbf{H}(\mathsf{P})$ is irreducible, hence $\mathbf{H}(\mathsf{V}) \cong \mathbf{H}(\mathsf{P})$ is also irreducible.

Now suppose conversely that $\mathbf{H}(\mathsf{V}) = \mathsf{V}/\mathsf{V}\mathbf{J}(\mathsf{K}\mathfrak{G})$ is irreducible. Let P be the indecomposable projective $\mathsf{K}\mathfrak{G}$-module for which $\mathbf{H}(\mathsf{P}) \cong \mathbf{H}(\mathsf{V})$ (10.9b)). Let v_{P} denote the natural epimorphism of P onto $\mathbf{H}(\mathsf{P})$, v_{V} the natural epimorphism of V onto $\mathbf{H}(\mathsf{V})$ and α an isomorphism of $\mathbf{H}(\mathsf{P})$ onto $\mathbf{H}(\mathsf{V})$.

$$\begin{array}{ccccc} \mathsf{P} & \xrightarrow{v_{\mathsf{P}}} & \mathbf{H}(\mathsf{P}) & & \\ {\scriptstyle\beta}\downarrow & & \downarrow{\scriptstyle\alpha} & & \\ \mathsf{V} & \xrightarrow[v_{\mathsf{V}}]{} & \mathbf{H}(\mathsf{V}) & \longrightarrow & 0 \end{array}$$

As P is projective, there exists a mapping $\beta \in \mathrm{Hom}_{\mathsf{K}\mathfrak{G}}(\mathsf{P}, \mathsf{V})$ such that

$$\beta v_{\mathsf{V}} = v_{\mathsf{P}}\alpha \neq 0.$$

Hence $\mathrm{im}\,\beta \subsetneqq \ker v_{\mathsf{V}}$. As $\ker v_{\mathsf{V}}$ is the only maximal submodule of V, this implies that $\mathrm{im}\,\beta = \mathsf{V}$.

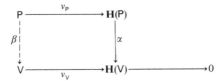

b) As the socle of P is irreducible (10.13a)), the existence of a monomorphism of V into P implies that $\mathbf{S}(V) = \mathbf{S}(P)$ is irreducible.

Suppose conversely that $\mathbf{S}(V)$ is irreducible. By 7.8, $K\mathfrak{G}$ is injective. Hence by 10.13c), there exists an indecomposable projective module P for which there is an isomorphism α of $\mathbf{S}(V)$ onto $\mathbf{S}(P)$.

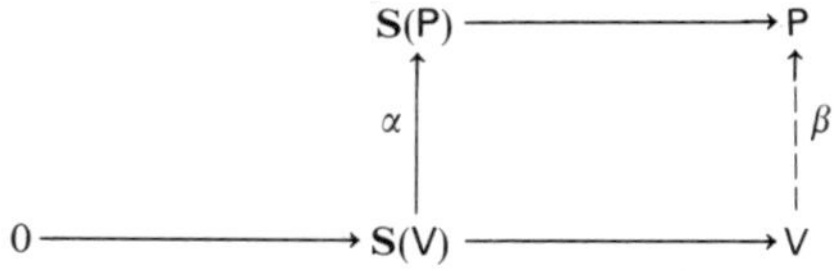

As P is injective (7.8), there exists a $\beta \in \mathrm{Hom}_{K\mathfrak{G}}(V, P)$ such that $v\alpha = v\beta$ for all $v \in \mathbf{S}(V)$. Hence

$$\ker \beta \cap \mathbf{S}(V) = 0$$

and thus $\ker \beta = 0$. Hence β is a monomorphism of V into P. **q.e.d.**

If K is a splitting field for $\mathfrak{G}$ and char K does not divide $|\mathfrak{G}|$, the structure of the free $K\mathfrak{G}$-module $K\mathfrak{G}$ is very simple, and we know from the Wedderburn theory that any irreducible module V occurs in $K\mathfrak{G}$ with multiplicity $\dim_K V$. If on the other hand char K divides $|\mathfrak{G}|$, this holds only for the multiplicity in $K\mathfrak{G}/\mathbf{J}(K\mathfrak{G})$. Our next steps are directed towards a determination of the multiplicities of the composition factors in $K\mathfrak{G}$ and in the indecomposable projective $K\mathfrak{G}$-modules.

10.15 Definition. Let $\mathfrak{A}$ be an algebra of finite dimension and let

$$\mathfrak{A} = e_1\mathfrak{A} \oplus \cdots \oplus e_n\mathfrak{A}$$

be a decomposition of $\mathfrak{A}$ into indecomposable right ideals $e_i\mathfrak{A}$, where the e_i are mutually orthogonal idempotents such that $1 = e_1 + \cdots + e_n$. We denote by c_{ij} the multiplicity of the irreducible $\mathfrak{A}$-module $\overline{e}_j\overline{\mathfrak{A}}$ as a composition factor of the indecomposable projective $\mathfrak{A}$-module $e_i\mathfrak{A}$. Since

$$e_i\mathfrak{A}/e_i\mathbf{J}(\mathfrak{A}) \cong \overline{e}_i\overline{\mathfrak{A}} \quad \text{(10.3c))},$$

certainly $c_{ii} \geq 1$. Suppose the numbering is so chosen that $\overline{e}_1\overline{\mathfrak{A}}, \ldots, \overline{e}_k\overline{\mathfrak{A}}$ are representatives of the isomorphism types of the irreducible $\mathfrak{A}$-modules. The $k \times k$ matrix $C = (c_{ij})$ is called the *Cartan matrix* of $\mathfrak{A}$.

Clearly C is the unit matrix if and only if each $e_i\mathfrak{A}$ is irreducible, that is, if $\mathfrak{A}$ is a completely reducible $\mathfrak{A}$-module and therefore a semisimple algebra.

10.16 Lemma. *Suppose that $\mathfrak{A}$ is a K-algebra, that e is an idempotent in $\mathfrak{A}$ and V an $\mathfrak{A}$-module. Then there is a K-isomorphism of* $\mathrm{Hom}_{\mathfrak{A}}(e\mathfrak{A}, \mathsf{V})$ *onto $\mathsf{V}e$.*

Proof. We define a mapping β of $\mathrm{Hom}_{\mathfrak{A}}(e\mathfrak{A}, \mathsf{V})$ into V by $\alpha\beta = e\alpha$ for $\alpha \in \mathrm{Hom}_{\mathfrak{A}}(e\mathfrak{A}, \mathsf{V})$. Since $e\alpha \in \mathsf{V}$, we have

$$e\alpha = e^2\alpha = (e\alpha)e \in Ve.$$

Thus β is a K-linear mapping of $\mathrm{Hom}_{\mathfrak{A}}(e\mathfrak{A}, \mathsf{V})$ into $\mathsf{V}e$. Given an element $ve \in \mathsf{V}e$, define the mapping α of $e\mathfrak{A}$ into V by

$$ea\alpha = vea \quad (a \in \mathfrak{A}).$$

Then $\alpha \in \mathrm{Hom}_{\mathfrak{A}}(e\mathfrak{A}, \mathsf{V})$ and $\alpha\beta = e\alpha = ve$. Thus β is an epimorphism of $\mathrm{Hom}_{\mathfrak{A}}(e\mathfrak{A}, \mathsf{V})$ onto $\mathsf{V}e$. If $\alpha\beta = 0$, then $e\alpha = 0$ and

$$(ea)\alpha = (e\alpha)a = 0$$

for all $a \in \mathfrak{A}$, so $\alpha = 0$. Thus β is an isomorphism. **q.e.d.**

10.17 Lemma. *Suppose that $\mathfrak{A}$ is an algebra of finite dimension over K. Let*

$$\mathfrak{A} = \bigoplus_{i=1}^{n} e_i\mathfrak{A}$$

be a decomposition of $\mathfrak{A}$ into indecomposable right ideals $e_i\mathfrak{A}$. If V is a finitely generated $\mathfrak{A}$-module, the multiplicity m of the irreducible module $\bar{e}_i\bar{\mathfrak{A}}$ (where $\bar{\mathfrak{A}} = \mathfrak{A}/\mathbf{J}(\mathfrak{A})$) as a composition factor of V is determined by

$$m \dim_{\mathsf{K}} \mathrm{Hom}_{\mathfrak{A}}(\bar{e}_i\bar{\mathfrak{A}}, \bar{e}_i\bar{\mathfrak{A}}) = \dim_{\mathsf{K}} \mathsf{V}e_i = \dim_{\mathsf{K}} \mathrm{Hom}_{\mathfrak{A}}(e_i\mathfrak{A}, \mathsf{V}).$$

Proof. We prove this by induction on dim V. Let V_1 be a maximal submodule of V. As $e_i\mathfrak{A}$ is projective, we obtain from the exact sequence

$$0 \to \mathsf{V}_1 \to \mathsf{V} \to \mathsf{V}/\mathsf{V}_1 \to 0$$

an exact sequence

$$0 \to \mathrm{Hom}_{\mathfrak{A}}(e_i\mathfrak{A}, \mathsf{V}_1) \to \mathrm{Hom}_{\mathfrak{A}}(e_i\mathfrak{A}, \mathsf{V}) \to \mathrm{Hom}_{\mathfrak{A}}(e_i\mathfrak{A}, \mathsf{V}/\mathsf{V}_1) \to 0.$$

This shows that

$$\dim_{\mathsf{K}}\mathrm{Hom}_{\mathfrak{A}}(e_i\mathfrak{A}, \mathsf{V}) = \dim_{\mathsf{K}}\mathrm{Hom}_{\mathfrak{A}}(e_i\mathfrak{A}, \mathsf{V}_1) + \dim_{\mathsf{K}}\mathrm{Hom}_{\mathfrak{A}}(e_i\mathfrak{A}, \mathsf{V}/\mathsf{V}_1).$$

We have

$$\dim_{\mathsf{K}}\mathrm{Hom}_{\mathfrak{A}}(e_i\mathfrak{A}, \mathsf{V}/\mathsf{V}_1) = \begin{cases} 0 & \text{if } \mathsf{V}/\mathsf{V}_1 \not\cong \bar{e}_i\bar{\mathfrak{A}}, \\ \dim_{\mathsf{K}}\mathrm{Hom}_{\mathfrak{A}}(\bar{e}_i\bar{\mathfrak{A}}, \bar{e}_i\bar{\mathfrak{A}}) & \text{if } \mathsf{V}/\mathsf{V}_1 \cong \bar{e}_i\bar{\mathfrak{A}}. \end{cases}$$

If m_1 is the multiplicity of $\bar{e}_i\bar{\mathfrak{A}}$ as a composition factor of V_1, then by the inductive hypothesis,

$$m_1 \dim_{\mathsf{K}}\mathrm{Hom}_{\mathfrak{A}}(\bar{e}_i\bar{\mathfrak{A}}, \bar{e}_i\bar{\mathfrak{A}}) = \dim_{\mathsf{K}}\mathrm{Hom}_{\mathfrak{A}}(e_i\mathfrak{A}, \mathsf{V}_1).$$

This proves the first part of the assertion; the second part follows by 10.16. **q.e.d.**

Lemma 10.17 gives us a formula for the Cartan numbers c_{ij}.

10.18 Theorem. *Suppose that $\mathfrak{A}$ is a K-algebra of finite dimension and that K is a splitting field for $\bar{\mathfrak{A}} = \mathfrak{A}/\mathbf{J}(\mathfrak{A})$. Suppose further that $\mathfrak{A} = \bigoplus_{i=1}^{n} e_i\mathfrak{A}$ is a direct decomposition of $\mathfrak{A}$ into indecomposable $\mathfrak{A}$-modules $e_i\mathfrak{A}$ as in 10.3.*

a) $c_{ij} = \dim_{\mathsf{K}} e_i\mathfrak{A}e_j$.

b) *The multiplicity of $e_i\mathfrak{A}$ as a direct summand of $\mathfrak{A}$ in the sense of the Krull-Schmidt theorem is* $\dim_{\mathsf{K}}\bar{e}_i\bar{\mathfrak{A}}$.

c) *The multiplicity of $\bar{e}_i\bar{\mathfrak{A}}$ as a composition factor of $\mathfrak{A}$ is* $\dim_{\mathsf{K}} \mathfrak{A}e_i$.

d) *If $c_{ij} = c_{ji}$ for all i, j, then* $\dim_{\mathsf{K}} e_i\mathfrak{A} = \dim_{\mathsf{K}} \mathfrak{A}e_i$. *(This hypothesis about the symmetry of the Cartan matrix is always satisfied in group-rings, as we shall see in 11.6d).)*

Proof. a) By Definition 10.15, c_{ij} is the multiplicity of $\bar{e}_j\bar{\mathfrak{A}}$ as a composition factor of $e_i\mathfrak{A}$, and by 10.17 this is $\dim_{\mathsf{K}} e_i\mathfrak{A}e_j$.

b) By the Wedderburn theory, the multiplicity of $\bar{e}_i\bar{\mathfrak{A}}$ in $\bar{\mathfrak{A}}$ is $\dim_{\mathsf{K}}\bar{e}_i\bar{\mathfrak{A}}$. By 10.9a), the indecomposable projective $\mathfrak{A}$-module $e_i\mathfrak{A}$ is determined to within isomorphism by its head $\bar{e}_i\bar{\mathfrak{A}}$, so $\dim_{\mathsf{K}}\bar{e}_i\bar{\mathfrak{A}}$ is also the multiplicity of $e_i\mathfrak{A}$ as a direct summand of $\mathfrak{A}$.

c) This follows at once from 10.17.

d) We count the multiplicity of $\bar{e}_i\bar{\mathfrak{A}}$ as a composition factor of $\mathfrak{A}$ in two ways.

On the one hand this multiplicity is $\dim_{\mathsf{K}} \mathfrak{A}e_i$, by c). On the other hand, the multiplicity of $e_j\mathfrak{A}$ as a direct summand of $\mathfrak{A}$ is $\dim_{\mathsf{K}} \overline{e}_j\overline{\mathfrak{A}}$ by b), and the multiplicity of $\overline{e}_i\overline{\mathfrak{A}}$ as a composition factor of $e_j\mathfrak{A}$ is c_{ji}. Thus if as in 10.15 $\overline{e}_1\overline{\mathfrak{A}}, \ldots, \overline{e}_k\overline{\mathfrak{A}}$ are the isomorphism types of irreducible $\mathfrak{A}$-modules, then

$$\dim_{\mathsf{K}} \mathfrak{A}e_i = \sum_{j=1}^{k} c_{ji} \dim_{\mathsf{K}} \overline{e}_j\overline{\mathfrak{A}} = \sum_{j=1}^{k} c_{ij} \dim_{\mathsf{K}} \overline{e}_j\overline{\mathfrak{A}} = \dim_{\mathsf{K}} e_i\mathfrak{A}. \qquad \textbf{q.e.d.}$$

Exercises

45) Let $\mathfrak{A}$ be the ring of all $n \times n$ triangular matrices with coefficients in a field K. Find $\mathbf{J}(\mathfrak{A})$, $\mathbf{S}_r(\mathfrak{A})$ and $\mathbf{S}_l(\mathfrak{A})$. Compare the right $\mathfrak{A}$-modules $\mathfrak{A}/\mathbf{J}(\mathfrak{A})$ and $\mathbf{S}_r(\mathfrak{A})$ and the left modules $\mathfrak{A}/\mathbf{J}(\mathfrak{A})$ and $\mathbf{S}_l(\mathfrak{A})$.

46) Let $\mathfrak{A}$ be an algebra of finite dimension and P a projective finitely generated $\mathfrak{A}$-module. We put $\overline{\mathsf{P}} = \mathsf{P}/\mathsf{P}\mathbf{J}(\mathfrak{A})$. If $\alpha \in \mathrm{Hom}_{\mathfrak{A}}(\mathsf{P}, \mathsf{P})$, then the mapping $\overline{\alpha}$ defined by

$$(x + \mathsf{P}\mathbf{J}(\mathfrak{A}))\overline{\alpha} = x\alpha + \mathsf{P}\mathbf{J}(\mathfrak{A}) \quad (x \in \mathsf{P})$$

is an element of $\mathrm{Hom}_{\overline{\mathfrak{A}}}(\overline{\mathsf{P}}, \overline{\mathsf{P}})$. Show that the mapping $\alpha \to \overline{\alpha}$ is an algebra-epimorphism of $\mathrm{Hom}_{\mathfrak{A}}(\mathsf{P}, \mathsf{P})$ onto $\mathrm{Hom}_{\overline{\mathfrak{A}}}(\overline{\mathsf{P}}, \overline{\mathsf{P}})$ with kernel $\mathrm{Hom}_{\mathfrak{A}}(\mathsf{P}, \mathsf{P}\mathbf{J}(\mathfrak{A})) = \mathbf{J}(\mathrm{Hom}_{\mathfrak{A}}(\mathsf{P}, \mathsf{P}))$.

47) Let $\mathfrak{A}$ be an algebra of finite dimension and let $\mathfrak{R}$ be a right ideal of $\mathfrak{A}$ such that $\mathfrak{R} \subsetneqq \mathbf{J}(\mathfrak{A})$. Show that $\mathfrak{R}$ contains an idempotent $e \neq 0$.

§ 11. Frobenius Algebras and Symmetric Algebras

In several places we have already encountered a certain duality of group-rings. We shall now study this duality in the more general context of Frobenius algebras and symmetric algebras.

11.1 Definition. Let $\mathfrak{A}$ be a K-algebra of finite dimension.

a) $\mathfrak{A}$ is called a *Frobenius algebra* if there exists a non-singular K-bilinear form (,) on $\mathfrak{A}$ such that for all $a, b, c \in \mathfrak{A}$ we have

$$(ab, c) = (a, bc).$$

b) $\mathfrak{A}$ is called a *symmetric algebra* if there exists a symmetric non-singular K-bilinear form $(,)$ on $\mathfrak{A}$ such that for all $a, b, c \in \mathfrak{A}$

$$(ab, c) = (a, bc).$$

For Lie algebras over a field of characteristic 0, the existence of a bilinear form with the properties introduced in a) is equivalent to semi-simplicity; this is an essential point of departure in E. Cartan and Killing's determination of all simple Lie algebras over the field of complex numbers. For associative algebras, however, the class of Frobenius algebras is larger than the class of semisimple algebras even in characteristic 0 (see Exercise 50).

11.2 Theorem. *If K is an arbitrary field, $\mathsf{K}\mathfrak{G}$ is a symmetric algebra.*

Proof. For $g_1, g_2 \in \mathfrak{G}$, we put

$$(g_1, g_2) = \begin{cases} 0 & \text{if } g_1 g_2 \neq 1 \\ 1 & \text{if } g_1 g_2 = 1 \end{cases}$$

and extend this to a bilinear form on $\mathsf{K}\mathfrak{G}$. Obviously this is a symmetric K-bilinear form. From

$$(g_1 g_2, g_3) = (g_1, g_2 g_3) \quad (g_i \in \mathfrak{G})$$

it follows easily that

$$(ab, c) = (a, bc)$$

for all $a, b, c \in \mathsf{K}\mathfrak{G}$. It remains only to show that $(,)$ is non-singular. Suppose that $a = \sum_{g \in \mathfrak{G}} a_g g$ and $(a, b) = 0$ for all $b \in \mathsf{K}\mathfrak{G}$. Then for all $h \in \mathfrak{G}$ we have

$$0 = (a, h^{-1}) = a_h.$$

Thus $a = 0$ and $(,)$ is non-singular. **q.e.d.**

In the following theorem we collect some basic properties of Frobenius algebras and use the notation of 10.7 for annihilators.

11.3 Theorem. *Let $\mathfrak{A}$ be a Frobenius algebra with bilinear form $(,)$.*

a) *If $\mathfrak{R}$ is a right ideal of $\mathfrak{A}$, then*

$$\mathbf{L}(\mathfrak{R}) = \{a | a \in \mathfrak{A}, (a, r) = 0 \quad \textit{for all } r \in \mathfrak{R}\}.$$

If $\mathfrak{L}$ is a left ideal of $\mathfrak{A}$, then

$$\mathbf{R}(\mathfrak{L}) = \{a | a \in \mathfrak{A}, (l, a) = 0 \quad \textit{for all } l \in \mathfrak{L}\}.$$

b) *For every right ideal $\mathfrak{R}$ of $\mathfrak{A}$ we have*

$$\dim_K \mathfrak{R} + \dim_K \mathbf{L}(\mathfrak{R}) = \dim_K \mathfrak{A} \tag{1}$$

and

$$\mathbf{R}(\mathbf{L}(\mathfrak{R})) = \mathfrak{R}. \tag{2}$$

Similarly, if $\mathfrak{L}$ is a left ideal of $\mathfrak{A}$, we have

$$\dim_K \mathfrak{L} + \dim_K \mathbf{R}(\mathfrak{L}) = \dim_K \mathfrak{A} \tag{1'}$$

and

$$\mathbf{L}(\mathbf{R}(\mathfrak{L})) = \mathfrak{L}. \tag{2'}$$

c) *The mapping $\mathfrak{R} \to \mathbf{L}(\mathfrak{R})$ is a duality of the lattice of right ideals of $\mathfrak{A}$ onto the lattice of left ideals of $\mathfrak{A}$, that is,*

$$\mathbf{L}(\mathfrak{R}_1 + \mathfrak{R}_2) = \mathbf{L}(\mathfrak{R}_1) \cap \mathbf{L}(\mathfrak{R}_2)$$

and

$$\mathbf{L}(\mathfrak{R}_1 \cap \mathfrak{R}_2) = \mathbf{L}(\mathfrak{R}_1) + \mathbf{L}(\mathfrak{R}_2)$$

for all right ideals $\mathfrak{R}_1$, $\mathfrak{R}_2$ of $\mathfrak{A}$.
Similarly $\mathfrak{L} \to \mathbf{R}(\mathfrak{L})$ is a duality of the lattice of left ideals of $\mathfrak{A}$ onto the lattice of right ideals of $\mathfrak{A}$.

Proof. a) Let $\mathfrak{R}$ be a right ideal of $\mathfrak{A}$ and put

$$\mathfrak{R}^\perp = \{a | a \in \mathfrak{A}, (a, r) = 0 \quad \text{for all } r \in \mathfrak{R}\}.$$

Suppose $x \in \mathfrak{R}^\perp$. Then for all $r \in \mathfrak{R}$ and $a \in \mathfrak{A}$ we have

$$0 = (x, ra) = (xr, a).$$

Since $(,)$ is non-singular, it follows that $xr = 0$ for all $r \in \mathfrak{R}$. Thus $x \in \mathbf{L}(\mathfrak{R})$ and hence $\mathfrak{R}^\perp \subseteq \mathbf{L}(\mathfrak{R})$. If conversely $x \in \mathbf{L}(\mathfrak{R})$, then for all $r \in \mathfrak{R}$,

$$0 = (xr, 1) = (x, r).$$

This shows that $\mathbf{L}(\mathfrak{R}) \subseteq \mathfrak{R}^\perp$. Together we obtain $\mathfrak{R}^\perp = \mathbf{L}(\mathfrak{R})$. The assertion for left ideals follows similarly.

b) Since $(,)$ is non-singular, it follows from well-known assertions about bilinear forms that

$$\dim_K \mathfrak{A} = \dim_K \mathfrak{R} + \dim_K \mathfrak{R}^\perp = \dim_K \mathfrak{R} + \dim_K \mathbf{L}(\mathfrak{R}).$$

The assertion (1′) is proved similarly. Since $\mathbf{L}(\mathfrak{R})\mathfrak{R} = 0$, we have certainly $\mathfrak{R} \subseteq \mathbf{R}(\mathbf{L}(\mathfrak{R}))$. We obtain from (1) and (1′)

$$\begin{aligned}\dim_K \mathbf{R}(\mathbf{L}(\mathfrak{R})) &= \dim_K \mathfrak{A} - \dim_K \mathbf{L}(\mathfrak{R}) \\ &= \dim_K \mathfrak{A} - (\dim_K \mathfrak{A} - \dim_K \mathfrak{R}) = \dim_K \mathfrak{R}.\end{aligned}$$

This shows that $\mathbf{R}(\mathbf{L}(\mathfrak{R})) = \mathfrak{R}$ for all right ideals $\mathfrak{R}$ of $\mathfrak{A}$. Similarly $\mathbf{L}(\mathbf{R}(\mathfrak{L})) = \mathfrak{L}$ for all left ideals $\mathfrak{L}$ of $\mathfrak{A}$.

c) Trivially we have

$$\mathbf{L}(\mathfrak{R}_1 + \mathfrak{R}_2) = \mathbf{L}(\mathfrak{R}_1) \cap \mathbf{L}(\mathfrak{R}_2)$$

and

$$\mathbf{R}(\mathfrak{L}_1 + \mathfrak{L}_2) = \mathbf{R}(\mathfrak{L}_1) \cap \mathbf{R}(\mathfrak{L}_2).$$

With b), it follows that

$$\begin{aligned}\mathbf{L}(\mathfrak{R}_1 \cap \mathfrak{R}_2) &= \mathbf{L}(\mathbf{R}(\mathbf{L}(\mathfrak{R}_1)) \cap \mathbf{R}(\mathbf{L}(\mathfrak{R}_2))) = \mathbf{L}(\mathbf{R}(\mathbf{L}(\mathfrak{R}_1) + \mathbf{L}(\mathfrak{R}_2))) \\ &= \mathbf{L}(\mathfrak{R}_1) + \mathbf{L}(\mathfrak{R}_2).\end{aligned}$$

The corresponding assertion for left ideals follows similarly. **q.e.d.**

11.4 Remarks. Here we cannot go into the many relations between the assertions made in 11.3; for this we must refer to the theory of Frobenius and quasi-Frobenius algebras (see CURTIS and REINER [1], chap. IX.) We confine ourselves to the following remarks.

a) In 7.9, quasi-Frobenius algebras were defined to be those algebras

$\mathfrak{A}$ of finite dimension for which $\mathfrak{A}$ is an injective right $\mathfrak{A}$-module. It can be shown that this is equivalent to the assertions (2) and (2′) in 11.3b) for all left ideals $\mathfrak{L}$ and all right ideals $\mathfrak{R}$. Also the assertions of 11.3c) characterize quasi-Frobenius algebras; indeed, any algebra of finite dimension which admits any duality of the lattice of right ideals onto the lattice of left ideals is a quasi-Frobenius algebra.

b) The assertions of 11.3b) characterize Frobenius algebras. Another characterization in terms of dual modules can be given. If V is a right $\mathfrak{A}$-module, then $\mathsf{V}^* = \operatorname{Hom}_{\mathsf{K}}(\mathsf{V}, \mathsf{K})$ can be made into a left $\mathfrak{A}$-module by putting

$$v(af) = (va)f$$

for $v \in \mathsf{V}$, $f \in \mathsf{V}^*$ and $a \in \mathfrak{A}$. Then Frobenius algebras are precisely those algebras $\mathfrak{A}$ of finite dimension for which the dual module $\mathfrak{A}^*$ of the free $\mathfrak{A}$-module $\mathfrak{A}$ on one generator is a free left $\mathfrak{A}$-module with one generator.

We prove only one of the facts mentioned in 11.4.

11.5 Theorem. *If $\mathfrak{A}$ is a Frobenius algebra, then $\mathfrak{A}$ is an injective right $\mathfrak{A}$-module.*

Proof. By Baer's lemma, it suffices to prove that if $\mathfrak{R}$ is a right ideal of $\mathfrak{A}$ and $\alpha \in \operatorname{Hom}_{\mathfrak{A}}(\mathfrak{R}, \mathfrak{A})$, then α can be extended to an element of $\operatorname{Hom}_{\mathfrak{A}}(\mathfrak{A}, \mathfrak{A})$.

For each $a \in \mathfrak{A}$, the mapping τ_a of $\mathfrak{R}$ into $\mathfrak{A}$ given by $r\tau_a = ar$ for $r \in \mathfrak{R}$ is an $\mathfrak{A}$-homomorphism of $\mathfrak{R}$ into $\mathfrak{A}$, and the mapping $a \to \tau_a$ is a K-homomorphism of $\mathfrak{A}$ into $\operatorname{Hom}_{\mathfrak{A}}(\mathfrak{R}, \mathfrak{A})$ with kernel $\mathbf{L}(\mathfrak{R})$. Obviously, each τ_a can be extended to an element τ'_a of $\operatorname{Hom}_{\mathfrak{A}}(\mathfrak{A}, \mathfrak{A})$ simply by putting $x\tau'_a = ax$ for all $x \in \mathfrak{A}$. We show that

$$\operatorname{Hom}_{\mathfrak{A}}(\mathfrak{R}, \mathfrak{A}) = \{\tau_a | a \in \mathfrak{A}\}.$$

Let (,) be a bilinear form defined on the Frobenius algebra according to Definition 11.1a). We then define $\phi \in \operatorname{Hom}_{\mathsf{K}}(\mathfrak{A}, \mathsf{K})$ by putting $a\phi = (a, 1)$ for $a \in \mathfrak{A}$. For $\alpha \in \operatorname{Hom}_{\mathfrak{A}}(\mathfrak{R}, \mathfrak{A})$ we put $\alpha' = \alpha\phi$. Thus $\alpha' \in \operatorname{Hom}_{\mathsf{K}}(\mathfrak{R}, \mathsf{K})$. If $\alpha' = 0$, then for all $r \in \mathfrak{R}$ and all $a \in \mathfrak{A}$ we have

$$0 = (ra)\alpha' = (ra)\alpha\phi = ((r\alpha)a)\phi = ((r\alpha)a, 1) = (r\alpha, a).$$

As (,) is non-singular, this implies $r\alpha = 0$ and thus $\alpha = 0$. Hence $\alpha \to \alpha'$ is a K-monomorphism of $\operatorname{Hom}_{\mathfrak{A}}(\mathfrak{R}, \mathfrak{A})$ into $\operatorname{Hom}_{\mathsf{K}}(\mathfrak{R}, \mathsf{K})$. Thus

$$\dim_{\mathsf{K}} \operatorname{Hom}_{\mathfrak{A}}(\mathfrak{R}, \mathfrak{A}) \leq \dim_{\mathsf{K}} \operatorname{Hom}_{\mathsf{K}}(\mathfrak{R}, \mathsf{K}) = \dim_{\mathsf{K}} \mathfrak{R}.$$

By 11.3b),

$$\begin{aligned}\dim_{\mathsf{K}} \mathfrak{R} &= \dim_{\mathsf{K}} \mathfrak{A} - \dim_{\mathsf{K}} \mathbf{L}(\mathfrak{R}) = \dim_{\mathsf{K}} \mathfrak{A}/\mathbf{L}(\mathfrak{R}) \\ &= \dim_{\mathsf{K}} \{\tau_a \mid a \in \mathfrak{A}\}.\end{aligned}$$

Hence $\operatorname{Hom}_{\mathfrak{A}}(\mathfrak{R}, \mathfrak{A}) = \{\tau_a \mid a \in \mathfrak{A}\}$, and so each α in $\operatorname{Hom}_{\mathfrak{A}}(\mathfrak{R}, \mathfrak{A})$ can be extended to an $\bar{\alpha}$ in $\operatorname{Hom}_{\mathfrak{A}}(\mathfrak{A}, \mathfrak{A})$. **q.e.d.**

We now collect those properties of symmetric algebras which are of interest for group-rings.

11.6 Theorem. *Suppose that $\mathfrak{A}$ is a symmetric algebra.*

a) *If $\mathfrak{J}$ is a two-sided ideal in $\mathfrak{A}$, then $\mathbf{L}(\mathfrak{J}) = \mathbf{R}(\mathfrak{J})$.*

b) $\mathbf{S}_l(\mathfrak{A}) = \mathbf{S}_r(\mathfrak{A})$.

c) *If $\mathfrak{A} = \bigoplus_{i=1}^n e_i\mathfrak{A}$ is a direct decomposition of $\mathfrak{A}$ into indecomposable right ideals $e_i\mathfrak{A}$ according to 10.3, then $e_i\mathbf{S}_r(\mathfrak{A})$ is the socle of $e_i\mathfrak{A}$ and*

$$e_i\mathbf{S}_r(\mathfrak{A}) \cong e_i\mathfrak{A}/e_i\mathbf{J}(\mathfrak{A}).$$

Thus head and socle of $e_i\mathfrak{A}$ are isomorphic.

d) *Suppose that K is a splitting field for $\mathfrak{A}/\mathbf{J}(\mathfrak{A})$. Then $c_{ij} = c_{ji}$. The Cartan matrix is thus symmetric.*

e) *Either*

(i) *$e_i\mathfrak{A}$ is irreducible, $c_{ii} = 1$ and $c_{ij} = 0$ for all $j \neq i$*

or

(ii) $c_{ii} \geq 2$.

Proof. a) On account of the symmetry of the bilinear form, it follows from 11.3a) that

$$\begin{aligned}\mathbf{L}(\mathfrak{J}) &= \{a \mid a \in \mathfrak{A}, (a, i) = 0 \quad \text{for all } i \in \mathfrak{J}\} \\ &= \{a \mid a \in \mathfrak{A}, (i, a) = 0 \quad \text{for all } i \in \mathfrak{J}\} = \mathbf{R}(\mathfrak{J}).\end{aligned}$$

b) By 10.7c), $\mathbf{S}_r(\mathfrak{A}) = \mathbf{L}(\mathbf{J}(\mathfrak{A}))$ and $\mathbf{S}_l(\mathfrak{A}) = \mathbf{R}(\mathbf{J}(\mathfrak{A}))$. Thus $\mathbf{S}_r(\mathfrak{A}) = \mathbf{S}_l(\mathfrak{A})$, by a).

c) By 11.5, $\mathfrak{A}$ is an injective $\mathfrak{A}$-module. Hence by 10.13, $\mathbf{S}(e_i\mathfrak{A}) = e_i\mathbf{S}_r(\mathfrak{A})$ is an irreducible $\mathfrak{A}$-module.

Put $\overline{\mathfrak{A}} = \mathfrak{A}/\mathbf{J}(\mathfrak{A})$ and let $\overline{\mathfrak{A}} = \bigoplus_{i=1}^k \bar{f}_i\overline{\mathfrak{A}}$ be the direct decomposition of $\overline{\mathfrak{A}}$ into simple two-sided ideals $\bar{f}_i\overline{\mathfrak{A}}$, where $\bar{1} = \bar{f}_1 + \cdots + \bar{f}_k$ is a decomposition of the unit element into mutually orthogonal central idempotents $\bar{f}_i$ of $\bar{\mathfrak{A}}$. We decompose the $\bar{f}_i\bar{\mathfrak{A}}$ into irreducible right ideals, say

$$\bar{f}_i\bar{\mathfrak{A}} = \bigoplus_j \bar{e}_{ij}\bar{\mathfrak{A}},$$

where $\bar{f}_i = \sum_j \bar{e}_{ij}$ is a decomposition of the unit element $\bar{f}_i$ of the algebra $\bar{f}_i\bar{\mathfrak{A}}$ into mutually orthogonal idempotents $\bar{e}_{ij}$. By 10.2b), there exist mutually orthogonal idempotents e_{ij} in $\mathfrak{A}$ such that $1 = \sum_{i,j} e_{ij}$ and $e_{ij} + \mathbf{J}(\mathfrak{A}) = \bar{e}_{ij}$. Put

$$f_i = \sum_j e_{ij}.$$

The f_i are then mutually orthogonal idempotents in $\mathfrak{A}$, but they are not necessarily central. If V is an irreducible $\mathfrak{A}$-module and $\mathsf{V}f_i \neq 0$, then $\mathsf{V}f_i = \mathsf{V}\bar{f}_i = \mathsf{V}$ and $\mathsf{V}\bar{f}_j = \mathsf{V}\bar{f}_i\bar{f}_j = 0$ for all $j \neq i$; thus V may be regarded as an irreducible $\bar{\mathfrak{A}}\bar{f}_i$-module and is therefore isomorphic to $\bar{e}_{i1}\bar{\mathfrak{A}}$, by V, 4.1. Thus to prove that $e_{i1}\mathbf{S}_r(\mathfrak{A}) \cong \bar{e}_{i1}\bar{\mathfrak{A}}$, we show that $e_{i1}\mathbf{S}_r(\mathfrak{A})f_i \neq 0$.

Suppose that $e_{i1}\mathbf{S}_r(\mathfrak{A})f_i = 0$. We observe first that $\mathbf{S}_r(\mathfrak{A})f_i$ is a two-sided ideal of $\mathfrak{A}$. To see this, observe that $f_i x - x f_i \in \mathbf{J}(\mathfrak{A})$ for all $x \in \mathfrak{A}$, since $\bar{f}_i$ is central in $\bar{\mathfrak{A}} = \mathfrak{A}/\mathbf{J}(\mathfrak{A})$. Since also $\mathbf{S}_r(\mathfrak{A})\mathbf{J}(\mathfrak{A}) = 0$,

$$\mathbf{S}_r(\mathfrak{A})f_i x = \mathbf{S}_r(\mathfrak{A})x f_i \subseteq \mathbf{S}_r(\mathfrak{A})f_i.$$

Since $\mathbf{S}_r(\mathfrak{A})$ is a left ideal, it follows that $\mathbf{S}_r(\mathfrak{A})f_i$ is a two-sided ideal. Hence by a),

$$e_{i1} \in \mathbf{L}(\mathbf{S}_r(\mathfrak{A})f_i) = \mathbf{R}(\mathbf{S}_r(\mathfrak{A})f_i),$$

which shows that

$$0 = \mathbf{S}_r(\mathfrak{A})f_i e_{i1} = \mathbf{S}_r(\mathfrak{A})e_{i1} = \mathbf{S}_l(\mathfrak{A})e_{i1}.$$

But $\mathfrak{A}e_{i1}$ is a non-zero left ideal of $\mathfrak{A}$ and therefore contains a minimal left ideal $\mathfrak{L}$. But then $\mathbf{J}(\mathfrak{A})\mathfrak{L} = 0$ and

$$0 \neq \mathfrak{L} \subseteq \mathfrak{A}e_{i1} \cap \mathbf{S}_l(\mathfrak{A}) = \mathbf{S}_l(\mathfrak{A})e_{i1}.$$

This is a contradiction, so c) is proved.

d) From the decomposition

$$\mathfrak{A} = \bigoplus_{i=1}^{n} e_i\mathfrak{A} = \bigoplus_{j=1}^{n} \mathfrak{A}e_j$$

we obtain

$$\mathfrak{A} = \bigoplus_{i,j=1}^{n} e_i\mathfrak{A}e_j,$$

which is a direct decomposition of $\mathfrak{A}$ as a vector space over K. For $j \neq k$ and all $a, b \in \mathfrak{A}$, we have

$$(e_iae_j, e_kbe_l) = (e_iae_je_k, be_l) = 0,$$

and from the symmetry of the form $(,)$, it follows that for $i \neq l$,

$$(e_iae_j, e_kbe_l) = (e_kbe_l, e_iae_j) = (e_kbe_le_i, ae_j) = 0.$$

The K-subspace $(e_i\mathfrak{A}e_j)^{\perp}$ orthogonal to $e_i\mathfrak{A}e_j$ therefore contains

$$\bigoplus_{(k,l)\neq(j,i)} e_k\mathfrak{A}e_l.$$

It now follows from 10.18a) that

$$\begin{aligned} c_{ij} &= \dim_{\mathsf{K}} e_i\mathfrak{A}e_j = \dim_{\mathsf{K}} \mathfrak{A} - \dim_{\mathsf{K}}(e_i\mathfrak{A}e_j)^{\perp} \\ &\leq \dim_{\mathsf{K}} \mathfrak{A} - \sum_{(k,l)\neq(j,i)} \dim_{\mathsf{K}} e_k\mathfrak{A}e_l = \dim_{\mathsf{K}} e_j\mathfrak{A}e_i = c_{ji}. \end{aligned}$$

Similarly we get $c_{ji} \leq c_{ij}$, so $c_{ij} = c_{ji}$.

e) If $e_i\mathfrak{A}$ is not irreducible, then $e_i\mathbf{S}_r(\mathfrak{A}) \subseteq e_i\mathbf{J}(\mathfrak{A})$ by 10.3c). But also $e_i\mathbf{S}_r(\mathfrak{A}) \cong e_i\mathfrak{A}/e_i\mathbf{J}(\mathfrak{A})$ by c), and so $c_{ii} \geq 2$. **q.e.d.**

11.7 Remarks. a) It can be shown that the two socles $\mathbf{S}_r(\mathfrak{A})$ and $\mathbf{S}_l(\mathfrak{A})$ coincide even for quasi-Frobenius algebras.

b) In VIII, 2.13 it will be proved that if $\mathfrak{I}$ is the augmentation ideal of the group-ring of a p-group over a field of characteristic p and $\mathfrak{I}^s \supset \mathfrak{I}^{s+1} = 0$, then $\mathfrak{I}^{s+1-i} = \mathbf{L}(\mathfrak{I}^i)$. The proof uses the symmetry of the group-ring. By 10.11b), this shows that

$$0 \subset \mathfrak{I}^s \subset \mathfrak{I}^{s-1} \subset \cdots \subset \mathfrak{I}^0 = \mathsf{K}\mathfrak{G}$$

is the upper Loewy series of $\mathsf{K}\mathfrak{G}$.

Exercises

48) If $\mathfrak{A}$ is a symmetric algebra, then the full matrix ring $(\mathfrak{A})_m$ is also a symmetric algebra.

49) Show that the tensor product of two Frobenius (or symmetric) algebras is again a Frobenius (or symmetric) algebra.

50) Let V be a finite dimensional vector space. Show that the exterior algebra of V is a Frobenius algebra. When is it a symmetric algebra?

51) If $0 \neq f \in \mathsf{K}[t]$, then $\mathsf{K}[t]/f\,\mathsf{K}[t]$ is a symmetric algebra.

52) Give an example of a symmetric algebra $\mathfrak{A}$ and a two-sided ideal $\mathfrak{I}$ in $\mathfrak{A}$ such that $\mathfrak{A}/\mathfrak{I}$ is not a Frobenius algebra.

53) Let $\mathfrak{A}$ be a K-algebra of finite dimension. Show that the following assertions are equivalent.

a) $\mathfrak{A}$ is a Frobenius algebra.

b) There exists $f \in \mathrm{Hom}_{\mathsf{K}}(\mathfrak{A}, \mathsf{K})$ such that the kernel of f contains no proper right ideal of $\mathfrak{A}$.

54) Let $\mathfrak{A}$ be a K-algebra of finite dimension. Show that the following assertions are equivalent.

a) $\mathfrak{A}$ is a symmetric algebra.

b) There exists $f \in \mathrm{Hom}_{\mathsf{K}}(\mathfrak{A}, \mathsf{K})$ such that the kernel of f contains no proper right ideal of $\mathfrak{A}$ and $f(ab) = f(ba)$ for all $a, b \in \mathfrak{A}$.

55) If an algebra $\mathfrak{A}$ of finite dimension contains only one irreducible right ideal, then $\mathfrak{A}$ is a Frobenius algebra.

56) (TSUSHIMA [1]) Let $\mathfrak{A}$ be a symmetric algebra and choose $f \in \mathrm{Hom}_{\mathsf{K}}(\mathfrak{A}, \mathsf{K})$ as in exercise 54).

a) For any $g \in \mathrm{Hom}_{\mathsf{K}}(\mathfrak{A}, \mathsf{K})$, there exists $a \in \mathfrak{A}$ such that $g(x) = f(ax)$ for all $x \in \mathfrak{A}$.

b) Let $\mathfrak{I}$ be a two-sided ideal in $\mathfrak{A}$ such that $\mathfrak{A}/\mathfrak{I}$ is a symmetric algebra and let g be an element of $\mathrm{Hom}_{\mathsf{K}}(\mathfrak{A}/\mathfrak{I}, \mathsf{K})$ as in exercise 54. Then there exists an element a in the centre of $\mathfrak{A}$ such that $g(x + \mathfrak{I}) = f(ax)$ for all $x \in \mathfrak{A}$. Further $\mathbf{L}(\mathfrak{I}) = \mathbf{R}(\mathfrak{I}) = a\mathfrak{A}$.

§ 12. Two-Sided Decompositions of Algebras

In § 10 we considered the direct decomposition of the group-ring $\mathfrak{A} = \mathsf{K}\mathfrak{G}$ into indecomposable right ideals. By lifting idempotents it was shown that these decompositions correspond exactly to the direct decompositions of $\overline{\mathfrak{A}} = \mathfrak{A}/\mathbf{J}(\mathfrak{A})$ into irreducible right ideals. Unfortunately the situation is not so simple for decompositions of $\mathfrak{A}$ into two-sided ideals, for a central idempotent of $\overline{\mathfrak{A}}$ cannot in general be lifted to a central idempotent of $\mathfrak{A}$.

First we prove a generalization of V, 3.8 about direct decompositions of algebras into two-sided ideals.

12.1 Theorem. *Let $\mathfrak{A}$ be an algebra of finite dimension over a field* K.

a) *There exists a direct decomposition*

$$\mathfrak{A} = \mathfrak{B}_1 \oplus \cdots \oplus \mathfrak{B}_t$$

of $\mathfrak{A}$ into two-sided ideals $\mathfrak{B}_i \neq 0$, which are indecomposable as two-sided ideals. For $i \neq j$ we have $\mathfrak{B}_i\mathfrak{B}_j = 0$.

b) *Suppose that $1 = f_1 + \cdots + f_t$, where $f_i \in \mathfrak{B}_i$. Then the f_i are mutually orthogonal idempotents in the centre $\mathbf{Z}(\mathfrak{A})$ of $\mathfrak{A}$ and $\mathfrak{B}_i = f_i\mathfrak{A} = \mathfrak{A}f_i$.*

c) *$\mathbf{Z}(\mathfrak{A}) = f_1\mathbf{Z}(\mathfrak{A}) \oplus \cdots \oplus f_t\mathbf{Z}(\mathfrak{A})$ is a direct decomposition of the commutative algebra $\mathbf{Z}(\mathfrak{A})$ into indecomposable ideals $f_i\mathbf{Z}(\mathfrak{A}) = \mathfrak{B}_i \cap \mathbf{Z}(\mathfrak{A})$. Each $f_i\mathbf{Z}(\mathfrak{A})$ is a local ring and so $f_i\mathbf{Z}(\mathfrak{A})/\mathbf{J}(f_i\mathbf{Z}(\mathfrak{A}))$ is a field.*

d) *For every $\mathfrak{A}$-module V,*

$$\mathsf{V} = \bigoplus_{i=1}^{t} \mathsf{V}f_i$$

is a direct decomposition of V into $\mathfrak{A}$-modules. If in particular V is indecomposable, then $\mathsf{V} = \mathsf{V}f_j$ for some j.

e) *For each right ideal $\mathfrak{R}$ of $\mathfrak{A}$ we have*

$$\mathfrak{R} = \bigoplus_{i=1}^{t} (\mathfrak{R} \cap \mathfrak{B}_i).$$

In particular, each indecomposable right ideal of $\mathfrak{A}$ lies in one of the $\mathfrak{B}_i$.

f) *If $\mathfrak{A}_1$ and $\mathfrak{A}_2$ are two-sided ideals of $\mathfrak{A}$ and $\mathfrak{A} = \mathfrak{A}_1 \oplus \mathfrak{A}_2$, then $\mathfrak{A}_1$ and $\mathfrak{A}_2$ are both direct sums of certain $\mathfrak{B}_i$. In particular, the $\mathfrak{B}_i$ in a) are uniquely determined.*

Proof. a) The existence of a direct decomposition of the algebra $\mathfrak{A}$ into indecomposable two-sided ideals is clear, since $\mathfrak{A}$ has finite dimension over K. For $i \neq j$ we have

$$\mathfrak{B}_i\mathfrak{B}_j \subseteq \mathfrak{B}_i \cap \mathfrak{B}_j = 0.$$

b) For $i \neq j$ we have $f_i f_j \in \mathfrak{B}_i\mathfrak{B}_j = 0$. Hence

$$f_i = 1f_i = \sum_{j=1}^{t} f_j f_i = f_i^2.$$

Further for all $a \in \mathfrak{A}$ we have

$$\sum_{i=1}^{t} af_i = a1 = 1a = \sum_{i=1}^{t} f_i a.$$

Since af_i and $f_i a$ lie in $\mathfrak{B}_i$, we see by comparing components that $af_i = f_i a$. Thus $f_i \in \mathbf{Z}(\mathfrak{A})$. Since $f_j\mathfrak{B}_i \subseteq \mathfrak{B}_j\mathfrak{B}_i = 0$ for $i \neq j$, we obtain

$$\mathfrak{B}_i = \left(\sum_{j=1}^{t} f_j\right)\mathfrak{B}_i = f_i\mathfrak{B}_i \subseteq f_i\mathfrak{A} \subseteq \mathfrak{B}_i.$$

Hence $\mathfrak{B}_i = f_i\mathfrak{A} = \mathfrak{A}f_i$.

c) Obviously $\mathbf{Z}(\mathfrak{A}) = \bigoplus_{i=1}^{t} f_i\mathbf{Z}(\mathfrak{A})$ is a direct decomposition of $\mathbf{Z}(\mathfrak{A})$. Suppose that $f_i = f' + f''$, where f', f'' are non-zero idempotents of $\mathbf{Z}(\mathfrak{A})$. Then

$$\mathfrak{B}_i = f_i\mathfrak{A} = f'\mathfrak{A} \oplus f''\mathfrak{A},$$

and since f', f'' lie in $\mathbf{Z}(\mathfrak{A})$, $f'\mathfrak{A}$ and $f''\mathfrak{A}$ are two-sided ideals of $\mathfrak{A}$. However this contradicts the indecomposability of $\mathfrak{B}_i = f_i\mathfrak{A}$ as two-sided ideal.

$f_i\mathbf{Z}(\mathfrak{A})/\mathbf{J}(f_i\mathbf{Z}(\mathfrak{A}))$ is a commutative semisimple algebra. Hence by Wedderburn's theorem

$$f_i\mathbf{Z}(\mathfrak{A})/\mathbf{J}(f_i\mathbf{Z}(\mathfrak{A})) \cong \mathsf{K}_1 \oplus \cdots \oplus \mathsf{K}_n$$

for certain fields $\mathsf{K}_1, \ldots, \mathsf{K}_n$. The K_j are minimal (right) ideals, so by 10.3, there exists a decomposition of $f_i\mathbf{Z}(\mathfrak{A})$ as the direct sum of n ideals. Since $f_i\mathbf{Z}(\mathfrak{A})$ is indecomposable, $n = 1$ and $f_i\mathbf{Z}(\mathfrak{A})/\mathbf{J}(f_i\mathbf{Z}(\mathfrak{A}))$ is a field. Thus $f_i\mathbf{Z}(\mathfrak{A})$ is a local ring.

d) We have

$$\mathsf{V} = \mathsf{V}1 = \mathsf{V}\sum_{i=1}^{t} f_i \subseteq \sum_{i=1}^{t} \mathsf{V}f_i \subseteq \mathsf{V}.$$

Hence $\mathsf{V} = \bigoplus_{i=1}^{t} \mathsf{V}f_i$. As f_i is central, $\mathsf{V}f_i$ is an $\mathfrak{A}$-module.

e) By d), $\mathfrak{R} = \bigoplus_{i=1}^{t} \mathfrak{R}f_i$ and

$$\mathfrak{R}f_i \subseteq \mathfrak{R} \cap \mathfrak{B}_i = \mathfrak{R} \cap \mathfrak{A}f_i \subseteq \mathfrak{R}f_i.$$

Thus

$$\mathfrak{R} = \bigoplus_{i=1}^{t} (\mathfrak{R} \cap \mathfrak{B}_i).$$

f) By e),

$$\mathfrak{A} = \mathfrak{A}_1 \oplus \mathfrak{A}_2 = \bigoplus_{i=1}^{t} (\mathfrak{A}_1 \cap \mathfrak{B}_i) \oplus \bigoplus_{i=1}^{t} (\mathfrak{A}_2 \cap \mathfrak{B}_i).$$

Hence for all i,

$$\mathfrak{B}_i = (\mathfrak{A}_1 \cap \mathfrak{B}_i) \oplus (\mathfrak{A}_2 \cap \mathfrak{B}_i).$$

The indecomposability of $\mathfrak{B}_i$ forces $\mathfrak{B}_i \subseteq \mathfrak{A}_1$ or $\mathfrak{B}_i \subseteq \mathfrak{A}_2$. Thus

$$\mathfrak{A}_j = \bigoplus_{\mathfrak{B}_i \subseteq \mathfrak{A}_j} \mathfrak{B}_i \quad (j = 1, 2).$$ **q.e.d.**

A different aspect of two-sided decompositions is often useful.

12.2 Theorem. *Suppose that $\mathfrak{A}$ is a K-algebra of finite dimension and that K is a splitting field for $\mathfrak{A}/\mathbf{J}(\mathfrak{A})$. With the notation of 12.1, the following assertions hold.*

a) $\mathbf{J}(\mathbf{Z}(\mathfrak{A})) = \mathbf{J}(\mathfrak{A}) \cap \mathbf{Z}(\mathfrak{A})$.

b) *The ring $f_i\mathbf{Z}(\mathfrak{A})/\mathbf{J}(f_i\mathbf{Z}(\mathfrak{A}))$ is isomorphic to K. Thus $\mathbf{Z}(\mathfrak{A})/\mathbf{J}(\mathbf{Z}(\mathfrak{A}))$ is the direct sum of t ideals each isomorphic to K.*

c) *There are precisely t algebra homomorphisms $\alpha_1, \ldots, \alpha_t$ of $\mathbf{Z}(\mathfrak{A})$ onto K. With appropriate numbering we have*

$$f_i\alpha_i = 1 \quad \textit{and} \quad f_j\alpha_i = 0 \quad (j \neq i).$$

Further

$$\ker \alpha_i = \mathbf{J}(f_i\mathbf{Z}(\mathfrak{A})) \oplus \bigoplus_{j \neq i} f_j\mathbf{Z}(\mathfrak{A}).$$

Proof. a) $\mathbf{J}(\mathfrak{A}) \cap \mathbf{Z}(\mathfrak{A})$ is a nilpotent ideal of $\mathbf{Z}(\mathfrak{A})$ and is therefore contained in $\mathbf{J}(\mathbf{Z}(\mathfrak{A}))$. Conversely, if $x \in \mathbf{J}(\mathbf{Z}(\mathfrak{A}))$, then x is nilpotent. Since $x \in \mathbf{Z}(\mathfrak{A})$, $x\mathfrak{A}$ is a nilpotent ideal of $\mathfrak{A}$, and it follows that $x \in \mathbf{J}(\mathfrak{A}) \cap \mathbf{Z}(\mathfrak{A})$.

b) Put $\mathsf{L}_i = f_i\mathbf{Z}(\mathfrak{A})/\mathbf{J}(f_i\mathbf{Z}(\mathfrak{A}))$. By 12.1c), L_i is a field. Since $\mathbf{Z}(\mathfrak{A}) = \bigoplus_{i=1}^{t} f_i\mathbf{Z}(\mathfrak{A})$, we have

$$\mathbf{Z}(\mathfrak{A})/\mathbf{J}(\mathbf{Z}(\mathfrak{A})) \cong \mathsf{L}_1 \oplus \cdots \oplus \mathsf{L}_t.$$

Using a), we obtain

$$\mathsf{L}_1 \oplus \cdots \oplus \mathsf{L}_t \cong \mathbf{Z}(\mathfrak{A})/(\mathbf{J}(\mathfrak{A}) \cap \mathbf{Z}(\mathfrak{A})) \cong (\mathbf{Z}(\mathfrak{A}) + \mathbf{J}(\mathfrak{A}))/\mathbf{J}(\mathfrak{A}).$$

Now $(\mathbf{Z}(\mathfrak{A}) + \mathbf{J}(\mathfrak{A}))/\mathbf{J}(\mathfrak{A})$ is a subalgebra of $\mathbf{Z}(\mathfrak{A}/\mathbf{J}(\mathfrak{A}))$, and since K is a splitting field for $\mathfrak{A}/\mathbf{J}(\mathfrak{A})$, we conclude that $\mathfrak{A}/\mathbf{J}(\mathfrak{A})$ is a direct sum of complete matrix rings over K. Hence $\mathbf{Z}(\mathfrak{A}/\mathbf{J}(\mathfrak{A}))$ is isomorphic to a direct sum of fields each isomorphic to K. Thus there is a K-algebra monomorphism σ_i of L_i $(i = 1, \ldots, t)$ into a direct sum $\mathsf{K} \oplus \cdots \oplus \mathsf{K}$. We denote by π_j the projection of $\mathsf{K} \oplus \cdots \oplus \mathsf{K}$ onto the j-th summand. Then there exists j such that $\sigma_i\pi_j \neq 0$. Since $\ker \sigma_i\pi_j$ is an ideal in L_i, $\sigma_i\pi_j$ is a K-linear isomorphism of L_i onto K. Thus $\mathbf{Z}(\mathfrak{A})/\mathbf{J}(\mathbf{Z}(\mathfrak{A}))$ is a direct sum of t ideals each isomorphic to K.

c) By b), $f_i\mathbf{Z}(\mathfrak{A})$ is spanned over K by $\mathbf{J}(f_i\mathbf{Z}(\mathfrak{A}))$ and f_i. Thus $\mathbf{Z}(\mathfrak{A})$ is spanned over K by $f_1, \ldots, f_t$ and $\mathbf{J}(\mathbf{Z}(\mathfrak{A}))$. Also, $f_1, \ldots, f_t$ are linearly independent modulo $\mathbf{J}(\mathbf{Z}(\mathfrak{A}))$. Thus for $1 \leq i \leq t$, there exists a K-linear mapping α_i of $\mathbf{Z}(\mathfrak{A})$ onto K such that $f_j\alpha_i = 0$ for $j \neq i$, $f_i\alpha_i = 1$ and $\mathbf{J}(\mathbf{Z}(\mathfrak{A}))\alpha_i = 0$. Clearly α_i is an algebra homomorphism and $\ker \alpha_i$ is as stated.

Conversely, suppose that α is an arbitrary algebra homomorphism of $\mathbf{Z}(\mathfrak{A})$ onto K. Since $1 = f_1 + \cdots + f_t$, there exists an i such that $f_i\alpha \neq 0$. Since for $j \neq i$ we have

$$0 = (f_if_j)\alpha = (f_i\alpha)(f_j\alpha),$$

we obtain $f_j\alpha = 0$. Obviously $\mathbf{J}(\mathbf{Z}(\mathfrak{A})) \subseteq \ker \alpha$, so $\alpha = \alpha_i$. **q.e.d.**

12.3 Remark. If $\mathsf{K}\mathfrak{G}$ is semisimple and K is a splitting field for $\mathsf{K}\mathfrak{G}$, the number t in 12.1 is the class-number $h(\mathfrak{G})$ of $\mathfrak{G}$ (V, 5.1). A corresponding description of t is unfortunately completely lacking in the case when char K divides $|\mathfrak{G}|$. It happens sometimes that $t = 1$, which means that $\mathsf{K}\mathfrak{G}$ is an indecomposable algebra. This seems to be rare for simple groups, but if $\mathfrak{G}$ is isomorphic to the Mathieu group $\mathfrak{M}_{22}$ and K is an algebraically closed field of characteristic 2, then $\mathsf{K}\mathfrak{G}$ is indecomposable (JAMES [1]). A necessary and sufficient condition for $t = 1$ is known only for p-constrained groups; this condition is that $\mathfrak{G}$ has no non-identity normal p'-subgroups, that is, $\mathbf{O}_{p'}(\mathfrak{G}) = 1$ (see 13.5). The existence of group-rings $\mathsf{K}\mathfrak{G}$ with $t = 1$ shows that central idempotents of $\mathsf{K}\mathfrak{G}/\mathbf{J}(\mathsf{K}\mathfrak{G})$ cannot in general be lifted to central idempotents of $\mathsf{K}\mathfrak{G}$.

If $\mathfrak{A}$ is a semisimple algebra, the two-sided ideals $\mathfrak{B}_i$ of 12.1 are just the sums of all the irreducible right ideals of a given isomorphism type. In general the connection between the two-sided decomposition in 12.1 and the decomposition of $\mathfrak{A}$ into right ideals in 10.3 is not so simple.

12.4 Theorem. *Let $\mathfrak{A}$ be an algebra of finite dimension over a field K. Let $\mathfrak{A} = \bigoplus_{i=1}^{n} e_i\mathfrak{A}$ be a decomposition of $\mathfrak{A}$ into indecomposable right ideals. Put $e_i\mathfrak{A} \sim e_j\mathfrak{A}$ if there exists a sequence*

$$e_i\mathfrak{A} = e_{i_1}\mathfrak{A}, e_{i_2}\mathfrak{A}, \ldots, e_{i_m}\mathfrak{A} = e_j\mathfrak{A},$$

such that any two neighbouring right ideals in this sequence have at least one composition factor in common. Obviously $\sim$ is an equivalence relation. If $\mathscr{A}_1, \ldots, \mathscr{A}_s$ are the equivalence classes of $\sim$, then $s = t$. If we put

$$\mathfrak{A}_j = \bigoplus_{e_i\mathfrak{A} \in \mathscr{A}_j} e_i\mathfrak{A},$$

then $\mathfrak{A} = \bigoplus_{j=1}^{t} \mathfrak{A}_j$ is the decomposition of $\mathfrak{A}$ into indecomposable two-sided ideals of 12.1.

Proof. If $e_i\mathfrak{A}$ and $e_j\mathfrak{A}$ have the composition factor $\overline{e}_k\overline{\mathfrak{A}}$ in common, then by 10.17, $e_i\mathfrak{A}e_k \neq 0$ and $e_j\mathfrak{A}e_k \neq 0$. By 12.1e), there exist ideals $\mathfrak{B}, \mathfrak{B}', \mathfrak{B}''$ in the set $\{\mathfrak{B}_1, \ldots, \mathfrak{B}_t\}$ such that $e_i\mathfrak{A} \subseteq \mathfrak{B}$, $e_j\mathfrak{A} \subseteq \mathfrak{B}'$ and $e_k\mathfrak{A} \subseteq \mathfrak{B}''$. It follows that $0 \neq e_i\mathfrak{A}e_k \subseteq \mathfrak{B}\mathfrak{B}''$, so $\mathfrak{B} = \mathfrak{B}''$; also $0 \neq e_j\mathfrak{A}e_k \subseteq \mathfrak{B}'\mathfrak{B}''$, so $\mathfrak{B}' = \mathfrak{B}''$. Hence $e_i\mathfrak{A} + e_j\mathfrak{A} \subseteq \mathfrak{B}$. Repeated application of this argument shows that the right ideal

$$\mathfrak{A}_j = \bigoplus_{e_i\mathfrak{A} \in \mathscr{A}_j} e_i\mathfrak{A}$$

is contained in a certain $\mathfrak{B}_{j'}$. Obviously $\mathfrak{A} = \bigoplus_{j=1}^{s} \mathfrak{A}_j$.

We show now that $\mathfrak{A}_j$ is a two-sided ideal of $\mathfrak{A}$. To do this, suppose that $e_i\mathfrak{A} \subseteq \mathfrak{A}_j$ and $e_k\mathfrak{A} \subseteq \mathfrak{A}_l$, where $j \neq l$. Then $e_i\mathfrak{A}$ and $e_k\mathfrak{A}$ have no composition factor in common. In particular, $e_k\mathfrak{A}$ has no composition factor isomorphic to $\overline{e_i\mathfrak{A}}$, so it follows from 10.17 that $e_k\mathfrak{A}e_i = 0$. This shows that $\mathfrak{A}_l\mathfrak{A}_j = 0$ for $j \neq l$. Thus

$$\mathfrak{A}\mathfrak{A}_j = \left(\bigoplus_{l=1}^{s} \mathfrak{A}_l\right)\mathfrak{A}_j = \mathfrak{A}_j^2 \subseteq \mathfrak{A}_j,$$

and so $\mathfrak{A}_j$ is a two-sided ideal in $\mathfrak{A}$. Since $\mathfrak{A}_j$ is also a direct summand of $\mathfrak{A}$, it follows from 12.1f) that $\mathfrak{A}_j$ is a direct sum of certain $\mathfrak{B}_i$. Since $\mathfrak{A}_j \subseteq \mathfrak{B}_{j'}$, we obtain $\mathfrak{A}_j = \mathfrak{B}_{j'}$. **q.e.d.**

12.5 Theorem. *Let $\mathfrak{A}$ be a K-algebra of finite dimension and suppose that K is a splitting field for $\mathfrak{A}/\mathbf{J}(\mathfrak{A})$. Further let*

$$\mathfrak{A} = \bigoplus_{i=1}^{n} e_i\mathfrak{A}$$

be a direct decomposition of $\mathfrak{A}$ into indecomposable right ideals.

a) *If $e_i\mathfrak{A}$ and $e_j\mathfrak{A}$ do not lie in the same two-sided indecomposable direct summand $\mathfrak{B}_k$ of $\mathfrak{A}$, then $c_{ij} = 0$. Thus the Cartan matrix can be written in the form*

$$C = \begin{pmatrix} C_1 & 0 & 0 & \cdots & 0 \\ 0 & C_2 & 0 & \cdots & 0 \\ \vdots & \vdots & & & \vdots \\ 0 & 0 & 0 & \cdots & C_t \end{pmatrix},$$

where C_i corresponds to the two-sided ideal $\mathfrak{B}_i$ ($i = 1, \ldots, t$).

b) *The matrices C_i in a) are indecomposable in the sense that the right ideals $e_j\mathfrak{A}$ in $\mathfrak{B}_i$ cannot be ordered such that C_i has the form*

$$\begin{pmatrix} C_{i1} & 0 \\ 0 & C_{i2} \end{pmatrix}$$

with square matrices C_{i1}, C_{i2}.

Proof. a) Suppose that $e_i\mathfrak{A} \subseteq \mathfrak{B}_{i'}$ and $e_j\mathfrak{A} \subseteq \mathfrak{B}_{j'}$, where $i' \neq j'$. Then

$$e_j\mathfrak{A}e_i \subseteq \mathfrak{B}_{j'}\mathfrak{B}_{i'} \subseteq \mathfrak{B}_{j'} \cap \mathfrak{B}_{i'} = 0.$$

Hence by 10.17

$$c_{ij}\dim_{\mathsf{K}}\operatorname{Hom}_{\mathfrak{A}}(\overline{e_j\mathfrak{A}}, \overline{e_j\mathfrak{A}}) = \dim_{\mathsf{K}} e_i\mathfrak{A}e_j = 0.$$

b) Suppose that an ordering is possible for which C_i is decomposable. Then the $e_j\mathfrak{A}$ which belong to C_{i1} form a union of equivalence classes for the relation $\sim$ of 12.4. But this contradicts 12.4. **q.e.d.**

We now introduce the important concept of blocks.

12.6 Definition. Let K be a field of characteristic p. Let

$$\mathsf{K}\mathfrak{G} = \bigoplus_{i=1}^{t} \mathfrak{B}_i = \bigoplus_{i=1}^{t} f_i\mathsf{K}\mathfrak{G}$$

be the decomposition of $\mathsf{K}\mathfrak{G}$ defined in 12.1.

a) $\mathfrak{B}_i$ is called a *block ideal* and f_i a *block idempotent* of $\mathsf{K}\mathfrak{G}$. If K is a splitting field for $\mathsf{K}\mathfrak{G}/\mathbf{J}(\mathsf{K}\mathfrak{G})$, we call the algebra homomorphism α_i of $\mathbf{Z}(\mathsf{K}\mathfrak{G})$ into K for which $f_i\alpha_i = 1$ (cf. 12.2c)) the *block character* of $\mathfrak{B}_i$.

b) For $i = 1, \ldots, t$, the block $\mathscr{B}_i$ is defined to be the class of all $\mathsf{K}\mathfrak{G}$-modules V for which $\mathsf{V}f_i = \mathsf{V}$. Since $f_if_j = 0$ for $i \neq j$, every non-zero $\mathsf{K}\mathfrak{G}$-module lies in at most one block. If V is indecomposable, then V belongs to a block by 12.1d).

A $\mathsf{K}\mathfrak{G}$-module belongs to a block if and only if all its indecomposable direct summands belong to the same block. A $\mathsf{K}\mathfrak{G}$-module belongs to a block if and only if all its composition factors belong to the same block.

Every block contains irreducible modules, for $\mathscr{B}_i$ contains the composition factors of $\mathfrak{B}_i$.

c) Let V be a $\mathsf{K}\mathfrak{G}$-module for the trivial representation of $\mathfrak{G}$ of degree 1. By b), V belongs to a block, and we choose the numbering so that this block is $\mathscr{B}_1$. Thus $\mathsf{V}f_1 = \mathsf{V}$. We call $\mathscr{B}_1$ the *principal block* of $\mathsf{K}\mathfrak{G}$.

For the proof of an important property of central idempotents in group-rings, we need a lemma which is central for some deeper results in modular representation theory.

12.7 Lemma (R. BRAUER [6]). *Suppose that K is a field of characteristic p, that $\mathfrak{P}$ is a p-subgroup of $\mathfrak{G}$ and that $\mathfrak{H}$ is a subgroup for which $\mathbf{C}_{\mathfrak{G}}(\mathfrak{P}) \leq \mathfrak{H} \leq \mathbf{N}_{\mathfrak{G}}(\mathfrak{P})$. We define a K-linear mapping γ of $\mathbf{Z}(\mathsf{K}\mathfrak{G})$ into $\mathsf{K}\mathfrak{H}$ by putting*

$$\left(\sum_{x \in \mathsf{C}} x\right)\gamma = \sum_{x \in \mathsf{C} \cap \mathbf{C}_{\mathfrak{G}}(\mathfrak{P})} x$$

for each conjugacy class C *of* $\mathfrak{G}$. *Then* γ *is an algebra homomorphism of* $\mathbf{Z}(\mathsf{K}\mathfrak{G})$ *into* $\mathbf{Z}(\mathsf{K}\mathfrak{H})$, *and the kernel of* γ *is spanned by the class-sums of those conjugacy classes* C *of* $\mathfrak{G}$ *for which* $\mathsf{C} \cap \mathbf{C}_{\mathfrak{G}}(\mathfrak{P}) = \varnothing$.

Proof. Since $\mathbf{C}_{\mathfrak{G}}(\mathfrak{P}) \leqq \mathfrak{H}$, the set $\mathsf{C} \cap \mathbf{C}_{\mathfrak{G}}(\mathfrak{P})$ is $\mathfrak{H}$-invariant for any conjugacy class C of $\mathfrak{G}$. Thus $\mathsf{C} \cap \mathbf{C}_{\mathfrak{G}}(\mathfrak{P})$ is a union of conjugacy classes of $\mathfrak{H}$. Hence γ is indeed a K-linear mapping of $\mathbf{Z}(\mathsf{K}\mathfrak{G})$ into $\mathbf{Z}(\mathsf{K}\mathfrak{H})$. For $\mathfrak{P} = 1$, we have $\mathfrak{H} = \mathfrak{G}$ and the assertion is trivial. Suppose henceforth that $\mathfrak{P} \neq 1$.

Let $\mathsf{C}_1, \dots, \mathsf{C}_h$ be the conjugacy classes of $\mathfrak{G}$. We put $\mathsf{C}_i' = \mathsf{C}_i - \mathsf{C}_i \cap \mathbf{C}_{\mathfrak{G}}(\mathfrak{P})$, $c_i = \sum_{x \in \mathsf{C}_i} x$ and $c_i' = \sum_{x \in \mathsf{C}_i'} x$. As $\{c_1, \dots, c_h\}$ is a K-basis of $\mathbf{Z}(\mathsf{K}\mathfrak{G})$, we have

$$(c_i\gamma + c_i')(c_j\gamma + c_j') = c_ic_j = \sum_{l=1}^{h} a_{ijl}c_l = \sum_{l=1}^{h} a_{ijl}(c_l\gamma + c_l')$$

for certain elements a_{ijl} in K. The elements of $\mathfrak{G}$ appearing in $(c_i\gamma)c_j'$ and in $c_i'(c_j\gamma)$ obviously all lie outside $\mathbf{C}_{\mathfrak{G}}(\mathfrak{P})$, and all elements appearing in $(c_i\gamma)(c_j\gamma)$ lie in $\mathbf{C}_{\mathfrak{G}}(\mathfrak{P})$. We show that no element of $\mathbf{C}_{\mathfrak{G}}(\mathfrak{P})$ appears in $c_i'c_j'$; it will then follow by comparing coefficients that

$$(c_i\gamma)(c_j\gamma) = \sum_{l=1}^{h} a_{ijl}(c_l\gamma) = (c_ic_j)\gamma.$$

Suppose that $g \in \mathbf{C}_{\mathfrak{G}}(\mathfrak{P})$, and for fixed i, j write

$$\mathfrak{A} = \{(x, y) \mid x \in \mathsf{C}_i', y \in \mathsf{C}_j', xy = g\}.$$

We have to show that $|\mathfrak{A}| \equiv 0 \ (p)$, for g appears in $c_i'c_j'$ with coefficient $|\mathfrak{A}|$. Since $g \in \mathbf{C}_{\mathfrak{G}}(\mathfrak{P})$, a permutation representation of $\mathfrak{P}$ on $\mathfrak{A}$ is defined by putting $(x, y) \to (x^u, y^u)$ for $u \in \mathfrak{P}$. Since $\mathsf{C}_i' \cap \mathbf{C}_{\mathfrak{G}}(\mathfrak{P}) = \varnothing$, there is no orbit of length one. The length of any orbit of $\mathfrak{P}$ on $\mathfrak{A}$ is thus a power of p larger than 1. Hence $|\mathfrak{A}| \equiv 0 \ (p)$.

We still have to prove the assertion about the kernel of γ. If C is a conjugacy class in $\mathfrak{G}$ for which $\mathsf{C} \cap \mathbf{C}_{\mathfrak{G}}(\mathfrak{P}) = \varnothing$, then obviously $(\sum_{x \in \mathsf{C}} x)\gamma = 0$. Since no element of $\mathbf{C}_{\mathfrak{G}}(\mathfrak{P})$ appears in more than one $c_i\gamma$, it follows from

$$\left(\sum_{i=1}^{h} a_ic_i\right)\gamma = 0 \quad (a_i \in \mathsf{K})$$

that each $a_ic_i\gamma = 0$. Thus $a_i = 0$ unless $\mathsf{C}_i \cap \mathbf{C}_{\mathfrak{G}}(\mathfrak{P}) = \varnothing$. **q.e.d.**

12.8 Theorem (OSIMA [1]). *Suppose that* char $\mathsf{K} = p$ *and that* $f = \sum_{g \in \mathfrak{G}} a_g g$ *is an idempotent in* $\mathbf{Z}(\mathsf{K}\mathfrak{G})$. *If* $a_g \neq 0$, *then* g *is a* p'*-element.*

Proof (PASSMAN [5]). Suppose that $a_g \neq 0$, but that p divides the order of g. Suppose that $g = uv = vu$, where u is a p-element and v is a p'-element. We apply 12.7 with $\mathfrak{P} = \langle u \rangle$ and $\mathfrak{H} = \mathbf{C}_{\mathfrak{G}}(u)$. Then $f' = f\gamma$ is an idempotent in $\mathbf{Z}(\mathsf{K}\mathfrak{H})$. If $f\gamma = \sum_{h \in \mathfrak{H}} b_h h$ with $b_h \in \mathsf{K}$, then from $g \in \mathbf{C}_{\mathfrak{G}}(u)$ it follows that $b_g = a_g \neq 0$.

Let n be a natural number such that $p^n \geq |\mathfrak{H}|$ and $p^n - 1$ is divisible by the order of v. Put $c = v^{-1} f'$ and write $c = \sum_{h \in \mathfrak{H}} c_h h$. Then since $f \in \mathbf{Z}(\mathsf{K}\mathfrak{H})$, we have

$$c^{p^n} = v^{-p^n} f'^{p^n} = v^{-1} f' = c.$$

We use the subspace $\mathbf{Q}(\mathsf{K}\mathfrak{H})$ of $\mathsf{K}\mathfrak{H}$ defined as in 3.1 by

$$\mathbf{Q}(\mathsf{K}\mathfrak{H}) = \langle ab - ba \mid a, b \in \mathsf{K}\mathfrak{H} \rangle.$$

By 3.3,

$$c = c^{p^n} = \sum_{h \in \mathfrak{H}} c_h^{p^n} h^{p^n} + r$$

for some $r \in \mathbf{Q}(\mathsf{K}\mathfrak{H})$. Since $p^n \geq |\mathfrak{H}|$, all the h^{p^n} are p'-elements. Since $f' = vc$, we get

$$0 \neq a_g = b_g = c_{v^{-1}g} = c_u.$$

Hence if we write $r = \sum_{h \in \mathfrak{H}} r_h h$, then $r_u \neq 0$. But by definition of $\mathbf{Q}(\mathsf{K}\mathfrak{H})$, r is a linear combination of elements of the form $xy - yx$ with $x, y \in \mathfrak{H}$. Thus there exist $x, y \in \mathfrak{H}$ such that $xy = u$, but $yx \neq u$. However this is impossible since u lies in the centre of $\mathfrak{H}$. **q.e.d.**

12.9 Theorem. *Let* $\mathfrak{N}$ *be a normal subgroup of* $\mathfrak{G}$ *and let* K *be a field of characteristic* p.

a) *Suppose that* $\mathfrak{N}$ *is a* p'*-group and put*

$$f = |\mathfrak{N}|^{-1} \sum_{x \in \mathfrak{N}} x.$$

Then f *is a central idempotent of* $\mathsf{K}\mathfrak{G}$ *and* $fx = f$ *for all* $x \in \mathfrak{N}$. *Further,*

$$\mathsf{K}\mathfrak{G} = f\mathsf{K}\mathfrak{G} \oplus (1 - f)\mathsf{K}\mathfrak{G}$$

is a direct decomposition of $\mathsf{K}\mathfrak{G}$ *into two-sided ideals. The mapping* α *of* $f\mathsf{K}\mathfrak{G}$ *onto* $\mathsf{K}(\mathfrak{G}/\mathfrak{N})$ *defined by* $(fg)\alpha = \mathfrak{N}g$ $(g \in \mathfrak{G})$ *is a* K*-algebra isomorphism and also an isomorphism of* $\mathsf{K}\mathfrak{G}$*-modules.*

b) *The following assertions are equivalent.*

(i) $\mathfrak{N}$ *is a* p'*-group.*

(ii) *There is an idempotent* e *in* $\mathsf{K}\mathfrak{G}$ *such that* $e\mathsf{K}\mathfrak{G}$ *and* $\mathsf{K}(\mathfrak{G}/\mathfrak{N})$ *are isomorphic* $\mathsf{K}\mathfrak{G}$*-modules.*

(iii) *There is a right ideal* $\mathfrak{R}$ *in* $\mathsf{K}\mathfrak{G}$ *such that* $\mathsf{K}\mathfrak{G} = \ker\beta \oplus \mathfrak{R}$, *where* β *is the natural epimorphism of* $\mathsf{K}\mathfrak{G}$ *onto* $\mathsf{K}(\mathfrak{G}/\mathfrak{N})$ *for which* $g\beta = g\mathfrak{N}$ *for all* $g \in \mathfrak{G}$.

Proof. a) Obviously $fx = f$ for all $x \in \mathfrak{N}$, so f is an idempotent. Since $\mathfrak{N}$ is the union of certain conjugacy classes of $\mathfrak{G}$, f lies in the centre of $\mathsf{K}\mathfrak{G}$. If T is a transversal of $\mathfrak{N}$ in $\mathfrak{G}$, then since $fx = f$ for all $x \in \mathfrak{N}$, we obtain

$$f\mathsf{K}\mathfrak{G} = \bigoplus_{t \in \mathsf{T}} \mathsf{K}(ft).$$

If $g \in \mathfrak{G}$ and $tg = yt'$ $(t, t' \in \mathsf{T}, y \in \mathfrak{N})$, then

$$(ft)g = ft' \quad \text{and} \quad (\mathfrak{N}t)g = \mathfrak{N}t'.$$

Hence there is a $\mathsf{K}\mathfrak{G}$-isomorphism α of $f\mathsf{K}\mathfrak{G}$ onto $\mathsf{K}(\mathfrak{G}/\mathfrak{N})$ such that $(ft)\alpha = \mathfrak{N}t$. It follows that $(fg)\alpha = \mathfrak{N}g$ for all $g \in \mathfrak{G}$. From

$$(fg_1fg_2)\alpha = (fg_1g_2)\alpha = \mathfrak{N}g_1g_2 = (\mathfrak{N}g_1)(\mathfrak{N}g_2)$$

we conclude that α is also an algebra isomorphism.

b) By a), (i) implies (ii).

If (ii) holds, then since

$$\mathsf{K}\mathfrak{G} = e\mathsf{K}\mathfrak{G} \oplus (1 - e)\mathsf{K}\mathfrak{G}$$

is a decomposition of $\mathsf{K}\mathfrak{G}$ into right ideals, $e\mathsf{K}\mathfrak{G}$ is a projective $\mathsf{K}\mathfrak{G}$-module. Since

$$e\mathsf{K}\mathfrak{G} \cong \mathsf{K}(\mathfrak{G}/\mathfrak{N}) \cong \mathsf{K}\mathfrak{G}/\ker\beta,$$

it follows that $\mathsf{K}\mathfrak{G} = \ker\beta \oplus \mathfrak{R}$ for some right ideal $\mathfrak{R}$.

If (iii) holds, $\mathfrak{R}$ is a projective right ideal of $\mathsf{K}\mathfrak{G}$ and

$$\mathfrak{R} \cong \mathsf{K}\mathfrak{G}/\ker\beta \cong \mathsf{K}(\mathfrak{G}/\mathfrak{N}).$$

It follows from 7.18 that $\mathfrak{N}$ is a p'-group. **q.e.d.**

Exercise

57) Let V be an indecomposable $K\mathfrak{G}$-module. If V does not belong to the principal block $\mathscr{B}_1$, then $\mathbf{H}^i(\mathfrak{G}, V) = 0$ for all $i \geq 1$. (For $i = 1$ use Exercise 34. For $i > 1$ consider an exact sequence

$$0 \to V \to P \to W \to 0$$

where P is projective, and P and W belong to the same block as V. Then use $\mathbf{H}^i(\mathfrak{G}, V) \cong \mathbf{H}^{i-1}(\mathfrak{G}, W)$.)

§ 13. Blocks of p-Constrained Groups

13.1 Lemma. *Let* K *be a field of characteristic p.*

a) *Let f_1 be the block idempotent for the principal block of* $K\mathfrak{G}$. *Then* $f_1(1 - g) = 0$ *for all* $g \in \mathbf{O}_{p'}(\mathfrak{G})$.

b) *If* $K\mathfrak{G}$ *is an indecomposable algebra, then* $\mathbf{O}_{p'}(\mathfrak{G}) = 1$.

Proof. a) Let

$$f = |\mathbf{O}_{p'}(\mathfrak{G})|^{-1} \sum_{g \in \mathbf{O}_{p'}(\mathfrak{G})} g.$$

By 12.9a), f is a central idempotent of $K\mathfrak{G}$, so $f_1 f$ and $f_1(1 - f)$ are central idempotents of $K\mathfrak{G}$. Since $f_1 K\mathfrak{G}$ is an indecomposable two-sided ideal of $K\mathfrak{G}$, either $f_1 f = 0$ or $f_1(1 - f) = 0$. But if Kv is the module for the trivial representation of $\mathfrak{G}$, then $vf = v$, so $vf_1 f = vf_1 \neq 0$ and hence $f_1 f \neq 0$. Thus $f_1(1 - f) = 0$ and $f_1 f = f_1$. Since $fg = f$ for all $g \in \mathbf{O}_{p'}(\mathfrak{G})$, it follows that $f_1 g = f_1 fg = f_1 f = f_1$.

b) Since

$$K\mathfrak{G} = f_1 K\mathfrak{G} \oplus (1 - f_1)K\mathfrak{G}$$

is a decomposition of $K\mathfrak{G}$ into two-sided ideals and $K\mathfrak{G}$ is indecomposable, $1 - f_1 = 0$ and $f_1 = 1$. Hence by a), $\mathbf{O}_{p'}(\mathfrak{G}) = 1$. **q.e.d.**

The converse of 13.1b) is in general false, for it can be shown that if $\mathfrak{G}$ is the alternating group of degree 5 and $p = 3$, then $\mathbf{O}_{p'}(\mathfrak{G}) = 1$ but $\mathfrak{G}$ has 3 blocks (Exercise 69). However, the converse is true for p-constrained groups. To prove this, we need the following lemma.

13.2 Lemma (COSSEY, GASCHÜTZ [1]). *Let* K *be a field of characteristic p and let* V *be an irreducible* $\mathsf{K}\mathfrak{G}$*-module. Then the following assertions are equivalent.*

a) V *belongs to the principal block.*

b) *If* C *is any conjugacy class of* $\mathfrak{G}$ *and* $c = \sum_{g \in \mathsf{C}} g$, *then* $\mathsf{V}(c - |\mathsf{C}|) = 0$.

Proof. Let $\mathfrak{J}$ be the K-space spanned by all $c - |\mathsf{C}|$, where C runs through the conjugacy classes of $\mathfrak{G}$ and $c = \sum_{g \in \mathsf{C}} g$. Thus $\dim_{\mathsf{K}} \mathfrak{J} = h - 1$, where h is the class number of $\mathfrak{G}$. So $\mathfrak{J}$ is a subspace of $\mathbf{Z}(\mathsf{K}\mathfrak{G})$ of codimension 1. Condition b) is equivalent to $\mathsf{V}\mathfrak{J} = 0$.

Let

$$\mathfrak{A} = \left\{ \sum_{g \in \mathfrak{G}} a_g g \,\middle|\, \sum_{g \in \mathfrak{G}} a_g = 0 \right\}$$

be the augmentation ideal of $\mathsf{K}\mathfrak{G}$. Then $\mathfrak{J} \subseteq \mathfrak{A}$, so $\mathfrak{J} \subseteq \mathfrak{A} \cap \mathbf{Z}(\mathsf{K}\mathfrak{G})$. But $\mathbf{Z}(\mathsf{K}\mathfrak{G}) \subsetneq \mathfrak{A}$, since $1 \in \mathbf{Z}(\mathsf{K}\mathfrak{G})$ and $1 \notin \mathfrak{A}$. Thus $\mathfrak{J} = \mathbf{Z}(\mathsf{K}\mathfrak{G}) \cap \mathfrak{A}$. Suppose

$$\mathsf{K}\mathfrak{G} = \bigoplus_{i=1}^{t} f_i \mathsf{K}\mathfrak{G}$$

as in 12.1. Now $\mathsf{K}\mathfrak{G}/\mathfrak{A}$ is a module for the trivial representation of $\mathfrak{G}$, so $(\mathsf{K}\mathfrak{G}/\mathfrak{A}) f_i = 0$ for all $i > 1$. Thus $f_i \in \mathfrak{A}$ and so $f_i \in \mathfrak{J}$ $(i > 1)$. Since

$$\mathbf{Z}(\mathsf{K}\mathfrak{G}) = \bigoplus_{i=1}^{t} f_i \mathbf{Z}(\mathsf{K}\mathfrak{G}),$$

it follows that

$$\mathfrak{J} = \mathfrak{J}_1 \oplus \bigoplus_{i=2}^{t} f_i \mathbf{Z}(\mathsf{K}\mathfrak{G}),$$

where $\mathfrak{J}_1 = f_1 \mathbf{Z}(\mathsf{K}\mathfrak{G}) \cap \mathfrak{J} = f_1 \mathbf{Z}(\mathsf{K}\mathfrak{G}) \cap \mathfrak{A}$. Hence $\dim_{\mathsf{K}} f_1 \mathbf{Z}(\mathsf{K}\mathfrak{G})/\mathfrak{J}_1 = 1$ and $\mathfrak{J}_1$ is a maximal ideal of the ring $f_1 \mathbf{Z}(\mathsf{K}\mathfrak{G})$. But by 12.1c), $f_1 \mathbf{Z}(\mathsf{K}\mathfrak{G})$ is a local ring, so $\mathfrak{J}_1 = \mathbf{J}(f_1 \mathbf{Z}(\mathsf{K}\mathfrak{G}))$ and the elements of $\mathfrak{J}_1$ are nilpotent elements of $\mathbf{Z}(\mathsf{K}\mathfrak{G})$. Hence $\mathfrak{J}_1 \subseteq \mathbf{J}(\mathsf{K}\mathfrak{G})$, and since V is irreducible, $\mathsf{V}\mathfrak{J}_1 = 0$. Thus

$$\mathsf{V}\mathfrak{J} = \mathsf{V}\mathfrak{J}_1 + \sum_{i=2}^{t} \mathsf{V} f_i \mathbf{Z}(\mathsf{K}\mathfrak{G}) = \sum_{i=2}^{t} \mathsf{V} f_i .$$

Thus $\mathsf{V}\mathfrak{J} = 0$ if and only if $\mathsf{V}f_i = 0$ for all $i \geq 2$. Since V necessarily lies in a block, it follows that $\mathsf{V}\mathfrak{J} = 0$ if and only if $\mathsf{V}f_1 = \mathsf{V}$. **q.e.d.**

13.3 Definition. A group $\mathfrak{G}$ is called *p-constrained* if

$$\mathbf{C}_{\mathfrak{G}}(\mathbf{O}_{p',p}(\mathfrak{G})/\mathbf{O}_{p'}(\mathfrak{G})) \leq \mathbf{O}_{p',p}(\mathfrak{G}).$$

We remark that by VI, 6.5 every p-soluble group is p-constrained.

13.4 Theorem. *Let* K *be a field of characteristic* p. *Then the intersection of the kernels of all irreducible* $\mathsf{K}\mathfrak{G}$*-modules is the largest normal* p*-subgroup* $\mathbf{O}_p(\mathfrak{G})$ *of* $\mathfrak{G}$.

Proof. We denote by $\mathfrak{K}$ the intersection of all the kernels of irreducible $\mathsf{K}\mathfrak{G}$-modules. Then $\mathbf{O}_p(\mathfrak{G}) \leq \mathfrak{K}$ by V, 5.17. Conversely, it follows from VI, 7.20 that $\mathfrak{K}$ consists only of p-elements, hence $\mathfrak{K} \leq \mathbf{O}_p(\mathfrak{G})$. **q.e.d.**

13.5 Theorem (COSSEY, FONG, GASCHÜTZ). *Let* $\mathfrak{G}$ *be a p-constrained group and let* K *be a field of characteristic* p. *Then* $\mathsf{K}\mathfrak{G}$ *is an indecomposable algebra if and only if* $\mathbf{O}_{p'}(\mathfrak{G}) = 1$.

Proof. If $\mathsf{K}\mathfrak{G}$ is indecomposable, then $\mathbf{O}_{p'}(\mathfrak{G}) = 1$ by 13.1b).

Suppose that $\mathbf{O}_{p'}(\mathfrak{G}) = 1$. It has to be shown that $\mathfrak{G}$ has only one block. Since every block contains irreducible $\mathsf{K}\mathfrak{G}$-modules, it is sufficient to show that any irreducible $\mathsf{K}\mathfrak{G}$-module V belongs to the principal block. We shall do this by using 13.2. Thus we show that if C is any conjugacy class of $\mathfrak{G}$ and $c = \sum_{g \in \mathsf{C}} g$, then $\mathsf{V}(c - |\mathsf{C}|) = 0$.

If $\mathsf{C} \subseteq \mathbf{O}_p(\mathfrak{G})$, then $\mathsf{V}(1 - x) = 0$ for all $x \in \mathsf{C}$ by 13.4, so $\mathsf{V}(c - |\mathsf{C}|) = 0$. Suppose that $\mathsf{C} \nsubseteq \mathbf{O}_p(\mathfrak{G})$. Choose $g \in \mathsf{C}$, then $g \notin \mathbf{O}_p(\mathfrak{G})$. Since $\mathbf{O}_{p'}(\mathfrak{G}) = 1$, it follows from the assumption that $\mathfrak{G}$ is p-constrained that $\mathbf{C}_{\mathfrak{G}}(\mathbf{O}_p(\mathfrak{G})) \leq \mathbf{O}_p(\mathfrak{G})$. Thus $\mathbf{O}_p(\mathfrak{G}) \nleq \mathbf{C}_{\mathfrak{G}}(g)$. Let

$$p^n = |\mathbf{C}_{\mathfrak{G}}(g)\mathbf{O}_p(\mathfrak{G}) : \mathbf{C}_{\mathfrak{G}}(g)| = |\mathbf{O}_p(\mathfrak{G}) : \mathbf{O}_p(\mathfrak{G}) \cap \mathbf{C}_{\mathfrak{G}}(g)|,$$

so $p^n > 1$. Let S be a transversal of $\mathbf{O}_p(\mathfrak{G}) \cap \mathbf{C}_{\mathfrak{G}}(g)$ in $\mathbf{O}_p(\mathfrak{G})$ and let T be a transversal of $\mathbf{C}_{\mathfrak{G}}(g)\mathbf{O}_p(\mathfrak{G})$ in $\mathfrak{G}$. Thus $|\mathsf{S}| = p^n$ and ST is a transversal of $\mathbf{C}_{\mathfrak{G}}(g)$ in $\mathfrak{G}$. Hence

$$\mathsf{C} = \{g^{st} | s \in \mathsf{S}, t \in \mathsf{T}\}.$$

As

$$g^{st}g^{-t} = [s, g^{-1}]^t \in \mathbf{O}_p(\mathfrak{G})$$

and $\mathbf{O}_p(\mathfrak{G})$ operates trivially on the irreducible $\mathsf{K}\mathfrak{G}$-module V (by 13.4), we obtain

$$vc = \sum_{t \in \mathsf{T}} \sum_{s \in \mathsf{S}} vg^{st} = |\mathsf{S}| \sum_{t \in \mathsf{T}} vg^t = p^n \sum_{t \in \mathsf{T}} vg^t = 0.$$

But also $|\mathsf{C}| = |\mathfrak{G} : \mathbf{C}_{\mathfrak{G}}(g)| \equiv 0(p)$, so $\mathsf{V}(c - |\mathsf{C}|) = 0$. **q.e.d.**

13.6 Theorem. *Let $\mathfrak{G}$ be a p-constrained group and put $\mathfrak{N} = \mathbf{O}_{p'}(\mathfrak{G})$. Let K be a field of characteristic p. Then*

$$|\mathfrak{N}|^{-1} \sum_{g \in \mathfrak{N}} g$$

is the central idempotent f_1 of $\mathsf{K}\mathfrak{G}$ belonging to the principal block of $\mathfrak{G}$.

Proof. We write

$$f = |\mathfrak{N}|^{-1} \sum_{g \in \mathfrak{N}} g.$$

By 12.9, f is a central idempotent in $\mathsf{K}\mathfrak{G}$ and there is an algebra isomorphism β of $f\mathsf{K}\mathfrak{G}$ onto $\mathsf{K}(\mathfrak{G}/\mathfrak{N})$ such that $(fg)\beta = \mathfrak{N}g$ for all $g \in \mathfrak{G}$. But since $\mathbf{O}_{p'}(\mathfrak{G}/\mathfrak{N}) = 1$ and $\mathfrak{G}/\mathfrak{N}$ is also p-constrained, $\mathsf{K}(\mathfrak{G}/\mathfrak{N})$ is an indecomposable algebra by 13.5. Thus $f\mathsf{K}\mathfrak{G}$ is indecomposable, so $f\mathsf{K}\mathfrak{G}$ is a block ideal of $\mathsf{K}\mathfrak{G}$. If $\mathsf{K}v$ is the module for the trivial representation of $\mathfrak{G}$, then $vf = v$, so $\mathsf{K}v$ belongs to the block defined by f. Thus f belongs to the principal block of $\mathfrak{G}$. **q.e.d.**

13.7 Theorem (FONG, GASCHÜTZ [1]). *Let $\mathfrak{G}$ be a p-constrained group and let K be a field of characteristic p. Let V be an irreducible $\mathsf{K}\mathfrak{G}$-module. Then the following conditions are equivalent.*

a) *V belongs to the principal block of $\mathfrak{G}$.*
b) *$\mathbf{O}_{p'}(\mathfrak{G})$ operates trivially on V.*
c) *$\mathbf{O}_{p',p}(\mathfrak{G})$ operates trivially on V.*

Proof. Let f_1 be the central idempotent belonging to the principal block of $\mathfrak{G}$. If a) holds, then $\mathsf{V}f_1 = \mathsf{V}$, so $\mathbf{O}_{p'}(\mathfrak{G})$ operates trivially on V by 13.1. Thus a) implies b).

If b) holds, V can be regarded as an irreducible $\mathsf{K}(\mathfrak{G}/\mathbf{O}_{p'}(\mathfrak{G}))$-module. It follows from 13.4 that $\mathbf{O}_{p',p}(\mathfrak{G})$ operates trivially on V. Thus b) implies c).

And if c) holds, then by 13.6, $vf_1 = v$ for all $v \in \mathsf{V}$. Thus V belongs to the principal block. Hence c) implies a). **q.e.d.**

From 13.7 one of the rare statements about tensor products follows.

13.8 Theorem (FONG, GASCHÜTZ [1]). *Let $\mathfrak{G}$ be a p-constrained group and let K be a field of characteristic p. If $\mathsf{V}_1, \mathsf{V}_2$ are irreducible $\mathsf{K}\mathfrak{G}$-modules in the principal block $\mathscr{B}_1$ of $\mathfrak{G}$, then all the composition factors of $\mathsf{V}_1 \otimes_{\mathsf{K}} \mathsf{V}_2$ belong to $\mathscr{B}_1$.*

Proof. By 13.7, $\mathbf{O}_{p'}(\mathfrak{G})$ operates trivially on V_1 and on V_2, hence also on $\mathsf{V}_1 \otimes_{\mathsf{K}} \mathsf{V}_2$. Thus $\mathbf{O}_{p'}(\mathfrak{G})$ operates trivially on all the composition factors of $\mathsf{V}_1 \otimes_{\mathsf{K}} \mathsf{V}_2$ and these therefore belong to the principal block by 13.7.
q.e.d.

We derive another result from 13.2.

13.9 Theorem (COSSEY, FONG, GASCHÜTZ). *Let $\mathsf{K} = GF(p)$ and let V be an irreducible $\mathsf{K}\mathfrak{G}$-module isomorphic to some chief factor $\mathfrak{M}/\mathfrak{N}$ of $\mathfrak{G}$. Then V belongs to the principal block.*

Proof. By 13.2, we have to show that

$$(1) \qquad \prod_{g \in \mathsf{C}} x^g x^{-1} \in \mathfrak{N}$$

for all $x \in \mathfrak{M}$ and for every conjugacy class C of $\mathfrak{G}$. If $x^g x^{-1} \in \mathfrak{N}$ for every $x \in \mathfrak{M}$ and $g \in \mathsf{C}$, this is clear. Suppose then that there exists $g \in \mathsf{C}$ such that $\mathfrak{M}/\mathfrak{N}$ is not centralized by $g\mathfrak{N}$. Let $\mathfrak{D}/\mathfrak{N} = \mathbf{C}_{\mathfrak{G}/\mathfrak{N}}(g\mathfrak{N})$; thus $\mathfrak{M} \not\leq \mathfrak{D}$. Let S be a transversal of $\mathfrak{D} \cap \mathfrak{M}$ in $\mathfrak{M}$; thus $|\mathsf{S}|$ is a power of p greater than 1. Let R be a transversal of $\mathfrak{M}\mathfrak{D}$ in $\mathfrak{G}$. Then SR is a transversal of $\mathfrak{D}$ in $\mathfrak{G}$. Finally let T be a transversal of $\mathbf{C}_{\mathfrak{G}}(g)$ in $\mathfrak{D}$. Then TSR is a transversal of $\mathbf{C}_{\mathfrak{G}}(g)$ in $\mathfrak{G}$ and

$$\mathsf{C} = \{g^{tsr} \mid t \in \mathsf{T}, s \in \mathsf{S}, r \in \mathsf{R}\}.$$

Since $t \in \mathfrak{D}$, $g^t \equiv g \bmod \mathfrak{N}$. Since also $s \in \mathfrak{M}$ and $\mathfrak{M}/\mathfrak{N}$ is Abelian,

$$x^{r^{-1}s^{-1}t^{-1}gtsr} \equiv x^{r^{-1}gr} \mod \mathfrak{N}$$

for all $x \in \mathfrak{M}$, so

$$\prod_{t \in \mathsf{T}} \prod_{s \in \mathsf{S}} \prod_{r \in \mathsf{R}} x^{r^{-1}s^{-1}t^{-1}gtsr} \equiv \left(\prod_{r \in \mathsf{R}} x^{r^{-1}gr}\right)^{|\mathsf{S}|\,|\mathsf{T}|} \equiv 1 \mod \mathfrak{N},$$

since $|\mathsf{S}|$ is a power of p greater than 1. And again, since $|\mathsf{C}| = |\mathsf{T}|\,|\mathsf{S}|\,|\mathsf{R}|$, also $x^{|\mathsf{C}|} \equiv 1 \bmod \mathfrak{N}$. Thus (1) is proved. **q.e.d.**

13.10 Remarks. a) Theorem 13.8 is not true for insoluble groups.

Let $\mathfrak{G}$ be the alternating group $\mathfrak{A}_5$ and K an algebraically closed field of characteristic 2. Then $\mathsf{K}\mathfrak{G}$ has two blocks. The principal block $\mathscr{B}_1$ contains the trivial module V_1 and two irreducible modules V_2, V_3 each of dimension 2, while $\mathscr{B}_2$ contains only one irreducible module, namely, $\mathsf{V}_4 \cong \mathsf{V}_2 \otimes_{\mathsf{K}} \mathsf{V}_3$ (see Exercises 64, 65). Hence the statement of 13.8 is not true for $\mathsf{K}\mathfrak{G}$.

b) Let $\mathfrak{G}$ be the symmetric group $\mathfrak{S}_4$ and K an algebraically closed field of characteristic 3. Then $\mathsf{K}\mathfrak{G}$ has three blocks $\mathscr{B}_1$, $\mathscr{B}_2$ and $\mathscr{B}_3$. Here $\mathscr{B}_1$ contains the trivial module V_1 and the module V of dimension 1 corresponding to the sign representation of $\mathfrak{S}_4$. The blocks $\mathscr{B}_2$ and $\mathscr{B}_3$ each contain only one irreducible module, say V_2 and V_3. Then $\mathsf{V}_3 \cong \mathsf{V}_2 \otimes_{\mathsf{K}} \mathsf{V}$ (see 15.10b)). Hence tensoring the module V_2 with the module V in $\mathscr{B}_1$ does not leave V_2 in its block. So any extension of 13.8 beyond the principal block seems impossible, even for soluble groups.

Exercises

58) Find the number t of indecomposable two-sided direct summands of $\mathsf{K}\mathfrak{G}$ in the case when $\mathfrak{G}$ is the alternating group of degree 4 and char K is 2 or 3.

59) Let $\mathfrak{G} = SL(n, q)$, where $q = p^f$. Let $\mathsf{K} = GF(q)$ and let V be the natural module for $\mathfrak{G}$ of dimension n over K.

a) If $\mathsf{V} \in \mathscr{B}_1$, then $(n, q - 1) = 1$.

b) If $(n, q - 1) = 1$ and n is not divisible by p, then $\mathsf{V} \in \mathscr{B}_1$.

(Hint for the proof of b): From Schur's lemma, it follows that for every conjugacy class C of $\mathfrak{G}$, $\sum_{g \in \mathsf{C}} g = a1$, where $a \in \mathsf{K}$ and $na = |\mathsf{C}|\,\mathrm{tr}\, g$. Calculate the action of $\sum_{x \in \mathsf{C}}(x - 1)$ on V, distinguishing between the cases when p divides $|\mathsf{C}|$ and when it does not.)

§ 14. Kernels of Blocks

14.1 Lemma (WILLEMS [3]). *Suppose that $\mathfrak{N} \trianglelefteq \mathfrak{G}$. We put*

$$a = a(\mathfrak{N}) = \sum_{x \in \mathfrak{N}} x.$$

Let V *be an irreducible* $\mathsf{K}\mathfrak{G}$*-module and* $\mathsf{P}_{\mathfrak{G}}(\mathsf{V})$ *the uniquely determined projective* $\mathsf{K}\mathfrak{G}$*-module such that*

$$\mathsf{P}_{\mathfrak{G}}(\mathsf{V})/\mathsf{P}_{\mathfrak{G}}(\mathsf{V})\mathbf{J}(\mathsf{K}\mathfrak{G}) \cong \mathsf{V} \quad (cf.\ 10.9).$$

a) $\mathsf{P}_{\mathfrak{G}}(\mathsf{V})a \cong \begin{cases} \mathsf{P}_{\mathfrak{G}/\mathfrak{N}}(\mathsf{V}) & \text{if } \mathfrak{N} \text{ operates trivially on } \mathsf{V}, \\ 0 & \text{otherwise.} \end{cases}$

(Here $\mathsf{P}_{\mathfrak{G}/\mathfrak{N}}(\mathsf{V})$ *is the uniquely determined projective* $\mathsf{K}(\mathfrak{G}/\mathfrak{N})$*-module for which* $\mathsf{P}_{\mathfrak{G}/\mathfrak{N}}(\mathsf{V})/\mathsf{P}_{\mathfrak{G}/\mathfrak{N}}(\mathsf{V})\mathbf{J}(\mathsf{K}(\mathfrak{G}/\mathfrak{N})) \cong \mathsf{V}$.*)*

b) $\mathsf{P}_{\mathfrak{G}}(\mathsf{V})a$ *is the maximal submodule of* $\mathsf{P}_{\mathfrak{G}}(\mathsf{V})$ *on which* $\mathfrak{N}$ *operates trivially.*

c) $\mathsf{P}\ (\mathsf{V})/\mathbf{An}_{\mathsf{P}_{\mathfrak{G}}(\mathsf{V})}a$ *is the maximal factor module of* $\mathsf{P}_{\mathfrak{G}}(\mathsf{V})$ *on which* $\mathfrak{N}$ *operates trivially.*

d) *If* char K *divides* $|\mathfrak{N}|$, *then* $a^2 = 0$ *and* $\mathsf{P}_{\mathfrak{G}}(\mathsf{V})a \subseteq \mathbf{An}_{\mathsf{P}_{\mathfrak{G}}(\mathsf{V})}a$.

Proof. a) Obviously, $xa = a \in \mathbf{Z}(\mathsf{K}\mathfrak{G})$ for all $x \in \mathfrak{N}$. Hence the K-linear mapping α, defined by

$$(g\mathfrak{N})\alpha = ga \quad (g \in \mathfrak{G})$$

is well-defined and is a $\mathsf{K}\mathfrak{G}$-epimorphism of $\mathsf{K}(\mathfrak{G}/\mathfrak{N})$ onto $\mathsf{K}\mathfrak{G}a$. If $\mathfrak{G} = \bigcup_{t \in \mathsf{T}} \mathfrak{N}t$, where T is a transversal of $\mathfrak{N}$ in $\mathfrak{G}$, then the elements ta $(t \in \mathsf{T})$ are linearly independent over K. Hence we obtain

$$\dim_{\mathsf{K}} \mathsf{K}\mathfrak{G}a \geq |\mathfrak{G}/\mathfrak{N}| = \dim_{\mathsf{K}} \mathsf{K}(\mathfrak{G}/\mathfrak{N}).$$

This yields the $\mathsf{K}\mathfrak{G}$-isomorphism

$$\mathsf{K}(\mathfrak{G}/\mathfrak{N}) \cong \mathsf{K}\mathfrak{G}a.$$

From

$$\mathsf{K}\mathfrak{G} = \mathsf{P}_{\mathfrak{G}}(\mathsf{V}) \oplus \mathsf{W}$$

with a suitable $\mathsf{K}\mathfrak{G}$-module W we obtain

$$\mathsf{K}\mathfrak{G}a = \mathsf{P}_{\mathfrak{G}}(\mathsf{V})a \oplus \mathsf{W}a,$$

for

$$\mathsf{P}_{\mathfrak{G}}(\mathsf{V})a \cap \mathsf{W}a \subseteq \mathsf{P}_{\mathfrak{G}}(\mathsf{V}) \cap \mathsf{W} = 0.$$

Thus $\mathsf{P}_{\mathfrak{G}}(\mathsf{V})a$ is isomorphic to a direct summand of $\mathsf{K}(\mathfrak{G}/\mathfrak{N})$ and hence is a projective $\mathsf{K}(\mathfrak{G}/\mathfrak{N})$-module. As $\mathsf{P}_{\mathfrak{G}}(\mathsf{V})$ has the irreducible head V, either $\mathsf{P}_{\mathfrak{G}}(\mathsf{V})a = 0$ or else $\mathsf{P}_{\mathfrak{G}}(\mathsf{V})a$ has the irreducible head V and is hence indecomposable. If $\mathfrak{N}$ does not operate trivially on V, we conclude that $\mathsf{P}_{\mathfrak{G}}(\mathsf{V})a = 0$. Hence we obtain

$$\bigoplus_{\mathsf{V}} d(\mathsf{V})\mathsf{P}_{\mathfrak{G}/\mathfrak{N}}(\mathsf{V}) \cong \mathsf{K}(\mathfrak{G}/\mathfrak{N}) \cong \mathsf{K}\mathfrak{G}a = \bigoplus_{\mathsf{V}} e(\mathsf{V})\mathsf{P}_{\mathfrak{G}}(\mathsf{V})a,$$

where $d(\mathsf{V})$ and $e(\mathsf{V})$ are certain multiplicities and both sums run over the irreducible $\mathsf{K}(\mathfrak{G}/\mathfrak{N})$-modules V, considered in the right-hand sum as $\mathsf{K}\mathfrak{G}$-modules. On the left side all these V appear as heads. On the right side V appears as a head if and only if $\mathsf{P}_{\mathfrak{G}}(\mathsf{V})a \neq 0$. The Krull-Schmidt theorem now shows that

$$\mathsf{P}_{\mathfrak{G}}(\mathsf{V})a \cong \mathsf{P}_{\mathfrak{G}/\mathfrak{N}}(\mathsf{V}).$$

b) We put

$$\mathsf{U} = \{u \mid u \in \mathsf{P}_{\mathfrak{G}}(\mathsf{V}),\ ux = u \quad \text{for all } x \in \mathfrak{N}\}.$$

As $\mathfrak{N} \trianglelefteq \mathfrak{G}$, certainly U is a $\mathsf{K}\mathfrak{G}$-submodule of $\mathsf{P}_{\mathfrak{G}}(\mathsf{V})$. From $ax = a$ for all $x \in \mathfrak{N}$ we conclude that $\mathsf{P}_{\mathfrak{G}}(\mathsf{V})a \subseteq \mathsf{U}$.

If $\mathfrak{N}$ does not operate trivially on V, we have by a) $\mathsf{P}_{\mathfrak{G}}(\mathsf{V})a = 0$. As the socle of $\mathsf{P}_{\mathfrak{G}}(\mathsf{V})$ is isomorphic to V (see 11.6c)), we obtain $\mathsf{U} \cap \mathbf{S}(\mathsf{P}_{\mathfrak{G}}(\mathsf{V})) = 0$, hence $\mathsf{U} = 0$.

Now suppose that $\mathfrak{N}$ operates trivially on V. Then by 10.14 we conclude from $\mathbf{S}(\mathsf{U}) \cong \mathsf{V}$ that U is isomorphic to a submodule of the uniquely determined projective (and injective) $\mathsf{K}(\mathfrak{G}/\mathfrak{N})$-module $\mathsf{P}_{\mathfrak{G}/\mathfrak{N}}(\mathsf{V})$ with socle V. By a) we have

$$\dim_{\mathsf{K}} \mathsf{U} \leq \dim_{\mathsf{K}} \mathsf{P}_{\mathfrak{G}/\mathfrak{N}}(\mathsf{V}) = \dim_{\mathsf{K}} \mathsf{P}_{\mathfrak{G}}(\mathsf{V})a \leq \dim_{\mathsf{K}} \mathsf{U}.$$

Hence $\mathsf{P}_{\mathfrak{G}}(\mathsf{V})a = \mathsf{U}$ in this case also.

c) Now let $\mathsf{P}_{\mathfrak{G}}(\mathsf{V})/\mathsf{W}$ be the largest factor module of $\mathsf{P}_{\mathfrak{G}}(\mathsf{V})$ on which $\mathfrak{N}$ operates trivially. As

$$z(x - 1) \in \mathbf{An}_{\mathsf{P}_{\mathfrak{G}}(\mathsf{V})}\, a$$

for all $z \in \mathsf{P}_{\mathfrak{G}}(\mathsf{V})$, $x \in \mathfrak{N}$, we certainly have

$$\mathbf{An}_{\mathsf{P}_{\mathfrak{G}}(\mathsf{V})}\, a \supseteq \mathsf{W}.$$

If $\mathfrak{N}$ does not operate trivially on V, then $P_{\mathfrak{G}}(V)/W = 0$, for V is the head of $P_{\mathfrak{G}}(V)$. In this case we have

$$W = \mathbf{An}_{P_{\mathfrak{G}}(V)}\, a = P_{\mathfrak{G}}(V).$$

Suppose now that $\mathfrak{N}$ does operate trivially on V. Then $P_{\mathfrak{G}}(V)/W$ has the irreducible head V. Thus by 10.14a), $P_{\mathfrak{G}}(V)/W$ is an epimorphic image of $P_{\mathfrak{G}/\mathfrak{N}}(V)$. Using a), we obtain

$$\begin{aligned}\dim_K P_{\mathfrak{G}/\mathfrak{N}}(V) &\geq \dim_K P_{\mathfrak{G}}(V)/W \geq \dim_K P_{\mathfrak{G}}(V)/\mathbf{An}_{P_{\mathfrak{G}}(V)}\, a\\ &= \dim_K P_{\mathfrak{G}}(V)a = \dim_K P_{\mathfrak{G}/\mathfrak{N}}(V).\end{aligned}$$

This proves $W = \mathbf{An}_{P_{\mathfrak{G}}(V)}\, a$.

d) As $xa = a$ for all $x \in \mathfrak{N}$, we have $a^2 = |\mathfrak{N}|a$. If char K divides $|\mathfrak{N}|$, then $a^2 = 0$ and so $P_{\mathfrak{G}}(V)a \subseteq \mathbf{An}_{P_{\mathfrak{G}}(V)}\, a$. **q.e.d.**

14.2 Lemma (WILLEMS [3]). *Let V be an irreducible $K(\mathfrak{G}/\mathfrak{N})$-module. Then*

$$\dim_K P_{\mathfrak{G}}(V) = \dim_K P_{\mathfrak{N}}(K) \dim_K P_{\mathfrak{G}/\mathfrak{N}}(V),$$

where K is to be considered as the trivial $K\mathfrak{G}$-module.

Proof. By 9.3,

$$P_{\mathfrak{G}}(V)_{\mathfrak{N}} = e(P_1 \oplus \cdots \oplus P_s),$$

where the P_j are projective, indecomposable $\mathfrak{G}$-conjugate $K\mathfrak{N}$-modules. As $\mathfrak{N}$ operates trivially on V, we must have $\mathbf{S}(P_j) = K$ for some j. But then P_j is $\mathfrak{G}$-conjugate only to P_j itself, so

$$P_{\mathfrak{G}}(V)_{\mathfrak{N}} = eP_{\mathfrak{N}}(K)$$

for some natural number e.

As $\mathfrak{N}$ operates trivially on K, we have by 14.1a) (with $(\mathfrak{N}, \mathfrak{N})$ in place of $(\mathfrak{G}, \mathfrak{N})$)

$$\dim_K P_{\mathfrak{G}}(V)a = e \dim_K P_{\mathfrak{N}}(K)a = e \dim_K P_{\mathfrak{N}/\mathfrak{N}}(K) = e \dim_K K = e.$$

Hence

$$\dim_K \mathbf{An}_{P_{\mathfrak{G}}(V)}\, a = \dim_K P_{\mathfrak{G}}(V) - \dim_K P_{\mathfrak{G}}(V)a = e(\dim_K P_{\mathfrak{N}}(K) - 1).$$

Therefore by 14.1,

$$e \dim_K \mathsf{P}_{\mathfrak{N}}(\mathsf{K}) = \dim_K \mathsf{P}_{\mathfrak{G}}(\mathsf{V}) = \dim_K \mathbf{An}_{\mathsf{P}_{\mathfrak{G}}(\mathsf{V})} a + \dim_K \mathsf{P}_{\mathfrak{G}}(\mathsf{V})a$$
$$= e(\dim_K \mathsf{P}_{\mathfrak{N}}(\mathsf{K}) - 1) + \dim_K \mathsf{P}_{\mathfrak{G}/\mathfrak{N}}(\mathsf{V}).$$

This shows that $e = \dim_K \mathsf{P}_{\mathfrak{G}/\mathfrak{N}}(\mathsf{V})$. **q.e.d.**

Let $\mathsf{P}_1 = \mathsf{P}_{\mathfrak{G}}(\mathsf{K})$ denote the indecomposable projective $\mathsf{K}\mathfrak{G}$-module for which $\mathsf{P}_1/\mathsf{P}_1\mathbf{J}(\mathsf{K}\mathfrak{G}) \cong \mathsf{K}$. By Dickson's theorem (7.16), we know that the highest power $|\mathfrak{G}|_p$ of $p = \operatorname{char} \mathsf{K}$ which divides $|\mathfrak{G}|$ also divides $\dim_K \mathsf{P}_1$. If $\mathfrak{G}$ has a p-complement, then $|\mathfrak{G}|_p = \dim_K \mathsf{P}_1$ by 10.12b). We show

14.3 Theorem (WILLEMS [3]). *The following assertions are equivalent.*
a) $\dim_K \mathsf{P}_{\mathfrak{G}}(\mathsf{K}) = |\mathfrak{G}|_p$.
b) *For every* $\mathfrak{N} \trianglelefteq \mathfrak{G}$, *we have*

$$\dim_K \mathsf{P}_{\mathfrak{N}}(\mathsf{K}) = |\mathfrak{N}|_p \quad \textit{and} \quad \dim_K \mathsf{P}_{\mathfrak{G}/\mathfrak{N}}(\mathsf{K}) = |\mathfrak{G}/\mathfrak{N}|_p.$$

c) *For every composition factor* $\mathfrak{K}$ *of* $\mathfrak{G}$ *we have* $\dim_K \mathsf{P}_{\mathfrak{K}}(\mathsf{K}) = |\mathfrak{K}|_p$.

Proof. a) ⇒ b): By 14.2, we have

$$\dim_K \mathsf{P}_{\mathfrak{N}}(\mathsf{K}) \dim_K \mathsf{P}_{\mathfrak{G}/\mathfrak{N}}(\mathsf{K}) = \dim_K \mathsf{P}_{\mathfrak{G}}(\mathsf{K}) = |\mathfrak{G}|_p.$$

As $|\mathfrak{N}|_p$ divides $\dim_K \mathsf{P}_{\mathfrak{N}}(\mathsf{K})$ and $|\mathfrak{G}/\mathfrak{N}|_p$ divides $\dim_K \mathsf{P}_{\mathfrak{G}/\mathfrak{N}}(\mathsf{K})$, we conclude that

$$|\mathfrak{N}|_p = \dim_K \mathsf{P}_{\mathfrak{N}}(\mathsf{K}) \quad \text{and} \quad |\mathfrak{G}/\mathfrak{N}|_p = \dim_K \mathsf{P}_{\mathfrak{G}/\mathfrak{N}}(\mathsf{K}).$$

b) ⇒ a): This follows immediately.
The equivalence of b) and c) is obvious. **q.e.d.**

We remark that the class of groups $\mathfrak{G}$ for which $\dim_K \mathsf{P}_{\mathfrak{G}}(\mathsf{K}) = |\mathfrak{G}|_p$ is definitely larger than the class of groups with a p-complement. The simple groups $PSL(2, p)$ have no p-complement for $p \geq 13$ (see II, 8.27), but it can be shown that $\dim_K \mathsf{P}_1 = p$ in all groups $PSL(2, p)$ (BURKHARDT [1]).

14.4 Definition. a) Let V be a $\mathsf{K}\mathfrak{G}$-module. We put

$$\mathfrak{K}(\mathsf{V}) = \{g \mid g \in \mathfrak{G} \text{ and } vg = v \text{ for all } v \in \mathsf{V}\}$$

and

$$\mathfrak{N}(\mathsf{V}) = \bigcap_{\mathsf{W}} \mathfrak{K}(\mathsf{W}),$$

where the intersection runs through all composition factors W of V.

b) Let $\mathscr{B}$ be a block of $\mathfrak{G}$ and f the corresponding idempotent in $\mathbf{Z}(\mathsf{K}\mathfrak{G})$. We put

$$\mathfrak{K}(\mathscr{B}) = \mathfrak{K}(f\mathsf{K}\mathfrak{G}) = \{g \mid g \in \mathfrak{G}, f(1-g) = 0\}$$

and

$$\mathfrak{N}(\mathscr{B}) = \mathfrak{N}(f\mathsf{K}\mathfrak{G}).$$

14.5 Lemma. a) *For every* $\mathsf{K}\mathfrak{G}$*-module* V *in the block* $\mathscr{B}$ *and every* $g \in \mathfrak{K}(\mathscr{B})$ *we have* $\mathsf{V}(1-g) = 0$.

b) *Let f be the idempotent in* $\mathbf{Z}(\mathsf{K}\mathfrak{G})$ *belonging to* $\mathscr{B}$. *Then* $\mathfrak{N}(\mathscr{B}) = \bigcap_{\mathsf{V}\,\mathrm{irr}}^{\mathsf{V}\in\mathscr{B}} \mathfrak{K}(\mathsf{V}) = \{g \mid g \in \mathfrak{G}, f(1-g) \in f\mathbf{J}(\mathsf{K}\mathfrak{G})\}$.

Proof. a) If V is a module in $\mathscr{B}$, then $\mathsf{V} = \mathsf{V}f$ and hence

$$\mathsf{V}(1-g) = \mathsf{V}f(1-g) = 0$$

for every $g \in \mathfrak{K}(\mathscr{B})$.

b) The first statement is obvious.

Suppose that $f(1-g) \in f\mathbf{J}(\mathsf{K}\mathfrak{G})$. If V is a composition factor of $f\mathsf{K}\mathfrak{G}$, then $\mathsf{V} = \mathsf{V}f$ and hence

$$\mathsf{V}(1-g) = \mathsf{V}f(1-g) \subseteq \mathsf{V}f\mathbf{J}(\mathsf{K}\mathfrak{G}) = 0.$$

This implies $g \in \mathfrak{N}(\mathscr{B})$.

Suppose conversely that $g \in \mathfrak{N}(\mathscr{B})$ and that

$$f\mathsf{K}\mathfrak{G} = \bigoplus_{i=1}^{s} e_i\mathsf{K}\mathfrak{G}$$

is a decomposition of $f\mathsf{K}\mathfrak{G}$ into indecomposable right ideals $e_i\mathsf{K}\mathfrak{G}$ and $f = \sum_{i=1}^{s} e_i$. Then by 10.9c), $e_i\mathsf{K}\mathfrak{G}/e_i\mathbf{J}(\mathsf{K}\mathfrak{G})$ is an irreducible module in $\mathscr{B}$. Thus

$$e_i(1-g) \in e_i\mathbf{J}(\mathsf{K}\mathfrak{G}) \subseteq \mathbf{J}(\mathsf{K}\mathfrak{G})$$

and

$$f(1 - g) = f \sum_{i=1}^{s} e_i(1 - g) \in f\mathbf{J}(\mathsf{K}\mathfrak{G}). \qquad \textbf{q.e.d.}$$

14.6 Theorem (WILLEMS [3]). *Let* V *be an irreducible* $\mathsf{K}\mathfrak{G}$*-module and* char $\mathsf{K} = p$.

a) $\mathfrak{K}(\mathsf{P}_{\mathfrak{G}}(\mathsf{V})) = \mathbf{O}_{p'}(\mathfrak{N}(\mathsf{P}_{\mathfrak{G}}(\mathsf{V}))) = \mathbf{O}_{p'}(\mathfrak{K}(\mathsf{V}))$.

b) $\mathfrak{N}(\mathsf{P}_{\mathfrak{G}}(\mathsf{V})) = \mathbf{O}_{p',p}(\mathfrak{K}(\mathsf{V}))$.

c) *If* K *is the trivial* $\mathsf{K}\mathfrak{G}$*-module, then* $\mathfrak{K}(\mathsf{P}_{\mathfrak{G}}(\mathsf{K})) = \mathbf{O}_{p'}(\mathfrak{G})$ *and* $\mathfrak{N}(\mathsf{P}_{\mathfrak{G}}(\mathsf{K})) = \mathbf{O}_{p',p}(\mathfrak{G})$.

Proof. a) We put $\mathfrak{N} = \mathbf{O}_{p'}(\mathfrak{K}(\mathsf{V}))$. As p does not divide $|\mathfrak{N}|$, the trivial $\mathsf{K}\mathfrak{N}$-module K is projective and hence $\mathsf{P}_{\mathfrak{N}}(\mathsf{K}) = \mathsf{K}$. We conclude from 14.2 that

$$\dim_{\mathsf{K}} \mathsf{P}_{\mathfrak{G}}(\mathsf{V}) = \dim_{\mathsf{K}} \mathsf{P}_{\mathfrak{G}/\mathfrak{N}}(\mathsf{V}).$$

Lemma 14.1a) now implies

$$\dim_{\mathsf{K}} \mathsf{P}_{\mathfrak{G}}(\mathsf{V}) = \dim_{\mathsf{K}} \mathsf{P}_{\mathfrak{G}/\mathfrak{N}}(\mathsf{V}) = \dim_{\mathsf{K}} \mathsf{P}_{\mathfrak{G}}(\mathsf{V})a;$$

thus $\mathsf{P}_{\mathfrak{G}}(\mathsf{V}) = \mathsf{P}_{\mathfrak{G}}(\mathsf{V})a$ and hence $\mathfrak{N}$ operates trivially on $\mathsf{P}_{\mathfrak{G}}(\mathsf{V})$. This shows $\mathfrak{N} \leq \mathfrak{K}(\mathsf{P}_{\mathfrak{G}}(\mathsf{V}))$.

We put $\mathfrak{M} = \mathfrak{K}(\mathsf{P}_{\mathfrak{G}}(\mathsf{V}))$. Then $\mathfrak{M}$ operates trivially on $\mathsf{P}_{\mathfrak{G}}(\mathsf{V})$. Hence

$$\mathsf{P}_{\mathfrak{G}}(\mathsf{V})_{\mathfrak{M}} \cong \mathsf{K} \oplus \cdots \oplus \mathsf{K},$$

and so K is a projective $\mathsf{K}\mathfrak{M}$-module. By Dickson's theorem (7.16), $\mathfrak{M}$ must be a p'-group. So we have proved

$$\mathfrak{N} = \mathbf{O}_{p'}(\mathfrak{K}(\mathsf{V})) \leq \mathfrak{K}(\mathsf{P}_{\mathfrak{G}}(\mathsf{V})) = \mathfrak{M} \leq \mathfrak{K}(\mathsf{V}) \cap \mathbf{O}_{p'}(\mathfrak{G}) = \mathbf{O}_{p'}(\mathfrak{K}(\mathsf{V})).$$

This shows

$$\mathfrak{K}(\mathsf{P}_{\mathfrak{G}}(\mathsf{V})) = \mathbf{O}_{p'}(\mathfrak{K}(\mathsf{V})).$$

We also have obviously

$$\mathfrak{K}(\mathsf{P}_{\mathfrak{G}}(\mathsf{V})) \leq \mathfrak{N}(\mathsf{P}_{\mathfrak{G}}(\mathsf{V})) \leq \mathfrak{K}(\mathsf{V}),$$

hence

$$\mathfrak{K}(\mathsf{P}_{\mathfrak{G}}(\mathsf{V})) \leq \mathbf{O}_{p'}(\mathfrak{N}(\mathsf{P}_{\mathfrak{G}}(\mathsf{V}))) \leq \mathbf{O}_{p'}(\mathfrak{K}(\mathsf{V})) = \mathfrak{K}(\mathsf{P}_{\mathfrak{G}}(\mathsf{V})).$$

So

$$\mathfrak{K}(\mathsf{P}_{\mathfrak{G}}(\mathsf{V})) = \mathbf{O}_{p'}(\mathfrak{N}(\mathsf{P}_{\mathfrak{G}}(\mathsf{V}))).$$

b) $\mathfrak{N}(\mathsf{P}_{\mathfrak{G}}(\mathsf{V}))/\mathfrak{K}(\mathsf{P}_{\mathfrak{G}}(\mathsf{V}))$ is faithfully represented on $\mathsf{P}_{\mathfrak{G}}(\mathsf{V})$ and operates trivially on all the $\mathfrak{G}$-composition factors. Hence it is a p-group. This shows that

$$\mathfrak{N}(\mathsf{P}_{\mathfrak{G}}(\mathsf{V}))/\mathfrak{K}(\mathsf{P}_{\mathfrak{G}}(\mathsf{V})) \leq \mathbf{O}_p(\mathfrak{K}(\mathsf{V})/\mathfrak{K}(\mathsf{P}_{\mathfrak{G}}(\mathsf{V}))) = \mathbf{O}_p(\mathfrak{K}(\mathsf{V})/\mathbf{O}_{p'}(\mathfrak{K}(\mathsf{V}))),$$

thus

$$\mathfrak{N}(\mathsf{P}_{\mathfrak{G}}(\mathsf{V})) \leq \mathbf{O}_{p',p}(\mathfrak{K}(\mathsf{V})).$$

As the normal p-subgroup

$$\mathbf{O}_{p',p}(\mathfrak{K}(\mathsf{V}))/\mathbf{O}_{p'}(\mathfrak{K}(\mathsf{V})) = \mathbf{O}_p(\mathfrak{K}(\mathsf{V})/\mathbf{O}_{p'}(\mathfrak{K}(\mathsf{V}))) = \mathbf{O}_p(\mathfrak{K}(\mathsf{V})/\mathfrak{K}(\mathsf{P}_{\mathfrak{G}}(\mathsf{V})))$$

of $\mathfrak{G}/\mathfrak{K}(\mathsf{P}_{\mathfrak{G}}(\mathsf{V}))$ operates trivially on every composition factor of $\mathsf{P}_{\mathfrak{G}}(\mathsf{V})$, we obtain

$$\mathbf{O}_{p',p}(\mathfrak{K}(\mathsf{V})) \leq \mathfrak{N}(\mathsf{P}_{\mathfrak{G}}(\mathsf{V})).$$

c) This is the special case $\mathsf{V} = \mathsf{K}$. **q.e.d.**

14.7 Theorem. *Let* V *be an irreducible* $\mathsf{K}\mathfrak{G}$*-module in the block* $\mathscr{B}$.

a) $\mathfrak{K}(\mathsf{P}_{\mathfrak{G}}(\mathsf{V})) = \mathfrak{K}(\mathscr{B})$. *So all the projective modules belonging to the block* $\mathscr{B}$ *have the same kernel.*

b) (BRAUER [8], MICHLER [1], WILLEMS [3])

$$\mathfrak{N}(\mathsf{P}_{\mathfrak{G}}(\mathsf{V})) = \mathfrak{N}(\mathscr{B}), \quad \mathfrak{K}(\mathscr{B}) = \mathbf{O}_{p'}(\mathfrak{N}(\mathscr{B}))$$

and

$$\mathfrak{N}(\mathscr{B})/\mathfrak{K}(\mathscr{B}) = \mathbf{O}_p(\mathfrak{G}/\mathfrak{K}(\mathscr{B})).$$

In particular, $\mathfrak{N}(\mathscr{B})$ *is p-nilpotent.*

Proof. a) If f is the idempotent corresponding to $\mathscr{B}$ in $\mathbf{Z}(\mathsf{K}\mathfrak{G})$, then

$$f\mathsf{K}\mathfrak{G} = \bigoplus_{i=1}^{s} e_i\mathsf{K}\mathfrak{G},$$

where the $e_i \mathsf{K}\mathfrak{G}$ are just the indecomposable projective $\mathsf{K}\mathfrak{G}$-modules in $\mathscr{B}$. Thus they are the $\mathsf{P}_{\mathfrak{G}}(\mathsf{V})$, where V runs through the irreducible modules in $\mathscr{B}$. Thus

$$\mathfrak{K}(\mathscr{B}) = \bigcap_{\substack{\mathsf{V} \in \mathscr{B} \\ \mathsf{V}\ \mathrm{irr}}} \mathfrak{K}(\mathsf{P}_{\mathfrak{G}}(\mathsf{V})).$$

Suppose $g \in \mathfrak{K}(\mathsf{P}_{\mathfrak{G}}(\mathsf{V}))$. Then by 14.6a),

$$g \in \mathfrak{K}(\mathsf{P}_{\mathfrak{G}}(\mathsf{V})) = \mathbf{O}_{p'}(\mathfrak{K}(\mathsf{V})) \leq \mathbf{O}_{p'}(\mathfrak{G}).$$

Thus for every composition factor W of $\mathsf{P}_{\mathfrak{G}}(\mathsf{V})$,

$$g \in \mathfrak{K}(\mathsf{W}) \cap \mathbf{O}_{p'}(\mathfrak{G}) = \mathbf{O}_{p'}(\mathfrak{K}(\mathsf{W})) = \mathfrak{K}(\mathsf{P}_{\mathfrak{G}}(\mathsf{W})).$$

Repeating this procedure, we see from 12.4 that $g \in \mathfrak{K}(\mathsf{P}_{\mathfrak{G}}(\mathsf{X}))$ for every irreducible module X in $\mathscr{B}$. Hence

$$\mathfrak{K}(\mathsf{P}_{\mathfrak{G}}(\mathsf{V})) \subseteq \bigcap_{\substack{\mathsf{W} \in \mathscr{B} \\ \mathsf{W}\ \mathrm{irr}}} \mathfrak{K}(\mathsf{P}_{\mathfrak{G}}(\mathsf{W})) = \mathfrak{K}(\mathscr{B}) \quad \text{and} \quad \mathfrak{K}(\mathsf{P}_{\mathfrak{G}}(\mathsf{V})) = \mathfrak{K}(\mathscr{B}).$$

b) We have $\mathfrak{N}(\mathscr{B}) \leq \mathfrak{N}(\mathsf{P}_{\mathfrak{G}}(\mathsf{V}))$ for all irreducible modules V in $\mathscr{B}$. As normal p-subgroups operate trivially on all irreducible modules, we also have

$$\begin{aligned}
\mathfrak{N}(\mathscr{B})/\mathfrak{K}(\mathscr{B}) &\leq \mathfrak{N}(\mathsf{P}_{\mathfrak{G}}(\mathsf{V}))/\mathfrak{K}(\mathsf{P}_{\mathfrak{G}}(\mathsf{V})) && \text{(by 14.7a))} \\
&= \mathbf{O}_{p',p}(\mathfrak{K}(\mathsf{V}))/\mathbf{O}_{p'}(\mathfrak{K}(\mathsf{V})) && \text{(by 14.6a), b))} \\
&= \mathbf{O}_{p}(\mathfrak{K}(\mathsf{V})/\mathbf{O}_{p'}(\mathfrak{K}(\mathsf{V}))) \\
&= \mathbf{O}_{p}(\mathfrak{K}(\mathsf{V})/\mathfrak{K}(\mathsf{P}_{\mathfrak{G}}(\mathsf{V}))) && \text{(by 14.6a))} \\
&= \mathbf{O}_{p}(\mathfrak{K}(\mathsf{V})/\mathfrak{K}(\mathscr{B})) \\
&\leq \mathbf{O}_{p}(\mathfrak{G}/\mathfrak{K}(\mathscr{B})) \leq \mathfrak{N}(\mathscr{B})/\mathfrak{K}(\mathscr{B}).
\end{aligned}$$

This implies

$$\mathfrak{N}(\mathsf{P}_{\mathfrak{G}}(\mathsf{V})) = \mathfrak{N}(\mathscr{B})$$

and

$$\mathbf{O}_{p}(\mathfrak{G}/\mathfrak{K}(\mathscr{B})) = \mathfrak{N}(\mathscr{B})/\mathfrak{K}(\mathscr{B}).$$

Finally, for every irreducible module V in $\mathscr{B}$ we have

$$\begin{aligned}\mathfrak{K}(\mathscr{B}) &= \mathfrak{K}(\mathsf{P}_{\mathfrak{G}}(\mathsf{V})) && (14.7\text{a})) \\ &= \mathbf{O}_{p'}(\mathfrak{N}(\mathsf{P}_{\mathfrak{G}}(\mathsf{V}))) && (14.6\text{a})) \\ &= \mathbf{O}_{p'}(\mathfrak{N}(\mathscr{B})).\end{aligned}$$

q.e.d.

We mention an important special case.

14.8 Theorem (BRAUER [7]. *If* $\mathscr{B}_1$ *is the principal block of* $\mathsf{K}\mathfrak{G}$, *then*

$$\mathfrak{K}(\mathscr{B}_1) = \mathbf{O}_{p'}(\mathfrak{G}) \quad \text{and} \quad \mathfrak{N}(\mathscr{B}_1) = \mathbf{O}_{p',p}(\mathfrak{G}).$$

Proof. If K denotes the trivial $\mathsf{K}\mathfrak{G}$-module, then

$$\begin{aligned}\mathfrak{K}(\mathscr{B}_1) &= \mathfrak{K}(\mathsf{P}_{\mathfrak{G}}(\mathsf{K})) && (14.7\text{a})) \\ &= \mathbf{O}_{p'}(\mathfrak{G}) && (14.6\text{c})).\end{aligned}$$

Also

$$\begin{aligned}\mathfrak{N}(\mathscr{B}_1) &= \mathfrak{N}(\mathsf{P}_{\mathfrak{G}}(\mathsf{K})) && (14.7\text{b})) \\ &= \mathbf{O}_{p',p}(\mathfrak{G}) && (14.6\text{c})).\end{aligned}$$

q.e.d.

As an application, we derive a representation-theoretic characterization of p-nilpotent groups.

14.9 Theorem. *Let* K *be a field of characteristic p. The following assertions are equivalent.*

a) *The principal block* $\mathscr{B}_1$ *of* $\mathfrak{G}$ *contains just one irreducible* $\mathsf{K}\mathfrak{G}$*-module.*
b) $\mathfrak{G}$ *is p-nilpotent.*
c) *Every block of* $\mathfrak{G}$ *contains just one irreducible* $\mathsf{K}\mathfrak{G}$*-module.*
d) *The Cartan matrix of* $\mathsf{K}\mathfrak{G}$ *is a diagonal matrix.*

Proof. a) ⇒ b): If $\mathscr{B}_1$ contains only one irreducible $\mathsf{K}\mathfrak{G}$-module, this is necessarily the module for the trivial representation of $\mathfrak{G}$. So by 14.5, $\mathfrak{N}(\mathscr{B}_1) = \mathfrak{G}$. Hence $\mathbf{O}_{p',p}(\mathfrak{G}) = \mathfrak{G}$ by 14.8, and $\mathfrak{G}$ is p-nilpotent.

b) ⇒ c): Suppose now that $\mathfrak{G}$ is p-nilpotent. By 14.8, $\mathfrak{N}(\mathscr{B}_1) = \mathfrak{G}$, so by 14.5, the only irreducible $\mathsf{K}\mathfrak{G}$-module in $\mathscr{B}_1$ is the module V_1 for the trivial representation of $\mathfrak{G}$.

Let $\mathsf{V}_1, \ldots, \mathsf{V}_k$ be all the irreducible $\mathsf{K}\mathfrak{G}$-modules to within isomorphism. By 10.9, there exists a projective indecomposable $\mathsf{K}\mathfrak{G}$-module P_i

such that $\mathsf{P}_i/\mathsf{P}_i\mathbf{J}(\mathsf{K}\mathfrak{G}) \cong \mathsf{V}_i$. We show that all the $\mathsf{K}\mathfrak{G}$-composition factors of P_i are isomorphic. This is clearly the case for P_1, since all composition factors of P_1 belong to $\mathscr{B}_1$ and are therefore isomorphic to V_1. If

$$0 = \mathsf{W}_0 \subset \mathsf{W}_1 \subset \cdots \subset \mathsf{W}_m = \mathsf{P}_1$$

is a $\mathsf{K}\mathfrak{G}$-composition series of P_1, then $\mathsf{W}_j/\mathsf{W}_{j-1} \cong \mathsf{V}_1$ $(j = 1, \ldots, m)$. Since

$$(\mathsf{W}_j \otimes_{\mathsf{K}} \mathsf{V}_i)/(\mathsf{W}_{j-1} \otimes_{\mathsf{K}} \mathsf{V}_i) \cong (\mathsf{W}_j/\mathsf{W}_{j-1}) \otimes_{\mathsf{K}} \mathsf{V}_i \cong \mathsf{V}_1 \otimes_{\mathsf{K}} \mathsf{V}_i \cong \mathsf{V}_i,$$

all the composition factors of $\mathsf{P}_1 \otimes_{\mathsf{K}} \mathsf{V}_i$ are isomorphic to V_i. By 7.19, $\mathsf{P}_1 \otimes_{\mathsf{K}} \mathsf{V}_i$ is projective, so by 10.9, $\mathsf{P}_1 \otimes_{\mathsf{K}} \mathsf{V}_i$ is the direct sum of certain P_j. Thus any composition factor of any of these P_j is isomorphic to V_i. It follows that $j = i$ and that any composition factor of P_i is isomorphic to V_i.

We conclude from 12.4 that P_i and P_j lie in distinct blocks for $i \neq j$. Hence each block contains only one V_i.

c) ⇒ d): This is true by definition of the Cartan matrix and 12.4.

d) ⇒ a): This is trivial. **q.e.d.**

We apply the preceding results in the following example.

14.10 Example. Let q be a prime and let $\mathfrak{G} = \langle x, y \rangle$ be a metacyclic group of order $q^n r$, defined by the relations

$$x^{q^n} = y^r = 1, x^y = x^k,$$

where $k^r \equiv 1$ (q^n) and $(r, q) = 1$. We make the assumption that $\mathbf{C}_{\mathfrak{G}}(x) = \langle x \rangle$. This means that $k^i \not\equiv 1$ (q^n) for $0 < i < r$. Observe that r divides $q - 1$.

We shall describe the irreducible modules, the indecomposable projective modules and the Cartan matrix of $\mathfrak{G}$ over a field K of characteristic p.

a) Suppose first that $p = q$.

We construct the principal indecomposable projective module P_0 by using 10.12b). (For technical reasons we start the numbering of the indecomposable projective modules with P_0 rather than the usual P_1.) Thus $\mathsf{P}_0 = \mathsf{K}^{\mathfrak{G}}$, where K is the trivial module for the p-complement $\langle y \rangle$ of $\mathfrak{G}$. As the Sylow p-subgroup $\mathfrak{P} = \langle x \rangle$ of $\mathfrak{G}$ is a transversal of $\langle y \rangle$ in $\mathfrak{G}$,

$$\mathsf{P}_0 = \langle 1 \otimes x^i | 0 \leq i < p^n \rangle.$$

$(P_0)_{\mathfrak{P}}$ is isomorphic to the indecomposable $K\mathfrak{P}$-module $K\mathfrak{P}$. Hence by 5.3 its only submodules are

$$V_i = P_0 \mathbf{J}(K\mathfrak{P})^i = 1 \otimes (1 - x)^i K\mathfrak{P} \quad (0 \leq i < p^n).$$

The action of y on these vector spaces is described by

$$\begin{aligned}(1 \otimes (1 - x)^i)y &= 1 \otimes (1 - x)^i y = 1 \otimes y(1 - x^y)^i = 1y \otimes (1 - x^k)^i \\ &= 1 \otimes ((1 - x)(1 + x + \cdots + x^{k-1}))^i \\ &= 1 \otimes (1 - x)^i \{k + (x - 1) + (x^2 - 1) + \cdots \\ &\quad + (x^{k-1} - 1)\}^i \\ &\equiv 1 \otimes (1 - x)^i k^i \mod V_{i+1}.\end{aligned}$$

Hence V_i/V_{i+1} is the 1-dimensional irreducible $K\mathfrak{G}$-module on which x operates trivially and y induces multiplication by k^i.

Let α denote the character of $\mathfrak{G}$ given by $\alpha(x) = 1$ and $\alpha(y^i) = k^i$. Let M_j be an irreducible $K\mathfrak{G}$-module corresponding to α^j. By the assumption that $\mathbf{C}_{\mathfrak{G}}(x) = \langle x \rangle$, it follows that α is faithful on $\langle y \rangle$. As $\langle x \rangle = \mathbf{O}_p(\mathfrak{G})$, $M_0, M_1, \ldots, M_{r-1}$ are all the irreducible $K\mathfrak{G}$-modules.

We determine the multiplicity of M_j as a composition factor in P_0. The calculation above shows that

$$V_i/V_{i+1} \cong M_i \quad (i \text{ modulo } r).$$

Hence we obtain the composition series of P_0 illustrated in the following diagram.

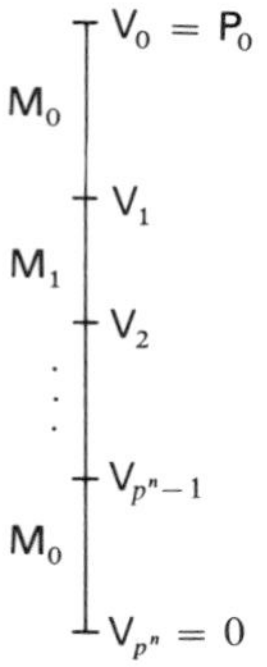

If $p^n - 1 = rs$, then the multiplicity of M_j in P_0 is s for $j \neq 0$ and $s + 1$ for $j = 0$.

We still have to find the other indecomposable projective $K\mathfrak{G}$-modules. In this example it is very simple. Put
$P_i = P_0 \otimes_K M_i (i = 0, \dots, r-1)$. Then by 7.19, P_i is projective. As

$$\dim_K P_i = \dim_K P_0 \dim_K M_i = \dim_K P_0 = p^n,$$

P_i is indecomposable by Dickson's theorem (7.16). The epimorphism α of P_0 onto M_0 gives rise to an epimorphism $\alpha \otimes 1$ of $P_i = P_0 \otimes_K M_i$ onto $M_0 \otimes_K M_i \cong M_i$. Hence P_i is the indecomposable projective $K\mathfrak{G}$-module with head M_i. We see immediately that P_i has a composition series described by the following diagram

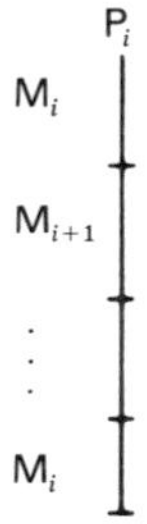

Thus the multiplicity of M_j in P_i is s for $i \neq j$ and $s+1$ for $i = j$. Hence the Cartan matrix of $K\mathfrak{G}$ is the $r \times r$ matrix

$$C = \begin{pmatrix} s+1 & s & \cdots & s \\ s & s+1 & \cdots & s \\ \vdots & \vdots & & \vdots \\ s & s & \cdots & s+1 \end{pmatrix}$$

As this matrix is obviously indecomposable in the sense of 12.5, $\mathfrak{G}$ has only one block for the prime p. Since $\mathbf{O}_{p'}(\mathfrak{G}) = 1$, this fact also follows from 13.5.

b) Now we assume that p divides r. Then $\mathfrak{G}$ is p-nilpotent and we can apply Theorem 14.9. Hence every block contains exactly one irreducible $K\mathfrak{G}$-module. We assume now that K is algebraically closed.

Let V be an irreducible $K\langle x\rangle$-module with character λ. We denote by $\mathfrak{I} = \mathfrak{I}(\lambda)$ the inertia subgroup of V in $\mathfrak{G}$. Let $\mathfrak{R}/\langle x\rangle$ be the normal p-complement of $\mathfrak{I}/\langle x\rangle$ and write $t = |\mathfrak{R}/\langle x\rangle|$. Then λ has exactly t extensions $\lambda_1, \dots, \lambda_t$ to $\mathfrak{R}$, which correspond to irreducible $K\mathfrak{R}$-modules $V_1, \dots, V_t$ of K-dimension 1. If K denotes the trivial $K\langle x\rangle$-module, then

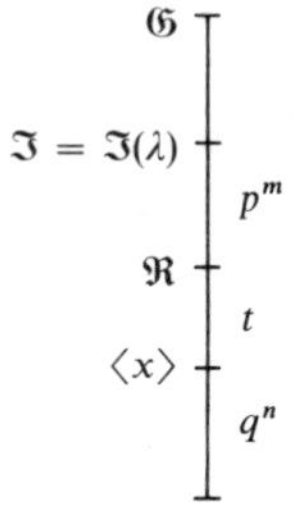

$$\mathsf{V}^{\mathfrak{N}} \cong ((\mathsf{V}_1)_{\langle x\rangle})^{\mathfrak{N}} \cong \mathsf{V}_1 \otimes_{\mathsf{K}} \mathsf{K}(\mathfrak{N}/\langle x\rangle) \quad (4.15b))$$
$$\cong \mathsf{V}_1 \oplus \cdots \oplus \mathsf{V}_t,$$

as $\mathsf{K}(\mathfrak{N}/\langle x\rangle)$ is a direct sum of modules of dimension 1.

We first construct indecomposable projective $\mathsf{K}\mathfrak{G}$-modules. We obtain

$$\mathsf{V}^{\mathfrak{I}} = \mathsf{V}_1^{\mathfrak{I}} \oplus \cdots \oplus \mathsf{V}_t^{\mathfrak{I}}.$$

As the Sylow p-subgroup $\mathfrak{S}$ of $\mathfrak{I}$ is a transversal of $\mathfrak{N}$ in $\mathfrak{I}$, we have

$$(\mathsf{V}_i^{\mathfrak{I}})_{\mathfrak{S}} \cong \mathsf{K}\mathfrak{S}.$$

Hence $\mathsf{V}_i^{\mathfrak{I}}$ is indecomposable, and thus by 9.6 the $\mathsf{V}_i^{\mathfrak{G}}$ are indecomposable. As $p \nmid |\mathfrak{N}|$, certainly V_i is a projective $\mathsf{K}\mathfrak{N}$-module, and hence $\mathsf{P}_i = \mathsf{V}_i^{\mathfrak{G}}$ is a projective $\mathsf{K}\mathfrak{G}$-module by 7.17.

Next we determine the head of P_i. As $\mathfrak{I}/\mathfrak{N}$ is a cyclic p-group, each of the characters λ_i of $\mathfrak{N}$ has a unique extension μ_i to $\mathfrak{I}$. As $\mathfrak{I} \trianglelefteq \mathfrak{G}$, we certainly have $\mathfrak{I}(\mu_i) \geq \mathfrak{I}$. But as $(\mu_i)_{\langle x\rangle} = \lambda$, we see that $\mathfrak{I}$ is the inertia group of μ_i. Let M_i be a $\mathsf{K}\mathfrak{I}$-module belonging to μ_i. By 9.6b) (with $\mathfrak{I}$ in the place of $\mathfrak{N}$ *and* $\mathfrak{I}$ of 9.6), $\mathsf{N}_i = \mathsf{M}_i^{\mathfrak{G}}$ is an irreducible $\mathsf{K}\mathfrak{G}$-module. Hence we obtain

$$\dim_{\mathsf{K}} \operatorname{Hom}_{\mathsf{K}\mathfrak{G}}(\mathsf{P}_i, \mathsf{N}_i) = \dim_{\mathsf{K}} \operatorname{Hom}_{\mathsf{K}\mathfrak{G}}(\mathsf{V}_i^{\mathfrak{G}}, \mathsf{N}_i) = \dim_{\mathsf{K}} \operatorname{Hom}_{\mathsf{K}\mathfrak{N}}(\mathsf{V}_i, (\mathsf{N}_i)_{\mathfrak{N}})$$
$$= \dim_{\mathsf{K}} \operatorname{Hom}_{\mathsf{K}\mathfrak{N}}(\mathsf{V}_i, \mathsf{V}_i \oplus \cdots) \geq 1.$$

Thus N_i is the head of P_i. As by 14.9 all composition factors of P_i are isomorphic to N_i, the multiplicity of N_i in P_i is

$$\dim_{\mathsf{K}} \mathsf{P}_i / \dim_{\mathsf{K}} \mathsf{N}_i = |\mathfrak{G} : \mathfrak{N}| / |\mathfrak{G} : \mathfrak{I}| = |\mathfrak{I}/\mathfrak{N}|.$$

Finally we have to show that the N_i are all the irreducible $\mathsf{K}\mathfrak{G}$-modules (to within isomorphism). Let W be any irreducible $\mathsf{K}\mathfrak{G}$-module and let λ be a character of $\langle x\rangle$ such that the $\mathsf{K}\langle x\rangle$-module V_λ belonging to λ is a direct summand of $\mathsf{W}_{\langle x\rangle}$. If we apply our construction to this λ, we obtain

$$\mathsf{V}_\lambda^{\mathfrak{G}} \cong (\mathsf{V}_\lambda^{\mathfrak{R}})^{\mathfrak{G}} \cong (\mathsf{V}_1 \oplus \cdots \oplus \mathsf{V}_t)^{\mathfrak{G}} \cong \mathsf{P}_1 \oplus \cdots \oplus \mathsf{P}_t.$$

Hence

$$\begin{aligned} 0 \neq \dim_{\mathsf{K}} \mathrm{Hom}_{\mathsf{K}\langle x\rangle}(\mathsf{W}_{\langle x\rangle}, \mathsf{V}_\lambda) &= \dim_{\mathsf{K}} \mathrm{Hom}_{\mathsf{K}\mathfrak{G}}(\mathsf{W}, \mathsf{V}_\lambda^{\mathfrak{G}}) \\ &= \dim_{\mathsf{K}} \mathrm{Hom}_{\mathsf{K}\mathfrak{G}}(\mathsf{W}, \mathsf{P}_1 \oplus \cdots \oplus \mathsf{P}_t). \end{aligned}$$

Thus W is isomorphic to an irreducible submodule of some P_i and hence to N_i.

Exercise

60) (WILLEMS) Suppose char $\mathsf{K} = p$. Let $\mathfrak{N}$ be a normal p-subgroup of $\mathfrak{G}$ and V an irreducible $\mathsf{K}\mathfrak{G}$-module. Then

$$L(\mathsf{P}_{\mathfrak{G}/\mathfrak{N}}(\mathsf{V})) + L(\mathsf{K}\mathfrak{N}) - 1 \leq L(\mathsf{P}_{\mathfrak{G}}(\mathsf{V})),$$

where L denotes the Loewy length.

§ 15. p-Chief Factors of p-Soluble Groups

If $\mathfrak{G}$ is a p-soluble group, any p-chief factor $\mathfrak{M}/\mathfrak{N}$ of $\mathfrak{G}$ may be regarded as an irreducible $\mathsf{K}\mathfrak{G}$-module, where $\mathsf{K} = GF(p)$, by putting

$$x\mathfrak{N} \to x^g\mathfrak{N} \quad (x \in \mathfrak{M}, g \in \mathfrak{G}).$$

Clearly, $\mathbf{O}_{p'}(\mathfrak{G})$ operates trivially on such a factor, so by 13.7, this module lies in the principal block. In this section we study relations between the Loewy series of the principal indecomposable projective $\mathsf{K}\mathfrak{G}$-module P_1 and the p-chief factors of $\mathfrak{G}$.

15.1 Definitions. Throughout this section we use the following notation.

Let $\mathfrak{U}$ be a subgroup of $\mathfrak{G}$ and $\mathsf{K} = GF(p)$. The $\mathsf{K}\mathfrak{G}$-module induced from the trivial $\mathsf{K}\mathfrak{U}$-module K is denoted by $\mathsf{K}_{\mathfrak{U}}^{\mathfrak{G}}$. We write

$$\mathsf{A}_{\mathfrak{U}}^{\mathfrak{G}} = \langle x(g-1) \mid x \in \mathsf{K}_{\mathfrak{U}}^{\mathfrak{G}}, g \in \mathfrak{G} \rangle = \mathsf{K}_{\mathfrak{U}}^{\mathfrak{G}} \mathfrak{J},$$

where $\mathfrak{J}$ is the augmentation ideal of $\mathsf{K}\mathfrak{G}$. If we regard $\mathsf{K}_{\mathfrak{U}}^{\mathfrak{G}}$ as being the K-vector space with K-basis the set of right cosets of $\mathfrak{U}$ in $\mathfrak{G}$, then $\{\mathfrak{U}g - \mathfrak{U} \mid g \in \mathfrak{G} - \mathfrak{U}\}$ is a K-basis of $\mathsf{A}_{\mathfrak{U}}^{\mathfrak{G}}$.

15.2 Lemma (GASCHÜTZ). *Suppose that $\mathfrak{G} = \mathfrak{C}\mathfrak{H}$ and $\mathfrak{C} \cap \mathfrak{H} = \mathfrak{D}$, where $\mathfrak{C}$ and $\mathfrak{D}$ are normal in $\mathfrak{G}$ and $\mathfrak{C}/\mathfrak{D}$ is an elementary Abelian p-group; thus $\mathfrak{C}/\mathfrak{D}$ may be regarded as a $\mathsf{K}\mathfrak{G}$-module. Let V be a $\mathsf{K}\mathfrak{G}$-module. If $\mathfrak{U} \leq \mathfrak{H}$, there exists a monomorphism σ of* $\operatorname{Hom}_{\mathsf{K}\mathfrak{G}}(\mathfrak{C}/\mathfrak{D}, \mathsf{V})$ *into* $\operatorname{Hom}_{\mathsf{K}\mathfrak{G}}(\mathsf{A}_{\mathfrak{U}}^{\mathfrak{G}}, \mathsf{V})$ *such that*

$$(\mathfrak{U}hc - \mathfrak{U})(\alpha\sigma) = (\mathfrak{D}c)\alpha$$

for all $\alpha \in$ $\operatorname{Hom}_{\mathsf{K}\mathfrak{G}}(\mathfrak{C}/\mathfrak{D}, \mathsf{V})$, *$h \in \mathfrak{H}$ and $c \in \mathfrak{C}$.*

Proof. Suppose that $\mathfrak{U}h_1c_1 = \mathfrak{U}h_2c_2$ with $h_i \in \mathfrak{H}$, $c_i \in \mathfrak{C}$. Then $\mathfrak{H}c_1 = \mathfrak{H}c_2$ since $\mathfrak{U} \leq \mathfrak{H}$, so $c_1c_2^{-1} \in \mathfrak{H} \cap \mathfrak{C} = \mathfrak{D}$ and $\mathfrak{D}c_1 = \mathfrak{D}c_2$. Thus if $\alpha \in \operatorname{Hom}_{\mathsf{K}\mathfrak{G}}(\mathfrak{C}/\mathfrak{D}, \mathsf{V})$, there exists a K-linear mapping $\alpha\sigma$ of $\mathsf{A}_{\mathfrak{U}}^{\mathfrak{G}}$ into V such that

$$(\mathfrak{U}hc - \mathfrak{U})(\alpha\sigma) = (\mathfrak{D}c)\alpha,$$

since $\mathfrak{G} = \mathfrak{H}\mathfrak{C}$ and $\{\mathfrak{U}g - \mathfrak{U} \mid g \in \mathfrak{G} - \mathfrak{U}\}$ is a K-basis of $\mathsf{A}_{\mathfrak{U}}^{\mathfrak{G}}$. $\alpha\sigma$ is a $\mathsf{K}\mathfrak{G}$-homomorphism, for if $g = h'c'$ with $h' \in \mathfrak{H}$, $c' \in \mathfrak{C}$,

$$\begin{aligned}((\mathfrak{U}hc - \mathfrak{U})g)(\alpha\sigma) &= ((\mathfrak{U}hh'c^{h'}c' - \mathfrak{U}) - (\mathfrak{U}h'c' - \mathfrak{U}))(\alpha\sigma)\\ &= (\mathfrak{D}c^{h'}c')\alpha - (\mathfrak{D}c')\alpha = (\mathfrak{D}c^{h'})\alpha\\ &= (\mathfrak{D}c^{g})\alpha = ((\mathfrak{D}c)\alpha)g\\ &= ((\mathfrak{U}hc - \mathfrak{U})(\alpha\sigma))g.\end{aligned}$$

Now σ is a homomorphism of $\operatorname{Hom}_{\mathsf{K}\mathfrak{G}}(\mathfrak{C}/\mathfrak{D}, \mathsf{V})$ into $\operatorname{Hom}_{\mathsf{K}\mathfrak{G}}(\mathsf{A}_{\mathfrak{U}}^{\mathfrak{G}}, \mathsf{V})$, since if $\alpha_i \in \operatorname{Hom}_{\mathsf{K}\mathfrak{G}}(\mathfrak{C}/\mathfrak{D}, \mathsf{V})$ $(i = 1, 2)$, then

$$\begin{aligned}(\mathfrak{U}hc - \mathfrak{U})((\alpha_1 + \alpha_2)\sigma) &= (\mathfrak{D}c)(\alpha_1 + \alpha_2)\\ &= (\mathfrak{D}c)\alpha_1 + (\mathfrak{D}c)\alpha_2\\ &= (\mathfrak{U}hc - \mathfrak{U})(\alpha_1\sigma + \alpha_2\sigma).\end{aligned}$$

Finally σ is a monomorphism, since if $\alpha\sigma = 0$, then $(\mathfrak{D}c)\alpha = 0$ for all $c \in \mathfrak{C}$ and so $\alpha = 0$. **q.e.d.**

15.3 Lemma (GASCHÜTZ). *Suppose that $\mathfrak{G}$ is p-soluble and $|\mathfrak{G}:\mathfrak{U}|$ is a power of p. Let* V *be an irreducible* $\mathsf{K}\mathfrak{G}$*-module and let*

$$\mathfrak{C} = \mathbf{C}_{\mathfrak{G}}(\mathsf{V}) = \{g \mid g \in \mathfrak{G}, vg = v \quad \textit{for all } v \in \mathsf{V}\}.$$

If we put $\mathfrak{D} = \mathfrak{C}'\mathfrak{C}^p$, there is a monomorphism τ of $\mathrm{Hom}_{\mathsf{K}\mathfrak{G}}(\mathsf{A}_{\mathfrak{U}}^{\mathfrak{G}}, \mathsf{V})$ *into* $\mathrm{Hom}_{\mathsf{K}\mathfrak{G}}(\mathfrak{C}/\mathfrak{D}, \mathsf{V})$ *such that*

$$(\mathfrak{D}c)(\beta\tau) = (\mathfrak{U}c - \mathfrak{U})\beta$$

for all $\beta \in \mathrm{Hom}_{\mathsf{K}\mathfrak{G}}(\mathsf{A}_{\mathfrak{U}}^{\mathfrak{G}}, \mathsf{V})$, $c \in \mathfrak{C}$.

Proof. Given $\beta \in \mathrm{Hom}_{\mathsf{K}\mathfrak{G}}(\mathsf{A}_{\mathfrak{U}}^{\mathfrak{G}}, \mathsf{V})$, put $c\bar{\beta} = (\mathfrak{U}c - \mathfrak{U})\beta$ for all $c \in \mathfrak{C}$. Then if $c_i \in \mathfrak{C}$ $(i = 1, 2)$,

$$\begin{aligned}(c_1c_2)\bar{\beta} &= (\mathfrak{U}c_1c_2 - \mathfrak{U})\beta = ((\mathfrak{U}c_1 - \mathfrak{U})c_2 + (\mathfrak{U}c_2 - \mathfrak{U}))\beta\\ &= ((\mathfrak{U}c_1 - \mathfrak{U})\beta)c_2 + (\mathfrak{U}c_2 - \mathfrak{U})\beta\\ &= (\mathfrak{U}c_1 - \mathfrak{U})\beta + (\mathfrak{U}c_2 - \mathfrak{U})\beta,\end{aligned}$$

since $\mathfrak{C} = \mathbf{C}_{\mathfrak{G}}(\mathsf{V})$. Hence

$$(c_1c_2)\bar{\beta} = c_1\bar{\beta} + c_2\bar{\beta}$$

and $\bar{\beta}$ is a homomorphism of $\mathfrak{C}$ into V. Since V is an elementary Abelian p-group, $\mathfrak{D} \leq \ker\bar{\beta}$. So there exists $\beta\tau \in \mathrm{Hom}_{\mathsf{K}}(\mathfrak{C}/\mathfrak{D}, \mathsf{V})$ such that

$$(\mathfrak{D}c)(\beta\tau) = (\mathfrak{U}c - \mathfrak{U})\beta$$

for all $c \in \mathfrak{C}$. In fact, $\beta\tau$ is a $\mathsf{K}\mathfrak{G}$-homomorphism, for if $c \in \mathfrak{C}$ and $g \in \mathfrak{G}$, then

$$\begin{aligned}((\mathfrak{D}c)g)(\beta\tau) &= (\mathfrak{D}c^g)(\beta\tau) = (\mathfrak{U}g^{-1}cg - \mathfrak{U})\beta\\ &= ((\mathfrak{U}g^{-1} - \mathfrak{U})cg + (\mathfrak{U}c - \mathfrak{U})g + (\mathfrak{U}g - \mathfrak{U}))\beta\\ &= ((\mathfrak{U}g^{-1} - \mathfrak{U})\beta)cg + ((\mathfrak{U}c - \mathfrak{U})\beta)g + (\mathfrak{U}g - \mathfrak{U})\beta\\ &= ((\mathfrak{U}g^{-1} - \mathfrak{U})g)\beta + ((\mathfrak{D}c)(\beta\tau))g + (\mathfrak{U}g - \mathfrak{U})\beta\\ &= ((\mathfrak{D}c)(\beta\tau))g.\end{aligned}$$

It is easy to see that τ is a homomorphism. To verify that it is a monomorphism, suppose that $0 \neq \beta \in \mathrm{Hom}_{\mathsf{K}\mathfrak{G}}(\mathsf{A}_{\mathfrak{U}}^{\mathfrak{G}}, \mathsf{V})$ and let

$$\mathfrak{K} = \{y | y \in \mathfrak{G}, (\mathfrak{U}y - \mathfrak{U})\beta = 0\}.$$

For $y_1, y_2 \in \mathfrak{K}$, we have

$$\begin{aligned}(\mathfrak{U}y_1 y_2 - \mathfrak{U})\beta &= (\mathfrak{U}y_1 - \mathfrak{U})y_2\beta + (\mathfrak{U}y_2 - \mathfrak{U})\beta \\ &= (\mathfrak{U}y_1 - \mathfrak{U})\beta y_2 = 0.\end{aligned}$$

Thus $\mathfrak{U} \leq \mathfrak{K} \leq \mathfrak{G}$ and so $|\mathfrak{G} : \mathfrak{K}|$ is a power of p. We show that

$$\bigcap_{g \in \mathfrak{G}} \mathfrak{K}^g \leq \mathfrak{C}.$$

Suppose that $y \in \bigcap_{g \in \mathfrak{G}} \mathfrak{K}^g$. Then for all $g \in \mathfrak{G}$,

$$\begin{aligned}(\mathfrak{U}g - \mathfrak{U})\beta y &= (\mathfrak{U}gy - \mathfrak{U})\beta - (\mathfrak{U}y - \mathfrak{U})\beta = (\mathfrak{U}gyg^{-1}g - \mathfrak{U})\beta \\ &= ((\mathfrak{U}gyg^{-1} - \mathfrak{U})g + (\mathfrak{U}g - \mathfrak{U}))\beta \\ &= (\mathfrak{U}gyg^{-1} - \mathfrak{U})\beta g + (\mathfrak{U}g - \mathfrak{U})\beta = (\mathfrak{U}g - \mathfrak{U})\beta,\end{aligned}$$

since $gyg^{-1} \in \mathfrak{K}$. As $\beta \neq 0$ and V is irreducible,

$$\langle (\mathfrak{U}g - \mathfrak{U})\beta | g \in \mathfrak{G} \rangle = \mathsf{V}.$$

Thus $y \in \mathbf{C}_{\mathfrak{G}}(\mathsf{V}) = \mathfrak{C}$.

Since $\mathfrak{G}/\mathfrak{C}$ is faithfully represented on the irreducible $\mathsf{K}\mathfrak{G}$-module V, we have $\mathbf{O}_p(\mathfrak{G}/\mathfrak{C}) = 1$. Now suppose that $\beta\tau = 0$. Then $(\mathfrak{U}c - \mathfrak{U})\beta = 0$ for all $c \in \mathfrak{C}$, so $\mathfrak{C} \leq \mathfrak{K}$. Since $\bigcap_{g \in \mathfrak{G}} \mathfrak{K}^g \leq \mathfrak{C}$, this implies that $\mathfrak{C} = \bigcap_{g \in \mathfrak{G}} \mathfrak{K}^g$. If $\mathfrak{N}/\mathfrak{C}$ is a minimal normal subgroup of $\mathfrak{G}/\mathfrak{C}$, then $\mathfrak{N}/\mathfrak{C}$ is a p'-group (as $\mathbf{O}_p(\mathfrak{G}/\mathfrak{C}) = 1$), hence $\mathfrak{N} \leq \mathfrak{K}$ and thus $\mathfrak{N} \leq \bigcap_{g \in \mathfrak{G}} \mathfrak{K}^g = \mathfrak{C}$. This proves $\mathfrak{G} = \mathfrak{C} = \mathfrak{K}$. Hence $(\mathfrak{U}g - \mathfrak{U})\beta = 0$ for all $g \in \mathfrak{G}$, and $\beta = 0$, a contradiction. **q.e.d.**

15.4 Lemma. *Let $\mathfrak{G}$ be a p-soluble group.*

a) *If $\mathfrak{M}$ is a minimal normal p-subgroup of $\mathfrak{G}$ and $\mathfrak{M} = \mathbf{C}_{\mathfrak{G}}(\mathfrak{M})$, then $\mathfrak{M}$ has a complement in $\mathfrak{G}$ (cf. II, 3.3).*

b) *Suppose that V is an irreducible $\mathsf{K}\mathfrak{G}$-module, where $\mathsf{K} = GF(p)$ and put $\mathfrak{C} = \mathbf{C}_{\mathfrak{G}}(\mathsf{V})$. Suppose also that $\mathfrak{D}$ is a normal subgroup of $\mathfrak{G}$ such that $\mathfrak{D} \leq \mathfrak{C}$, that $\mathfrak{C}/\mathfrak{D}$ is an elementary Abelian p-group and, as a $\mathsf{K}\mathfrak{G}$-module, $\mathfrak{C}/\mathfrak{D}$ is the direct sum of submodules isomorphic to V. Then there exists a subgroup $\mathfrak{H}$ of $\mathfrak{G}$ such that $\mathfrak{G} = \mathfrak{C}\mathfrak{H}$ and $\mathfrak{C} \cap \mathfrak{H} = \mathfrak{D}$.*

Proof. a) We may assume that $\mathfrak{M} < \mathfrak{G}$. Let $\mathfrak{N}/\mathfrak{M}$ be a minimal normal subgroup of $\mathfrak{G}/\mathfrak{M}$. As $\mathfrak{G}/\mathfrak{M}$ is faithfully and irreducibly represented on $\mathfrak{M}$, certainly $\mathfrak{N}/\mathfrak{M}$ is not a p-group. Thus $\mathfrak{N}/\mathfrak{M}$ is a p'-group. By I, 18.1, $\mathfrak{M}$ has a complement $\mathfrak{Q}$ in $\mathfrak{N}$, and by I, 18.2, all such complements are conjugate in $\mathfrak{N}$. Hence by the Frattini argument, $\mathfrak{G} = \mathfrak{N}\mathbf{N}_{\mathfrak{G}}(\mathfrak{Q}) = \mathfrak{M}\mathbf{N}_{\mathfrak{G}}(\mathfrak{Q})$. Thus $\mathfrak{M} \cap \mathbf{N}_{\mathfrak{G}}(\mathfrak{Q}) \trianglelefteq \mathfrak{G}$. But since $\mathfrak{M}$ is minimal and $\mathfrak{Q}$ does not centralize $\mathfrak{M}$, we have $\mathfrak{M} \cap \mathbf{N}_{\mathfrak{G}}(\mathfrak{Q}) = 1$ and $\mathbf{N}_{\mathfrak{G}}(\mathfrak{Q})$ is therefore a complement of $\mathfrak{M}$ in $\mathfrak{G}$.

b) If V is the trivial module, $\mathfrak{C} = \mathfrak{G}$ and we may take $\mathfrak{H} = \mathfrak{D}$. Otherwise $\mathfrak{C} < \mathfrak{G}$. Choose a minimal subgroup $\mathfrak{H}$ of $\mathfrak{G}$ for which $\mathfrak{G} = \mathfrak{C}\mathfrak{H}$ and $\mathfrak{C} \cap \mathfrak{H} \geq \mathfrak{D}$. Then $\mathfrak{C} \cap \mathfrak{H} \trianglelefteq \mathfrak{G}$, so if $\mathfrak{C} \cap \mathfrak{H} > \mathfrak{D}$, we may choose a normal subgroup $\mathfrak{N}$ of $\mathfrak{G}$, maximal with respect to $\mathfrak{D} \leq \mathfrak{N} < \mathfrak{C} \cap \mathfrak{H}$. Then by hypothesis, the $\mathsf{K}\mathfrak{G}$-module $(\mathfrak{C} \cap \mathfrak{H})/\mathfrak{N}$ is the direct sum of $\mathsf{K}\mathfrak{G}$-modules isomorphic to V.
Hence

$$\mathbf{C}_{\mathfrak{H}/\mathfrak{N}}((\mathfrak{C} \cap \mathfrak{H})/\mathfrak{N}) = (\mathfrak{C} \cap \mathfrak{H})/\mathfrak{N}.$$

Thus by a), $(\mathfrak{C} \cap \mathfrak{H})/\mathfrak{N}$ has a complement $\mathfrak{K}/\mathfrak{N}$ in $\mathfrak{H}/\mathfrak{N}$. Hence

$$\mathfrak{G} = \mathfrak{C}\mathfrak{H} = \mathfrak{C}(\mathfrak{C} \cap \mathfrak{H})\mathfrak{K} = \mathfrak{C}\mathfrak{K}$$

and $\mathfrak{C} \cap \mathfrak{K} \geq \mathfrak{N} \geq \mathfrak{D}$. This contradicts the minimality of $\mathfrak{H}$, so $\mathfrak{C} \cap \mathfrak{H} = \mathfrak{D}$. **q.e.d.**

15.5 Theorem (GASCHÜTZ). *Let $\mathfrak{G}$ be a p-soluble group, $\mathsf{K} = GF(p)$, let V be an irreducible $\mathsf{K}\mathfrak{G}$-module and put $\mathfrak{C} = \mathbf{C}_{\mathfrak{G}}(\mathsf{V})$.*

a) *For any p-complement $\mathfrak{U}$ of $\mathfrak{G}$, $\mathrm{Hom}_{\mathsf{K}\mathfrak{G}}(\mathsf{A}_{\mathfrak{U}}^{\mathfrak{G}}, \mathsf{V})$ and $\mathrm{Hom}_{\mathsf{K}\mathfrak{G}}(\mathfrak{C}/\mathfrak{C}'\mathfrak{C}^p, \mathsf{V})$ are isomorphic.*

b) *Let P_1 be the principal indecomposable projective $\mathsf{K}\mathfrak{G}$-module. Then V occurs as a composition factor in the completely reducible $\mathsf{K}\mathfrak{G}$-module $\mathsf{P}_1\mathbf{J}(\mathsf{K}\mathfrak{G})/\mathsf{P}_1\mathbf{J}(\mathsf{K}\mathfrak{G})^2$ if and only if V is isomorphic to a complemented p-chief factor of $\mathfrak{G}$, and the multiplicity of V in $\mathsf{P}_1\mathbf{J}(\mathsf{K}\mathfrak{G})/\mathsf{P}_1\mathbf{J}(\mathsf{K}\mathfrak{G})^2$ is equal to its multiplicity in the maximal completely reducible factor module of $\mathfrak{C}/\mathfrak{C}'\mathfrak{C}^p$.*

Proof. a) Let $\mathfrak{D}$ be the intersection of all normal subgroups $\mathfrak{X}$ of $\mathfrak{G}$ for which $\mathfrak{C}'\mathfrak{C}^p \leq \mathfrak{X} < \mathfrak{C}$ and $\mathfrak{C}/\mathfrak{X}$ is $\mathsf{K}\mathfrak{G}$-isomorphic to V. Then $\mathrm{Hom}_{\mathsf{K}\mathfrak{G}}(\mathfrak{C}/\mathfrak{C}'\mathfrak{C}^p, \mathsf{V})$ is isomorphic to $\mathrm{Hom}_{\mathsf{K}\mathfrak{G}}(\mathfrak{C}/\mathfrak{D}, \mathsf{V})$ and $\mathfrak{C}/\mathfrak{D}$ is the direct sum of $\mathsf{K}\mathfrak{G}$-submodules isomorphic to V. By 15.4, there exists a subgroup $\mathfrak{H}$ of $\mathfrak{G}$ such that $\mathfrak{G} = \mathfrak{C}\mathfrak{H}$ and $\mathfrak{C} \cap \mathfrak{H} = \mathfrak{D}$. Hence $|\mathfrak{G} : \mathfrak{H}| = |\mathfrak{C} : \mathfrak{D}|$ is a power of p. Thus there exists a p-complement $\mathfrak{U}$ of $\mathfrak{G}$ such that $\mathfrak{U} \leq \mathfrak{H}$. By 15.3, there is a monomorphism τ of $\mathrm{Hom}_{\mathsf{K}\mathfrak{G}}(\mathsf{A}_{\mathfrak{U}}^{\mathfrak{G}}, \mathsf{V})$ into $\mathrm{Hom}_{\mathsf{K}\mathfrak{G}}(\mathfrak{C}/\mathfrak{D}, \mathsf{V})$ such that

$$(\mathfrak{D}c)(\beta\tau) = (\mathfrak{U}c - \mathfrak{U})\beta$$

for all $c \in \mathfrak{C}$, $\beta \in \mathrm{Hom}_{\mathsf{K}\mathfrak{G}}(\mathsf{A}_{\mathfrak{U}}^{\mathfrak{G}}, \mathsf{V})$. And by 15.2, there is a monomorphism σ of $\mathrm{Hom}_{\mathsf{K}\mathfrak{G}}(\mathfrak{C}/\mathfrak{D}, \mathsf{V})$ into $\mathrm{Hom}_{\mathsf{K}\mathfrak{G}}(\mathsf{A}_{\mathfrak{U}}^{\mathfrak{G}}, \mathsf{V})$ such that $\sigma\tau$ is the identity mapping. Hence τ is an isomorphism.

Since the p-complements of $\mathfrak{G}$ are conjugate (VI, 1.7), a) holds for any p-complement $\mathfrak{U}$ of $\mathfrak{G}$.

b) Let r_1, r_2 be the multiplicities of V in $\mathsf{P}_1\mathbf{J}(\mathsf{K}\mathfrak{G})/\mathsf{P}_1\mathbf{J}(\mathsf{K}\mathfrak{G})^2$ and in the maximal completely reducible factor module of $\mathfrak{C}/\mathfrak{C}'\mathfrak{C}^p$ respectively. By 4.12,

$$r_1 \dim_{\mathsf{K}} \mathrm{Hom}_{\mathsf{K}\mathfrak{G}}(\mathsf{V}, \mathsf{V}) = \dim_{\mathsf{K}} \mathrm{Hom}_{\mathsf{K}\mathfrak{G}}(\mathsf{P}_1\mathbf{J}(\mathsf{K}\mathfrak{G}), \mathsf{V})$$

and

$$r_2 \dim_{\mathsf{K}} \mathrm{Hom}_{\mathsf{K}\mathfrak{G}}(\mathsf{V}, \mathsf{V}) = \dim_{\mathsf{K}} \mathrm{Hom}_{\mathsf{K}\mathfrak{G}}(\mathfrak{C}/\mathfrak{C}'\mathfrak{C}^p, \mathsf{V}).$$

By 10.12, $\mathsf{P}_1 = \mathsf{K}_{\mathfrak{U}}^{\mathfrak{G}}$, so $\mathsf{P}_1\mathbf{J}(\mathsf{K}\mathfrak{G}) = \mathsf{A}_{\mathfrak{U}}^{\mathfrak{G}}$. Hence by a), $\mathrm{Hom}_{\mathsf{K}\mathfrak{G}}(\mathsf{P}_1\mathbf{J}(\mathsf{K}\mathfrak{G}), \mathsf{V})$ and $\mathrm{Hom}_{\mathsf{K}\mathfrak{G}}(\mathfrak{C}/\mathfrak{C}'\mathfrak{C}^p, \mathsf{V})$ are isomorphic and $r_1 = r_2$.

Let $\mathfrak{A}/\mathfrak{B}$ be a p-chief factor of $\mathfrak{G}$ complemented by $\mathfrak{H}_1$. If $\mathfrak{C}_1 = \mathbf{C}_{\mathfrak{G}}(\mathfrak{A}/\mathfrak{B})$, then $\mathfrak{C}_1/(\mathfrak{C}_1 \cap \mathfrak{H}_1)$ is a self-centralizing complemented p-chief factor of $\mathfrak{G}$ $\mathsf{K}\mathfrak{G}$-isomorphic to $\mathfrak{A}/\mathfrak{B}$. Thus $\mathfrak{C}_1/(\mathfrak{C}_1 \cap \mathfrak{H}_1)$ is an irreducible $\mathsf{K}\mathfrak{G}$-module which occurs as a composition factor in the maximal completely reducible factor module of $\mathfrak{C}_1/\mathfrak{C}_1'\mathfrak{C}_1^p$ and hence in $\mathsf{P}_1\mathbf{J}(\mathsf{K}\mathfrak{G})/\mathsf{P}_1\mathbf{J}(\mathsf{K}\mathfrak{G})^2$.

Conversely, by a), a composition factor of $\mathsf{P}_1\mathbf{J}(\mathsf{K}\mathfrak{G})/\mathsf{P}_1\mathbf{J}(\mathsf{K}\mathfrak{G})^2$ is of the form $\mathfrak{C}/\mathfrak{D}$, where $\mathfrak{D} \trianglelefteq \mathfrak{G}$ and $\mathfrak{C}/\mathfrak{D}$ is self-centralizing. By 15.4a), $\mathfrak{C}/\mathfrak{D}$ is complemented. **q.e.d.**

15.6 Remark. The multiplicity r of V in $\mathsf{P}_1\mathbf{J}(\mathsf{K}\mathfrak{G})/\mathsf{P}_1\mathbf{J}(\mathsf{K}\mathfrak{G})^2$ can also be found in the following way. In any chief series of $\mathfrak{G}$, r is the number of complemented p-chief factors in this series which are $\mathsf{K}\mathfrak{G}$-isomorphic to V. (This number is indeed the same for every chief series.)

Theorem 15.5 gives a complete description of the second Loewy factor $\mathsf{P}_1\mathbf{J}(\mathsf{K}\mathfrak{G})/\mathsf{P}_1\mathbf{J}(\mathsf{K}\mathfrak{G})^2$ of P_1. Much less is known about the composition factors of $\mathsf{P}_1\mathbf{J}(\mathsf{K}\mathfrak{G})^2$. We shall however show that every p-chief factor of $\mathfrak{G}$ is isomorphic to some composition factor of a certain factor module of P_1. We first make the following elementary remarks.

15.7 Lemma. *Suppose that $\mathfrak{U} \leq \mathfrak{G}$.*

a) *If $\mathfrak{N} \trianglelefteq \mathfrak{G}$, then $\mathsf{K}_{\mathfrak{U}\mathfrak{N}/\mathfrak{N}}^{\mathfrak{G}/\mathfrak{N}}$ (regarded by means of inflation as a $\mathsf{K}\mathfrak{G}$-module) is isomorphic to $\mathsf{K}_{\mathfrak{U}\mathfrak{N}}^{\mathfrak{G}}$.*

b) *If* $\mathfrak{U} \leq \mathfrak{P} \leq \mathfrak{G}$ *the* K*-linear mapping* β *of* $\mathsf{K}_{\mathfrak{U}}^{\mathfrak{G}}$ *onto* $\mathsf{K}_{\mathfrak{P}}^{\mathfrak{G}}$ *given by* $(\mathfrak{U}g)\beta = \mathfrak{P}g$ $(g \in \mathfrak{G})$ *is a* $\mathsf{K}\mathfrak{G}$*-epimorphism of* $\mathsf{K}_{\mathfrak{U}}^{\mathfrak{G}}$ *onto* $\mathsf{K}_{\mathfrak{P}}^{\mathfrak{G}}$.

Proof. a) If v is the natural homomorphism of $\mathfrak{G}$ onto $\mathfrak{G}/\mathfrak{N}$, there exists a $\mathsf{K}\mathfrak{G}$-isomorphism α of $\mathsf{K}_{\mathfrak{U}\mathfrak{N}}^{\mathfrak{G}}$ onto $\mathsf{K}_{\mathfrak{U}\mathfrak{N}/\mathfrak{N}}^{\mathfrak{G}/\mathfrak{N}}$ such that

$$((\mathfrak{U}\mathfrak{N})g)\alpha = (\mathfrak{U}v)(gv)$$

for all $g \in \mathfrak{G}$.

b) This is obvious. **q.e.d.**

15.8 Theorem (GREEN, HILL [1]). *Suppose that* $\mathfrak{G}$ *is a p-soluble group,* $\mathfrak{U}$ *is a p-complement of* $\mathfrak{G}$ *and* $\mathfrak{S} = \mathbf{N}_{\mathfrak{G}}(\mathfrak{U})$. *Then* $\mathsf{K}_{\mathfrak{S}}^{\mathfrak{G}}$ *is an indecomposable factor module of the principal indecomposable projective* $\mathsf{K}\mathfrak{G}$*-module* $\mathsf{P}_1 = \mathsf{K}_{\mathfrak{U}}^{\mathfrak{G}}$. *Every p-chief factor of* $\mathfrak{G}$ *is isomorphic to a composition factor of* $\mathsf{K}_{\mathfrak{S}}^{\mathfrak{G}}$.

Proof. By 15.7b), $\mathsf{K}_{\mathfrak{S}}^{\mathfrak{G}}$ is an epimorphic image of the $\mathsf{K}\mathfrak{G}$-module $\mathsf{P}_1 = \mathsf{K}_{\mathfrak{U}}^{\mathfrak{G}}$. By 10.12b), P_1 is the projective $\mathsf{K}\mathfrak{G}$-module for which $\mathsf{P}_1/\mathsf{P}_1\mathbf{J}(\mathsf{K}\mathfrak{G})$ is the module for the trivial representation of $\mathfrak{G}$. By 10.9a), $\mathsf{K}_{\mathfrak{S}}^{\mathfrak{G}}$ has only one maximal submodule and is therefore indecomposable.

Suppose that $\mathfrak{N}/\mathfrak{M}$ is a p-chief factor of $\mathfrak{G}$. If $\mathfrak{N}/\mathfrak{M}$ is central in $\mathfrak{G}$, then $\mathfrak{N}/\mathfrak{M}$ is $\mathsf{K}\mathfrak{G}$-isomorphic to $\mathsf{K}_{\mathfrak{S}}^{\mathfrak{G}}/\mathsf{K}_{\mathfrak{S}}^{\mathfrak{G}}\mathbf{J}(\mathsf{K}\mathfrak{G})$. Hence we may assume that $\mathfrak{N}/\mathfrak{M}$ is not central.

We show first that it is sufficient to consider the case $\mathfrak{M} = 1$. For if the theorem is proved in this case, then the $\mathsf{K}(\mathfrak{G}/\mathfrak{M})$-module $\mathfrak{N}/\mathfrak{M}$ is isomorphic to a composition factor of $\mathsf{K}_{\mathfrak{T}/\mathfrak{M}}^{\mathfrak{G}/\mathfrak{M}}$, where $\mathfrak{T}/\mathfrak{M} = \mathbf{N}_{\mathfrak{G}/\mathfrak{M}}(\mathfrak{U}\mathfrak{M}/\mathfrak{M})$. Hence by 15.7a), the $\mathsf{K}\mathfrak{G}$-module $\mathfrak{N}/\mathfrak{M}$ is isomorphic to a composition factor of $\mathsf{K}_{\mathfrak{T}}^{\mathfrak{G}}$. Now $\mathfrak{S} \leq \mathfrak{T}$ and hence $\mathfrak{N}/\mathfrak{M}$ is isomorphic to a composition factor of $\mathsf{K}_{\mathfrak{S}}^{\mathfrak{G}}$ by 15.7b).

Suppose then that $\mathfrak{M} = 1$. Then $\mathfrak{N}$ is a minimal normal non-central p-subgroup of $\mathfrak{G}$. Let $\mathfrak{P}$ be a Sylow p-subgroup of $\mathfrak{G}$. Then $\mathfrak{N} \leq \mathfrak{P}$ and $\mathfrak{G} = \mathfrak{U}\mathfrak{P}$. Now $[\mathfrak{N}, \mathfrak{U}] \leq \mathfrak{N}$ and $[\mathfrak{S}, \mathfrak{U}] \leq \mathfrak{U}$ since $\mathfrak{S} = \mathbf{N}_{\mathfrak{G}}(\mathfrak{U})$. Thus

$$[\mathfrak{N} \cap \mathfrak{S}, \mathfrak{U}] \leq \mathfrak{N} \cap \mathfrak{U} = 1.$$

Hence

$$[\mathfrak{N} \cap \mathfrak{S}, \mathfrak{G}] = [\mathfrak{N} \cap \mathfrak{S}, \mathfrak{U}\mathfrak{P}] = [\mathfrak{N} \cap \mathfrak{S}, \mathfrak{P}] < \mathfrak{N}$$

by III, 1.2 and 2.6. But by III, 1.6b), $[\mathfrak{N} \cap \mathfrak{S}, \mathfrak{G}] \trianglelefteq \mathfrak{G}$, so $[\mathfrak{N} \cap \mathfrak{S}, \mathfrak{G}] = 1$. Thus $\mathfrak{N} \cap \mathfrak{S} \leq \mathfrak{N} \cap \mathbf{Z}(\mathfrak{G}) = 1$.

By 15.7b), the K-linear mapping β of $\mathsf{K}_{\mathfrak{S}}^{\mathfrak{G}}$ onto $\mathsf{K}_{\mathfrak{S}\mathfrak{N}}^{\mathfrak{G}}$, given by $(\mathfrak{S}g)\beta = \mathfrak{S}\mathfrak{N}g$ $(g \in \mathfrak{G})$ is a $\mathsf{K}\mathfrak{G}$-epimorphism of $\mathsf{K}_{\mathfrak{S}}^{\mathfrak{G}}$ onto $\mathsf{K}_{\mathfrak{S}\mathfrak{N}}^{\mathfrak{G}}$. We have

$$\begin{aligned}\dim_{\mathsf{K}} \ker \beta &= \dim_{\mathsf{K}} \mathsf{K}_{\mathfrak{S}}^{\mathfrak{G}} - \dim_{\mathsf{K}} \mathsf{K}_{\mathfrak{S}\mathfrak{N}}^{\mathfrak{G}} \\ &= |\mathfrak{G} : \mathfrak{S}| - |\mathfrak{G} : \mathfrak{S}\mathfrak{N}| = (|\mathfrak{N}| - 1)\,|\mathfrak{G} : \mathfrak{S}\mathfrak{N}|,\end{aligned}$$

since $\mathfrak{S} \cap \mathfrak{N} = 1$. But if T is a transversal of $\mathfrak{S}\mathfrak{N}$ in $\mathfrak{G}$, then $\mathfrak{S}yt - \mathfrak{S}t \in \ker \beta$ for all $y \in \mathfrak{N} - \{1\}$, $t \in \mathsf{T}$, and these elements $\mathfrak{S}yt - \mathfrak{S}t$ are linearly independent. Since the number of them is $\dim_{\mathsf{K}} \ker \beta$, it follows that they form a K-basis of $\ker \beta$. Hence there exists a K-linear mapping γ of $\ker \beta$ into $\mathfrak{N}$ given by

$$((\mathfrak{S}y - \mathfrak{S})t)\gamma = y^t \quad (y \in \mathfrak{N},\ t \in \mathsf{T}).$$

This is a $\mathsf{K}\mathfrak{G}$-homomorphism, since if $g \in \mathfrak{G}$, $y \in \mathfrak{N}$, $t \in \mathsf{T}$ and $tg = sy't'$ with $s \in \mathfrak{S}$, $y' \in \mathfrak{N}$ and $t' \in \mathsf{T}$, then

$$\begin{aligned}((\mathfrak{S}y - \mathfrak{S})tg)\gamma &= ((\mathfrak{S}ysy' - \mathfrak{S}y')t')\gamma = ((\mathfrak{S}y^sy' - \mathfrak{S}y')t')\gamma \\ &= (y^sy')^{t'}(y'^{t'})^{-1} = y^{st'} \\ &= y^{sy't'} \qquad \text{(since } \mathfrak{N} \text{ is Abelian)} \\ &= y^{tg} = (((\mathfrak{S}y - \mathfrak{S})t)\gamma)^g.\end{aligned}$$

Thus γ is a $\mathsf{K}\mathfrak{G}$-epimorphism of $\ker \beta$ onto $\mathfrak{N}$, and so $\mathfrak{N}$ is isomorphic to a composition factor of $\mathsf{K}_{\mathfrak{S}}^{\mathfrak{G}}$. **q.e.d.**

15.9 Examples. a) Suppose $\mathfrak{G} = \mathfrak{Q}\mathfrak{Z}$ is the split extension of the quaternion group $\mathfrak{Q}$ of order 8 by a cyclic group $\mathfrak{Z}$ of order 3, where $\mathfrak{Z}$ operates non-trivially on $\mathfrak{Q}$. Let $\mathsf{K} = GF(2)$ and $\mathsf{L} = GF(4)$.

By 10.12, the principal indecomposable projective $\mathsf{K}\mathfrak{G}$-module P_1' is induced from the trivial module K for the 2-complement $\mathfrak{Z}$ of $\mathfrak{G}$. Hence $\dim_{\mathsf{K}} \mathsf{P}_1' = |\mathfrak{G} : \mathfrak{Z}| = 8$.

Let V be the irreducible $\mathsf{K}\mathfrak{G}$-module isomorphic to the complemented 2-chief factor $\mathfrak{Q}/\mathbf{Z}(\mathfrak{Q})$ of $\mathfrak{G}$. Then $\mathbf{C}_{\mathfrak{G}}(\mathsf{V}) = \mathfrak{Q}$, and by 15.5, V appears as a composition factor of $\mathsf{P}_1'\mathbf{J}(\mathsf{K}\mathfrak{G})/\mathsf{P}_1'\mathbf{J}(\mathsf{K}\mathfrak{G})^2$ with multiplicity 1. As $\mathfrak{Q}/\mathbf{Z}(\mathfrak{Q})$ is the only complemented 2-chief factor of $\mathfrak{G}$, by 15.5,

$$\mathsf{P}_1'\mathbf{J}(\mathsf{K}\mathfrak{G})/\mathsf{P}_1'\mathbf{J}(\mathsf{K}\mathfrak{G})^2 \cong \mathsf{V}.$$

By 11.6c) the socle S of P_1' is isomorphic to the module K for the trivial representation of $\mathsf{K}\mathfrak{G}$. Thus by 8.3a),

$$\mathsf{K} \cong \mathsf{S}^* \cong \mathsf{P}_1'^*/\mathsf{S}^{\perp},$$

and $\mathsf{P}_1'^*$ is indecomposable by 8.3f). By 8.23b), $\mathsf{P}_1'^*$ is also projective, so $\mathsf{P}_1'^* \cong \mathsf{P}_1'$. From $\dim_{\mathsf{K}}(\mathsf{P}_1'\mathbf{J}(\mathsf{K}\mathfrak{G}))^{\perp} = 1$ we conclude that

$$\begin{aligned}(\mathsf{P}_1'\mathbf{J}(\mathsf{K}\mathfrak{G}))^* &\cong \mathsf{P}_1'^*/(\mathsf{P}_1'\mathbf{J}(\mathsf{K}\mathfrak{G}))^{\perp} \quad \text{(8.3a))}\\ &\cong \mathsf{P}_1'/\mathsf{S}.\end{aligned}$$

As $\mathsf{P}_1'\mathbf{J}(\mathsf{K}\mathfrak{G})$ has only one irreducible factor module and this is isomorphic to V, P_1'/S has only one irreducible submodule U/S, and this is isomorphic to V^*. Hence $\mathsf{U} \subseteq \mathsf{P}_1'\mathbf{J}(\mathsf{K}\mathfrak{G})^2$. But as $\mathbf{O}_2(\mathfrak{G}) = \mathfrak{Q}$ centralizes all irreducible $\mathsf{K}\mathfrak{G}$-modules, V is the only irreducible $\mathsf{K}\mathfrak{G}$-module of dimension 2 and hence $\mathsf{V}^* \cong \mathsf{V}$. Therefore we have the following diagram of submodules of P_1'.

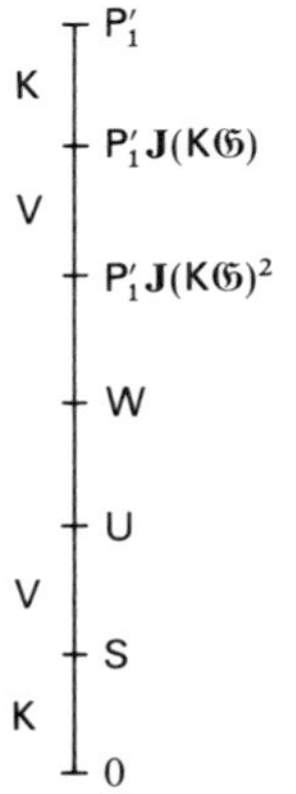

We still have to determine the 2-dimensional module

$$\mathsf{P}_1'\mathbf{J}(\mathsf{K}\mathfrak{G})^2/\mathsf{U}.$$

As $\mathfrak{Z} < \mathfrak{Z} \times \mathbf{Z}(\mathfrak{Q}) < \mathfrak{G}$ and $|\mathfrak{G} : \mathfrak{Z} \times \mathbf{Z}(\mathfrak{Q})| = 4$, P_1' has a factor module P_1'/W of dimension 4 by 15.7. Hence

$$\mathsf{U} \subset \mathsf{W} \subset \mathsf{P}_1'\mathbf{J}(\mathsf{K}\mathfrak{G})^2$$

and

$$\mathsf{W}/\mathsf{U} \cong \mathsf{P}_1'\mathbf{J}(\mathsf{K}\mathfrak{G})^2/\mathsf{W} \cong \mathsf{K}.$$

We claim that $P_1' \mathbf{J}(K\mathfrak{G})^2/U$ is isomorphic to $K \oplus K$. Otherwise by 10.14a) it would be an epimorphic image of P_1'. But this is impossible, as $P_1'/P_1'\mathbf{J}(K\mathfrak{G})^2$ has only one composition factor isomorphic to K. Thus the lattice of submodules of P_1' is the following.

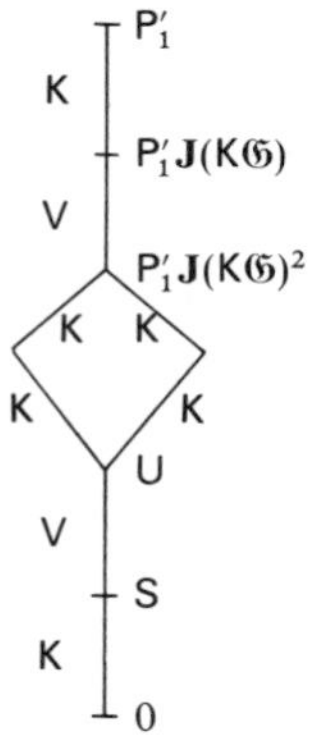

Obviously $V \otimes_K L \cong V_2 \oplus V_3$, where V_2, V_3 are irreducible non-isomorphic $L\mathfrak{G}$-modules, and L, V_2, V_3 represent all types of irreducible $L\mathfrak{G}$-modules. As complete reducibility is not affected by extension of the ground-field (1.8), we see that the factor modules in the Loewy series of $P_1 = P_1' \otimes_K L$ are

$$L,\ V_2 \oplus V_3,\ L \oplus L,\ V_2 \oplus V_3,\ L.$$

As in Example 14.10, we see that $P_i = P_1 \otimes_L V_i$ $(i = 2, 3)$ is the indecomposable projective $L\mathfrak{G}$-module with head V_i. As $V_2 \otimes_L V_2 \cong V_3$ and $V_2 \otimes_L V_3 \cong L$, we see that P_2 has a descending series of submodules with factor modules (from above) isomorphic to

$$V_2,\ L \oplus V_3,\ V_2 \oplus V_2,\ L \oplus V_3,\ V_2.$$

To show that these factors are the Loewy factors of P_2, we have only to observe that

$$P_2 \otimes_L V_3 \cong (P_1 \otimes_L V_2) \otimes_L V_3 \cong P_1 \otimes_L (V_2 \otimes_L V_3) \cong P_1 \otimes_L L \cong P_1$$

and that every completely reducible $K\mathfrak{G}$-module remains completely reducible by tensoring with the module V_3 of dimension 1.

Similarly, P_3 has the Loewy factors

$$\mathsf{V}_3,\ \mathsf{L} \oplus \mathsf{V}_2,\ \mathsf{V}_3 \oplus \mathsf{V}_3,\ \mathsf{L} \oplus \mathsf{V}_2,\ \mathsf{V}_3.$$

Hence the Cartan matrix is

$$C = \begin{pmatrix} 4 & 2 & 2 \\ 2 & 4 & 2 \\ 2 & 2 & 4 \end{pmatrix}.$$

b) It is much simpler to study the group-ring $\mathsf{K}\mathfrak{G}$, where $\mathfrak{G} = \mathfrak{Q}\mathfrak{Z}$ is the same group as in a), but K is an algebraically closed field of characteristic 3. Now $\mathfrak{G}$ is 3-nilpotent and has three 3′-classes, hence there are three irreducible $\mathsf{K}\mathfrak{G}$-modules $\mathsf{V}_1, \mathsf{V}_2, \mathsf{V}_3$. We assume $\mathsf{V}_1 = \mathsf{K}$.

As $\mathfrak{Q}$ is a 3-complement of $\mathfrak{G}$, by 10.12b) $\mathsf{P}_1 = \mathsf{W}_1^{\mathfrak{G}}$ is the principal indecomposable projective $\mathsf{K}\mathfrak{G}$-module, where W_1 denotes the trivial $\mathsf{K}\mathfrak{Q}$-module. The Cartan matrix of $\mathsf{K}\mathfrak{G}$ is a diagonal matrix by 14.9. Hence P_1 has the Loewy factors $\mathsf{V}_1, \mathsf{V}_1, \mathsf{V}_1$. So $c_{11} = 3$ and $c_{1i} = 0$ for $i = 2, 3$.

Let W_2 be an irreducible, non-trivial $\mathsf{K}\mathfrak{Q}$-module of dimension 1. Then $\mathfrak{Q}$ is the inertia group of W_2, hence $\mathsf{P}_2 = \mathsf{W}_2^{\mathfrak{G}}$ is irreducible, by 9.6. As W_2 is a projective $\mathsf{K}\mathfrak{Q}$-module, P_2 is a projective $\mathsf{K}\mathfrak{G}$-module, by 7.17. We put $\mathsf{V}_2 = \mathsf{P}_2$.

Let W_3 be the irreducible $\mathsf{K}\mathfrak{Q}$-module of dimension 2. Its inertia group is $\mathfrak{G}$. Hence by 9.9a) there exists an irreducible $\mathsf{K}\mathfrak{G}$-module V_3 of dimension 2 such that $(\mathsf{V}_3)_{\mathfrak{Q}} \cong \mathsf{W}_3$. We put $\mathsf{P}_3 = \mathsf{W}_3^{\mathfrak{G}}$. Then P_3 is projective by 7.17. We have

$$\dim_{\mathsf{K}} \operatorname{Hom}_{\mathsf{K}\mathfrak{G}}(\mathsf{P}_3, \mathsf{V}_i) = \dim_{\mathsf{K}} \operatorname{Hom}_{\mathsf{K}\mathfrak{Q}}(\mathsf{W}_3, (\mathsf{V}_i)_{\mathfrak{Q}}).$$

Now $\mathbf{Z}(\mathfrak{Q})$ is represented faithfully on W_3, but $\mathbf{Z}(\mathfrak{Q})$ is represented trivially on $(\mathsf{V}_i)_{\mathfrak{Q}}$ for $i = 1, 2$. Thus

$$\dim_{\mathsf{K}} \operatorname{Hom}_{\mathsf{K}\mathfrak{G}}(\mathsf{P}_3, \mathsf{V}_i) = \begin{cases} 0 & \text{for } i = 1, 2, \\ 1 & \text{for } i = 3. \end{cases}$$

Hence $\mathsf{P}_3 = \mathsf{W}_3^{\mathfrak{G}}$ is the projective indecomposable $\mathsf{K}\mathfrak{G}$-module with head V_3. As

$$\dim_{\mathsf{K}} \mathsf{P}_3 = 6 = 3 \dim_{\mathsf{K}} \mathsf{V}_3,$$

the Loewy factors of P_3 are $\mathsf{V}_3, \mathsf{V}_3, \mathsf{V}_3$. Hence the Cartan matrix of $\mathfrak{G}$ is

$$C = \begin{pmatrix} 3 & 0 & 0 \\ 0 & 1 & 0 \\ 0 & 0 & 3 \end{pmatrix}.$$

15.10 Examples. a) We now consider the case $\mathfrak{G} = \mathfrak{S}_4$ and K any field of characteristic 2. As $\mathfrak{G}/\mathbf{O}_2(\mathfrak{G}) \cong \mathfrak{S}_3$, $\mathsf{K}\mathfrak{G}$ has two irreducible modules K and V, where V is absolutely irreducible of dimension 2. Thus K is a splitting field for $\mathsf{K}\mathfrak{G}/\mathbf{J}(\mathsf{K}\mathfrak{G})$. We denote by P_1, P_2 the indecomposable projective $\mathsf{K}\mathfrak{G}$-modules for which

$$\mathsf{P}_1/\mathsf{P}_1\mathbf{J}(\mathsf{K}\mathfrak{G}) \cong \mathsf{K}, \mathsf{P}_2/\mathsf{P}_2\mathbf{J}(\mathsf{K}\mathfrak{G}) \cong \mathsf{V}.$$

As P_1 is induced from the trivial module for a 2-complement of $\mathfrak{G}$, $\dim_{\mathsf{K}}\mathsf{P}_1 = 8$. By 10.18b),

$$\mathsf{K}\mathfrak{G} \cong \mathsf{P}_1 \oplus \mathsf{P}_2 \oplus \mathsf{P}_2;$$

hence $\dim_{\mathsf{K}}\mathsf{P}_2 = 8$.

Next we determine the Cartan matrix of $\mathfrak{G}$. Using the symmetry of C, we obtain

$$8 = \dim_{\mathsf{K}}\mathsf{P}_1 = c_{11} + c_{12}\dim_{\mathsf{K}}\mathsf{V} = c_{11} + 2c_{12},$$
$$8 = \dim_{\mathsf{K}}\mathsf{P}_2 = c_{12} + 2c_{22}.$$

Thus c_{12} is even. By 15.5, $\mathsf{P}_1\mathbf{J}(\mathsf{K}\mathfrak{G})/\mathsf{P}_1\mathbf{J}(\mathsf{K}\mathfrak{G})^2$ is isomorphic to $\mathsf{K} \oplus \mathsf{V}$. As head and socle of P_1 are isomorphic to K, we conclude that $c_{11} \geq 3$ and $c_{12} \geq 1$. As c_{12} is even, this forces $c_{11} = 4, c_{12} = c_{21} = 2, c_{22} = 3$. Thus

$$C = \begin{pmatrix} 4 & 2 \\ 2 & 3 \end{pmatrix}.$$

(As $\mathfrak{G}$ is not 2-nilpotent, we could have deduced $c_{12} \neq 0$ from 14.9.)

It is considerably more difficult to obtain the Loewy series of P_1 and P_2. We start by looking at several $\mathsf{K}\mathfrak{G}$-modules.

(1) $\mathsf{V} \otimes_{\mathsf{K}} \mathsf{V} \cong \mathsf{P}_1' \oplus \mathsf{V}$, where P_1' is indecomposable and has two composition factors isomorphic to K.

As the Klein subgroup operates trivially on V, we can regard V as an irreducible $\mathsf{K}\mathfrak{S}_3$-module. If P_1', P_2' are the indecomposable projective $\mathsf{K}\mathfrak{S}_3$-modules with heads K, V respectively, then

$$6 = \dim_K K\mathfrak{S}_3 = \dim_K P_1' + 2\dim_K P_2'.$$

This implies that $\dim_K P_1' = 2$ and $P_2' = V$. Hence V is a projective $K\mathfrak{S}_3$-module. (It is *not* projective as a $K\mathfrak{G}$-module!). By 7.19), $V \otimes_K V$ is a projective $K\mathfrak{S}_3$-module. We consider the operation of an element g of order 3 on $V \otimes_K V$. The eigen-values of g on V are ε, ε^{-1}, where $\varepsilon^3 = 1 \neq \varepsilon$. Hence on $V \otimes_K V$ the eigen-values of g are 1, 1, ε, ε^{-1}. This rules out the possibilities $V \otimes_K V \cong P_1' \oplus P_1'$ and $V \otimes_K V \cong V \oplus V$. Hence

$$V \otimes_K V \cong P_1' \oplus V.$$

(2) Let W be the $K\mathfrak{G}$-module of dimension 4 induced from the trivial $K\mathfrak{S}_3$-module. Then the head and socle of W are isomorphic to K, and the composition factors of W are K, V, K.

By Nakayama's theorem, the multiplicity of K in the head of W is

$$\dim_K \mathrm{Hom}_{K\mathfrak{G}}(W, K) = \dim_K \mathrm{Hom}_{K\mathfrak{S}_3}(K, K) = 1.$$

The multiplicity of V in the head of W is

$$\dim_K \mathrm{Hom}_{K\mathfrak{G}}(W, V) = \dim_K \mathrm{Hom}_{K\mathfrak{S}_3}(K, V_{\mathfrak{S}_3}) = 0.$$

The statement about the socle of W follows similarly. If all composition factors of W are isomorphic to K, the elements of order 3 of $\mathfrak{G}$ operate trivially on W, by Maschke's theorem. But this is not true, for W is the module for the natural permutation representation of $\mathfrak{G} = \mathfrak{S}_4$ of degree 4.

(3) There exists a $K\mathfrak{G}$-module U_1 such that $\dim_K U_1 = 4$, the head of U_1 is isomorphic to V, the socle of U_1 is isomorphic to K and the composition factors of U_1 are V, K, K.

Let

$$0 = W_0 \subset W_1 \subset W_2 \subset W_3 = W$$

be a composition series of W. Then by (2),

$$W_1/W_0 \cong W_3/W_2 \cong K \quad \text{and} \quad W_2/W_1 \cong V.$$

We consider $T = W \otimes_K V$ and put $T_i = W_i \otimes_K V$. Then

$$T_1/T_0 \cong (W_1/W_0) \otimes_K V \cong V \cong T_3/T_2$$

and by (1),

$$T_2/T_1 \cong (W_2/W_1) \otimes_K V \cong V \otimes_K V \cong P_1' \oplus V.$$

We have

$$T \cong K^{\mathfrak{G}} \otimes_K V \simeq (K \otimes_K V_{\mathfrak{S}_3})^{\mathfrak{G}} \cong (V_{\mathfrak{S}_3})^{\mathfrak{G}}.$$

As

$$\dim_K \operatorname{Hom}_{K\mathfrak{G}}(T, K) = \dim_K \operatorname{Hom}_{K\mathfrak{S}_3}(V_{\mathfrak{S}_3}, K) = 0,$$

T has no factor module isomorphic to K. Thus T has a factor module U_1 with the required properties.

(4) The dual U_1^* of U_1 is a module U_2 with only one composition series

$$0 = Y_0 \subset Y_1 \subset Y_2 \subset Y_3 = U_2,$$

where

$$Y_1/Y_0 \cong V^* \cong V \quad \text{and} \quad Y_2/Y_1 \cong Y_3/Y_2 \cong K.$$

(5) Now we can determine the Loewy series of P_1. By 15.5,

$$P_1\mathbf{J}(K\mathfrak{G})/P_1\mathbf{J}(K\mathfrak{G})^2 \cong K \oplus V.$$

As W and U_2 are isomorphic to factor modules of P_1 (10.14a)), there are submodules Z_1 and Z_2 of P_1 for which

$$P_1/Z_1 \cong W \quad \text{and} \quad P_1/Z_2 \cong U_2.$$

Hence

$$K \cong W\mathbf{J}(K\mathfrak{G})^2 \cong P_1\mathbf{J}(K\mathfrak{G})^2/(P_1\mathbf{J}(K\mathfrak{G})^2 \cap Z_1)$$

and

$$V \cong U_2\mathbf{J}(K\mathfrak{G})^2 \cong P_1\mathbf{J}(K\mathfrak{G})^2/(P_1\mathbf{J}(K\mathfrak{G})^2 \cap Z_2).$$

Therefore $P_1\mathbf{J}(K\mathfrak{G})^2/P_1\mathbf{J}(K\mathfrak{G})^3$ has composition factors K and V. As $\mathbf{S}(P_1) \cong K$, this implies

$$P_1\mathbf{J}(K\mathfrak{G})^2/P_1\mathbf{J}(K\mathfrak{G})^3 \cong K \oplus V$$

and

$$P_1\mathbf{J}(K\mathfrak{G})^3 = \mathbf{S}(P_1) \cong K.$$

To determine the Loewy series of the indecomposable projective $K\mathfrak{G}$-module P_2 with head V, we show first that

(6) $P_1 \otimes_K V \cong P_2 \oplus P_2$.

By 7.19c), $P_1 \otimes_K V$ is projective and hence is isomorphic to a direct sum of copies of P_1 and P_2. The composition factors of $P_1 \otimes_K V$ are those of

$$K \otimes_K V = V, \text{ taken 4 times,}$$

and those of

$$V \otimes_K V \cong P_1' \oplus V \text{ (cf. (1)), taken twice.}$$

Hence the multiplicities of K, V in $P_1 \otimes_K V$ are 4, 6 respectively. Using the Cartan matrix, this implies that

$$P_1 \otimes_K V \cong P_2 \oplus P_2.$$

(7) From the Loewy series of P_1 we see that $P_1 \otimes_K V$ has submodules R_1, R_2 such that

$$(P_1 \otimes_K V)/R_1 \cong V \quad \text{and} \quad R_1/R_2 \cong (K \oplus V) \otimes_K V \cong V \oplus (V \oplus P_1').$$

Hence $(P_2 \oplus P_2)/(P_2 \oplus P_2)\mathbf{J}(K\mathfrak{G})^2$ has at least the composition factors V, V, V, K. Thus $P_2\mathbf{J}(K\mathfrak{G})/P_2\mathbf{J}(K\mathfrak{G})^2$ must have composition factors V and K. Hence $\dim_K P_2\mathbf{J}(K\mathfrak{G})^2 \leq 3$. This still allows the possibilities that

(i) $$\dim_K P_2\mathbf{J}(K\mathfrak{G})^2 = 3,\ P_2\mathbf{J}(K\mathfrak{G})^3 = \mathbf{S}(P_2) = V$$

or

(ii) $$P_2\mathbf{J}(K\mathfrak{G})^2 = \mathbf{S}(P_2) = V.$$

We show that case (ii) is not possible.

In (3) we constructed a $K\mathfrak{G}$-module U_1 with head V, socle K and of dimension 4. By 10.14a), this module is isomorphic to a factor module of P_2, say $\mathsf{P}_2/\mathsf{Y} \cong \mathsf{U}_1$. But this implies

$$\mathsf{K} \cong \mathsf{U}_1\mathbf{J}(\mathsf{K}\mathfrak{G})^2 \cong (\mathsf{P}_2\mathbf{J}(\mathsf{K}\mathfrak{G})^2 + \mathsf{Y})/\mathsf{Y} \cong \mathsf{P}_2\mathbf{J}(\mathsf{K}\mathfrak{G})^2/(\mathsf{P}_2\mathbf{J}(\mathsf{K}\mathfrak{G})^2 \cap \mathsf{Y}).$$

Thus case (i) holds. The structure of P_2 is illustrated in the following diagram.

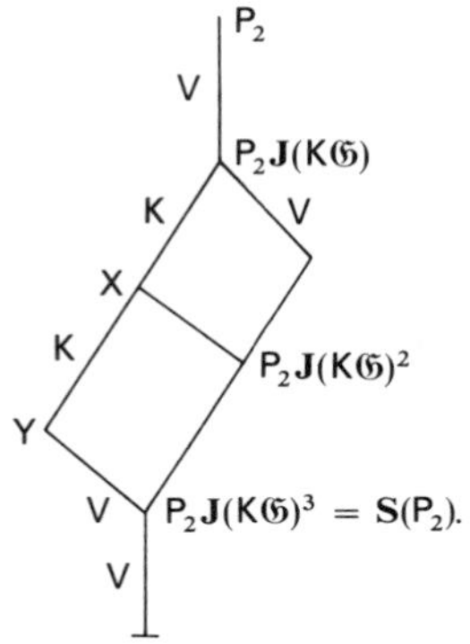

Hence

$$\mathsf{P}_2\mathbf{J}(\mathsf{K}\mathfrak{G})^i/\mathsf{P}_2\mathbf{J}(\mathsf{K}\mathfrak{G})^{i+1} \cong \begin{cases} \mathsf{V} & \text{for } i = 0, \\ \mathsf{K} \oplus \mathsf{V} & \text{for } i = 1, \\ \mathsf{K} & \text{for } i = 2, \\ \mathsf{V} & \text{for } i = 3. \end{cases}$$

The upper Loewy series of P_2 is

$$0 \subset \mathbf{S}(\mathsf{P}_2) \subset \mathsf{X} \subset \mathsf{P}_2\mathbf{J}(\mathsf{K}\mathfrak{G}) \subset \mathsf{P}_2;$$

the terms of the upper and lower Loewy series are therefore not the same.

b) We consider $\mathsf{K}\mathfrak{G}$, where again $\mathfrak{G} = \mathfrak{S}_4$, but K is an algebraically closed field of characteristic 3.

Now $\mathfrak{G}$ has four 3′-classes. Hence there are four irreducible $\mathsf{K}\mathfrak{G}$-modules $\mathsf{V}_1, \mathsf{V}_2, \mathsf{V}_3, \mathsf{V}_4$. We assume that V_1 is the trivial module and V_2 the module corresponding to the sign representation of $\mathfrak{G}$. $\mathfrak{G}$ has only one 3-chief factor and this is complemented. V_2 is the module obtained from this by extension of the ground-field.

Let W_1 be the module for the trivial representation of a Sylow 2-subgroup $\mathfrak{P}$ of $\mathfrak{G}$. Then $\mathsf{P}_1 = \mathsf{W}_1^{\mathfrak{G}}$ is by 10.12b) the principal indecomposable projective $\mathsf{K}\mathfrak{G}$-module. As

$$\mathsf{P}_1\mathbf{J}(\mathsf{K}\mathfrak{G})/\mathsf{P}_1\mathbf{J}(\mathsf{K}\mathfrak{G})^2 \cong \mathsf{V}_2$$

by 15.5, the Loewy factors of P_1 are $\mathsf{V}_1, \mathsf{V}_2, \mathsf{V}_1$.

Certainly $\mathsf{P}_2 = \mathsf{P}_1 \otimes_{\mathsf{K}} \mathsf{V}_2$ is projective and indecomposable. As P_2 has a factor module isomorphic to $\mathsf{V}_1 \otimes_{\mathsf{K}} \mathsf{V}_2 \cong \mathsf{V}_2$, P_2 is the indecomposable projective $\mathsf{K}\mathfrak{G}$-module with head V_2. From $\mathsf{V}_2 \otimes_{\mathsf{K}} \mathsf{V}_2 \cong \mathsf{V}_1$ we conclude easily that P_2 has the Loewy factors $\mathsf{V}_2, \mathsf{V}_1, \mathsf{V}_2$.

Let T be the trivial module for $\mathsf{K}\mathfrak{S}_3$ and consider $\mathsf{T}^{\mathfrak{G}}$, on which $\mathfrak{G}$ induces the natural permutation representation of degree 4. If $\{v_1, v_2, v_3, v_4\}$ is a K-basis of $\mathsf{T}^{\mathfrak{G}}$ the elements of which are permuted by $\mathfrak{G}$, then as char $\mathsf{K} = 3$, we have the direct decomposition

$$\mathsf{T}^{\mathfrak{G}} = \mathsf{K}(v_1 + v_2 + v_3 + v_4) \oplus \mathsf{W},$$

where

$$\mathsf{W} = \left\{ \sum_{i=1}^{4} a_i v_i \,\middle|\, \sum_{i=1}^{4} a_i = 0 \right\}.$$

This is a decomposition into $\mathsf{K}\mathfrak{G}$-submodules. By Nakayama's lemma,

$$\begin{aligned} \dim_{\mathsf{K}} \operatorname{Hom}_{\mathsf{K}\mathfrak{G}}(\mathsf{T}^{\mathfrak{G}}, \mathsf{V}_i) &= \dim_{\mathsf{K}} \operatorname{Hom}_{\mathsf{K}\mathfrak{S}_3}(\mathsf{T}, (\mathsf{V}_i)_{\mathfrak{S}_3}) \\ &= \begin{cases} 1 & \text{for } i = 1, \\ 0 & \text{for } i = 2. \end{cases} \end{aligned}$$

Hence W has no factor module isomorphic to V_1 or V_2. A similar calculation shows that W has no submodule isomorphic to V_1 or V_2. As V_1 and V_2 are obviously all the irreducible $\mathsf{K}\mathfrak{G}$-modules of dimension 1, we see that W must be irreducible. We put $\mathsf{W} = \mathsf{V}_3$.

Finally put $\mathsf{V}_4 = \mathsf{V}_3 \otimes_{\mathsf{K}} \mathsf{V}_2$. As V_2 has dimension 1, V_4 is irreducible. If $g = (1, 2) \in \mathfrak{S}_4$ and χ_i denotes the character of $\mathfrak{G}$ belonging to V_i, then $\chi_3(g) = 1$, hence

$$\chi_4(g) = \chi_3(g)\chi_2(g) = -\chi_3(g) \neq \chi_3(g).$$

This proves that $\mathsf{V}_3 \ncong \mathsf{V}_4$.

From 10.18b), we conclude that

$$\begin{aligned} 24 = \dim_{\mathsf{K}} \mathsf{K}\mathfrak{G} &= \sum_{i=1}^{4} \dim_{\mathsf{K}} \mathsf{V}_i \dim_{\mathsf{K}} \mathsf{P}_i \\ &\geq 3 + 3 + 3\dim_{\mathsf{K}} \mathsf{V}_3 + 3\dim_{\mathsf{K}} \mathsf{V}_4 = 24. \end{aligned}$$

This shows that $V_i \cong P_i$ for $i = 3, 4$. The Cartan matrix is hence

$$C = \begin{pmatrix} 2 & 1 & 0 & 0 \\ 1 & 2 & 0 & 0 \\ 0 & 0 & 1 & 0 \\ 0 & 0 & 0 & 1 \end{pmatrix}$$

Thus $K\mathfrak{G}$ has three blocks.

We remark that V_3 and V_4 can also be constructed in a different way. Let $\mathfrak{P}$ be a Sylow 2-subgroup of $\mathfrak{G}$ and U an irreducible $K\mathfrak{P}$-module of dimension 1. Then

$$\mathrm{Hom}_{K\mathfrak{G}}(U^{\mathfrak{G}}, V_3) \cong \mathrm{Hom}_{K\mathfrak{P}}(U, (V_3)_{\mathfrak{P}}).$$

As $(V_3)_{\mathfrak{P}}$ is a direct sum of irreducible $K\mathfrak{P}$-modules of dimensions 1 or 2, there exists a U such that $\mathrm{Hom}_{K\mathfrak{G}}(U^{\mathfrak{G}}, V_3) \neq 0$. Thus $U^{\mathfrak{G}}$ has a factor module isomorphic to V_3, hence $U^{\mathfrak{G}} \cong V_3$. As U is a projective $K\mathfrak{P}$-module, this construction once more makes it evident, by 7.17, that V_3 is a projective $K\mathfrak{G}$-module.

Exercises

61) Let K be a field of characteristic p and P_1 the principal indecomposable projective $K\mathfrak{G}$-module. Then the following assertions are equivalent.

a) $\mathfrak{G}$ has a p-factor group different from 1.

b) $P_1\mathbf{J}(K\mathfrak{G})/P_1\mathbf{J}(K\mathfrak{G})^2$ has a composition factor isomorphic to the trivial module.

62) In examples 15.9a) and b), determine

a) the composition factors of all $V_i \otimes_K V_j$,

b) the decomposition of the projective modules $P_i \otimes_K V_j$ and $P_i \otimes_K P_j$ into indecomposable projective modules.

63) In example 15.10a), prove the following decompositions.

$P_1 \otimes_K V \cong P_2 \oplus P_2$;
$P_2 \otimes_K V \cong P_1 \oplus P_2$;
$P_1 \otimes_K P_1 \cong 4P_1 \oplus 4P_2$;
$P_1 \otimes_K P_2 \cong 2P_1 \oplus 6P_2$;
$P_2 \otimes_K P_2 \cong 3P_1 \oplus 5P_2$.
(Here, $4P_1$ stands for $P_1 \oplus P_1 \oplus P_1 \oplus P_1$, etc.)

64) Let $\mathfrak{G}$ be the alternating group $\mathfrak{A}_5$ and K an algebraically closed field of characteristic 2. Determine all irreducible $\mathsf{K}\mathfrak{G}$-modules by the following procedure. Let $\mathsf{V}_1 = \mathsf{K}$ be the module for the trivial representation of $\mathfrak{G}$.

a) As $\mathfrak{G} \cong SL(2, 4)$ (II, 6.14), there are irreducible non-isomorphic modules V_2, V_3 of dimension 2. Show that $\mathsf{V}_i^* \cong \mathsf{V}_i$ $(i = 2, 3)$.

b) Show that $\mathsf{V}_4 = \mathsf{V}_2 \otimes_{\mathsf{K}} \mathsf{V}_3$ is an indecomposable projective $\mathsf{K}\mathfrak{G}$-module by considering $(\mathsf{V}_4)_{\mathfrak{P}}$ for a Sylow 2-subgroup $\mathfrak{P}$ of $\mathfrak{G}$.

c) Show that V_4 is irreducible.

d) $\mathsf{V}_1, \mathsf{V}_2, \mathsf{V}_3, \mathsf{V}_4$ are all irreducible $\mathsf{K}\mathfrak{G}$-modules (to within isomorphism).

e) Let W be the trivial module for the subgroup $\mathfrak{A}_4$ of $\mathfrak{G}$. Show that $\mathsf{W}^{\mathfrak{G}} \cong \mathsf{V}_1 \oplus \mathsf{V}_4$. Use $\mathsf{W}^{\mathfrak{G}}$ to construct V_4 independently of b) and c).

65) In the case of Exercise 64, determine the Cartan matrix by the following arguments.

a) Let K be the module for the trivial representation of a Sylow 5-subgroup of $\mathfrak{G}$. Show by Nakayama's reciprocity theorem that $\mathsf{K}^{\mathfrak{G}}$ is the principal indecomposable projective $\mathsf{K}\mathfrak{G}$-module P_1.

b) Use Nakayama's reciprocity theorem and Mackey's theorem to prove that $c_{11} = 4$.

c) Show that the Cartan matrix is of the form

$$C = \begin{pmatrix} 4 & 2 & 2 & 0 \\ 2 & c_{22} & c_{23} & 0 \\ 2 & c_{23} & c_{33} & 0 \\ 0 & 0 & 0 & 1 \end{pmatrix}$$

where either $c_{22} = c_{33} = 2$, $c_{23} = 1$ or $c_{22} = c_{33} = 3$, $c_{23} = 0$.

d) Show that $\mathsf{V}_2 \otimes_{\mathsf{K}} \mathsf{V}_2$ has the composition factors $\mathsf{V}_1, \mathsf{V}_1, \mathsf{V}_3$.

e) If P_i denotes the indecomposable projective $\mathsf{K}\mathfrak{G}$-module with head V_i, show that $\mathsf{P}_1 \otimes_{\mathsf{K}} \mathsf{V}_2 \cong 2\mathsf{P}_2 \oplus 2\mathsf{P}_4$.

f) Conclude that $c_{22} = c_{33} = 2$ and $c_{23} = c_{32} = 1$.

66) In the case of Exercise 64, determine

a) the composition factors of all the tensor products $\mathsf{V}_i \otimes_{\mathsf{K}} \mathsf{V}_j$.

b) the decomposition of all the modules $\mathsf{P}_i \otimes_{\mathsf{K}} \mathsf{V}_j$ and $\mathsf{P}_i \otimes_{\mathsf{K}} \mathsf{P}_j$ as direct sums of indecomposable projective modules.

67) Consider the example of exercise 64 and show the following.

a) P_1 has Loewy factors V_1, $\mathsf{V}_2 \oplus \mathsf{V}_3$, $\mathsf{V}_1 \oplus \mathsf{V}_1$, $\mathsf{V}_2 \oplus \mathsf{V}_3$, V_1. (Use Exercise 61.)

b) Show that there does not exist an indecomposable $K\mathfrak{G}$-module W with socle $S \cong V_1 \oplus V_1$ and $W/S \cong V_2$.
(Suppose that $W/S \cong V_2$ and $S \cong V_1 \oplus V_1$. Let $\mathfrak{P}$ be a Sylow 2-subgroup of $\mathfrak{G}$. Show first that if U is a K-subspace of S such that $U \cap W\mathbf{J}(K\mathfrak{P}) = 0$, then U is a $K\mathfrak{P}$-direct summand of $W_{\mathfrak{P}}$ and a $K\mathfrak{G}$-submodule, hence a $K\mathfrak{G}$-direct summand of W, a contradiction. Hence $S \subseteq W\mathbf{J}(K\mathfrak{P})$. Show that $\dim_K W\mathbf{J}(K\mathfrak{P}) = 3$ and conclude that $W_{\mathfrak{P}}$ is projective. Hence W is a projective $K\mathfrak{G}$-module, a contradiction.)

c) P_2 has Loewy factors V_2, V_1, V_3, V_1, V_2. (Use the fact that P_2, and hence also $P_2\mathbf{J}(K\mathfrak{G})/\mathbf{S}(P_2)$, is self-dual.)

d) Show that there does not exist an indecomposable $K\mathfrak{G}$-module W with socle S, where $W/S \cong V_2$ and $S \cong V_2$ or $S \cong V_3$.

68) Consider $K\mathfrak{G}$, where $\mathfrak{G} = \mathfrak{A}_5$ and K is an algebraically closed field of characteristic 5.

a) As $\mathfrak{G} \cong PSL(2, 5)$ (II, 6.14), there exist irreducible $K\mathfrak{G}$-modules V_1, V_2, V_3 of dimensions 1, 3, 5, and no others.

b) Let P_i again denote the projective $K\mathfrak{G}$-module with head V_i. Show that $\dim_K P_1 = 5$, therefore $c_{11} = 2$ and $c_{12} = 1$. Conclude that $\dim_K P_2 = 10$ and $P_3 = V_3$.

c) Show that the Cartan matrix is

$$C = \begin{pmatrix} 2 & 1 & 0 \\ 1 & 3 & 0 \\ 0 & 0 & 1 \end{pmatrix}.$$

d) The Loewy factors of P_2 are $V_2, V_1 \oplus V_2, V_2$. (Use the fact that P_2 is self-dual.)

69) Finally consider $K\mathfrak{G}$, where $\mathfrak{G} = \mathfrak{A}_5$ and K is an algebraically closed field of characteristic 3.

a) There is no irreducible $K\mathfrak{G}$-module of dimension 2. (Consider the restriction to a Sylow 2-subgroup.)

b) Let $\mathfrak{N}$ denote the normalizer of a Sylow 5-subgroup of $\mathfrak{G}$ and consider $K^{\mathfrak{N}}$, where K denotes the trivial $K\mathfrak{N}$-module. Show that $K^{\mathfrak{N}}$ is projective and has a factor module isomorphic to the module V_1 for the trivial representation of $\mathfrak{G}$. Show finally that $K^{\mathfrak{G}} = P_1$ and that P_1 has Loewy factors V_1, V_2, V_1, where V_2 is an irreducible $K\mathfrak{G}$-module of dimension 4.

c) Show that $\dim_K P_2 = 9$ and that there are irreducible projective $K\mathfrak{G}$-modules $V_i = P_i$ $(i = 3, 4)$ of dimension 3.

d) The Cartan matrix of $K\mathfrak{G}$ is

$$C = \begin{pmatrix} 2 & 1 & 0 & 0 \\ 1 & 2 & 0 & 0 \\ 0 & 0 & 1 & 0 \\ 0 & 0 & 0 & 1 \end{pmatrix}$$

and $\mathsf{K}\mathfrak{G}$ has three blocks.

e) Let K be the trivial module for the subgroup $\mathfrak{A}_4$ of $\mathfrak{G}$. Show that $\mathsf{K}^{\mathfrak{G}} \cong \mathsf{V}_1 \oplus \mathsf{V}_2$.

f) Let W be the non-trivial $\mathsf{K}\mathfrak{N}$-module of dimension 1. Show that $\mathsf{W}^{\mathfrak{G}} \cong \mathsf{V}_3 \oplus \mathsf{V}_4$.
(Exclude the possibility that $\mathsf{W}^{\mathfrak{G}} \cong \mathsf{V}_3 \oplus \mathsf{V}_3$ by using $\mathsf{K}\mathfrak{G} \cong (\mathsf{K}\mathfrak{N})^{\mathfrak{G}}$ and the decomposition of $\mathsf{K}\mathfrak{N}$ into irreducible modules.)

§ 16. Green's Indecomposability Theorem

16.1 Definition. A subgroup $\mathfrak{N}$ of a group $\mathfrak{G}$ is called *subnormal* if there exists a series

$$\mathfrak{N} = \mathfrak{N}_1 \leq \mathfrak{N}_2 \leq \cdots \leq \mathfrak{N}_n = \mathfrak{G}$$

of subgroups such that $\mathfrak{N}_i \trianglelefteq \mathfrak{N}_{i+1}$ $(1 \leq i < n)$.

The following theorem plays a part in many investigations and is the main theorem of this section.

16.2 Theorem (GREEN [1]). *Let K be a perfect field of characteristic p and let $\mathfrak{N}$ be a subnormal subgroup of $\mathfrak{G}$ of index a power of p. If V is an absolutely indecomposable $\mathsf{K}\mathfrak{N}$-module, then $\mathsf{V}^{\mathfrak{G}}$ is an absolutely indecomposable $\mathsf{K}\mathfrak{G}$-module.*

For the proof we need a lemma and some preliminary remarks.

16.3 Remarks. a) We may suppose that K is algebraically closed.

For let $\hat{\mathsf{K}}$ be the algebraic closure of K. By 6.7, $\mathsf{V}_{\hat{\mathsf{K}}}$ is absolutely indecomposable. Thus, assuming the theorem in the algebraically closed case, $(\mathsf{V}_{\hat{\mathsf{K}}})^{\mathfrak{G}}$ is absolutely indecomposable. Thus $(\mathsf{V}^{\mathfrak{G}})_{\hat{\mathsf{K}}}$ is indecomposable and by 6.9, $\mathsf{V}^{\mathfrak{G}}$ is absolutely indecomposable.

b) We may suppose that $|\mathfrak{G} : \mathfrak{N}| = p$.

For by 16.1, there is a series

$$\mathfrak{N} = \mathfrak{N}_1 < \mathfrak{N}_2 < \cdots < \mathfrak{N}_k = \mathfrak{G}$$

of subgroups such that $\mathfrak{N}_i \trianglelefteq \mathfrak{N}_{i+1}$ $(1 \leq i < k)$. By adding more terms if necessary, we may suppose that $|\mathfrak{N}_{i+1} : \mathfrak{N}_i| = p$ $(i = 1, 2, \ldots, k - 1)$. Thus if the result is known for index p, the absolute indecomposability of $\mathsf{V}^{\mathfrak{N}_{i+1}}$ follows from that of $\mathsf{V}^{\mathfrak{N}_i}$, since

$$\mathsf{V}^{\mathfrak{N}_{i+1}} = (\mathsf{V}^{\mathfrak{N}_i})^{\mathfrak{N}_{i+1}}.$$

c) We may suppose that the inertia subgroup of V in $\mathfrak{G}$ is $\mathfrak{G}$ itself.

For otherwise, by b), the inertia subgroup is $\mathfrak{N}$. Then $\mathsf{V}^{\mathfrak{G}}$ is an indecomposable $\mathsf{K}\mathfrak{G}$-module by 9.6a). As K is algebraically closed, $\mathsf{V}^{\mathfrak{G}}$ is absolutely indecomposable.

Next we calculate $\mathrm{Hom}_{\mathsf{K}\mathfrak{G}}(\mathsf{V}^{\mathfrak{G}}, \mathsf{V}^{\mathfrak{G}})$.

16.4 Lemma. *Suppose that* $\mathfrak{N} \trianglelefteq \mathfrak{G}$ *and that* V *is a* $\mathsf{K}\mathfrak{N}$*-module such that the inertia group of* V *in* $\mathfrak{G}$ *is* $\mathfrak{G}$ *itself. Write* $\mathfrak{S} = \mathrm{Hom}_{\mathsf{K}\mathfrak{G}}(\mathsf{V}^{\mathfrak{G}}, \mathsf{V}^{\mathfrak{G}})$ *and* $\overline{\mathfrak{G}} = \mathfrak{G}/\mathfrak{N}$. *Then, corresponding to each* $h \in \overline{\mathfrak{G}}$, *there is a subspace* $\mathfrak{S}_h$ *of* $\mathfrak{S}$ *with the following properties.*

a) *If* $g \in \mathfrak{G}$ *and* $\overline{g} = g\mathfrak{N}$,

$$\mathfrak{S}_{\overline{g}} = \{\alpha \mid \alpha \in \mathfrak{S}, (\mathsf{V} \otimes 1)\alpha \subseteq \mathsf{V} \otimes g^{-1}\}.$$

b) $\mathfrak{S} = \bigoplus_{h \in \overline{\mathfrak{G}}} \mathfrak{S}_h$.

c) $\mathfrak{S}_{h_1}\mathfrak{S}_{h_2} = \mathfrak{S}_{h_1 h_2}$ *for all* h_1, h_2 *in* $\overline{\mathfrak{G}}$. *Further, there exist units* ψ_h *in* $\mathfrak{S}$ *such that*

$$\mathfrak{S}_h = \psi_h \mathfrak{S}_1 = \mathfrak{S}_1 \psi_h.$$

(Thus $\mathfrak{S}$ *is an algebra graded by* $\mathfrak{G}/\mathfrak{N}$*).*

d) $\mathfrak{S}_1$ *and* $\mathrm{Hom}_{\mathsf{K}\mathfrak{N}}(\mathsf{V}, \mathsf{V})$ *are isomorphic* K*-algebras.*

Proof. a) Let T be a transversal of $\mathfrak{N}$ in $\mathfrak{G}$. For each $t \in \mathsf{T}$, define

$$\mathfrak{S}_{\overline{t}} = \{\alpha \mid \alpha \in \mathfrak{S}, (\mathsf{V} \otimes 1)\alpha \subseteq \mathsf{V} \otimes t^{-1}\}.$$

Then $\mathfrak{S}_{\overline{t}}$ is a subspace of $\mathfrak{S}$. If $g \in \mathfrak{G}$, $\overline{g} = g\mathfrak{N}$ and $g = xt$ with $x \in \mathfrak{N}$, $t \in \mathsf{T}$, then $\mathsf{V} \otimes g^{-1} = \mathsf{V} \otimes t^{-1}$; hence $\mathfrak{S}_{\overline{g}}$ is well-defined.

To prove the remaining assertions, observe that for any $g \in \mathfrak{G}$, $\mathsf{V} \otimes g$ is a $\mathsf{K}\mathfrak{N}$-submodule of $\mathsf{V}^{\mathfrak{G}}$. We shall use the following.

(*) If $\xi \in \mathrm{Hom}_{\mathsf{K}\mathfrak{N}}(\mathsf{V} \otimes 1, \mathsf{V} \otimes g^{-1})$, there exists $\xi' \in \mathfrak{S}_{\overline{g}}$ such that

$$(v \otimes g')\xi' = (v \otimes 1)\xi g' \quad (v \in \mathsf{V}, g' \in \mathfrak{G}).$$

To see this, observe that we may define $\xi' \in \mathrm{Hom}_{\mathsf{K}}(\mathsf{V}^{\mathfrak{G}}, \mathsf{V}^{\mathfrak{G}})$ by

$$(v \otimes t)\xi' = (v \otimes 1)\xi t \quad (v \in \mathsf{V}, t \in \mathsf{T}).$$

Then if $g' \in \mathfrak{G}$ and $g' = xt$ with $x \in \mathfrak{N}$, $t \in \mathsf{T}$,

$$(v \otimes g')\xi' = (vx \otimes t)\xi' = (vx \otimes 1)\xi t = (v \otimes 1)\xi xt = (v \otimes 1)\xi g'.$$

It follows at once that $\xi' \in \mathrm{Hom}_{\mathsf{K}\mathfrak{G}}(\mathsf{V}^{\mathfrak{G}}, \mathsf{V}^{\mathfrak{G}})$. And since $(\mathsf{V} \otimes 1)\xi' = (\mathsf{V} \otimes 1)\xi \subseteq \mathsf{V} \otimes g^{-1}$, we obtain $\xi' \in \mathfrak{S}_{\bar{g}}$.

b) It is to be proved that

$$\mathfrak{S} = \bigoplus_{t \in \mathsf{T}} \mathfrak{S}_{\bar{t}}.$$

First suppose that $0 = \sum_{t \in \mathsf{T}} \gamma_t$, with $\gamma_t \in \mathfrak{S}_{\bar{t}}$. Then for any $v \in \mathsf{V}$, $g \in \mathfrak{G}$,

$$\sum_{t \in \mathsf{T}} (v \otimes g)\gamma_t = 0.$$

But $(v \otimes g)\gamma_t = (v \otimes 1)\gamma_t g \subseteq \mathsf{V} \otimes t^{-1}g$ and $\mathsf{V}^{\mathfrak{G}} = \bigoplus_{t \in \mathsf{T}} \mathsf{V} \otimes t^{-1}g$. Hence $(v \otimes g)\gamma_t = 0$ for all $v \in \mathsf{V}$, $g \in \mathfrak{G}$, and $\gamma_t = 0$.

Next suppose that $\alpha \in \mathfrak{S}$. Let π_t be the projection of $\mathsf{V}^{\mathfrak{G}}$ onto $\mathsf{V} \otimes t^{-1}$; thus $\pi_t \in \mathrm{Hom}_{\mathsf{K}\mathfrak{N}}(\mathsf{V}^{\mathfrak{G}}, \mathsf{V} \otimes t^{-1})$ and

$$\sum_{t \in \mathsf{T}} \pi_t = 1.$$

By (*), applied to the restriction of $\alpha\pi_t$ to $\mathsf{V} \otimes 1$, there exists $\beta_t \in \mathfrak{S}_{\bar{t}}$ such that

$$(v \otimes g)\beta_t = (v \otimes 1)\alpha\pi_t g \quad (v \in \mathsf{V}, g \in \mathfrak{G}).$$

Hence

$$(v \otimes g) \sum_{t \in \mathsf{T}} \beta_t = \sum_{t \in \mathsf{T}} (v \otimes 1)\alpha\pi_t g = (v \otimes 1)\alpha g = (v \otimes g)\alpha,$$

so

$$\alpha = \sum_{t \in \mathsf{T}} \beta_t.$$

Thus

$$\mathfrak{S} = \bigoplus_{t \in \mathsf{T}}' \mathfrak{S}_{\bar{t}}.$$

c) If $h_i = g_i \mathfrak{N}$ and $\alpha_i \in \mathfrak{S}_{h_i}$ $(i = 1, 2)$ then by a),

$$\begin{aligned}(\mathsf{V} \otimes 1)\alpha_1\alpha_2 &\subseteq (\mathsf{V} \otimes g_1^{-1})\alpha_2 = ((\mathsf{V} \otimes 1)\alpha_2)g_1^{-1} \\ &\subseteq \mathsf{V} \otimes g_2^{-1}g_1^{-1} = \mathsf{V} \otimes (g_1g_2)^{-1},\end{aligned}$$

so $\alpha_1\alpha_2 \in \mathfrak{S}_{h_1h_2}$. Thus $\mathfrak{S}_{h_1}\mathfrak{S}_{h_2} \subseteq \mathfrak{S}_{h_1h_2}$.

Suppose that $t \in \mathsf{T}$. By hypothesis, there exists a $\mathsf{K}\mathfrak{N}$-isomorphism ρ_t of $\mathsf{V} \otimes 1$ onto $\mathsf{V} \otimes t^{-1}$. By (*), there exists $\psi_{\bar{t}} \in \mathfrak{S}_{\bar{t}}$ such that

$$(v \otimes g)\psi_{\bar{t}} = (v \otimes 1)\rho_t g.$$

It is clear that $\psi_{\bar{t}}$ is surjective and hence non-singular. Thus $\psi_{\bar{t}}^{-1} \in \mathfrak{S}$ and $\psi_{\bar{t}}^{-1} \in \mathfrak{S}_{\bar{t}^{-1}}$, since $(\mathsf{V} \otimes t)\psi_{\bar{t}} = \mathsf{V} \otimes 1$. In general, we have

$$\psi_h \in \mathfrak{S}_h, \ \psi_h^{-1} \in \mathfrak{S}_{h^{-1}} \quad (h \in \overline{\mathfrak{G}}).$$

Thus

$$\mathfrak{S}_{h_1h_2} = \mathfrak{S}_{h_1h_2}\psi_{h_2}^{-1}\psi_{h_2} \subseteq \mathfrak{S}_{h_1h_2}\mathfrak{S}_{h_2^{-1}}\psi_{h_2} \subseteq \mathfrak{S}_{h_1}\psi_{h_2} \subseteq \mathfrak{S}_{h_1}\mathfrak{S}_{h_2} \subseteq \mathfrak{S}_{h_1h_2},$$

so

$$\mathfrak{S}_{h_1h_2} = \mathfrak{S}_{h_1}\mathfrak{S}_{h_2} = \mathfrak{S}_{h_1}\psi_{h_2}.$$

In particular, $\mathfrak{S}_h = \mathfrak{S}_1\psi_h$, and similarly, $\mathfrak{S}_h = \psi_h\mathfrak{S}_1$.

d) If $\alpha \in \mathrm{Hom}_{\mathsf{K}\mathfrak{N}}(\mathsf{V}, \mathsf{V})$, then by (*), there exists $\beta \in \mathfrak{S}_1$ such that

$$(v \otimes g)\beta = (v\alpha \otimes 1)g = v\alpha \otimes g.$$

The mapping $\alpha \to \beta$ is an algebra monomorphism. It is in fact an isomorphism, since if $\beta' \in \mathfrak{S}_1$,

$$(v \otimes 1)\beta' = v\alpha' \otimes 1$$

for some $\alpha' \in \mathrm{Hom}_{\mathsf{K}\mathfrak{N}}(\mathsf{V}, \mathsf{V})$. **q.e.d.**

16.5 *Proof of 16.2*. By the remarks in 16.3, we may suppose that $\mathfrak{N} \trianglelefteq \mathfrak{G}$, $\mathfrak{G}/\mathfrak{N}$ is a p-group and the inertia subgroup of V in $\mathfrak{G}$ is $\mathfrak{G}$ itself. Put

$\overline{\mathfrak{G}} = \mathfrak{G}/\mathfrak{N}$ and $\mathfrak{S} = \mathrm{Hom}_{\mathsf{K}\mathfrak{G}}(\mathsf{V}^{\mathfrak{G}}, \mathsf{V}^{\mathfrak{G}})$. By 16.4, $\mathfrak{S}$ has a subalgebra $\mathfrak{S}_1$ isomorphic to $\mathrm{Hom}_{\mathsf{K}\mathfrak{N}}(\mathsf{V}, \mathsf{V})$, and there exist units ψ_h ($h \in \overline{\mathfrak{G}}$) of $\mathfrak{S}$ such that $\psi_h \mathfrak{S}_1 = \mathfrak{S}_1 \psi_h = \mathfrak{S}_h$, $\mathfrak{S}_{h_1}\mathfrak{S}_{h_2} = \mathfrak{S}_{h_1 h_2}$ and

$$\mathfrak{S} = \bigoplus_{h \in \overline{\mathfrak{G}}} \mathfrak{S}_h.$$

Thus $\psi_h^{-1}\mathfrak{S}_1\psi_h = \mathfrak{S}_1$, whence $\psi_h^{-1}\mathbf{J}(\mathfrak{S}_1)\psi_h = \mathbf{J}(\mathfrak{S}_1)$. Thus

$$\mathbf{J}(\mathfrak{S}_1)\mathfrak{S} = \sum_{h \in \overline{\mathfrak{G}}} \mathbf{J}(\mathfrak{S}_1)\psi_h\mathfrak{S}_1 = \sum_{h \in \overline{\mathfrak{G}}} \psi_h\mathbf{J}(\mathfrak{S}_1)\mathfrak{S}_1 = \sum_{h \in \overline{\mathfrak{G}}} \psi_h\mathbf{J}(\mathfrak{S}_1) = \mathfrak{S}\mathbf{J}(\mathfrak{S}_1).$$

Hence $\mathfrak{J} = \mathbf{J}(\mathfrak{S}_1)\mathfrak{S}$ is a two-sided ideal of $\mathfrak{S}$ and

$$\mathfrak{J} = \bigoplus_{h \in \overline{\mathfrak{G}}} \psi_h\mathbf{J}(\mathfrak{S}_1), \tag{1}$$

since $\psi_h\mathbf{J}(\mathfrak{S}_1) \subseteq \mathfrak{S}_h\mathfrak{S}_1 = \mathfrak{S}_h$. If $\mathbf{J}(\mathfrak{S}_1)^k = 0$,

$$\mathfrak{J}^k = \sum \psi_{h_1}\mathbf{J}(\mathfrak{S}_1) \cdots \psi_{h_k}\mathbf{J}(\mathfrak{S}_1) = \sum \psi_{h_1} \cdots \psi_{h_k}\mathbf{J}(\mathfrak{S}_1)^k = 0.$$

Thus $\mathfrak{J} \subseteq \mathbf{J}(\mathfrak{S})$.

Since V is an absolutely indecomposable $\mathsf{K}\mathfrak{N}$-module,

$$\mathfrak{S}_1/\mathbf{J}(\mathfrak{S}_1) \cong \mathsf{K}.$$

Thus the elements of $\mathfrak{S}_1$ are expressible as $\lambda 1 + a$ with $\lambda \in \mathsf{K}$, $a \in \mathbf{J}(\mathfrak{S}_1)$. Hence the elements of $\mathfrak{S}$ are all of the form

$$\sum_{h \in \overline{\mathfrak{G}}} \lambda_h\psi_h + b$$

with $b \in \mathfrak{J}$. It follows from this and (1) that if $e_h = \psi_h + \mathfrak{J}$, then $\{e_h | h \in \overline{\mathfrak{G}}\}$ is a K-basis of the algebra $\mathfrak{S}/\mathfrak{J}$. Now

$$\psi_{h_1}\psi_{h_2} \in \mathfrak{S}_{h_1}\mathfrak{S}_{h_2} = \mathfrak{S}_{h_1 h_2} = \mathfrak{S}_1\psi_{h_1 h_2},$$

so

$$\psi_{h_1}\psi_{h_2} = s_{h_1,h_2}\psi_{h_1 h_2}$$

for some unit s_{h_1,h_2} in $\mathfrak{S}_1$. Thus $s_{h_1,h_2} = \lambda_{h_1,h_2} + c_{h_1,h_2}$ with $0 \neq \lambda_{h_1,h_2} \in \mathsf{K}$, $c_{h_1,h_2} \in \mathfrak{J}$. Hence

$$e_{h_1}e_{h_2} = \lambda_{h_1,h_2}e_{h_1h_2}.$$

Then

$$\lambda_{h_1,h_2}\lambda_{h_1h_2,h_3}e_{h_1h_2h_3} = (e_{h_1}e_{h_2})e_{h_3}$$

$$= e_{h_1}(e_{h_2}e_{h_3}) = \lambda_{h_2,h_3}\lambda_{h_1,h_2h_3}e_{h_1h_2h_3},$$

so

$$(2) \qquad \lambda_{h_1,h_2}\lambda_{h_1h_2,h_3} = \lambda_{h_2,h_3}\lambda_{h_1,h_2h_3}.$$

(Thus we have an element of $\mathbf{Z}^2(\overline{\mathfrak{G}}, \mathsf{K}^\times)$.) Put

$$\mu_h = \prod_{h' \in \overline{\mathfrak{G}}} \lambda_{h,h'}.$$

It follows from (2) that

$$\lambda_{h_1,h_2}^{|\overline{\mathfrak{G}}|}\mu_{h_1h_2} = \mu_{h_2}\mu_{h_1}.$$

Since K is perfect and $|\overline{\mathfrak{G}}|$ is a power of p, there is an element $v_h \in \mathsf{K}^\times$ such that $v_h^{|\overline{\mathfrak{G}}|} = \mu_h$ ($h \in \overline{\mathfrak{G}}$). Since char $\mathsf{K} = p$, it follows that

$$\lambda_{h_1,h_2}v_{h_1h_2} = v_{h_2}v_{h_1}.$$

Put $f_h = v_h^{-1}e_h$; then

$$f_{h_1}f_{h_2} = v_{h_1}^{-1}v_{h_2}^{-1}e_{h_1}e_{h_2} = v_{h_1h_2}^{-1}e_{h_1h_2} = f_{h_1h_2}.$$

Also $\{f_h | h \in \overline{\mathfrak{G}}\}$ is a K-basis of $\mathfrak{S}/\mathfrak{J}$, so the algebra $\mathfrak{S}/\mathfrak{J}$ is isomorphic to $\mathsf{K}\overline{\mathfrak{G}}$. Since $\overline{\mathfrak{G}}$ is a p-group,

$$\mathsf{K}\overline{\mathfrak{G}}/\mathbf{J}(\mathsf{K}\overline{\mathfrak{G}}) \cong \mathsf{K}.$$

Since $\mathfrak{J} \subseteq \mathbf{J}(\mathfrak{S})$, $\mathbf{J}(\mathfrak{S}/\mathfrak{J}) = \mathbf{J}(\mathfrak{S})/\mathfrak{J}$. It follows that

$$\mathfrak{S}/\mathbf{J}(\mathfrak{S}) \cong (\mathfrak{S}/\mathfrak{J})/\mathbf{J}(\mathfrak{S}/\mathfrak{J}) \cong \mathsf{K}\overline{\mathfrak{G}}/\mathbf{J}(\mathsf{K}\overline{\mathfrak{G}}) \cong \mathsf{K}.$$

Hence $\mathsf{V}^{\mathfrak{G}}$ is absolutely indecomposable. **q.e.d.**

We mention a special case of 16.2.

16.6 Theorem. *Suppose that $\mathfrak{G}$ is a p-group and that $\mathfrak{U} \leq \mathfrak{G}$. Let K be a perfect field of characteristic p and let V be an absolutely indecomposable $\mathsf{K}\mathfrak{U}$-module. Then $\mathsf{V}^{\mathfrak{G}}$ is absolutely indecomposable.*

Proof. This follows at once from 16.2, since $\mathfrak{U}$ is subnormal in $\mathfrak{G}$ (see III, 2.3). **q.e.d.**

16.7 Theorem (WILLEMS [1]). *Let K be an algebraically closed field of characteristic p. Suppose that $\mathfrak{N} \trianglelefteq \mathfrak{G}$ and let V be an indecomposable $\mathsf{K}\mathfrak{N}$-module with inertia subgroup $\mathfrak{I}$. Then the following are equivalent.*

a) *$\mathsf{V}^{\mathfrak{G}}$ is an indecomposable $\mathsf{K}\mathfrak{G}$-module.*

b) *$\mathsf{V}^{\mathfrak{I}}$ is an indecomposable $\mathsf{K}\mathfrak{I}$-module.*

c) *$\mathfrak{I}/\mathfrak{N}$ is a p-group.*

Proof. a) ⇒ b): Since $\mathsf{V}^{\mathfrak{G}} = (\mathsf{V}^{\mathfrak{I}})^{\mathfrak{G}}$, this is trivial.

b) ⇒ c): If $\mathfrak{I}/\mathfrak{N}$ is not a p-group, it contains a non-identity cyclic p'-subgroup $\mathfrak{U}/\mathfrak{N}$. By 9.9b), there exists a $\mathsf{K}\mathfrak{U}$-module W such that $\mathsf{W}_{\mathfrak{N}} \cong \mathsf{V}$. Hence by 4.15b),

$$\mathsf{V}^{\mathfrak{U}} \cong \mathsf{W} \otimes_{\mathsf{K}} \mathsf{K}(\mathfrak{U}/\mathfrak{N}).$$

But $\mathfrak{U}/\mathfrak{N}$ is a non-identity p'-group, so

$$\mathsf{K}(\mathfrak{U}/\mathfrak{N}) \cong \mathsf{K} \oplus \mathsf{U}$$

where K is the trivial $\mathsf{K}\mathfrak{U}$-module and U is a non-zero $\mathsf{K}\mathfrak{U}$-module. Hence

$$\mathsf{V}^{\mathfrak{U}} \cong \mathsf{W} \oplus (\mathsf{W} \otimes_{\mathsf{K}} \mathsf{U})$$

and

$$\mathsf{V}^{\mathfrak{I}} \cong \mathsf{W}^{\mathfrak{I}} \oplus (\mathsf{W} \otimes_{\mathsf{K}} \mathsf{U})^{\mathfrak{I}}.$$

Thus $\mathsf{V}^{\mathfrak{I}}$ is decomposable, contrary to b).

c) ⇒ a): As $\mathfrak{I}/\mathfrak{N}$ is a p-group, $\mathsf{V}^{\mathfrak{I}}$ is indecomposable, by 16.2. Hence $\mathsf{V}^{\mathfrak{G}}$ is indecomposable by 9.6. **q.e.d.**

In preparation for the proof of a theorem of Fong, we introduce the projective envelope.

16.8 Lemma. *Let* V *be a* $\mathsf{K}\mathfrak{G}$*-module.*

a) *To within isomorphism, there is exactly one projective* $\mathsf{K}\mathfrak{G}$*-module* P *for which* $\mathsf{P}/\mathsf{P}\mathbf{J}(\mathsf{K}\mathfrak{G}) \cong \mathsf{V}/\mathsf{V}\mathbf{J}(\mathsf{K}\mathfrak{G})$. *We put* $\mathsf{P} = \mathsf{P}(\mathsf{V})$; $\mathsf{P}(\mathsf{V})$ *is called the projective envelope of* V.

b) *There exists an epimorphism* λ_{V} *of* $\mathsf{P}(\mathsf{V})$ *onto* V *such that* $\ker \lambda_{\mathsf{V}} \subseteq \mathsf{P}(\mathsf{V})\mathbf{J}(\mathsf{K}\mathfrak{G})$.

c) *Suppose that* V, W *are* $\mathsf{K}\mathfrak{G}$*-modules and* α *is an epimorphism of* V *onto* W. *Then* $\mathsf{P}(\mathsf{W})$ *is isomorphic to a direct summand of* $\mathsf{P}(\mathsf{V})$.

Proof. a) By 10.9b), there exists a projective $\mathsf{K}\mathfrak{G}$-module P such that $\mathsf{P}/\mathsf{P}\mathbf{J}(\mathsf{K}\mathfrak{G}) \cong \mathsf{V}/\mathsf{V}\mathbf{J}(\mathsf{K}\mathfrak{G})$, since $\mathsf{V}/\mathsf{V}\mathbf{J}(\mathsf{K}\mathfrak{G})$ is completely reducible. By 10.6, P is uniquely determined by this condition.

b) Let α be an isomorphism of $\mathsf{P}/\mathsf{P}\mathbf{J}(\mathsf{K}\mathfrak{G})$ onto $\mathsf{V}/\mathsf{V}\mathbf{J}(\mathsf{K}\mathfrak{G})$. By 10.5a), there is an epimorphism λ of P onto V for which

$$x\lambda + \mathsf{V}\mathbf{J}(\mathsf{K}\mathfrak{G}) = (x + \mathsf{P}\mathbf{J}(\mathsf{K}\mathfrak{G}))\alpha.$$

If $x \in \mathsf{P}$ and $x\lambda = 0$, then $(x + \mathsf{P}\mathbf{J}(\mathsf{K}\mathfrak{G}))\alpha = 0$ and $x \in \mathsf{P}\mathbf{J}(\mathsf{K}\mathfrak{G})$.

c) Consider the commutative diagram

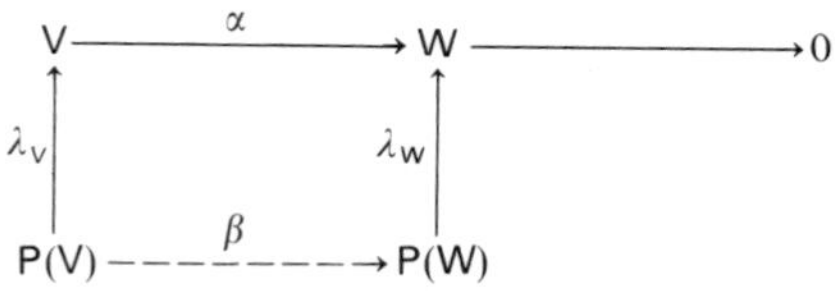

Since λ_{W} is an epimorphism, there exists $\beta \in \mathrm{Hom}_{\mathsf{K}\mathfrak{G}}(\mathsf{P}(\mathsf{V}), \mathsf{P}(\mathsf{W}))$ such that $\beta\lambda_{\mathsf{W}} = \lambda_{\mathsf{V}}\alpha$. Since $\lambda_{\mathsf{V}}\alpha$ is an epimorphism,

$$\begin{aligned} \mathsf{P}(\mathsf{W}) &= \mathsf{P}(\mathsf{V})\beta + \ker \lambda_{\mathsf{W}} \\ &= \mathsf{P}(\mathsf{V})\beta + \mathsf{P}(\mathsf{W})\mathbf{J}(\mathsf{K}\mathfrak{G}). \end{aligned}$$

Hence by nilpotency of $\mathbf{J}(\mathsf{K}\mathfrak{G})$, $\mathsf{P}(\mathsf{W}) = \mathsf{P}(\mathsf{V})\beta$ and β is an epimorphism. Since $\mathsf{P}(\mathsf{W})$ is projective, $\mathsf{P}(\mathsf{W})$ is isomorphic to a direct summand of $\mathsf{P}(\mathsf{V})$. **q.e.d.**

Now we can prove an interesting theorem about the projective indecomposable modules for soluble groups.

16.9 Theorem (FONG [1]). *Suppose that* $\mathfrak{G}$ *is soluble and that* K *is an algebraically closed field of characteristic p. Let* V *be an irreducible* $\mathsf{K}\mathfrak{G}$*-module and let* $\mathsf{P}(\mathsf{V})$ *be its projective envelope. Then*

$$\dim_K P(V) = |\mathfrak{G}|_p(\dim_K V)_{p'}.$$

Proof (W. SCHWARZ [1]). This is proved by induction on $|\mathfrak{G}|$. Let $\mathfrak{N}$ be a maximal normal subgroup of $\mathfrak{G}$. Then either $|\mathfrak{G}/\mathfrak{N}| = p$ or $|\mathfrak{G}/\mathfrak{N}|$ is not divisible by p.

a) Suppose first that $|\mathfrak{G}/\mathfrak{N}| = p$. By Clifford's theorem, there exists a decomposition

$$V_{\mathfrak{N}} = W_1 \oplus \cdots \oplus W_r,$$

where $W_1, \ldots, W_r$ are irreducible $K\mathfrak{N}$-modules of equal dimensions. By 9.20, r divides $|\mathfrak{G}/\mathfrak{N}| = p$. Thus

$$(\dim_K V)_{p'} = (\dim_K W_1)_{p'}.$$

By Nakayama's reciprocity theorem, the multiplicity of V in the head of $W_1^{\mathfrak{G}}$ is

$$\dim_K \operatorname{Hom}_{K\mathfrak{G}}(W_1^{\mathfrak{G}}, V) = \dim_K \operatorname{Hom}_{K\mathfrak{N}}(W_1, V_{\mathfrak{N}}) > 0.$$

Thus there is an epimorphism of $W_1^{\mathfrak{G}}$ onto V. Hence by 16.8c), $P(V)$ is isomorphic to a direct summand of $P(W_1^{\mathfrak{G}})$. Since W_1 is irreducible, $P(W_1)$ is indecomposable; hence by 16.2, $P(W_1)^{\mathfrak{G}}$ is indecomposable. By 16.8b), there is an epimorphism of $P(W_1)$ onto W_1, so by 4.2, there is an epimorphism of $P(W_1)^{\mathfrak{G}}$ onto $W_1^{\mathfrak{G}}$. By 16.8c), $P(W_1^{\mathfrak{G}})$ is a direct summand of $P(P(W_1)^{\mathfrak{G}})$. By 7.17, $P(W_1)^{\mathfrak{G}}$ is projective, so $P(P(W_1)^{\mathfrak{G}}) = P(W_1)^{\mathfrak{G}}$ and $P(W_1^{\mathfrak{G}})$ is a direct summand of $P(W_1)^{\mathfrak{G}}$. Since $P(W_1)^{\mathfrak{G}}$ is indecomposable, so is $P(W_1^{\mathfrak{G}})$, so $P(V) \cong P(W_1^{\mathfrak{G}}) = P(W_1)^{\mathfrak{G}}$. Thus

$$\begin{aligned}\dim_K P(V) &= |\mathfrak{G}/\mathfrak{N}| \dim_K P(W_1)\\ &= |\mathfrak{G}/\mathfrak{N}|\,|\mathfrak{N}|_p(\dim_K W_1)_{p'},\end{aligned}$$

by the inductive hypothesis. Hence

$$\dim_K P(V) = |\mathfrak{G}|_p(\dim_K V)_{p'},$$

as required.

b) Now suppose that $|\mathfrak{G}/\mathfrak{N}|$ is prime to p. By 7.21b), $\mathbf{J}(K\mathfrak{G}) = \mathbf{J}(K\mathfrak{N})K\mathfrak{G}$. If $g \in \mathfrak{G}$, $g^{-1}(K\mathfrak{N})g = K\mathfrak{N}$, so $g^{-1}\mathbf{J}(K\mathfrak{N})g = \mathbf{J}(K\mathfrak{N})$. Thus

$$\mathbf{J}(K\mathfrak{G}) = K\mathfrak{G}\mathbf{J}(K\mathfrak{N}).$$

Consider $\mathsf{K}\mathfrak{G}$ as a $\mathsf{K}\mathfrak{N}$-module; then the greatest completely reducible factor module of $(\mathsf{K}\mathfrak{G})_{\mathfrak{N}}$ is

$$(\mathsf{K}\mathfrak{G})_{\mathfrak{N}}/(\mathsf{K}\mathfrak{G})_{\mathfrak{N}}\mathbf{J}(\mathsf{K}\mathfrak{N}) = (\mathsf{K}\mathfrak{G}/\mathbf{J}(\mathsf{K}\mathfrak{G}))_{\mathfrak{N}}.$$

Write

$$\mathsf{K}\mathfrak{G}/\mathbf{J}(\mathsf{K}\mathfrak{G}) = \bigoplus_i d_i \mathsf{V}_i$$

with non-isomorphic irreducible $\mathsf{K}\mathfrak{G}$-modules V_i. Then

$$\mathsf{K}\mathfrak{G} = \bigoplus_i d_i \mathsf{P}(\mathsf{V}_i)$$

and

$$(\mathsf{K}\mathfrak{G})_{\mathfrak{N}} = \bigoplus_i d_i \mathsf{P}(\mathsf{V}_i)_{\mathfrak{N}}.$$

But since $\mathsf{K}\mathfrak{G}$ is a projective $\mathsf{K}\mathfrak{N}$-module,

$$\begin{aligned}(\mathsf{K}\mathfrak{G})_{\mathfrak{N}} &= \mathsf{P}((\mathsf{K}\mathfrak{G})_{\mathfrak{N}})\\ &= \mathsf{P}((\mathsf{K}\mathfrak{G})_{\mathfrak{N}}/(\mathsf{K}\mathfrak{G})_{\mathfrak{N}}\mathbf{J}(\mathsf{K}\mathfrak{N}))\\ &= \mathsf{P}((\mathsf{K}\mathfrak{G}/\mathbf{J}(\mathsf{K}\mathfrak{G}))_{\mathfrak{N}})\\ &= \bigoplus_i d_i \mathsf{P}((\mathsf{V}_i)_{\mathfrak{N}}).\end{aligned}$$

But since there is an epimorphism of $\mathsf{P}(\mathsf{V}_i)$ onto V_i and hence of $\mathsf{P}(\mathsf{V}_i)_{\mathfrak{N}}$ onto $(\mathsf{V}_i)_{\mathfrak{N}}$, $\mathsf{P}((\mathsf{V}_i)_{\mathfrak{N}})$ is a direct summand of $\mathsf{P}(\mathsf{P}(\mathsf{V}_i)_{\mathfrak{N}}) = \mathsf{P}(\mathsf{V}_i)_{\mathfrak{N}}$. Hence

$$\mathsf{P}((\mathsf{V}_i)_{\mathfrak{N}}) = \mathsf{P}(\mathsf{V}_i)_{\mathfrak{N}}$$

and

$$\dim_{\mathsf{K}} \mathsf{P}(\mathsf{V}_i) = \dim_{\mathsf{K}} \mathsf{P}((\mathsf{V}_i)_{\mathfrak{N}}).$$

But by Clifford's theorem,

$$(\mathsf{V}_i)_{\mathfrak{N}} = \bigoplus_j c_{ij} \mathsf{W}_{ij}$$

with non-isomorphic irreducible $\mathsf{K}\mathfrak{N}$-modules W_{ij} of equal dimension. Thus

$$\mathsf{P}((\mathsf{V}_i)_{\mathfrak{N}}) = \bigoplus_j c_{ij}\mathsf{P}(\mathsf{W}_{ij})$$

and

$$\dim_{\mathsf{K}} \mathsf{P}(\mathsf{V}_i) = \sum_j c_{ij} \dim_{\mathsf{K}} \mathsf{P}(\mathsf{W}_{ij}).$$

By the inductive hypothesis,

$$\dim_{\mathsf{K}} \mathsf{P}(\mathsf{W}_{ij}) = |\mathfrak{N}|_p(\dim_{\mathsf{K}} \mathsf{W}_{ij})_{p'},$$

so

$$\dim_{\mathsf{K}} \mathsf{P}(\mathsf{V}_i) = \left(\sum_j c_{ij}\right)|\mathfrak{N}|_p(\dim_{\mathsf{K}} \mathsf{W}_{i1})_{p'}.$$

By 9.20, $\sum_j c_{ij}$ divides $|\mathfrak{G}/\mathfrak{N}|$ and is therefore a p'-number. Thus

$$\begin{aligned}\dim_{\mathsf{K}} \mathsf{P}(\mathsf{V}_i) &= |\mathfrak{N}|_p\left(\sum c_{ij} \dim_{\mathsf{K}} \mathsf{W}_{ij}\right)_{p'}\\ &= |\mathfrak{G}|_p(\dim_{\mathsf{K}} \mathsf{V}_i)_{p'}.\end{aligned}$$ **q.e.d.**

Theorem 16.9 holds also for p-soluble groups; the proof is similar and uses the extension of 9.20 to p-soluble groups.

With our methods, we can deal with p-nilpotent groups.

16.10 Theorem (MICHLER [2], SRINIVASAN [1]). *Suppose that $\mathfrak{G}$ is p-nilpotent and $\mathfrak{N} = \mathbf{O}_{p'}(\mathfrak{G})$. Let K be a perfect splitting field for $\mathfrak{N}$ of characteristic p. Let $f_1, \ldots, f_m$ be a complete set of representatives of the $\mathfrak{G}$-conjugacy classes of indecomposable central idempotents in $\mathbf{Z}(\mathsf{K}\mathfrak{N})$. Let*

$$\mathfrak{T}_i = \{g | g \in \mathfrak{G}, g^{-1}f_i g = f_i\}$$

be the inertia group of f_i, and let

$$\mathfrak{G} = \bigcup_{j=1}^{p^{a_i}} \mathfrak{T}_i g_{ij}$$

be a decomposition of $\mathfrak{G}$ into right cosets of $\mathfrak{T}_i$.

a) *The elements*

$$f_i' = \sum_{j=1}^{p^{a_i}} g_{ij}^{-1} f_i g_{ij} \quad (i = 1, \ldots, m)$$

are the indecomposable central idempotents of $\mathsf{K}\mathfrak{G}$. *In particular,* $\mathfrak{G}$ *has exactly m blocks.*

b) *Let* $\mathsf{V}_1, \ldots, \mathsf{V}_m$ *be a complete set of representatives of the* $\mathfrak{G}$-*conjugacy classes of irreducible* $\mathsf{K}\mathfrak{N}$-*modules. Write* $\mathsf{P}_i = \mathsf{V}_i^{\mathfrak{G}}$. *Then to within isomorphism,* $\mathsf{P}_1, \ldots, \mathsf{P}_m$ *are all the indecomposable projective* $\mathsf{K}\mathfrak{G}$-*modules.*

c) $\mathsf{W}_i = \mathsf{P}_i/\mathsf{P}_i\mathbf{J}(\mathsf{K}\mathfrak{G})$ *is an absolutely irreducible* $\mathsf{K}\mathfrak{G}$-*module. In particular,* K *is a splitting field for* $\mathfrak{G}$.

d) $(\mathsf{W}_i)_{\mathfrak{N}} \cong \bigoplus_{j=1}^{p^{a_i}} \mathsf{V}_i \otimes g_{ij}$.
In particular, $\dim_{\mathsf{K}} \mathsf{W}_i = |\mathfrak{G} : \mathfrak{T}_i| \dim_{\mathsf{K}} \mathsf{V}_i$. *Also, the composition length of* P_i *is* $|\mathfrak{T}_i/\mathfrak{N}|$.

e) *If* $|\mathfrak{G}/\mathfrak{N}| = p^a$,

$$\dim_{\mathsf{K}} f_i' \, \mathsf{K}\mathfrak{G} = (\dim_{\mathsf{K}} \mathsf{V}_i)^2 p^{a+a_i}$$

and

$$\dim_{\mathsf{K}} \mathbf{J}(f_i' \mathsf{K}\mathfrak{G}) = (\dim_{\mathsf{K}} \mathsf{V}_i)^2 p^{2a_i}(p^{a-a_i} - 1).$$

Proof. (1) $\mathfrak{G}$ has at least m blocks.

The indecomposable central indempotents of $\mathsf{K}\mathfrak{N}$ are the elements $g_{ij}^{-1} f_i g_{ij}$, so

$$1 = \sum_{i=1}^{m} \sum_{j=1}^{p^{a_i}} g_{ij}^{-1} f_i g_{ij} = \sum_{i=1}^{m} f_i'.$$

Clearly $f_i' \in \mathbf{Z}(\mathsf{K}\mathfrak{G})$, so $\mathfrak{G}$ has at least m blocks.

(2) As K is a splitting field of $\mathfrak{N}$ and $\mathfrak{N}$ is a p'-group,

$$\mathsf{K}\mathfrak{N} = \bigoplus_{i=1}^{r} \mathsf{V}_i$$

for absolutely irreducible $\mathsf{K}\mathfrak{N}$-modules V_i. Then

$$\mathsf{K}\mathfrak{G} \cong (\mathsf{K}\mathfrak{N})^{\mathfrak{G}} = \bigoplus_{i=1}^{r} \mathsf{V}_i^{\mathfrak{G}},$$

and by 16.2, $V_i^{\mathfrak{G}}$ is absolutely indecomposable. If V_i is $\mathfrak{G}$-conjugate to V_j, $V_i^{\mathfrak{G}} \cong V_j^{\mathfrak{G}}$. Thus the decomposition of $K\mathfrak{G}$ contains at most m isomorphism types of indecomposable direct summands. Being direct summands of $K\mathfrak{G}$, the $V_i^{\mathfrak{G}}$ are of course projective.

Combining (1) and (2), we obtain the following.

(3) $\mathfrak{G}$ has exactly m blocks and the idempotents f_i' are the block idempotents of $K\mathfrak{G}$. Supposing (as we may) that no two of $V_1, \ldots, V_m$ are isomorphic, $P_1, \ldots, P_m$ are all the types of projective indecomposable $K\mathfrak{G}$-modules. In particular, all the indecomposable projective modules in a given block are isomorphic. The Cartan matrix is thus a diagonal matrix (cf. 14.9).

Thus a), b) are proved.

(4) Since the P_i are absolutely indecomposable, it follows that for any extension L of K,

$$\begin{aligned}(P_i/P_i\mathbf{J}(K\mathfrak{G})) \otimes_K L &\cong (P_i \otimes_K L)/(P_i \otimes_K L)(\mathbf{J}(K\mathfrak{G}) \otimes_K L)\\ &\cong (P_i \otimes_K L)/(P_i \otimes_K L)\mathbf{J}(L\mathfrak{G}) \qquad \text{(by 1.5)}\end{aligned}$$

is an irreducible $L\mathfrak{G}$-module. Hence W_i is absolutely irreducible.

(5) By Clifford's theorem, $(W_i)_{\mathfrak{N}}$ is a completely reducible $K\mathfrak{N}$-module. It follows from Nakayama's reciprocity theorem that

$$\dim_K \operatorname{Hom}_{K\mathfrak{N}}(V_j, (W_i)_{\mathfrak{N}}) = \dim_K \operatorname{Hom}_{K\mathfrak{G}}(V_j^{\mathfrak{G}}, W_i).$$

This is 1 if $V_j^{\mathfrak{G}} \cong V_i^{\mathfrak{G}}$ and 0 otherwise. But $V_j^{\mathfrak{G}} \cong V_i^{\mathfrak{G}}$ if and only if V_j, V_i are $\mathfrak{G}$-conjugate. Thus

$$(W_i)_{\mathfrak{N}} \cong \bigoplus_{j=1}^{p^{a_i}} V_i \otimes g_{ij}$$

and

$$\dim_K W_i = p^{a_i} \dim_K V_i = |\mathfrak{G} : \mathfrak{T}_i| \dim_K V_i.$$

Since the Cartan matrix is diagonal, all the composition factors of P_i are isomorphic to W_i. The composition length of P_i is thus

$$\frac{\dim_K P_i}{\dim_K W_i} = \frac{|\mathfrak{G}/\mathfrak{N}| \dim_K V_i}{|\mathfrak{G}/\mathfrak{T}_i| \dim_K V_i} = |\mathfrak{T}_i/\mathfrak{N}|.$$

(6) Since the block ideal $f_i' K\mathfrak{G}$ is the direct sum of $\dim_K W_i$ indecomposable $K\mathfrak{G}$-modules each isomorphic to P_i,

$$\begin{aligned}\dim_K f_i' K\mathfrak{G} &= (\dim_K W_i)(\dim_K P_i) \\ &= p^{a_i}(\dim_K V_i) p^a (\dim_K V_i) \\ &= p^{a+a_i}(\dim_K V_i)^2.\end{aligned}$$

Further, $f_i' K\mathfrak{G}/\mathbf{J}(f_i' K\mathfrak{G})$ is isomorphic to the direct sum of $\dim_K W_i$ modules isomorphic to W_i, so

$$\dim_K f_i' K\mathfrak{G} - \dim_K \mathbf{J}(f_i' K\mathfrak{G}) = (\dim_K W_i)^2 = p^{2a_i}(\dim_K V_i)^2.$$

Thus

$$\dim_K \mathbf{J}(f_i' K\mathfrak{G}) = (p^{a+a_i} - p^{2a_i})(\dim_K V_i)^2. \qquad \textbf{q.e.d.}$$

A number of similar facts can also be proved for p-soluble groups (see SCHWARZ [1]).

Exercises

70) Suppose that $K = GF(2)$, $\mathfrak{G} = \mathfrak{S}_3$ and $\mathfrak{N} = \mathfrak{A}_3$. Then there is an irreducible $K\mathfrak{N}$-module V of dimension 2 such that $V^{\mathfrak{G}}$ is decomposable. (Thus in 16.2, absolute indecomposability cannot be replaced by indecomposability.)

71) Suppose that K is algebraically closed of characteristic 5, $\mathfrak{G} = \mathfrak{A}_5$ and $\mathfrak{U} = \mathfrak{A}_4$. Then there is an irreducible $K\mathfrak{U}$-module V of dimension 3 for which $V^{\mathfrak{G}}$ is decomposable. (Thus in 16.2, the condition that $\mathfrak{N}$ be subnormal cannot be replaced by $\mathfrak{N} \leq \mathfrak{G}$).

72) Is Theorem 16.9 valid for arbitrary groups?

73) (WILLEMS [1]) Suppose that K is an algebraically closed field of characteristic p. Suppose that $\mathfrak{N} \trianglelefteq \mathfrak{G}$ and that V is an irreducible $K\mathfrak{N}$-module with inertia group $\mathfrak{I}$. Then the following are equivalent.

a) $\mathfrak{I}/\mathfrak{N}$ is a p'-group.

b) $V^{\mathfrak{G}}$ is completely reducible.

c) If $\mathfrak{P}/\mathfrak{N} \in \mathbf{S}_p(\mathfrak{G}/\mathfrak{N})$, $V^{\mathfrak{P}}$ is completely reducible.

74) a) Formulate and prove the assertions about the injective envelope dual to 16.8.

b) Give an example of a module for which the injective and projective envelopes are not isomorphic.

Notes on Chapter VII

§ 1: 1.17 is stated in BRAUER [4], but essentially, it is already contained in BRAUER [1] on page 101. The proof of 1.20 was communicated to us by A. Dress. 1.22 was first proved by E. Noether and published in DEURING [1].
§ 3: 3.9 was first proved in BRAUER [2]. The proof given here is taken from BRAUER [6]. For an interesting different treatment of 3.10, we refer to GLOVER [1].
§ 4: This section follows mainly two fundamental papers of D. G. HIGMAN [2, 3].
§ 5: The central result 5.4 of this section comes from D. G. HIGMAN [3], with some improvements from KASCH, KNESER and KUPISCH [1].
§ 6: After 6.6, the treatment follows HUPPERT [4]. The result in 6.11 is already contained in the unpublished thesis of D. VOIGT (Kiel 1965).
§ 7: The use of Gaschütz operators to characterize projective and injective modules first appears in GASCHÜTZ [1]. Relative projective and relative injective modules are introduced in D. G. HIGMAN [4].
§ 8: 8.11 was communicated to us in a letter by R. Gow.
§ 9: 9.18 was proved for characteristic 0 in ROTH [1] by character calculations, and the general case is from SCHWARZ [1]. The Exercises 40, 41, 43 and 44 are taken from WILLEMS [1].
§ 10: Most of the content of this section appears in the work of Brauer, Nakayama and Nesbitt on algebras. In NAKAYAMA [1], 10.13–10.18, among other results, will be found. The concepts of projective and injective modules, only introduced some years later, did much to clarify the situation. In particular, compare 10.13 with 11.5. The basic result 10.12 is essentially in BRAUER and NESBITT [1].
§ 11: Frobenius and symmetric algebras are introduced in NESBITT [1]; for instance, most of 11.6 is there. 11.3 is from NAKAYAMA [2].
§ 12: Many results in this section are from the basic paper BRAUER [6].
§ 13: 13.5, 13.7 and 13.8 were first proved using character theory in FONG and GASCHÜTZ [1]. The treatment in this section follows COSSEY and GASCHÜTZ [1].
§ 14: Most of this section follows WILLEMS [3]. Most of 14.9 is already in BRAUER and NESBITT [1].
§ 15: 15.1–15.6 are due to oral communications by Gaschütz.
§ 16: The proof of Lemma 16.4 is strongly influenced by a lecture of E. Dade at Aarhus in 1978.

Chapter VIII

Linear Methods in Nilpotent Groups

The subject of this chapter is commutator calculation. It will be recalled that the commutator $[x, y]$ of two elements x, y of a group is defined by the relation

$$[x, y] = x^{-1}y^{-1}xy.$$

We then have

$$[xy, z] = [x, z]^y[y, z], \quad [x, yz] = [x, z][x, y]^z.$$

These relations are rather similar to the conditions for bilinearity of forms, and there are a number of ways of formalizing this similarity. Once this is done, commutator calculations can be done by linear methods. Several examples of theorems proved by this method will be given in this chapter.

One of these is the determination of the Suzuki 2-groups (see 7.1). This is accomplished by applying the idea sketched above to express the relevant commutation in terms of bilinear forms. No other structure of linear algebra is used in this problem. But the *Witt identity*

$$[x, y^{-1}, z]^y[y, z^{-1}, x]^z[z, x^{-1}, y]^x = 1$$

is remarkably like the *Jacobi identity*

$$[x, y, z] + [y, z, x] + [z, x, y] = 0$$

of Lie rings. This can be formalized by associating a Lie ring with any nilpotent group, and this will be used in the discussion of other problems.

One of the best-known such problems is the following. Suppose that $\mathsf{A}\mathfrak{G}$ is the group-ring of the group $\mathfrak{G}$ over the commutative ring A, and let $\mathfrak{J}$ be the augmentation ideal of $\mathsf{A}\mathfrak{G}$. If n is a positive integer, define $\mathfrak{G}_n$ to be the set of elements x of $\mathfrak{G}$ such that $x - 1 \in \mathfrak{J}^n$. It is easy to see that $\mathfrak{G}_n$ is a normal subgroup of $\mathfrak{G}$. The *dimension subgroup problem* is to characterize $\mathfrak{G}_n$ within $\mathfrak{G}$ itself.

Magnus and Witt used the procedure mentioned above to solve this problem in the case when A is the ring $\mathbb{Z}$ of integers and $\mathfrak{G}$ is a free group; this is presented in § 11. No definitive answer is known for a general group $\mathfrak{G}$ when A is the ring of integers. On the other hand, the problem was solved by Jennings for any group $\mathfrak{G}$ in the case when A is a field. This is presented in § 2 in the case of prime characteristic. The Lie ring method is not needed for this.

The simplification effected when commutation is replaced by a bilinear form is, of course, not very great in itself, but an additional advantage of doing so is that operations on the linear structures involved may be used, even when no corresponding operation exists for the group. The operation most successfully used in this connection is extension of the ground-ring. An application involving this is given in §§ 9–10, where it is shown that the class of a nilpotent group having a fixed point free automorphism of prime order p is bounded by a function of p.

Another method for applying linear methods to nilpotent groups is the use of the Baker-Hausdorff formula and the corresponding inversion formula (see, for example, AMAYO and STEWART [1]). This, however, will not be described in this chapter.

§ 1. Central Series with Elementary Abelian Factors

Throughout this section let p be a fixed prime. We shall give a construction of two central series $\kappa_n(\mathfrak{G})$, $\lambda_n(\mathfrak{G})$ $(n \geq 1)$ of a group $\mathfrak{G}$, for which $\kappa_n(\mathfrak{G})/\kappa_{n+1}(\mathfrak{G})$ and $\lambda_n(\mathfrak{G})/\lambda_{n+1}(\mathfrak{G})$ are elementary Abelian p-groups. The *lower central series* of $\mathfrak{G}$ will be denoted by

$$\mathfrak{G} = \gamma_1(\mathfrak{G}) \geq \gamma_2(\mathfrak{G}) \geq \cdots \geq \gamma_n(\mathfrak{G}) \geq \cdots;$$

here $\gamma_n(\mathfrak{G}) = [\gamma_{n-1}(\mathfrak{G}), \mathfrak{G}]$ for $n > 1$. If m is a positive integer, $\mathfrak{G}^m = \langle x^m | x \in \mathfrak{G} \rangle$. Thus $\mathfrak{G}^{mn} \leq (\mathfrak{G}^m)^n$ and $(\mathfrak{G}/\mathfrak{N})^n = \mathfrak{G}^n\mathfrak{N}/\mathfrak{N}$ for $\mathfrak{N} \trianglelefteq \mathfrak{G}$.

1.1 Lemma. a) *If x, y are elements of a group $\mathfrak{G}$,*

$$(xy)^{p^n} \equiv x^{p^n} y^{p^n} \mod \gamma_2(\mathfrak{G})^{p^n} \prod_{r=1}^{n} \gamma_{p^r}(\mathfrak{G})^{p^{n-r}}$$

for all $n \geq 1$.

b) *If* x, y *are elements of a group* $\mathfrak{G}$, $\mathfrak{H} \leq \mathfrak{G}$ *and* $x, [x, y]$ *belong to* $\mathfrak{H}$,

$$[x^{p^n}, y] \equiv [x, y]^{p^n} \mod \gamma_2(\mathfrak{H})^{p^n} \prod_{r=1}^{n} \gamma_{p^r}(\mathfrak{H})^{p^{n-r}}$$

for all $n \geq 1$.

Proof. a) Let

$$\mathfrak{N} = \gamma_2(\mathfrak{G})^{p^n} \prod_{r=1}^{n} \gamma_{p^r}(\mathfrak{G})^{p^{n-r}}.$$

By the Hall-Petrescu formula (III, 9.4), there exist elements $c_i \in \gamma_i(\mathfrak{G})$ $(i = 2, \ldots, p^n)$ such that

$$x^{p^n} y^{p^n} = (xy)^{p^n} c_2^{\binom{p^n}{2}} \cdots c_{p^n}.$$

If $(i, p) = 1$, then $\binom{p^n}{i}$ is divisible by p^n, and for $i > 1$, $c_i^{\binom{p^n}{i}} \in \gamma_2(\mathfrak{G})^{p^n} \leq \mathfrak{N}$. And if $i = p^r j$, where $r \geq 1$ and $(p, j) = 1$, then $\binom{p^n}{i}$ is divisible by p^{n-r} and $i \geq p^r$, so $c_i^{\binom{p^n}{i}} \in \gamma_{p^r}(\mathfrak{G})^{p^{n-r}} \leq \mathfrak{N}$.
Hence

$$x^{p^n} y^{p^n} \equiv (xy)^{p^n} \mod \mathfrak{N}.$$

b) By a),

$$(x[x, y])^{p^n} \equiv x^{p^n}[x, y]^{p^n} \mod \gamma_2(\mathfrak{H})^{p^n} \prod_{r=1}^{n} \gamma_{p^r}(\mathfrak{H})^{p^{n-r}}.$$

But

$$(x[x, y])^{p^n} = (x^y)^{p^n} = (x^{p^n})^y = x^{p^n}[x^{p^n}, y],$$

so the assertion follows at once. **q.e.d.**

1.2 Lemma. *If* $n \geq 0$,

$$[\gamma_i(\mathfrak{G})^{p^n}, \gamma_j(\mathfrak{G})] \leq \prod_{r=0}^{n} \gamma_{j+ip^r}(\mathfrak{G})^{p^{n-r}}.$$

Proof. Let $\mathfrak{N} = \prod_{r=0}^{n} \gamma_{j+ip^r}(\mathfrak{G})^{p^{n-r}}$. Since $\mathfrak{N} \trianglelefteq \mathfrak{G}$, it is sufficient to show that $[x^{p^n}, y] \in \mathfrak{N}$ for any $x \in \gamma_i(\mathfrak{G})$, $y \in \gamma_j(\mathfrak{G})$. By 1.1b),

$$[x^{p^n}, y] \equiv [x, y]^{p^n} \quad \bmod \gamma_2(\mathfrak{H})^{p^n} \prod_{r=1}^{n} \gamma_{p^r}(\mathfrak{H})^{p^{n-r}},$$

where $\mathfrak{H} = \langle x, [x, y]\rangle$. Then $\gamma_2(\mathfrak{H})$ is the normal closure of $[x, y, x]$ in $\mathfrak{H}$, by III, 1.11, so $\gamma_2(\mathfrak{H}) \leq \gamma_{2i+j}(\mathfrak{G})$. Since $\mathfrak{H} \leq \gamma_i(\mathfrak{G})$, it follows that $\gamma_m(\mathfrak{H}) \leq \gamma_{mi+j}(\mathfrak{G})$ for all $m \geq 2$. Thus $\gamma_{p^r}(\mathfrak{H})^{p^{n-r}} \leq \gamma_{p^r i+j}(\mathfrak{G})^{p^{n-r}} \leq \mathfrak{N}$ for $r = 1, \ldots, n$. Also $\gamma_2(\mathfrak{H})^{p^n} \leq \gamma_{i+j}(\mathfrak{G})^{p^n} \leq \mathfrak{N}$ and $[x, y]^{p^n} \in \gamma_{i+j}(\mathfrak{G})^{p^n} \leq \mathfrak{N}$. Hence $[x^{p^n}, y] \in \mathfrak{N}$, as required. **q.e.d.**

1.3 Lemma. *For $i \geq 1$, $k \geq 0$, put*

$$\mathfrak{N}_{ik} = \gamma_i(\mathfrak{G})^{p^k}\gamma_{i+1}(\mathfrak{G})^{p^{k-1}} \cdots \gamma_{i+k}(\mathfrak{G}).$$

Then $\mathfrak{N}_{ik}^{p^m} \leq \mathfrak{N}_{i,k+m}$ and

$$[\mathfrak{N}_{ik}, \mathfrak{N}_{jl}] \leq \mathfrak{N}_{i+j,k+l}.$$

Proof. By 1.2,

$$[\gamma_{i+h}(\mathfrak{G})^{p^{k-h}}, \gamma_j(\mathfrak{G})] \leq \prod_{r=0}^{k-h} \gamma_{j+(i+h)p^r}(\mathfrak{G})^{p^{k-h-r}} \leq \mathfrak{N}_{i+j,k},$$

since $j + (i + h)p^r + (k - h - r) \geq j + ip^r + k - r \geq j + i + k$. Hence

$$[\mathfrak{N}_{ik}, \gamma_j(\mathfrak{G})] \leq \mathfrak{N}_{i+j,k}. \tag{1}$$

In particular, $\mathfrak{N}_{ik}' \leq [\mathfrak{N}_{ik}, \gamma_i(\mathfrak{G})] \leq \mathfrak{N}_{2i,k} \leq \mathfrak{N}_{i,k+1}$. Thus

$$\mathfrak{N}_{ik}^p \leq \mathfrak{N}_{ik}'\gamma_i(\mathfrak{G})^{p^{k+1}} \cdots \gamma_{i+k}(\mathfrak{G})^p \leq \mathfrak{N}_{i,k+1}.$$

It follows at once that $\mathfrak{N}_{ik}^{p^m} \leq \mathfrak{N}_{i,k+m}$.

Now suppose that $x \in \gamma_i(\mathfrak{G})$, $y \in \mathfrak{N}_{jl}$. By (1), $[x, y] \in \mathfrak{N}_{i+j,l}$ and $[x, y, x] \in \mathfrak{N}_{2i+j,l}$. Thus if $\mathfrak{H} = \langle x, [x, y]\rangle$, $\gamma_2(\mathfrak{H}) \leq \mathfrak{N}_{2i+j,l}$. Again by (1), it follows that $\gamma_n(\mathfrak{H}) \leq \mathfrak{N}_{ni+j,l}$. Thus for $1 \leq r \leq k$,

$$\gamma_{p^r}(\mathfrak{H})^{p^{k-r}} \leq \mathfrak{N}_{p^r i+j,l}^{p^{k-r}} \leq \mathfrak{N}_{p^r i+j,l+k-r} \leq \mathfrak{N}_{i+j,k+l}.$$

Also, $[x, y]^{p^k}$ and $\gamma_2(\mathfrak{H})^{p^k}$ are both contained in

$$\mathfrak{N}_{i+j,l}^{p^k} \leq \mathfrak{N}_{i+j,k+l}.$$

Hence by 1.1b),

$$[x^{p^k}, y] \in \mathfrak{N}_{i+j,k+l}.$$

It follows that

$$[\gamma_i(\mathfrak{G})^{p^k}, \mathfrak{N}_{jl}] \leq \mathfrak{N}_{i+j,k+l}$$

and

$$[\gamma_{i+h}(\mathfrak{G})^{p^{k-h}}, \mathfrak{N}_{jl}] \leq \mathfrak{N}_{i+j+h,k+l-h} \leq \mathfrak{N}_{i+j,k+l}.$$

Hence

$$[\mathfrak{N}_{ik}, \mathfrak{N}_{jl}] \leq \mathfrak{N}_{i+j,k+l}.$$ **q.e.d.**

1.4 Definition. For any group $\mathfrak{G}$, put

$$\lambda_n(\mathfrak{G}) = \gamma_1(\mathfrak{G})^{p^{n-1}}\gamma_2(\mathfrak{G})^{p^{n-2}} \cdots \gamma_n(\mathfrak{G}) \quad (n \geq 1).$$

Thus $\lambda_n(\mathfrak{G})$ is a characteristic subgroup of $\mathfrak{G}$ and

$$\mathfrak{G} = \lambda_1(\mathfrak{G}) \geq \lambda_2(\mathfrak{G}) \geq \cdots \geq \lambda_n(\mathfrak{G}) \geq \cdots.$$

1.5 Theorem. a) $[\lambda_m(\mathfrak{G}), \lambda_n(\mathfrak{G})] \leq \lambda_{m+n}(\mathfrak{G})$.
b) $\lambda_n(\mathfrak{G})^{p^j} \leq \lambda_{n+j}(\mathfrak{G})$.
c) *For* $n > 1$, $\lambda_n(\mathfrak{G}) = [\lambda_{n-1}(\mathfrak{G}), \mathfrak{G}]\lambda_{n-1}(\mathfrak{G})^p$.

Proof. In the notation of 1.3, $\lambda_n(\mathfrak{G}) = \mathfrak{N}_{1,n-1}$, so a), b) follow at once. It follows from a) and b) that

$$[\lambda_{n-1}(\mathfrak{G}), \mathfrak{G}]\lambda_{n-1}(\mathfrak{G})^p \leq \lambda_n(\mathfrak{G}),$$

so it only remains to prove that

$$\gamma_i(\mathfrak{G})^{p^{n-i}} \leq [\lambda_{n-1}(\mathfrak{G}), \mathfrak{G}]\lambda_{n-1}(\mathfrak{G})^p$$

for $i = 1, \ldots, n$. But for $i = 1, \ldots, n-1$,

$$\gamma_i(\mathfrak{G})^{p^{n-i}} \leq (\gamma_i(\mathfrak{G})^{p^{n-i-1}})^p \leq \lambda_{n-1}(\mathfrak{G})^p,$$

and

$$\gamma_n(\mathfrak{G}) = [\gamma_{n-1}(\mathfrak{G}), \mathfrak{G}] \leq [\lambda_{n-1}(\mathfrak{G}), \mathfrak{G}].$$ **q.e.d.**

1.6 Corollary. *Suppose that*

$$\mathfrak{G} = \mathfrak{G}_1 \geq \mathfrak{G}_2 \geq \cdots \geq \mathfrak{G}_n \geq \mathfrak{G}_{n+1} \geq \cdots,$$

where $\mathfrak{G}_i \trianglelefteq \mathfrak{G}$, $\mathfrak{G}_i/\mathfrak{G}_{i+1} \leq \mathbf{Z}(\mathfrak{G}/\mathfrak{G}_{i+1})$ *and* $\mathfrak{G}_i/\mathfrak{G}_{i+1}$ *is an elementary Abelian p-group* ($i = 1, 2, \ldots$). *Then* $\lambda_n(\mathfrak{G}) \leq \mathfrak{G}_n$.

Proof. We use induction on n. For $n > 1$, $\lambda_{n-1}(\mathfrak{G}) \leq \mathfrak{G}_{n-1}$, by the inductive hypothesis. Hence by 1.5,

$$\lambda_n(\mathfrak{G}) = [\lambda_{n-1}(\mathfrak{G}), \mathfrak{G}]\lambda_{n-1}(\mathfrak{G})^p \leq [\mathfrak{G}_{n-1}, \mathfrak{G}]\mathfrak{G}_{n-1}^p \leq \mathfrak{G}_n.$$ **q.e.d.**

1.7 Theorem. a) *Suppose that* α, β *are homomorphisms of* $\mathfrak{G}$ *into* $\tilde{\mathfrak{G}}$ *which induce the same homomorphism on* $\mathfrak{G}/\lambda_2(\mathfrak{G})$. *Then* α, β *induce the same homomorphism on* $\lambda_n(\mathfrak{G})/\lambda_{n+1}(\mathfrak{G})$ *for all* $n \geq 1$.

b) *Suppose that* $\mathfrak{H} \leq \mathfrak{G}$ *and* $\mathfrak{H}\lambda_2(\mathfrak{G}) = \mathfrak{G}$. *Then* $\mathfrak{H}\lambda_{n+1}(\mathfrak{G}) = \mathfrak{G}$ *for all* $n \geq 1$.

Proof. a) This is proved by induction on n and is trivial for $n = 1$. For $n > 1$, it follows from the inductive hypothesis that if $u \in \lambda_{n-1}(\mathfrak{G})$, then $u\alpha = (u\beta)v$ for some $v \in \lambda_n(\tilde{\mathfrak{G}})$. Thus

$$u^p\alpha = ((u\beta)v)^p \equiv (u\beta)^p v^p \equiv (u\beta)^p = u^p\beta \mod \lambda_{n+1}(\tilde{\mathfrak{G}}),$$

since by 1.5, $\lambda_n(\tilde{\mathfrak{G}})/\lambda_{n+1}(\tilde{\mathfrak{G}})$ is central and elementary Abelian. Also, if $x \in \mathfrak{G}$,

$$[u, x]\alpha = [u\alpha, x\alpha] = [(u\beta)v, x\alpha] \equiv [u\beta, x\alpha] \mod \lambda_{n+1}(\tilde{\mathfrak{G}}).$$

But by hypothesis, $x\alpha = (x\beta)y$ for some $y \in \lambda_2(\tilde{\mathfrak{G}})$, so

$$[u, x]\alpha \equiv [u\beta, (x\beta)y] \equiv [u\beta, y][u\beta, x\beta]^y \equiv [u\beta, y][u\beta, x\beta] \mod \lambda_{n+1}(\tilde{\mathfrak{G}}).$$

Using 1.5a),

$$[u\beta, y] \in [\lambda_{n-1}(\tilde{\mathfrak{G}}), \lambda_2(\tilde{\mathfrak{G}})] \leq \lambda_{n+1}(\tilde{\mathfrak{G}}),$$

so

$$[u, x]\alpha \equiv [u, x]\beta \mod \lambda_{n+1}(\tilde{\mathfrak{G}}).$$

By 1.5c),

$$\lambda_n(\mathfrak{G}) = \langle u^p, [u, x] \mid u \in \lambda_{n-1}(\mathfrak{G}), x \in \mathfrak{G} \rangle,$$

so α, β induce the same homomorphism on $\lambda_n(\mathfrak{G})/\lambda_{n+1}(\mathfrak{G})$.

b) Again, we use induction on n. For $n > 1$, the inductive hypothesis yields $\mathfrak{G} = \mathfrak{H}\lambda_n(\mathfrak{G})$. Thus by 1.5,

$$\begin{aligned}\lambda_n(\mathfrak{G}) &= [\lambda_{n-1}(\mathfrak{G}), \mathfrak{G}]\lambda_{n-1}(\mathfrak{G})^p \\ &\leq [\mathfrak{H}\lambda_n(\mathfrak{G}), \mathfrak{H}\lambda_n(\mathfrak{G})](\mathfrak{H}\lambda_n(\mathfrak{G}))^p \\ &\leq \mathfrak{H}\lambda_{n+1}(\mathfrak{G})\end{aligned}$$

and $\mathfrak{G} \leq \mathfrak{H}\lambda_{n+1}(\mathfrak{G})$. **q.e.d.**

To investigate further the factors of this series, we prove the following.

1.8 Theorem. *For $i \geq 1$, $k \geq 0$, let*

$$\mathfrak{N}_{ik} = \gamma_i(\mathfrak{G})^{p^k}\gamma_{i+1}(\mathfrak{G})^{p^{k-1}} \cdots \gamma_{i+k}(\mathfrak{G}).$$

a) *Any element of $\mathfrak{N}_{ik}$ can be written in the form*

$$a_i^{p^k} a_{i+1}^{p^{k-1}} \cdots a_{i+k}$$

for suitable $a_j \in \gamma_j(\mathfrak{G})$.

b) *If $x \in \gamma_i(\mathfrak{G})$ and $y \in \gamma_i(\mathfrak{G})^p\gamma_{i+1}(\mathfrak{G})$, then*

$$(xy)^{p^k} \equiv x^{p^k} \mod \mathfrak{N}_{i,k+1}.$$

c) *There is a surjective mapping β of the direct product*

$$(\gamma_i(\mathfrak{G})/\gamma_i(\mathfrak{G})^p\gamma_{i+1}(\mathfrak{G})) \times \cdots \times (\gamma_{i+k}(\mathfrak{G})/\gamma_{i+k}(\mathfrak{G})^p\gamma_{i+k+1}(\mathfrak{G}))$$

onto $\mathfrak{N}_{ik}/\mathfrak{N}_{i,k+1}$ given by

$$(\bar{a}_i, \bar{a}_{i+1}, \ldots, \bar{a}_{i+k})\beta = a_i^{p^k} a_{i+1}^{p^{k-1}} \cdots a_{i+k}\mathfrak{N}_{i,k+1}, \tag{2}$$

where $a_{i+j} \in \gamma_{i+j}(\mathfrak{G})$ and $\bar{a}_{i+j} = a_{i+j}\gamma_{i+j}(\mathfrak{G})^p\gamma_{i+j+1}(\mathfrak{G})$. Except in the case when $p = 2$ and $i = 1$, β is an epimorphism.

Proof. Note first that $\mathfrak{N}_{ik} \leq \mathfrak{N}_{jl}$ whenever $i \geq j$ and $i + k \geq j + l$.

Suppose that x, y are elements of $\gamma_i(\mathfrak{G})$ and $\mathfrak{H} = \langle x, y \rangle$. Thus $\mathfrak{H} \leq \gamma_i(\mathfrak{G})$ and $\gamma_m(\mathfrak{H}) \leq \gamma_{im}(\mathfrak{G})$. Hence by 1.1,

$$(3) \qquad (xy)^{p^k} \equiv x^{p^k} y^{p^k} \mod \gamma_{2i}(\mathfrak{G})^{p^k} \gamma_p(\mathfrak{H})^{p^{k-1}} \prod_{r=2}^{k} \gamma_{ip^r}(\mathfrak{G})^{p^{k-r}}.$$

a) This is proved by induction on k. For $k = 0$, it is trivial. For $k > 0$, observe that

$$\gamma_{2i}(\mathfrak{G})^{p^k} \leq \mathfrak{N}_{i+1,k-1},$$
$$\gamma_p(\mathfrak{H})^{p^{k-1}} \leq \gamma_{pi}(\mathfrak{G})^{p^{k-1}} \leq \mathfrak{N}_{i+1,k-1},$$
$$\gamma_{ip^r}(\mathfrak{G})^{p^{k-r}} \leq \mathfrak{N}_{i+1,k-1},$$

since $ip^r + k - r \geq i(r + 1) + k - r \geq i + k$. Thus by (3), $x \rightarrow x^{p^k}\mathfrak{N}_{i+1,k-1}$ is a homomorphism of $\gamma_i(\mathfrak{G})$.

Now suppose that $u \in \mathfrak{N}_{ik}$. Then there exist elements $x_1, \ldots, x_m$ in $\gamma_i(\mathfrak{G})$ and $v \in \mathfrak{N}_{i+1,k-1}$ such that

$$u = x_1^{p^k} \cdots x_m^{p^k} v.$$

Since $x \rightarrow x^{p^k}\mathfrak{N}_{i+1,k-1}$ is a homomorphism,

$$x_1^{p^k} \cdots x_m^{p^k} = a_i^{p^k} v'$$

for $a_i = x_1 \cdots x_m$, $v' \in \mathfrak{N}_{i+1,k-1}$. Thus $u = a_i^{p^k} v'v$, and the assertion follows by applying the inductive hypothesis to $v'v$.

b) By (3),

$$(xy)^{p^k} \equiv x^{p^k} y^{p^k} \mod \gamma_{2i}(\mathfrak{G})^{p^k} \gamma_p(\mathfrak{H})^{p^{k-1}} \prod_{r=2}^{k} \gamma_{ip^r}(\mathfrak{G})^{p^{k-r}},$$

where $\mathfrak{H} = \langle x, y \rangle$. Now $y \in \mathfrak{N}_{i1}$, so by 1.3, $y^{p^k} \in \mathfrak{N}_{i,k+1}$. Also $\gamma_{2i}(\mathfrak{G})^{p^k} \leq \mathfrak{N}_{i,k+1}$ and, for $r \geq 2$, $\gamma_{ip^r}(\mathfrak{G})^{p^{k-r}} \leq \mathfrak{N}_{i,k+1}$, since

$$\begin{aligned} ip^r + k - r &\geq i + k + i(p^r - 1) - r \geq i + k + (p^r - r - 1) \\ &\geq i + k + 1. \end{aligned}$$

Finally,

$$\gamma_2(\mathfrak{H}) \leq [\gamma_i(\mathfrak{G}), \gamma_i(\mathfrak{G})^p \gamma_{i+1}(\mathfrak{G})] = [\mathfrak{N}_{i0}, \mathfrak{N}_{i1}] \leq \mathfrak{N}_{2i,1}$$

by 1.3. Thus

$$\gamma_p(\mathfrak{H})^{p^{k-1}} \leq \mathfrak{N}_{2i,1}^{p^{k-1}} \leq \mathfrak{N}_{2i,k} \leq \mathfrak{N}_{i,k+1}.$$

Hence

$$(xy)^{p^k} \equiv x^{p^k} \mod \mathfrak{N}_{i,k+1}.$$

c) It follows from b) that there exists a mapping β given by (2). For if $a_{i+j} \in \gamma_{i+j}(\mathfrak{G})$ and $x \in \gamma_{i+j}(\mathfrak{G})^p \gamma_{i+j+1}(\mathfrak{G})$,

$$(a_{i+j}x)^{p^{k-j}} \equiv a_{i+j}^{p^{k-j}} \mod \mathfrak{N}_{i+j,k-j+1};$$

also $\mathfrak{N}_{i+j,k-j+1} \leq \mathfrak{N}_{i,k+1}$.

By a), β is surjective.

Suppose that x, y are elements of $\gamma_i(\mathfrak{G})$. Since $\mathfrak{H} = \langle x, y \rangle \leq \gamma_i(\mathfrak{G})$, $\gamma_p(\mathfrak{H}) \leq \gamma_{ip}(\mathfrak{G})$. Thus by (3),

$$(xy)^{p^k} \equiv x^{p^k} y^{p^k} \mod \gamma_{2i}(\mathfrak{G})^{p^k} \prod_{r=1}^{k} \gamma_{ip^r}(\mathfrak{G})^{p^{k-r}}.$$

Now $\gamma_{2i}(\mathfrak{G})^{p^k} \leq \mathfrak{N}_{i,k+1}$. And except in the case when $i = 1$, $p = 2$ and $r = 1$, $i(p^r - 1) \geq r + 1$, so

$$ip^r + k - r \geq i + k + 1$$

and $\gamma_{ip^r}(\mathfrak{G})^{p^{k-r}} \leq \mathfrak{N}_{i,k+1}$. Thus except when $i = 1$ and $p = 2$,

$$(xy)^{p^k} \equiv x^{p^k} y^{p^k} \mod \mathfrak{N}_{i,k+1}.$$

Since $\mathfrak{N}_{i+j,k-j+1} \leq \mathfrak{N}_{i,k+1}$, it follows that β is an epimorphism. **q.e.d.**

In § 13, we shall apply the following theorem to free groups.

1.9 Theorem. *Suppose that $\mathfrak{G}$ is a group and that for $i = 1, \ldots, n+1$, no non-identity element of $\mathfrak{G}/\gamma_i(\mathfrak{G})$ is of finite order; (that is, $\mathfrak{G}/\gamma_i(\mathfrak{G})$ is torsion-free).*

a) *For $i = 1, \ldots, n+1$,*

$$\lambda_{n+1}(\mathfrak{G}) \cap \gamma_i(\mathfrak{G}) = \gamma_i(\mathfrak{G})^{p^{n-i+1}} \gamma_{i+1}(\mathfrak{G})^{p^{n-i}} \cdots \gamma_{n+1}(\mathfrak{G}).$$

b) *There is a bijective mapping α of*

$$(\gamma_1(\mathfrak{G})/\gamma_1(\mathfrak{G})^p\gamma_2(\mathfrak{G})) \times \cdots \times (\gamma_n(\mathfrak{G})/\gamma_n(\mathfrak{G})^p\gamma_{n+1}(\mathfrak{G}))$$

onto $\lambda_n(\mathfrak{G})/\lambda_{n+1}(\mathfrak{G})$ *given by*

$$(\bar{a}_1, \ldots, \bar{a}_n)\alpha = a_1^{p^{n-1}} \cdots a_n\lambda_{n+1}(\mathfrak{G}),$$

where $a_i \in \gamma_i(\mathfrak{G})$ *and* $\bar{a}_i = a_i\gamma_i(\mathfrak{G})^p\gamma_{i+1}(\mathfrak{G})$.

c) *For p odd,* α *is an isomorphism. For* $p = 2$, *the restriction of* α *to*

$$(\gamma_2(\mathfrak{G})/\gamma_2(\mathfrak{G})^p\gamma_3(\mathfrak{G})) \times \cdots \times (\gamma_n(\mathfrak{G})/\gamma_n(\mathfrak{G})^p\gamma_{n+1}(\mathfrak{G}))$$

is an isomorphism of this group onto $(\lambda_n(\mathfrak{G}) \cap \gamma_2(\mathfrak{G}))\lambda_{n+1}(\mathfrak{G})/\lambda_{n+1}(\mathfrak{G})$.

Proof. a) This is proved by induction on i and is trivial for $i = 1$. Suppose that $i > 1$ and put

$$\mathfrak{N}_{i,n-i+1} = \gamma_i(\mathfrak{G})^{p^{n-i+1}} \cdots \gamma_{n+1}(\mathfrak{G}).$$

Clearly $\mathfrak{N}_{i,n-i+1} \leq \lambda_{n+1}(\mathfrak{G}) \cap \gamma_i(\mathfrak{G})$. Suppose that $x \in \lambda_{n+1}(\mathfrak{G}) \cap \gamma_i(\mathfrak{G})$. Then

$$x \in \lambda_{n+1}(\mathfrak{G}) \cap \gamma_{i-1}(\mathfrak{G}) = \gamma_{i-1}(\mathfrak{G})^{p^{n-i+2}} \cdots \gamma_{n+1}(\mathfrak{G}),$$

by the inductive hypothesis. It follows from 1.8a) that $x = a^{p^{n-i+2}}y$, where $a \in \gamma_{i-1}(\mathfrak{G})$ and $y \in \mathfrak{N}_{i,n-i+1}$. Thus $a^{p^{n-i+2}} = xy^{-1} \in \gamma_i(\mathfrak{G})$. Since $\mathfrak{G}/\gamma_i(\mathfrak{G})$ is torsion-free, $a \in \gamma_i(\mathfrak{G})$. Hence $a^{p^{n-i+2}} \in \mathfrak{N}_{i,n-i+1}$ and $x \in \mathfrak{N}_{i,n-i+1}$. Therefore

$$\lambda_{n+1}(\mathfrak{G}) \cap \gamma_i(\mathfrak{G}) = \mathfrak{N}_{i,n-i+1}.$$

b) By 1.8c), α is surjective. To show that α is injective, suppose that a_i, b_i are elements of $\gamma_i(\mathfrak{G})$ for which

$$a_1^{p^{n-1}} \cdots a_n \equiv b_1^{p^{n-1}} \cdots b_n \mod \lambda_{n+1}(\mathfrak{G}).$$

We prove that $a_i \equiv b_i \bmod \gamma_i(\mathfrak{G})^p\gamma_{i+1}(\mathfrak{G})$ $(i = 1, \ldots, n)$ by induction on i. If $i > 1$, the inductive hypothesis gives $a_j \equiv b_j \bmod \gamma_j(\mathfrak{G})^p\gamma_{j+1}(\mathfrak{G})$ for $j = 1, \ldots, i-1$. Hence by 1.8b), $a_j^{p^{n-j}} \equiv b_j^{p^{n-j}} \bmod \lambda_{n+1}(\mathfrak{G})$. Thus

$$a_i^{p^{n-i}} \cdots a_n \equiv b_i^{p^{n-i}} \cdots b_n \mod \lambda_{n+1}(\mathfrak{G}) \cap \gamma_i(\mathfrak{G});$$

this also holds for $i = 1$. By a),

$$a_i^{p^{n-i}} \cdots a_n \equiv b_i^{p^{n-i}} \cdots b_n \quad \bmod \gamma_i(\mathfrak{G})^{p^{n-i+1}} \cdots \gamma_{n+1}(\mathfrak{G}).$$

Hence

$$a_i^{p^{n-i}} \equiv b_i^{p^{n-i}} \quad \bmod \gamma_i(\mathfrak{G})^{p^{n-i+1}} \gamma_{i+1}(\mathfrak{G})$$

and

$$(a_i \gamma_{i+1}(\mathfrak{G}))^{p^{n-i}} = (b_i c^p \gamma_{i+1}(\mathfrak{G}))^{p^{n-i}}$$

for some $c \in \gamma_i(\mathfrak{G})$. Since $\gamma_i(\mathfrak{G})/\gamma_{i+1}(\mathfrak{G})$ is a torsion-free Abelian group, it follows that

$$a_i \gamma_{i+1}(\mathfrak{G}) = b_i c^p \gamma_{i+1}(\mathfrak{G})$$

and $a_i \equiv b_i \bmod \gamma_i(\mathfrak{G})^p \gamma_{i+1}(\mathfrak{G})$. Thus α is injective.

c) By b) and 1.8c), α is an isomorphism for p odd. For $p = 2$, the restriction α_1 of α to

$$\mathfrak{F} = (\gamma_2(\mathfrak{G})/\gamma_2(\mathfrak{G})^p \gamma_3(\mathfrak{G})) \times \cdots \times (\gamma_n(\mathfrak{G})/\gamma_n(\mathfrak{G})^p \gamma_{n+1}(\mathfrak{G}))$$

is $\beta\gamma$, where β is the epimorphism of $\mathfrak{F}$ onto $\mathfrak{N}_{2,n-2}/\mathfrak{N}_{2,n-1}$ given in 1.8c) and γ is the natural epimorphism of $\mathfrak{N}_{2,n-2}/\mathfrak{N}_{2,n-1}$ onto $\mathfrak{N}_{2,n-2}\lambda_{n+1}(\mathfrak{G})/\lambda_{n+1}(\mathfrak{G})$. By a) and b), α_1 is an isomorphism onto $(\lambda_n(\mathfrak{G}) \cap \gamma_2(\mathfrak{G}))\lambda_{n+1}(\mathfrak{G})/\lambda_{n+1}(\mathfrak{G})$. **q.e.d.**

We turn now to our second central series.

1.10 Definition. For any positive integer n, write

$$\kappa_n(\mathfrak{G}) = \prod_{ip^k \geq n} \gamma_i(\mathfrak{G})^{p^k}.$$

Then $\kappa_n(\mathfrak{G})$ is a characteristic subgroup of $\mathfrak{G}$, and

$$\mathfrak{G} = \kappa_1(\mathfrak{G}) \geq \kappa_2(\mathfrak{G}) \geq \cdots \geq \kappa_n(\mathfrak{G}) \geq \cdots.$$

Note that it is possible to have $\kappa_{n-1}(\mathfrak{G}) = \kappa_n(\mathfrak{G}) > \kappa_{n+1}(\mathfrak{G})$. For if $\mathfrak{G}$ is Abelian, $\kappa_n(\mathfrak{G}) = \mathfrak{G}^{p^k}$, where k is the smallest integer for which $p^k \geq n$. Thus $\kappa_n(\mathfrak{G}) = \mathfrak{G}^{p^k}$ if $p^{k-1} < n \leq p^k$.

If $\mathfrak{G}$ is a finite p-group, $\kappa_2(\mathfrak{G}) = \lambda_2(\mathfrak{G})$ is the Frattini subgroup of $\mathfrak{G}$. If $\mathfrak{G}$ is a group of exponent p, $\kappa_n(\mathfrak{G}) = \gamma_n(\mathfrak{G})$ for all n.

To prove that this series is central, we need the following lemmas.

1.11 Lemma. *Suppose that* $i \geq 1, j \geq 1, h \geq 0$. *Let*

$$\mathfrak{N} = \prod_{r=0}^{h} \gamma_{i+jp^r}(\mathfrak{G})^{p^{h-r}}.$$

If $x \in \gamma_i(\mathfrak{G})$ *and* $n \geq 2$,

$$\gamma_n(\langle x, \mathfrak{N} \rangle) \leq \prod_{r=0}^{h} \gamma_{in+jp^r}(\mathfrak{G})^{p^{h-r}}.$$

Proof. This is proved by induction on n. For $n = 2$, we have

$$\gamma_2(\langle x, \mathfrak{N} \rangle) = [\mathfrak{N}, \langle x, \mathfrak{N} \rangle]$$

by III, 1.11; since $\langle x, \mathfrak{N} \rangle \leq \gamma_i(\mathfrak{G})$, it follows that

$$\gamma_2(\langle x, \mathfrak{N} \rangle) \leq [\mathfrak{N}, \gamma_i(\mathfrak{G})].$$

For $n > 2$, we have

$$\gamma_n(\langle x, \mathfrak{N} \rangle) \leq [\gamma_{n-1}(\langle x, \mathfrak{N} \rangle), \gamma_i(\mathfrak{G})].$$

Using the definition of $\mathfrak{N}$ for $n = 2$ and the inductive hypothesis for $n > 2$, it follows that for $n \geq 2$,

$$\gamma_n(\langle x, \mathfrak{N} \rangle) \leq \left[\prod_{r=0}^{h} \gamma_{in-i+jp^r}(\mathfrak{G})^{p^{h-r}}, \gamma_i(\mathfrak{G}) \right].$$

Applying III, 1.10a) and 1.2,

$$\begin{aligned} \gamma_n(\langle x, \mathfrak{N} \rangle) &\leq \prod_{r=0}^{h} [\gamma_{in-i+jp^r}(\mathfrak{G})^{p^{h-r}}, \gamma_i(\mathfrak{G})] \\ &\leq \prod_{r=0}^{h} \prod_{s=0}^{h-r} \gamma_{i+p^s(in-i+jp^r)}(\mathfrak{G})^{p^{h-r-s}} \\ &\leq \prod_{r+s \leq h} \gamma_{in+jp^{r+s}}(\mathfrak{G})^{p^{h-r-s}}, \end{aligned}$$

since $i + p^s in - p^s i \geq in$. The lemma follows at once. **q.e.d.**

1.12 Lemma. *If* $i \geq 1, j \geq 1, h \geq 0, k \geq 0$,

$$[\gamma_i(\mathfrak{G})^{p^k}, \gamma_j(\mathfrak{G})^{p^h}] \leq \kappa_{ip^k+jp^h}(\mathfrak{G}).$$

Proof. Suppose that $x \in \gamma_i(\mathfrak{G})$ and $y \in \gamma_j(\mathfrak{G})$. Let $z = [x, y^{p^h}]$ and $\mathfrak{H} = \langle x, z \rangle$. By 1.1b),

$$[x^{p^k}, y^{p^h}] \equiv [x, y^{p^h}]^{p^k} = z^{p^k} \mod \gamma_2(\mathfrak{H})^{p^k} \prod_{m=1}^{k} \gamma_{p^m}(\mathfrak{H})^{p^{k-m}}.$$

For $n = 1, \ldots, p^k$, define

$$\mathfrak{H}_n = \prod_{r=0}^{h} \gamma_{in+jp^r}(\mathfrak{G})^{p^{h-r}}.$$

By 1.2, $z \in \mathfrak{H}_1$. Thus $\mathfrak{H} \leq \langle x, \mathfrak{H}_1 \rangle$, and by 1.11, $\gamma_n(\mathfrak{H}) \leq \mathfrak{H}_n$. Hence

$$[x^{p^k}, y^{p^h}] \in \prod_{m=0}^{n} \mathfrak{H}_{p^m}^{p^{k-m}}.$$

Now let $\mathfrak{N} = \kappa_{ip^k+jp^h}(\mathfrak{G})$. If $m + h - r \geq k$, then $\gamma_{ip^m+jp^r}(\mathfrak{G})^{p^{h-r}} \leq \mathfrak{N}$, by 1.10. Thus if $s = \max(m + h - k + 1, 0)$,

$$\mathfrak{H}_{p^m} \leq \mathfrak{N} \prod_{r=s}^{h} \gamma_{ip^m+jp^r}(\mathfrak{G})^{p^{h-r}} \leq \mathfrak{N}\gamma_{ip^m+jp^s}(\mathfrak{G}).$$

Hence

$$\mathfrak{H}_{p^m}^{p^{k-m}} \leq \mathfrak{N}\gamma_{ip^m+jp^s}(\mathfrak{G})^{p^{k-m}} = \mathfrak{N},$$

since $ip^k + jp^{s+k-m} \geq ip^k + jp^h$.

Thus $[x^{p^k}, y^{p^h}] \in \mathfrak{N}$ and $x^{p^k}\mathfrak{N}$ commutes with $y^{p^h}\mathfrak{N}$. Hence each element of $\gamma_i(\mathfrak{G})^{p^k}\mathfrak{N}/\mathfrak{N}$ commutes with each element of $\gamma_j(\mathfrak{G})^{p^h}\mathfrak{N}/\mathfrak{N}$, and

$$[\gamma_i(\mathfrak{G})^{p^k}, \gamma_j(\mathfrak{G})^{p^h}] \leq \mathfrak{N}.$$ **q.e.d.**

1.13 Theorem. a) $[\kappa_m(\mathfrak{G}), \kappa_n(\mathfrak{G})] \leq \kappa_{m+n}(\mathfrak{G})$.

b) $\kappa_n(\mathfrak{G})^p \leq \kappa_{pn}(\mathfrak{G})$.

c) *For* $n > 1$, $\kappa_n(\mathfrak{G}) = [\kappa_{n-1}(\mathfrak{G}), \mathfrak{G}]\kappa_m(\mathfrak{G})^p$, *where* m *is the least integer for which* $pm \geq n$.

Proof. a) follows at once from 1.10 and 1.12.

b) By a), $\gamma_p(\kappa_n(\mathfrak{G})) \leq \kappa_{pn}(\mathfrak{G})$, so $\kappa_n(\mathfrak{G})/\kappa_{pn}(\mathfrak{G})$ is regular (III, 10.2). But $\kappa_n(\mathfrak{G})/\kappa_{pn}(\mathfrak{G})$ is generated by elements of order p since if $ip^k \geq n$ and $x \in \gamma_i(\mathfrak{G})$, $(x^{p^k})^p \in \kappa_{ip^{k+1}}(\mathfrak{G}) \leq \kappa_{pn}(\mathfrak{G})$. Hence $(\kappa_n(\mathfrak{G})/\kappa_{pn}(\mathfrak{G}))^p = 1$ by III, 10.5, and $\kappa_n(\mathfrak{G})^p \leq \kappa_{pn}(\mathfrak{G})$.

c) By a) and b), $[\kappa_{n-1}(\mathfrak{G}), \mathfrak{G}]\kappa_m(\mathfrak{G})^p \leq \kappa_n(\mathfrak{G})$. Suppose that $ip^k \geq n$. If $k = 0$, $\gamma_i(\mathfrak{G})^{p^k} = \gamma_i(\mathfrak{G}) \leq \gamma_n(\mathfrak{G}) = [\gamma_{n-1}(\mathfrak{G}), \mathfrak{G}] \leq [\kappa_{n-1}(\mathfrak{G}), \mathfrak{G}]$. If $k > 0$, then $ip^{k-1} \geq m$ by definition of m, and $\gamma_i(\mathfrak{G})^{p^k} \leq (\gamma_i(\mathfrak{G})^{p^{k-1}})^p \leq \kappa_{ip^{k-1}}(\mathfrak{G})^p \leq \kappa_m(\mathfrak{G})^p$. Thus $\gamma_i(\mathfrak{G})^{p^k} \leq [\kappa_{n-1}(\mathfrak{G}), \mathfrak{G}]\kappa_m(\mathfrak{G})^p$ in any case, and the assertion follows from 1.10. **q.e.d.**

1.14 Corollary. *Suppose that*

$$\mathfrak{G} = \mathfrak{K}_1 \geq \mathfrak{K}_2 \geq \mathfrak{K}_3 \geq \cdots,$$

where $\mathfrak{K}_n \trianglelefteq \mathfrak{G}$, $[\mathfrak{K}_n, \mathfrak{G}] \leq \mathfrak{K}_{n+1}$ *and* $\mathfrak{K}_n^p \leq \mathfrak{K}_{np}$ *for all* $n \geq 1$.
Then $\mathfrak{K}_n \geq \kappa_n(\mathfrak{G})$ *for all* $n \geq 1$.

Proof. This follows from 1.13c) by induction on n (cf. 1.6). **q.e.d.**

1.15 Example. Let $\mathfrak{G}$ be a p-group of maximal class of order p^m, where $m \geq p + 1$. For $n > 1$, define the integers l, a as follows:

$$p^l \leq n - 1 < p^{l+1}, \quad 0 \leq n - 1 - ap^l < p^l, \quad 0 < a < p.$$

Then $\kappa_n(\mathfrak{G}) = \gamma_{a+1+l(p-1)}(\mathfrak{G})$. If $m - 1 = (p - 1)q + r$ and $2 \leq r \leq p$, $\kappa_{rp^q}(\mathfrak{G}) > \kappa_{1+rp^q}(\mathfrak{G}) = 1$.

For such a group $\gamma_i(\mathfrak{G})^{p^k} = \gamma_{i+k(p-1)}(\mathfrak{G})$ (III, 14.16).
Hence $\gamma_{a+1+l(p-1)}(\mathfrak{G}) = \gamma_{a+1}(\mathfrak{G})^{p^l} \leq \kappa_{(a+1)p^l}(\mathfrak{G}) \leq \kappa_n(\mathfrak{G})$, as $(a + 1)p^l \geq n$.
Conversely, if $ip^k \geq n$ and $k \leq l$, then $i \geq 1 + ap^{l-k}$, so that

$$\begin{aligned} i + k(p - 1) &\geq 1 + k(p - 1) + ap^{l-k} \geq 1 + k(p - 1) + a + p^{l-k} - 1 \\ &\geq 1 + k(p - 1) + a + (l - k)(p - 1) = 1 + l(p - 1) + a. \end{aligned}$$

Also if $k > l$, $1 + k(p - 1) \geq a + 1 + l(p - 1)$. Thus
$\kappa_n(\mathfrak{G}) \leq \gamma_{a+1+l(p-1)}(\mathfrak{G})$.

Of course $\lambda_n(\mathfrak{G}) = \gamma_n(\mathfrak{G})$, by 1.6.

Exercise

1) Prove that for any group $\mathfrak{G}$,

$$[\gamma_i(\mathfrak{G})^{p^k}\gamma_{i+1}(\mathfrak{G})^{p^{k-1}} \cdots \gamma_{i+k}(\mathfrak{G}), \mathfrak{G}] = \gamma_{i+1}(\mathfrak{G})^{p^k}\gamma_{i+2}(\mathfrak{G})^{p^{k-1}} \cdots \gamma_{i+k+1}(\mathfrak{G}).$$

Deduce that

a) $[\lambda_n(\mathfrak{G}), \mathfrak{G}] = \gamma_2(\mathfrak{G})^{p^{n-1}} \cdots \gamma_{n+1}(\mathfrak{G})$; and

b) for $n > 1$, $\lambda_n(\mathfrak{G}) = [\lambda_{n-1}(\mathfrak{G}), \mathfrak{G}]\mathfrak{G}^{p^{n-1}}$.

§ 2. Jennings' Theorem

There is a relation between the central series $\kappa_n(\mathfrak{G})$ introduced in § 1 and the group-ring over a field of characteristic p. This was first proved by JENNINGS [1]. Jennings proved his theorem only for finite p-groups; however it is true for all groups. For the rather easy extension to the general case we need the following lemma.

2.1 Lemma. *Suppose that $\mathfrak{G}$ is a nilpotent group, $\mathfrak{G} = \langle x_1, \ldots, x_n\rangle$ and each x_i is of finite order. Then $\mathfrak{G}$ is finite.*

Proof. We use induction on the class c of $\mathfrak{G}$. By III, 1.11, $\gamma_c(\mathfrak{G}) = \langle X\rangle$, where $X = \{[x_{i_1}, \ldots, x_{i_c}] \mid 1 \leq i_j \leq n\}$. X is finite and by III, 6.8, each element of X is of finite order. Hence the Abelian group $\gamma_c(\mathfrak{G})$ is finite. This is the assertion if $c = 1$. If $c > 1$, $\mathfrak{G}/\gamma_c(\mathfrak{G})$ is finite by the inductive hypothesis, so $\mathfrak{G}$ is finite. **q.e.d.**

2.2 Definitions. If A is any commutative ring with identity and $\mathfrak{G}$ is a group, the *group-ring* (I, 16.6) of $\mathfrak{G}$ over A will be denoted by $\mathsf{A}\mathfrak{G}$. If $\mathsf{J}_1, \mathsf{J}_2$ are A-submodules of $\mathsf{A}\mathfrak{G}$, $\mathsf{J}_1\mathsf{J}_2$ denotes the A-module spanned by all products ab, with $a \in \mathsf{J}_1$, $b \in \mathsf{J}_2$. Thus $(\mathsf{J}_1\mathsf{J}_2)\mathsf{J}_3 = \mathsf{J}_1(\mathsf{J}_2\mathsf{J}_3)$, and we can form the powers J_1^n of J_1. If $\mathfrak{J}_1$, $\mathfrak{J}_2$ are 2-sided ideals of the ring $\mathsf{A}\mathfrak{G}$, so are $\mathfrak{J}_1\mathfrak{J}_2$ and $\mathfrak{J}_1^n$. The *augmentation ideal* $\mathfrak{J}$ of $\mathsf{A}\mathfrak{G}$ (III, 18.3) is defined by

$$\mathfrak{J} = \left\{ \sum_{g \in \mathfrak{G}} a_g g \,\middle|\, \sum_{g \in \mathfrak{G}} a_g = 0 \right\}.$$

$\mathfrak{J}$ has A-basis $\{g - 1 \mid g \in \mathfrak{G}, g \neq 1\}$.

2.3 Lemma. *Let $\mathfrak{J}_1$, $\mathfrak{J}_2$ be two-sided ideals of $\mathsf{A}\mathfrak{G}$ and let*

$$\mathfrak{K}_i = \{x \mid x \in \mathfrak{G}, x - 1 \in \mathfrak{J}_i\} \quad (i = 1, 2).$$

a) $\mathfrak{K}_i \trianglelefteq \mathfrak{G}$.

b) *If* $\mathfrak{K} = \{x \mid x \in \mathfrak{G}, x - 1 \in \mathfrak{J}_1\mathfrak{J}_2 + \mathfrak{J}_2\mathfrak{J}_1\}$, *then* $[\mathfrak{K}_1, \mathfrak{K}_2] \leq \mathfrak{K}$.

c) *If* $x \in \mathfrak{K}_1$, $y \in \mathfrak{K}_2$, *then*

$$[x, y] - 1 - \{(x - 1)(y - 1) - (y - 1)(x - 1)\} \in \mathfrak{J}(\mathfrak{J}_1\mathfrak{J}_2 + \mathfrak{J}_2\mathfrak{J}_1),$$

where $\mathfrak{J}$ *is the augmentation ideal of* $\mathsf{A}\mathfrak{G}$.

Proof. If ν_i is the natural homomorphism of the ring $\mathsf{A}\mathfrak{G}$ onto $\mathsf{A}\mathfrak{G}/\mathfrak{J}_i$, the restriction μ_i of ν_i to $\mathfrak{G}$ is a group-homomorphism of $\mathfrak{G}$ into the group of units of $\mathsf{A}\mathfrak{G}/\mathfrak{J}_i$. Since $\mathfrak{K}_i = \ker \mu_i$, a) is clear. Suppose that $x \in \mathfrak{K}_1$ and $y \in \mathfrak{K}_2$. Then

$$[x, y] - 1 = x^{-1}y^{-1}xy - 1 = x^{-1}y^{-1}(xy - yx) = x^{-1}y^{-1}z,$$

where $z = (x - 1)(y - 1) - (y - 1)(x - 1)$. But $x - 1 \in \mathfrak{J}_1$ and $y - 1 \in \mathfrak{J}_2$, so $z \in \mathfrak{J}_1\mathfrak{J}_2 + \mathfrak{J}_2\mathfrak{J}_1$. Hence $[x, y] - 1 \in \mathfrak{J}_1\mathfrak{J}_2 + \mathfrak{J}_2\mathfrak{J}_1$ and $[x, y] \in \mathfrak{K}$. By a), $\mathfrak{K}$ is a subgroup, so b) is clear. Also

$$[x, y] - 1 - z = (x^{-1}y^{-1} - 1)z.$$

Since $x^{-1}y^{-1} - 1 \in \mathfrak{J}$, c) is proved. **q.e.d.**

2.4 Definition. For any group $\mathfrak{G}$ and any prime p, $\kappa_n(\mathfrak{G})/\kappa_{n+1}(\mathfrak{G})$ is an elementary Abelian p-group by 1.13. We may therefore think of $\kappa_n(\mathfrak{G})/\kappa_{n+1}(\mathfrak{G})$ as a vector space over $GF(p)$ and choose a $GF(p)$-basis $\overline{\mathsf{B}}_n$ of it. For each $\bar{b} \in \overline{\mathsf{B}}_n$, choose a fixed $b \in \kappa_n(\mathfrak{G})$ such that $\bar{b} = b\kappa_{n+1}(\mathfrak{G})$, and denote by B_n the set of all such elements b. Let $\mathsf{B} = \bigcup_{n \geq 1} \mathsf{B}_n$. Such a set B will be called a *κ-net* on $\mathfrak{G}$. Clearly $\mathsf{B}_n = (\mathsf{B} \cap \kappa_n(\mathfrak{G})) - (\mathsf{B} \cap \kappa_{n+1}(\mathfrak{G}))$.

By an *ordered κ-net* is meant a κ-net B fully ordered in such a way that (i) $a < b$ if $a \in \mathsf{B}_m$, $b \in \mathsf{B}_n$, $m < n$; and (ii) every non-empty subset of B_n has a greatest element.

By the well-ordering principle, every group $\mathfrak{G}$ possesses an ordered κ-net.

2.5 Lemma. *Let* B *be a κ-net on* $\mathfrak{G}$ *and let n be a positive integer. For each* $d = 1, \ldots, n$, *let* $\hat{\mathsf{B}}_d$ *be a finite subset of* $(\mathsf{B} \cap \kappa_d(\mathfrak{G})) - (\mathsf{B} \cap \kappa_{d+1}(\mathfrak{G}))$. *Let* F *be a finite subset of* $\mathfrak{G}$. *Then there exists a subgroup* $\mathfrak{H}$ *of* $\mathfrak{G}$ *and a normal subgroup* $\mathfrak{K}$ *of* $\mathfrak{H}$ *with the following properties.*

a) $\mathfrak{G}_1 = \mathfrak{H}/\mathfrak{K}$ *is a finite p-group and* $\kappa_{n+1}(\mathfrak{G}_1) = 1$.

b) $\hat{\mathsf{B}}_d \subseteq \kappa_d(\mathfrak{H})$ *and* $\mathsf{F} \subseteq \mathfrak{H}$.

c) *There exists a* κ-*net* C *on* $\mathfrak{G}_1$ *such that for* $d = 1, \ldots, n$, $(\mathsf{C} \cap \kappa_d(\mathfrak{G}_1)) - (\mathsf{C} \cap \kappa_{d+1}(\mathfrak{G}_1)) \supseteq \{b\mathfrak{K} | b \in \hat{\mathsf{B}}_d\}$.

Proof. If $x \in \kappa_d(\mathfrak{G})$, there is a finitely generated subgroup $\mathfrak{G}^*$ of $\mathfrak{G}$ such that $x \in \kappa_d(\mathfrak{G}^*)$; this is clear from 1.13c). Hence there exists a finitely generated subgroup $\mathfrak{H}$ of $\mathfrak{G}$ containing F such that $b \in \kappa_d(\mathfrak{H})$ for all $b \in \hat{\mathsf{B}}_d$ $(d = 1, \ldots, n)$, since F and the $\hat{\mathsf{B}}_d$ are finite. Let $\mathfrak{K} = \kappa_{n+1}(\mathfrak{H})$. By 1.10, $\mathfrak{K} \geq \mathfrak{H}^{p^{n+1}}$, so every element of $\mathfrak{G}_1 = \mathfrak{H}/\mathfrak{K}$ is of finite order. Also $\mathfrak{K} \geq \gamma_{n+1}(\mathfrak{H})$, so $\mathfrak{G}_1$ is finitely generated and nilpotent. By 2.1, $\mathfrak{G}_1$ is finite. Since $\mathfrak{G}_1^{p^{n+1}} = 1$, $\mathfrak{G}_1$ is a finite p-group. Thus a) and b) are clear.

Now $\{b\kappa_{d+1}(\mathfrak{G}) | b \in \hat{\mathsf{B}}_d\}$ is a linearly independent set of elements of $\kappa_d(\mathfrak{G})/\kappa_{d+1}(\mathfrak{G})$. Hence $\{b\kappa_{d+1}(\mathfrak{H}) | b \in \hat{\mathsf{B}}_d\}$ is a linearly independent set of elements of $\kappa_d(\mathfrak{H})/\kappa_{d+1}(\mathfrak{H})$. Hence $\{b\kappa_{d+1}(\mathfrak{H}) | b \in \hat{\mathsf{B}}_d\}$ may be embedded in a $GF(p)$-basis $\overline{\mathsf{C}}_d$ of $\kappa_d(\mathfrak{H})/\kappa_{d+1}(\mathfrak{H})$. Clearly a κ-net C on $\mathfrak{G}_1$ may be formed from the $\overline{\mathsf{C}}_d$ so that c) holds. **q.e.d.**

2.6 Theorem (JENNINGS [1]). *Let* $\mathfrak{G}$ *be a group and let* p *be a prime. Let* B *be an ordered* κ-*net on* $\mathfrak{G}$, *and* $\mathsf{B}_n = (\mathsf{B} \cap \kappa_n(\mathfrak{G})) - (\mathsf{B} \cap \kappa_{n+1}(\mathfrak{G}))$ $(n \geq 1)$. *Let* K *be a field of characteristic* p *and let* $\mathfrak{J}$ *be the augmentation ideal of* $\mathsf{K}\mathfrak{G}$. *If* $x \in \kappa_n(\mathfrak{G})$, *then* $x - 1 \in \mathfrak{J}^n$. *For each* $n \geq 1$, *define*

$$\mathsf{B}_n^* = \left\{(b_1 - 1)^{e_1} \cdots (b_r - 1)^{e_r} \middle| \; b_i \in \mathsf{B}_{d_i}, \right.$$

$$\left. 1 \leq e_i < p, b_1 < \cdots < b_r, \sum_{i=1}^{r} e_i d_i = n \right\}.$$

Then $\{b^* + \mathfrak{J}^{n+1} | b^* \in \mathsf{B}_n^*\}$ *is a* K-*basis of the vector space* $\mathfrak{J}^n/\mathfrak{J}^{n+1}$.

Proof. The proof will be carried out in several steps.

a) If $x \in \kappa_n(\mathfrak{G})$, then $x - 1 \in \mathfrak{J}^n$. Also $\mathsf{B}_n^* \subseteq \mathfrak{J}^n$.

Let

$$\mathfrak{G}_n = \{x | x \in \mathfrak{G}, x - 1 \in \mathfrak{J}^n\} \quad (n \geq 1).$$

Thus $\mathfrak{G}_1 = \mathfrak{G}$. By 2.3, $\mathfrak{G}_n \trianglelefteq \mathfrak{G}$ and $[\mathfrak{G}_n, \mathfrak{G}] = [\mathfrak{G}_n, \mathfrak{G}_1] \leq \mathfrak{G}_{n+1}$. Thus

$$\mathfrak{G} = \mathfrak{G}_1 \geq \mathfrak{G}_2 \geq \cdots$$

is a central series of $\mathfrak{G}$. If $x \in \mathfrak{G}_n$, $x^p - 1 = (x - 1)^p \in \mathfrak{J}^{pn}$, since K is of characteristic p, whence $x^p \in \mathfrak{G}_{pn}$. Thus $\mathfrak{G}_n^p \leq \mathfrak{G}_{pn}$. By 1.14, $\mathfrak{G}_n \geq \kappa_n(\mathfrak{G})$

for all $n \geq 1$. Thus if $x \in \kappa_n(\mathfrak{G})$, $x - 1 \in \mathfrak{J}^n$. In particular $b - 1 \in \mathfrak{J}^n$ if $b \in \mathsf{B}_n$. It follows that $\mathsf{B}_n^* \subseteq \mathfrak{J}^n$.

b) If $b_1, \ldots, b_m$ are elements of B_n and $x \in \kappa_{n+1}(\mathfrak{G})$,

$$b_1 \cdots b_m x - 1 \equiv \sum_{i=1}^{m} (b_i - 1) \mod \mathfrak{J}^{n+1}.$$

To see this, expand

$$b_1 \cdots b_m x = \{1 + (b_1 - 1)\} \cdots \{1 + (b_m - 1)\}\{1 + (x - 1)\}.$$

By a), $b_i - 1 \in \mathfrak{J}^n$, so any product $(b_i - 1)(b_j - 1)$ lies in $\mathfrak{J}^{n+1}$. Also $x - 1 \in \mathfrak{J}^{n+1}$ by a); hence the assertion.

c) $\mathfrak{J}$ is spanned over K by $\mathfrak{J}^2$ and all $b - 1$ $(b \in \mathsf{B}_1)$.

The augmentation ideal $\mathfrak{J}$ is spanned over K by all $x - 1$ with $x \in \mathfrak{G}$. By definition of B_1, $x = b_1 \cdots b_m y$, where $b_i \in \mathsf{B}_1$, $y \in \kappa_2(\mathfrak{G})$. By b), $x - 1$ lies in the space spanned by $\mathfrak{J}^2$ and all $b - 1$ $(b \in \mathsf{B}_1)$.

d) $\mathfrak{J}^n$ is spanned over K by B_n^* and $\mathfrak{J}^{n+1}$.

Suppose that this is false. Let V_n be the vector space spanned over K by B_n^* and $\mathfrak{J}^{n+1}$. By a), $\mathsf{V}_n \subseteq \mathfrak{J}^n$. Also $\mathsf{V}_1 = \mathfrak{J}$ by c), for $\mathsf{B}_1^* = \{b - 1 \mid b \in \mathsf{B}_1\}$. Let n be the smallest integer for which $\mathsf{V}_n \neq \mathfrak{J}^n$. Thus $n > 1$. It follows from c) that $\mathfrak{J}^n$ is spanned over K by $\mathfrak{J}^{n+1}$ and all $(b_1 - 1) \cdots (b_n - 1)$ with $b_i \in \mathsf{B}_1$. Since $\mathsf{V}_n \neq \mathfrak{J}^n$, some such element does not lie in V_n. Let X be the set of finite sequences $(b_1, \ldots, b_m)$ for which $b_i \in \mathsf{B}_{d_i}$, $d_1 + \cdots + d_m = n$ and $(b_1 - 1) \cdots (b_m - 1) \notin \mathsf{V}_n$. Thus X is non-empty.

Choose $(b_1, \ldots, b_m) \in \mathsf{X}$ with $b_i \in \mathsf{B}_{d_i}$, $d_1 + \cdots + d_m = n$ and b_m as large as possible in the ordering of B; this is possible since $d_m \leq n$ and every non-empty subset of B_d has a greatest element. If $d_m = n$, we have $b_m - 1 \in \mathsf{B}_n^* \subseteq \mathsf{V}_n$, a contradiction. Thus $d_m < n$. Write $d = d_m$, $b = b_m$. Now $(b_1 - 1) \cdots (b_{m-1} - 1) \in \mathfrak{J}^{n-d}$ by a). By choice of n, $\mathsf{V}_{n-d} = \mathfrak{J}^{n-d}$; it follows that $(b_1 - 1) \cdots (b_m - 1)$ lies in the vector space spanned over K by $\mathfrak{J}^{n+1}$ and $\{u(b - 1) \mid u \in \mathsf{B}_{n-d}^*\}$. Hence there exists $u \in \mathsf{B}_{n-d}^*$ such that $u(b - 1) \notin \mathsf{V}_n$. Write

$$u = (a_1 - 1)^{e_1} \cdots (a_r - 1)^{e_r},$$

where $a_i \in \mathsf{B}_{c_i}$, $1 \leq e_i < p$, $a_1 < \cdots < a_r$ and $e_1 c_1 + \cdots + e_r c_r = n - d$. Since $u(b - 1) \notin \mathsf{V}_n$, we have $u(b - 1) \notin \mathsf{B}_n^*$. Hence either $a_r > b$, or $a_r = b$ and $e_r = p - 1$.

First suppose that $a_r > b$. Since $a_r - 1 \in \mathfrak{J}^{c_r}$ and $b - 1 \in \mathfrak{J}^d$,

$$[a_r, b] - 1 \equiv (a_r - 1)(b - 1) - (b - 1)(a_r - 1) \mod \mathfrak{J}^{c+1},$$

where $c = c_r + d$, by 2.3c). By 1.13a), $[a_r, b] \in \kappa_c(\mathfrak{G})$. Hence we may write

$$[a_r, b] = a'_1 \cdots a'_s a',$$

where $a'_j \in \mathsf{B}_c$ $(j = 1, \ldots, s)$ and $a' \in \kappa_{c+1}(\mathfrak{G})$. By b),

$$[a_r, b] - 1 \equiv \sum_{j=1}^{s} (a'_j - 1) \quad \operatorname{mod} \mathfrak{J}^{c+1}.$$

Hence

$$(a_r - 1)(b - 1) \equiv (b - 1)(a_r - 1) + \sum_{j=1}^{s} (a'_j - 1) \quad \operatorname{mod} \mathfrak{J}^{c+1}.$$

Multiplying on the left by $u' = (a_1 - 1)^{e_1} \cdots (a_r - 1)^{e_r - 1}$,

$$u(b - 1) \equiv u'(b - 1)(a_r - 1) + \sum_{j=1}^{s} u'(a'_j - 1) \quad \operatorname{mod} \mathfrak{J}^{n+1},$$

since $u' \in \mathfrak{J}^{n-c}$. But now $(\underbrace{a_1, \ldots, a_1}_{e_1}, \ldots, \underbrace{a_r, \ldots, a_r}_{e_r - 1}, b, a_r)$ and $(\underbrace{a_1, \ldots, a_1}_{e_1}, \ldots, \underbrace{a_r, \ldots, a_r}_{e_r - 1}, a'_j)$ cannot lie in X on account of the definition of b, for $a_r > b$ and $a'_j > b$. Hence $u'(b - 1)(a_r - 1) \in \mathsf{V}_n$ and $u'(a'_j - 1) \in \mathsf{V}_n$. Since $\mathfrak{J}^{n+1} \subseteq \mathsf{V}_n$, it follows that $u(b - 1) \in \mathsf{V}_n$, a contradiction.

Thus $a_r = b$ and $e_r = p - 1$. Hence $u(b - 1) = v(b - 1)^p$, where $v = (a_1 - 1)^{e_1} \cdots (a_{r-1} - 1)^{e_{r-1}} \in \mathfrak{J}^{n-pd}$ (since $c_r = d$). Since K is of characteristic p, $(b - 1)^p = b^p - 1$. By 1.13b), $\kappa_d(\mathfrak{G})^p \leq \kappa_{pd}(\mathfrak{G})$, so $b^p \in \kappa_{pd}(\mathfrak{G})$. Hence we may write

$$b^p = b'_1 \cdots b'_t b'$$

with $b'_j \in \mathsf{B}_{pd}$ $(j = 1, \ldots, t)$ and $b' \in \kappa_{pd+1}(\mathfrak{G})$. By b),

$$b^p - 1 \equiv \sum_{j=1}^{t} (b'_j - 1) \quad \operatorname{mod} \mathfrak{J}^{pd+1}.$$

Since $v \in \mathfrak{J}^{n-pd}$,

$$u(b - 1) = v(b^p - 1) \equiv \sum_{j=1}^{t} v(b'_j - 1) \quad \operatorname{mod} \mathfrak{J}^{n+1}.$$

Since $b'_j > b$, $(a_1, \underset{e_1}{\ldots}, a_1, \ldots, a_{r-1}, \underset{e_{r-1}}{\ldots}, a_{r-1}, b'_j) \notin \mathsf{X}$. Thus $v(b'_j - 1) \in \mathsf{V}_n$. Hence $u(b - 1) \in \mathsf{V}_n$, a contradiction. Thus d) is proved.

e) $\{b^* + \mathfrak{J}^{n+1} | b^* \in \mathsf{B}_n^*\}$ is linearly independent over K.

Suppose that this is false. Then there is a non-trivial relation $\sum \lambda_k b_k^* = u \in \mathfrak{J}^{n+1}$. Write each $b_k^* \in \mathsf{B}_n^*$ in this relation as $(b_1 - 1)^{e_1} \cdots (b_r - 1)^{e_r}$ with $b_i \in \mathsf{B}$, $b_1 < \cdots < b_r$, $1 \leq e_i < p$; write the element u of $\mathfrak{J}^{n+1}$ as a K-linear combination of elements $(x_1 - 1) \cdots (x_{n+1} - 1)$ with $x_i \in \mathfrak{G}$. The set $\hat{\mathsf{B}}_d$ of elements of B_d occurring in this relation is finite $(d = 1, \ldots, n)$; also the set F of elements x_i of $\mathfrak{G}$ occurring in it is finite. By 2.5, there exists a subgroup $\mathfrak{H}$ and a normal subgroup $\mathfrak{K}$ of $\mathfrak{H}$ such that $\mathfrak{G}_1 = \mathfrak{H}/\mathfrak{K}$ is a finite p-group; further there exists a κ-net C on $\mathfrak{G}_1$ such that $\mathsf{C}_d = (\mathsf{C} \cap \kappa_d(\mathfrak{G}_1)) - (\mathsf{C} \cap \kappa_{d+1}(\mathfrak{G}_1)) \supseteq \{b\mathfrak{K} | b \in \hat{\mathsf{B}}_d\}$. We may extend the ordering on $\hat{\mathsf{B}}_d$ to an ordering of C_d and thus make C into an ordered κ-net. Let $\mathsf{C} = \{a_1, \ldots, a_m\}$, $a_1 < a_2 < \cdots < a_m$. Since $|\kappa_d(\mathfrak{G}_1) : \kappa_{d+1}(\mathfrak{G}_1)| = p^{|\mathsf{C}_d|}$, $|\mathfrak{G}_1| = p^m$.

We wish to apply d) to the ordered κ-net C on $\mathfrak{G}_1$. Thus for each $d \geq 1$, let

$$\mathsf{C}_d^* = \{(a_1 - 1)^{e_1} \cdots (a_m - 1)^{e_m} | 0 \leq e_i < p, \quad \textstyle\sum e_i c_i = d\},$$

where $a_i \in \mathsf{C}_{c_i}$. If $\mathfrak{J}$ is the augmentation ideal of $\mathsf{K}\mathfrak{G}_1$, $\mathfrak{J}^d$ is spanned by C_d^* and $\mathfrak{J}^{d+1}$; thus

$$|\mathsf{C}_d^*| \geq \dim_{\mathsf{K}}(\mathfrak{J}^d/\mathfrak{J}^{d+1}) \quad (d \geq 1).$$

But the above relation implies a non-trivial relation between the elements of C_n^* modulo $\mathfrak{J}^{n+1}$, so

$$|\mathsf{C}_n^*| > \dim_{\mathsf{K}}(\mathfrak{J}^n/\mathfrak{J}^{n+1}).$$

Hence if $\mathsf{C}^* = \bigcup_{d \geq 1} \mathsf{C}_d^*$,

$$|\mathsf{C}^*| > \sum_{d \geq 1} \dim_{\mathsf{K}}(\mathfrak{J}^d/\mathfrak{J}^{d+1}).$$

But

$$\mathsf{C}^* = \{(a_1 - 1)^{e_1} \cdots (a_m - 1)^{e_m} | 0 \leq e_i < p, \quad \text{not all } e_i = 0\};$$

thus $|\mathsf{C}^*| \leq p^m - 1$. Now by V, 5.16, $\mathfrak{J}$ is the radical of $\mathsf{K}\mathfrak{G}_1$. By V, 2.4, $\mathfrak{J}$ is nilpotent. Hence

$$p^m - 1 \geq |\mathsf{C}^*| > \sum_{d \geq 1} \dim_{\mathsf{K}}(\mathfrak{g}^d/\mathfrak{g}^{d+1}) = \dim_{\mathsf{K}} \mathfrak{g}.$$

But $\dim_{\mathsf{K}} \mathfrak{g} = |\mathfrak{G}_1| - 1 = p^m - 1$, a contradiction.

With d) and e), 2.6 is proved. **q.e.d.**

2.7 Theorem (JENNINGS [1]). *Suppose that $\mathfrak{G}$ is a group, K is a field of characteristic p, and $\mathfrak{J}$ is the augmentation ideal of $\mathsf{K}\mathfrak{G}$. Suppose that $x \in \mathfrak{G}$. Then $x \in \kappa_n(\mathfrak{G})$ if and only if $x - 1 \in \mathfrak{J}^n$.*

Proof. If $x \in \kappa_n(\mathfrak{G})$, then $x - 1 \in \mathfrak{J}^n$ by 2.6. If $x \notin \kappa_n(\mathfrak{G})$, there is a least integer m such that $x \notin \kappa_m(\mathfrak{G})$ and $1 < m \leq n$. Thus $x \in \kappa_{m-1}(\mathfrak{G})$. There exists a κ-net B on $\mathfrak{G}$ such that $x \in \mathsf{B}$. By 2.6, there exists a K-basis of $\mathfrak{J}^{m-1}/\mathfrak{J}^m$ containing $x - 1$. Thus $x - 1 \notin \mathfrak{J}^m$. Since $n \geq m$, $x - 1 \notin \mathfrak{J}^n$. **q.e.d.**

2.8 Corollary. *Let $\mathfrak{G}$ be a finite p-group, K a field of characteristic p and $\mathfrak{J}$ the augmentation ideal of $\mathsf{K}\mathfrak{G}$. Suppose that $\kappa_l(\mathfrak{G}) > \kappa_{l+1}(\mathfrak{G}) = 1$,*

$$|\kappa_n(\mathfrak{G}) : \kappa_{n+1}(\mathfrak{G})| = p^{d_n} \quad (n = 1, \ldots, l), \quad \textit{and}$$

$$s = (p - 1) \sum_{n=1}^{l} n d_n.$$

Then $\mathfrak{J}^{s+1} = 0$. $\mathfrak{J}^s$ is of dimension 1 over K and is spanned by $\sum_{x \in \mathfrak{G}} x$.

Proof. Let B be an ordered κ-net on $\mathfrak{G}$. By 2.6, $\mathfrak{J}^{s+1} = 0$ and $\mathfrak{J}^s$ is spanned by $\prod_{b \in \mathsf{B}} (b - 1)^{p-1}$, where the factors are taken in increasing order of $b \in \mathsf{B}$. But

$$\prod_{b \in \mathsf{B}} (b - 1)^{p-1} = \prod_{b \in \mathsf{B}} (1 + b + \cdots + b^{p-1}) = \sum_{x \in \mathfrak{G}} x. \qquad \textbf{q.e.d.}$$

2.9 Examples. a) Suppose that $\mathfrak{G}$ is an Abelian p-group of exponent p^m and that $|\mathfrak{G}^{p^{i-1}} : \mathfrak{G}^{p^i}| = p^{\omega_i}$ $(i = 1, \ldots, m)$. By 1.10, $\kappa_{n+1}(\mathfrak{G}) = \kappa_n(\mathfrak{G})$ if n is not a power of p, and $\kappa_{p^i}(\mathfrak{G}) = \mathfrak{G}^{p^i}$, $\kappa_{p^i+1}(\mathfrak{G}) = \mathfrak{G}^{p^{i+1}}$. Thus $d_{p^{i-1}} = \omega_i$ $(i = 1, \ldots, m)$, and

$$s = (p - 1) \sum_{i=1}^{m} p^{i-1} \omega_i.$$

b) Let $\mathfrak{G}$ be a p-group of maximal class of order p^m, where $m \geq p + 1$, $m = (p - 1)q + r + 1$ and $1 \leq r \leq p - 1$. It follows from 1.15 that

$$|\kappa_n(\mathfrak{G}):\kappa_{n+1}(\mathfrak{G})| = \begin{cases} p^2 & \text{if } n = 1; \\ p & \text{if } n = ap^l \text{ with } 1 \le a \le p-1, 0 \le l \le q, \\ & (a, l) \ne (1, 0) \text{ and } a \le r \text{ if } l = q; \\ 1 & \text{otherwise.} \end{cases}$$

The integer s of 2.8 is given by

$$s = (p-1) + \tfrac{1}{2}p(p-1)(p^q - 1) + \tfrac{1}{2}(p-1)p^q r(r+1).$$

For $p = 2$, this reduces to $s = 2^{m-1}$.

2.10 Theorem *(Jennings' formula). Suppose that $\mathfrak{G}$ is a finite p-group, K is a field of characteristic p and $\mathfrak{J}$ is the augmentation ideal of $\mathsf{K}\mathfrak{G}$. Suppose further that $\kappa_l(\mathfrak{G}) > \kappa_{l+1}(\mathfrak{G}) = 1$ and that $|\kappa_n(\mathfrak{G}):\kappa_{n+1}(\mathfrak{G})| = p^{d_n}$ $(n = 1, \ldots, l)$. Let $s = (p-1)\sum_{n=1}^{l} nd_n$, and define the integers c_n by*

$$\prod_{i=1}^{l} (1 + t^i + t^{2i} + \cdots + t^{(p-1)i})^{d_i} = \sum_{n=0}^{s} c_n t^n.$$

(In particular $c_1 = d_1$). Then $c_n = \dim_{\mathsf{K}} \mathfrak{J}^n/\mathfrak{J}^{n+1}$ $(1 \le n \le s)$.

Proof. Let B be a κ-net on $\mathfrak{G}$, and let $\mathsf{B}_n = (\mathsf{B} \cap \kappa_n(\mathfrak{G})) - (\mathsf{B} \cap \kappa_{n+1}(\mathfrak{G}))$. Thus $|\mathsf{B}_n| = d_n$. Write $\mathsf{B}_n = \{b_{n1}, \ldots, b_{nd_n}\}$. By 2.6, $\dim_{\mathsf{K}} \mathfrak{J}^n/\mathfrak{J}^{n+1}$ is the number of lexicographically ordered products

$$\prod_{i,j} (b_{ij} - 1)^{e_{ij}}$$

with $0 \le e_{ij} < p$ and $\sum_{i,j} ie_{ij} = n$. This number is obviously c_n. **q.e.d.**

2.11 Remarks. a) If $\mathsf{K} = GF(p)$, $\mathfrak{J}/\mathfrak{J}^2$ is isomorphic to $\mathfrak{G}/\kappa_2(\mathfrak{G})$ since $c_1 = d_1$. More explicitly, there is an isomorphism between them in which $x - 1 + \mathfrak{J}^2$ corresponds to $x\kappa_2(\mathfrak{G})$ for all $x \in \mathfrak{G}$.

b) If $\mathfrak{G}$ is a 2-group of maximal class of order 2^m, 2.10 yields $c_n = 2$ for $1 \le n < 2^{m-1}$ and $c_{2^{m-1}} = 1$.

A remarkable symmetry of the c_n follows from 2.10.

2.12 Theorem. *Suppose that $\mathfrak{G}$ is a finite p-group, K is a field of characteristic p and $\mathfrak{J}$ is the augmentation ideal of $\mathsf{K}\mathfrak{G}$. If $\mathfrak{J}^s \supset \mathfrak{J}^{s+1} = 0$,*

$$\dim_{\mathsf{K}} \mathfrak{J}^n/\mathfrak{J}^{n+1} = \dim_{\mathsf{K}} \mathfrak{J}^{s-n}/\mathfrak{J}^{s-n+1} \quad (n = 0, \ldots, s),$$

and

$$\dim_{\mathsf{K}} \mathfrak{J}^n + \dim_{\mathsf{K}} \mathfrak{J}^{s-n+1} = |\mathfrak{G}| \quad (n = 1, \dots, s).$$

Proof. In the notation of 2.10, $s = (p-1)\sum_{i=1}^{l} id_i$ and

$$\begin{aligned}
\sum_{n=0}^{s} c_n t^n &= \prod_{i=1}^{l} (1 + t^i + t^{2i} + \cdots + t^{(p-1)i})^{d_i} \\
&= \prod_{i=1}^{l} t^{(p-1)id_i}(t^{-(p-1)i} + t^{-(p-2)i} + \cdots + 1)^{d_i} \\
&= t^s \prod_{i=1}^{l} (1 + t^{-i} + \cdots + t^{-(p-1)i})^{d_i} \\
&= t^s \sum_{n=0}^{s} c_n t^{-n} \\
&= \sum_{n=0}^{s} c_{s-n} t^n.
\end{aligned}$$

Comparing coefficients, $c_n = c_{s-n}$ $(n = 0, \dots, s)$. Thus if $1 \leq n \leq s-1$,

$$\dim_{\mathsf{K}} \mathfrak{J}^n/\mathfrak{J}^{n+1} = c_n = c_{s-n} = \dim_{\mathsf{K}} \mathfrak{J}^{s-n}/\dim_{\mathsf{K}} \mathfrak{J}^{s-n+1}.$$

Also $c_s = c_0 = 1$ (cf. 2.8), so

$$\dim_{\mathsf{K}} \mathsf{K}\mathfrak{G}/\mathfrak{J} = 1 = c_s = \dim_{\mathsf{K}} \mathfrak{J}^s.$$

Hence if $1 \leq n \leq s$,

$$\begin{aligned}
\dim_{\mathsf{K}} \mathsf{K}\mathfrak{G}/\mathfrak{J}^n &= \dim_{\mathsf{K}} \mathsf{K}\mathfrak{G}/\mathfrak{J} + \sum_{m=1}^{n-1} \dim_{\mathsf{K}} \mathfrak{J}^m/\mathfrak{J}^{m+1} \\
&= \dim_{\mathsf{K}} \mathfrak{J}^s + \sum_{m=1}^{n-1} \dim_{\mathsf{K}} \mathfrak{J}^{s-m}/\mathfrak{J}^{s-m+1} \\
&= \dim_{\mathsf{K}} \mathfrak{J}^{s-n+1}.
\end{aligned}$$

Thus

$$\dim_{\mathsf{K}} \mathfrak{J}^n + \dim_{\mathsf{K}} \mathfrak{J}^{s-n+1} = \dim_{\mathsf{K}} \mathsf{K}\mathfrak{G} = |\mathfrak{G}|.$$ **q.e.d.**

The following theorem shows that the right and left annihilators of powers of $\mathfrak{J}$ are powers of $\mathfrak{J}$. In the terminology of *Loewy series* (see VII, 10.10), the upper and lower Loewy series have the same terms.

2.13 Theorem (E. T. HILL [1]). *Suppose that $\mathfrak{G}$ is a finite p-group, K is a field of characteristic p and $\mathfrak{J}$ is the augmentation ideal of $\mathsf{K}\mathfrak{G}$. Suppose $\mathfrak{J}^s \supset \mathfrak{J}^{s+1} = 0$. For any element a of $\mathsf{K}\mathfrak{G}$ and any integer n such that $1 \leq n \leq s$, the following statements are equivalent:*

a) $a \in \mathfrak{J}^{s+1-n}$.

b) *The coefficient of* $1_{\mathfrak{G}}$ *in ab is* 0 *for every* $b \in \mathfrak{J}^n$.

c) $ab = 0$ *for every* $b \in \mathfrak{J}^n$.

d) *The coefficient of* $1_{\mathfrak{G}}$ *in ba is* 0 *for every* $b \in \mathfrak{J}^n$.

e) $ba = 0$ *for every* $b \in \mathfrak{J}^n$.

Proof. For $1 \leq n \leq s$, we define

$$\mathsf{U} = \{a \mid a \in \mathsf{K}\mathfrak{G}, \quad ab = 0 \quad \text{for all } b \in \mathfrak{J}^n\},$$

$$\mathsf{V} = \{a \mid a \in \mathsf{K}\mathfrak{G}, \quad \text{the coefficient of } 1_{\mathfrak{G}} \text{ in } ab \text{ is } 0 \quad \text{for all } b \in \mathfrak{J}^n\}.$$

Clearly, $\mathfrak{J}^{s+1-n} \subseteq \mathsf{U} \subseteq \mathsf{V}$.

For any elements a, b of $\mathsf{K}\mathfrak{G}$, denote by (a, b) the coefficient of $1_{\mathfrak{G}}$ in ab. Then $(\ ,\ )$ is a K-bilinear form on $\mathsf{K}\mathfrak{G}$. We show that this form is non-singular. Suppose that $a \in \mathsf{K}\mathfrak{G}$ and $a \neq 0$. Then $a = \sum_{x \in \mathfrak{G}} \lambda_x x$, where $\lambda_x \in \mathsf{K}$ and $\lambda_y \neq 0$ for some $y \in \mathfrak{G}$, and $(a, y^{-1}) = (y^{-1}, a) = \lambda_y \neq 0$.

Now

$$\mathsf{V} = \{a \mid a \in \mathsf{K}\mathfrak{G}, \quad (a, b) = 0 \quad \text{for all } b \in \mathfrak{J}^n\}.$$

Since $(\ ,\ )$ is non-singular,

$$\dim_{\mathsf{K}} \mathsf{V} = \dim_{\mathsf{K}} \mathsf{K}\mathfrak{G} - \dim_{\mathsf{K}} \mathfrak{J}^n.$$

Using 2.12, it follows that $\dim_{\mathsf{K}} \mathsf{V} = \dim_{\mathsf{K}} \mathfrak{J}^{s-n+1}$. Since $\mathfrak{J}^{s+1-n} \subseteq \mathsf{U} \subseteq \mathsf{V}$, it follows that $\mathfrak{J}^{s+1-n} = \mathsf{U} = \mathsf{V}$. This establishes the equivalence of a), b), c). Similarly a), d), e) are equivalent. **q.e.d.**

It follows from 2.13 that, in the notation of VII, 10.7b),

$$\mathbf{R}(\mathfrak{J}^n) = \mathbf{L}(\mathfrak{J}^n) = \mathfrak{J}^{s-n+1};$$

thus $\mathfrak{J}^{s-n+1}$ is the right and left annihilator of $\mathfrak{J}^n$. The lower Loewy series (VII, 10.10) of the $\mathsf{K}\mathfrak{G}$-module $\mathsf{K}\mathfrak{G}$ is

$$\mathsf{K}\mathfrak{G} \supset \mathfrak{J} \supset \mathfrak{J}^2 \supset \cdots \supset \mathfrak{J}^s \supset \mathfrak{J}^{s+1} = 0.$$

It follows from VII, 10.11b) that the upper Loewy series is the same series in the reverse order.

We conclude this section by showing how 2.7 can be used to deduce a slight generalization of III, 9.7 from the following elementary ring-theoretical formula.

2.14 Lemma. *If a, b are elements of an associative algebra over a field of odd characteristic p,*

$$\sum_{i=0}^{p-1} (a + ib)^p + \sum_{j=0}^{p-1} b^j a b^{p-1-j} = 0.$$

Proof. We have

$$(a + b)^p = a^p + b^p + \sum_{j=1}^{p-1} s_j,$$

where s_j is the sum of all products $c_1 c_2 \cdots c_p$ in which j of the c_i are b and the remainder are a. Replacing b by ib ($i = 0, 1, \ldots$), we see that

$$(a + ib)^p = a^p + i^p b^p + \sum_{j=1}^{p-1} i^j s_j.$$

Summing over $i = 0, 1, \ldots, p - 1$ and using Fermat's theorem,

$$\sum_{i=0}^{p-1} (a + ib)^p = \left(\sum_{i=1}^{p-1} i\right) b^p + \sum_{j=1}^{p-2} \left(\sum_{i=1}^{p-1} i^j\right) s_j + (p - 1) s_{p-1}.$$

But for $1 \leq j \leq p - 2$, $\sum_{i=1}^{p-1} i^j \equiv 0\,(p)$, so since p is odd,

$$\sum_{i=0}^{p-1} (a + ib)^p = -s_{p-1}.$$

The assertion follows from this since by definition of s_{p-1},

$$s_{p-1} = \sum_{j=0}^{p-1} b^j a b^{p-1-j}.$$

q.e.d.

To interpret the second of the two sums in 2.14 group-theoretically, we need the following lemma.

2.15 Lemma. *Let* $\mathfrak{G}$ *be a group,* K *a field of characteristic* p *and* $\mathfrak{J}$ *the augmentation ideal of* $\mathsf{K}\mathfrak{G}$. *Let* x, y *be elements of* $\mathfrak{G}$ *and write* $a = x - 1$, $b = y - 1$. *Define the mapping* ξ *of* $\mathsf{K}\mathfrak{G}$ *into itself by*

$$c\xi = cb - bc \quad (c \in \mathsf{K}\mathfrak{G}).$$

a) *For* $n \geq 0$,

$$a\xi^n \equiv [x, y, \underset{n}{\ldots}, y] - 1 \mod \mathfrak{J}^{n+2}.$$

b) *For all* $c \in \mathsf{K}\mathfrak{G}$,

$$c\xi^{p-1} = \sum_{i=0}^{p-1} b^i c b^{p-1-i}.$$

Proof. a) This is proved by induction on n. It is trivial for $n = 0$. If $n > 0$, there exists $c \in \mathfrak{J}^{n+1}$ such that

$$a\xi^{n-1} = [x, y, \underset{n-1}{\ldots}, y] - 1 + c,$$

by the inductive hypothesis. Now since $b \in \mathfrak{J}$, ξ carries $\mathfrak{J}^m$ into $\mathfrak{J}^{m+1}$; hence ξ^m carries $\mathfrak{J}$ into $\mathfrak{J}^{m+1}$. Thus $a\xi^{n-1} \in \mathfrak{J}^n$ and

$$[x, y, \underset{n-1}{\ldots}, y] - 1 \in \mathfrak{J}^n.$$

By 2.3c),

$$[[x, y, \underset{n-1}{\ldots}, y], y] - 1 - ([x, y, \underset{n-1}{\ldots}, y] - 1)\xi \in \mathfrak{J}^{n+2}.$$

Hence

$$[x, y, \underset{n}{\ldots}, y] - 1 - (a\xi^{n-1} - c)\xi \in \mathfrak{J}^{n+2}.$$

Since ξ is linear and $c\xi \in \mathfrak{J}^{n+2}$, the assertion follows.

b) Define the mappings ξ_1, ξ_2 of $\mathsf{K}\mathfrak{G}$ into itself by

$$c\xi_1 = cb, \quad c\xi_2 = bc \quad (c \in \mathsf{K}\mathfrak{G}).$$

Thus $\xi = \xi_1 - \xi_2$ and $\xi_1\xi_2 = \xi_2\xi_1$. Since char $\mathsf{K} = p$, it follows that

$$\xi^{p-1} = \sum_{i=0}^{p-1} \xi_2^i \xi_1^{p-1-i},$$

which is the assertion. **q.e.d.**

2.16 Theorem (ZASSENHAUS). *Suppose that p is an odd prime. If x, y are elements of a group* $\mathfrak{G}$,

$$[x, y, \underset{p-1}{\ldots}, y] \prod_{i=0}^{p-1} (xy^i)^p \in \mathfrak{G}^{p^2}\gamma_2(\mathfrak{G})^p\gamma_{p+1}(\mathfrak{G}).$$

(This holds for any order of the factors in the product).

Proof. Let $\mathsf{K} = GF(p)$ and let $a = x - 1$, $b = y - 1$ in the group-ring $\mathsf{K}\mathfrak{G}$. If $\mathfrak{J}$ is the augmentation ideal of $\mathsf{K}\mathfrak{G}$, we have

$$xy^i = (1 + a)(1 + b)^i = 1 + a + ib + c_i,$$

where $c_i \in \mathfrak{J}^2$. Therefore

$$(xy^i)^p = 1 + (a + ib + c_i)^p \equiv 1 + (a + ib)^p \mod \mathfrak{J}^{p+1}.$$

Hence

$$\prod_{i=0}^{p-1} (xy^i)^p \equiv 1 + \sum_{i=0}^{p-1} (a + ib)^p \mod \mathfrak{J}^{p+1}.$$

But also by 2.15,

$$[x, y, \underset{p-1}{\ldots}, y] \equiv 1 + a\xi^{p-1} \equiv 1 + \sum_{i=0}^{p-1} b^i a b^{p-1-i} \mod \mathfrak{J}^{p+1}.$$

Multiplying,

$$[x, y, \underset{p-1}{\ldots}, y] \prod_{i=0}^{p-1} (xy^i)^p \equiv 1 + \sum_{i=0}^{p-1} (a + ib)^p + \sum_{i=0}^{p-1} b^i a b^{p-1-i} \mod \mathfrak{J}^{p+1},$$

since $a \in \mathfrak{J}$, $b \in \mathfrak{J}$. It follows from 2.14 that

$$[x, y, \underset{p-1}{\ldots}, y] \prod_{i=0}^{p-1} (xy^i)^p - 1 \in \mathfrak{J}^{p+1}.$$

By 2.7,

$$[x, y, \underset{p-1}{\ldots}, y] \prod_{i=0}^{p-1} (xy^i)^p \in \kappa_{p+1}(\mathfrak{G}).$$

By 1.10, $\kappa_{p+1}(\mathfrak{G}) = \mathfrak{G}^{p^2}\gamma_2(\mathfrak{G})^p\gamma_{p+1}(\mathfrak{G})$, which yields the assertion. **q.e.d.**

The following can be used to prove III, 14.21.

2.17 Theorem. *If p is a prime and x, y are elements of a group $\mathfrak{G}$, $[x, y, \underset{p-1}{\ldots}, y] \in \mathfrak{G}^{(p)}\mathfrak{G}'^p\gamma_{p+1}(\mathfrak{G})$, where $\mathfrak{G}^{(p)}$ is the group generated by all elements of the form $(uv)^{-p}u^pv^p$ with u, v in $\mathfrak{G}$.*

Proof. Let $\mathfrak{F}$ be a free group for which there is an epimorphism of $\mathfrak{F}$ onto $\mathfrak{G}$ and let a, b be preimages of x, y respectively. It is sufficient to prove that $c = [a, b, \underset{p-1}{\ldots}, b] \in \mathfrak{F}^{(p)}\mathfrak{F}'^p\gamma_{p+1}(\mathfrak{F})$. By 2.16, $c \in \mathfrak{F}^p\gamma_{p+1}(\mathfrak{F})$, so

$$c = f_1^p \cdots f_k^p g$$

for certain $f_i \in \mathfrak{F}$ $(i = 1, \ldots, k)$, $g \in \gamma_{p+1}(\mathfrak{F})$. Thus $f_1^p \cdots f_k^p = cg^{-1} \in \mathfrak{F}'$. Since $\mathfrak{F}/\mathfrak{F}'$ is Abelian, $(f_1 \cdots f_k)^p \in \mathfrak{F}'$. But $\mathfrak{F}/\mathfrak{F}'$ is a free Abelian group and is therefore torsion-free. Hence $f_1 \cdots f_k \in \mathfrak{F}'$. Now $(uv)^p \equiv u^pv^p$ mod $\mathfrak{F}^{(p)}$ for all u, v in $\mathfrak{F}$, so

$$f_1^p \cdots f_k^p \equiv (f_1 \cdots f_k)^p \bmod \mathfrak{F}^{(p)}.$$

Therefore $f_1^p \cdots f_k^p \in \mathfrak{F}^{(p)}\mathfrak{F}'^p$ and $c \in \mathfrak{F}^{(p)}\mathfrak{F}'^p\gamma_{p+1}(\mathfrak{F})$. **q.e.d.**

Exercises

2) For any finite p-group $\mathfrak{G}$, write $c_n(\mathfrak{G})$ for the integer denoted in 2.10 by c_n. Prove that if $\mathfrak{G}$, $\mathfrak{H}$ are finite p-groups,

$$\sum c_n(\mathfrak{G} \times \mathfrak{H})t^n = \left(\sum c_n(\mathfrak{G})t^n\right)\left(\sum c_n(\mathfrak{H})t^n\right).$$

3) Let K be a field of characteristic p. Let $\mathfrak{A}$ be a K-algebra having K-basis $\{a_{i_1} \cdots a_{i_n} | 0 \leq n < q\}$ for elements $a_1, \ldots, a_d$, in which any product of q of the a_i is zero. Show that $1 + a_i$ is a unit of $\mathfrak{A}$, and let $\mathfrak{G}$ be the group generated by $1 + a_1, \ldots, 1 + a_d$. Let $\mathfrak{F}$ be a free group of rank d. Prove that $\mathfrak{A} \cong \mathsf{K}\mathfrak{F}/\mathfrak{J}^q$, where $\mathfrak{J}$ is the augmentation ideal of $\mathsf{K}\mathfrak{F}$, and deduce that $\mathfrak{G} \cong \mathfrak{F}/\kappa_q(\mathfrak{F})$.

§ 3. Transitive Linear Groups

Our usual assumption that all groups mentioned are finite unless otherwise stated comes into force again here.

In § 3–§ 7, the structure of a 2-group $\mathfrak{G}$ in which the involutions are permuted transitively by a soluble subgroup $\mathfrak{A}$ of **Aut** $\mathfrak{G}$ will be discussed. In this situation, $\Omega_1(\mathbf{Z}(\mathfrak{G}))$ can be regarded as a vector space over $GF(2)$ and $\mathfrak{A}$ induces a soluble group of linear transformations of this vector space which permutes the one-dimensional subspaces transitively. The corresponding condition over $GF(p)$ will be studied in general in XII, § 7; here we shall only deal with the case $p = 2$. We say that a group $\mathfrak{A}$ of operators on a group $\mathfrak{W}$ is *irreducible* if $\mathfrak{W} \neq 1$ and there is no $\mathfrak{A}$-invariant subgroup $\mathfrak{U}$ of $\mathfrak{W}$ such that $1 < \mathfrak{U} < \mathfrak{W}$. Otherwise $\mathfrak{A}$ is *reducible*. If $\mathfrak{W}$ is soluble, it is of course necessary for irreducibility that $\mathfrak{W}$ be elementary Abelian.

The first theorem is concerned with $\frac{1}{2}n$-transitivity (II, 1.14).

3.1 Theorem (WIELANDT [4]). a) *If $\mathfrak{G}$ is a $\frac{3}{2}$-transitive permutation group, then either $\mathfrak{G}$ is primitive or $\mathfrak{G}$ is a Frobenius group.*

b) *Let $\mathfrak{A}$ be a reducible group of automorphisms of a finite group $\mathfrak{W}$. Suppose that the orbits of $\mathfrak{A}$ on $\mathfrak{W} - \{1\}$ are all of the same length. Then the stabiliser of any non-identity element of $\mathfrak{W}$ is $\{1\}$.*

Proof. a) Suppose that $\mathfrak{G}$ is $\frac{3}{2}$-transitive and imprimitive on Ω. Then $\Omega = \Psi_1 \cup \cdots \cup \Psi_r$ and $\Psi_i \cap \Psi_j = \varnothing$ $(i \neq j)$, where $1 < r < |\Omega|$ and $\mathfrak{G}$ permutes the Ψ_i. Write

$$\Psi_i = \{a_{ij} | j = 1, \ldots, s\}$$

and put $a_{11} = 1$. Then the orbits of the stabiliser $\mathfrak{G}_1$ on $\Omega - \{1\}$ are all of the same length $m > 1$.

Since $1 \in \Psi_1$, Ψ_1 remains fixed under $\mathfrak{G}_1$. Hence Ψ_1 is a union of orbits of $\mathfrak{G}_1$, among them $\{1\}$. Hence $s = |\Psi_1| \equiv 1$ (m); in particular, $(m, s) = 1$. For $i \neq 1$, $\Psi_i\mathfrak{G}_1$ remains fixed under $\mathfrak{G}_1$ but does not contain 1; hence $\Psi_i\mathfrak{G}_1$ is a union of orbits of $\mathfrak{G}_1$ of length m. Hence $|\Psi_i\mathfrak{G}_1| \equiv 0\,(m)$. On the other hand, $\Psi_i\mathfrak{G}_1$ is the disjoint union of certain

Ψ_j, so $|\Psi_i \mathfrak{G}_1| \equiv 0$ (s). Since $(m, s) = 1$, it follows that $|\Psi_i \mathfrak{G}_1| \equiv 0$ (ms). Further

$$(1) \qquad \Psi_i \mathfrak{G}_1 = \bigcup_{j=1}^{s} a_{ij} \mathfrak{G}_1,$$

and $|a_{ij}\mathfrak{G}_1| = m$ since $a_{ij} \neq 1$. Therefore (1) is a disjoint union. In particular,

$$(2) \qquad a_{ij}\mathfrak{G}_1 \cap \Psi_i = \{a_{ij}\}$$

for $i \neq 1$ and all j.

We show that if $a \in \Psi_i$ with $i \neq 1$, then $\mathfrak{G}_{1,a}$ leaves invariant every element of Ψ_1. For there exists $h \in \mathfrak{G}$ such that $ah = 1$. Then $\Psi_i h = \Psi_1$ and $\Psi_1 h = \Psi_j$ for some $j \neq 1$. Suppose that $g \in \mathfrak{G}_{1,a}$. Then $g^h \in \mathfrak{G}_{1h,1}$ and $\Psi_j g^h = \Psi_j$. Also, if $b \in \Psi_1$, then $bh \in \Psi_j$ and by (2),

$$bh\mathfrak{G}_1 \cap \Psi_j = \{bh\}.$$

Hence

$$bg = bhh^{-1}g = (bh\mathfrak{G}_1 \cap \Psi_j)g^h h^{-1} = (bh\mathfrak{G}_1 \cap \Psi_j)h^{-1} = bhh^{-1} = b.$$

It follows that if $a \notin \Psi_1$ and $1 \neq b \in \Psi_1$, then $\mathfrak{G}_{1,a} = \mathfrak{G}_{1,b}$. For $\mathfrak{G}_{1,a} \leq \mathfrak{G}_{1,b}$ and $|\mathfrak{G}_1 : \mathfrak{G}_{1,a}| = m = |\mathfrak{G}_1 : \mathfrak{G}_{1,b}|$. Hence if $1 \neq b \in \Psi_1$, $\mathfrak{G}_{1,b} = \mathfrak{G}_{1,c}$ for all $1 \neq c \in \Omega$ and $\mathfrak{G}_{1,b} = 1$. Thus $\mathfrak{G}$ is a Frobenius group.

b) Let ρ be the regular representation of $\mathfrak{W}$; thus $\mathfrak{A}$ normalises $\rho(\mathfrak{W})$. Let $\mathfrak{G} = \mathfrak{A}\rho(\mathfrak{W})$. Then $\mathfrak{G}$ is transitive on $\mathfrak{W}$ and the stabiliser of 1 is $\mathfrak{A}$. It follows from the hypothesis that $\mathfrak{G}$ is $\frac{3}{2}$-transitive. But since $\mathfrak{A}$ is reducible, $\mathfrak{A}$ is not a maximal subgroup of $\mathfrak{G}$ and $\mathfrak{G}$ is not primitive. By a), $\mathfrak{G}$ is a Frobenius group. Thus if $v \in \mathfrak{W} - \{1\}$, $\mathfrak{A}_v = \mathfrak{G}_{1,v} = 1$.

q.e.d.

3.2 Theorem. *If the group $\mathfrak{G}$ possesses a faithful irreducible representation, the centre $\mathfrak{Z}$ of $\mathfrak{G}$ is cyclic.*

Proof. Let ρ be a faithful, irreducible representation of $\mathfrak{G}$ on a vector space V. Since $\mathfrak{Z}$ is the centre of $\mathfrak{G}$, $\rho(\mathfrak{Z}) \subseteq \mathrm{Hom}_{K\mathfrak{G}}(V, V)$. By I, 10.5, $\mathrm{Hom}_{K\mathfrak{G}}(V, V)$ is a division algebra. Thus $\rho(\mathfrak{Z})$ generates a field contained in $\mathrm{Hom}_{K\mathfrak{G}}(V, V)$. Since finite multiplicative subgroups of fields are cyclic, $\rho(\mathfrak{Z})$ is cyclic. Since ρ is faithful, $\mathfrak{Z}$ is cyclic. **q.e.d.**

3.3 Lemma. *Let $\mathfrak{A}$ be a p-group and let V be a faithful, irreducible $\mathsf{K}\mathfrak{A}$-module for any field K. Suppose that $\mathfrak{A}$ has a normal subgroup $\mathfrak{N}$ which is elementary Abelian of order p^2. Let $\mathfrak{B} = \mathbf{C}_{\mathfrak{A}}(\mathfrak{N})$. Then $|\mathfrak{A} : \mathfrak{B}| = p$ and $\mathsf{V} = \mathsf{V}_1 \oplus \cdots \oplus \mathsf{V}_p$, where each V_i is $\mathfrak{B}$-invariant. For any $\alpha \in \mathfrak{A}$, there exists a permutation π of $\{1, \ldots, p\}$ such that $\mathsf{V}_i\alpha = \mathsf{V}_{i\pi}$.*

Proof. Since $\mathfrak{N} \leq \mathfrak{B}$, the centre of $\mathfrak{B}$ is not cyclic. Hence by 3.2, V is a reducible $\mathsf{K}\mathfrak{B}$-module. Thus $\mathfrak{B} \neq \mathfrak{A}$. Since p^2 does not divide the order of the group of automorphisms of $\mathfrak{N}$, $|\mathfrak{A} : \mathfrak{B}| = p$. By V, 17.3,

$$\mathsf{V}_{\mathfrak{B}} = \mathsf{V}_1 \oplus \cdots \oplus \mathsf{V}_t,$$

where each V_i is a completely reducible $\mathsf{K}\mathfrak{B}$-module, all irreducible $\mathsf{K}\mathfrak{B}$-submodules of V_i are isomorphic and an irreducible $\mathsf{K}\mathfrak{B}$-submodule of V_i is not isomorphic to an irreducible $\mathsf{K}\mathfrak{B}$-submodule of V_j if $i \neq j$. It follows from 3.2 that no V_i is a faithful $\mathsf{K}\mathfrak{B}$-module, so $t > 1$. By V, 17.3d), there exists a transitive permutation representation ρ of $\mathfrak{A}$ on $\{\mathsf{V}_1, \ldots, \mathsf{V}_t\}$ such that $\mathsf{V}_i\rho(\alpha) = \mathsf{V}_i\alpha$ for all $\alpha \in \mathfrak{A}$. Thus $\mathfrak{B}$ is contained in the kernel of ρ and $t = p$, by I, 5.11. **q.e.d.**

3.4 Lemma. *Let V be an elementary Abelian q-group and let $\mathfrak{A}$ be a p-group of automorphisms of V, where p is an odd prime. Suppose that all the orbits of $\mathfrak{A}$ on $\mathsf{V} - \{1\}$ are of the same length m. Then $\mathfrak{A}$ is cyclic and the stabiliser of any non-identity element of V is 1.*

Proof. We observe first that the two conclusions of the lemma are equivalent. For if $\mathfrak{A}$ is cyclic, the stabiliser of any non-identity element of V is the unique subgroup of $\mathfrak{A}$ of index m, which is thus 1. And if $\mathfrak{A}$ is not cyclic, then by V, 8.7, the semidirect product $\mathsf{V}\mathfrak{A}$ is not a Frobenius group, since p is odd. Hence the stabiliser in $\mathfrak{A}$ of some non-identity element of V is not 1.

Suppose that the lemma is false. Then the stabiliser of some non-identity element of V is not 1, so $\mathfrak{A}$ is irreducible by 3.1. Also $\mathfrak{A}$ is not cyclic and p is odd, so by III, 7.5, $\mathfrak{A}$ possesses a normal subgroup $\mathfrak{N}$ which is elementary Abelian of order p^2. Let $\mathfrak{B} = \mathbf{C}_{\mathfrak{A}}(\mathfrak{N})$. By 3.3, $|\mathfrak{A} : \mathfrak{B}| = p$ and V is the direct product of $\mathfrak{B}$-invariant subgroups $\mathsf{V}_1, \ldots, \mathsf{V}_p$. Suppose $1 \leq j < k \leq p$ and choose non-identity elements u, v in V_j, V_k respectively. Let $\mathfrak{U}, \mathfrak{V}, \mathfrak{W}$ be the stabilisers of u, v, uv respectively in $\mathfrak{A}$. If $\mathfrak{W} \not\leq \mathfrak{B}$, choose $\alpha \in \mathfrak{W}$, $\alpha \notin \mathfrak{B}$; by 3.3, $\mathsf{V}_j\alpha = \mathsf{V}_r$, $\mathsf{V}_k\alpha = \mathsf{V}_s$ for suitable r, s. Since $(u\alpha)(v\alpha) = uv$, $\{r, s\} = \{j, k\}$. Thus $\mathsf{V}_j\alpha^2 = \mathsf{V}_j$ and V_j is invariant under $\langle \alpha^2, \mathfrak{B} \rangle = \mathfrak{A}$, contrary to the irreducibility of $\mathfrak{A}$. Hence $\mathfrak{W} \leq \mathfrak{B}$. If $\alpha \in \mathfrak{W}$, the equation $(u\alpha)(v\alpha) = uv$ now yields $u\alpha = u$, $v\alpha = v$, so $\alpha \in \mathfrak{U} \cap \mathfrak{V}$. Thus $\mathfrak{W} \leq \mathfrak{U} \cap \mathfrak{V}$. But by hypothesis, $\mathfrak{U}, \mathfrak{V}, \mathfrak{W}$ are all of the same order. Hence $\mathfrak{U} = \mathfrak{V}$. Thus all the non-identity elements of all the V_i have the same stabiliser, which is thus 1. This is a contradiction. **q.e.d.**

3.5 Theorem. *Let* V *be an elementary Abelian 2-group of order* 2^n, *and let* $\mathfrak{A}$ *be a soluble group of automorphisms of* V *of odd order. Suppose that* $\mathfrak{A}$ *permutes the set of non-identity elements of* V *transitively. Then* $\mathfrak{A}$ *is similar to a subgroup of the group of permutations of* $K = GF(2^n)$ *of the form* $x \to a(x\theta)$ $(a \in K^\times, \theta \in \mathbf{Aut}\, K)$. *If* $\mathfrak{B}$ *is the subgroup of* $\mathfrak{A}$ *consisting of elements which correspond to permutations of* K *of the form* $x \to ax$, *then* $\mathfrak{B} \trianglelefteq \mathfrak{A}$ *and* $\mathfrak{B}$ *is cyclic of order* k, *where* k *divides* $2^n - 1$ *and* $2^n - 1$ *divides* kn. *Also* $\mathfrak{B}$ *is irreducible on* V, *and* $\mathfrak{B}$ *permutes the set of non-identity elements of* V *in orbits of length* k.

Proof. Let $\mathfrak{F}$ be the Fitting subgroup of $\mathfrak{A}$. Each Sylow subgroup $\mathfrak{S}$ of $\mathfrak{F}$ is normal in $\mathfrak{A}$ and thus permutes the set of non-identity elements of V in orbits of equal length (II, 1.5). Since $|\mathfrak{A}|$ is odd, it follows from 3.4 that $\mathfrak{S}$ is cyclic and that the stabiliser in $\mathfrak{S}$ of any non-identity element of V is $\{1\}$. Since $\mathfrak{F}$ is the direct product of its Sylow subgroups, $\mathfrak{F}$ is cyclic. Also the stabiliser in $\mathfrak{F}$ of any non-identity element of V is $\{1\}$, since such a stabiliser would otherwise contain an element of prime order. Thus $\mathfrak{F}$ permutes the set of non-identity elements of V in orbits of length $|\mathfrak{F}|$.

Since an automorphism of a cyclic group is determined by the image of a generator, $|\mathbf{Aut}\,\mathfrak{F}| < |\mathfrak{F}|$. Hence $|\mathfrak{A} : \mathbf{C}_{\mathfrak{A}}(\mathfrak{F})| < |\mathfrak{F}|$. But $\mathbf{C}_{\mathfrak{A}}(\mathfrak{F}) = \mathfrak{F}$, by III, 4.2b). Thus $|\mathfrak{A}| \leq |\mathfrak{F}|^2 - 1$. Since $\mathfrak{A}$ permutes the set of non-identity elements of V transitively, $|V| - 1$ divides $|\mathfrak{A}|$. Thus $|V| \leq |\mathfrak{F}|^2$.

Now suppose that U is an irreducible $\mathfrak{F}$-invariant subgroup of V of minimal order. Since the orbits of $\mathfrak{F}$ on $V - \{1\}$ are of length $|\mathfrak{F}|$, $|U| - 1$ is divisible by $|\mathfrak{F}|$. Thus $|U| > |\mathfrak{F}|$. Thus $|U|^2 > |\mathfrak{F}|^2 \geq |V|$. By the Maschke-Schur theorem (I, 17.7), $V = U \times W$ for some $\mathfrak{F}$-invariant subgroup W of V. Hence $|W| < |U|$. By minimality of $|U|$, $W = 1$. Thus $V = U$ and V is an irreducible $\mathfrak{F}$-group. By II, 3.11, $\mathfrak{A}$ is similar to a subgroup of the group of permutations of $K = GF(2^n)$ of the form $x \to a(x\theta)$ $(a \in K^\times, \theta \in \mathbf{Aut}\, K)$; also, if $\mathfrak{B}$ is the subgroup of $\mathfrak{A}$ consisting of elements which correspond to permutations of K of the form $x \to ax$, $\mathfrak{B} = \mathbf{C}_{\mathfrak{A}}(\mathfrak{F})$. Hence $\mathfrak{B} = \mathfrak{F} \trianglelefteq \mathfrak{A}$, $\mathfrak{B}$ is cyclic and $\mathfrak{B}$ is irreducible on V. Let $k = |\mathfrak{B}|$. Since $\mathfrak{A}/\mathfrak{B}$ is isomorphic to a subgroup of $\mathbf{Aut}\, K$, $|\mathfrak{A} : \mathfrak{B}|$ divides n. Hence $|\mathfrak{A}|$ divides kn. Since $|V| - 1$ divides $|\mathfrak{A}|$, $2^n - 1$ divides kn. Since $\mathfrak{B}$ is isomorphic to a subgroup of $K^\times$, k divides $2^n - 1$ and $\mathfrak{B}$ permutes the set of non-identity elements of V in orbits of length k. **q.e.d.**

3.6 Lemma. *Let* V *be an elementary Abelian q-group of order* q^n, *and let* α *be an automorphism of* V *of order k, where* $(k, q - 1) = 1$ *and* $q^n - 1 \leq kn$. *Suppose that* $\langle\alpha\rangle$ *is irreducible, that* $\mathsf{V} = \mathsf{V}_1 \times \cdots \times \mathsf{V}_r$ *and that* α *permutes the* V_i. *Then* $r = 1$.

Proof. By II, 3.10, k divides $q^n - 1$. Hence $(k, q(q - 1)) = 1$ and k is odd. Since $\langle\alpha\rangle$ is irreducible, $\langle\alpha\rangle$ permutes the V_i transitively. Hence r divides k, and the V_i are all of the same order q^m, where $rm = n$. Further, the stabiliser of V_i under the permutation group induced by $\langle\alpha\rangle$ is $\langle\alpha^r\rangle$, and $\langle\alpha^r\rangle$ operates faithfully and irreducibly on each V_i. Hence $q^m - 1$ is divisible by $\frac{k}{r}$, by II, 3.10. Hence

$$q^m - 1 \geq \frac{k}{r} \geq \frac{q^n - 1}{rn} = \frac{q^{rm} - 1}{r^2 m}.$$

Thus

$$\begin{aligned} r^2 m &\geq 1 + q^m + \cdots + q^{m(r-1)} \\ &\geq 1 + 2^m + \cdots + 2^{m(r-1)}. \end{aligned}$$

Since $2^{m-1} \geq m$, it follows that

$$mr^2 > 2m + 2^2 m + \cdots + 2^{r-1} m$$

and $r^2 > 2^r - 2$. Hence $r \leq 4$. Since $mr^2 \geq (2^{mr} - 1)/(2^m - 1)$, it is easy to check that if $r = 4$ or $r = 3$, then $m = 1$. Since r divides k and k is odd, $r \neq 2$. Hence either $r = 1$ or $m = 1$. But if $m = 1$, $|\mathsf{V}_i| = q$. Since $(k, q - 1) = 1$, α^r is identity on each V_i. Thus $\alpha^r = 1$. If $v \in \mathsf{V}_1$ and $v \neq 1$, $v(v\alpha) \cdots (v\alpha^{r-1})$ is then invariant under α. Since $\langle\alpha\rangle$ is irreducible, $n = 1$; hence $r = 1$ in any case. **q.e.d.**

§ 4. Some Number-Theoretical Lemmas

We shall need some number-theoretical properties of the integers k, n occurring in 3.5.

As far as 4.5 inclusive, let k, n denote positive integers such that

(1) $n > 1$, k divides $2^n - 1$ and $2^n - 1$ divides nk.

Thus if $d = (n, 2^n - 1)$, the possible values of k are $c\dfrac{2^n - 1}{d}$, where c is any divisor of d. Hence if $d = 1$, $k = 2^n - 1$. If $d > 1$, there is an odd prime divisor p of n such that n is divisible by the order of 2 modulo p. Thus it follows, for instance, that if $d > 1$, n cannot be a power of a prime; again, if $n = 2p$ for some prime p, then $p = 3$. Using these facts it is easy to check that the following table gives all values of $n \leq 55$ for which $d > 1$.

n	6	12	18	20	21	24	30	36	40	42	48	54
d	3	3	9	5	7	3	3	9	5	21	3	27

We need to solve certain congruences of the following kind.

4.1 Lemma. *Suppose that* $i_1, \ldots, i_r$ $(r \geq 2)$ *are non-negative integers,* $0 \leq l \leq r$ *and*

$$2^{i_1} + \cdots + 2^{i_l} \equiv 2^{i_{l+1}} + \cdots + 2^{i_r} \quad (k).$$

If the i_j *are distinct modulo n, then*

$$k + 1 \leq 2^{1+n\left(1-\frac{1}{r}\right)}$$

and $2^{\frac{n}{r}} < 2n$.

Proof. Since $2^n \equiv 1(k)$, we may suppose that $0 \leq i_j < n$ for all j. Thus the i_j are distinct. Suppose that the i_j, when written in increasing order, are $h_1, \ldots, h_r$; thus

$$h_1 < h_2 < \cdots < h_r.$$

If the given congruence is multiplied by 2^{n-h_j}, a congruence is obtained which involves

$$2^{h_1+n-h_j}, \ldots, 2^{h_{j-1}+n-h_j}, 1, 2^{h_{j+1}-h_j}, \ldots, 2^{h_r-h_j}.$$

Denote the largest exponent occurring here by d_j; thus
$d_j = h_{j-1} + n - h_j$ if $j > 1$ and $d_1 = h_r - h_1$. If $d = \min(d_1, \ldots, d_r)$, then

$$rd \leq d_1 + \cdots + d_r = (r - 1)n. \tag{2}$$

Now one of these congruences is of the form $L \equiv R\ (k)$, where L, R are sums of powers of 2 such that 2^d appears in L, and all the other exponents which occur in either L or R are distinct and less than d.

Thus

$$R \leq 2^{d-1} + 2^{d-2} + \cdots + 1 < 2^d \leq L.$$

Hence $L \geq R + k$. Thus

$$k \leq L \leq 2^d + 2^{d-1} + \cdots + 1 \leq 2^{d+1} - 1.$$

Combined with (2), this is the first of the stated inequalities.

By (1), $nk \geq 2^n - 1$, so

$$n \cdot 2^{1+n\left(1-\frac{1}{r}\right)} \geq n(k+1) \geq 2^n - 1 + n > 2^n.$$

Hence $2n > 2^{\frac{n}{r}}$. **q.e.d.**

This gives a bound for n in terms of r, which we give explicitly for $1 < r \leq 8$.

4.2 Lemma. *Suppose that n, r are integers greater than* 1 *and that* $2^{n/r} < 2n$. *Then* $n \leq f(r)$, *where* $f(2) = 7$, $f(3) = 14$, $f(4) = 21$, $f(5) = 29$, $f(6) = 37$, $f(7) = 45$ *and* $f(8) = 54$.

Thus in solving the congruence of 4.1 for a particular value of r, only a finite number of cases need be considered. We consider first some cases with $r = 2$ and $r = 3$.

4.3 Lemma. a) *If* $2^i \equiv 1\ (k)$, *then* $i \equiv 0\ (n)$.

b) *If* $2^i \equiv -1\ (k)$, *then* $n = 2$ *and* $i \equiv 1\ (2)$.

c) *If* $2^i + 2^j \equiv 1\ (k)$, *then* $i \equiv j \equiv n - 1\ (n)$.

d) *If* $2^i + 2^j \equiv -1\ (k)$, *then either* (i) $n = 2$, $i \equiv j \equiv 0\ (2)$, (ii) $n = 3$, $\{i, j\} \equiv \{1, 2\}\ (3)$, *or* (iii) $n = 6$, $k = 21$ *and* $\{i, j\} \equiv \{2, 4\}\ (6)$.

(A statement such as $\{i, j\} \equiv \{1, 2\}\ (n)$ *means of course that either* $i \equiv 1$, $j \equiv 2\ (n)$ *or* $i \equiv 2$, $j \equiv 1\ (n)$*).*

Proof. In a) and b) we may suppose $0 < i < n$. In c) it is clear that $i \not\equiv 0\ (n)$ and $j \not\equiv 0\ (n)$; thus we may suppose that $0 < i < j < n$ since a) may be used in the case $i = j$. In d), if $i \equiv 0\ (n)$, $j \equiv 0\ (n)$ or $i \equiv j\ (n)$, we may use b), and this gives the solution (i); thus we may suppose that $0 < i < j < n$ in this case also.

Under these suppositions the conditions of 4.1 are satisfied with $r = 2$ in a), b) and $r = 3$ in c), d). By 4.1, $2^{n/r} < 2n$. By 4.2, $n \leq 7$ in a), b) and $n \leq 14$ in c), d).

In a), b) the situation is clear if $n = 2$. If $n > 2$, suppose $2^i \mp 1 = kl$. Then $kl \leq 2^{n-1} + 1 < 2^n - 1$. Hence $k < 2^n - 1$ and from the above table $n = 6$, $k = 21$. Then $21l < 63$, $l < 3$, $l = 1$ and $2^i \mp 1 = k = 21$, which is false.

In c), d) the situation is clear if $n = 3$, since $0 < i < j < n$. If $n > 3$, suppose $2^i + 2^j \mp 1 = kl$. Then $kl \leq 2^{n-1} + 2^{n-2} + 1 < 2^n - 1$. From the list $n = 6$ or $n = 12$ and $2^n - 1 = 3k$. Hence $l = 1$, and $2^i + 2^j \mp 1 = k$. We use the fact that every non-negative integer is uniquely expressible as a sum of distinct powers of 2. For $n = 12$,

$$1 + 2^2 + 2^4 + \cdots + 2^{10} = k = 2^i + 2^j \mp 1$$

is impossible. For $n = 6$, $1 + 2^2 + 2^4 = k = 2^i + 2^j \mp 1$ implies that $i = 2$, $j = 4$ in d).

Next we handle some cases when the number of powers of 2 involved is 4.

4.4 Lemma. a) *If* $2^h + 2^i + 2^j \equiv 1$ (k), *then either* (i) $\{h, i, j\} \equiv \{-1, -2, -2\}$ (n), *or* (ii) $n = 6$, $k = 21$, $\{h, i, j\} \equiv \{1, 2, 4\}$ (6).

b) *If* $2^h + 2^i + 2^j \equiv -1$ (k), *then either* (i) $n = 2$, $\{h, i, j\} \equiv \{0, 1, 1\}$ (2); (ii) $n = 3$, $\{h, i, j\} \equiv \{0, 0, 2\}$ *or* $\{1, 1, 1\}$ (3); (iii) $n = 4$, $\{h, i, j\} \equiv \{1, 2, 3\}$ (4); *or* (iv) $n = 6$, $k = 21$, $\{h, i, j\} \equiv \{0, 3, 5\}$, $\{1, 1, 4\}$ *or* $\{2, 3, 3\}$ (6).

c) *If* $2^i + 2^j \equiv 1 + 2^h$ (k), *either* (i) $\{i, j\} \equiv \{0, h\}$ (n), (ii) $n = 6$, $k = 21$, $\{i, j\} \equiv \{3, 4\}$ (6), $h \equiv 1$ (6), *or* (iii) $n = 6$, $k = 21$, $\{i, j\} \equiv \{2, 3\}$ (6), $h \equiv 5$ (6).

Proof. These results are easily deduced from 4.3 except in the case when the exponents are distinct and non-zero modulo n. In this case $2^{n/4} < 2n$ by 4.1 and $n \leq 21$ by 4.2. Clearly we may suppose $n > 4$. We also suppose that h, i, j lie between 0 and n, and write

$$2^h + 2^i + 2^j \mp 1 = kl, \quad \text{or} \quad 2^i + 2^j = 1 + 2^h + kl.$$

Then $|kl| \leq 2^{n-1} + 2^{n-2} + 2^{n-3} + 1 < 2^n - 1$. Thus $k < 2^n - 1$ and the possible values of n are 6, 12, 18, 20, 21. Of course l is odd and $|l| < \dfrac{2^n - 1}{k}$.

If $k = (2^n - 1)/3$, then $l = \pm 1$. Also n is even and $k = 1 + 2^2 + \cdots + 2^{n-2}$. In a), b), $l = 1$. Now $k = 2^h + 2^i + 2^j + 1$ implies $\{h, i, j\} = \{2, 4, 6\}$ and $n = 8$, a contradiction. And $k = 2^h + 2^i + 2^j - 1$ implies $\{h, i, j\} = \{1, 2, 4\}$ and $n = 6$, which is one of the stated solutions. In c), $l = \pm 1$. But $2^i + 2^j = 1 + 2^h + k$ implies $\{i, j\} \equiv \{3, 4\}$ (6), $h \equiv 1$, $n = 6$, $k = 21$; and $2^i + 2^j + k = 1 + 2^h$ implies $\{i, j\} \equiv \{2, 3\}$ (6), $h \equiv 5$, $n = 6$, $k = 21$.

If $k \neq (2^n - 1)/3$, $n = 18$, 20 or 21. These cases have to be treated individually. One writes out all the possible values of kl as sums of powers of 2, and it is very easy to convince oneself that no such kl can satisfy the above equations. **q.e.d.**

4.5 Lemma. a) *If* $2^g + 2^h + 2^i + 2^j \equiv 1$ (k), *either* (i) $\{g, h, i, j\} \equiv \{-2, -2, -2, -2\}$ *or* $\{-3, -3, -2, -1\}$ (n), *or* (ii) $n = 6$, $k = 21$ *and* $\{g, h, i, j\} \equiv \{0, 0, 2, 4\}, \{0, 1, 3, 5\}, \{1, 1, 1, 4\}$ *or* $\{1, 2, 3, 3\}$ (n).

b) *If* $2^h + 2^i + 2^j \equiv 1 + 2^g$ (k), *either* (i) $\{h, i, j\} \equiv \{0, g-1, g-1\}$ *or* $\{-1, -1, g\}$ (n), *or* (ii) $n = 6$, $k = 21$ *and one of the following holds:*

$$g \equiv 0, \quad \{h, i, j\} \equiv \{2, 3, 5\}\ (6);$$

$$g \equiv 1, \quad \{h, i, j\} \equiv \{2, 2, 4\} \text{ or } \{3, 3, 3\}\ (6);$$

$$g \equiv 2, \quad \{h, i, j\} \equiv \{1, 3, 4\}\ (6);$$

$$g \equiv 4, \quad \{h, i, j\} \equiv \{1, 2, 5\}\ (6);$$

$$g \equiv 5, \quad \{h, i, j\} \equiv \{1, 1, 3\} \text{ or } \{2, 2, 2\}\ (6).$$

Proof. This is proved along the same lines. First we prove the results in the case when some of the exponents are congruent modulo n or when one of them is zero modulo n by appealing to 4.3 and 4.4. In the remaining cases $n \leq 29$ by 4.1 and 4.2. There are no solutions for $n = 5$, so n is 6, 12, 18, 20, 21 or 24. The case $k = (2^n - 1)/3$ can be handled rather easily as in 4.4; the other cases must be treated individually. **q.e.d.**

On account of the exceptional role of $n = 6$, we shall need a certain fact about $GF(2^6)$.

4.6 Lemma. *Let* $\mathsf{F} = GF(2^6)$. *If* ε *is any non-zero element of* F, *there exists elements* a, b *in* F, *not both zero, such that*

$$a^5 + b^2 + \varepsilon a^{16} b^4 = 0.$$

Proof. Let β be the mapping of F into itself for which $x\beta = x(1 + \varepsilon x)$. Then β is an additive homomorphism, and $|\ker \beta| = 2$ since $\varepsilon \neq 0$. Hence $|\operatorname{im} \beta| = 2^5$. If K is the subfield of F having 4 elements, $\mathsf{K} \cap (\operatorname{im} \beta) \neq 0$. Thus there exists $x \in \mathsf{F}$ such that $x(1 + \varepsilon x)$ is a non-zero element of K. Now K consists of the 21st powers of the elements of F; thus $x(1 + \varepsilon x) = a^{21}$ for some $a \in \mathsf{F}$, $a \neq 0$. Define $b \in \mathsf{F}$ by $x = a^{16}b^2$. Then $b^2(1 + \varepsilon a^{16}b^2) = b^2a^{21}x^{-1} = a^5$, and $a^5 + b^2 + \varepsilon a^{16}b^4 = 0$.

q.e.d.

§ 5. Lemmas on 2-Groups

In this section, we present some results on groups of prime-power order which are useful for the determination of the Suzuki 2-groups.

We begin with a counting argument. In 5.1 and 5.2, p is an arbitrary prime, and if $\mathfrak{X}$ is a finite p-group, $j(\mathfrak{X})$ is the number of elements of $\mathfrak{X}$ of order at most p.

5.1 Theorem (BLACKBURN [1]). *Let $\mathfrak{G}$ be a p-group and let $\mathfrak{N}$ be a non-identity subgroup of exponent p in the centre of $\mathfrak{G}$. If $\mathscr{H}$ is the set of maximal subgroups of $\mathfrak{N}$,*

$$\sum_{\mathfrak{M} \in \mathscr{H}} j(\mathfrak{G}/\mathfrak{M}) = (|\mathscr{H}| - 1)j(\mathfrak{G}/\mathfrak{N}) + j(\mathfrak{G}).$$

Proof. Let $k = |\mathscr{H}|$. Thus $|\mathfrak{N}| = (p - 1)k + 1$. Note that if $\mathfrak{Z}$ is a subgroup of $\mathfrak{N}$ of order p, the number of maximal subgroups of $\mathfrak{N}/\mathfrak{Z}$ is $(k - 1)/p$.

Let

$$\mathscr{Y} = \{(x, \mathfrak{M}) | x \in \mathfrak{G} - \mathfrak{N}, \mathfrak{M} \in \mathscr{H}, x^p \in \mathfrak{M}\}.$$

If $x \in \mathfrak{G} - \mathfrak{N}$ and $x^p \in \mathfrak{N}$, let

$$\mathscr{Y}_x = \{\mathfrak{M} | \mathfrak{M} \in \mathscr{H}, x^p \in \mathfrak{M}\}.$$

Clearly, $|\mathscr{Y}_x|$ is the number of maximal subgroups of $\mathfrak{N}/\langle x^p \rangle$. Thus $|\mathscr{Y}_x| = k$ if $x^p = 1$ and $|\mathscr{Y}_x| = (k - 1)/p$ if $x^p \neq 1$. Hence

$$|\mathscr{Y}| = \sum_x |\mathscr{Y}_x| = \sum_{x^p = 1} k + \sum_{x^p \neq 1} \frac{k - 1}{p} = \sum_{x^p = 1} \left(k - \frac{k - 1}{p}\right) + \sum_x \frac{k - 1}{p},$$

where x runs through the set of elements of $\mathfrak{G} - \mathfrak{N}$ for which $x^p \in \mathfrak{N}$. Now $x^p \in \mathfrak{N}$ if and only if $(x\mathfrak{N})^p = 1$; thus the number of $x \in \mathfrak{G} - \mathfrak{N}$ for which $x^p \in \mathfrak{N}$ is $(j(\mathfrak{G}/\mathfrak{N}) - 1)|\mathfrak{N}|$. Hence

$$|\mathscr{Y}| = (j(\mathfrak{G}) - |\mathfrak{N}|)\left(k - \frac{k-1}{p}\right) + (j(\mathfrak{G}/\mathfrak{N}) - 1)|\mathfrak{N}|\frac{k-1}{p}$$

$$= j(\mathfrak{G})\left(k - \frac{k-1}{p}\right) + j(\mathfrak{G}/\mathfrak{N})|\mathfrak{N}|\frac{k-1}{p} - k|\mathfrak{N}|.$$

On the other hand, suppose that $\mathfrak{M} \in \mathscr{H}$. The number of elements $x \in \mathfrak{G} - \mathfrak{N}$ for which $x^p \in \mathfrak{M}$ is $(j(\mathfrak{G}/\mathfrak{M}) - p)|\mathfrak{M}|$, so

$$|\mathscr{Y}| = \sum_{\mathfrak{M} \in \mathscr{H}} (j(\mathfrak{G}/\mathfrak{M}) - p)\frac{|\mathfrak{N}|}{p}.$$

Comparing the two expressions for $|\mathscr{Y}|$,

$$\frac{j(\mathfrak{G})}{p} + j(\mathfrak{G}/\mathfrak{N})\frac{k-1}{p} - k = \sum_{\mathfrak{M} \in \mathscr{H}} (j(\mathfrak{G}/\mathfrak{M}) - p)\frac{1}{p}.$$

Multiplication by p yields the stated result. **q.e.d.**

To apply this, we shall need to know $j(\mathfrak{X})$ if $\mathfrak{X}$ is a 2-group and $|\Phi(\mathfrak{X})| = 2$. First we prove the following.

5.2 Lemma. *Suppose that $\mathfrak{G}$ is a p-group and that $\mathfrak{G} = \mathfrak{A}\mathfrak{B}$, where $[\mathfrak{A}, \mathfrak{B}] = 1$. Let $\mathfrak{D} = \mathfrak{A} \cap \mathfrak{B}$. Suppose that $|\mathfrak{D}| = p$ and that $\mathfrak{A}/\mathfrak{D}$, $\mathfrak{B}/\mathfrak{D}$ are of exponent p. Then*

$$j(\mathfrak{G}) = \frac{1}{p}\left(j(\mathfrak{A})j(\mathfrak{B}) + \frac{(|\mathfrak{A}| - j(\mathfrak{A}))(|\mathfrak{B}| - j(\mathfrak{B}))}{p-1}\right).$$

Proof. We have $\mathfrak{G} \cong (\mathfrak{A} \times \mathfrak{B})/\langle(x, x^{-1})\rangle$ for a generator x of $\mathfrak{D}$. Thus

$$pj(\mathfrak{G}) = |\{(a, b) | a \in \mathfrak{A}, b \in \mathfrak{B}, a^p = x^i, b^p = x^{-i} \quad \text{for some } i\}|.$$

If $i = 0$, the number of solutions of $a^p = x^i$ is $j(\mathfrak{A})$; otherwise it is $(|\mathfrak{A}| - j(\mathfrak{A}))/(p-1)$, since $\mathfrak{A}/\mathfrak{D}$ is of exponent p. Similar statements hold in $\mathfrak{B}$, and the assertion follows at once. **q.e.d.**

5.3 Lemma. *Suppose that $\mathfrak{G}$ is a 2-group, $|\mathfrak{G} : \Phi(\mathfrak{G})| = 2^d$ and $|\Phi(\mathfrak{G})| = 2$. If $j = |\{x | x \in \mathfrak{G}, x^2 = 1\}|$, j is either 2^d, $2^d + 2^{d-r}$ or $2^d - 2^{d-r}$, where r is a positive integer satisfying $2r \leq d$. Further, if j is $2^d + 2^{d-\frac{1}{2}d}$ or $2^d - 2^{d-\frac{1}{2}d}$, $\mathfrak{G}$ is extraspecial.*

Proof. This is proved by induction on $|\mathfrak{G}|$. If $\mathfrak{G}$ is Abelian, $\Phi(\mathfrak{G}) = \mathfrak{G}^2$ and $j = |\mathfrak{G} : \mathfrak{G}^2| = 2^d$. If $\mathfrak{G}$ is non-Abelian, there exist non-commuting elements x, y of $\mathfrak{G}$. If $\mathfrak{H} = \langle x, y \rangle$, $\mathfrak{H}$ is non-Abelian of order 8, so the number i of elements of $\mathfrak{H}$ of order at most 2 is either 2 or 6. Since $\langle x, \Phi(\mathfrak{G}) \rangle$ and $\langle y, \Phi(\mathfrak{G}) \rangle$ are non-central normal subgroups of $\mathfrak{G}$ of order 4, their centralizers are of index 2, for the order of the automorphism group of a group of order 4 is not divisible by 4. Hence if $\mathfrak{C} = \mathbf{C}_{\mathfrak{G}}(\mathfrak{H})$, $|\mathfrak{G} : \mathfrak{C}| \leq 4$. But $\mathfrak{C} \cap \mathfrak{H} = \Phi(\mathfrak{H}) = \Phi(\mathfrak{G})$, so $|\mathfrak{C}\mathfrak{H}| = 4|\mathfrak{C}| \geq |\mathfrak{G}|$ and $\mathfrak{G} = \mathfrak{C}\mathfrak{H}$. Thus by 5.2,

$$j = \tfrac{1}{2}(il + (8 - i)(2^{d-1} - l)) = il + 2^{d+1} - 4l - 2^{d-2}i,$$

where l is the number of elements of $\mathfrak{C}$ of order at most 2. If $\mathfrak{C}$ is elementary Abelian, $l = 2^{d-1}$, so $j = 2^{d-2}i$. If $i = 2$, $j = 2^d - 2^{d-1}$, and if $i = 6$, $j = 2^d + 2^{d-1}$. If $\mathfrak{C}$ is not elementary Abelian, then $\Phi(\mathfrak{C}) = \Phi(\mathfrak{G})$ and we can apply the inductive hypothesis to $\mathfrak{C}$. Thus l is either 2^{d-2}, $2^{d-2} + 2^{d-s-2}$ or $2^{d-2} - 2^{d-s-2}$, where s is a positive integer satisfying $2s \leq d - 2$. If $l = 2^{d-2}$, $j = 2^d$, and in the other cases, j is $2^d - 2^{d-r}$ or $2^d + 2^{d-r}$, where $r = s + 1$, so $2r \leq d$.

Suppose that $\mathfrak{G}$ is non-Abelian and is not extraspecial. If $\mathfrak{C}$ is elementary Abelian, (so that $r = 1$), then $|\mathfrak{C}| > 2$, so $d > 2 = 2r$. In the remaining case, $\mathfrak{C}$ is not extraspecial, so the inductive hypothesis gives $2s < d - 2$. Thus $2r < d$ in any case. **q.e.d.**

The following theorem is an application of this counting argument.

5.4 Theorem. *Suppose that $\mathfrak{G}$ is a 2-group of class at most 2 and exponent at most 4. Suppose that $\{x | x \in \mathfrak{G}, x^2 = 1\}$ is a subgroup $\mathfrak{N}$ of $\mathfrak{G}$. Then $|\mathfrak{G}| \leq |\mathfrak{N}|^3$; indeed $|\mathfrak{G}| < |\mathfrak{N}|^3$ if $\mathfrak{G}$ is not special.*

Proof. We prove this by induction on $|\mathfrak{N}|$. If $|\mathfrak{N}| = 2$, $\mathfrak{G}$ is either cyclic of order at most 4 or the quaternion group of order 8, and the assertion is clear. Suppose then that $|\mathfrak{N}| > 2$. Clearly $\Phi(\mathfrak{G}) \leq \mathfrak{N}$.

First suppose that $\mathfrak{G}' < \mathfrak{N}$. We have $\mathfrak{N} \leq \mathfrak{L}$, where $\mathfrak{L}/\mathfrak{G}' = \Omega_1(\mathfrak{G}/\mathfrak{G}')$. Since $\mathfrak{L}/\mathfrak{G}'$ is elementary Abelian,

$$\mathfrak{L}/\mathfrak{G}' = (\mathfrak{N}/\mathfrak{G}') \times (\mathfrak{K}/\mathfrak{G}')$$

for some subgroup $\mathfrak{K} \leq \mathfrak{L}$. Since $\mathfrak{G}' < \mathfrak{N}$, $\mathfrak{K} < \mathfrak{L}$, so the inductive hypothesis may be applied to $\mathfrak{K}$. Thus $|\mathfrak{K}| \leq |\mathfrak{N} \cap \mathfrak{K}|^3 = |\mathfrak{G}'|^3$, so $|\mathfrak{L}| \leq |\mathfrak{N}||\mathfrak{G}'|^2$. But

$$|\mathfrak{L} : \mathfrak{G}'| = |\Omega_1(\mathfrak{G}/\mathfrak{G}')| = |\mathfrak{G}/\mathfrak{G}' : (\mathfrak{G}/\mathfrak{G}')^2| = |\mathfrak{G} : \Phi(\mathfrak{G})|.$$

Hence

$$|\mathfrak{G}| = |\mathfrak{G} : \mathfrak{L}||\mathfrak{L}| = |\Phi(\mathfrak{G}) : \mathfrak{G}'||\mathfrak{L}| \leq |\mathfrak{N}|^2|\mathfrak{G}'| < |\mathfrak{N}|^3.$$

Suppose then that $\mathfrak{G}' = \mathfrak{N}$; thus $\Phi(\mathfrak{G}) = \mathfrak{N}$ also. Since $\mathfrak{G}$ is of class 2, $\mathfrak{N} \leq \mathbf{Z}(\mathfrak{G})$. Suppose next that $\mathfrak{N} < \mathbf{Z}(\mathfrak{G})$. Choose $z \in \mathbf{Z}(\mathfrak{G})$, $z \notin \mathfrak{N}$. Let $\mathfrak{M}$ be a maximal subgroup of $\mathfrak{G}$ such that $z \notin \mathfrak{M}$. Thus

$$\mathfrak{G} = \mathfrak{M}\langle z\rangle, \quad \mathfrak{M} \cap \langle z\rangle = \mathfrak{Z},$$

where $\mathfrak{Z} = \langle z^2\rangle \neq 1$. We apply the inductive hypothesis to $\mathfrak{M}/\mathfrak{Z}$. If $x \in \mathfrak{M}$ and $x^2 \in \mathfrak{Z}$, we have $x^2 = z^{2m}$ for some m and $xz^{-m} \in \mathfrak{N}$. Hence $z^m \in \mathfrak{M} \cap \langle z\rangle = \mathfrak{Z}$ and $x\mathfrak{Z} \in \mathfrak{N}/\mathfrak{Z}$. Thus

$$|\mathfrak{M}/\mathfrak{Z}| \leq |\mathfrak{N}/\mathfrak{Z}|^3$$

and

$$|\mathfrak{G}| = 2|\mathfrak{M}| \leq \frac{2|\mathfrak{N}|^3}{|\mathfrak{Z}|^2} = \tfrac{1}{2}|\mathfrak{N}|^3 < |\mathfrak{N}|^3.$$

Thus we may suppose that $\mathfrak{G}' = \Phi(\mathfrak{G}) = \mathbf{Z}(\mathfrak{G}) = \mathfrak{N}$. Write $j(\mathfrak{X})$ for the number of elements of order at most 2 in the 2-group $\mathfrak{X}$. Then $j(\mathfrak{G}) = |\mathfrak{N}| = 2^n$, and $j(\mathfrak{G}/\mathfrak{N}) = |\mathfrak{G}/\mathfrak{N}| = 2^d$, say. Let $\mathscr{H}$ denote the set of maximal subgroups of $\mathfrak{N}$. Thus $|\mathscr{H}| = 2^n - 1$. Hence by 5.1,

$$\text{(1)} \qquad \sum_{\mathfrak{M} \in \mathscr{H}} j(\mathfrak{G}/\mathfrak{M}) = (|\mathscr{H}| - 1)j(\mathfrak{G}/\mathfrak{N}) + j(\mathfrak{G})$$
$$= (2^n - 2)2^d + 2^n.$$

By 5.3, $j(\mathfrak{G}/\mathfrak{M})$ is an integer of the form 2^d, $2^d + 2^{d-r}$ or $2^d - 2^{d-r}$, where r is a positive integer such that $2r \leq d$. Substituting in (1), we obtain an equation of the form

$$2^{n+d} - 2^{d+1} + 2^n = \sum_i (\pm 2^{s_i}),$$

where $s_i \geq \frac{1}{2}d$. Suppose now that $n < \frac{1}{2}d$. Then $s_i > n$ for all i, so the right-hand side is divisible by 2^{n+1}. But also, the terms 2^{n+d}, 2^{d+1} of the left-hand side are divisible by 2^{n+1}, whereas 2^n is not. This is a contradiction. Hence $n \geq \frac{1}{2}d$ and $|\mathfrak{G}| = 2^{d+n} \leq 2^{3n} = |\mathfrak{N}|^3$. **q.e.d.**

5.5 Theorem. *Suppose that $\mathfrak{G}$ is a non-identity 2-group and that* **Aut** $\mathfrak{G}$ *permutes the set* I *of involutions in $\mathfrak{G}$ transitively. Let $|\Omega_1(\mathbf{Z}(\mathfrak{G}))| = q$.*

a) $\Omega_1(\mathbf{Z}(\mathfrak{G})) = \mathsf{I} \cup \{1\}$.

b) (Gross[5]) $|\mathfrak{G}| = q^m$ *for some m.*

c) *If $\mathfrak{G}$ is non-Abelian and of exponent 4, then $\mathfrak{G}' = \mathbf{Z}(\mathfrak{G}) = \Phi(\mathfrak{G})$ and $|\mathfrak{G}|$ is q^2 or q^3.*

Proof. a) Since $\mathfrak{G} \neq 1$, $\Omega_1(\mathbf{Z}(\mathfrak{G}))$ contains an element $t \in \mathsf{I}$. Thus $\mathsf{I} = \{t\alpha | \alpha \in \mathbf{Aut}\,\mathfrak{G}\} \subseteq \Omega_1(\mathbf{Z}(\mathfrak{G}))$ and $\mathsf{I} \cup \{1\} = \Omega_1(\mathbf{Z}(\mathfrak{G}))$.

b) Given $t \in \mathsf{I}$, write $m_k = |\{y | y \in \mathfrak{G}, y^{2^{k-1}} = t\}|$ $(k \geq 1)$. Since **Aut** $\mathfrak{G}$ permutes I transitively, m_k is independent of t, and the number of elements of order 2^k in $\mathfrak{G}$ is $(q-1)m_k$ $(k \geq 1)$. Hence

$$|\mathfrak{G}| - 1 = (q-1)(m_1 + m_2 + \cdots),$$

and $q - 1$ divides $|\mathfrak{G}| - 1$. Since q and $|\mathfrak{G}|$ are both powers of 2, it follows that $|\mathfrak{G}|$ is a power of q.

c) By a), $\mathsf{I} \subseteq \Omega_1(\mathbf{Z}(\mathfrak{G}))$, so $\mathfrak{G}/\Omega_1(\mathbf{Z}(\mathfrak{G}))$ is of exponent 2 and hence elementary Abelian. Thus $\mathfrak{G}' \leq \Omega_1(\mathbf{Z}(\mathfrak{G}))$. Since $\mathfrak{G}$ is non-Abelian, $1 \neq \mathfrak{G}'$. But then, if $\mathfrak{G}' < \Omega_1(\mathbf{Z}(\mathfrak{G}))$, there are involutions both inside and outside $\mathfrak{G}'$, which is not possible on account of the transitivity of **Aut** $\mathfrak{G}$ on I. Hence $\mathfrak{G}' = \Omega_1(\mathbf{Z}(\mathfrak{G}))$. Therefore $\mathfrak{G}/\mathfrak{G}'$ is elementary Abelian and $\Phi(\mathfrak{G}) = \mathfrak{G}'$. If $\mathfrak{G}' < \mathbf{Z}(\mathfrak{G})$, $\mathbf{Z}(\mathfrak{G})$ contains an element z of order 4. Now if $y \in \mathfrak{G} - \mathbf{Z}(\mathfrak{G})$, $y^2 \in \mathsf{I}$. Since also $z^2 \in \mathsf{I}$, $y^2 = (z^2)\alpha$ for some $\alpha \in \mathbf{Aut}\,\mathfrak{G}$. But then $y^2 = (z\alpha)^2$ and $y(z\alpha)^{-1} \in \mathsf{I} \cup \{1\} \subseteq \mathbf{Z}(\mathfrak{G})$. Thus $y \in \mathbf{Z}(\mathfrak{G})$, a contradiction. Hence $\mathfrak{G}' = \mathbf{Z}(\mathfrak{G}) = \Omega_1(\mathbf{Z}(\mathfrak{G}))$.

By b), $|\mathfrak{G}| = q^m$ for some m. By 5.4, $m \leq 3$. Since $\mathfrak{G}$ is non-Abelian, $m > 1$. **q.e.d.**

5.6 Definition. An Abelian p-group $\mathfrak{A}$ is called *homocyclic* if $\mathfrak{A}$ is the direct product of cyclic groups of the same order.

5.7 Lemma. *Let $\mathfrak{A}$ be a homocyclic p-group of exponent p^n.*

a) *Suppose that $\mathfrak{A} = \langle a_1, \ldots, a_d \rangle$, where $p^d = |\mathfrak{A} : \Phi(\mathfrak{A})|$. Given elements $x_1, \ldots, x_d$ in any Abelian p-group $\mathfrak{X}$ of exponent at most p^n, there exists a unique homomorphism α of $\mathfrak{A}$ into $\mathfrak{X}$ for which $a_i\alpha = x_i$ $(i = 1, \ldots, d)$.*

b) *Suppose that* β *is a homomorphism of* $\mathfrak{A}$ *into* $\mathfrak{B}/\mathfrak{C}$, *where* $\mathfrak{B}$ *is an Abelian p-group of exponent at most* p^n *and* $\mathfrak{C} \leq \mathfrak{B}$. *Then there exists a homomorphism* ξ *of* $\mathfrak{A}$ *into* $\mathfrak{B}$ *such that* $\beta = \xi\nu$, *where* ν *is the natural epimorphism of* $\mathfrak{B}$ *onto* $\mathfrak{B}/\mathfrak{C}$.

Proof. a) This is obvious since

$$\mathfrak{A} = \langle a_1 \rangle \times \cdots \times \langle a_d \rangle.$$

b) If $a_i\beta = b_i\mathfrak{C}$, the homorphism ξ for which $a_i\xi = b_i$ has the required property. **q.e.d.**

5.8 Theorem. a) *If* $\mathfrak{A}$ *is an Abelian p-group and* **Aut** $\mathfrak{A}$ *induces an irreducible group on* $\Omega_1(\mathfrak{A})$, $\mathfrak{A}$ *is homocyclic.*

b) *If* $\mathfrak{A}$ *is an Abelian p-group and* **Aut** $\mathfrak{A}$ *permutes the set of subgroups of* $\mathfrak{A}$ *of order p transitively, then* $\mathfrak{A}$ *is homocyclic.*

c) *A homocyclic p-group* $\mathfrak{A}$ *possesses an automorphism* α *such that* $\langle\alpha\rangle$ *permutes the set of elements of* $\mathfrak{A}$ *of order p transitively.*

Proof. a) Suppose that $\mathfrak{A}$ is not homocyclic and let p^n be the exponent of $\mathfrak{A}$. Then $1 < \mathfrak{A}^{p^{n-1}} < \Omega_1(\mathfrak{A})$, so **Aut** $\mathfrak{A}$ induces a reducible group on $\Omega_1(\mathfrak{A})$.

b) This follows at once from a).

c) $\mathfrak{A}/\Phi(\mathfrak{A})$ is isomorphic to the additive group of some Galois field $\mathsf{K} = GF(p^k)$. If $\mathsf{K}^\times = \langle\omega\rangle$, the mapping $\lambda \to \lambda\omega$ ($\lambda \in \mathsf{K}$) is an additive automorphism γ of K such that $\gamma^{p^k-1} = 1$ and γ permutes transitively the set of elements of order p. Thus $\mathfrak{A}/\Phi(\mathfrak{A})$ possesses such an automorphism β. Let ν be the natural epimorphism of $\mathfrak{A}$ onto $\mathfrak{A}/\Phi(\mathfrak{A})$. Then by 5.7b), $\nu\beta = \alpha\nu$ for some endomorphism α of $\mathfrak{A}$. Then $\langle\alpha\rangle$ induces on $\mathfrak{A}/\Phi(\mathfrak{A})$ a group of automorphisms which permutes the non-identity elements transitively. If the exponent of $\mathfrak{A}$ is p^n, $x\Phi(\mathfrak{A}) \to x^{p^{n-1}}$ is an (**End** $\mathfrak{A}$)-isomorphism of $\mathfrak{A}/\Phi(\mathfrak{A})$ onto $\Omega_1(\mathfrak{A})$, so $\langle\alpha\rangle$ induces on $\Omega_1(\mathfrak{A})$ a group of automorphisms which permutes the non-identity elements transitively. Thus α is an automorphism of $\mathfrak{A}$. **q.e.d.**

5.9 Theorem. *Suppose that* $\mathfrak{A}$ *is an Abelian p-group and that* $\mathfrak{X}$ *is a* p'-*group of automorphisms of* $\mathfrak{A}$. *Then* $\mathfrak{A}$ *is the direct product of* $\mathfrak{X}$-*invariant homocyclic subgroups* $\mathfrak{A}_1, \ldots, \mathfrak{A}_r$, *and* $\mathfrak{A}_i/\Phi(\mathfrak{A}_i)$ *is an irreducible* $\mathfrak{X}$-*group.*

Proof. This is proved by induction on $|\mathfrak{A}|$. Let p^n be the exponent of $\mathfrak{A}$.

Suppose first that $\Omega_1(\mathfrak{A})$ is an irreducible $\mathfrak{X}$-group. Then $\Omega_1(\mathfrak{A})$ is irreducible under **Aut** $\mathfrak{A}$, so by 5.8a), $\mathfrak{A}$ is homocyclic. Hence $\mathfrak{A}/\Phi(\mathfrak{A})$ is $\mathfrak{X}$-isomorphic to $\Omega_1(\mathfrak{A})$ and is therefore an irreducible $\mathfrak{X}$-group. The theorem is thus proved in this case.

Now suppose that $\Omega_1(\mathfrak{A})$ is a reducible $\mathfrak{X}$-group. Since $\mathfrak{A}^{p^{n-1}} \neq 1$, there is a minimal non-identity $\mathfrak{X}$-invariant subgroup $\mathfrak{B}$ of $\mathfrak{A}^{p^{n-1}}$. Since $\Omega_1(\mathfrak{A})$ is reducible, $\mathfrak{B} < \Omega_1(\mathfrak{A})$. By the Maschke-Schur theorem (I, 17.7), $\Omega_1(\mathfrak{A}) = \mathfrak{B} \times \mathfrak{U}$ for some $\mathfrak{X}$-invariant subgroup $\mathfrak{U}$ of $\Omega_1(\mathfrak{A})$. Thus $\mathfrak{U} \neq 1$. If $v \in \mathfrak{B} - \{1\}$, there exists $x \in \mathfrak{A}$ such that $x^{p^{n-1}} = v$. Then the order of $x\mathfrak{U}$ is p^n, since $v \notin \mathfrak{U}$. Thus the exponent of $\mathfrak{A}/\mathfrak{U}$ is p^n.

Since $\mathfrak{U} \neq 1$, it follows from the inductive hypothesis that $\mathfrak{A}/\mathfrak{U}$ is the direct product of $\mathfrak{X}$-invariant homocyclic subgroups with $\mathfrak{X}$-irreducible Frattini factor groups. Since the exponent of $\mathfrak{A}/\mathfrak{U}$ is p^n, at least one of these factors, say $\mathfrak{H}/\mathfrak{U}$, is of exponent p^n. We write

$$\mathfrak{A}/\mathfrak{U} = (\mathfrak{H}/\mathfrak{U}) \times (\mathfrak{B}/\mathfrak{U}),$$

where $\mathfrak{H}$, $\mathfrak{B}$ are $\mathfrak{X}$-invariant subgroups of $\mathfrak{A}$. Also $\mathfrak{H}/\mathfrak{U}$ is homocyclic of exponent p^n, and its Frattini factor group is $\mathfrak{X}$-irreducible.

We observe that $\mathfrak{H}^p \cap \mathfrak{U} = 1$. For if $z \in \mathfrak{H}^p \cap \mathfrak{U}$, there exists $y \in \mathfrak{H}$ such that $z = y^p$. Then $y\mathfrak{U}$ is an element of $\mathfrak{H}/\mathfrak{U}$ of order at most p. Since $\mathfrak{H}/\mathfrak{U}$ is homocyclic of exponent p^n,

$$y\mathfrak{U} \in (\mathfrak{H}/\mathfrak{U})^{p^{n-1}} = \mathfrak{H}^{p^{n-1}}\mathfrak{U}/\mathfrak{U}$$

and $y \in \mathfrak{H}^{p^{n-1}}\mathfrak{U}$. Hence $z = y^p \in \mathfrak{H}^{p^n}\mathfrak{U}^p = 1$. Thus $\mathfrak{H}^p \cap \mathfrak{U} = 1$.

Now $\mathfrak{U}\mathfrak{H}^p/\mathfrak{H}^p$ is an $\mathfrak{X}$-invariant subgroup of $\mathfrak{H}/\mathfrak{H}^p$. Hence by the Maschke-Schur theorem,

$$\mathfrak{H}/\mathfrak{H}^p = (\mathfrak{U}\mathfrak{H}^p/\mathfrak{H}^p) \times (\mathfrak{D}/\mathfrak{H}^p)$$

for some $\mathfrak{X}$-invariant subgroup $\mathfrak{D}$ of $\mathfrak{H}$. Thus $\mathfrak{H} = \mathfrak{D}\mathfrak{U}\mathfrak{H}^p$, and $\mathfrak{H} = \mathfrak{D}\mathfrak{U}$. Since $\mathfrak{A}/\mathfrak{U} = (\mathfrak{H}/\mathfrak{U})(\mathfrak{B}/\mathfrak{U})$, it follows that

$$\mathfrak{A} = \mathfrak{H}\mathfrak{B} = \mathfrak{D}\mathfrak{U}\mathfrak{B} = \mathfrak{D}\mathfrak{B}.$$

Also $\mathfrak{H} \cap \mathfrak{B} = \mathfrak{U}$ and $\mathfrak{U}\mathfrak{H}^p \cap \mathfrak{D} = \mathfrak{H}^p$, so

$$\mathfrak{D} \cap \mathfrak{B} = \mathfrak{D} \cap \mathfrak{H} \cap \mathfrak{B} = \mathfrak{D} \cap \mathfrak{U} = \mathfrak{D} \cap \mathfrak{U}\mathfrak{H}^p \cap \mathfrak{U} = \mathfrak{H}^p \cap \mathfrak{U} = 1.$$

Thus $\mathfrak{A} = \mathfrak{D} \times \mathfrak{B}$. Then $\mathfrak{H} = \mathfrak{D} \times (\mathfrak{B} \cap \mathfrak{H}) = \mathfrak{D} \times \mathfrak{U}$, so $\mathfrak{D} \cong \mathfrak{H}/\mathfrak{U}$ is homocyclic and $\mathfrak{D}/\Phi(\mathfrak{D})$ is $\mathfrak{X}$-irreducible. Also $\mathfrak{D} \neq 1$, so the inductive hypothesis may be applied to $\mathfrak{B}$, and the assertion of the theorem follows at once. **q.e.d.**

5.10 Theorem. *Suppose that $\mathfrak{A}$ is an Abelian p-group and that $\mathfrak{X}$ is a p'-group of automorphisms of $\mathfrak{A}$. The following conditions are equivalent.*

a) *$\mathfrak{A}$ is $\mathfrak{X}$-indecomposable.*

b) *$\mathfrak{A}/\Phi(\mathfrak{A})$ is $\mathfrak{X}$-irreducible.*

c) *The only $\mathfrak{X}$-subgroups of $\mathfrak{A}$ are $\Omega_i(\mathfrak{A})$ $(i = 0, 1, \ldots)$.*

d) *$\Omega_1(\mathfrak{A})$ is $\mathfrak{X}$-irreducible.*

Proof. First we deduce b), c) and d) from a). Suppose that $\mathfrak{A}$ is $\mathfrak{X}$-indecomposable. By 5.9, $\mathfrak{A}$ is homocyclic and b) holds. Therefore $\Omega_1(\mathfrak{A})$ is $\mathfrak{X}$-isomorphic to $\mathfrak{A}/\Phi(\mathfrak{A})$. Hence d) holds. To prove c), let $\mathfrak{B}$ be a non-identity $\mathfrak{X}$-subgroup of $\mathfrak{A}$. If the exponent of $\mathfrak{B}$ is p^m, $\mathfrak{B}^{p^{m-1}}$ is a non-identity $\mathfrak{X}$-invariant subgroup of $\Omega_1(\mathfrak{A})$, so by d), $\Omega_1(\mathfrak{A}) = \mathfrak{B}^{p^{m-1}} = \Omega_1(\mathfrak{B})$. Thus $\mathfrak{B}$ is homocyclic and $\mathfrak{B} = \Omega_i(\mathfrak{A})$ for some i.

The assertions c) $\Rightarrow$ d), d) $\Rightarrow$ a), b) $\Rightarrow$ a) are all trivial. **q.e.d.**

5.11 Corollary. *Suppose that $\mathfrak{A}$ is an Abelian p-group, that $\mathfrak{X}$ is a p'-group of automorphisms of $\mathfrak{A}$ and that $\mathfrak{A}$ is $\mathfrak{X}$-indecomposable.*

a) *Every $\mathfrak{X}$-subgroup of $\mathfrak{A}$ is $\mathfrak{X}$-indecomposable.*

b) *Every $\mathfrak{X}$-factor group of $\mathfrak{A}$ is $\mathfrak{X}$-indecomposable.*

Proof. This follows at once from 5.10. **q.e.d.**

5.12 Examples. a) It is possible to have a homocyclic p-group $\mathfrak{A}$ with a p-group $\mathfrak{X}$ of automorphisms for which $\mathfrak{A}$ is $\mathfrak{X}$-indecomposable but $\mathfrak{A}/\Phi(\mathfrak{A})$ is reducible.

For let $\mathfrak{A} = \langle a_1 \rangle \oplus \langle a_2 \rangle$, where a_1, a_2 are of order p^2, and let $\mathfrak{X}$ be the group of all automorphisms α which induce the identity automorphism of $\mathfrak{A}/\Phi(\mathfrak{A})$. Then

$$a_1\alpha = (1 + pk)a_1 + pla_2, \quad a_2\alpha = pma_1 + (1 + pn)a_2,$$

so $\mathfrak{X}$ is elementary Abelian of order p^4. Also $\mathfrak{X}$ induces the identity automorphism on $\Omega_1(\mathfrak{A})$. But $\mathfrak{A}$ is an indecomposable $\mathfrak{X}$-group. For otherwise $\mathfrak{A} = \langle b_1 \rangle \oplus \langle b_2 \rangle$ with $b_i\alpha \in \langle b_i \rangle$ for all $\alpha \in \mathfrak{X}$. But then

$$b_i\alpha - b_i \in \langle b_i \rangle \cap \Phi(\mathfrak{A}) = \langle pb_i \rangle,$$

so $|\mathfrak{X}| \leq p^2$, a contradiction.

b) There exist non-homocyclic Abelian p-groups which are $\mathfrak{X}$-indecomposable for some automorphism group $\mathfrak{X}$.

For let $\mathfrak{A} = \langle a_1 \rangle \oplus \langle a_2 \rangle$, where a_1 is of order p^2 and a_2 is of order p. Let α be the automorphism of $\mathfrak{A}$ for which

$$a_1\alpha = (p + 1)a_1, \quad a_2\alpha = pa_1 + a_2.$$

If $\mathfrak{A}$ is $\langle \alpha \rangle$-decomposable, $\mathfrak{A} = \langle b_1 \rangle \oplus \langle b_2 \rangle$, where $p^2 b_1 = pb_2 = 0$ and $b_i\alpha \in \langle b_i \rangle$ $(i = 1, 2)$. Since α induces the identity mapping on $\mathfrak{A}/\Phi(\mathfrak{A})$, $b_2\alpha - b_2 \in \langle b_2 \rangle \cap \langle pb_1 \rangle = 0$, so $b_2\alpha = b_2$. Since $\mathbf{C}_{\mathfrak{A}}(\alpha) \cap \Omega_1(\mathfrak{A}) = \langle pa_1 \rangle = p\mathfrak{A}$, $b_2 \in p\mathfrak{A}$, a contradiction.

5.13 Theorem. *Suppose that $\mathfrak{G}$ is a 2-group, $\mathfrak{A}$ is an Abelian subgroup of $\mathfrak{G}$ and $x \in \mathfrak{G}$. If $x^2 \in \mathfrak{A}^2$ and $[\mathfrak{A}, x] \leq \mathfrak{A}^4$, there exists $a \in \mathfrak{A}$ such that $(xa)^2 = 1$.*

Proof. If this is false, there exists a least positive integer l such that $(xa)^2 \notin \mathfrak{A}^{2^l}$ for all $a \in \mathfrak{A}$. Since $x^2 \in \mathfrak{A}^2$, $l > 1$. Thus there exists $a \in \mathfrak{A}$ and $b \in \mathfrak{A}$ such that $(xa)^2 = b^{2^{l-1}}$. Let $c = b^{2^{l-2}}$. Then

$$(xac^{-1})^2 = (xa)^2[xa, c]c^{-2} = [xa, c].$$

Now $[\mathfrak{A}, x] \leq \mathfrak{A}^4$, so $[xa, b] \in \mathfrak{A}$ and

$$[xa, c] = [xa, b]^{2^{l-2}} = [x, b]^{2^{l-2}} \in \mathfrak{A}^{2^l}.$$

Thus $(xac^{-1})^2 \in \mathfrak{A}^{2^l}$, contrary to the definition of l. **q.e.d.**

5.14 Lemma. *Let $\xi_1, \ldots, \xi_n$ be linear transformations of a finite-dimensional vector space V. Suppose that $\xi_i\xi_j = \xi_j\xi_i$ $(i \neq j)$ and $\xi_i^2 = \xi_i$ $(i = 1, \ldots, n)$. Then $\xi_1, \ldots, \xi_n$ have a common eigen-vector.*

Proof. Let U be a non-zero subspace of minimal dimension for which $\mathsf{U}\xi_i \subseteq \mathsf{U}$ for all $i = 1, \ldots, n$. Suppose that $\dim \mathsf{U} > 1$. Let u be a non-zero element of U. Then U is not spanned by u, so by minimality of $\dim \mathsf{U}$, $u\xi_i$ does not lie in the space spanned by u for some i. Hence the restriction ξ of ξ_i to U is neither the zero nor the identity mapping. Since $\xi^2 = \xi$, it follows that $\ker \xi$ is a proper non-zero subspace of U. Since $\xi_i\xi_j = \xi_j\xi_i$, $(\ker \xi)\xi_j \subseteq \ker \xi$ for all $j = 1, \ldots, n$. This contradicts the minimality of U. Hence $\dim \mathsf{U} = 1$, and the non-zero elements of U are eigen-vectors of all the ξ_i. **q.e.d.**

5.15 Theorem. *Suppose that $\mathfrak{A}$ is a normal homocyclic Abelian subgroup of the 2-group $\mathfrak{G}$ such that $\mathfrak{A}^4 \neq 1$, $[\Omega_1(\mathfrak{A}), \mathfrak{G}] = 1$ and $[\mathfrak{A}, \Phi(\mathfrak{G})] = 1$. Then $\mathfrak{A}$ contains a cyclic normal subgroup of $\mathfrak{G}$ of order 4.*

Proof. We show first that given $x \in \mathfrak{G}$, there is an endomorphism $\eta(x)$ of $\mathfrak{A}$ such that

$$a^x = a(1 - 2\eta(x)) \tag{2}$$

for all $a \in \mathfrak{A}$. To do this, write $\mathfrak{A} = \langle a_1 \rangle \times \cdots \times \langle a_r \rangle$. Since $\mathfrak{A}$ is homocyclic, $\mathfrak{A}/\mathfrak{A}^2$ is $\mathfrak{G}$-isomorphic to $\Omega_1(\mathfrak{A})$. Hence $[\mathfrak{A}, \mathfrak{G}] \leq \mathfrak{A}^2$ and $[a_i, x] = a_i'^2$ for some $a_i' \in \mathfrak{A}$. By 5.7a), there is an endomorphism $\eta(x)$ of $\mathfrak{A}$ such that $a_i\eta(x) = a_i'^{-1}$ $(i = 1, \ldots, r)$. Clearly $\eta(x)$ satisfies (2).

If x, y are in $\mathfrak{G}$ and $a \in \mathfrak{A}$,

$$a(1 - 2\eta(xy)) = a^{xy} = (a^x)(1 - 2\eta(y)) = a(1 - 2\eta(x))(1 - 2\eta(y)).$$

Hence

$$2\eta(xy) = 2\eta(x) + 2\eta(y) - 4\eta(x)\eta(y) \tag{3}$$

for all x, y in $\mathfrak{G}$. Of course if $z \in \mathbf{C}_{\mathfrak{G}}(\mathfrak{A})$, then $2\eta(z) = 0$ by (2), and $2\eta(xz) = 2\eta(x)$ for all $x \in \mathfrak{G}$ by (3). Thus for any x, y in $\mathfrak{G}$, $2\eta(xy) = 2\eta(yx)$, since $[x, y] \in \Phi(\mathfrak{G}) \leq \mathbf{C}_{\mathfrak{G}}(\mathfrak{A})$. Hence by (3),

$$4\eta(x)\eta(y) = 4\eta(y)\eta(x). \tag{4}$$

Also, putting $x = y$ in (3), we get

$$4\eta(x) = 4\eta(x)^2, \tag{5}$$

as $x^2 \in \Phi(\mathfrak{G}) \leq \mathbf{C}_{\mathfrak{G}}(\mathfrak{A})$.

Let $\bar{\eta}(x)$ be the restriction of $\eta(x)$ to $\mathfrak{B} = \Omega_1(\mathfrak{A})$. Thus $\bar{\eta}(x)$ is an endomorphism of $\mathfrak{B}$. Since $\mathfrak{A}^4 \neq 1$ and $\mathfrak{A}$ is homocyclic, $\mathfrak{B} \leq \mathfrak{A}^4$. Thus (4) and (5) give

$$\bar{\eta}(x)\bar{\eta}(y) = \bar{\eta}(y)\bar{\eta}(x),$$
$$\bar{\eta}(x)^2 = \bar{\eta}(x).$$

Since $\mathfrak{B}$ may be regarded as a vector space over $GF(2)$, it follows from 5.14 that $\mathfrak{B}$ contains an element b such that $b\bar{\eta}(x) \in \langle b \rangle$ for all $x \in \mathfrak{G}$. Then $b = a^2$ for some a and $(a^2)\bar{\eta}(x) = a^{2j}$ for some j. Thus

$$a^x = a((a^{-2})\eta(x)) = a^{1-2j}.$$

Hence $\langle a \rangle$ is a cyclic normal subgroup of $\mathfrak{G}$ of order 4. **q.e.d.**

In the application of this, we shall need the following elementary fact.

5.16 Lemma. *Let* V *be a vector space over a field* K, *let* S *be a set of linear transformations of* V, *and let* $\phi_1, \ldots, \phi_n$ *be distinct mappings of* S *into* K. *Let*

$$\mathsf{V}_i = \{v \mid v \in \mathsf{V}, \quad vx = \phi_i(x)v \quad \textit{for all } x \in \mathsf{S}\} \quad (i = 1, \ldots, n).$$

If U *is the* K*-subspace of* V *spanned by* $\mathsf{V}_1, \ldots, \mathsf{V}_n$,

$$\mathsf{U} = \mathsf{V}_1 \oplus \cdots \oplus \mathsf{V}_n.$$

Proof. V_i is a K-subspace of V, so

$$\mathsf{U} = \mathsf{V}_1 + \cdots + \mathsf{V}_n.$$

Suppose that the sum is not direct. Let d be the smallest positive integer for which there exist non-zero elements $v_1, \ldots, v_d$ such that

$$v_1 + \cdots + v_d = 0$$

and $v_1 \in \mathsf{V}_{i_1}, \ldots, v_d \in \mathsf{V}_{i_d}$ with $i_1 < i_2 < \cdots < i_d$. Then $d > 1$, so there exists $x \in \mathsf{S}$ such that $\phi_{i_1}(x) \neq \phi_{i_2}(x)$. But

$$\begin{aligned} 0 &= v_1 x + \cdots + v_d x \\ &= \phi_{i_1}(x)v_1 + \cdots + \phi_{i_d}(x)v_d. \end{aligned}$$

Hence

$$(\phi_{i_2}(x) - \phi_{i_1}(x))v_2 + \cdots + (\phi_{i_d}(x) - \phi_{i_1}(x))v_d = 0,$$

contrary to the definition of d. **q.e.d.**

Exercise

4) Let $\mathfrak{G}$ be a special non-Abelian 2-group in which every involution lies in $\mathbf{Z}(\mathfrak{G})$. Suppose that the number of involutions in $\mathfrak{G}/\mathfrak{M}$ is the same for all maximal subgroups $\mathfrak{M}$ of $\mathbf{Z}(\mathfrak{G})$. Prove that $|\mathfrak{G}|$ is $|\mathbf{Z}(\mathfrak{G})|^2$ or $|\mathbf{Z}(\mathfrak{G})|^3$.

§ 6. Commutators and Bilinear Mappings

The general formulae

$$[xy, z] = [x, z]^y[y, z], \tag{1}$$

$$[x, yz] = [x, z][x, y]^z \tag{2}$$

closely resemble the conditions for a mapping to be bilinear over $\mathbb{Z}$. We make this more precise.

6.1 Lemma. *Suppose that $\mathfrak{H}_1, \mathfrak{H}_2, \overline{\mathfrak{H}}_1, \overline{\mathfrak{H}}_2, \mathfrak{K}, \overline{\mathfrak{K}}$ are subgroups of a group $\mathfrak{G}$. Suppose that $\overline{\mathfrak{K}} \trianglelefteq \mathfrak{K}$, $\overline{\mathfrak{H}}_i \trianglelefteq \mathfrak{H}_i$ and that $\mathfrak{H}_i/\overline{\mathfrak{H}}_i$ is Abelian ($i = 1, 2$). Suppose also that*

$$[\mathfrak{H}_1, \mathfrak{H}_2] \leq \mathfrak{K}, \quad [\mathfrak{H}_1, \overline{\mathfrak{H}}_2] \leq \overline{\mathfrak{K}}, \quad [\overline{\mathfrak{H}}_1, \mathfrak{H}_2] \leq \overline{\mathfrak{K}}, \quad [\mathfrak{K}, \mathfrak{G}] \leq \overline{\mathfrak{K}}.$$

Then there exists a $\mathbb{Z}$-bilinear mapping γ of $(\mathfrak{H}_1/\overline{\mathfrak{H}}_1) \times (\mathfrak{H}_2/\overline{\mathfrak{H}}_2)$ into $\mathfrak{K}/\overline{\mathfrak{K}}$ such that if $x \in \mathfrak{H}_1$ and $y \in \mathfrak{H}_2$,

$$(x\overline{\mathfrak{H}}_1, y\overline{\mathfrak{H}}_2)\gamma = [x, y]\overline{\mathfrak{K}}.$$

Proof. Let g be the mapping of $\mathfrak{H}_1 \times \mathfrak{H}_2$ into $\mathfrak{K}/\overline{\mathfrak{K}}$ for which

$$(x, y)g = [x, y]\overline{\mathfrak{K}} \quad (x \in \mathfrak{H}_1, y \in \mathfrak{H}_2).$$

We verify that if $u \in \overline{\mathfrak{H}}_1$ and $v \in \overline{\mathfrak{H}}_2$, then

$$(xu, yv)g = (x, y)g.$$

Indeed

$$\begin{aligned}(xu, yv)g &= [xu, yv]\overline{\mathfrak{K}}\\ &= [x, v]^u[x, y]^{vu}[u, v][u, y]^v\overline{\mathfrak{K}},\end{aligned}$$

by (1) and (2). From the hypotheses we see that all these terms lie in $\overline{\mathfrak{K}}$ except $[x, y]^{vu}$. But also $[x, y, vu] \in \overline{\mathfrak{K}}$ since $[\mathfrak{K}, \mathfrak{G}] \leq \overline{\mathfrak{K}}$, so

$$(xu, yv)g = [x, y]\overline{\mathfrak{K}} = (x, y)g,$$

as asserted.

It follows that there exists a mapping γ of $(\mathfrak{H}_1/\overline{\mathfrak{H}}_1) \times (\mathfrak{H}_2/\overline{\mathfrak{H}}_2)$ into $\mathfrak{K}/\overline{\mathfrak{K}}$ such that if $x \in \mathfrak{H}_1$, $y \in \mathfrak{H}_2$,

$$(x\overline{\mathfrak{H}}_1, y\overline{\mathfrak{H}}_2)\gamma = (x, y)g = [x, y]\overline{\mathfrak{K}}.$$

If also $x' \in \mathfrak{H}_1$,

$$\begin{aligned}(xx'\overline{\mathfrak{H}}_1, y\overline{\mathfrak{H}}_2)\gamma &= [xx', y]\overline{\mathfrak{K}}\\ &= [x, y]^{x'}[x', y]\overline{\mathfrak{K}}\end{aligned}$$

by (1). Since $[x, y, x'] \in [\mathfrak{K}, \mathfrak{G}] \leq \overline{\mathfrak{K}}$,

$$\begin{aligned}(xx'\overline{\mathfrak{H}}_1, y\overline{\mathfrak{H}}_2)\gamma &= [x, y][x', y]\overline{\mathfrak{K}}\\ &= ((x\overline{\mathfrak{H}}_1, y\overline{\mathfrak{H}}_2)\gamma)((x'\overline{\mathfrak{H}}_1, y\overline{\mathfrak{H}}_2)\gamma).\end{aligned}$$

This and the similar equation in the second variable show that γ is $\mathbb{Z}$-bilinear. **q.e.d.**

In our first applications of 6.1, $\mathfrak{G}$ has an automorphism ξ which induces irreducible automorphisms on the $\mathfrak{H}_i/\overline{\mathfrak{H}}_i$. It follows from II, 3.10 that under these circumstances, there are isomorphisms of these groups onto the additive groups of fields. Thus we examine bilinear mappings on fields.

6.2 Lemma. *Suppose that* K, F *are fields and that* F *is a Galois extension of* K. *Let* $\mathfrak{G}$ *be the Galois group of* F *over* K *and let* V *be a finite dimensional vector space over* F.

a) *If* α *is a* K*-linear mapping of* F *into* V, *there exists a unique set of elements* $v_\xi \in \mathsf{V}$ $(\xi \in \mathfrak{G})$ *such that*

$$a\alpha = \sum_{\xi \in \mathfrak{G}} (a\xi)v_\xi \quad (a \in \mathsf{F}).$$

b) *Suppose that* $\mathsf{K} \subseteq \mathsf{L}_i \subseteq \mathsf{F}$, L_i *is a Galois extension of* K *and* $\mathfrak{H}_i$ *is the Galois group of* L_i *over* K $(i = 1, 2)$. *If* β *is a* K*-bilinear mapping of* $\mathsf{L}_1 \times \mathsf{L}_2$ *into* V, *there exists a unique set of elements* $v_{\xi\eta} \in \mathsf{V}$ $(\xi \in \mathfrak{H}_1, \eta \in \mathfrak{H}_2)$ *such that*

$$(a, b)\beta = \sum_{\xi, \eta} (a\xi)(b\eta)v_{\xi\eta} \quad (a \in \mathsf{L}_1, b \in \mathsf{L}_2).$$

c) *Let* γ *be the* K*-bilinear mapping of* $\mathsf{F} \times \mathsf{F}$ *into* V *given by*

$$(a, b)\gamma = \sum_{\xi,\eta} (a\xi)(b\eta)v_{\xi\eta}.$$

If $(a, a)\gamma = 0$ *for all* $a \in \mathsf{F}$, *then* $v_{\xi\xi} = v_{\xi\eta} + v_{\eta\xi} = 0$ *for all* $\xi \in \mathfrak{G}$, $\eta \in \mathfrak{G}$.

Proof. a) Let M be the K-space of all row vectors (v_ξ) with $v_\xi \in \mathsf{V}$ and let N be the K-space of all K-linear mappings of F into V. Thus

$$\dim_{\mathsf{K}} \mathsf{M} = |\mathfrak{G}| \dim_{\mathsf{K}} \mathsf{V} = [\mathsf{F} : \mathsf{K}] \dim_{\mathsf{K}} \mathsf{V} = \dim_{\mathsf{K}} \mathsf{N}.$$

Define a mapping μ of M into N by $(v_\xi)\mu = \alpha$, where

$$a\alpha = \sum_{\xi} (a\xi)v_\xi \quad (a \in \mathsf{F}).$$

Thus μ is K-linear. Suppose that $(v_\xi)\mu = 0$. Then

$$\sum_{\xi} (a\xi)v_\xi = 0$$

for all $a \in \mathsf{F}$. If $u_1, \ldots, u_m$ is an F-basis of V and $v_\xi = \sum_{i=1}^{m} \lambda_{\xi,i}u_i$, $(\lambda_{\xi,i} \in \mathsf{F})$, then for any $a \in \mathsf{F}$,

$$a(\sum \lambda_{\xi,i}\xi) = \sum_{\xi} (a\xi)\lambda_{\xi,i} = 0.$$

Since distinct homomorphisms of a group into the multiplicative group of non-zero elements of a field are linearly independent (Dedekind's lemma), it follows that all $\lambda_{\xi,i} = 0$ and $(v_\xi) = 0$. Thus μ is a monomorphism. Since $\dim_{\mathsf{K}} \mathsf{M} = \dim_{\mathsf{K}} \mathsf{N}$, it follows that μ is an isomorphism. Thus a) is proved.

b) For each $b \in \mathsf{L}_2$, the mapping $a \to (a, b)\beta$ is a K-linear mapping of L_1 into V, so by a), there exists a unique set of elements $v_\xi(b) \in \mathsf{V}$ $(\xi \in \mathfrak{H}_1)$ such that

$$(a, b)\beta = \sum_{\xi \in \mathfrak{H}_1} (a\xi)v_\xi(b) \quad (a \in \mathsf{L}_1, b \in \mathsf{L}_2).$$

But then v_ξ is a K-linear mapping of L_2 into V, so again by a), there is a unique set of elements $v_{\xi\eta} \in \mathsf{V}$ $(\eta \in \mathfrak{H}_2)$ such that

$$v_\xi(b) = \sum_{\eta \in \mathfrak{H}_2} (b\eta)v_{\xi\eta} \quad (b \in \mathsf{L}_2).$$

The assertion now follows easily.

c) By b), the K-dimension of the space N′ of mappings

$$(a, b) \to \sum_{\xi, \eta \in \mathfrak{G}} (a\xi)(b\eta)v_{\xi\eta},$$

where $v_{\xi\xi} = v_{\xi\eta} + v_{\eta\xi} = 0$, is $\frac{1}{2}|\mathfrak{G}|(|\mathfrak{G}| - 1)\dim_K V$, which is also the dimension of the space M′ of bilinear mappings γ of $F \times F$ into V for which $(a, a)\gamma = 0$. Since $N' \subseteq M'$, it follows that $N' = M'$. **q.e.d.**

6.3 Corollary. *Let F be a Galois extension of K. Suppose that $F = K(\lambda)$, $\mu \in F$ and α is a non-zero K-linear mapping of F into a vector space V over F such that $(\lambda a)\alpha = \mu(a\alpha)$ for all $a \in F$. Then there is a field automorphism θ of F over K such that $\mu = \lambda\theta$ and $a\alpha = (a\theta)v$ for some $v \in V$.*

Proof. Let $\mathfrak{G}$ be the Galois group of F over K. By 6.2, there exist $v_\xi \in V$ $(\xi \in \mathfrak{G})$ such that

$$a\alpha = \sum_{\xi \in \mathfrak{G}} (a\xi)v_\xi$$

for all $a \in F$. Thus

$$\mu \sum_\xi (a\xi)v_\xi = \mu(a\alpha) = (\lambda a)\alpha = \sum_\xi (a\xi)(\lambda\xi)v_\xi.$$

By the uniqueness assertion of 6.2, $(\mu - \lambda\xi)v_\xi = 0$ for all $\xi \in \mathfrak{G}$. Since $\alpha \neq 0$, $v_\theta \neq 0$ for some $\theta \in \mathfrak{G}$. Thus $\mu = \lambda\theta$. If $\xi \neq \theta$, then $\mu = \lambda\theta \neq \lambda\xi$, since $F = K(\lambda)$. Hence $v_\xi = 0$ for all $\xi \neq \theta$, and $a\alpha = (a\theta)v_\theta$. **q.e.d.**

6.4 Lemma. *Let $K = GF(2)$, $F = GF(2^n)$. Suppose that $\lambda \in F^\times$ and that the multiplicative order of λ is k, where $2^n - 1$ divides kn. Suppose that $\mu \in F$ and that β is a K-bilinear mapping of $F \times F$ into F such that $(a, a)\beta = 0$ and $(\lambda a, \lambda b)\beta = \mu((a, b)\beta)$ for all, a, b in F. Then there is a K-linear transformation α of F and a field automorphism θ such that*

$$(a, b)\beta = (a(b\theta) + (a\theta)b)\alpha.$$

Proof. If $n = 1$, β is the zero mapping since β is symplectic. If $\beta = 0$, the assertion is trivial with $\theta = 1$. Suppose that $\beta \neq 0$. Thus $n > 1$.

F is a Galois extension of K and the Galois group is generated by the automorphism $x \to x^2$. Thus by 6.2,

$$(a, b)\beta = \sum_{i,j=1}^{n} \varepsilon_{ij} a^{2^{i-1}} b^{2^{j-1}}$$

for suitable $\varepsilon_{ij} \in \mathsf{F}$ with $\varepsilon_{ij} + \varepsilon_{ji} = \varepsilon_{ii} = 0$. However, not all ε_{ij} are zero; suppose $\varepsilon_{h,h+r} \neq 0$, where $0 < r < n$. Then

$$\sum_{i,j=1}^{n} \mu\varepsilon_{ij} a^{2^{i-1}} b^{2^{j-1}} = \mu((a, b)\beta) = (\lambda a, \lambda b)\beta$$

$$= \sum_{i,j=1}^{n} \varepsilon_{ij} \lambda^{2^{i-1}+2^{j-1}} a^{2^{i-1}} b^{2^{j-1}},$$

and by the uniqueness assertion of 6.2,

$$\varepsilon_{ij}(\mu - \lambda^{2^{i-1}+2^{j-1}}) = 0$$

for all i, j. Since $\varepsilon_{h,h+r} \neq 0$,

$$\mu = (\lambda^{1+2^r})^{2^{h-1}}.$$

Thus

$$\varepsilon_{ij}(\lambda^{(1+2^r)2^{h-1}} - \lambda^{2^{i-1}+2^{j-1}}) = 0$$

for all i, j. Hence if $\varepsilon_{ij} \neq 0$,

$$\lambda^{1+2^r} = \lambda^{2^{i-h+n}+2^{j-h+n}}$$

and

$$1 + 2^r \equiv 2^{i-h+n} + 2^{j-h+n} \quad (k).$$

But k, n satisfy the conditions (1) of § 4. Hence by 4.4c), $j - i \equiv \pm r\ (n)$. Hence, since $\varepsilon_{ij} + \varepsilon_{ji} = 0$,

$$(a, b)\beta = \sum_{i=1}^{n} \varepsilon_i (ab^{2^r} - a^{2^r} b)^{2^{i-1}}$$

for some ε_i. If α is the K-linear transformation of F for which

$$c\alpha = \sum_{i=1}^{n} \varepsilon_i c^{2^{i-1}},$$

the assertion is clear, θ being the automorphism $a \to a^{2^r}$. **q.e.d.**

6.5 Theorem. *Let $\mathfrak{G}$ be a 2-group, ξ an automorphism of $\mathfrak{G}$, $\mathfrak{H}$ a ξ-invariant normal subgroup of $\mathfrak{G}$. Suppose the following hold.*

(i) $|\mathfrak{G} : \mathfrak{H}| = |\mathfrak{H}| = 2^n$.

(ii) *$\mathfrak{G}$ is not elementary Abelian.*

(iii) *ξ induces irreducible automorphisms on $\mathfrak{G}/\mathfrak{H}$ and $\mathfrak{H}$, and if k is the order of that induced on $\mathfrak{G}/\mathfrak{H}$, $2^n - 1$ divides nk. Let $\mathsf{K} = GF(2)$, $\mathsf{F} = GF(2^n)$.*

a) *There exist isomorphisms ρ, σ of $\mathfrak{G}/\mathfrak{H}$, $\mathfrak{H}$ respectively onto the additive group of F, and there exists a field automorphism θ of F such that if $x \in \mathfrak{G}$ and $(x\mathfrak{H})\rho = a$, then $(x^2)\sigma = a(a\theta)$. Also θ is not of order 2, and $\theta = 1$ if and only if $\mathfrak{G}$ is Abelian.*

b) *There exists $\lambda \in \mathsf{F}$ such that $\mathsf{F} = \mathsf{K}(\lambda(\lambda\theta)) = \mathsf{K}(\lambda)$ and*

$$((x\xi)\mathfrak{H})\rho = \lambda((x\mathfrak{H})\rho), \quad y\xi\sigma = \lambda(\lambda\theta)(y\sigma)$$

for $x \in \mathfrak{G}$, $y \in \mathfrak{H}$.

Proof. First suppose that $\mathfrak{G}$ is Abelian. By (iii), $\mathfrak{H}$ is elementary Abelian, so $\Omega_1(\mathfrak{G}) \geq \mathfrak{H}$. Since $\Omega_1(\mathfrak{G})$ is ξ-invariant and ξ is irreducible on $\mathfrak{G}/\mathfrak{H}$, $\Omega_1(\mathfrak{G}) = \mathfrak{H}$ or $\mathfrak{G}$. By (ii), $\Omega_1(\mathfrak{G}) = \mathfrak{H}$. By 5.8, $\mathfrak{G}$ is homocyclic, so there exists an isomorphism χ of $\mathfrak{G}/\mathfrak{H}$ onto $\mathfrak{H}$ such that

$$(x\mathfrak{H})\chi = x^2 \quad (x \in \mathfrak{G}).$$

By (iii) and II, 3.10, there is an isomorphism ρ of $\mathfrak{G}/\mathfrak{H}$ onto the additive group of F and there is an element λ of F such that $\mathsf{F} = \mathsf{K}(\lambda)$ and

$$((x\xi)\mathfrak{H})\rho = \lambda((x\mathfrak{H})\rho)$$

for all $x \in \mathfrak{G}$. For $y \in \mathfrak{H}$, let $y\sigma = (y\chi^{-1}\rho)^2$; thus σ is an isomorphism of $\mathfrak{H}$ onto F. If $(x\mathfrak{H})\rho = a$, then $(x^2)\sigma = a^2 = a(a\theta)$, where $\theta = 1_{\mathsf{F}}$. Finally, if $y \in \mathfrak{H}$, then $y = x^2$ for some $x \in \mathfrak{G}$ and

$$y\xi\sigma = (y\xi\chi^{-1}\rho)^2 = ((x\xi)^2\chi^{-1}\rho)^2 = (((x\xi)\mathfrak{H})\rho)^2$$

$$= \lambda^2((x\mathfrak{H})\rho)^2 = \lambda^2(y\sigma).$$

Thus 6.5 is proved in this case, except for the assertion $\mathsf{F} = \mathsf{K}(\lambda(\lambda\theta))$.

Now suppose that $\mathfrak{G}$ is non-Abelian. Thus $n > 1$. Since $\mathfrak{G}'$, $[\mathfrak{G}, \mathfrak{H}]$ are ξ-invariant and $1 \neq \mathfrak{G}' \leq \mathfrak{H}$, $[\mathfrak{G}, \mathfrak{H}] < \mathfrak{H}$, it follows from (iii) that $\mathfrak{G}' = \mathfrak{H}$ and $[\mathfrak{G}, \mathfrak{H}] = 1$. Hence by 6.1, there exists a $\mathbb{Z}$-bilinear mapping γ of $(\mathfrak{G}/\mathfrak{H}) \times (\mathfrak{G}/\mathfrak{H})$ into $\mathfrak{H}$ such that if x, y are in $\mathfrak{G}$,

(3) $$(x\mathfrak{H}, y\mathfrak{H})\gamma = [x, y].$$

But by (i), (iii) and II, 3.10, there exist isomorphisms ρ, $\tilde{\rho}$ of $\mathfrak{G}/\mathfrak{H}$, $\mathfrak{H}$ onto the additive group of F, and there exist elements λ, μ in F such that $\mathsf{F} = \mathsf{K}(\lambda) = \mathsf{K}(\mu)$ and

(4) $$((x\xi)\mathfrak{H})\rho = \lambda((x\mathfrak{H})\rho), \quad y\xi\tilde{\rho} = \mu(y\tilde{\rho})$$

for all $x \in \mathfrak{G}$, $y \in \mathfrak{H}$. For a, b in F, write

(5) $$(a, b)\beta = ((a\rho^{-1}, b\rho^{-1})\gamma)\tilde{\rho}.$$

Then β is a K-bilinear mapping of $\mathsf{F} \times \mathsf{F}$ into F. If $a = (x\mathfrak{H})\rho, b = (y\mathfrak{H})\rho$,

$$\begin{aligned}
(\lambda a, \lambda b)\beta &= (((x\xi)\mathfrak{H})\rho, ((y\xi)\mathfrak{H})\rho)\beta && \text{(by (4))}\\
&= [x\xi, y\xi]\tilde{\rho} && \text{(by (5) and (3))}\\
&= [x, y]\xi\tilde{\rho} &&\\
&= \mu([x, y]\tilde{\rho}) && \text{(by (4))}\\
&= \mu((a, b)\beta) && \text{(by (5) and (3)).}
\end{aligned}$$

Clearly, $(a, a)\beta = 0$ for all $a \in \mathsf{F}$, and k is the order of λ. Hence by 6.4, there exists a K-linear transformation α of F and a field automorphism θ of F such that

(6) $$(a, b)\beta = (a(b\theta) + (a\theta)b)\alpha.$$

Since $\mathfrak{H} = \mathfrak{G}'$, $\mathfrak{H}$ is generated by $[x, y]$ with x, y in $\mathfrak{G}$. Hence F is the additive group generated by all $[x, y]\tilde{\rho}$; hence F is generated by $\operatorname{im}\beta$. Thus α is an epimorphism and F is generated by all $a(b\theta) + (a\theta)b$.

It follows that α is non-singular. Let $\sigma = \tilde{\rho}\alpha^{-1}$. If x, y are in $\mathfrak{G}$ and $a = (x\mathfrak{H})\rho$, $b = (y\mathfrak{H})\rho$, then by (3) and (5),

$$[x, y]\sigma = ((x\mathfrak{H}, y\mathfrak{H})\gamma)\sigma = (a, b)\beta\alpha^{-1}.$$

Hence by (6),

(7) $$[x, y]\sigma = a(b\theta) + (a\theta)b.$$

Since $((x\xi)\mathfrak{H})\rho = \lambda a$ and $((y\xi)\mathfrak{H})\rho = \lambda b$, (7) gives

$$[x\xi, y\xi]\sigma = \lambda(\lambda\theta)(a(b\theta) + (a\theta)b) = \lambda(\lambda\theta)([x, y]\sigma).$$

Hence $z\xi\sigma = \lambda(\lambda\theta)(z\sigma)$ for all $z \in \mathfrak{H} = \mathfrak{G}'$. Thus the equations of b) are proved. It follows also from (7) that $\theta \neq 1_{\mathsf{F}}$, since $\mathfrak{G}$ is non-Abelian.

To prove a), observe that if $a \in \mathsf{F}$, then $a = (x\mathfrak{H})\rho$ for some $x \in \mathfrak{G}$ and x^2 depends only on a. Write $a\tau = (x^2)\sigma$. If $b = (y\mathfrak{H})\rho$, then $a + b = (xy\mathfrak{H})\rho$ and by (7),

$$\begin{aligned}(a + b)\tau &= ((xy)^2)\sigma \\ &= (x^2y^2[x, y])\sigma \\ &= (x^2)\sigma + (y^2)\sigma + [x, y]\sigma \\ &= a\tau + b\tau + a(b\theta) + (a\theta)b.\end{aligned}$$

Hence if $a\tau' = a\tau + a(a\theta)$,

$$(a + b)\tau' = a\tau + b\tau + a(a\theta) + b(b\theta) = a\tau' + b\tau'.$$

Thus τ' is a K-linear mapping of F into F. By 6.2,

$$a\tau' = \sum_{i=1}^{n} \varepsilon_i a^{2^{i-1}}$$

for suitable ε_i, or

$$a\tau = a(a\theta) + \sum_{i=1}^{n} \varepsilon_i a^{2^{i-1}}.$$

Thus

$$(\lambda a)\tau = \lambda(\lambda\theta)a(a\theta) + \sum_{i=1}^{n} \varepsilon_i \lambda^{2^{i-1}} a^{2^{i-1}}.$$

But

$$\begin{aligned}(\lambda a)\tau &= ((x\xi)\mathfrak{H})\rho\tau = ((x\xi)^2)\sigma = (x^2)\xi\sigma \\ &= (\lambda(\lambda\theta))(x^2)\sigma = \lambda(\lambda\theta)(a\tau) = \lambda(\lambda\theta)a(a\theta) + \sum_{i=1}^{n} \varepsilon_i \lambda(\lambda\theta) a^{2^{i-1}}.\end{aligned}$$

By the uniqueness assertion of 6.2,

$$\varepsilon_i(\lambda^{2^{i-1}} - \lambda(\lambda\theta)) = 0$$

for all i. But $\theta \neq 1$; hence if $a\theta = a^{2^r}$, $r \not\equiv 0$ (n). Hence by 4.3c), $1 + 2^r \not\equiv 2^{i-1}$ (k), and $\lambda(\lambda\theta) \neq \lambda^{2^{i-1}}$ for all i. Thus $\varepsilon_i = 0$ for all i, and $(x^2)\sigma = a\tau = a(a\theta)$.

Whether or not $\mathfrak{G}$ is Abelian, let $\mathsf{F}_1 = \mathsf{K}(\lambda(\lambda\theta))$. If $y\sigma \in \mathsf{F}_1$ $(y \in \mathfrak{H})$, then $y\xi\sigma \in \mathsf{F}_1$. Hence $\mathsf{F}_1\sigma^{-1}$ is a ξ-invariant subgroup of $\mathfrak{H}$. Since $\lambda \neq 0$, it follows that $\mathsf{F}_1\sigma^{-1} = \mathfrak{H}$ and $\mathsf{F}_1 = \mathfrak{H}\sigma = \mathsf{F}$. Thus $\mathsf{F} = \mathsf{K}(\lambda(\lambda\theta))$. It follows in particular that θ is not of order 2, since otherwise $\lambda(\lambda\theta)$ lies in the field of elements fixed by θ. **q.e.d.**

We show next that the group $\mathfrak{G}$ in 6.5 is determined to within isomorphism by θ.

6.6 Theorem. *Suppose that $\mathfrak{G}_1$, $\mathfrak{G}_2$ are 2-groups. Suppose that $\mathfrak{H}_i$ is an elementary Abelian subgroup of the centre of $\mathfrak{G}_i$ and that $\mathfrak{G}_i/\mathfrak{H}_i$ is elementary Abelian $(i = 1, 2)$. Suppose that ρ is an isomorphism of $\mathfrak{G}_1/\mathfrak{H}_1$ onto $\mathfrak{G}_2/\mathfrak{H}_2$, that σ is an isomorphism of $\mathfrak{H}_1$ onto $\mathfrak{H}_2$ and that if $(x\mathfrak{H}_1)\rho = y\mathfrak{H}_2$, then $(x^2)\sigma = y^2$ $(x \in \mathfrak{G}_1, y \in \mathfrak{G}_2)$. Then there is an isomorphism of $\mathfrak{G}_1$ onto $\mathfrak{G}_2$ which carries $\mathfrak{H}_1$ onto $\mathfrak{H}_2$ and induces ρ, σ on $\mathfrak{G}_1/\mathfrak{H}_1$, $\mathfrak{H}_1$ respectively.*

Proof. Let $x_1\mathfrak{H}_1, \ldots, x_m\mathfrak{H}_1$ be a basis of the elementary Abelian group $\mathfrak{G}_1/\mathfrak{H}_1$. For $i = 1, \ldots, m$, choose $y_i \in \mathfrak{G}_2$ such that $(x_i\mathfrak{H}_1)\rho = y_i\mathfrak{H}_2$. Then $(x_i^2)\sigma = y_i^2$ and $((x_ix_j)^2)\sigma = (y_iy_j)^2$. Since $(x_ix_j)^2 = x_i^2x_j^2[x_j, x_i]$, it follows that $[x_j, x_i]\sigma = [y_j, y_i]$. Now if $x \in \mathfrak{G}_1$, x can be written uniquely as $x = x_1^{l_1} \cdots x_m^{l_m}y$ with $0 \leq l_i \leq 1$ and $y \in \mathfrak{H}_1$, so we may put

$$x\chi = y_1^{l_1} \cdots y_m^{l_m}(y\sigma).$$

It is easy to verify that χ is a homomorphism of $\mathfrak{G}_1$ into $\mathfrak{G}_2$. Also χ carries $\mathfrak{H}_1$ onto $\mathfrak{H}_2$, χ induces σ on $\mathfrak{H}_1$ and $x_i\chi = y_i$ $(i = 1, \ldots, m)$. Since ρ, σ are isomorphisms, so is χ, and χ induces ρ on $\mathfrak{G}_1/\mathfrak{H}_1$. **q.e.d.**

To establish the existence of groups satisfying the conclusion of 6.5, we give matrix representations of them.

6.7 Example. Suppose that $\mathsf{F} = GF(2^n)$ and θ is an automorphism of F. We denote by $A(n, \theta)$ the set of all matrices of the form

$$u(a, b) = \begin{pmatrix} 1 & 0 & 0 \\ a & 1 & 0 \\ b & a\theta & 1 \end{pmatrix},$$

with a, b in F. Then

$$u(a, b)u(a', b') = u(a + a', b + b' + a'(a\theta)),$$

and

$$u(a, b)^{-1} = u(a, b + a(a\theta)).$$

Then $A(n, \theta)$ is a group of order 2^{2n} with unit element $u(0, 0)$.

The mapping $u(a, b) \to a$ is a homomorphism of $A(n, \theta)$ onto the additive group of F with kernel

$$\mathfrak{N} = \{u(0, b) | b \in \mathsf{F}\}.$$

Thus there is an isomorphism $\tilde{\rho}$ of $A(n, \theta)/\mathfrak{N}$ onto F such that

$$(u(a, b)\mathfrak{N})\tilde{\rho} = a.$$

Also, there is an isomorphism $\tilde{\sigma}$ of $\mathfrak{N}$ onto F given by

$$u(0, b)\tilde{\sigma} = b,$$

and

$$(u(a, b)^2)\tilde{\sigma} = a(a\theta).$$

Note that $\mathfrak{N} - \{1\}$ is the set of involutions in $A(n, \theta)$, and that $A(n, \theta)$ is Abelian if and only if $\theta = 1$.

Note also that if $\lambda \in \mathsf{F}^{\times}$, there is an automorphism ξ_λ of $A(n, \theta)$ given by

$$u(a, b)\xi_\lambda = u(\lambda a, \lambda(\lambda\theta)b).$$

Theorem 6.5 may now be restated as follows.

6.8 Theorem. *Let $\mathfrak{G}$ be a 2-group, ξ an automorphism of $\mathfrak{G}$, $\mathfrak{H}$ a ξ-invariant normal subgroup of $\mathfrak{G}$. Suppose that the following hold.*

(i) $|\mathfrak{G} : \mathfrak{H}| = |\mathfrak{H}| = 2^n$.

(ii) *$\mathfrak{G}$ is not elementary Abelian.*

(iii) *ξ induces irreducible automorphisms on $\mathfrak{G}/\mathfrak{H}$ and $\mathfrak{H}$, and if k is the order of that induced on $\mathfrak{G}/\mathfrak{H}$, $2^n - 1$ divides nk.*

Then $\mathfrak{G} \cong A(n, \theta)$, where θ is some automorphism of $\mathsf{F} = GF(2^n)$ not *of order 2. Also, if* $\bar{\xi}$ *is the automorphism of* $A(n, \theta)$ *corresponding to* ξ, *there exists* $\lambda \in \mathsf{F}$ *such that* $\mathsf{F} = GF(2)(\lambda(\lambda\theta))$ *and* $\bar{\xi}$, ξ_λ *induce the same automorphisms on* $\mathfrak{N}$ *and on* $A(n, \theta)/\mathfrak{N}$.

Proof. An isomorphism ψ of $\mathfrak{G}$ onto $A(n, \theta)$ is obtained by applying 6.6 to the isomorphisms $\rho\tilde{\rho}^{-1}$, $\sigma\tilde{\sigma}^{-1}$, where ρ, σ are as in 6.5 and $\tilde{\rho}$, $\tilde{\sigma}$ are as in 6.7. The assertion about $\bar{\xi}$ is equivalent to 6.5b). **q.e.d.**

Not all the groups $A(n, \theta)$ are non-isomorphic, and not all λ can arise in 6.8.

6.9 Theorem. *Let* $\mathsf{F} = GF(2^n)$, *and let* θ *be an automorphism of* F.

a) $A(n, \theta) \cong A(n, \theta^{-1})$.

b) *The mapping* $a \to a(a\theta)$ *of* F *into* F *is injective if and only if* θ *is of odd order.*

c) *If* θ *is of odd order, there exists* $\lambda \in \mathsf{F}$ *such that the set of involutions of* $A(n, \theta)$ *is transitively permuted by* $\langle \xi_\lambda \rangle$.

d) *If* θ *is of even order, the set of involutions of* $A(n, \theta)$ *is intransitively permuted by the group of automorphisms of* $A(n, \theta)$.

Proof. a) Let $\mathfrak{M}$ be the set of elements of order at most 2 in $A(n, \theta^{-1})$, and let ρ', σ' be isomorphisms of $A(n, \theta^{-1})/\mathfrak{M}$, $\mathfrak{M}$ respectively onto F such that

$$(x^2)\sigma' = ((x\mathfrak{M})\rho')((x\mathfrak{M})\rho'\theta^{-1})$$

(see 6.7). The isomorphisms $\tilde{\rho}\rho'^{-1}$, $\tilde{\sigma}\theta^{-1}\sigma'^{-1}$ (where $\tilde{\rho}$, $\tilde{\sigma}$ are as in 6.7) satisfy the conditions of 6.6. Hence $A(n, \theta) \cong A(n, \theta^{-1})$.

b) Put $a\chi = a(a\theta)$ ($a \in \mathsf{F}^\times$). Then χ is an endomorphism of $\mathsf{F}^\times$. If $\ker \chi \neq 1$, there exists $a \neq 1$ such that $a\theta = a^{-1}$, so the order of θ is even. If the order of θ is even, the subfield of elements fixed by θ^2 is different from the subfield F_1 of elements fixed by θ. Thus there exists $a \in \mathsf{F}$ such that $a\theta \neq a$, $a\theta^2 = a$. Then $a(a\theta) \in \mathsf{F}_1^\times$, so, since $|\mathsf{F}_1^\times|$ is odd, there exists $b \in \mathsf{F}_1$ such that $b^2 = a(a\theta)$. Let $c = ab^{-1}$; $c \neq 1$ since $a \notin \mathsf{F}_1$. Thus $c\theta = (a\theta)(b\theta)^{-1} = (a\theta)b^{-1} = ba^{-1}$, by definition of b. Hence $c\theta = c^{-1}$ and $\ker \chi \neq 1$.

c) Suppose that θ is of odd order. Let ω be a generator of $\mathsf{F}^\times$. By b), there exists $\lambda \in \mathsf{F}$ such that $\lambda(\lambda\theta) = \omega$. Then $u(0, b)\xi_\lambda = u(\lambda 0, \omega b)$, so ξ_λ has the stated property.

d) Suppose that θ is of even order. By b), χ is not injective and thus not surjective. Hence there exists $a \in \mathsf{F}$ such that a is not of the form $b(b\theta)$ with $b \in \mathsf{F}$; thus $a\tilde{\sigma}^{-1}$ is an involution in $A(n, \theta)$ but is not a square.

Since $A(n, \theta)$ is of exponent 4, some involutions are squares. The assertion follows at once. **q.e.d.**

$A(n, \theta)$ was constructed above by a rather special representation. We give a more theoretical way of constructing it.

6.10 Theorem. *Let* $\mathsf{K} = GF(2)$ *and suppose that* U, V *are finite-dimensional vector spaces over* K. *Suppose that* τ *is a mapping of* U *into* V *such that the mapping* β *of* $\mathsf{U} \times \mathsf{U}$ *into* V, *defined by*

$$(u, u')\beta = (u + u')\tau + u\tau + u'\tau \quad (u \in \mathsf{U}, u' \in \mathsf{U}),$$

is bilinear. Then there exists a 2-group $\mathfrak{G}$, *a subgroup* $\mathfrak{H}$ *of the centre of* $\mathfrak{G}$ *and isomorphisms* ρ, σ *of* $\mathfrak{G}/\mathfrak{H}$, $\mathfrak{H}$ *onto* U, V *respectively such that if* $(x\mathfrak{H})\rho = u$, *then* $(x^2)\sigma = u\tau$.

Proof. Let $u_1, \ldots, u_m$ be a K-basis of U. Put

$$(u_i, u_i)\xi = u_i\tau,$$

$$(u_i, u_j)\xi = 0 \qquad (i < j),$$

$$(u_i, u_j)\xi = (u_j, u_i)\beta \quad (i > j),$$

and extend ξ to a bilinear mapping of $\mathsf{U} \times \mathsf{U}$ into V. Then

$$(u', u'')\xi - (u + u', u'')\xi + (u, u' + u'')\xi - (u, u')\xi = 0$$

and $(u, 0)\xi = (0, u)\xi = 0$. Let $\mathfrak{G} = \{(u, v) | u \in \mathsf{U}, v \in \mathsf{V}\}$, and define multiplication in $\mathfrak{G}$ by the rule

$$(u, v)(u', v') = (u + u', v + v' + (u, u')\xi).$$

By I, 14.2, $\mathfrak{G}$ is a group with unit element $(0, 0)$, and

$$(u, v)^{-1} = (u, v + (u, u)\xi).$$

Obviously, the mapping $(u, v) \to u$ is a homomorphism of $\mathfrak{G}$ onto U with kernel $\mathfrak{H} = \{(0, v) | v \in \mathsf{V}\}$. Thus there is an isomorphism ρ of $\mathfrak{G}/\mathfrak{H}$ onto U such that $((u, v)\mathfrak{H})\rho = u$. $\mathfrak{H}$ is contained in the centre of $\mathfrak{G}$, and the mapping $(0, v) \to v$ is an isomorphism σ of $\mathfrak{H}$ onto V. It remains to verify that $((u, v)^2)\sigma = u\tau$, that is, $(u, u)\xi = u\tau$. Suppose that

$u = u_{i_1} + \cdots + u_{i_r}$ with $i_1 < \cdots < i_r$; we use induction on r. It is clear for $r = 0$ and $r = 1$. If $r > 1$, the inductive hypothesis gives $(x, x)\xi = x\tau$, where $x = u_{i_1} + \cdots + u_{i_{r-1}}$. But by definition of β,

$$(x, u_{i_r})\beta = u\tau + x\tau + u_{i_r}\tau.$$

Since β is bilinear, it follows that

$$(x, u_{i_r})\beta = \sum_{j<r} (u_{i_j}, u_{i_r})\beta = \sum_{j<r} (u_{i_r}, u_{i_j})\xi = (u_{i_r}, x)\xi.$$

Thus

$$\begin{aligned} u\tau &= (u_{i_r}, x)\xi + (x, x)\xi + (u_{i_r}, u_{i_r})\xi \\ &= (u, u)\xi - (x, u_{i_r})\xi. \end{aligned}$$

By definition of ξ, $(x, u_{i_r})\xi = 0$, so $u\tau = (u, u)\xi$. **q.e.d.**

We remark that the groups $A(n, \theta)$ of 6.7 may be constructed from 6.10 by taking $\mathsf{U} = \mathsf{V} = GF(2^n)$ and $u\tau = u(u\theta)$, for if

$$\begin{aligned} (u, u')\beta &= (u + u')\tau + u\tau + u'\tau \\ &= u(u'\theta) + (u\theta)u', \end{aligned}$$

β is bilinear.

Exercises

5) Show that for any group $\mathfrak{G}$ and any $m \geq 1$, there exists a $\mathbb{Z}$-multilinear mapping f of $\mathfrak{G}/\mathfrak{G}'$ onto $\gamma_m(\mathfrak{G})/\gamma_{m+1}(\mathfrak{G})$ given by

$$(g_1\mathfrak{G}', \ldots, g_m\mathfrak{G}')f = [g_1, \ldots, g_m]\gamma_{m+1}(\mathfrak{G}).$$

6) Let F be a Galois extension of K with Galois group $\mathfrak{G}$. Let ε_ξ $(\xi \in \mathfrak{G})$ be elements of F. Prove that the K-linear mapping

$$a \to \sum_{\xi \in \mathfrak{G}} \varepsilon_\xi(a\xi) \quad (a \in \mathsf{F})$$

is non-singular if and only if the matrix $(\varepsilon_{\xi\eta^{-1}}\eta)$ is non-singular.

7) Formulate and prove the analogues of 6.6 and 6.10 for an arbitrary prime p.

8) (G. HIGMAN [2]). Prove that if $A(n, \theta) \cong A(n, \phi)$, then ϕ is θ or θ^{-1}.

9) Let θ be an automorphism of $\mathsf{F} = GF(2^n)$. Suppose that θ is not of order 2.

a) If ω is a generator of $\mathsf{F}^\times$, $\omega(\omega\theta)$ generates F over $GF(2)$.

b) The only characteristic subgroup of $A(n, \theta)$ other than $A(n, \theta)$ and 1 is the group generated by the involutions.

c) $A(n, \theta)$ has a cyclic group of automorphisms which permutes transitively the set of all maximal subgroups of $A(n, \theta)$.

10) (ALPERIN and GORENSTEIN [1]). Suppose that $\mathfrak{G}$ is a p-group of class at most $p + 1$, that $\mathfrak{G}/\mathfrak{G}'$ is elementary Abelian of order q and that α is an automorphism of $\mathfrak{G}$ of order $q - 1$ which induces an irreducible linear transformation on $\mathfrak{G}/\mathfrak{G}'$. If $\mathbf{C}_{\mathfrak{G}}(\alpha) \neq 1$, q is 2, 4, 8 or 9.

§ 7. Suzuki 2-Groups

Suppose that $\mathfrak{G}$ is an Abelian p-group and that **Aut** $\mathfrak{G}$ permutes transitively the set of subgroups of $\mathfrak{G}$ of order p. Then by 5.8b), $\mathfrak{G}$ is homocyclic. The corresponding question when $\mathfrak{G}$ is possibly non-Abelian has been considered, and it has been proved by SHULT [2, 3] that if p is odd, $\mathfrak{G}$ must be Abelian. This is not the case for $p = 2$, since by 6.9, if θ is an automorphism of $GF(2^n)$ of odd order, $A(n, \theta)$ possesses an automorphism ξ such that $\langle \xi \rangle$ permutes the set of subgroups of order 2 transitively. All 2-groups having such an automorphism were determined by G. HIGMAN [2], and his results were extended by Goldschmidt, Shaw and Gross. In this section, one of these results, on 2-groups having a soluble automorphism group which permutes the set of involutions transitively, will be proved (Theorem 7.9).

If the 2-group $\mathfrak{G}$ has only one involution, the identity automorphism permutes the set of involutions in $\mathfrak{G}$ transitively, but the only non-Abelian 2-groups with just one involution are the generalized quaternion groups (III, 8.2). We therefore exclude this case and make the following definition.

7.1 Definition. A *Suzuki 2-group* is a group $\mathfrak{G}$ which has the following properties.

a) $\mathfrak{G}$ is a non-Abelian 2-group.

b) $\mathfrak{G}$ has more than one involution.

c) There exists a soluble group of automorphisms of $\mathfrak{G}$ which permutes the set of involutions in $\mathfrak{G}$ transitively.

Throughout this section, $\mathfrak{G}$ denotes a Suzuki 2-group.

7.2 Lemma. a) *If* $\mathfrak{J} = \Omega_1(\mathbf{Z}(\mathfrak{G}))$, $\mathfrak{J}$ *is the set of elements of* $\mathfrak{G}$ *of order at most 2. Let* $|\mathfrak{J}| = 2^n$.

b) *There exists a soluble group* $\mathfrak{Y}$ *of automorphisms of* $\mathfrak{G}$ *such that* $\mathfrak{Y}$ *permutes the set of involutions in* $\mathfrak{G}$ *transitively, and any prime divisor of* $|\mathfrak{Y}|$ *divides* $2^n - 1$. *In particular* $|\mathfrak{Y}|$ *is odd.*

c) *If* $\mathfrak{V}$ *is the kernel of the representation of* $\mathfrak{Y}$ *on* $\mathfrak{J}$, $\mathfrak{Y}/\mathfrak{V}$ *has a cyclic normal subgroup* $\mathfrak{Z}/\mathfrak{V}$ *of order k, where k divides* $2^n - 1$ *and* $2^n - 1$ *divides nk.* $\mathfrak{Z}$ *permutes the set of involutions in* $\mathfrak{G}$ *in orbits of length k, and the representation of* $\mathfrak{Z}$ *on* $\mathfrak{J}$ *is irreducible.*

Proof. a) This follows from 5.5a).

b) By 7.1, there exists a soluble group $\mathfrak{X}$ of automorphisms of $\mathfrak{G}$ which permutes the set of involutions of $\mathfrak{G}$ transitively. Let π be the set of prime divisors of $2^n - 1$ and let $\mathfrak{Y}$ be a Hall π-subgroup of $\mathfrak{X}$. If $\mathfrak{S}$ is the stabiliser in $\mathfrak{X}$ of an involution t, $|\mathfrak{X}:\mathfrak{S}| = 2^n - 1$. Hence $|\mathfrak{X}:\mathfrak{S}|$ and $|\mathfrak{X}:\mathfrak{Y}|$ are coprime. By I, 2.13, $\mathfrak{S}\mathfrak{Y} = \mathfrak{X}$. Hence the orbit of t under $\mathfrak{Y}$ is the same as that under $\mathfrak{X}$, that is, the set of all involutions of $\mathfrak{G}$. Hence $\mathfrak{Y}$ permutes the set of involutions in $\mathfrak{G}$ transitively.

c) If $\mathfrak{Y}/\mathfrak{V}$ is regarded as a group of automorphisms of $\mathfrak{J}$, the conditions of 3.5 are satisfied, and the assertion is clear. **q.e.d.**

As far as Theorem 7.9, the symbols $\mathfrak{J}$, n, $\mathfrak{Y}$, $\mathfrak{V}$, $\mathfrak{Z}$, k will have the same connotation as in 7.2. Write $q = 2^n$. Note that n, k satisfy the conditions (1) of § 4. F will denote the field $GF(2^n)$. Since $\mathfrak{Z}/\mathfrak{V}$ is cyclic, there exists $\xi \in \mathfrak{Z}$ such that $\mathfrak{Z} = \mathfrak{V}\langle\xi\rangle$.

7.3 Lemma. a) ξ *permutes the elements of* $\mathfrak{J} - \{1\}$ *in orbits of length k and induces an irreducible automorphism on* $\mathfrak{J}$. *Hence* $\mathfrak{J}$ *is contained in every non-identity* ξ*-invariant subgroup of* $\mathfrak{G}$.

b) *Suppose that* $\mathfrak{H}$ *is a* ξ*-invariant subgroup of* $\mathfrak{G}$ *such that* $\mathfrak{H} > \mathfrak{J}$ *and* $\mathfrak{H}/\mathfrak{J}$ *is elementary Abelian. Then* $\langle\xi\rangle$ *induces a group of permutations of the non-identity elements of* $\mathfrak{H}/\mathfrak{J}$ *in which the length of every orbit is divisible by k. Further* $\mathfrak{H}/\mathfrak{J}$ *is of order a power of q.*

Proof. a) Since $\mathfrak{Z}/\mathfrak{B}$ is cyclic and $\mathfrak{B}$ is represented trivially on $\mathfrak{I}$, ξ has the same orbits as $\mathfrak{Z}$ on $\mathfrak{I}$. Hence ξ permutes the elements of $\mathfrak{I} - \{1\}$ in orbits of length k, by 7.2c). Similarly, since the representation of $\mathfrak{Z}$ on $\mathfrak{I}$ is irreducible, ξ induces an irreducible automorphism on $\mathfrak{I}$. Any non-identity ξ-invariant subgroup of $\mathfrak{G}$ possesses an involution and therefore contains $\mathfrak{I}$.

b) Suppose that k_1 is the length of the orbit of ξ on $\mathfrak{H}/\mathfrak{I}$ containing $x\mathfrak{I}$ ($x \notin \mathfrak{I}$). Then $x\xi^{k_1} = xy$ for some $y \in \mathfrak{I}$. Since $y^2 = 1$ and $y \in \mathbf{Z}(\mathfrak{G})$, $(x^2)\xi^{k_1} = x^2$. But $x \notin \mathfrak{I}$, so by 7.2a), $x^2 \neq 1$; thus x^2 is an involution. By a), $k_1 \equiv 0\ (k)$. Hence $|\mathfrak{H}/\mathfrak{I}| - 1 \equiv 0\ (k)$. By 4.3a), $|\mathfrak{H}/\mathfrak{I}|$ is a power of $2^n = q$. **q.e.d.**

7.4 Lemma. *Suppose that $\mathfrak{A}$ is a normal Abelian ξ-invariant subgroup of $\mathfrak{G}$.*

a) *$\mathfrak{A}$ is homocyclic. The only ξ-invariant subgroups of $\mathfrak{A}$ are the $\mathfrak{A}^{2^i}$, and $[\mathfrak{A}^{2^i}, \mathfrak{G}] \leq \mathfrak{A}^{2^{i+1}}$.*

b) *If $i \geq 0$ and $\mathfrak{A}^{2^i} \neq 1$, $\mathfrak{A}^{2^i}/\mathfrak{A}^{2^{i+1}}$ is ξ-isomorphic to $\mathfrak{I}$.*

c) *If $\mathfrak{B}$ is a minimal normal ξ-invariant subgroup of $\mathfrak{G}$ such that $\mathfrak{B} > \mathfrak{A}$, ξ induces an irreducible automorphism on $\mathfrak{B}/\mathfrak{A}$.*

Proof. a) Let $\mathfrak{D}$ be a non-identity ξ-invariant subgroup of $\mathfrak{A}$. By 7.3a), $\mathfrak{I} = \Omega_1(\mathfrak{D})$, so ξ induces an irreducible automorphism on $\Omega_1(\mathfrak{D})$. By 5.8a), $\mathfrak{D}$ is homocyclic. In particular, $\mathfrak{A}$ is homocyclic and, since $\mathfrak{D} \geq \mathfrak{I}$, $\mathfrak{D} = \mathfrak{A}^{2^i}$ for some i. Since $[\mathfrak{A}^{2^i}, \mathfrak{G}]$ is a proper ξ-invariant subgroup of $\mathfrak{A}^{2^i}$, $[\mathfrak{A}^{2^i}, \mathfrak{G}] \leq \mathfrak{A}^{2^{i+1}}$.

b) If the exponent of $\mathfrak{A}$ is 2^m, $x\mathfrak{A}^{2^{i+1}} \to x^{2^{m-i-1}}$ is a ξ-isomorphism of $\mathfrak{A}^{2^i}/\mathfrak{A}^{2^{i+1}}$ onto $\mathfrak{I}$.

c) Since $\mathfrak{B} > \mathfrak{A}$, $\mathfrak{B} > [\mathfrak{B}, \mathfrak{G}]\mathfrak{A} \geq \mathfrak{A}$. Since $[\mathfrak{B}, \mathfrak{G}]$ is a normal ξ-invariant subgroup of $\mathfrak{G}$, it follows that $[\mathfrak{B}, \mathfrak{G}] \leq \mathfrak{A}$. Thus if $\mathfrak{B} \geq \mathfrak{B}_0 \geq \mathfrak{A}$, $\mathfrak{B}_0 \trianglelefteq \mathfrak{G}$. Hence $\mathfrak{B}$ is a minimal ξ-invariant subgroup of $\mathfrak{G}$ for which $\mathfrak{B} > \mathfrak{A}$. **q.e.d.**

7.5 Lemma. *Let $\mathfrak{A}$ be a normal Abelian ξ-invariant subgroup of $\mathfrak{G}$, and let $\mathfrak{B}$ be a minimal normal ξ-invariant subgroup of $\mathfrak{G}$ for which $\mathfrak{B} > \mathfrak{A}$. If $\mathfrak{B}$ is non-Abelian and $\mathfrak{B}' < \mathfrak{A}$, then $\Phi(\mathfrak{B}) = \Phi(\mathfrak{A})$.*

Proof. By 7.4a), $\mathfrak{B}' = \mathfrak{A}^{2^r}$; by hypothesis, $r > 0$ and $\mathfrak{A}^{2^r} \neq 1$. Hence $\Omega_r(\mathfrak{A}) \neq \mathfrak{A}$ and so $\Omega_r(\mathfrak{A}) \leq \mathfrak{A}^2$.

If $x \in \mathfrak{B}$, put $x\phi = x^{2^{r+1}}\mathfrak{A}^{2^{r+1}}$. Evidently $x^2 \in \mathfrak{A}$ and $x^{2^{r+1}} \in \mathfrak{A}^{2^r}$, so ϕ is a mapping of $\mathfrak{B}$ into $\mathfrak{A}^{2^r}/\mathfrak{A}^{2^{r+1}}$. We show that ϕ is a ξ-homomorphism. Clearly $\phi\xi = \xi\phi$. If x, y are in $\mathfrak{B}$, $(xy)^2 = x^2y^2[y, x]^y$, and since x^2, y^2 and $[y, x]$ all lie in $\mathfrak{A}$,

$$(xy)^{2^{r+1}} = x^{2^{r+1}} y^{2^{r+1}} ([y, x]^y)^{2^r}.$$

But $[y, x] \in \mathfrak{B}' \leq \mathfrak{A}^2$, so $(xy)\phi = (x\phi)(y\phi)$.

Thus $\operatorname{im}\phi$ is a ξ-subgroup of $\mathfrak{A}^{2^r}/\mathfrak{A}^{2^{r+1}}$. By 7.4b), $\mathfrak{A}^{2^r}/\mathfrak{A}^{2^{r+1}}$ is ξ-isomorphic to $\mathfrak{J}$, so ξ is irreducible on $\mathfrak{A}^{2^r}/\mathfrak{A}^{2^{r+1}}$. Thus either $\operatorname{im}\phi = 1$ or $\operatorname{im}\phi = \mathfrak{A}^{2^r}/\mathfrak{A}^{2^{r+1}}$. Suppose first that $\operatorname{im}\phi = 1$. Then given $x \in \mathfrak{B}$, $x^{2^{r+1}} \in \mathfrak{A}^{2^{r+1}}$, so $x^{2^{r+1}} = y^{2^{r+1}}$ for some $y \in \mathfrak{A}$. Thus $(x^2y^{-2})^{2^r} = 1$ and $x^2y^{-2} \in \Omega_r(\mathfrak{A})$. Thus $x^2y^{-2} \in \mathfrak{A}^2$ and $x^2 \in \mathfrak{A}^2$. Since $\Phi(\mathfrak{B}) = \langle x^2 | x \in \mathfrak{B} \rangle$, it follows that $\Phi(\mathfrak{B}) \leq \mathfrak{A}^2 = \Phi(\mathfrak{A})$ and $\Phi(\mathfrak{B}) = \Phi(\mathfrak{A})$, as asserted.

Suppose then that $\operatorname{im}\phi = \mathfrak{A}^{2^r}/\mathfrak{A}^{2^{r+1}}$. Thus $\ker\phi \neq \mathfrak{B}$. But obviously $\mathfrak{A} \leq \ker\phi$ and $\ker\phi$ is a ξ-invariant subgroup; thus by 7.4c), $\ker\phi = \mathfrak{A}$. Hence there is a ξ-isomorphism ψ of $\mathfrak{B}/\mathfrak{A}$ onto $\mathfrak{A}^{2^r}/\mathfrak{A}^{2^{r+1}}$ such that

$$(x\mathfrak{A})\psi = x^{2^{r+1}}\mathfrak{A}^{2^{r+1}} \quad (x \in \mathfrak{B}).$$

By 7.4, $\mathfrak{A}$ is homocyclic and $\mathfrak{A} \geq \mathfrak{J}$. Thus $|\mathfrak{A}^{2^r} : \mathfrak{A}^{2^{r+1}}| = 2^n$ and by II, 3.10, there exist an isomorphism σ of $\mathfrak{A}^{2^r}/\mathfrak{A}^{2^{r+1}}$ onto the additive group of F and an element λ of F such that

$$((y\xi)\mathfrak{A}^{2^{r+1}})\sigma = \lambda((y\mathfrak{A}^{2^{r+1}})\sigma)$$

for all $y \in \mathfrak{A}^{2^r}$. Put $\rho = \psi\sigma$; thus ρ is an isomorphism of $\mathfrak{B}/\mathfrak{A}$ onto F, and if $x \in \mathfrak{B}$,

$$((x\xi)\mathfrak{A})\rho = ((x\xi)^{2^{r+1}}\mathfrak{A}^{2^{r+1}})\sigma = \lambda((x^{2^{r+1}}\mathfrak{A}^{2^{r+1}})\sigma) = \lambda((x\mathfrak{A})\rho).$$

Suppose that $[\mathfrak{B}, \mathfrak{A}] \leq \mathfrak{A}^{2^{r+1}}$. Then, since $[\mathfrak{A}^{2^r}, \mathfrak{G}] \leq \mathfrak{A}^{2^{r+1}}$, we can apply 6.1. Thus there exists a bilinear mapping γ of $(\mathfrak{B}/\mathfrak{A}) \times (\mathfrak{B}/\mathfrak{A})$ into $\mathfrak{A}^{2^r}/\mathfrak{A}^{2^{r+1}}$ such that if $x \in \mathfrak{B}$ and $y \in \mathfrak{B}$,

$$(x\mathfrak{A}, y\mathfrak{A})\gamma = [x, y]\mathfrak{A}^{2^{r+1}}.$$

A bilinear mapping β of $\mathsf{F} \times \mathsf{F}$ into F is thus defined by

$$(a, b)\beta = ((a\rho^{-1}, b\rho^{-1})\gamma)\sigma.$$

Obviously $(a, a)\beta = 0$, so by 6.2c) there exist $\varepsilon_{ij} \in \mathsf{F}$ with $\varepsilon_{ii} = \varepsilon_{ij} + \varepsilon_{ji} = 0$, such that

$$(a, b)\beta = \sum_{i,j=1}^{n} \varepsilon_{ij} a^{2^{i-1}} b^{2^{j-1}}.$$

Now if $a = (x\mathfrak{A})\rho$, $b = (y\mathfrak{A})\rho$,

$$(a, b)\beta = ((x\mathfrak{A}, y\mathfrak{A})\gamma)\sigma = ([x, y]\mathfrak{A}^{2^{r+1}})\sigma.$$

Thus, since $\lambda a = ((x\xi)\mathfrak{A})\rho$ and $\lambda b = ((y\xi)\mathfrak{A})\rho$,

$$\begin{aligned}(\lambda a, \lambda b)\beta &= ([x\xi, y\xi]\mathfrak{A}^{2^{r+1}})\sigma = (([x, y]\xi)\mathfrak{A}^{2^{r+1}})\sigma \\ &= \lambda([x, y]\mathfrak{A}^{2^{r+1}})\sigma = \lambda(a, b)\beta.\end{aligned}$$

Hence

$$\sum_{i,j=1}^{n} \varepsilon_{ij}\lambda^{2^{i-1}+2^{j-1}}a^{2^{i-1}}b^{2^{j-1}} = \sum_{i,j=1}^{n} \lambda\varepsilon_{ij}a^{2^{i-1}}b^{2^{j-1}}$$

for all a, b in F. By the uniqueness assertion of 6.2b),

$$\varepsilon_{ij}(\lambda^{2^{i-1}+2^{j-1}} - \lambda) = 0$$

for all i, j. Since $\mathfrak{A}^{2^r}/\mathfrak{A}^{2^{r+1}}$ is ξ-isomorphic to $\mathfrak{J}$, the order of λ is k. Hence by 4.3c), $\lambda^{2^{i-1}+2^{j-1}} \neq \lambda$ for $i \neq j$. Hence $\varepsilon_{ij} = 0$ for all i, j. Thus $(a, b)\beta = 0$ for all a, b and $[x, y] \in \mathfrak{A}^{2^{r+1}}$ for all x, y in $\mathfrak{B}$. However, this implies that $\mathfrak{B}' \leq \mathfrak{A}^{2^{r+1}}$, whereas $\mathfrak{B}' = \mathfrak{A}^{2^r}$. This is a contradiction.

Hence $[\mathfrak{B}, \mathfrak{A}] \not\leq \mathfrak{A}^{2^{r+1}}$. We show now that $[\mathfrak{B}, \Phi(\mathfrak{B})] \leq \mathfrak{A}^{2^{r+1}}$. To do this, it suffices to prove that $[x, y^2] \in \mathfrak{A}^{2^{r+1}}$ for any x, y in $\mathfrak{B}$. We have

$$[x, y^2] = [x, y][x, y]^y = [x, y]^2[x, y, y].$$

But $[x, y] \in \mathfrak{B}' \leq \mathfrak{A}^{2^r}$ and $[\mathfrak{A}^{2^r}, \mathfrak{G}] \leq \mathfrak{A}^{2^{r+1}}$, so $[x, y, y] \in \mathfrak{A}^{2^{r+1}}$. Since also $[x, y]^2 \in \mathfrak{A}^{2^{r+1}}$, our assertion follows.

Thus $\Phi(\mathfrak{B}) \neq \mathfrak{A}$. Hence $\Phi(\mathfrak{B}) < \mathfrak{A}$, and since $\Phi(\mathfrak{B})$ is a ξ-invariant subgroup, $\Phi(\mathfrak{B}) \leq \mathfrak{A}^2 = \Phi(\mathfrak{A})$. Hence $\Phi(\mathfrak{B}) = \Phi(\mathfrak{A})$. **q.e.d.**

7.6 Lemma. *Let $\mathfrak{A}$ be a normal Abelian ξ-invariant subgroup of $\mathfrak{G}$, and let $\mathfrak{B}$ be a minimal normal ξ-invariant subgroup of $\mathfrak{G}$ for which $\mathfrak{B} > \mathfrak{A}$. If $\mathfrak{B}' = \mathfrak{A}$, then $\mathfrak{A}$ is elementary Abelian.*

Proof. Suppose that this is false. By 7.4, $\mathfrak{A}$ is homocyclic, $\mathfrak{A} \geq \mathfrak{J}$ and $[\mathfrak{A}^{2^i}, \mathfrak{B}] \leq \mathfrak{A}^{2^{i+1}}$ for all $i \geq 0$. On the other hand $\mathfrak{B}/\mathfrak{B}' = \mathfrak{B}/\mathfrak{A}$ is elementary Abelian, so by III, 2.13, $\gamma_j(\mathfrak{B})/\gamma_{j+1}(\mathfrak{B})$ is elementary Abelian for all $j \geq 1$. It follows by induction on j that $\gamma_j(\mathfrak{B}) = \mathfrak{A}^{2^{j-2}}$, for if $j > 2$, $\gamma_{j-1}(\mathfrak{B}) = \mathfrak{A}^{2^{j-3}}$ by the inductive hypothesis and

$$\gamma_j(\mathfrak{B}) = [\gamma_{j-1}(\mathfrak{B}), \mathfrak{B}] = [\mathfrak{A}^{2^{j-3}}, \mathfrak{B}] \leq \mathfrak{A}^{2^{j-2}} = (\gamma_{j-1}(\mathfrak{B}))^2 \leq \gamma_j(\mathfrak{B}).$$

Let $\mathfrak{H} = \mathfrak{H}_1 = \mathfrak{B}/\mathfrak{A}^4$, $\mathfrak{H}_2 = \mathfrak{A}/\mathfrak{A}^4$, $\mathfrak{H}_3 = \mathfrak{A}^2/\mathfrak{A}^4$. Thus the lower central series of $\mathfrak{H}$ is $\mathfrak{H} = \mathfrak{H}_1 > \mathfrak{H}_2 > \mathfrak{H}_3 > 1$. $\mathfrak{H}$ is a ξ-group, and $\mathfrak{H}_1/\mathfrak{H}_2$ is ξ-isomorphic to $\mathfrak{B}/\mathfrak{A}$. By 7.4c), ξ induces an irreducible automorphism on $\mathfrak{B}/\mathfrak{A}$ and hence on $\mathfrak{H}_1/\mathfrak{H}_2$. By hypothesis $\mathfrak{A}^2 \neq 1$, so by 7.4b), $\mathfrak{H}_3$ and $\mathfrak{H}_2/\mathfrak{H}_3$ are ξ-isomorphic to $\mathfrak{J}$. In particular ξ^k induces the identity automorphism on $\mathfrak{H}_2/\mathfrak{H}_3$ and $\mathfrak{H}_3$. Thus if $y \in \mathfrak{H}_2$, $y\xi^k = yz$ for some $z \in \mathfrak{H}_3$, and if $x \in \mathfrak{H}$, $[x, y] \in \mathfrak{H}_3$, so that

$$[x, y] = [x, y]\xi^k = [x\xi^k, yz] = [x\xi^k, y].$$

Thus $(x\xi^k)x^{-1}$ commutes with y. Hence $(x\xi^k)x^{-1} \in \mathbf{C}_{\mathfrak{H}}(\mathfrak{H}_2)$. But $\mathbf{C}_{\mathfrak{H}}(\mathfrak{H}_2)$ is ξ-invariant and $\mathfrak{H}_2 \leq \mathbf{C}_{\mathfrak{H}}(\mathfrak{H}_2) < \mathfrak{H}$; hence $\mathbf{C}_{\mathfrak{H}}(\mathfrak{H}_2) = \mathfrak{H}_2$. Thus $x\xi^k \equiv x \bmod \mathfrak{H}_2$ for all $x \in \mathfrak{H}$, and ξ^k induces the identity mapping on $\mathfrak{H}/\mathfrak{H}_2$. Hence the order k_1 of the automorphism induced on $\mathfrak{H}/\mathfrak{H}_2$ by ξ is a divisor of k. But if x, y are in $\mathfrak{H}$,

$$[x, y]\xi^{k_1} = [x\xi^{k_1}, y\xi^{k_1}] \equiv [x, y] \bmod \mathfrak{H}_3,$$

since $\mathfrak{H}_2/\mathfrak{H}_3$ is contained in the centre of $\mathfrak{H}/\mathfrak{H}_3$. Thus ξ^{k_1} induces the identity mapping on $\mathfrak{H}_2/\mathfrak{H}_3$. Hence $k_1 = k$. Thus $\mathfrak{H}/\mathfrak{H}_2$ has an irreducible automorphism of order k. By II, 3.10, $|\mathfrak{H}/\mathfrak{H}_2| = 2^{n_1}$, where n_1 is the smallest positive integer such that $2^{n_1} \equiv 1$ (k). By 4.3, $n_1 = n$; thus $|\mathfrak{H} : \mathfrak{H}_2| = 2^n$.

$\mathfrak{H}/\mathfrak{H}_3$ therefore satisfies the conditions of 6.5. It follows that there exist isomorphisms ρ, σ of $\mathfrak{H}/\mathfrak{H}_2$, $\mathfrak{H}_2/\mathfrak{H}_3$ respectively onto the additive group of F, and there exists a field automorphism θ of F such that if $x \in \mathfrak{H}$ and $(x\mathfrak{H}_2)\rho = a$, then $(x^2\mathfrak{H}_3)\sigma = a(a\theta)$. Also θ is not of order 2; and $\theta \neq 1$ since $\mathfrak{H}/\mathfrak{H}_3$ is non-Abelian. Further, there exists $\lambda \in \mathsf{F}$ such that $\mathsf{F} = \mathsf{K}(\lambda)$ and

$$((x\xi)\mathfrak{H}_2)\rho = \lambda((x\mathfrak{H}_2)\rho),$$

$$((y\xi)\mathfrak{H}_3)\sigma = \lambda(\lambda\theta)((y\mathfrak{H}_3)\sigma)$$

for all $x \in \mathfrak{H}$, $y \in \mathfrak{H}_2$.

If $z \in \mathfrak{H}_3$, then $z = t^2$ for some $t \in \mathfrak{H}_2$ and $t\mathfrak{H}_3$ depends only on z; thus we may put $z\tau = (t\mathfrak{H}_3)\sigma$. Then τ is an isomorphism of $\mathfrak{H}_3$ onto the additive group of F and

$$z\xi\tau = \lambda(\lambda\theta)(z\tau)$$

for all $z \in \mathfrak{H}_3$.

By 6.1, commutation induces a bilinear mapping γ of $(\mathfrak{H}/\mathfrak{H}_2) \times (\mathfrak{H}_2/\mathfrak{H}_3)$ into $\mathfrak{H}_3$. A bilinear mapping β of $\mathsf{F} \times \mathsf{F}$ into F is defined by putting

$$(a, b)\beta = ((a\rho^{-1}, b\sigma^{-1})\gamma)\tau,$$

and

$$(\lambda a, \lambda(\lambda\theta)b)\beta = \lambda(\lambda\theta)((a, b)\beta).$$

By 6.2,

$$(a, b)\beta = \sum_{i,j=1}^{n} \varepsilon_{ij} a^{2^{i-1}} b^{2^{j-1}}$$

for certain $\varepsilon_{ij} \in \mathsf{F}$, and by the uniqueness part of 6.2,

$$\lambda(\lambda\theta)\varepsilon_{ij} = \varepsilon_{ij}\lambda^{2^{i-1}}(\lambda(\lambda\theta))^{2^{j-1}}$$

for all i, j.

Suppose $\varepsilon_{ij} \neq 0$. If, then, $a\theta = a^{2^r}$ for all $a \in \mathsf{F}$, $2r \not\equiv 0\ (n)$ as $\theta^2 \neq 1$, and

$$1 + 2^r \equiv 2^{i-1} + 2^{j-1} + 2^{j+r-1} \quad (k).$$

The solutions to this may be read off from 4.5; they are (i) $i \equiv j \equiv r\ (n)$ if $2r \equiv 1\ (n)$, (ii) $i \equiv j - r - 1 \equiv 0\ (n)$ if $2r \equiv -1\ (n)$, (iii) $n = 6, k = 21$, $r \equiv j \equiv i - 3 \equiv 2\ (n)$ or (iv) $n = 6, k = 21, r \equiv j + 2 \equiv i + 1 \equiv 4\ (n)$. Thus for any given values of n, r, there is at most one non-zero ε_{ij}, and we may write

$$(a, b)\beta = \varepsilon a^{2^{i-1}} b^{2^{j-1}}.$$

Thus

$$(a, a(a\theta))\beta = \varepsilon a^{2^{i-1}+2^{j-1}+2^{j+r-1}}.$$

But if $a = (x\mathfrak{H}_2)\rho$, then $a(a\theta) = (x^2\mathfrak{H}_3)\sigma$ and

$$(a, a(a\theta))\beta = ((x\mathfrak{H}_2, x^2\mathfrak{H}_3)\gamma)\tau = [x, x^2]\tau = 1\tau = 0.$$

Thus $\varepsilon = 0$ and $(a, b)\beta = 0$ for all a, b in F. Thus $[\mathfrak{H}, \mathfrak{H}_2] = 1$, a contradiction. **q.e.d.**

7.7 Lemma. *Suppose that $\mathfrak{A}$ is a maximal normal Abelian ξ-invariant subgroup of $\mathfrak{G}$. Then $\mathfrak{A}^4 = 1$ and $\mathfrak{G}/\mathfrak{A}$ is elementary Abelian.*

Proof. Suppose that $\mathfrak{A}^4 \neq 1$. Since $\mathfrak{G}$ is non-Abelian, there exists a minimal normal ξ-invariant subgroup $\mathfrak{B}$ of $\mathfrak{G}$ such that $\mathfrak{B} > \mathfrak{A}$. Thus $\mathfrak{B}$ is non-Abelian. By 7.6, $\mathfrak{B}' < \mathfrak{A}$, and by 7.5, $\Phi(\mathfrak{B}) = \Phi(\mathfrak{A})$. Since there are no involutions in $\mathfrak{B} - \mathfrak{A}$, it follows from 5.13 that $[\mathfrak{A}, \mathfrak{B}] \nleq \mathfrak{A}^4$. Thus $[\Omega_2(\mathfrak{A}), \mathfrak{B}] \neq 1$ and if $\mathfrak{C} = \mathbf{C}_{\mathfrak{B}}(\Omega_2(\mathfrak{A}))$, $\mathfrak{A} \leq \mathfrak{C} < \mathfrak{B}$. Since $\mathfrak{C}$ is ξ-invariant, it follows from the minimality of $\mathfrak{B}$ that $\mathfrak{C} = \mathfrak{A}$.

Since $\Phi(\mathfrak{B}) = \Phi(\mathfrak{A})$, $[\mathfrak{A}, \Phi(\mathfrak{B})] = 1$. Hence by 5.15, $\mathfrak{A}$ contains a cyclic normal subgroup $\langle a \rangle$ of $\mathfrak{B}$ of order 4. Now if $v \in \mathfrak{J}$ and $x \in \mathfrak{B}$, write $v = u^2$ with $u \in \mathfrak{A}$ and

$$v\eta(x) = [u, x].$$

Then $\eta(x)$ is a mapping of $\mathfrak{J}$ into itself, and indeed $\eta(x)$ is an endomorphism of $\mathfrak{J}$. Since $[a, x] \in \langle a \rangle$ for all $x \in \mathfrak{B}$,

$$(a^2)\eta(x) = a^{2g(x)},$$

where $g(x) \in GF(2)$. Let $\mathscr{F}$ be the set of mappings of $\mathfrak{B}$ into $GF(2)$, and if $f \in \mathscr{F}$, define

$$\mathfrak{J}_f = \{v \mid v \in \mathfrak{J},\ v\eta(x) = v^{f(x)} \quad \text{for all } x \in \mathfrak{B}\}.$$

Thus $a^2 \in \mathfrak{J}_g$ and $\mathfrak{J}_g \neq 1$. Now ξ permutes the $\mathfrak{J}_f$, for $\mathfrak{J}_f\xi = \mathfrak{J}_{f'}$, where

$$f'(x) = f(x\xi^{-1}) \quad (x \in \mathfrak{B}).$$

Hence $\prod_{f \in \mathscr{F}} \mathfrak{J}_f$ is a non-identity ξ-invariant subgroup of $\mathfrak{J}$. By 7.3a),

$$\mathfrak{J} = \prod_{f \in \mathscr{F}} \mathfrak{J}_f.$$

By 5.16, $\mathfrak{J}$ is the direct product of the $\mathfrak{J}_f$. By 3.6, $\mathfrak{J} = \mathfrak{J}_g$, so

$$[u, x] = (u^2)\eta(x) = u^{2g(x)}$$

for all $u \in \Omega_2(\mathfrak{A})$, and $u^x = u^{1+2g(x)}$. Hence if $x \in \mathfrak{B} - \mathfrak{C}$, $u^x = u^{-1}$ for all $u \in \Omega_2(\mathfrak{A})$. Thus $|\mathfrak{B} : \mathfrak{C}| = 2$.

It follows that $|\mathfrak{B} : \mathfrak{A}| = 2$ and $\mathfrak{B} = \langle \mathfrak{A}, x \rangle$ for some x; also $x\xi \equiv x \bmod \mathfrak{A}$. Let

$$\mathfrak{A}_1 = \{a^x a \mid a \in \mathfrak{A}\}, \quad \mathfrak{A}_2 = \langle y^2 \mid y \in \mathfrak{B} - \mathfrak{A} \rangle.$$

Thus $\mathfrak{A}_1$, $\mathfrak{A}_2$ are ξ-invariant subgroups of $\mathfrak{A}$. If $y \in \mathfrak{B} - \mathfrak{A}$, then $y = xa$ for some $a \in \mathfrak{A}$ and $y^2 = x^2a^xa$. Hence $\mathfrak{A}_1 \le \mathfrak{A}_2$ and $\mathfrak{A}_2 = \langle \mathfrak{A}_1, x^2 \rangle$. By 7.4, $\mathfrak{A}_1 = \mathfrak{A}^{2^i}$ and $\mathfrak{A}_2 = \mathfrak{A}^{2^j}$ for some i, j, so $\mathfrak{A}^{2^j}/\mathfrak{A}^{2^i}$ is cyclic. But $\mathfrak{J}$ is not cyclic, so by 7.4b), $i = j$ and $\mathfrak{A}_1 = \mathfrak{A}_2$. Thus $x^{-2} \in \mathfrak{A}_1$ and $x^{-2} = a^xa$ for some $a \in \mathfrak{A}$. Therefore $(xa)^2 = 1$ and $xa \in \mathfrak{J} \le \mathfrak{A}$. This gives $x \in \mathfrak{A}$, a contradiction. Hence $\mathfrak{A}^4 = 1$.

Now suppose that $\mathfrak{G}/\mathfrak{A}$ is not elementary Abelian. Then $\mathfrak{A} \not\ge \Phi(\mathfrak{G})$, so $\mathfrak{A}\Phi(\mathfrak{G}) > \mathfrak{A}$. Let $\mathfrak{D}$ be a minimal normal ξ-invariant subgroup of $\mathfrak{G}$ such that $\mathfrak{A}\Phi(\mathfrak{G}) \ge \mathfrak{D} > \mathfrak{A}$. Then $\mathfrak{D} = \mathfrak{A}(\mathfrak{D} \cap \Phi(\mathfrak{G}))$ and $\mathfrak{D} \cap \Phi(\mathfrak{G}) \not\le \mathfrak{A}$. By 7.4a), $[\mathfrak{A}, \mathfrak{G}] \le \mathfrak{A}^2$ and $[\mathfrak{A}^2, \mathfrak{G}] = 1$; thus $[\mathfrak{A}, \mathfrak{G}, \mathfrak{G}] = 1$. Hence if $x \in \mathfrak{A}$ and $y \in \mathfrak{G}$,

$$[x, y^2] = [x, y][x, y]^y = [x, y]^2 = 1.$$

Thus $[\mathfrak{A}, \Phi(\mathfrak{G})] = 1$. If $z \in \mathfrak{D} \cap \Phi(\mathfrak{G})$ and $z \notin \mathfrak{A}$, it follows from 5.13 that $z^2 \notin \mathfrak{A}^2$. Hence $\Phi(\mathfrak{D}) \not\le \mathfrak{A}^2$. By 7.4a), $\Phi(\mathfrak{D}) = \mathfrak{A}$. By 7.5, $\mathfrak{D}' = \mathfrak{A}$, and by 7.6, $\mathfrak{A}$ is elementary Abelian. Hence $\mathfrak{A} = \mathfrak{J}$. In particular, $\mathfrak{A} \le \mathbf{Z}(\mathfrak{G})$. Thus $\mathfrak{A} = \mathbf{Z}(\mathfrak{G})$ on account of the maximality of $\mathfrak{A}$.

Let $1 < \mathbf{Z}_1(\mathfrak{G}) < \mathbf{Z}_2(\mathfrak{G}) \le \cdots$ be the upper central series of $\mathfrak{G}$. Then $\mathbf{Z}_2(\mathfrak{G}) \cap \mathfrak{G}'$ is Abelian, since $[\mathbf{Z}_2(\mathfrak{G}), \mathfrak{G}'] = 1$ (III, 2.11). Since $\mathfrak{A} = \mathfrak{D}'$, $\mathfrak{A} \le \mathbf{Z}_2(\mathfrak{G}) \cap \mathfrak{G}'$. Again, it follows from the maximality of $\mathfrak{A}$ that $\mathfrak{A} = \mathbf{Z}_2(\mathfrak{G}) \cap \mathfrak{G}'$. Hence $\mathbf{Z}_2(\mathfrak{G}) \cap \mathfrak{G}' = \mathbf{Z}_1(\mathfrak{G}) \cap \mathfrak{G}'$. By III, 2.6 applied to $\mathfrak{G}'\mathbf{Z}_1(\mathfrak{G})/\mathbf{Z}_1(\mathfrak{G})$, $\mathfrak{G}' \le \mathbf{Z}_1(\mathfrak{G})$. Hence $\mathbf{Z}_2(\mathfrak{G}) = \mathfrak{G}$. By III,2.13, $\mathfrak{G}/\mathbf{Z}_1(\mathfrak{G}) = \mathbf{Z}_2(\mathfrak{G})/\mathbf{Z}_1(\mathfrak{G})$ is elementary Abelian, since $\mathfrak{A} = \mathfrak{J} = \mathbf{Z}(\mathfrak{G})$. Thus $\Phi(\mathfrak{G}) \le \mathbf{Z}_1(\mathfrak{G}) = \mathfrak{A}$, a contradiction. **q.e.d.**

7.8 Lemma. *$\Phi(\mathfrak{G})$ is elementary Abelian.*

Proof. Suppose that this is false. The long argument needed to obtain a contradiction will be divided into 5 steps.

Step 1. Preliminaries

Let $\mathfrak{A}$ be a maximal normal ξ-invariant Abelian subgroup of $\mathfrak{G}$. By 7.7, $\mathfrak{A}^4 = 1$ and $\mathfrak{A} \ge \Phi(\mathfrak{G})$. Since $\Phi(\mathfrak{G})$ is not elementary Abelian, $\Phi(\mathfrak{G}) \not\le \mathfrak{A}^2$. But by 7.4, the only ξ-invariant subgroups of $\mathfrak{A}$ are $\mathfrak{A}$, $\mathfrak{A}^2$ and 1. Hence $\Phi(\mathfrak{G}) = \mathfrak{A}$. Thus $\Phi(\mathfrak{G})$ is homocyclic of order q^2 and is a maximal ξ-invariant Abelian subgroup of $\mathfrak{G}$. Also $[\Phi(\mathfrak{G}), \mathfrak{G}] \le \mathfrak{J}$, by 7.4.

We show that if $\mathfrak{H}$ is a minimal ξ-invariant subgroup of $\mathfrak{G}$ such that $\mathfrak{H} > \Phi(\mathfrak{G})$, then $\mathfrak{H}/\mathfrak{J}$ is elementary Abelian. Indeed, as $\mathfrak{A}^2 \ne 1$, $\mathfrak{H}' < \Phi(\mathfrak{G})$ by 7.6, so $\Phi(\mathfrak{H}) = \Phi(\mathfrak{A}) = \mathfrak{J}$ by 7.5.

We deduce that $\mathbf{C}_{\mathfrak{G}}(\Phi(\mathfrak{G})) = \Phi(\mathfrak{G})$. For otherwise, we can choose a minimal ξ-invariant subgroup $\mathfrak{H}$ of $\mathfrak{G}$ such that $\Phi(\mathfrak{G}) < \mathfrak{H} \le \mathbf{C}_{\mathfrak{G}}(\Phi(\mathfrak{G}))$.

Then $\mathfrak{H}/\mathfrak{J}$ is elementary Abelian. Hence if $x \in \mathfrak{H} - \Phi(\mathfrak{G})$, $x^2 \in \mathfrak{J} = (\Phi(\mathfrak{G}))^2$, so $x^2 = y^2$ for some $y \in \Phi(\mathfrak{G})$. But $[x, y] = 1$, so $xy^{-1} \in \mathfrak{J}$ and $x \in \langle \mathfrak{J}, y \rangle \leq \Phi(\mathfrak{G})$, a contradiction.

We prove next that ξ is of order k. Since ξ is of odd order and the order of the restriction of ξ to $\mathfrak{J}$ is k, it suffices to show that ξ^k induces the identity automorphism on $\mathfrak{G}/\Phi(\mathfrak{G})$, $\Phi(\mathfrak{G})/\mathfrak{J}$ and $\mathfrak{J}$, by I, 4.4. This is true for $\Phi(\mathfrak{G})/\mathfrak{J}$ by 7.4b). Suppose that $x \in \mathfrak{G}$. Then for any $y \in \Phi(\mathfrak{G})$, $[y, x] \in \mathfrak{J}$ and $y\xi^k \equiv y \bmod \mathfrak{J}$, so

$$[y, x] = [y, x]\xi^k = [y\xi^k, x\xi^k] = [y, x\xi^k].$$

Hence $y^x = y^{x\xi^k}$. Thus $(x\xi^k)x^{-1} \in \mathbf{C}_{\mathfrak{G}}(\Phi(\mathfrak{G})) = \Phi(\mathfrak{G})$, so ξ^k induces the identity automorphism on $\mathfrak{G}/\Phi(\mathfrak{G})$.

Since ξ is of odd order, it follows from the Maschke-Schur theorem that $\mathfrak{G}/\Phi(\mathfrak{G})$ is the direct product of groups $\mathfrak{H}_l/\Phi(\mathfrak{G})$, where $\mathfrak{H}_l$ is a minimal ξ-invariant subgroup of $\mathfrak{G}$ such that $\mathfrak{H}_l > \Phi(\mathfrak{G})$. From above, $\mathfrak{H}_l/\mathfrak{J}$ is elementary Abelian. But $\mathfrak{G}/\mathfrak{J}$ is not elementary Abelian, so $[\mathfrak{H}_1/\mathfrak{J}, \mathfrak{H}_2/\mathfrak{J}] \neq 1$ for appropriate $\mathfrak{H}_1$, $\mathfrak{H}_2$. Thus $\mathfrak{H}_1 \neq \mathfrak{H}_2$ and $[\mathfrak{H}_1, \mathfrak{H}_2] \not\leq \mathfrak{J}$. Since $[\mathfrak{H}_1, \mathfrak{H}_2]$ is a ξ-invariant subgroup of $\Phi(\mathfrak{G})$, $[\mathfrak{H}_1, \mathfrak{H}_2] = \Phi(\mathfrak{G})$. Note also that $\mathfrak{H}_l$ is non-Abelian on account of the maximality of $\Phi(\mathfrak{G})$. Hence $\mathfrak{H}_l' = \mathfrak{J}$. The remainder of the argument only concerns the groups $\mathfrak{H}_1$, $\mathfrak{H}_2$.

Step 2. Structure of $\Phi(\mathfrak{G})$, $\mathfrak{K}_1$, $\mathfrak{K}_2$

By the Maschke-Schur theorem, $\mathfrak{H}_l/\mathfrak{J} = (\Phi(\mathfrak{G})/\mathfrak{J}) \times (\mathfrak{K}_l/\mathfrak{J})$ for some ξ-invariant subgroup $\mathfrak{K}_l$. By 7.3b), ξ permutes the non-identity elements of $\mathfrak{K}_l/\mathfrak{J}$ in orbits the lengths of which are divisible by k. Hence the order of the automorphism induced by ξ on $\mathfrak{K}_l/\mathfrak{J}$ is k. By II, 3.10 and 4.3, $|\mathfrak{K}_l/\mathfrak{J}| = q$. Thus $|\mathfrak{H}_l| = q^3$ and $\mathfrak{H}_l' = \mathfrak{J}$ $(l = 1, 2)$, since $\mathfrak{H}_l'$ is a non-identity ξ-invariant subgroup of $\mathfrak{J}$.

We shall now apply 6.5 to each of the groups $\Phi(\mathfrak{G})$, $\mathfrak{K}_1$, $\mathfrak{K}_2$. First, consider $\Phi(\mathfrak{G})$, which is Abelian. There exist isomorphisms π, σ of $\Phi(\mathfrak{G})/\mathfrak{J}$, $\mathfrak{J}$ respectively onto the additive group of F such that

$$(1) \qquad (x^2)\sigma = ((x\mathfrak{J})\pi)^2 \quad (x \in \Phi(\mathfrak{G}));$$

further there exists $\lambda \in \mathsf{F}$ such that $\mathsf{F} = \mathsf{K}(\lambda^2)$, (where $\mathsf{K} = GF(2)$), and

$$(2) \qquad ((x\xi)\mathfrak{J})\pi = \lambda((x\mathfrak{J})\pi), \quad y\xi\sigma = \lambda^2(y\sigma)$$

for $x \in \Phi(\mathfrak{G})$, $y \in \mathfrak{J}$. (The relevant field automorphism is 1 in this case, since $\Phi(\mathfrak{G})$ is Abelian).

Next, consider $\mathfrak{K}_l$ ($l = 1, 2$). There exist isomorphisms ρ_l', σ_l of $\mathfrak{K}_l/\mathfrak{J}$, $\mathfrak{J}$ respectively onto the additive group of F and a field automorphism θ_l, not of order 2, such that if $x \in \mathfrak{K}_l$ and $(x\mathfrak{J})\rho_l' = a$, then

$$(3) \qquad (x^2)\sigma_l = a(a\theta_l).$$

Also there exists $\mu_l' \in \mathsf{F}$ such that $\mathsf{F} = \mathsf{K}(\mu_l'(\mu_l'\theta_l))$ and

$$(4) \qquad ((x\xi)\mathfrak{J})\rho_l' = \mu_l'((x\mathfrak{J})\rho_l'), \quad y\xi\sigma_l = \mu_l'(\mu_l'\theta_l)(y\sigma_l)$$

for all $x \in \mathfrak{K}_l$, $y \in \mathfrak{J}$. Note that replacing σ_l by $\sigma_l\theta_l^{-1}$ is equivalent to replacing θ_l by θ_l^{-1}. Thus if $a\theta_l = a^{2^{r_l}}$, we may suppose that $0 \leq r_l < \frac{1}{2}n$; ($r_l \neq \frac{1}{2}n$ since θ_l is not of order 2).

We now have 3 mappings of $\mathfrak{J}$ onto F, namely $\sigma, \sigma_1, \sigma_2$. Our next aim is to remove the necessity of considering the last two; thus we must derive equations corresponding to (3), (4) involving σ instead of σ_l. To do this, define $\alpha_l = \sigma_l^{-1}\sigma$. Then α_l is a K-linear mapping of F onto F, and by (2) and (4), $\lambda^2(a\alpha_l) = (\mu_l'(\mu_l'\theta_l)a)\alpha_l$ for all $a \in \mathsf{F}$. By 6.3, there is a field automorphism ψ_l of F such that $\lambda^2 = \mu_l(\mu_l\theta_l)$, where $\mu_l = \mu_l'\psi_l$ and $a\alpha_l = v_l'(a\psi_l)$ for some $v_l' \in \mathsf{F}$. Since $|\mathsf{F}| = 2^n$, we can write $v_l' = v_l^2$. Put $\rho_l = \rho_l'\psi_l$. We prove that if $x \in \mathfrak{K}_l$ and $(x\mathfrak{J})\rho_l = a$, then

$$(5) \qquad (x^2)\sigma = v_l^2 a(a\theta_l).$$

Indeed, $(x\mathfrak{J})\rho_l' = a\psi_l^{-1}$, so by (3),

$$(x^2)\sigma_l = (a\psi_l^{-1})(a\psi_l^{-1}\theta_l) = (a(a\theta_l))\psi_l^{-1};$$

hence $(x^2)\sigma = (x^2)\sigma_l\alpha_l = v_l^2(a(a\theta_l))$.

We observe also that if $x \in \mathfrak{K}_l$, then

$$(6) \qquad ((x\xi)\mathfrak{J})\rho_l = \mu_l((x\mathfrak{J})\rho_l).$$

For by (4),

$$((x\xi)\mathfrak{J})\rho_l = ((x\xi)\mathfrak{J})\rho_l'\psi_l = (\mu_l'((x\mathfrak{J})\rho_l'))\psi_l = \mu_l((x\mathfrak{J})\rho_l).$$

Step 3. Commutation between $\Phi(\mathfrak{G})$ and $\mathfrak{K}_l$

By 6.1, there is a $\mathbb{Z}$-bilinear mapping γ_l of $(\Phi(\mathfrak{G})/\mathfrak{J}) \times (\mathfrak{K}_l/\mathfrak{J})$ into $\mathfrak{J}$ such that if $x \in \Phi(\mathfrak{G})$ and $y \in \mathfrak{K}_l$,

$$(x\mathfrak{J}, y\mathfrak{J})\gamma_l = [x, y].$$

A K-bilinear mapping β_l of $\mathsf{F} \times \mathsf{F}$ into F is defined by putting

$$(a, b)\beta_l = v_l^{-2}(((v_l a)\pi^{-1}, b\rho_l^{-1})\gamma_l\sigma).$$

Since $\mathbf{C}_{\mathfrak{G}}(\Phi(\mathfrak{G})) = \Phi(\mathfrak{G})$, $\beta_l \neq 0$. Thus by 6.2, there exist $\varepsilon_{ij}^l \in \mathsf{F}$, not all zero, such that

$$(a, b)\beta_l = \sum_{i,j=1}^{n} \varepsilon_{ij}^l a^{2^{i-1}} b^{2^{j-1}} \tag{7}$$

for all a, b in F. Put $(v_l a)\pi^{-1} = x\mathfrak{J}$, $b\rho_l^{-1} = y\mathfrak{J}$, where $x \in \Phi(\mathfrak{G})$, $y \in \mathfrak{K}_l$. Then by (2) and (6),

$$((x\xi)\mathfrak{J})\pi = \lambda((x\mathfrak{J})\pi) = \lambda v_l a,$$

$$((y\xi)\mathfrak{J})\rho_l = \mu_l((y\mathfrak{J})\rho_l) = \mu_l b.$$

Thus, by definition of β_l,

$$\begin{aligned}(\lambda a, \mu_l b)\beta_l &= v_l^{-2}(((v_l \lambda a)\pi^{-1}, (\mu_l b)\rho_l^{-1})\gamma_l\sigma) \\ &= v_l^{-2}(((x\xi)\mathfrak{J}, (y\xi)\mathfrak{J})\gamma_l\sigma).\end{aligned}$$

Using the definition of γ_l and (2),

$$\begin{aligned}(\lambda a, \mu_l b)\beta_l &= v_l^{-2}([x, y]\xi\sigma) = v_l^{-2}\lambda^2([x, y]\sigma) \\ &= v_l^{-2}\lambda^2((x\mathfrak{J}, y\mathfrak{J})\gamma_l\sigma) \\ &= v_l^{-2}\lambda^2(((v_l a)\pi^{-1}, b\rho_l^{-1})\gamma_l\sigma).\end{aligned}$$

Hence

$$(\lambda a, \mu_l b)\beta_l = \lambda^2((a, b)\beta_l).$$

It follows in the usual way from 6.2 that

$$\varepsilon_{ij}^l(\lambda^2 - \lambda^{2^{i-1}}\mu_l^{2^{j-1}}) = 0$$

for all $i, j = 1, \ldots, n$. Since $\lambda^2 = \mu_l(\mu_l\theta_l) = \mu_l^{1+2^{r_l}}$,

$$\varepsilon_{ij}^l(\mu_l(\mu_l\theta_l) - \mu_l^{2^{i-2}}(\mu_l\theta_l)^{2^{i-2}}\mu_l^{2^{j-1}}) = 0.$$

Since k is the order of the automorphism induced on $\mathfrak{K}_l/\mathfrak{J}$ by ξ, it follows from (6) that k is the multiplicative order of μ_l. Hence if $\varepsilon_{ij}^l \neq 0$,

$$1 + 2^{r_l} \equiv 2^{i-2} + 2^{i-2+r_l} + 2^{j-1}\ (k).$$

By 4.5b), the only solutions of this satisfying $0 \leq r_l < \frac{1}{2}n$ are (i) $r_l = 0$, $i = j = 1$, (ii) $r_l = \frac{1}{2}(n-1)$, $i = r_l + 2$, $j = 0$, and (iii) $n = 6$, $k = 21$, $r_l = 2$, $i = 3$, $j = 5$. In case (iii), $(a, b)\beta_l = \varepsilon a^4 b^{16}$ for $\varepsilon \neq 0$. Now by 4.6, there exist a, b in F, not both zero, such that

$$a^2 + b^5 + \varepsilon a^4 b^{16} = 0.$$

Suppose that $(v_l a)\pi^{-1} = x\mathfrak{J}$ and $b\rho_l^{-1} = y\mathfrak{J}$; thus x, y are not both in $\mathfrak{J}$, and so $xy \notin \mathfrak{J}$. Then $a^2 = v_l^{-2}((x^2)\sigma)$ by (1) and $b^5 = b(b\theta_l) = v_l^{-2}((y^2)\sigma)$ by (5). Thus

$$\begin{aligned}(x^2y^2)\sigma &= (x^2)\sigma + (y^2)\sigma = v_l^2(a^2 + b^5) = v_l^2 \varepsilon a^4 b^{16}\\ &= v_l^2((a, b)\beta_l) = [x, y]\sigma.\end{aligned}$$

Thus $[x, y] = x^2y^2$ and $(xy)^2 = 1$. However $xy \notin \mathfrak{J}$, so this contradicts 7.2a), and (iii) cannot occur. Thus either $r_l = 0$ and $(a, b)\beta_l = \varepsilon_l ab$, or $r_l = \frac{1}{2}(n-1)$ and $(a, b)\beta_l = \varepsilon_l((a^2)\theta_l)(b\theta_l^2)$; in either case $\varepsilon_l \neq 0$ $(l = 1, 2)$.

Step 4. Commutation between $\mathfrak{K}_1$ and $\mathfrak{K}_2$

The subgroups $\mathfrak{H}_1$, $\mathfrak{H}_2$ were so chosen that $[\mathfrak{H}_1, \mathfrak{H}_2] \not\leq \mathfrak{J}$. But $\mathfrak{H}_l = \mathfrak{K}_l \Phi(\mathfrak{G})$ and $\Phi(\mathfrak{G})/\mathfrak{J}$ is a central factor of $\mathfrak{G}$. Hence $[\mathfrak{K}_1, \mathfrak{K}_2] \not\leq \mathfrak{J}$. But $[\mathfrak{K}_1, \mathfrak{K}_2] \leq \Phi(\mathfrak{G})$ and again $[\mathfrak{G}, \Phi(\mathfrak{G})] \leq \mathfrak{J}$. Hence by 6.1, there is a non-trivial bilinear mapping γ of $(\mathfrak{K}_1/\mathfrak{J}) \times (\mathfrak{K}_2/\mathfrak{J})$ into $\Phi(\mathfrak{G})/\mathfrak{J}$ such that if $x_i \in \mathfrak{K}_i$ $(i = 1, 2)$,

$$(x_1\mathfrak{J}, x_2\mathfrak{J})\gamma = [x_1, x_2]\mathfrak{J}.$$

We derive the non-zero bilinear mapping β of $\mathsf{F} \times \mathsf{F}$ into F by putting

$$(a, b)\beta = (a\rho_1^{-1}, b\rho_2^{-1})\gamma\pi.$$

By 6.2, there exist $\varepsilon_{ij} \in \mathsf{F}$, not all zero, such that

$$(a, b)\beta = \sum_{i,j=1}^{n} \varepsilon_{ij} a^{2^{i-1}} b^{2^{j-1}}$$

for all a, b in F. Put $a = (x\mathfrak{J})\rho_1$, $b = (y\mathfrak{J})\rho_2$, so $x \in \mathfrak{K}_1$, $y \in \mathfrak{K}_2$. By (6), $((x\xi)\mathfrak{J})\rho_1 = \mu_1 a$ and $((y\xi)\mathfrak{J})\rho_2 = \mu_2 b$, so by the definitions of β and γ,

$$(\mu_1 a, \mu_2 b)\beta = ((x\xi)\mathfrak{J}, (y\xi)\mathfrak{J})\gamma\pi = ([x\xi, y\xi]\mathfrak{J})\pi = (([x, y]\xi)\mathfrak{J})\pi,$$

and

$$(a, b)\beta = (x\mathfrak{J}, y\mathfrak{J})\gamma\pi = ([x, y]\mathfrak{J})\pi.$$

But by (2), $(([x, y]\xi)\mathfrak{J})\pi = \lambda(([x, y]\mathfrak{J})\pi)$, so

$$(\mu_1 a, \mu_2 b)\beta = \lambda((a, b)\beta).$$

Hence by 6.2,

$$\varepsilon_{ij}(\mu_1^{2^{i-1}}\mu_2^{2^{j-1}} - \lambda) = 0$$

for all i, j. If $\varepsilon_{ij} \neq 0$, it follows that $\lambda = \mu_1^{2^{i-1}}\mu_2^{2^{j-1}}$. Since $\lambda^2 = \mu_l^{1+2^{r_l}}$ (p. 309), we find by taking the $(1 + 2^{r_1})(1 + 2^{r_2})$-*th* power that

$$(8) \qquad (1 + 2^{r_1})(1 + 2^{r_2}) \equiv 2^i(1 + 2^{r_2}) + 2^j(1 + 2^{r_1}) \quad (k).$$

Step 5. The contradiction

We obtain a contradiction by using the Witt identity, which takes the form

$$[x, y, z][y, z, x][z, x, y] = 1,$$

since $\mathfrak{G}$ is metabelian (III, Aufg. 1)). To express this in the notation we have developed, suppose that a, a', b, b' are elements of F. Write

$$a = (x\mathfrak{J})\rho_1, a' = (x'\mathfrak{J})\rho_1, b = (y\mathfrak{J})\rho_2, b' = (y'\mathfrak{J})\rho_2.$$

Then $(a, b)\beta = ([x, y]\mathfrak{J})\pi$, so

$$\begin{aligned}(v_2^{-1}((a, b)\beta), b')\beta_2 &= v_2^{-2}(([x, y]\mathfrak{J}, y'\mathfrak{J})\gamma_2\sigma)\\ &= v_2^{-2}([x, y, y']\sigma).\end{aligned}$$

Since y, y' are in $\mathfrak{K}_2$, $[y, y'] \in \mathfrak{J}$, so the above form of the Witt identity reduces to $[x, y, y'][y', x, y] = 1$, or $[x, y, y'] = [x, y', y]$. Hence

$$(9) \qquad (v_2^{-1}((a, b)\beta), b')\beta_2 = (v_2^{-1}((a, b')\beta), b)\beta_2.$$

Similarly,

$$(10) \qquad (v_1^{-1}((a,b)\beta), a')\beta_1 = (v_1^{-1}((a',b)\beta), a)\beta_1.$$

First suppose that $r_1 = r_2 = 0$. Then (8) and 4.3c) imply $i = j = n$; hence $(a, b)\beta = \varepsilon a^{2^{n-1}} b^{2^{n-1}}$ and $(a, b)\beta_2 = \varepsilon_2 ab$. Thus by (9) $a^{2^{n-1}} b^{2^{n-1}} b' = a^{2^{n-1}} b'^{2^{n-1}} b$ for all a, b, b' in F, which is absurd, as $|\mathsf{F}| > 2$.

If $r_1 = 0$ and $r_2 = \frac{1}{2}(n-1)$, (8) and 4.5b) imply $i = \frac{1}{2}(n+1)$, $j = n - 1$, so $(a, b)\beta = \varepsilon a^{2^{\frac{1}{2}(n-1)}} b^{2^{n-2}}$ and $(a, b)\beta_1 = \varepsilon_1 ab$. Again (10) yields an absurdity. The case $r_1 = \frac{1}{2}(n-1)$, $r_2 = 0$ need not be considered on account of symmetry.

Finally, suppose $r_1 = r_2 = \frac{1}{2}(n-1) = m$, say. Since $(2^m + 1, k) = (2^m + 1, 2^n - 1) = 1$, (8) and 4.4c) imply that $\{i, j\} = \{0, m\}$. Thus

$$(a, b)\beta = \bar{\varepsilon}_1 a^{2^{n-1}} b^{2^{m-1}} + \bar{\varepsilon}_2 a^{2^{m-1}} b^{2^{n-1}},$$

where $\bar{\varepsilon}_1, \bar{\varepsilon}_2$ are not both 0. Substituting in (9) yields $\bar{\varepsilon}_2 = 0$; substitution in (10) then yields $\bar{\varepsilon}_1 = 0$. **q.e.d.**

7.9 Theorem. *Let $\mathfrak{G}$ be a Suzuki 2-group.*

a) $\mathfrak{G}' = \Phi(\mathfrak{G}) = \mathbf{Z}(\mathfrak{G}) = \{x \mid x \in \mathfrak{G}, x^2 = 1\}$.

b) *Either* (i) $\mathfrak{G} \cong A(n, \theta)$ *for some non-identity automorphism* θ *of* $GF(2^n)$ *of odd order, or* (ii) $|\mathfrak{G}| = |\mathbf{Z}(\mathfrak{G})|^3$.

Proof. a) By 7.8, $\Phi(\mathfrak{G}) \leq \mathfrak{J} \leq \mathbf{Z}(\mathfrak{G})$, so the exponent of $\mathfrak{G}$ is 4. Hence by 5.5c), $\mathfrak{G}' = \mathbf{Z}(\mathfrak{G}) = \Phi(\mathfrak{G}) = \mathfrak{J}$.

b) By 5.5, $|\mathfrak{G}|$ is q^2 or q^3. If $|\mathfrak{G}| = q^2$, $|\mathfrak{G} : \mathfrak{J}| = q$, so by 7.3b), ξ is irreducible on $\mathfrak{G}/\mathfrak{J}$, and $\mathfrak{G}$ satisfies all the conditions of 6.8. Hence $\mathfrak{G} \cong A(n, \theta)$ for some automorphism θ of $GF(2^n)$. Since **Aut** $\mathfrak{G}$ permutes the set of involutions of $\mathfrak{G}$ transitively, it follows from 6.9d) that θ is of odd order. Since $\mathfrak{G}$ is non-Abelian, $\theta \neq 1$, by 6.7. **q.e.d.**

7.10 Remark. G. HIGMAN [2] and SHAW [1] proved that if $\mathfrak{G}$ is a Suzuki 2-group of order $|\mathbf{Z}(\mathfrak{G})|^3$, there exist isomorphisms τ of $\mathfrak{G}/\mathfrak{J}$ onto $\mathsf{F} \oplus \mathsf{F}$ and σ of $\mathfrak{J}$ onto F such that one of the following holds.

a) If $x \in \mathfrak{G}$ and $(x\mathfrak{J})\tau = (a, b)$, then

$$(x^2)\sigma = a(a\theta) + \varepsilon a(b\theta) + b(b\theta);$$

here θ is an automorphism of F of odd order and ε is a non-zero element of F such that $\varepsilon \neq c^{-1} + (c\theta)$ for all $c \in \mathsf{F}$.

b) $n = 2m + 1$ and if $(x\mathfrak{J})\tau = (a, b)$, then

$$(x^2)\sigma = a(a\theta) + \varepsilon(a\theta^2)((b^2)\theta) + b^2;$$

here θ is the automorphism $a \to a^{2^m}$ and $\varepsilon \neq c + c((c^{-2})\theta)$ for all $c \in \mathsf{F}$.

c) $n = 5m$ and if $(x\mathfrak{J})\tau = (a, b)$, then

$$(x^2)\sigma = a(a\theta) + \varepsilon(a\theta^3)(b\theta) + b(b\theta^2);$$

here θ is the automorphism $a \to a^{2^m}$ and $\varepsilon \neq c^{-1}(1 + (c\theta)(c\theta^4))$ for all $c \in \mathsf{F}$.

This is proved by showing that $\mathfrak{G}/\mathfrak{J}$ is the direct product of two groups $\mathfrak{H}/\mathfrak{J}$, $\mathfrak{K}/\mathfrak{J}$, where $\mathfrak{H}$, $\mathfrak{K}$ both satisfy the conditions of 6.5. There thus arise two mappings of $\mathfrak{J}$ onto F, and these may be compared by using 6.3. Methods similar to those used in this section then lead to the congruence

$$(1 + 2^r)(1 + 2^s) \equiv (1 + 2^s)2^{i-1} + (1 + 2^r)2^{j-1} \quad (k).$$

The solution of this by the methods of § 4 is a long, wearisome process, but it leads to the various possibilities for the "square mapping" $x\mathfrak{J} \to x^2$.

By 6.6, the group $\mathfrak{G}$ is determined to within isomorphism by the conditions a), b), c). The existence of groups satisfying these conditions follows easily from 6.10; further, matrix representations in F are easily found. Each of these groups possesses an automorphism which permutes the set of involutions transitively.

Let $q = 2^n$. By II, 10.12b), the matrices

$$Q(x, y) = \begin{pmatrix} 1 & x & y \\ 0 & 1 & x^q \\ 0 & 0 & 1 \end{pmatrix},$$

where x, y are elements of $GF(q^2)$ and $y + y^q + xx^q = 0$, form a Sylow 2-subgroup $\mathfrak{Q}$ of $SU(3, q^2)$. Since

$$Q(x, y)^2 = Q(0, xx^q),$$

the involutions in $\mathfrak{Q}$ are the elements $Q(0, y)$ with $y \in \mathsf{F} = GF(q)$. Now for $0 \neq \lambda \in GF(q^2)$, put

$$H(\lambda) = \begin{pmatrix} \lambda^{-q} & & \\ & \lambda^{q-1} & \\ & & \lambda \end{pmatrix}.$$

Then

$$H(\lambda)^{-1}Q(x, y)H(\lambda) = Q(x\lambda^{2q-1}, y\lambda^{q+1}),$$

and $\mathfrak{Q}$ is a Suzuki 2-group of order q^3.

If η is an element of $GF(q^2)^\times$ of order $q + 1$, $\{1, \eta\}$ is a basis of $GF(q^2)$ over F. Thus there is an isomorphism τ of $\mathfrak{Q}/\mathbf{Z}(\mathfrak{Q})$ onto $\mathsf{F} \oplus \mathsf{F}$ given by

$$(Q(a + b\eta, 0)\mathbf{Z}(\mathfrak{Q}))\tau = (a, b) \quad (a \in \mathsf{F}, b \in \mathsf{F}).$$

Since

$$\begin{aligned} Q(a + b\eta, 0)^2 &= Q(0, (a + b\eta)(a + b\eta)^q) \\ &= Q(0, a^2 + ab(\eta + \eta^q) + b^2\eta^{q+1}) \\ &= Q(0, a^2 + \varepsilon ab + b^2), \end{aligned}$$

where $\varepsilon = \eta + \eta^q = \eta + \eta^{-1}$, $\mathfrak{Q}$ occurs above in a). Of course $\varepsilon \neq c + c^{-1}$ for $c \in \mathsf{F}$.

7.11 Remarks. a) It was proved by SHULT [2, 3] that if p is odd, $\mathfrak{G}$ is a p-group and $\mathbf{Aut}\,\mathfrak{G}$ permutes transitively the set of subgroups of order p, then $\mathfrak{G}$ is Abelian. The proof is also by linear methods, and the main lemma is the following.

Let $\mathsf{K} = GF(q)$, let $\mathfrak{G}$ be a finite group and let U, V be (finite) $\mathsf{K}\mathfrak{G}$-modules. Suppose that $\mathfrak{G}$ permutes the one-dimensional subspaces of V transitively. Let μ be a $\mathsf{K}\mathfrak{G}$-homomorphism of $\mathsf{U} \otimes_{\mathsf{K}} \mathsf{V}$ into V and suppose that for each $u \in \mathsf{U}$, the linear transformation $v \to (u \otimes v)\mu$ of V is nilpotent. Then $\mu = 0$.

The proof of this makes use of the Feit-Thompson theorem for $q = 2$ and a deep theorem of BRAUER [3] for $q > 2$.

b) The remaining problem is thus the complete determination of the 2-groups $\mathfrak{G}$ for which $\mathbf{Aut}\,\mathfrak{G}$ permutes the involutions transitively. Using a combination of the method of Shult with that of Higman presented in this section, GROSS [5] has proved that one of the following holds for such a group.

(i) $\mathfrak{G}$ is homocyclic.

(ii) $\mathfrak{G}$ is special and $|\mathfrak{G}|$ is $|\mathbf{Z}(\mathfrak{G})|^2$ or $|\mathbf{Z}(\mathfrak{G})|^3$.

(iii) $\mathfrak{G}$ has lower central series $\mathfrak{G} = \mathfrak{G}_1 > \mathfrak{G}_2 > \mathfrak{G}_3 > 1$, where $\mathfrak{G}_2 = \mathfrak{G}' = \Phi(\mathfrak{G}) = \mathbf{C}_{\mathfrak{G}}(\mathfrak{G}')$ and $\mathfrak{G}_3 = [\mathfrak{G}_2, \mathfrak{G}] = \mathfrak{G}_2^2 = \mathbf{Z}(\mathfrak{G})$.

c) Another result related to the determination of the Suzuki 2-groups is the following theorem of LANDROCK [1].

Suppose that $\mathfrak{G}$ is a non-Abelian 2-group, α is an automorphism of $\mathfrak{G}$ of order 2 and $\mathfrak{C} = \mathbf{C}_{\mathfrak{G}}(\alpha)$ is elementary Abelian of order 2^n. Suppose that $\mathfrak{G}$ has an automorphism ρ of order $2^n - 1$ which commutes with α. If ρ permutes transitively the set of non-identity elements of $\mathfrak{C}$, then $\mathfrak{G}$ is isomorphic to a Sylow 2-subgroup of $PSU(3, 2^n)$ or of $PSL(3, 2^n)$.

§ 8. Lie Algebras

In § 6 we considered commutation as a bilinear form. But we wish to study a whole series of commutations together, such as those which arise from the lower central series. In order to do this, we need the notion of a Lie ring. One of the advantages obtained by doing this is the possibility of extending the ring of coefficients. We therefore define Lie algebras over any commutative ring. Throughout this section A denotes a fixed commutative ring with identity.

8.1 Definitions. a) Let $\mathfrak{g}$ be an A-module, and suppose that a multiplication is defined in $\mathfrak{g}$ in which the product of a and b is denoted by $[a, b]$. $\mathfrak{g}$ is called a *Lie algebra* over A if the following hold for any a, b, c in $\mathfrak{g}$ and any λ, μ in A:

(i) $[\lambda a + \mu b, c] = \lambda[a, c] + \mu[b, c]$,

(ii) $[a, \lambda b + \mu c] = \lambda[a, b] + \mu[a, c]$,

(iii) $[[a, b], c] + [[b, c], a] + [[c, a], b] = 0$,

(iv) $[a, a] = 0$.

(iii) is called the *Jacobi identity*. A Lie algebra over $\mathbb{Z}$ is called a *Lie ring*. If a, b are elements of a Lie algebra, then

$$[a + b, a + b] = [a, a] + [a, b] + [b, a] + [b, b]$$

by (i) and (ii). It follows from (iv) that

$$[a, b] + [b, a] = 0$$

and hence

(iv′) $[a, b] = -[b, a]$.

If $a_1, \ldots, a_n$ $(n \geq 2)$ are elements of $\mathfrak{g}$, the n-fold product $[a_1, \ldots, a_n]$ is defined by induction on n; for $n > 2$,

$$[a_1, \ldots, a_n] = [[a_1, \ldots, a_{n-1}], a_n].$$

Thus the Jacobi identity may be written

(iii) $[a, b, c] + [b, c, a] + [c, a, b] = 0.$

b) If $\mathfrak{g}_1, \mathfrak{g}_2$ are Lie algebras over A, a (*Lie*) *homomorphism* of $\mathfrak{g}_1$ into $\mathfrak{g}_2$ is an A-module homomorphism ρ of $\mathfrak{g}_1$ into $\mathfrak{g}_2$ for which

$$[a, b]\rho = [a\rho, b\rho]$$

for all a, b in $\mathfrak{g}_1$. The homomorphism ρ is called an *epimorphism* if it is surjective and a *monomorphism* if it is injective. An *isomorphism* is a homomorphism which is both injective and surjective. An *automorphism* of a Lie algebra is an isomorphism of it onto itself. The product of two homomorphisms is a homomorphism, and the automorphisms of a Lie algebra form a group.

c) If $\mathfrak{g}$ is an A-module and U, V are A-submodules of $\mathfrak{g}$, write

$$\mathsf{U} + \mathsf{V} = \{a + b | a \in \mathsf{U}, v \in \mathsf{V}\}.$$

Thus $\mathsf{U} + \mathsf{V}$ is an A-submodule of $\mathfrak{g}$, and

$$\mathsf{U} + \mathsf{V} = \mathsf{V} + \mathsf{U}, \quad (\mathsf{U} + \mathsf{V}) + \mathsf{W} = \mathsf{U} + (\mathsf{V} + \mathsf{W}).$$

If also $\mathfrak{g}$ is a Lie algebra over A, denote by $[\mathsf{U}, \mathsf{V}]$ the A-submodule spanned by all products $[a, b]$ with $a \in \mathsf{U}, b \in \mathsf{V}$.

By (iv′), $[\mathsf{U}, \mathsf{V}] = [\mathsf{V}, \mathsf{U}]$. Clearly, if $\mathsf{U} \subseteq \mathsf{V}$, $[\mathsf{U}, \mathsf{W}] \subseteq [\mathsf{V}, \mathsf{W}]$ and by (i) and (ii),

$$[\mathsf{U} + \mathsf{V}, \mathsf{W}] = [\mathsf{U}, \mathsf{W}] + [\mathsf{V}, \mathsf{W}], \quad [\mathsf{U}, \mathsf{V} + \mathsf{W}] = [\mathsf{U}, \mathsf{V}] + [\mathsf{U}, \mathsf{W}].$$

d) An *ideal* of a Lie algebra $\mathfrak{g}$ is an A-submodule $\mathfrak{j}$ such that $[\mathfrak{j}, \mathfrak{g}] \subseteq \mathfrak{j}$. By c), this is equivalent to the condition $[\mathfrak{g}, \mathfrak{j}] \subseteq \mathfrak{j}$. The kernel of a homomorphism is an ideal.

If $\mathfrak{j}$ is an ideal of $\mathfrak{g}$, the A-module $\mathfrak{g}/\mathfrak{j}$ is a Lie algebra in which

$$[a + \mathfrak{j}, b + \mathfrak{j}] = [a, b] + \mathfrak{j},$$

as is easily verified. The mapping ν of $\mathfrak{g}$ onto $\mathfrak{g}/\mathfrak{j}$ defined by putting $a\nu = a + \mathfrak{j}$ $(a \in \mathfrak{g})$ is a Lie epimorphism, called the *natural homomorphism* ν of $\mathfrak{g}$ onto $\mathfrak{g}/\mathfrak{j}$. The kernel of ν is $\mathfrak{j}$.

The intersection of a set of ideals of $\mathfrak{g}$ is again an ideal of $\mathfrak{g}$. If X is a subset of $\mathfrak{g}$, the intersection of the (non-empty) set of ideals which contain X is an ideal, called the *ideal of* $\mathfrak{g}$ *generated* by X.

If $\mathfrak{i}_1, \mathfrak{i}_2$ are ideals of $\mathfrak{g}$, so is $\mathfrak{i}_1 + \mathfrak{i}_2$, for by c),

$$[\mathfrak{i}_1 + \mathfrak{i}_2, \mathfrak{g}] = [\mathfrak{i}_1, \mathfrak{g}] + [\mathfrak{i}_2, \mathfrak{g}] \subseteq \mathfrak{i}_1 + \mathfrak{i}_2.$$

e) A *subalgebra* of a Lie algebra $\mathfrak{g}$ is an A-submodule $\mathfrak{h}$ of $\mathfrak{g}$ such that $[\mathfrak{h}, \mathfrak{h}] \subseteq \mathfrak{h}$. The intersection of a set of subalgebras of $\mathfrak{g}$ is again a subalgebra. If X is a subset of $\mathfrak{g}$, the intersection of the (non-empty) set of subalgebras which contain X is a subalgebra, called the *subalgebra of* $\mathfrak{g}$ *generated* by X. An ideal of a Lie algebra is a subalgebra; hence the ideal of $\mathfrak{g}$ generated by X contains the subalgebra of $\mathfrak{g}$ generated by X.

If $\mathfrak{i}$ is an ideal of $\mathfrak{g}$ and $\mathfrak{h}$ is a subalgebra, $\mathfrak{i} + \mathfrak{h}$ is a subalgebra, for by c),

$$[\mathfrak{i} + \mathfrak{h}, \mathfrak{i} + \mathfrak{h}] \subseteq [\mathfrak{i}, \mathfrak{g}] + [\mathfrak{h}, \mathfrak{h}] \subseteq \mathfrak{i} + \mathfrak{h}.$$

8.2 Example. Let $\mathfrak{A}$ be an associative algebra over A. (By this is meant an A-module in which multiplication is defined satisfying the associative and distributive laws; also $\mathfrak{A}$ is to have an identity element. No assumption analogous to that of finite dimension in V, § 1 is made, but it will always be supposed that a homomorphism of associative algebras carries the identity element into the identity element.) Given elements a, b of $\mathfrak{A}$, write

$$[a, b] = ab - ba.$$

The axioms (i), (ii), (iv) of 8.1a) are obviously satisfied by this product. Further,

$$\begin{aligned}[][[a, b], c] &= [a, b]c - c[a, b] \\ &= abc - bac - c[a, b] \\ &= abc - bca + b[c, a] - c[a, b].\end{aligned}$$

Hence (iii) also holds. Thus we may define a Lie algebra $\mathfrak{l}(\mathfrak{A})$ as follows: the underlying A-module of $\mathfrak{l}(\mathfrak{A})$ is $\mathfrak{A}$, and the product of two elements a, b of $\mathfrak{l}(\mathfrak{A})$ is $[a, b]$.

Suppose that $\mathfrak{A}, \mathfrak{B}$ are associative algebras and ρ is an associative homomorphism of $\mathfrak{A}$ into $\mathfrak{B}$; thus $(ab)\rho = (a\rho)(b\rho)$ and $1\rho = 1$. Put $\mathfrak{l}(\rho) = \rho$; then $\mathfrak{l}(\rho)$ is a Lie homomorphism of $\mathfrak{l}(\mathfrak{A})$ into $\mathfrak{l}(\mathfrak{B})$.

If $\mathfrak{C}$ is a subalgebra of $\mathfrak{A}$, $\mathfrak{l}(\mathfrak{C})$ is a Lie subalgebra of $\mathfrak{l}(\mathfrak{A})$. If $\mathfrak{J}$ is a (two-sided) ideal of $\mathfrak{A}$, the set $\mathfrak{J}$ is an ideal of $\mathfrak{l}(\mathfrak{A})$ and $\mathfrak{l}(\mathfrak{A}/\mathfrak{J}) = \mathfrak{l}(\mathfrak{A})/\mathfrak{J}$.

8.3 Lemma. *Let $\mathfrak{g}$ be a Lie algebra over* A.

a) *For $n > 1$, let $\mathfrak{S}_n^*$ be the set of permutations α of $\{1, \ldots, n\}$ for which there exists an integer k $(0 \leq k < n)$ such that*

$$1\alpha > \cdots > k\alpha > (k+1)\alpha < (k+2)\alpha < \cdots < n\alpha,$$

and put $s(\alpha) = (-1)^k$. Then if $b, a_1, \ldots, a_n$ are elements of $\mathfrak{g}$,

$$[b, [a_1, \ldots, a_n]] = \sum_{\alpha \in \mathfrak{S}_n^*} s(\alpha)[b, a_{1\alpha}, \ldots, a_{n\alpha}].$$

b) *If X is a subset of $\mathfrak{g}$, the subalgebra of $\mathfrak{g}$ generated by X is the* A-*submodule spanned by all products $[a_1, \ldots, a_n]$ with $n > 0$ and $a_i \in X$. (The product $[a]$ of a single element a is understood to be a).*

c) *Suppose that X, Y are subsets of $\mathfrak{g}$ and that $\mathfrak{g}$ is the subalgebra generated by Y. The ideal of $\mathfrak{g}$ generated by X is the* A-*submodule spanned by all products $[b, a_1, \ldots, a_n]$ with $n \geq 0$, $b \in X$, $a_i \in Y$.*

Proof. a) is proved by induction on n. For $n = 2$, $\mathfrak{S}_2^* = \mathfrak{S}_2$, and the assertion is

$$[b, [a_1, a_2]] = [b, a_1, a_2] - [b, a_2, a_1]. \tag{1}$$

This is clear, however, since by (iv′), (i) and (iii),

$$\begin{aligned} [b, a_1, a_2] - [b, a_2, a_1] &= [b, a_1, a_2] - [-[a_2, b], a_1] \\ &= [b, a_1, a_2] + [a_2, b, a_1] \\ &= -[a_1, a_2, b] \\ &= [b, [a_1, a_2]]. \end{aligned}$$

For $n > 2$, observe that for $\alpha \in \mathfrak{S}_n^*$, either $n\alpha = n$ or $1\alpha = n$. Hence the right-hand side of the required equation is

$$\left[\sum_{\beta \in \mathfrak{S}_{n-1}^*} s(\beta)[b, a_{1\beta}, \ldots, a_{(n-1)\beta}], a_n\right] - \sum_{\beta \in \mathfrak{S}_{n-1}^*} s(\beta)[b, a_n, a_{1\beta}, \ldots, a_{(n-1)\beta}].$$

Applying the inductive hypothesis to both sums, this becomes

$$[b, [a_1, \ldots, a_{n-1}], a_n] - [b, a_n, [a_1, \ldots, a_{n-1}]].$$

By (1), this is $[b, [a_1, \ldots, a_n]]$.

b) It suffices to show that the A-submodule $\mathfrak{h}$ spanned by all products $[a_1, \ldots, a_n]$ $(n > 0, a_i \in X)$ is a subalgebra. Thus we must show that if $a_i \in X$ and $a'_j \in X$, then

$$[[a_1, \ldots, a_m], [a'_1, \ldots, a'_n]] \in \mathfrak{h}.$$

This is clear if $n = 1$ and follows at once from a) if $n > 1$.

c) Let $\mathfrak{j}$ be the A-module spanned by all products $[b, a_1, \ldots, a_n]$ with $n \geq 0$, $b \in X$, $a_i \in Y$. By a),

$$[[b, a_1, \ldots, a_m], [a'_1, \ldots, a'_n]] \in \mathfrak{j}$$

if $b \in X$, $a_i \in Y$, $a'_j \in Y$. Since $\mathfrak{g}$ is the Lie algebra generated by Y, it follows from b) that $\mathfrak{g}$ is the A-module spanned by all $[a'_1, \ldots, a'_n]$ with $a'_j \in Y$. Hence

$$[b, a_1, \ldots, a_m, c] \in \mathfrak{j}$$

if $b \in X$, $a_i \in Y$, $c \in \mathfrak{g}$. Thus $[\mathfrak{j}, \mathfrak{g}] \subseteq \mathfrak{j}$ and $\mathfrak{j}$ is an ideal of $\mathfrak{g}$. Hence $\mathfrak{j}$ is the ideal generated by X. **q.e.d.**

8.4 Definition. If $\mathsf{U}_1, \ldots, \mathsf{U}_n$ are A-submodules of the Lie algebra $\mathfrak{g}$, define $[\mathsf{U}_1, \ldots, \mathsf{U}_n]$ to be the A-submodule spanned by all $[x_1, \ldots, x_n]$ with $x_i \in \mathsf{U}_i$. By a trivial induction,
$[\mathsf{U}_1, \ldots, \mathsf{U}_n] = [[\mathsf{U}_1, \ldots, \mathsf{U}_{n-1}], \mathsf{U}_n]$ if $n > 1$.

8.5 Lemma. *Suppose that* U, V, W *are* A-*submodules of a Lie algebra* $\mathfrak{g}$.

a) *(cf. III, 1.10)* $[\mathsf{U}, \mathsf{V}, \mathsf{W}] \subseteq [\mathsf{V}, \mathsf{W}, \mathsf{U}] + [\mathsf{W}, \mathsf{U}, \mathsf{V}]$.

b) *If* ρ *is a homomorphism of* $\mathfrak{g}$, $[\mathsf{U}\rho, \mathsf{V}\rho] = [\mathsf{U}, \mathsf{V}]\rho$ *for any* A-*submodules* U, V. *If* $\mathfrak{j}$ *is an ideal of* $\mathfrak{g}$, $[\mathsf{U} + \mathfrak{j}/\mathfrak{j}, \mathsf{V} + \mathfrak{j}/\mathfrak{j}] = ([\mathsf{U}, \mathsf{V}] + \mathfrak{j})/\mathfrak{j}$.

c) *If* $\mathfrak{i}_1, \mathfrak{i}_2$ *are ideals of* $\mathfrak{g}$, *so is* $[\mathfrak{i}_1, \mathfrak{i}_2]$.

Proof. a) follows immediately from 8.4 and the Jacobi identity. The first part of b) is trivial, and the second part is the application of the first to the natural homomorphism of $\mathfrak{g}$ onto $\mathfrak{g}/\mathfrak{j}$. For c), we note that $[\mathfrak{i}_2, \mathfrak{g}, \mathfrak{i}_1] \subseteq [\mathfrak{i}_2, \mathfrak{i}_1] = [\mathfrak{i}_1, \mathfrak{i}_2]$ and $[\mathfrak{g}, \mathfrak{i}_1, \mathfrak{i}_2] \subseteq [\mathfrak{i}_1, \mathfrak{i}_2]$; thus $[\mathfrak{i}_1, \mathfrak{i}_2, \mathfrak{g}] \subseteq [\mathfrak{i}_1, \mathfrak{i}_2]$ by a). **q.e.d.**

8.6 Definition. If $\mathfrak{g}$ is a Lie algebra over A, write $\mathfrak{g}^n = [\mathfrak{g}, \underset{n}{\ldots}, \mathfrak{g}]$. Thus $\mathfrak{g}^1 = \mathfrak{g}$ and $\mathfrak{g}^n = [\mathfrak{g}^{n-1}, \mathfrak{g}]$ for $n > 1$. By 8.4, $\mathfrak{g}^n$ is the A-module spanned by all $[x_1, \ldots, x_n]$ with $x_i \in \mathfrak{g}$. By 8.5c), $\mathfrak{g}^n$ is an ideal of $\mathfrak{g}$. Hence $\mathfrak{g}^n \subseteq \mathfrak{g}^{n-1}$ $(n > 1)$; the series

$$\mathfrak{g} = \mathfrak{g}^1 \supseteq \mathfrak{g}^2 \supseteq \mathfrak{g}^3 \supseteq \cdots$$

is called the *lower central series* of $\mathfrak{g}$.

8.7 Lemma. *Let* $\mathfrak{g}$ *be a Lie algebra over* A.

a) *If* $\mathfrak{j}$ *is an* A*-submodule of* $\mathfrak{g}$ *and* $\mathfrak{g}^{n+1} \subseteq \mathfrak{j} \subseteq \mathfrak{g}^n$, *then* $\mathfrak{j}$ *is an ideal of* $\mathfrak{g}$.

b) *If* $\mathfrak{g}$ *is the subalgebra generated by* X, $\mathfrak{g}^n$ *is the* A*-module spanned by all* $[x_1, \ldots, x_k]$ *with* $x_i \in X$ *and* $k \geq n$.

c) *If* $\mathfrak{j}$ *is an ideal of* $\mathfrak{g}$, $(\mathfrak{g}/\mathfrak{j})^n = (\mathfrak{g}^n + \mathfrak{j})/\mathfrak{j}$.

d) *If* $\mathfrak{g} = \mathfrak{j}_1 \supseteq \mathfrak{j}_2 \supseteq \cdots$ *is a series of ideals of* $\mathfrak{g}$ *and* $[\mathfrak{j}_n, \mathfrak{g}] \subseteq \mathfrak{j}_{n+1}$ *for all* $n \geq 1$, *then* $\mathfrak{j}_n \supseteq \mathfrak{g}^n$.

e) $[\mathfrak{g}^m, \mathfrak{g}^n] \subseteq \mathfrak{g}^{m+n}$ *for all* $m \geq 1, n \geq 1$.

Proof. a) $[\mathfrak{j}, \mathfrak{g}] \subseteq [\mathfrak{g}^n, \mathfrak{g}] = \mathfrak{g}^{n+1} \subseteq \mathfrak{j}$.

b) is proved by induction on n. For $n = 1$, the assertion is the same as that of 8.3b). For $n > 1$, $\mathfrak{g}^{n-1}$ is the A-module generated by all $[x_1, \ldots, x_k]$ with $x_i \in X$ and $k \geq n - 1$, by the inductive hypothesis. Hence $\mathfrak{g}^n$ is the A-module generated by all

$$[x_1, \ldots, x_k, [x_{k+1}, \ldots, x_l]]$$

with $x_i \in X$, $l > k \geq n - 1$. The result follows by applying 8.3a).

c), d), e) are all proved by induction on n and are all trivial for $n = 1$. In c) for $n > 1$, we have

$$\begin{aligned}(\mathfrak{g}/\mathfrak{j})^n &= [(\mathfrak{g}/\mathfrak{j})^{n-1}, \mathfrak{g}/\mathfrak{j}] = [(\mathfrak{g}^{n-1} + \mathfrak{j})/\mathfrak{j}, \mathfrak{g}/\mathfrak{j}] \\ &= ([\mathfrak{g}^{n-1}, \mathfrak{g}] + \mathfrak{j})/\mathfrak{j} = (\mathfrak{g}^n + \mathfrak{j})/\mathfrak{j}\end{aligned}$$

by 8.6, the inductive hypothesis and 8.5b). In d) for $n > 1$,

$$\mathfrak{j}_n \supseteq [\mathfrak{j}_{n-1}, \mathfrak{g}] \supseteq [\mathfrak{g}^{n-1}, \mathfrak{g}] = \mathfrak{g}^n.$$

Finally, to prove e), observe that for $n > 1$,

$$[\mathfrak{g}, \mathfrak{g}^m, \mathfrak{g}^{n-1}] = [\mathfrak{g}^{m+1}, \mathfrak{g}^{n-1}] \subseteq \mathfrak{g}^{m+n}$$

and

$$[\mathfrak{g}^m, \mathfrak{g}^{n-1}, \mathfrak{g}] \subseteq [\mathfrak{g}^{m+n-1}, \mathfrak{g}] \subseteq \mathfrak{g}^{m+n}$$

by the inductive hypothesis and 8.6. By 8.5a), $[\mathfrak{g}^{n-1}, \mathfrak{g}, \mathfrak{g}^m] \subseteq \mathfrak{g}^{m+n}$, whence

$$[\mathfrak{g}^m, \mathfrak{g}^n] = [\mathfrak{g}^n, \mathfrak{g}^m] \subseteq \mathfrak{g}^{m+n}.$$

q.e.d.

8.8 Theorem. *Let* $\mathfrak{g}$ *be a Lie algebra over* A *and let* $\mathfrak{h}$ *be a Lie subalgebra such that* $\mathfrak{h}^i + \mathfrak{g}^{i+1} = \mathfrak{g}^i$ *for some* $i \geq 1$. *Then* $\mathfrak{h}^n + \mathfrak{g}^m = \mathfrak{g}^n$ *whenever* $m > n \geq i$.

Proof. This is proved by induction on $m - i$. If $m - i = 1$, the assertion is the hypothesis. Suppose that $m - i > 1$. We have

$$\begin{aligned} \mathfrak{g}^{i+1} &= [\mathfrak{g}^i, \mathfrak{g}] \\ &= [\mathfrak{h}^i + \mathfrak{g}^{i+1}, \mathfrak{g}] \\ &= [\mathfrak{h}^i, \mathfrak{g}] + \mathfrak{g}^{i+2}. \end{aligned}$$

Now $[\mathfrak{h}^i, \mathfrak{g}] = [\mathfrak{g}, \mathfrak{h}^i]$ is the A-module spanned by all $[b, [a_1, \ldots, a_i]]$ with $b \in \mathfrak{g}$, $a_j \in \mathfrak{h}$ $(1 \leq j \leq i)$. Hence by 8.3a),

$$[\mathfrak{h}^i, \mathfrak{g}] \subseteq [\mathfrak{g}, \underbrace{\mathfrak{h}, \ldots, \mathfrak{h}}_{i}].$$

But now

$$\begin{aligned} [\mathfrak{g}, \underbrace{\mathfrak{h}, \ldots, \mathfrak{h}}_{i}] &\subseteq [\mathfrak{g}^i, \mathfrak{h}] \\ &= [\mathfrak{h}^i + \mathfrak{g}^{i+1}, \mathfrak{h}] \\ &= [\mathfrak{h}^i, \mathfrak{h}] + [\mathfrak{g}^{i+1}, \mathfrak{h}] \\ &\subseteq \mathfrak{h}^{i+1} + \mathfrak{g}^{i+2}. \end{aligned}$$

Thus

$$\begin{aligned} \mathfrak{g}^{i+1} &= [\mathfrak{h}^i, \mathfrak{g}] + \mathfrak{g}^{i+2} \\ &\subseteq [\mathfrak{g}, \underbrace{\mathfrak{h}, \ldots, \mathfrak{h}}_{i}] + \mathfrak{g}^{i+2} \\ &\subseteq \mathfrak{h}^{i+1} + \mathfrak{g}^{i+2}. \end{aligned}$$

The assertion in the case when $n > i$ now follows at once from the inductive hypothesis; in particular

$$\mathfrak{h}^{i+1} + \mathfrak{g}^m = \mathfrak{g}^{i+1} \quad (m > i + 1).$$

Hence

$$\mathfrak{g}^i = \mathfrak{h}^i + \mathfrak{g}^{i+1} = \mathfrak{h}^i + \mathfrak{h}^{i+1} + \mathfrak{g}^m = \mathfrak{h}^i + \mathfrak{g}^m,$$

which is the assertion in the case $n = i$. **q.e.d.**

8.9 Definition. The Lie algebra $\mathfrak{g}$ is called *Abelian* if $\mathfrak{g}^2 = 0$, that is, if $[a, b] = 0$ for all $a \in \mathfrak{g}$, $b \in \mathfrak{g}$. Any A-module can be regarded as an Abelian Lie algebra by simply defining the product of two elements to be zero.

$\mathfrak{g}$ is called *nilpotent* if there exists an integer n such that $\mathfrak{g}^{n+1} = 0$. The smallest such integer n is called the *class* of $\mathfrak{g}$. Thus $\mathfrak{g}$ is Abelian if and only if $\mathfrak{g}$ is nilpotent of class at most 1.

8.10 Theorem. *Let $\mathfrak{g}$ be a Lie algebra over* A.

a) *If $\mathfrak{j}$ is an ideal of $\mathfrak{g}$, $\mathfrak{g}/\mathfrak{j}$ is nilpotent of class at most n if and only if $\mathfrak{j} \supseteq \mathfrak{g}^{n+1}$. In particular, $\mathfrak{g}/\mathfrak{j}$ is Abelian if and only if $\mathfrak{g}^2 \subseteq \mathfrak{j}$.*

b) *$\mathfrak{g}$ is nilpotent of class at most n if and only if there exists a series*

$$\mathfrak{g} = \mathfrak{j}_1 \supseteq \mathfrak{j}_2 \supseteq \cdots \supseteq \mathfrak{j}_{n+1} = 0$$

of ideals of $\mathfrak{g}$ such that $[\mathfrak{j}_i, \mathfrak{g}] \subseteq \mathfrak{j}_{i+1}$ $(i = 1, \ldots, n)$.

c) *If $\mathfrak{g}$ is nilpotent of class at most n, so is any subalgebra or quotient algebra of $\mathfrak{g}$.*

d) *If $\mathfrak{g}$ is nilpotent and $\mathfrak{h}$ is a subalgebra such that $\mathfrak{g}^i = \mathfrak{h}^i + \mathfrak{g}^{i+1}$ for some $i \geq 1$, then $\mathfrak{g}^j = \mathfrak{h}^j$ for all $j \geq i$.*

Proof. a) This follows from 8.7c).

b) This follows from 8.7d).

c) This follows from a) and the fact that $\mathfrak{h}^m \subseteq \mathfrak{g}^m$ for any subalgebra $\mathfrak{h}$ of $\mathfrak{g}$.

d) If m is the class of $\mathfrak{g}$, $\mathfrak{g}^j = \mathfrak{h}^j + \mathfrak{g}^{m+1} = \mathfrak{h}^j$, by 8.8. **q.e.d.**

8.11 Definitions. For any Lie algebra $\mathfrak{g}$, the A-submodule $\mathfrak{g}^{(n)}$ is defined inductively by

$$\mathfrak{g}^{(0)} = \mathfrak{g}, \quad \mathfrak{g}^{(n)} = [\mathfrak{g}^{(n-1)}, \mathfrak{g}^{(n-1)}] \quad (n \geq 1).$$

In particular, we write $\mathfrak{g}' = \mathfrak{g}^{(1)} = [\mathfrak{g}, \mathfrak{g}] = \mathfrak{g}^2$, $\mathfrak{g}'' = \mathfrak{g}^{(2)}$, etc. By 8.5c), $\mathfrak{g}^{(n)}$ is an ideal of $\mathfrak{g}$. By 8.10a), $\mathfrak{g}^{(n-1)}/\mathfrak{g}^{(n)}$ is Abelian for $n \geq 1$. The series

$$\mathfrak{g} \supseteq \mathfrak{g}' \supseteq \mathfrak{g}'' \supset \cdots$$

is called the *derived series* of $\mathfrak{g}$. By 8.7e), $\mathfrak{g}^{(n)} \subseteq \mathfrak{g}^{2^n}$ (cf. III, 2.12). $\mathfrak{g}$ is called *soluble* if there is an integer d such that $\mathfrak{g}^{(d)} = 0$, and the smallest such integer d is called the *derived length* of $\mathfrak{g}$. Thus $\mathfrak{g}$ is Abelian if and only if $\mathfrak{g}$ is soluble of derived length 1. A nilpotent Lie algebra is soluble.

8.12 Theorem. *Let* $\mathfrak{g}$ *be a Lie algebra over* A.

a) *If* $\mathfrak{j}$ *is an ideal of* $\mathfrak{g}$, $(\mathfrak{g}/\mathfrak{j})^{(n)} = (\mathfrak{g}^{(n)} + \mathfrak{j})/\mathfrak{j}$. *Thus* $\mathfrak{g}/\mathfrak{j}$ *is soluble of derived length at most d if and only if* $\mathfrak{j} \supseteq \mathfrak{g}^{(d)}$.

b) *If* $\mathfrak{g} = \mathfrak{j}_0 \supseteq \mathfrak{j}_1 \supseteq \mathfrak{j}_2 \supseteq \cdots$ *is a series of subalgebras of* $\mathfrak{g}$ *such that* $\mathfrak{j}_n$ *is an ideal of* $\mathfrak{j}_{n-1}$ *and* $\mathfrak{j}_{n-1}/\mathfrak{j}_n$ *is Abelian for each* $n > 0$, *then* $\mathfrak{j}_n \supseteq \mathfrak{g}^{(n)}$. *Thus* $\mathfrak{g}$ *is soluble of derived length at most d if and only if there exists such a series with* $\mathfrak{j}_d = 0$.

c) *If* $\mathfrak{g}$ *is soluble, so is any subalgebra or quotient algebra of* $\mathfrak{g}$.

d) *If* $\mathfrak{j}$ *is an ideal of* $\mathfrak{g}$ *and* $\mathfrak{j}$, $\mathfrak{g}/\mathfrak{j}$ *are both soluble, then* $\mathfrak{g}$ *is soluble.*

Proof. a) follows from 8.5b). b) is proved by induction on n. For $n > 0$, $\mathfrak{j}_{n-1}/\mathfrak{j}_n$ is Abelian, so $\mathfrak{j}_n \supseteq \mathfrak{j}'_{n-1}$ by 8.10a). Thus

$$\mathfrak{j}_n \supseteq [\mathfrak{j}_{n-1}, \mathfrak{j}_{n-1}] \supseteq [\mathfrak{g}^{(n-1)}, \mathfrak{g}^{(n-1)}] = \mathfrak{g}^{(n)}.$$

c) follows from a) and the obvious fact that $\mathfrak{h}^{(m)} \subseteq \mathfrak{g}^{(m)}$ for any subalgebra $\mathfrak{h}$ of $\mathfrak{g}$. To prove d), observe that $\mathfrak{g}^{(d)} \subseteq \mathfrak{j}$ for some integer d, by a). Hence $\mathfrak{g}^{(d+n)} \subseteq \mathfrak{j}^{(n)}$ for all n. Since there exists an integer n for which $\mathfrak{j}^{(n)} = 0$, $\mathfrak{g}$ is soluble. **q.e.d.**

These results are obvious analogues of the elementary theories of nilpotent and soluble groups. But in Lie algebras, one can extend the ground-ring; there is no analogue of this for groups. In the following, A is a subring of a commutative ring A^* containing the identity element of A^*.

8.13 Theorem *(cf. V, 11.1)*. *Suppose that* $\mathfrak{g}$ *is a Lie algebra over* A *and let* $\mathfrak{g}^* = \mathfrak{g} \otimes_{\mathsf{A}} \mathsf{A}^*$.

a) $\mathfrak{g}^*$ *has the structure of a Lie algebra over* A^* *with*

$$\lambda(a \otimes \lambda') = a \otimes \lambda\lambda', \tag{2}$$

$$[a \otimes \lambda, a' \otimes \lambda'] = [a, a'] \otimes \lambda\lambda' \tag{3}$$

for all λ, λ' *in* A^* *and* a, a' *in* $\mathfrak{g}$.

b) *If* α *is an automorphism of* $\mathfrak{g}$, $\alpha^* = \alpha \otimes 1$ *is an automorphism of* $\mathfrak{g}^*$.

c) *Suppose that* A^*, *regarded as an* A*-module, is free. If* U, V *are* A*-submodules of* $\mathfrak{g}$, $\mathsf{U} \otimes \mathsf{A}^*$ *and* $\mathsf{V} \otimes \mathsf{A}^*$ *can be regarded as* A^**-submodules of* $\mathfrak{g}^*$. *Then* $\mathsf{U} \otimes \mathsf{A}^* \subseteq \mathsf{V} \otimes \mathsf{A}^*$ *if and only if* $\mathsf{U} \subseteq \mathsf{V}$. *Also*

$$[\mathsf{U}, \mathsf{V}] \otimes \mathsf{A}^* = [\mathsf{U} \otimes \mathsf{A}^*, \mathsf{V} \otimes \mathsf{A}^*],$$

and if $\mathfrak{j}$ *is an ideal of* $\mathfrak{g}$, $\mathfrak{j} \otimes \mathsf{A}^*$ *is an ideal of* $\mathfrak{g}^*$.

d) *Suppose that* A^* *is a free* A*-module, that* α *is an automorphism of* $\mathfrak{g}$ *and* $\alpha^* = \alpha \otimes 1$. *If* U *is an* A*-submodule of* $\mathfrak{g}$ *which contains all elements of* $\mathfrak{g}$ *invariant under* α, *then* $\mathsf{U} \otimes \mathsf{A}^*$ *contains all elements of* $\mathfrak{g}^*$ *invariant under* α^*.

Proof. a) Certainly $\mathfrak{g}^*$ is an A^*-module in which (2) holds. Multiplication in $\mathfrak{g}$ is an A-bilinear mapping of $\mathfrak{g} \times \mathfrak{g}$ into $\mathfrak{g}$; this can therefore be extended to an A^*-bilinear mapping of $\mathfrak{g}^* \times \mathfrak{g}^*$ into $\mathfrak{g}^*$ satisfying (3). The verification of the Jacobi identity and the skew-symmetry of the multiplication in $\mathfrak{g}^*$ are tedious but routine matters.

b) If $\alpha^* = \alpha \otimes 1$ and $\beta^* = \alpha^{-1} \otimes 1$, then α^* and β^* are A^*-endomorphisms of $\mathfrak{g}^*$ and $\alpha^*\beta^* = \beta^*\alpha^* = 1$. Since also α^* preserves Lie products, α^* is an automorphism of $\mathfrak{g}^*$.

c) Since A^* is a free A-module, it follows from V, 9.5 that $\mathsf{U} \otimes \mathsf{A}^*$, $\mathsf{V} \otimes \mathsf{A}^*$ can be monomorphically embedded as A^*-modules in $\mathfrak{g} \otimes \mathsf{A}^*$ and may therefore be regarded as A^*-submodules of $\mathfrak{g}^*$. It is clear that $\mathsf{U} \otimes \mathsf{A}^* \subseteq \mathsf{V} \otimes \mathsf{A}^*$ if and only if $\mathsf{U} \subseteq \mathsf{V}$. By 8.1c), $[\mathsf{U} \otimes \mathsf{A}^*, \mathsf{V} \otimes \mathsf{A}^*]$ is spanned over A^* by all $[u \otimes \lambda, v \otimes \lambda'] = [u, v] \otimes \lambda\lambda' = \lambda\lambda'([u, v] \otimes 1)$ with λ, λ' in A^* and $u \in \mathsf{U}$, $v \in \mathsf{V}$, so

$$[\mathsf{U} \otimes \mathsf{A}^*, \mathsf{V} \otimes \mathsf{A}^*] = [\mathsf{U}, \mathsf{V}] \otimes \mathsf{A}^*.$$

If $\mathfrak{j}$ is an ideal of $\mathfrak{g}$,

$$[\mathfrak{j} \otimes \mathsf{A}^*, \mathfrak{g}^*] = [\mathfrak{j} \otimes \mathsf{A}^*, \mathfrak{g} \otimes \mathsf{A}^*] = [\mathfrak{j}, \mathfrak{g}] \otimes \mathsf{A}^* \subseteq \mathfrak{j} \otimes \mathsf{A}^*.$$

Thus $\mathfrak{j} \otimes \mathsf{A}^*$ is an ideal of $\mathfrak{g}^*$.

d) Suppose that b is an element of $\mathfrak{g}^*$ invariant under α^*. There are distinct elements $\lambda_1, \ldots, \lambda_m$ of an A-basis of A^* such that

$$b = \sum_{i=1}^{m} a_i \otimes \lambda_i,$$

where $a_i \in \mathfrak{g}$. Since $b\alpha^* = b$, $\sum_i (a_i - a_i\alpha) \otimes \lambda_i = 0$. By V, 9.5, $a_i = a_i\alpha$, so $a_i \in \mathsf{U}$. Thus $b \in \mathsf{U} \otimes \mathsf{A}^*$. **q.e.d.**

Exercises

11) Let $\mathfrak{g}$ be the Lie algebra generated by X and let $\mathfrak{j}$ be the ideal of $\mathfrak{g}$ generated by Y. Show that $[\mathfrak{j}, \mathfrak{g}]$ is the ideal of $\mathfrak{g}$ generated by $\{[a, b] \mid a \in X,\ b \in Y\}$. Show that $\mathfrak{g}^n$ is the ideal of $\mathfrak{g}$ generated by $\{[a_1, \ldots, a_n] \mid a_i \in X\}$.

12) Prove that a proper subalgebra of a nilpotent Lie algebra is an ideal of a strictly larger subalgebra.

13) Suppose that $n > 1$ and $\mathfrak{g}$ is the Lie algebra of all $n \times n$ matrices (λ_{ij}) with $\lambda_{ij} \in \mathsf{A}$ and $\lambda_{ij} = 0$ if $i > j$. Prove the following.

a) $\mathfrak{h} = \mathfrak{g}^2 = \{(\lambda_{ij}) | \lambda_{ij} = 0 \quad \text{if } i \geq j\}$.

b) $\mathfrak{g}^2 = \mathfrak{g}^3$.

c) $\mathfrak{h}^m = \{(\lambda_{ij}) | \lambda_{ij} = 0 \quad \text{if } i > j - m\}$ $(1 \leq m \leq n)$.

d) $\mathfrak{h}$ is nilpotent. $\mathfrak{g}$ is soluble but not nilpotent.

14) Formulate and prove the analogue of Theorem 8.8 for groups.

§ 9. The Lie Ring Method and an Application

In order to use Lie algebras for the study of nilpotent groups, we make a construction which enables one to associate a Lie ring with any strongly central series of a group.

9.1 Definition. A *strongly central series* of a group $\mathfrak{G}$ is a series

(S) $$\mathfrak{G} = \mathfrak{G}_1 \geq \mathfrak{G}_2 \geq \mathfrak{G}_3 \geq \cdots.$$

of subgroups of $\mathfrak{G}$ for which $[\mathfrak{G}_i, \mathfrak{G}_j] \leq \mathfrak{G}_{i+j}$ for all i, j. This implies that $[\mathfrak{G}_i, \mathfrak{G}] = [\mathfrak{G}_i, \mathfrak{G}_1] \leq \mathfrak{G}_{i+1}$ for all $i \geq 1$, so (S) is a central series in the sense of III, 2.1. Hence $\mathfrak{G}_i \trianglelefteqq \mathfrak{G}$ and $\mathfrak{G}_i/\mathfrak{G}_{i+1}$ is contained in the centre of $\mathfrak{G}/\mathfrak{G}_{i+1}$. In particular $\mathfrak{G}_i/\mathfrak{G}_{i+1}$ is Abelian.

For example, the lower central series of any group is strongly central (III, 2.11b)).

With a strongly central series we obtain something resembling the Jacobi identity.

9.2 Lemma. *Suppose that*

(S) $$\mathfrak{G} = \mathfrak{G}_1 \geq \mathfrak{G}_2 \geq \mathfrak{G}_3 \geq \cdots$$

is a strongly central series of a group $\mathfrak{G}$. *If* $x \in \mathfrak{G}_i$, $y \in \mathfrak{G}_j$, $z \in \mathfrak{G}_k$,

$$[x, y, z][y, z, x][z, x, y] \in \mathfrak{G}_{i+j+k+1}.$$

Proof. Since $[x, y] = x^{-1}(y^{-1}xyx^{-1})x = [y, x^{-1}]^x$, we have

$$\begin{aligned}[x, y, z] &= [[y, x^{-1}]^x, (xzx^{-1})^x] \\ &= [y, x^{-1}, (xzx^{-1}z^{-1})z]^x \\ &= [y, x^{-1}, z]^x[y, x^{-1}, [x^{-1}, z^{-1}]]^{zx}.\end{aligned}$$

Since (S) is strongly central,

$$[y, x^{-1}, [x^{-1}, z^{-1}]]^{zx} \in [\mathfrak{G}_{i+j}, \mathfrak{G}_{i+k}]^{zx} \leq \mathfrak{G}_{i+j+k+i} \leq \mathfrak{G}_{i+j+k+1}.$$

Since also $\mathfrak{G}_{i+j+k}/\mathfrak{G}_{i+j+k+1}$ is Abelian, it follows that

$$\begin{aligned}&[x, y, z][y, z, x][z, x, y]\mathfrak{G}_{i+j+k+1} \\ &= [y, x^{-1}, z]^x[x, z^{-1}, y]^z[z, y^{-1}, x]^y\mathfrak{G}_{i+j+k+1} = 1,\end{aligned}$$

by III, 1.4. **q.e.d.**

9.3 Theorem *(cf. III, Aufg. 8). Let*

(S) $$\mathfrak{G} = \mathfrak{G}_1 \geq \mathfrak{G}_2 \geq \mathfrak{G}_3 \geq \cdots$$

be a strongly central series of a group $\mathfrak{G}$.

a) *There exists a Lie ring* $\mathfrak{g}$, *unique to within isomorphism, having the following properties.*

(i) *For each* $i \geq 1$, *there exists a monomorphism* σ_i *of* $\mathfrak{G}_i/\mathfrak{G}_{i+1}$ *into the additive group of* $\mathfrak{g}$, *and if* $\mathsf{G}_i = \operatorname{im}\sigma_i$,

$$\mathfrak{g} = \mathsf{G}_1 \oplus \mathsf{G}_2 \oplus \cdots.$$

(ii) *If* $x \in \mathfrak{G}_i$ *and* $y \in \mathfrak{G}_j$,

$$[(x\mathfrak{G}_{i+1})\sigma_i, (y\mathfrak{G}_{j+1})\sigma_j] = ([x, y]\mathfrak{G}_{i+j+1})\sigma_{i+j}.$$

b) *Suppose that* $\mathfrak{h}$ *is a Lie ring and that for each* $i \geq 1$, *there is a homomorphism* ρ_i *of* $\mathfrak{G}_i/\mathfrak{G}_{i+1}$ *into* $\mathfrak{h}$ *such that if* $x \in \mathfrak{G}_i$ *and* $y \in \mathfrak{G}_j$,

$$[(x\mathfrak{G}_{i+1})\rho_i, (y\mathfrak{G}_{j+1})\rho_j] = ([x, y]\mathfrak{G}_{i+j+1})\rho_{i+j}.$$

Then there is a Lie ring homomorphism θ *of* $\mathfrak{g}$ *into* $\mathfrak{h}$ *such that* $\rho_i = \sigma_i\theta$ *for each* $i \geq 1$. *If each* ρ_i *is a monomorphism and the sum of the* $\operatorname{im}\rho_i$ *is direct,* θ *is a monomorphism. If* $\mathfrak{h}$ *is spanned by the* $\operatorname{im}\rho_i$, θ *is an epimorphism.*

c) *If* α *is an automorphism of* $\mathfrak{G}$ *and* $\mathfrak{G}_i\alpha = \mathfrak{G}_i$ *for all i, there exists an automorphism* ρ *of* $\mathfrak{g}$ *such that if* $x \in \mathfrak{G}_i$,

$$(x\mathfrak{G}_{i+1})\sigma_i\rho = ((x\alpha)\mathfrak{G}_{i+1})\sigma_i.$$

Proof. a) Let σ_i be an isomorphism of $\mathfrak{G}_i/\mathfrak{G}_{i+1}$ onto an additive group G_i ($i = 1, 2, \ldots$), and let $\mathfrak{g}$ be the direct sum of the G_i:

$$\mathfrak{g} = \mathsf{G}_1 \oplus \mathsf{G}_2 \oplus \mathsf{G}_3 \cdots.$$

Since (S) is strongly central, it follows from 6.1 that for $i \geq 1, j \geq 1$, there exists a $\mathbb{Z}$-bilinear mapping γ_{ij} of $(\mathfrak{G}_i/\mathfrak{G}_{i+1}) \times (\mathfrak{G}_j/\mathfrak{G}_{j+1})$ into $\mathfrak{G}_{i+j}/\mathfrak{G}_{i+j+1}$ such that if $x \in \mathfrak{G}_i$ and $y \in \mathfrak{G}_j$,

$$(x\mathfrak{G}_{i+1}, y\mathfrak{G}_{j+1})\gamma_{ij} = [x, y]\mathfrak{G}_{i+j+1}.$$

Thus there exists a $\mathbb{Z}$-bilinear mapping β_{ij} of $\mathsf{G}_i \times \mathsf{G}_j$ into G_{i+j} such that if $a \in \mathsf{G}_i$ and $b \in \mathsf{G}_j$,

$$(a, b)\beta_{ij} = ((a\sigma_i^{-1}, b\sigma_j^{-1})\gamma_{ij})\sigma_{i+j}.$$

Since $\mathfrak{g}$ is the direct sum of the G_i, there exists a bilinear mapping $(a, b) \to [a, b]$ of $\mathfrak{g} \times \mathfrak{g}$ into $\mathfrak{g}$, the restriction of which to $\mathsf{G}_i \times \mathsf{G}_j$ is β_{ij}. Thus if $x \in \mathfrak{G}_i$ and $y \in \mathfrak{G}_j$,

$$\begin{aligned}[(x\mathfrak{G}_{i+1})\sigma_i, (y\mathfrak{G}_{j+1})\sigma_j] &= ((x\mathfrak{G}_{i+1})\sigma_i, (y\mathfrak{G}_{j+1})\sigma_j)\beta_{ij} \\ &= ((x\mathfrak{G}_{i+1}, y\mathfrak{G}_{j+1})\gamma_{ij})\sigma_{i+j} \\ &= ([x, y]\mathfrak{G}_{i+j+1})\sigma_{i+j},\end{aligned}$$

as asserted in (ii). We show that $\mathfrak{g}$ is a Lie ring.

The distributive laws are equivalent to the bilinearity already established. To verify the skew-symmetry, suppose that $a \in \mathfrak{g}$. We can write

$$a = a_1 + \cdots + a_k$$

with $a_i \in \mathsf{G}_i$ and $a_i = (x_i\mathfrak{G}_{i+1})\sigma_i$ $(i = 1, \ldots, k)$. Now

$$\begin{aligned}[a, a] &= \sum_{i,j=1}^{k} [a_i, a_j] \\ &= \sum_{i=1}^{k} [a_i, a_i] + \sum_{i<j} ([a_i, a_j] + [a_j, a_i]).\end{aligned}$$

Thus, to prove $[a, a] = 0$, it suffices to prove that $[a_i, a_i] = 0$ and $[a_i, a_j] + [a_j, a_i] = 0$. But by (ii)

$$[a_i, a_i] = ([x_i, x_i]\mathfrak{G}_{2i+1})\sigma_{2i} = 1\sigma_{2i} = 0,$$

and

$$\begin{aligned}[a_i, a_j] + [a_j, a_i] &= ([x_i, x_j]\mathfrak{G}_{i+j+1})\sigma_{i+j} + ([x_j, x_i]\mathfrak{G}_{i+j+1})\sigma_{i+j} \\ &= ([x_i, x_j][x_j, x_i]\mathfrak{G}_{i+j+1})\sigma_{i+j} = 1\sigma_{i+j} = 0.\end{aligned}$$

As for the Jacobi identity, it is clear from the linearity that it is sufficient to prove that

$$[[a, b], c] + [[b, c], a] + [[c, a], b] = 0$$

if $a \in \mathsf{G}_i, b \in \mathsf{G}_j, c \in \mathsf{G}_k$. Let $a = (x\mathfrak{G}_{i+1})\sigma_i, b = (y\mathfrak{G}_{j+1})\sigma_j, c = (z\mathfrak{G}_{k+1})\sigma_k$. Then, by two applications of (ii),

$$[a, b] = ([x, y]\mathfrak{G}_{i+j+1})\sigma_{i+j}$$

and

$$[[a, b], c] = ([x, y, z]\mathfrak{G}_{i+j+k+1})\sigma_{i+j+k}.$$

Thus

$$\begin{aligned}&[[a, b], c] + [[b, c], a] + [[c, a], b] \\ &= ([x, y, z][y, z, x][z, x, y]\mathfrak{G}_{i+j+k+1})\sigma_{i+j+k} \\ &= 1\sigma_{i+j+k} = 0,\end{aligned}$$

by 9.2. Thus, in a), it only remains to prove the uniqueness assertion; first we prove b).

b) It is clear that for each $i \geq 1$, there is an additive homomorphism θ_i of G_i into $\mathfrak{h}$ such that $\sigma_i\theta_i = \rho_i$. Since $\mathfrak{g}$ is the direct sum of the G_i, it follows that there is an additive homomorphism θ of $\mathfrak{g}$ into $\mathfrak{h}$ such that θ_i is the restriction to G_i of θ. To show that θ is a Lie ring homomorphism, it suffices in view of the linearity to show that

$$[a, b]\theta = [a\theta, b\theta]$$

for $a \in \mathsf{G}_i$, $b \in \mathsf{G}_j$. But if $a = (x\mathfrak{G}_{i+1})\sigma_i$ and $b = (y\mathfrak{G}_{j+1})\sigma_j$, then $a\theta = a\theta_i = (x\mathfrak{G}_{i+1})\sigma_i\theta_i = (x\mathfrak{G}_{i+1})\rho_i$ and similarly $b\theta = (y\mathfrak{G}_{j+1})\rho_j$, whence

$$\begin{aligned}[a, b]\theta &= [(x\mathfrak{G}_{i+1})\sigma_i, (y\mathfrak{G}_{j+1})\sigma_j]\theta_{i+j} \\ &= ([x, y]\mathfrak{G}_{i+j+1})\sigma_{i+j}\theta_{i+j} \\ &= ([x, y]\mathfrak{G}_{i+j+1})\rho_{i+j} \\ &= [(x\mathfrak{G}_{i+1})\rho_i, (y\mathfrak{G}_{j+1})\rho_j] \\ &= [a\theta, b\theta],\end{aligned}$$

as stated. The assertions concerning monomorphisms and epimorphisms are easy to verify.

The uniqueness assertion of a) follows at once from this.

c) Suppose that α is an automorphism of $\mathfrak{G}$ and $\mathfrak{G}_i\alpha = \mathfrak{G}_i$ for all i. Then α induces automorphisms α_i of $\mathfrak{G}_i/\mathfrak{G}_{i+1}$:

$$(x\mathfrak{G}_{i+1})\alpha_i = (x\alpha)\mathfrak{G}_{i+1} \quad (x \in \mathfrak{G}_i).$$

We may apply b) to the isomorphisms $\alpha_i\sigma_i$ of $\mathfrak{G}_i/\mathfrak{G}_{i+1}$ onto G_i, since if $x \in \mathfrak{G}_i$ and $y \in \mathfrak{G}_j$,

$$\begin{aligned}[(x\mathfrak{G}_{i+1})\alpha_i\sigma_i, (y\mathfrak{G}_{j+1})\alpha_j\sigma_j] &= [((x\alpha)\mathfrak{G}_{i+1})\sigma_i, ((y\alpha)\mathfrak{G}_{j+1})\sigma_j] \\ &= ([x\alpha, y\alpha]\mathfrak{G}_{i+j+1})\sigma_{i+j} \\ &= ([x, y]\mathfrak{G}_{i+j+1})\alpha_{i+j}\sigma_{i+j}.\end{aligned}$$

Thus there exists an automorphism ρ of $\mathfrak{g}$ such that $\sigma_i\rho = \alpha_i\sigma_i$ for all i, so that if $x \in \mathfrak{G}_i$,

$$(x\mathfrak{G}_{i+1})\sigma_i\rho = (x\mathfrak{G}_{i+1})\alpha_i\sigma_i = ((x\alpha)\mathfrak{G}_{i+1})\sigma_i. \qquad \textbf{q.e.d.}$$

9.4 Lemma. *Suppose that $\mathfrak{G}$ is a group and that $\mathfrak{g}$ is a Lie ring. Suppose that for $i \geq 1$, σ_i is a homomorphism of $\gamma_i(\mathfrak{G})/\gamma_{i+1}(\mathfrak{G})$ onto an additive subgroup G_i of $\mathfrak{g}$ such that $\mathfrak{g} = \mathsf{G}_1 + \mathsf{G}_2 + \cdots$. Suppose further that if $x \in \gamma_i(\mathfrak{G})$, $y \in \gamma_j(\mathfrak{G})$,*

$$(1) \qquad [(x\gamma_{i+1}(\mathfrak{G}))\sigma_i, (y\gamma_{j+1}(\mathfrak{G}))\sigma_j] = ([x, y]\gamma_{i+j+1}(\mathfrak{G}))\sigma_{i+j}.$$

a) *If $\mathfrak{G} = \langle X \rangle$, $\mathfrak{g}$ is the Lie algebra generated by*

$$Y = \{(x\gamma_2(\mathfrak{G}))\sigma_1 | x \in X\}.$$

b) $\mathfrak{g}^n = \mathsf{G}_n + \mathsf{G}_{n+1} + \cdots \quad (n = 1, 2, \ldots)$.

c) *$\mathfrak{g}^n/\mathfrak{g}^{n+1}$ is isomorphic to $\mathsf{G}_n/(\mathsf{G}_n \cap \mathfrak{g}^{n+1})$.*

d) $\mathfrak{g}^{(d)} \subseteq \sum_{i\geq 1} ((\mathfrak{G}^{(d)} \cap \gamma_i(\mathfrak{G}))\gamma_{i+1}(\mathfrak{G})/\gamma_{i+1}(\mathfrak{G}))\sigma_i$ *for all $d \geq 0$.*

Proof. a) If $x \in X$, write $\bar{x} = (x\gamma_2(\mathfrak{G}))\sigma_1$. By a simple induction on n, it follows from (1) that

$$[\bar{x}_1, \ldots, \bar{x}_n] = ([x_1, \ldots, x_n]\gamma_{n+1}(\mathfrak{G}))\sigma_n.$$

By III, 1.11, $\gamma_n(\mathfrak{G})/\gamma_{n+1}(\mathfrak{G})$ is generated by all $[x_1, \ldots, x_n]\gamma_{n+1}(\mathfrak{G})$ with $x_i \in X$. Since $\mathsf{G}_n = \operatorname{im} \sigma_n$, G_n is generated by all $[\bar{x}_1, \ldots, \bar{x}_n]$. Hence $\mathfrak{g}$ is generated as an additive group by all $[\bar{x}_1, \ldots, \bar{x}_n]$ ($\bar{x}_i \in Y, n \geq 1$) and as a Lie algebra by Y.

b) By 8.7b), $\mathfrak{g}^n$ is the additive group generated by all $[\bar{x}_1, \ldots, \bar{x}_k]$ with $k \geq n$. Hence $\mathfrak{g}^n$ is generated by all G_k with $k \geq n$.

c) By b), $\mathfrak{g}^n = \mathsf{G}_n + \mathfrak{g}^{n+1}$ and $\mathfrak{g}^n/\mathfrak{g}^{n+1} \cong \mathsf{G}_n/(\mathsf{G}_n \cap \mathfrak{g}^{n+1})$.

d) This is proved by induction on d. For $d = 0$, it is clear. If $d > 0$, put $\mathfrak{H}_i = \mathfrak{G}^{(d-1)} \cap \gamma_i(\mathfrak{G})$. Then by the inductive hypothesis,

$$\mathfrak{g}^{(d-1)} \subseteq \sum_{i \geq 1} (\mathfrak{H}_i\gamma_{i+1}(\mathfrak{G})/\gamma_{i+1}(\mathfrak{G}))\sigma_i.$$

Hence

$$\begin{aligned} \mathfrak{g}^{(d)} &= [\mathfrak{g}^{(d-1)}, \mathfrak{g}^{(d-1)}] \\ &\subseteq \sum_{i,j} [(\mathfrak{H}_i\gamma_{i+1}(\mathfrak{G})/\gamma_{i+1}(\mathfrak{G}))\sigma_i, (\mathfrak{H}_j\gamma_{j+1}(\mathfrak{G})/\gamma_{j+1}(\mathfrak{G}))\sigma_j]. \end{aligned}$$

Using (1), it follows that

$$\mathfrak{g}^{(d)} \subseteq \sum_{i,j} ([\mathfrak{H}_i, \mathfrak{H}_j]\gamma_{i+j+1}(\mathfrak{G})/\gamma_{i+j+1}(\mathfrak{G}))\sigma_{i+j}.$$

But from the definition of the $\mathfrak{H}_i$, it is clear that

$$[\mathfrak{H}_i, \mathfrak{H}_j] \subseteq \mathfrak{G}^{(d)} \cap \gamma_{i+j}(\mathfrak{G}),$$

so d) is proved. **q.e.d.**

Theorem 9.3 has a number of applications in group-theory; the use of it is known as the *Lie ring method for nilpotent groups*. The idea is to prove a theorem about Lie rings and then use Theorem 9.3 to derive information about groups.

Even in simple cases when the group-theoretical result can be proved directly, some simplification is often effected by using the Lie ring method, since calculation is usually easier in the Lie ring than in the group. For example, put

$$f(a, b, c) = [a, b, c] - [b, c, a]$$

for any elements a, b, c of a Lie ring. Then it is very easy to verify the following identities:

$$f(b, c, a) + 2f(c, a, b) = -3[a, b, c];$$

$$f([a, b], c, d) - f(a, [b, c], d) - f(a, b, [c, d])$$
$$= 2[a, b, c, d] - [f(a, b, c), d] + [f(b, c, d), a];$$
$$f(a, b, c) = [b, c, c] + [b, a, a] - [b, c + a, c + a].$$

Suppose then that $\mathfrak{g}$ is a Lie ring in which $[a, b, b] = 0$ for all a, b in $\mathfrak{g}$. It follows at once that $3[a, b, c] = 2[a, b, c, d] = 0$, so $[a, b, c, d] = 0$. The proof of the corresponding result for groups (III, 6.5) was not as simple as this, but it can be deduced from these facts about Lie rings without much difficulty.

The first example of the Lie ring method that we give is a theorem of Vaughan-Lee on p-groups (9.12). We begin by proving the following simple lemma.

9.5 Lemma. *Let $t_1, \ldots, t_n$ be independent indeterminates over a commutative ring A. Suppose that $g \in \mathsf{A}[t_1, \ldots, t_n]$, that $g(0, \ldots, 0) = 1$ and that $g(\lambda_1, \ldots, \lambda_n) = 0$ whenever $\lambda_1, \ldots, \lambda_n$ are elements of $\{0, 1\}$ not all zero. Then the degree of g is at least n.*

Proof. We may write

$$g(t_1, \ldots, t_n) = 1 + \sum_{i_1 < \cdots < i_r} t_{i_1} \cdots t_{i_r} f_{i_1 \cdots i_r}(t_{i_1}, \ldots, t_{i_r}).$$

It will be proved by induction on r ($r = 1, \ldots, n$) that $f_{i_1 \cdots i_r}(1, \ldots, 1) = (-1)^r$. To do this, substitute 1 for $t_{i_1}, \ldots, t_{i_r}$ and 0 for all the other t_i; this gives

$$0 = 1 + \sum_{s=1}^{r-1} \binom{r}{s}(-1)^s + f_{i_1 \cdots i_r}(1, \ldots, 1),$$

the inductive hypothesis being used here if $r > 1$. The assertion thus follows from the binomial theorem.

In particular, $f_{12 \cdots n}(1, \ldots, 1) = (-1)^n$. Hence $f_{12 \cdots n} \neq 0$ and the degree of g is at least n. **q.e.d.**

9.6 Lemma. *Let* V, W *be finite-dimensional vector spaces over a field* K *such that* $\dim_{\mathsf{K}}\mathsf{V} > 1$, *and let* f *be a symplectic bilinear mapping of* $\mathsf{V} \times \mathsf{V}$ *into* W. *If there exists a maximal subspace* U *of* V *such that* $f(\mathsf{U}, \mathsf{U}) = f(\mathsf{V}, \mathsf{V})$, *there exist at least two such subspaces.*

(If X *is a subspace of* V, $f(\mathsf{X}, \mathsf{X})$ *denotes the subspace of* W *spanned by all* $f(x, y)$ *with* $x \in \mathsf{X}$, $y \in \mathsf{X}$.*)*

Proof. Suppose that this is false. Then $f \neq 0$, since $\dim_{\mathsf{K}}\mathsf{V} > 1$. Let $a_1, \ldots, a_n$ be a K-basis of U. Then $f(\mathsf{V}, \mathsf{V}) = f(\mathsf{U}, \mathsf{U})$ is spanned by all $f(a_i, a_j)$ $(1 \leq i < j \leq n)$. Let

$$\mathsf{S} = \{(i, j) \mid 1 \leq i < j \leq n\}.$$

Since $f \neq 0$, $n > 1$ and S is non-empty. If $s \in \mathsf{S}$, write $s = (i_s, j_s)$ and put $c_s = f(a_{i_s}, a_{j_s})$. Since U is maximal, $\mathsf{V} = \mathsf{U} + \mathsf{K}a$ for some $a \in \mathsf{V}$. For $i = 1, \ldots, n$, write

$$f(a, a_i) = \sum_{s \in \mathsf{S}} \mu_{is} c_s \quad (\mu_{is} \in \mathsf{K}).$$

Suppose that $\lambda_1, \ldots, \lambda_n$ are elements of K not all zero. Put $b_i = \lambda_i a + a_i$ $(i = 1, \ldots, n)$, and let U_1 be the subspace spanned by $b_1, \ldots, b_n$. Since $\lambda_1, \ldots, \lambda_n$ are not all zero, $\mathsf{U} \neq \mathsf{U}_1$. Since the assertion of the theorem is false, $f(\mathsf{U}_1, \mathsf{U}_1) \neq f(\mathsf{V}, \mathsf{V})$. But $f(\mathsf{U}_1, \mathsf{U}_1)$ is spanned by all

$$\begin{aligned} f(b_i, b_j) &= f(\lambda_i a + a_i, \lambda_j a + a_j) \\ &= \sum_{s \in \mathsf{S}} (\lambda_i \mu_{js} - \lambda_j \mu_{is}) c_s + f(a_i, a_j) \end{aligned}$$

with $i < j$. If r, s are elements of S, put

$$\alpha_{rs} = \lambda_{i_r} \mu_{j_r s} - \lambda_{j_r} \mu_{i_r s},$$

and let A be the $\mathsf{S} \times \mathsf{S}$ matrix the (r, s)-coefficient of which is α_{rs}. Then

$$f(b_{i_r}, b_{j_r}) = \sum_{s \in \mathsf{S}} (\alpha_{rs} + \delta_{rs}) c_s,$$

where δ_{rs} is the Kronecker δ. These elements do not span $f(\mathsf{V}, \mathsf{V})$, so $A + I$ is a singular matrix and

$$\det(A + I) = 0.$$

Now let $t_1, \ldots, t_n$ be independent indeterminates over K, and let B be the $\mathsf{S} \times \mathsf{S}$ matrix the (r, s)-coefficient of which is $t_{i_r}\mu_{j_r s} - t_{j_r}\mu_{i_r s}$. Let

$$g(t_1, \ldots, t_n) = \det(B + I).$$

Thus $g \in \mathsf{K}[t_1, \ldots, t_n]$ and $g(\lambda_1, \ldots, \lambda_n) = 0$ whenever $\lambda_1, \ldots, \lambda_n$ are elements of K not all 0. Also $g(0, \ldots, 0) = 1$. By 9.5, the degree of g is at least n.

Let B_r, I_r be the r-rows of B, I respectively ($r \in \mathsf{S}$). Since the determinant is multilinear in its rows, $g(t_1, \ldots, t_n)$ is the sum of the determinants of certain matrices the rows of which are all taken from the B_r or the I_r. Since the degree of g is at least n, it follows that there exists a non-singular $\mathsf{S} \times \mathsf{S}$ matrix C, n rows of which are taken from the B_r. Thus there exist n of the B_r which are linearly independent over the field of rational functions $\mathsf{K}(t_1, \ldots, t_n)$. But if C_i is the row-vector (μ_{is}) $(i = 1, \ldots, n)$,

$$\begin{aligned} B_{(i,j)} &= t_i C_j - t_j C_i \\ &= t_i t_j (t_j^{-1} C_j - t_i^{-1} C_i), \end{aligned}$$

so all the B_r are linear combinations of the $n - 1$ row-vectors $t_i^{-1}C_i - t_n^{-1}C_n$ $(i = 1, \ldots, n - 1)$. Hence any n of the B_r are linearly dependent. This is a contradiction. **q.e.d.**

9.7 Definition. Let V, W be finite-dimensional vector spaces over a field K and let f be a symplectic bilinear mapping of $\mathsf{V} \times \mathsf{V}$ into W. If $v \in \mathsf{V}$, we put

$$v^\perp = \{u \mid u \in \mathsf{V}, f(u, v) = 0\}.$$

Thus $v^\perp$ is a K-subspace of V, and since f is symplectic, $\mathsf{K}v \subseteq v^\perp$.

9.8 Theorem. *Let* V, W *be finite-dimensional vector spaces over a field* K *and let* f *be a symplectic bilinear mapping of* $\mathsf{V} \times \mathsf{V}$ *into* W. *Suppose that* $f(\mathsf{X}, \mathsf{X}) \subset f(\mathsf{V}, \mathsf{V})$ *for every proper subspace* X *of* V. *Then* V *is spanned by*

$$\{v \mid v \in \mathsf{V}, v^\perp = \mathsf{K}v\}.$$

Proof. Suppose that this is false and that we have a counterexample for which $\dim_\mathsf{K} \mathsf{V}$ is minimal. Then $f \neq 0$; in particular, $\dim_\mathsf{K} \mathsf{V} > 1$.

(1) There exists a maximal subspace U of V such that if $v \in \mathsf{V}$ and $v \notin \mathsf{U}$, then $v^\perp \cap \mathsf{U} \neq 0$.

Since the conclusion of the theorem is false, there exists a maximal subspace U of V containing all $v \in \mathsf{V}$ for which $v^\perp = \mathsf{K}v$. Thus if $v \notin \mathsf{U}$, $v^\perp \supset \mathsf{K}v$ and $v^\perp \cap \mathsf{U} \neq 0$.

(2) There exists a maximal subspace A of U such that if $v \in \mathsf{V}$ and $v \notin \mathsf{U}$, then $v^\perp \cap \mathsf{A} \neq 0$.

Since U is a maximal subspace of V, $\mathsf{V} = \mathsf{U} \oplus \mathsf{K}g$ for some $g \in \mathsf{V}$. Thus

$$f(\mathsf{V}, \mathsf{V}) = f(\mathsf{U} \oplus \mathsf{K}g, \mathsf{U} \oplus \mathsf{K}g) = f(\mathsf{U}, \mathsf{U}) + f(\mathsf{U}, \mathsf{K}g) = f(\mathsf{U}, \mathsf{V}).$$

Hence $f(\mathsf{U}, \mathsf{V}) \neq f(\mathsf{U}, \mathsf{U})$. But if $v \in \mathsf{V}$ and $v \notin \mathsf{U}$, then

$$\begin{aligned} f(v^\perp \cap \mathsf{U}, \mathsf{V}) &= f(v^\perp \cap \mathsf{U}, \mathsf{K}v \oplus \mathsf{U}) \\ &= f(v^\perp \cap \mathsf{U}, \mathsf{K}v) + f(v^\perp \cap \mathsf{U}, \mathsf{U}) \subseteq f(\mathsf{U}, \mathsf{U}). \end{aligned}$$

Hence U is not spanned by the $v^\perp \cap \mathsf{U}$ for all $v \in \mathsf{V}$, $v \notin \mathsf{U}$. Hence there exists a maximal subspace A of U such that $v^\perp \cap \mathsf{U} \subseteq \mathsf{A}$ whenever $v \in \mathsf{V}$, $v \notin \mathsf{U}$. By (1), $v^\perp \cap \mathsf{A} \neq 0$.

(3) Suppose that Y is a proper subspace of V and X is a maximal subspace of Y. Suppose also that $y^\perp \cap \mathsf{X} \neq 0$ whenever $y \in \mathsf{Y}$, $y \notin \mathsf{X}$. Then there exists a maximal subspace X_1 of Y such that $\mathsf{X}_1 \neq \mathsf{X}$ and $f(\mathsf{X}_1, \mathsf{X}_1) = f(\mathsf{Y}, \mathsf{Y})$.

Let Y_1 be the subspace of Y spanned by all $y \in \mathsf{Y}$ for which $y^\perp \cap \mathsf{Y} = \mathsf{K}y$. If $y \in \mathsf{Y}$ and $y \notin \mathsf{X}$, $y^\perp \cap \mathsf{X} \neq 0$, whence $y^\perp \cap \mathsf{Y} \neq \mathsf{K}y$. Hence $\mathsf{Y}_1 \neq \mathsf{Y}$. But then, by minimality of $\dim_\mathsf{K} \mathsf{V}$, there exists a maximal subspace X_1 of Y such that $f(\mathsf{X}_1, \mathsf{X}_1) = f(\mathsf{Y}, \mathsf{Y})$. Since $\mathsf{X} \neq 0$, we have $\dim_\mathsf{K} \mathsf{Y} > 1$, so by 9.6, we may suppose that $\mathsf{X}_1 \neq \mathsf{X}$.

Choose $g \in \mathsf{V}$ such that $\mathsf{V} = \mathsf{U} \oplus \mathsf{K}g$, and let $\mathsf{L} = \mathsf{A} \oplus \mathsf{K}g$. By (2), A and L satisfy the conditions of (3). Hence there exists a maximal subspace M of L such that $\mathsf{M} \neq \mathsf{A}$, and

(4) $f(\mathsf{M}, \mathsf{M}) = f(\mathsf{L}, \mathsf{L})$.

Let $\mathsf{B} = \mathsf{M} \cap \mathsf{A}$ and write $\mathsf{A} = \mathsf{B} \oplus \mathsf{K}a$. Then $a \notin \mathsf{M}$, so $\mathsf{L} = \mathsf{M} \oplus \mathsf{K}a$ and $g = v + \lambda a$ for some $v \in \mathsf{M}$, $\lambda \in \mathsf{K}$. Since $g \notin \mathsf{A}$, $v \notin \mathsf{B}$ and $\mathsf{M} = \mathsf{B} \oplus \mathsf{K}v$. Write $\mathsf{U} = \mathsf{A} \oplus \mathsf{K}z$. Thus $\mathsf{V} = \mathsf{U} + \mathsf{L} = \mathsf{L} \oplus \mathsf{K}z$. Let $\mathsf{P} = \mathsf{B} \oplus \mathsf{K}(v + z)$.

(5) If $x \in \mathsf{P}$ and $x \notin \mathsf{B}$, $x^\perp \cap \mathsf{B} \neq 0$.

Since $x \notin \mathsf{B}$, $x \notin \mathsf{U}$, so by (2), $\mathsf{Y} = x^\perp \cap \mathsf{A} \neq 0$. Also $x = \mu z + y$, where $\mu \neq 0$ and $y \in \mathsf{L}$, so $f(\mathsf{Y}, \mathsf{K}z) = f(\mathsf{Y}, \mathsf{K}(x - y)) = f(\mathsf{Y}, \mathsf{K}y) \subseteq f(\mathsf{L}, \mathsf{L})$. Thus

$$f(\mathsf{M} + \mathsf{K}z + \mathsf{Y}, \mathsf{Y}) = f(\mathsf{M}, \mathsf{Y}) + f(\mathsf{K}z, \mathsf{Y}) + f(\mathsf{Y}, \mathsf{Y})$$
$$\subseteq f(\mathsf{L}, \mathsf{L}) = f(\mathsf{M}, \mathsf{M}),$$

by (4). Hence

$$f(\mathsf{M} + \mathsf{K}z + \mathsf{Y}, \mathsf{M} + \mathsf{K}z + \mathsf{Y}) \subseteq f(\mathsf{M} + \mathsf{K}z, \mathsf{M} + \mathsf{K}z) \subset f(\mathsf{V}, \mathsf{V}),$$

so $\mathsf{M} + \mathsf{K}z + \mathsf{Y} \subset \mathsf{V}$. Since $\mathsf{M} \oplus \mathsf{K}z$ is a maximal subspace of V, $\mathsf{Y} \subseteq \mathsf{M} \oplus \mathsf{K}z$. Hence

$$\mathsf{Y} \subseteq (\mathsf{M} \oplus \mathsf{K}z) \cap \mathsf{A} = (\mathsf{M} \oplus \mathsf{K}z) \cap \mathsf{L} \cap \mathsf{A} = \mathsf{M} \cap \mathsf{A} = \mathsf{B},$$

and $\mathsf{Y} = x^{\perp} \cap \mathsf{B} \neq 0$.

By (3) and (5), there is a maximal subspace N of P such that $\mathsf{N} \neq \mathsf{B}$ and

(6) $f(\mathsf{N}, \mathsf{N}) = f(\mathsf{P}, \mathsf{P})$.

Let $\mathsf{C} = \mathsf{B} \cap \mathsf{N}$ and write $\mathsf{B} = \mathsf{C} \oplus \mathsf{K}b$. Since $\mathsf{P} = \mathsf{N} + \mathsf{B}$, there exists $b' \in \mathsf{B}$ such that $v + z - b' \in \mathsf{N}$. Put $w = v - b'$, so that $\mathsf{M} = \mathsf{B} \oplus \mathsf{K}w$ and $w + z \in \mathsf{N}$. Finally, let $\mathsf{Q} = \mathsf{N} \oplus \mathsf{K}(b + z)$.

(7) $f(\mathsf{P}, \mathsf{K}b) \subseteq f(\mathsf{Q}, \mathsf{Q})$.

For $f(\mathsf{P}, \mathsf{K}b) \subseteq f(\mathsf{P}, \mathsf{P}) = f(\mathsf{N}, \mathsf{N}) \subseteq f(\mathsf{Q}, \mathsf{Q})$.

(8) $f(\mathsf{K}w \oplus \mathsf{N}, \mathsf{K}z) \subseteq f(\mathsf{Q}, \mathsf{Q})$.

For

$$f(\mathsf{K}w \oplus \mathsf{N}, \mathsf{K}z) \subseteq f(\mathsf{K}(w + z) + \mathsf{K}z + \mathsf{N}, \mathsf{K}z)$$
$$\subseteq f(\mathsf{N} + \mathsf{K}z, \mathsf{K}z)$$
$$\subseteq f(\mathsf{N}, \mathsf{K}z)$$
$$\subseteq f(\mathsf{N}, \mathsf{K}(z + b)) + f(\mathsf{N}, \mathsf{K}b)$$
$$\subseteq f(\mathsf{Q}, \mathsf{Q}),$$

by (7).

(9) $f(\mathsf{M}, \mathsf{C}) \subseteq f(\mathsf{Q}, \mathsf{Q})$.

For $\mathsf{M} = \mathsf{B} \oplus \mathsf{K}w = \mathsf{C} \oplus \mathsf{K}b \oplus \mathsf{K}w$. Since $\mathsf{C} \subseteq \mathsf{Q}$, $f(\mathsf{C}, \mathsf{C}) \subseteq f(\mathsf{Q}, \mathsf{Q})$. By (7), $f(\mathsf{K}b, \mathsf{C}) \subseteq f(\mathsf{Q}, \mathsf{Q})$. And

$$f(\mathsf{K}w, \mathsf{C}) \subseteq f(\mathsf{K}(w + z), \mathsf{C}) + f(\mathsf{K}z, \mathsf{C})$$
$$\subseteq f(\mathsf{N}, \mathsf{C}) + f(\mathsf{K}z, \mathsf{C})$$
$$\subseteq f(\mathsf{Q}, \mathsf{Q}),$$

by (8). Thus $f(\mathsf{M}, \mathsf{C}) \subseteq f(\mathsf{Q}, \mathsf{Q})$.

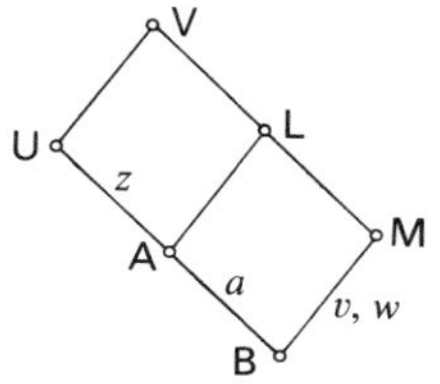

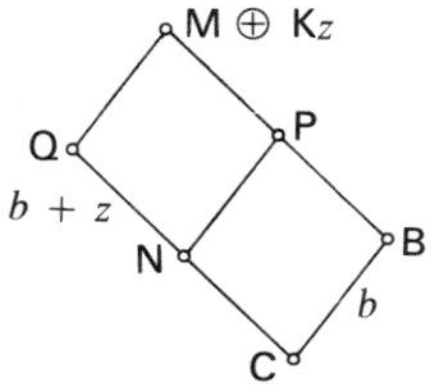

Now $\mathsf{M} = \mathsf{C} \oplus \mathsf{K}b \oplus \mathsf{K}w$, so

$$f(\mathsf{M}, \mathsf{M}) = f(\mathsf{M}, \mathsf{C}) + \mathsf{K}f(b, w).$$

But by (4), $f(\mathsf{M}, \mathsf{M}) = f(\mathsf{L}, \mathsf{L})$, so

$$f(\mathsf{L}, \mathsf{L}) \subseteq f(\mathsf{Q}, \mathsf{Q}) + \mathsf{K}f(b, w)$$

by (9). Hence there exist β, γ in K such that

(10) $$f(a, w) - \beta f(b, w) \in f(\mathsf{Q}, \mathsf{Q}),$$

(11) $$f(a, b) - \gamma f(b, w) \in f(\mathsf{Q}, \mathsf{Q}).$$

For $\alpha \in \mathsf{K}$, put $\mathsf{R} = \mathsf{Q} + \mathsf{K}(a + \alpha z)$. Then $\mathsf{R} + \mathsf{K}z$ contains z, a and Q. Since $b + z \in \mathsf{Q}$ and $\mathsf{C} \subseteq \mathsf{Q}$,

$$\mathsf{R} + \mathsf{K}z \supseteq \mathsf{C} + \mathsf{K}b + \mathsf{K}z + \mathsf{K}a = \mathsf{B} + \mathsf{K}a + \mathsf{K}z = \mathsf{A} + \mathsf{K}z = \mathsf{U}.$$

But also $w + z \in \mathsf{N} \subseteq \mathsf{Q}$, so $w \in \mathsf{R} + \mathsf{K}z$ and $\mathsf{R} + \mathsf{K}z \supseteq \mathsf{U} + \mathsf{K}w = \mathsf{V}$. Hence $\dim_{\mathsf{K}} \mathsf{R} = \dim_{\mathsf{K}} \mathsf{V} - 1$, so $f(\mathsf{R}, \mathsf{R}) \neq f(\mathsf{V}, \mathsf{V})$. Since $\mathsf{R} + \mathsf{K}z = \mathsf{V}$, then, $f(\mathsf{R}, \mathsf{K}z) \nsubseteq f(\mathsf{R}, \mathsf{R})$. Since $\mathsf{R} = \mathsf{N} + \mathsf{K}(b + z) + \mathsf{K}(a + \alpha z)$, it follows from (8) that

(12) $$f(\mathsf{K}b + \mathsf{K}a, \mathsf{K}z) \nsubseteq f(\mathsf{R}, \mathsf{R}).$$

On the other hand,

$$f(b + z, w + z) = f(b, w) + f(b, z) - f(w, z)$$

$$f(a + \alpha z, w + z) = f(a, w) + f(a, z) - \alpha f(w, z)$$

$$f(a + \alpha z, b + z) = f(a, b) + f(a, z) - \alpha f(b, z)$$

all lie in $f(\mathsf{R}, \mathsf{R})$. By (8), $f(w, z) \in f(\mathsf{R}, \mathsf{R})$. Using also (10) and (11) to eliminate $f(a, w), f(a, b)$, it follows that

$$f(b, w) + f(b, z), \quad \beta f(b, w) + f(a, z), \quad \gamma f(b, w) + f(a, z) - \alpha f(b, z)$$

all lie in $f(\mathsf{R}, \mathsf{R})$. Hence, eliminating $f(b, w)$,

$$f(a, z) - \beta f(b, z), \quad f(a, z) - (\alpha + \gamma) f(b, z)$$

also lie in $f(\mathsf{R}, \mathsf{R})$. But then, if α is chosen to be different from $\beta - \gamma$, it follows that $f(a, z), f(b, z)$ lie in $f(\mathsf{R}, \mathsf{R})$, contrary to (12). **q.e.d.**

The main theorem of this section concerns the number of elements in conjugacy classes, or equivalently, the index of centralizers. We shall need similar notions for Lie algebras.

9.9 Definition. Let $\mathfrak{g}$ be a Lie algebra over A. If $x \in \mathfrak{g}$, put

$$g(\operatorname{ad} x) = [g, x]$$

for all $g \in \mathfrak{g}$; thus $\operatorname{ad} x$ is an A-homomorphism of $\mathfrak{g}$ into $\mathfrak{g}$. Put

$$\mathbf{C}_{\mathfrak{g}}(x) = \ker(\operatorname{ad} x);$$

thus $\mathbf{C}_{\mathfrak{g}}(x)$ is an A-submodule of $\mathfrak{g}$ and $\mathfrak{g}/\mathbf{C}_{\mathfrak{g}}(x)$ is A-isomorphic to the A-submodule

$$\{[g, x] \mid g \in \mathfrak{g}\}.$$

In particular, if $\mathfrak{g}$ is finite, $|\mathfrak{g} : \mathbf{C}_{\mathfrak{g}}(x)|$ is the number of $[g, x]$ $(g \in \mathfrak{g})$.

We now apply 9.8 to Lie algebras.

9.10 Theorem. *Let p be a prime and let $\mathfrak{g}$ be a nilpotent Lie ring for which $|\mathfrak{g}|$ is a power of p and $p\mathfrak{g} \cap \mathfrak{g}^2 = p\mathfrak{g}^2$. Let $\mathfrak{h}$ be a minimal element of the set of subrings of $\mathfrak{g}$ for which $\mathfrak{h}^2 = \mathfrak{g}^2$. Put $|\mathfrak{h} : \mathfrak{h}^2 + p\mathfrak{h}| = p^d$, $\mathfrak{i} = p\mathfrak{g} + \mathfrak{g}^3$, $\overline{\mathfrak{g}} = \mathfrak{g}/\mathfrak{i}$ and $\overline{\mathfrak{h}} = (\mathfrak{h} + \mathfrak{i})/\mathfrak{i}$.*

a) *$\mathfrak{h}$ is a minimal element of the set of subrings of $\mathfrak{g}$ for which $\mathfrak{h}^2 + \mathfrak{i} = \mathfrak{g}^2 + \mathfrak{i}$.*

b) *$\overline{\mathfrak{h}}$ is a minimal element of the set of subrings of $\overline{\mathfrak{g}}$ for which $\overline{\mathfrak{h}}^2 = \overline{\mathfrak{g}}^2$.*

c) *$\mathfrak{h}$ is generated by d elements $x_1, \ldots, x_d$ for which $|\overline{\mathfrak{h}} : \mathbf{C}_{\overline{\mathfrak{h}}}(\overline{x}_i)| = p^{d-1}$, where $\overline{x}_i = x_i + \mathfrak{i}$ $(i = 1, \ldots, d)$.*

Proof. a) Clearly $\mathfrak{i}$ is an ideal of $\mathfrak{g}$. Suppose that $\mathfrak{f}$ is a subring of $\mathfrak{h}$ and $\mathfrak{f}^2 + \mathfrak{i} = \mathfrak{g}^2 + \mathfrak{i}$. Then $\mathfrak{g}^2 = \mathfrak{f}^2 + (\mathfrak{i} \cap \mathfrak{g}^2)$. But

$$\mathfrak{i} \cap \mathfrak{g}^2 = (p\mathfrak{g} \cap \mathfrak{g}^2) + \mathfrak{g}^3 = p\mathfrak{g}^2 + \mathfrak{g}^3,$$

so $\mathfrak{g}^2 = \mathfrak{f}^2 + p\mathfrak{g}^2 + \mathfrak{g}^3$. Hence $\mathsf{U} = \mathfrak{g}^2/(\mathfrak{f}^2 + \mathfrak{g}^3)$ is an additive p-group for which $p\mathsf{U} = \mathsf{U}$. Thus $\mathsf{U} = 0$ and $\mathfrak{g}^2 = \mathfrak{f}^2 + \mathfrak{g}^3$. By 8.10d), $\mathfrak{g}^2 = \mathfrak{f}^2$. By minimality of $\mathfrak{h}$, $\mathfrak{f} = \mathfrak{h}$.

b) We have

$$\bar{\mathfrak{h}}^2 = (\mathfrak{h}^2 + \mathfrak{i})/\mathfrak{i} = (\mathfrak{g}^2 + \mathfrak{i})/\mathfrak{i} = \bar{\mathfrak{g}}^2.$$

Suppose that $\bar{\mathfrak{f}}$ is a subring of $\bar{\mathfrak{h}}$ for which $\bar{\mathfrak{f}}^2 = \bar{\mathfrak{h}}^2$. Then $\bar{\mathfrak{f}} = (\mathfrak{f} + \mathfrak{i})/\mathfrak{i}$ for some subring $\mathfrak{f}$ of $\mathfrak{h}$, and $\mathfrak{f}^2 + \mathfrak{i} = \mathfrak{h}^2 + \mathfrak{i} = \mathfrak{g}^2 + \mathfrak{i}$. By a), $\mathfrak{f} = \mathfrak{h}$. Thus $\bar{\mathfrak{f}} = \bar{\mathfrak{h}}$.

c) $\bar{\mathfrak{h}}/\bar{\mathfrak{h}}^2$ and $\bar{\mathfrak{h}}^2$ are vector spaces over $\mathsf{K} = GF(p)$, and since $\bar{\mathfrak{h}}^3 = 0$, there is a symplectic K-bilinear mapping f of $\bar{\mathfrak{h}}/\bar{\mathfrak{h}}^2$ into $\bar{\mathfrak{h}}^2$ given by

$$f(\bar{u} + \bar{\mathfrak{h}}^2, \bar{v} + \bar{\mathfrak{h}}^2) = [\bar{u}, \bar{v}] \quad (\bar{u} \in \bar{\mathfrak{h}}, \bar{v} \in \bar{\mathfrak{h}}).$$

Any proper subspace of $\bar{\mathfrak{h}}/\bar{\mathfrak{h}}^2$ is of the form $\bar{\mathfrak{f}}/\bar{\mathfrak{h}}^2$ for some proper ideal $\bar{\mathfrak{f}}$ of $\bar{\mathfrak{h}}$, and by b),

$$f(\bar{\mathfrak{f}}/\bar{\mathfrak{h}}^2, \bar{\mathfrak{f}}/\bar{\mathfrak{h}}^2) = \bar{\mathfrak{f}}^2 \subset \bar{\mathfrak{h}}^2 = f(\bar{\mathfrak{h}}/\bar{\mathfrak{h}}^2, \bar{\mathfrak{h}}/\bar{\mathfrak{h}}^2).$$

Hence by 9.8, $\bar{\mathfrak{h}}/\bar{\mathfrak{h}}^2$ is spanned by the set X of those $\bar{x} + \bar{\mathfrak{h}}^2$ ($\bar{x} \in \bar{\mathfrak{h}}$) for which $(\bar{x} + \bar{\mathfrak{h}}^2)^\perp = \mathsf{K}(\bar{x} + \bar{\mathfrak{h}}^2)$. But

$$\begin{aligned}(\bar{x} + \bar{\mathfrak{h}}^2)^\perp &= \{\bar{u} + \bar{\mathfrak{h}}^2 \mid \bar{u} \in \bar{\mathfrak{h}}, [\bar{x}, \bar{u}] = 0\} \\ &= \mathbf{C}_{\bar{\mathfrak{h}}}(\bar{x})/\bar{\mathfrak{h}}^2.\end{aligned}$$

Thus, putting $\dim_{\mathsf{K}}(\bar{\mathfrak{h}}/\bar{\mathfrak{h}}^2) = n$, $\bar{x} + \bar{\mathfrak{h}}^2 \in X$ if and only if $\dim_{\mathsf{K}}(\bar{\mathfrak{h}}/\mathbf{C}_{\bar{\mathfrak{h}}}(\bar{x})) = n - 1$. Also $\bar{\mathfrak{h}}/\bar{\mathfrak{h}}^2$ is spanned by linearly independent elements $\bar{x}_1 + \bar{\mathfrak{h}}^2, \dots, \bar{x}_n + \bar{\mathfrak{h}}^2$ of X. Hence $|\bar{\mathfrak{h}} : \mathbf{C}_{\bar{\mathfrak{h}}}(\bar{x}_i)| = p^{n-1}$ $(i = 1, \dots, n)$, and $\bar{\mathfrak{h}}$ is spanned by $\bar{x}_1, \dots, \bar{x}_n$ and $\bar{\mathfrak{h}}^2$.

Put $\bar{x}_i = x_i + \mathfrak{i}$ with $x_i \in \mathfrak{h}$, and let $\mathfrak{h}_1$ be the subring generated by $x_1, \dots, x_n$. Then $(\mathfrak{h}_1 + \mathfrak{i}/\mathfrak{i}) + \bar{\mathfrak{h}}^2 = \bar{\mathfrak{h}}$. Squaring, $\bar{\mathfrak{h}}^2 = (\mathfrak{h}_1^2 + \mathfrak{i})/\mathfrak{i}$, so $\mathfrak{h}_1^2 + \mathfrak{i} = \mathfrak{h}^2 + \mathfrak{i}$. By a), $\mathfrak{h}_1 = \mathfrak{h}$, so $\mathfrak{h}$ is generated by $x_1, \dots, x_n$. Since $\mathfrak{h}^2 + \mathfrak{i} \supseteq \mathfrak{h}^2 + p\mathfrak{h}$, it follows that $\mathfrak{h}/(\mathfrak{h}^2 + p\mathfrak{h})$ is spanned over K by the linearly independent elements $x_i + (\mathfrak{h}^2 + p\mathfrak{h})$ $(i = 1, \dots, n)$, and $n = d$. **q.e.d.**

9.11 Lemma. *Let $\mathfrak{h}$ be a nilpotent Lie ring for which $|\mathfrak{h}|$ is a power of the prime p. Suppose $\mathfrak{h}$ is generated by $y_1, \ldots, y_n$, where $|\mathfrak{h} : \mathbf{C}_{\mathfrak{h}}(y_i)| \leq p^b$ $(i = 1, \ldots, n)$. Let $\tilde{\mathfrak{h}} = \mathfrak{h}/(p\mathfrak{h} + \mathfrak{h}^3)$, $\tilde{y}_i = y_i + (p\mathfrak{h} + \mathfrak{h}^3)$, and suppose that $|\tilde{\mathfrak{h}} : \mathbf{C}_{\tilde{\mathfrak{h}}}(\tilde{y}_i)| \geq p^c$ $(i = 1, \ldots, n)$. Put $\mathfrak{j} = p\mathfrak{h}^2 + \mathfrak{h}^3$. Suppose also that $\mathfrak{l}$ is a subring of $\mathfrak{h}$ and*

$$|\mathfrak{h} : \mathfrak{l} + p\mathfrak{h} + \mathfrak{h}^2| = p^a, \quad |\mathfrak{l} + p\mathfrak{h} + \mathfrak{h}^2 : \mathfrak{l}| = p^f.$$

a) $|\mathfrak{h}^2 : \mathfrak{l}^2 + \mathfrak{j}| \leq p^{\frac{1}{2}a(a-1)+a(n-a)}$.
b) $|\mathfrak{l}^2 + \mathfrak{j} : \mathfrak{l}^2| \leq p^{a(b-c)+(n-a)f}$.
c) $|\mathfrak{j}| \leq p^{n(b-c)}$.

Proof. a) Put $\mathfrak{i} = p\mathfrak{h} + \mathfrak{h}^2$. Then $\mathfrak{h}/\mathfrak{i}$ is a vector space over $\mathsf{K} = GF(p)$ spanned by $y_1 + \mathfrak{i}, \ldots, y_n + \mathfrak{i}$. Renumbering the y_i if necessary, we may suppose that $\mathfrak{h}$ is spanned by $\mathfrak{l} + \mathfrak{i}$ and $y_1, \ldots, y_a$. Then $\mathfrak{l} + \mathfrak{i}$ is spanned by $\mathfrak{i}$ and $n - a$ elements $z_1, \ldots, z_{n-a}$ of $\mathfrak{l}$. Hence $\mathfrak{h}^2$ is spanned by the $\frac{1}{2}a(a - 1)$ elements $[y_i, y_j]$ $(1 \leq i < j \leq a)$, the $a(n - a)$ elements $[y_i, z_j]$ and $\mathfrak{l}^2 + p\mathfrak{h}^2 + \mathfrak{h}^3 = \mathfrak{l}^2 + \mathfrak{j}$. Thus

$$|\mathfrak{h}^2 : \mathfrak{l}^2 + \mathfrak{j}| \leq p^{\frac{1}{2}a(a-1)+a(n-a)}.$$

b) If $\mathfrak{h}_1$ is the subalgebra generated by $X = \{y_1, \ldots, y_a, z_1, \ldots, z_{n-a}\}$, $\mathfrak{h} = \mathfrak{h}_1 + \mathfrak{i} = \mathfrak{h}_1 + p\mathfrak{h} + \mathfrak{h}^2$. Thus $\mathsf{A} = \mathfrak{h}/(\mathfrak{h}_1 + \mathfrak{h}^2)$ is an additive p-group for which $p\mathsf{A} = \mathsf{A}$ and $\mathfrak{h} = \mathfrak{h}_1 + \mathfrak{h}^2$. By 8.10, $\mathfrak{h} = \mathfrak{h}_1$. Thus $\mathfrak{h}$ is generated by X. By 8.7, $\mathfrak{j} = p\mathfrak{h}^2 + \mathfrak{h}^3$ is the additive group generated by all $p[x_1, x_2]$ and all $[x_1, \ldots, x_m]$ for which $m \geq 3$ $(x_i \in X)$. Hence if

$$\mathsf{U}_i = \{[u, y_i] | u \in \mathfrak{i}\} \quad (i = 1, \ldots, a),$$
$$\mathsf{V}_j = \{[u, z_j] | u \in \mathfrak{i}\} \quad (j = 1, \ldots, n - a),$$

then

$$\mathfrak{j} = \mathsf{U}_1 + \cdots + \mathsf{U}_a + \mathsf{V}_1 + \cdots + \mathsf{V}_{n-a}.$$

Since $|\mathfrak{l} + \mathfrak{i} : \mathfrak{l}| = p^f$, we see from the mapping $u + \mathfrak{l} \to [u, z_j] + \mathfrak{l}^2$ $(u \in \mathfrak{i})$ that

$$(13) \qquad |\mathsf{V}_j + \mathfrak{l}^2 : \mathfrak{l}^2| \leq p^f \quad (j = 1, \ldots, n - a).$$

To estimate $|\mathsf{U}_i|$, let

$$\mathsf{T}_i = \{[v, y_i] | v \in \mathfrak{h}\} \quad (i = 1, \ldots, a).$$

Thus $\mathsf{U}_i \subseteq \mathsf{T}_i \cap \mathfrak{j} \subseteq \mathsf{T}_i \cap (p\mathfrak{h} + \mathfrak{h}^3)$. Since T_i is an additive subgroup of $\mathfrak{h}$,

$$|\mathsf{T}_i \cap (p\mathfrak{h} + \mathfrak{h}^3)| = |\mathsf{T}_i|/|(\mathsf{T}_i + p\mathfrak{h} + \mathfrak{h}^3)/(p\mathfrak{h} + \mathfrak{h}^3)|.$$

Now by 9.9, $|\mathsf{T}_i| = |\mathfrak{h} : \mathbf{C}_{\mathfrak{h}}(y_i)| \leq p^b$. Similarly

$$\begin{aligned} |(\mathsf{T}_i + p\mathfrak{h} + \mathfrak{h}^3)/(p\mathfrak{h} + \mathfrak{h}^3)| &= |\{[\tilde{v}, \tilde{y}_i] \mid \tilde{v} \in \tilde{\mathfrak{h}}\}| \\ &= |\tilde{\mathfrak{h}} : \mathbf{C}_{\tilde{\mathfrak{h}}}(\tilde{y}_i)| \geq p^c. \end{aligned}$$

Thus

$$|\mathsf{T}_i \cap (p\mathfrak{h} + \mathfrak{h}^3)| \leq p^{b-c},$$

and

$$(14) \qquad |\mathsf{U}_i| \leq p^{b-c} \quad (i = 1, \ldots, a).$$

By (13) and (14),

$$\begin{aligned} |\mathfrak{l}^2 + \mathfrak{j} : \mathfrak{l}^2| &\leq \prod_{i=1}^{a} |\mathsf{U}_i| \cdot \prod_{j=1}^{n-a} |\mathsf{V}_j + \mathfrak{l}^2 : \mathfrak{l}^2| \\ &\leq p^{a(b-c)+(n-a)f}. \end{aligned}$$

c) This follows from b) by taking $\mathfrak{l} = 0$, in which case we may suppose that $n = a$. **q.e.d.**

9.12 Theorem (VAUGHAN-LEE [2]). *Suppose that $\mathfrak{G}$ is a finite p-group and that every conjugacy class of $\mathfrak{G}$ contains at most p^b elements. Then*

$$|\mathfrak{G}'| \leq p^{\frac{1}{2}b(b+1)}.$$

More generally, if $\mathfrak{H} \leq \mathfrak{G}$ and $|\mathfrak{G} : \mathfrak{H}| = p^r$, then

$$|\mathfrak{G}' : \mathfrak{H}'| \leq p^b p^{b-1} \cdots p^v,$$

where $v = \max(0, b - r + 1)$.

Proof. Let c be the class of $\mathfrak{G}$ and let $\mathfrak{G}_i = \gamma_i(\mathfrak{G})$ $(i = 1, \ldots, c + 1)$. By 9.3 there exists a Lie ring $\mathfrak{g} = \mathsf{G}_1 \oplus \cdots \oplus \mathsf{G}_c$ and, for $i = 1, \ldots, c$, an isomorphism σ_i of $\mathfrak{G}_i/\mathfrak{G}_{i+1}$ onto G_i such that for any $x \in \mathfrak{G}_i$, $y \in \mathfrak{G}_j$,

$$[(x\mathfrak{G}_{i+1})\sigma_i, (y\mathfrak{G}_{j+1})\sigma_j] = ([x, y]\mathfrak{G}_{i+j+1})\sigma_{i+j}.$$

Thus $[\mathsf{G}_i, \mathsf{G}_j] \subseteq \mathsf{G}_{i+j}$ and, by 9.4,

$$\mathfrak{g}^m = \mathsf{G}_m \oplus \mathsf{G}_{m+1} \oplus \cdots \oplus \mathsf{G}_c \quad (m = 1, \ldots, c + 1).$$

Thus $p\mathfrak{g} \cap \mathfrak{g}^2 = p\mathfrak{g}^2$. Let

$$\mathfrak{i} = p\mathsf{G}_1 \oplus p\mathsf{G}_2 \oplus \mathsf{G}_3 \oplus \cdots \oplus \mathsf{G}_c,$$

$$\mathfrak{j} = p\mathsf{G}_2 \oplus \mathsf{G}_3 \oplus \cdots \oplus \mathsf{G}_c.$$

Thus $\mathfrak{i}$, $\mathfrak{j}$ are ideals of $\mathfrak{g}$, $\mathfrak{i} = p\mathfrak{g} + \mathfrak{g}^3$ and $\mathfrak{j} = p\mathfrak{g}^2 + \mathfrak{g}^3$. Let $\mathfrak{h}$ be a minimal element of the set of subrings of $\mathfrak{g}$ for which $\mathfrak{h}^2 = \mathfrak{g}^2$. Put $|\mathfrak{h} : \mathfrak{h}^2 + p\mathfrak{h}| = p^d$ and $\bar{\mathfrak{h}} = (\mathfrak{h} + \mathfrak{i})/\mathfrak{i}$. We show that $\mathfrak{h}$ satisfies the conditions of 9.11.

a) $\mathfrak{h}$ is generated by d elements $y_1, \ldots, y_d$ for which $|\mathfrak{h} : \mathbf{C}_{\mathfrak{h}}(y_i)| \leq p^b$ $(i = 1, \ldots, d)$. Also $|\bar{\mathfrak{h}} : \mathbf{C}_{\bar{\mathfrak{h}}}(\bar{y}_i)| = p^{d-1}$, where $\bar{y}_i = y_i + \mathfrak{i}$, so $d - 1 \leq b$.

Since $p\mathfrak{g} \cap \mathfrak{g}^2 = p\mathfrak{g}^2$, it follows from 9.10c) that $\mathfrak{h}$ is generated by d elements $x_1, \ldots, x_d$ for which $|\bar{\mathfrak{h}} : \mathbf{C}_{\bar{\mathfrak{h}}}(\bar{x}_i)| = p^{d-1}$, where $\bar{x}_i = x_i + \mathfrak{i}$. Write $x_i = y_i + z_i$, where $y_i \in \mathsf{G}_1$, $z_i \in \mathfrak{g}^2 = \mathfrak{h}^2$. Thus $y_i \in \mathfrak{h}$ and if $\mathfrak{h}_1$ is the subring generated by $y_1, \ldots, y_d$, $\mathfrak{h}_1 + \mathfrak{h}^2 = \mathfrak{h}$. By 8.10, $\mathfrak{h} = \mathfrak{h}_1$. Also $\bar{x}_i - \bar{y}_i \in \bar{\mathfrak{h}}^2$, so $\mathbf{C}_{\bar{\mathfrak{h}}}(\bar{y}_i) = \mathbf{C}_{\bar{\mathfrak{h}}}(\bar{x}_i)$ and $|\bar{\mathfrak{h}} : \mathbf{C}_{\bar{\mathfrak{h}}}(\bar{y}_i)| = p^{d-1}$.

Since $y_i \in \mathsf{G}_1$, $y_i = (g_i\mathfrak{G}_2)\sigma_1$ for some $g_i \in \mathfrak{G}$. Since $[\mathsf{G}_1, \mathsf{G}_j] \subseteq \mathsf{G}_{j+1}$,

$$\mathbf{C}_{\mathfrak{g}}(y_i) = (\mathsf{G}_1 \cap \mathbf{C}_{\mathfrak{g}}(y_i)) \oplus \cdots \oplus (\mathsf{G}_c \cap \mathbf{C}_{\mathfrak{g}}(y_i)).$$

But for $j = 1, \ldots, c$,

$$\begin{aligned} \mathsf{G}_j \cap \mathbf{C}_{\mathfrak{g}}(y_i) &= \{(g\mathfrak{G}_{j+1})\sigma_j \mid g \in \mathfrak{G}_j, [g_i, g] \in \mathfrak{G}_{j+2}\} \\ &\supseteq \{(g\mathfrak{G}_{j+1})\sigma_j \mid g \in \mathfrak{G}_j \cap \mathbf{C}_{\mathfrak{G}}(g_i)\}. \end{aligned}$$

Since σ_j is an isomorphism, it follows that

$$\begin{aligned} |\mathsf{G}_j \cap \mathbf{C}_{\mathfrak{g}}(y_i)| &\geq |(\mathfrak{G}_j \cap \mathbf{C}_{\mathfrak{G}}(g_i))\mathfrak{G}_{j+1}/\mathfrak{G}_{j+1}| \\ &= |\mathfrak{G}_j \cap \mathbf{C}_{\mathfrak{G}}(g_i) : \mathfrak{G}_{j+1} \cap \mathbf{C}_{\mathfrak{G}}(g_i)|. \end{aligned}$$

Thus

$$|\mathbf{C}_{\mathfrak{g}}(y_i)| = \prod_{j=1}^{c} |\mathsf{G}_j \cap \mathbf{C}_{\mathfrak{g}}(y_i)| \geq |\mathbf{C}_{\mathfrak{G}}(g_i)|,$$

and since $|\mathfrak{g}| = |\mathfrak{G}|$,

$$|\mathfrak{g} : \mathbf{C}_{\mathfrak{g}}(y_i)| \leq |\mathfrak{G} : \mathbf{C}_{\mathfrak{G}}(g_i)| \leq p^b.$$

Hence $|\mathfrak{h} : \mathbf{C}_{\mathfrak{h}}(y_i)| \leq p^b$ $(i = 1, \ldots, d)$.

Let $\tilde{\mathfrak{h}} = \mathfrak{h}/(p\mathfrak{h} + \mathfrak{h}^3)$, $\tilde{y}_i = y_i + (p\mathfrak{h} + \mathfrak{h}^3)$.

b) $|\tilde{\mathfrak{h}} : \mathbf{C}_{\tilde{\mathfrak{h}}}(\tilde{y}_i)| \geq p^{d-1}$ $(i = 1, \ldots, d)$.

For, using 9.9,

$$\begin{aligned} |\tilde{\mathfrak{h}} : \mathbf{C}_{\tilde{\mathfrak{h}}}(\tilde{y}_i)| &= |\{[z, y_i] + (p\mathfrak{h} + \mathfrak{h}^3) | z \in \mathfrak{h}\}| \\ &\geq |\{[z, y_i] + \mathfrak{i} | z \in \mathfrak{h}\}| \\ &= |\bar{\mathfrak{h}} : \mathbf{C}_{\bar{\mathfrak{h}}}(\bar{y}_i)| \\ &= p^{d-1}, \end{aligned}$$

by a).

c) $\mathfrak{j} = p\mathfrak{h}^2 + \mathfrak{h}^3$.

By 8.10d), $\mathfrak{g}^i = \mathfrak{h}^i$ for each $i \geq 2$, so $\mathfrak{j} = p\mathfrak{g}^2 + \mathfrak{g}^3 = p\mathfrak{h}^2 + \mathfrak{h}^3$.

For $i = 1, \ldots, c$, put $\mathsf{L}_i = ((\mathfrak{H} \cap \mathfrak{G}_i)\mathfrak{G}_{i+1}/\mathfrak{G}_{i+1})\sigma_i$. Let

$$\mathfrak{l}^* = \mathsf{L}_1 \oplus \cdots \oplus \mathsf{L}_c$$

and $\mathfrak{l} = \mathfrak{l}^* \cap \mathfrak{h}$. Put

$$|\mathfrak{h} : \mathfrak{l} + p\mathfrak{h} + \mathfrak{h}^2| = p^a, \quad |\mathfrak{l} + p\mathfrak{h} + \mathfrak{h}^2 : \mathfrak{l}| = p^f.$$

d) $\mathfrak{l}$ is a subring of $\mathfrak{h}$ and $|\mathfrak{G}' : \mathfrak{H}'| \leq |\mathfrak{h}^2 : \mathfrak{l}^2|$. Also $a + f \leq r$.

$[\mathsf{L}_i, \mathsf{L}_j]$ is the additive group spanned by all

$$[(x\mathfrak{G}_{i+1})\sigma_i, (y\mathfrak{G}_{j+1})\sigma_j] = ([x, y]\mathfrak{G}_{i+j+1})\sigma_{i+j}$$

with $x \in \mathfrak{H} \cap \mathfrak{G}_i$, $y \in \mathfrak{H} \cap \mathfrak{G}_j$, so

$$[\mathsf{L}_i, \mathsf{L}_j] \subseteq ((\mathfrak{H}' \cap \mathfrak{G}_{i+j})\mathfrak{G}_{i+j+1}/\mathfrak{G}_{i+j+1})\sigma_{i+j} \subseteq \mathsf{L}_{i+j}.$$

Thus $\mathfrak{l}^*$ is a Lie subring of $\mathfrak{g}$ and $\mathfrak{l}^{*2} \subseteq \mathsf{L}'_2 \oplus \cdots \oplus \mathsf{L}'_c$, where

$$\mathsf{L}'_i = ((\mathfrak{H}' \cap \mathfrak{G}_i)\mathfrak{G}_{i+1}/\mathfrak{G}_{i+1})\sigma_i \quad (i = 2, \ldots, c).$$

Hence $\mathfrak{l}$ is a subring of $\mathfrak{h}$. Now

$$|\mathsf{L}_i| = |\mathfrak{H} \cap \mathfrak{G}_i : \mathfrak{H} \cap \mathfrak{G}_{i+1}|, \quad |\mathsf{L}'_i| = |\mathfrak{H}' \cap \mathfrak{G}_i : \mathfrak{H}' \cap \mathfrak{G}_{i+1}|.$$

Thus $|\mathfrak{l}^*| = |\mathfrak{H}|$ and $|\mathfrak{l}^{*^2}| \leq |\mathfrak{H}'|$.

Since

$$p^{a+f} = |\mathfrak{h} : \mathfrak{l}| = |\mathfrak{h} : \mathfrak{h} \cap \mathfrak{l}^*| = |\mathfrak{h} + \mathfrak{l}^* : \mathfrak{l}^*| \leq |\mathfrak{g} : \mathfrak{l}^*| = |\mathfrak{G} : \mathfrak{H}| = p^r,$$

$a + f \leq r$. Also $|\mathfrak{G}' : \mathfrak{H}'| \leq |\mathfrak{g}^2 : \mathfrak{l}^{*^2}| \leq |\mathfrak{h}^2 : \mathfrak{l}^2|$. Thus d) is proved.

By 9.11c), $|\mathfrak{j}| \leq p^{d(b-d+1)}$, so

$$|\mathfrak{l}^2 + \mathfrak{j} : \mathfrak{l}^2| \leq |\mathfrak{j}| \leq p^{d(b-d+1)}.$$

Combining with 9.11a), we find that $|\mathfrak{h}^2 : \mathfrak{l}^2| \leq p^x$, where

$$x = d(b - d + 1) + \tfrac{1}{2}a(a - 1) + a(d - a).$$

Now

$$x = \tfrac{1}{2}b(b + 1) - \tfrac{1}{2}v(v - 1) - x',$$

where

$$x' = \tfrac{1}{2}(b - d)(b - d + 1) + \tfrac{1}{2}(d - a)(d - a - 1) - \tfrac{1}{2}v(v - 1).$$

Thus if $x' \geq 0$, we have

$$|\mathfrak{G}' : \mathfrak{H}'| \leq |\mathfrak{h}^2 : \mathfrak{l}^2| \leq p^x \leq p^b p^{b-1} \cdots p^v,$$

as required. Suppose then that $x' < 0$. Then $v > d - a$. Since $d - a \geq 0$, it follows that $v > 0$, so $v = b - r + 1$ and $b - r + 1 > d - a$. Hence $b - d - r + a \geq 0$. By a), $d - 1 \leq b$, so

$$2b - 2d - r + a + 1 \geq 0.$$

In this case we use 9.11a) and b), from which it follows that $|\mathfrak{h}^2 : \mathfrak{l}^2| \leq p^y$, where

$$y = a(b - d + 1) + (d - a)f + \tfrac{1}{2}a(a - 1) + a(d - a).$$

Now

$$\tfrac{1}{2}b(b + 1) - \tfrac{1}{2}v(v - 1) - y$$
$$= \tfrac{1}{2}(r - a)(2b - 2d - r + a + 1) + (d - a)(r - a - f) \geq 0,$$

since by d), $r - a \geq r - a - f \geq 0$. Hence

$$|\mathfrak{G}' : \mathfrak{H}'| \leq |\mathfrak{h}^2 : \mathfrak{l}^2| \leq p^y \leq p^b p^{b-1} \cdots p^v.$$

Taking $\mathfrak{H} = 1$, we obtain $|\mathfrak{G}'| \leq p^{\frac{1}{2}b(b+1)}$. **q.e.d.**

9.13 Remarks. a) The inequality for $|\mathfrak{G}'|$ in 9.12 is best possible, as is shown by the group generated by $x_0, \ldots, x_b$ with defining relations

$$x_i^p = [x_i, x_j]^p = [x_i, x_j, x_k] = 1.$$

b) VAUGHAN-LEE [1] proved the analogue of Theorem 9.12 for Lie rings.

c) It is not difficult to prove by slightly strengthening IV, 2.3 that there is a function $f(n)$ such that whenever every conjugacy class of $\mathfrak{G}$ contains at most n elements, $|\mathfrak{G}'| \leq f(n)$. It has been proved by P. M. NEUMANN and M. R. VAUGHAN-LEE [1] that we may take

$$f(n) \leq n^{\frac{1}{2}(3 + 5 \log n)},$$

where the logarithm is taken to base 2. In the same paper it is shown that for soluble groups

$$f(n) \leq n^{\frac{1}{2}(5 + \log n)}.$$

The proof utilizes Theorem 9.12 in the case when $|\mathfrak{G} : \mathfrak{H}| < p^b$.

It was conjectured for some years that if every conjugacy class of the p-group $\mathfrak{G}$ contains at most p^b elements, the class c of $\mathfrak{G}$ is at most $b + 1$ (see I. D. MACDONALD [1]), but a counterexample to this was given recently[1]. However, it is known that $c \leq 2b$, and this will now be proved.

9.14 Lemma. *Suppose that*

$$\mathfrak{G} = \mathfrak{G}_1 \geq \mathfrak{G}_2 \geq \cdots \mathfrak{G}_k \geq \mathfrak{G}_{k+1} = 1$$

is a central series of the p-group $\mathfrak{G}$. If $x \in \mathfrak{G}$ and there exist s suffices i for which $1 \leq i \leq k - 1$ and $x \notin \mathbf{C}_{\mathfrak{G}}(\mathfrak{G}_i/\mathfrak{G}_{i+2})$, then the number of conjugates of x in $\mathfrak{G}$ is at least p^s.

1 See FELSCH, NEUBÜSER and PLESKEN [1].

Proof. This is proved by induction on k and is trivial for $k = 1$. If $k > 1$ and $x \in \mathbf{C}_{\mathfrak{G}}(\mathfrak{G}_{k-1})$, it follows from the inductive hypothesis that $x\mathfrak{G}_k$ has at least p^s conjugates in $\mathfrak{G}/\mathfrak{G}_k$ and the assertion follows trivially. If $k > 1$ and $x \notin \mathbf{C}_{\mathfrak{G}}(\mathfrak{G}_{k-1})$, the inductive hypothesis shows that $x\mathfrak{G}_k$ has at least p^{s-1} conjugates in $\mathfrak{G}/\mathfrak{G}_k$, whence $|\mathfrak{G} : \mathfrak{C}| \geq p^{s-1}$, where $\mathfrak{C}/\mathfrak{G}_k = \mathbf{C}_{\mathfrak{G}/\mathfrak{G}_k}(x\mathfrak{G}_k)$. Now $\mathbf{C}_{\mathfrak{G}}(x) \leq \mathfrak{C}$. But $\mathfrak{G}_{k-1} \leq \mathfrak{C}$ since the given series is central, and $\mathfrak{G}_{k-1} \not\leq \mathbf{C}_{\mathfrak{G}}(x)$. Hence $\mathbf{C}_{\mathfrak{G}}(x) < \mathfrak{C}$ and $|\mathfrak{G} : \mathbf{C}_{\mathfrak{G}}(x)| \geq p^s$.

q.e.d.

9.15 Theorem. (LEEDHAM-GREEN, NEUMANN and WIEGOLD [1]). *If $\mathfrak{G}$ is a p-group and the number of elements in any conjugacy class of $\mathfrak{G}$ is at most p^b, then the class c of $\mathfrak{G}$ is less than* $\dfrac{bp}{p-1} + 1$ *and is thus at most 2b.*

Proof. For each element $x \in \mathfrak{G}$ and for $1 \leq i \leq c - 1$, write

$$t_i(x) = \begin{cases} 1 & \text{if } \; x \in \mathbf{C}_{\mathfrak{G}}(\gamma_i(\mathfrak{G})/\gamma_{i+2}(\mathfrak{G})), \\ 0 & \text{otherwise.} \end{cases}$$

Thus for $i = 1, \ldots, c - 1$,

$$\sum_{x \in \mathfrak{G}} t_i(x) = |\mathbf{C}_{\mathfrak{G}}(\gamma_i(\mathfrak{G})/\gamma_{i+2}(\mathfrak{G}))| \leq \frac{|\mathfrak{G}|}{p},$$

since $\gamma_i(\mathfrak{G})/\gamma_{i+2}(\mathfrak{G})$ is not a central factor. Hence

$$\sum_{i=1}^{c-1} \sum_{x \in \mathfrak{G}} t_i(x) \leq \frac{c-1}{p}|\mathfrak{G}|.$$

But $\sum_{i=1}^{c-1} t_i(x) = c - 1 - s(x)$, where $s(x)$ denotes the number of suffices i for which $1 \leq i \leq c - 1$ and $x \notin \mathbf{C}_{\mathfrak{G}}(\gamma_i(\mathfrak{G})/\gamma_{i+2}(\mathfrak{G}))$. Thus

$$\sum_{x \in \mathfrak{G}} (c - 1 - s(x)) \leq \frac{c-1}{p}|\mathfrak{G}|.$$

But $s(1) = 0$, so

$$\sum_{1 \neq x \in \mathfrak{G}} (c - 1 - s(x)) \leq \frac{c-1}{p}|\mathfrak{G}| - (c - 1) < \frac{c-1}{p}(|\mathfrak{G}| - 1).$$

Hence there exists $x \in \mathfrak{G}$ such that

$$c - 1 - s(x) < \frac{c-1}{p}.$$

By 9.14, $s(x) \leq b$, so

$$b > (c - 1) - \frac{c-1}{p} = \frac{(c-1)(p-1)}{p}.$$

This gives the stated inequalities. **q.e.d.**

Under special conditions, there are other ways for forming a Lie ring from a group. The following is a very simple example, which will be used in Chapter X.

9.16 Lemma. *Suppose that $\mathfrak{G}$ is a (not necessarily finite) nilpotent group of class at most 2 in which every element x has a unique square root $x^{\frac{1}{2}}$. For any elements x, y of $\mathfrak{G}$, put*

$$x + y = xy[y, x]^{\frac{1}{2}}.$$

Then $\mathfrak{G}$ is a Lie ring with respect to addition and commutation. Every element of $\mathfrak{G}$ has the same order with respect to the two group operations on $\mathfrak{G}$, and the automorphisms of the multiplicative group $\mathfrak{G}$ are the same as the automorphisms of the Lie ring $\mathfrak{G}$.

Proof. First observe that since $\mathfrak{G}$ is of class at most 2,

$$[x^{\frac{1}{2}}, y]^2 = [x, y^{\frac{1}{2}}]^2 = [x, y],$$

so $[x, y]^{\frac{1}{2}} = [x^{\frac{1}{2}}, y] = [x, y^{\frac{1}{2}}] \in \mathbf{Z}(\mathfrak{G})$. Thus

$$[x, y]^{\frac{1}{2}}[y, x]^{\frac{1}{2}} = [x^{\frac{1}{2}}, y][y, x^{\frac{1}{2}}] = 1.$$

Hence

$$xy[y, x]^{\frac{1}{2}} = yx[x, y][y, x]^{\frac{1}{2}} = yx[x, y]^{\frac{1}{2}}$$

and $x + y = y + x$. Again,

$$\begin{aligned}(x + y) + z &= (x + y)z[z, x + y]^{\frac{1}{2}} \\ &= xy[y, x]^{\frac{1}{2}} z[z, xy]^{\frac{1}{2}} \\ &= xyz[y, x]^{\frac{1}{2}}[z, y]^{\frac{1}{2}}[z, x]^{\frac{1}{2}} \\ &= x(y + z)[yz, x]^{\frac{1}{2}} \\ &= x + (y + z).\end{aligned}$$

Thus $\mathfrak{G}$ is an additive group with zero element 1, and $-x = x^{-1}$.

We have

$$[x + y, z] = [xy, z] = [x, z][y, z] = [x, z] + [y, z].$$

Also $[x, x] = 1$ and $[x, y, z] = 1$. Thus $\mathfrak{G}$ is a Lie ring.

If $n \in \mathbb{Z}$, $nx = x^n$. Thus x has the same order with respect to the two group operations on $\mathfrak{G}$. Also $\frac{1}{2}x = x^{\frac{1}{2}}$.

Finally, note that

$$xy = x + y + \tfrac{1}{2}[x, y],$$

so the automorphisms of $\mathfrak{G}$, regarded as a multiplicative group or as a Lie ring, are the same. **q.e.d.**

9.17 Remark. The procedure in 9.16 is the first step in the inversion of the *Baker-Hausdorff* formula. The Baker-Hausdorff formula (see, for example, JACOBSON [1], page 173) is

$$x * y = \sum_{m \geq 1} \sum_{p_i + q_i > 0} \frac{(-1)^{m-1}[x, \overset{p_1}{\ldots}, x, y, \overset{q_1}{\ldots}, y, x, \overset{p_2}{\ldots}, x, \ldots, y, \overset{q_m}{\ldots}, y]}{m\Sigma(p_i + q_i)p_1!\, q_1! \cdots p_m!\, q_m!}$$

$$= x + y + \tfrac{1}{2}[x, y] + \tfrac{1}{12}[x, y, y] - \tfrac{1}{12}[x, y, x] + \cdots.$$

If $\mathfrak{g}$ is a Lie ring and the right-hand side of this formula has a meaning for all x, y in $\mathfrak{g}$, then $\mathfrak{g}$ becomes a group with respect to $*$, the identity element being 0 and the inverse of x being $-x$. The commutator $[x, y]^*$ of x and y is

$$[x, y]^* = [x, y] + \tfrac{1}{2}[x, y, x] + \tfrac{1}{2}[x, y, y] + \cdots.$$

Conversely, the formula can be inverted, so that $x + y$ and $[x, y]$ are expressible in terms of the $*$-operation. Thus, under appropriate conditions, a group can be regarded as a Lie ring (see, for example, AMAYO and STEWART [1], Chapter 5).

The above special case arises in a quite different way in a paper of C. HOPKINS [1].

§ 10. Regular Automorphisms

Our next aim is to use Theorem 9.3 to prove some theorems about groups having fixed point free automorphisms. To do this, it is necessary to obtain first the corresponding results for Lie rings. We therefore wish to study Lie rings having an automorphism α which leaves only the zero element fixed, that is, for which $\ker(\alpha - 1) = 0$. The results obtained are more general, and the hypothesis on α can be replaced by assertions about the ideal generated by the elements fixed by α. Consider, for example, the following theorem about automorphisms of order 2.

10.1 Theorem. *Suppose that $\mathfrak{g}$ is a Lie ring, that α is an automorphism of $\mathfrak{g}$ of order 2 and that $\mathfrak{j}$ is an ideal of $\mathfrak{g}$ which contains all elements of $\mathfrak{g}$ left fixed by α.*

a) $a\alpha \equiv -a \bmod \mathfrak{j}$, *for all* $a \in \mathfrak{g}$.

b) *If* $\mathsf{G}_1 = \{a | a \in \mathfrak{g}, a\alpha = -a\}$, G_1 *is an additive subgroup of* $\mathfrak{g}$, $[\mathsf{G}_1, \mathsf{G}_1] \subseteq \mathfrak{j}$ *and* $2\mathfrak{g} \subseteq \mathfrak{j} + \mathsf{G}_1$.

c) α *induces the identity automorphism on* $(\mathfrak{g}^2 + \mathfrak{j})/\mathfrak{j}$.

d) $4\mathfrak{g}^2 \subseteq \mathfrak{j}$.

Proof. a) If $a \in \mathfrak{g}$, $(a\alpha + a)\alpha = a\alpha^2 + a\alpha = a\alpha + a$. Hence $a\alpha + a$ is fixed by α and lies in $\mathfrak{j}$.

b) G_1 is clearly an additive subgroup. If $a \in \mathsf{G}_1$ and $b \in \mathsf{G}_1$,

$$[a, b]\alpha = [a\alpha, b\alpha] = [-a, -b] = [a, b],$$

so $[a, b] \in \mathfrak{j}$. Thus $[\mathsf{G}_1, \mathsf{G}_1] \subseteq \mathfrak{j}$. If $a \in \mathfrak{g}$, $2a = (a + a\alpha) + (a - a\alpha)$. By a), $a + a\alpha \in \mathfrak{j}$. And $a - a\alpha \in \mathsf{G}_1$, since

$$(a - a\alpha)\alpha = a\alpha - a\alpha^2 = a\alpha - a = -(a - a\alpha).$$

c) If a, b are in $\mathfrak{g}$, $[a\alpha, b\alpha] \equiv [-a, -b] \bmod \mathfrak{j}$, by a). Hence $[a, b]\alpha \equiv [a, b] \bmod \mathfrak{j}$. Thus α induces the identity automorphism on $(\mathfrak{g}^2 + \mathfrak{j})/\mathfrak{j}$.

d) $4\mathfrak{g}^2 = [2\mathfrak{g}, 2\mathfrak{g}] \subseteq [\mathfrak{j} + \mathsf{G}_1, \mathfrak{j} + \mathsf{G}_1]$ by b). Using b) again,

$$4\mathfrak{g}^2 \subseteq [\mathfrak{j}, \mathfrak{g}] + [\mathsf{G}_1, \mathsf{G}_1] \subseteq \mathfrak{j}.$$

q.e.d.

Suppose now that $\mathfrak{g}$ has an automorphism of order 2 which leaves only the zero element fixed; then $\mathfrak{g}$ is Abelian (cf. V, 8.18). This can be deduced from 10.1 in two ways. In both, we take $\mathfrak{j} = 0$. On the one hand, it follows from c) that every element of $\mathfrak{g}^2$ is fixed by α, so $\mathfrak{g}^2 = 0$. On the other hand, it follows from d) that $4\mathfrak{g}^2 = 0$; hence $\mathfrak{g}^2 = 0$ since an automorphism of order p of a non-zero Abelian p-group always has a non-zero fixed point (see 10.6 below). The first of these procedures seems difficult to generalize to automorphisms of other orders, though this can be done for automorphisms of order 3 (see Exercise 16). The second procedure can, however, be generalized. Before doing so, we shall consider it for automorphisms of order 3.

10.2 Theorem. *Suppose that $\mathfrak{g}$ is a Lie ring, that α is an automorphism of $\mathfrak{g}$ of order 3 and that $\mathfrak{j}$ is an ideal of $\mathfrak{g}$ which contains all elements of $\mathfrak{g}$ left fixed by α. Then $27\mathfrak{g}^3 \subseteq \mathfrak{j}$.*

Proof. In this case we are unable to make use immediately of the decomposition corresponding to that of $2a$ in 10.1b); first we must extend the ground-ring. Thus let ω be a complex cube root of unity and let A be the ring of all numbers $x + y\omega$ with x, y in $\mathbb{Z}$. Thus $\{1, \omega\}$ is a $\mathbb{Z}$-basis of A. Let $\mathfrak{g}^* = \mathfrak{g} \otimes_{\mathbb{Z}} \mathsf{A}$, $\alpha^* = \alpha \otimes 1$, $\mathfrak{j}^* = \mathfrak{j} \otimes_{\mathbb{Z}} \mathsf{A}$. By 8.13, $\mathfrak{g}^*$ is a Lie algebra over A, α^* is an automorphism of $\mathfrak{g}^*$, $\mathfrak{g}^3 \otimes_{\mathbb{Z}} \mathsf{A} = (\mathfrak{g}^*)^3$, and $\mathfrak{j}^*$ is an ideal of $\mathfrak{g}^*$ which contains all elements of $\mathfrak{g}^*$ invariant under α^*. Let

$$\mathsf{G}_1 = \{a \mid a \in \mathfrak{g}^*, a\alpha^* = \omega a\}, \quad \mathsf{G}_{-1} = \{a \mid a \in \mathfrak{g}^*, a\alpha^* = \omega^{-1} a\}.$$

Thus G_1, G_{-1} are A-submodules of $\mathfrak{g}^*$. We have

$$[\mathsf{G}_1, \mathsf{G}_{-1}] \subseteq \mathfrak{j}^*,$$

since if $a \in \mathsf{G}_1$ and $b \in \mathsf{G}_{-1}$,

$$[a, b]\alpha^* = [a\alpha^*, b\alpha^*] = [\omega a, \omega^{-1} b] = [a, b].$$

Also

$$3\mathfrak{g}^* \subseteq \mathfrak{j}^* + \mathsf{G}_1 + \mathsf{G}_{-1}.$$

For if $a \in \mathfrak{g}^*$, write

$$a_1 = a + a\alpha^* + a\alpha^{*2}$$
$$a_2 = a + \omega a\alpha^* + \omega^2 a\alpha^{*2}$$
$$a_3 = a + \omega^2 a\alpha^* + \omega a\alpha^{*2}.$$

Then clearly, $a_1 \in \mathfrak{j}^*$, $a_2 \in \mathsf{G}_{-1}$, $a_3 \in \mathsf{G}_1$ and $a_1 + a_2 + a_3 = 3a$.
Now

$$\begin{aligned}(3\mathfrak{g}^*)^2 &\subseteq [\mathfrak{j}^* + \mathsf{G}_1 + \mathsf{G}_{-1}, \mathfrak{j}^* + \mathsf{G}_1 + \mathsf{G}_{-1}] \\ &\subseteq [\mathfrak{j}^*, \mathfrak{g}^*] + [\mathsf{G}_1, \mathsf{G}_1] + [\mathsf{G}_1, \mathsf{G}_{-1}] + [\mathsf{G}_{-1}, \mathsf{G}_{-1}] \\ &\subseteq \mathfrak{j}^* + [\mathsf{G}_1, \mathsf{G}_1] + [\mathsf{G}_{-1}, \mathsf{G}_{-1}].\end{aligned}$$

Hence

$$\begin{aligned}(3\mathfrak{g}^*)^3 &\subseteq [\mathfrak{j}^* + [\mathsf{G}_1, \mathsf{G}_1] + [\mathsf{G}_{-1}, \mathsf{G}_{-1}], \mathfrak{j}^* + \mathsf{G}_1 + \mathsf{G}_{-1}] \\ &\subseteq [\mathfrak{j}^*, \mathfrak{g}^*] + [\mathsf{G}_1, \mathsf{G}_1, \mathsf{G}_1] + [\mathsf{G}_1, \mathsf{G}_1, \mathsf{G}_{-1}] \\ &\quad + [\mathsf{G}_{-1}, \mathsf{G}_{-1}, \mathsf{G}_1] + [\mathsf{G}_{-1}, \mathsf{G}_{-1}, \mathsf{G}_{-1}].\end{aligned}$$

But all these summands are contained in $\mathfrak{j}^*$. For by 8.4, $[\mathsf{G}_1, \mathsf{G}_1, \mathsf{G}_1]$ is spanned over A by all $[a, b, c]$ with a, b, c in G_1, and

$$[a, b, c]\alpha^* = [a\alpha^*, b\alpha^*, c\alpha^*] = [\omega a, \omega b, \omega c] = [a, b, c].$$

A similar reason holds for $[\mathsf{G}_{-1}, \mathsf{G}_{-1}, \mathsf{G}_{-1}]$. Finally, by 8.5a),

$$[\mathsf{G}_1, \mathsf{G}_1, \mathsf{G}_{-1}] \subseteq [\mathsf{G}_1, \mathsf{G}_{-1}, \mathsf{G}_1] + [\mathsf{G}_{-1}, \mathsf{G}_1, \mathsf{G}_1] \subseteq \mathfrak{j}^*,$$

since $[\mathsf{G}_1, \mathsf{G}_{-1}] \subseteq \mathfrak{j}^*$, and a similar argument holds for $[\mathsf{G}_{-1}, \mathsf{G}_{-1}, \mathsf{G}_1]$.
Thus $27(\mathfrak{g}^*)^3 \subseteq \mathfrak{j}^*$. Hence $(27\mathfrak{g}^3) \otimes \mathsf{A} \subseteq \mathfrak{j} \otimes \mathsf{A}$. By 8.13, $27\mathfrak{g}^3 \subseteq \mathfrak{j}$.

q.e.d.

10.3 Lemma. *Suppose that* A *is a commutative ring which contains a primitive n-th root of unity* ω. *Suppose that* $\mathfrak{g}$ *is a Lie algebra over* A *and that* α *is an automorphism of* $\mathfrak{g}$ *of order n. For each integer j, write*

$$\mathsf{G}_j = \{a \mid a \in \mathfrak{g},\ a\alpha = \omega^j a\}.$$

a) G_j *is an* A*-submodule of* $\mathfrak{g}$, *and* $\mathsf{G}_i = \mathsf{G}_j$ *if* $i \equiv j\ (n)$.

b) $[\mathsf{G}_i, \mathsf{G}_j] \subseteq \mathsf{G}_{i+j}$ *for any integers i, j.*

c) *If* $\mathfrak{h} = \mathsf{G}_0 + \cdots + \mathsf{G}_{n-1}$, *then* $\mathfrak{h}$ *is a subalgebra of* $\mathfrak{g}$ *invariant under* α *and* $n\mathfrak{g} \subseteq \mathfrak{h}$. *Also if* $b_i \in \mathsf{G}_i$ $(i = 0, \ldots, n-1)$ *and* $b_0 + \cdots + b_{n-1} = 0$, *then* $nb_i = 0$.

d) *If d is a non-negative integer,*

$$\mathfrak{h}^{(d)} = \sum_{k=0}^{n-1} (\mathfrak{h}^{(d)} \cap \mathsf{G}_k),$$

and if $d > 0$,

$$n(\mathfrak{h}^{(d)} \cap \mathsf{G}_k) = n \sum_{i=0}^{n-1} [\mathfrak{h}^{(d-1)} \cap \mathsf{G}_i, \mathfrak{h}^{(d-1)} \cap \mathsf{G}_{k-i}]$$

for any integer k.

Proof. a) This is trivial.

b) If $u \in \mathsf{G}_i$ and $v \in \mathsf{G}_j$,

$$[u, v]\alpha = [u\alpha, v\alpha] = [\omega^i u, \omega^j v] = \omega^{i+j}[u, v].$$

Thus $[u, v] \in \mathsf{G}_{i+j}$, as required.

c) $\mathfrak{h}$ is a subalgebra on account of b). Since $\mathsf{G}_i\alpha = \mathsf{G}_i$, $\mathfrak{h}$ is invariant under α. If $a \in \mathfrak{g}$, write

$$a_i = a + \omega^{-i}(a\alpha) + \cdots + \omega^{-(n-1)i}(a\alpha^{n-1})$$

for $i = 0, \ldots, n-1$. Then $a_i\alpha = \omega^i a_i$, so $a_i \in \mathsf{G}_i$. Since ω is a primitive n-th root of unity, $a_0 + a_1 + \cdots + a_{n-1} = na$. Hence $na \in \mathfrak{h}$. Thus $n\mathfrak{g} \subseteq \mathfrak{h}$. Apply $\omega^{-ij}\alpha^j$ to the equation $b_0 + \cdots + b_{n-1} = 0$. We find that for $i, j = 0, \ldots, n-1$,

$$\omega^{-ij}b_0 + \omega^{-(i-1)j}b_1 + \cdots + b_i + \cdots + \omega^{(n-i-1)j}b_{n-1} = 0.$$

Summing over $j = 0, 1, \ldots, n-1$, $nb_i = 0$.

d) The first assertion is proved by induction on d. It is trivial for $d = 0$. If $d > 0$,

$$\mathfrak{h}^{(d)} = [\mathfrak{h}^{(d-1)}, \mathfrak{h}^{(d-1)}] = \sum_{i,j} [\mathfrak{h}^{(d-1)} \cap \mathsf{G}_i, \mathfrak{h}^{(d-1)} \cap \mathsf{G}_j],$$

by the inductive hypothesis.
By b), $[\mathfrak{h}^{(d-1)} \cap \mathsf{G}_i, \mathfrak{h}^{(d-1)} \cap \mathsf{G}_j] \subseteq \mathfrak{h}^{(d)} \cap \mathsf{G}_{i+j}$, so the assertion is clear.

Suppose that $a \in \mathfrak{h}^{(d)} \cap \mathsf{G}_k$ ($d > 0$). By definition of $\mathfrak{h}^{(d)}$ and the previous assertion, a is a sum of elements of the form $[x, y]$, with $x \in \mathfrak{h}^{(d-1)} \cap \mathsf{G}_i$, $y \in \mathfrak{h}^{(d-1)} \cap \mathsf{G}_j$. By b), $[x, y] \in \mathsf{G}_{i+j}$. Hence by c), $n(a - s) = 0$, where s is the sum of those $[x, y]$ for which $i + j \equiv k$ (n). The assertion follows at once from this. **q.e.d.**

10.4 Theorem. *Suppose that* A *is a commutative ring which contains a primitive* n*-th root of unity* ω. *Suppose that* $\mathfrak{g}$ *is a Lie algebra over* A *and* *that α is an automorphism of $\mathfrak{g}$ of order n. For each integer j, write*

$$\mathsf{G}_j = \{a \mid a \in \mathfrak{g},\ a\alpha = \omega^j a\}.$$

Let $\mathfrak{h} = \mathsf{G}_0 + \cdots + \mathsf{G}_{n-1}$, let $\mathfrak{j}$ be the ideal of $\mathfrak{h}$ generated by G_0 and let $\mathfrak{h}_k$ be the subalgebra of $\mathfrak{g}$ generated by $\mathsf{G}_{k+1}, \ldots, \mathsf{G}_{n-1}$ $(k = 0, \ldots, n-1)$; in particular $\mathfrak{h}_{n-1} = 0$.

a) *For $k = 1, \ldots, n-1$,*

$$n^k(\mathfrak{h}^{(2^{k-1})} \cap \mathsf{G}_k) \subseteq n^k(\mathfrak{h}_k + \mathfrak{j}).$$

b) *For $k = 1, \ldots, n-1$ and $i = 1, \ldots, k$,*

$$n^k(\mathfrak{h}^{(2^k-1)} \cap \mathsf{G}_i) \subseteq n^k(\mathfrak{h}_k + \mathfrak{j}).$$

c) $n^{n-1}\mathfrak{h}^{(2^{n-1}-1)} \subseteq n^{n-1}\mathfrak{j}$.

Proof. a) and b) are proved together by induction on k. For $k = 1$, both state that $n(\mathfrak{h}' \cap \mathsf{G}_1) \subseteq n(\mathfrak{h}_1 + \mathfrak{j})$. By 10.3d), it suffices to prove that $n[\mathsf{G}_j, \mathsf{G}_{1-j}] \subseteq n(\mathfrak{h}_1 + \mathfrak{j})$ for $j = 0, \ldots, n-1$. For $j = 0$ or $j = 1$, $[\mathsf{G}_j, \mathsf{G}_{1-j}] \subseteq \mathfrak{j}$ since $\mathsf{G}_0 \subseteq \mathfrak{j}$. If $2 \leq j \leq n-1$, then $\mathsf{G}_j \subseteq \mathfrak{h}_1$ and $\mathsf{G}_{1-j} = \mathsf{G}_{n+1-j} \subseteq \mathfrak{h}_1$, since $2 \leq n+1-j \leq n-1$. Thus the result is clear.

Now suppose that $1 < k \leq n-1$. Suppose that a) is false. By 10.3d), there exist integers i, j such that $0 \leq i < n$, $0 \leq j < n$, $i + j \equiv k\ (n)$ and

$$n^k[\mathfrak{h}^{(2^{k-1}-1)} \cap \mathsf{G}_i, \mathfrak{h}^{(2^{k-1}-1)} \cap \mathsf{G}_j] \nsubseteq n^k(\mathfrak{h}_k + \mathfrak{j}).$$

Since $\mathsf{G}_0 \subseteq \mathfrak{j}$, $i > 0$ and $j > 0$. Also either $i \leq k$ or $j \leq k$; otherwise $[\mathsf{G}_i, \mathsf{G}_j]$ would be contained in the algebra generated by $\mathsf{G}_{k+1}, \ldots, \mathsf{G}_{n-1}$, which is $\mathfrak{h}_k$. Hence $i + j < k + n$, but also $i + j \equiv k\ (n)$, so $i + j = k$. Hence $0 < i < k$ and $0 < j < k$. Thus by the inductive hypothesis of b),

$$n^{k-1}(\mathfrak{h}^{(2^{k-1}-1)} \cap \mathsf{G}_j) \subseteq n^{k-1}(\mathfrak{h}_{k-1} + \mathfrak{j}).$$

Now there exist $u \in \mathfrak{h}^{(2^{k-1}-1)} \cap \mathsf{G}_i$ and $v \in \mathfrak{h}^{(2^{k-1}-1)} \cap \mathsf{G}_j$ such that $n^k[u, v] \notin n^k(\mathfrak{h}_k + \mathfrak{j})$. But $n^{k-1}v \in n^{k-1}(\mathfrak{h}_{k-1} + \mathfrak{j})$, so there exist $v_1 \in \mathfrak{h}_{k-1}$ and $v_2 \in \mathfrak{j}$ such that $n^{k-1}(v - v_1 - v_2) = 0$. By 8.3b), v_1 is a sum of products $[u_1, \ldots, u_h]$, where $h > 0$, $u_r \in \mathsf{G}_{q_r}$ and $k \leq q_r < n$; note that this product lies in $\mathsf{G}_{q_1 + \cdots + q_h}$ by 10.3b). Similarly by 8.3c), v_2 is a sum of products $[w_1, \ldots, w_f]$, where $f > 0$, $w_r \in \mathsf{G}_{p_r}$ and $p_1 = 0$; this product lies in $\mathsf{G}_l \cap \mathfrak{j}$ for some l. By substituting these sums in $n^{k-1}(v - v_1 - v_2) = 0$ and applying 10.3c), we find that $n^k(v - v_1' - v_2') = 0$, where v_1' is the sum of those $[u_1, \ldots, u_h]$ for which $q_1 + \cdots + q_h \equiv j\ (n)$ and $v_2' \in \mathsf{G}_j \cap \mathfrak{j}$. Now

$$n^k[u, v] = n^k[u, v_1'] + n^k[u, v_2'].$$

But $n^k[u, v] \notin n^k(\mathfrak{h}_k + \mathfrak{j})$ and $n^k[u, v_2'] \in n^k\mathfrak{j}$. Hence there is a product $[u_1, \ldots, u_h]$ for which $q_1 + \cdots + q_h \equiv j\ (n)$ and

$$n^k[u, [u_1, \ldots, u_h]] \notin n^k(\mathfrak{h}_k + \mathfrak{j}).$$

By 8.3a), there is a permutation π of $\{1, \ldots, h\}$ such that

$$[u, u_{1\pi}, \ldots, u_{h\pi}] \notin \mathfrak{h}_k + \mathfrak{j}.$$

Write $a = [u, u_{1\pi}, \ldots, u_{(h-1)\pi}]$, $b = u_{h\pi}$, $q = q_{h\pi}$. Since

$$\begin{aligned} i + q_{1\pi} + \cdots + q_{(h-1)\pi} &= i + (q_1 + \cdots + q_h) - q \\ &\equiv i + j - q \equiv k - q\ (n), \end{aligned}$$

it follows from 10.3b) that $a \in \mathsf{G}_{k-q}$. Also $b \in \mathsf{G}_q$, $k \leq q < n$, and $[a, b] \notin \mathfrak{h}_k + \mathfrak{j}$. Thus $a \notin \mathfrak{j}$, so $k \neq q$. Hence $k < q < n$, so $a \in \mathsf{G}_{n+k-q}$ and $k < n + k - q < n$. Thus $a \in \mathfrak{h}_k$ and $b \in \mathfrak{h}_k$, a contradiction. Thus a) is proved.

Since $2^{k-1} \leq 2^k - 1$, it follows from a) that

$$n^k(\mathfrak{h}^{(2^k-1)} \cap \mathsf{G}_k) \subseteq n^k(\mathfrak{h}_k + \mathfrak{j}),$$

which is the case $i = k$ of b). Suppose now that $1 \leq i < k$. We apply the inductive hypothesis to $\mathfrak{h}^{(2^{k-1})}$. To do this, note that

$$\{a \mid a \in \mathfrak{h}^{(2^{k-1})}, a\alpha = \omega^j a\} = \mathsf{G}_j \cap \mathfrak{h}^{(2^{k-1})}$$

and, by 10.3d),

$$\mathfrak{h}^{(2^{k-1})} = \sum_{j=0}^{n-1} (\mathfrak{h}^{(2^{k-1})} \cap \mathsf{G}_j).$$

Note also that $(\mathfrak{h}^{(2^{k-1})})^{(2^{k-1}-1)} = \mathfrak{h}^{(2^k-1)}$. Thus b) (for $k-1$) gives

$$n^{k-1}(\mathfrak{h}^{(2^k-1)} \cap \mathsf{G}_i) \subseteq n^{k-1}(\mathfrak{h}^* + \mathfrak{j}),$$

where $\mathfrak{h}^*$ is the subalgebra generated by $\mathsf{G}_k \cap \mathfrak{h}^{(2^{k-1})}, \ldots, \mathsf{G}_{n-1} \cap \mathfrak{h}^{(2^{k-1})}$. All these subspaces are contained in $\mathfrak{h}_k + \mathfrak{j}$, except possibly $\mathsf{G}_k \cap \mathfrak{h}^{(2^{k-1})}$. But by a), $\mathsf{G}_k \cap \mathfrak{h}^{(2^{k-1})} \subseteq \mathfrak{h}_k + \mathfrak{j} + \mathfrak{t}$, where $\mathfrak{t} = \{b \mid b \in \mathfrak{g},\ n^k b = 0\}$. Since $\mathfrak{t}$ is an ideal of $\mathfrak{g}$, $\mathfrak{h}_k + \mathfrak{j} + \mathfrak{t}$ is a subalgebra. Hence $\mathfrak{h}^* + \mathfrak{j} \subseteq \mathfrak{h}_k + \mathfrak{j} + \mathfrak{t}$ and $n^k(\mathfrak{h}^* + \mathfrak{j}) \subseteq n^k(\mathfrak{h}_k + \mathfrak{j})$. Combined with the above, this gives b).

To prove c), put $k = n-1$ in b). This gives

$$n^{n-1}(\mathfrak{h}^{(2^{n-1}-1)} \cap \mathsf{G}_i) \subseteq n^{n-1}\mathfrak{j} \quad (i = 1, \ldots, n-1),$$

since $\mathfrak{h}_{n-1} = 0$. This is also true for $i = 0$, since $\mathsf{G}_0 \subseteq \mathfrak{j}$. By 10.3d),

$$n^{n-1}\mathfrak{h}^{(2^{n-1}-1)} = n^{n-1}\sum_{i=0}^{n-1}(\mathfrak{h}^{(2^{n-1}-1)} \cap \mathsf{G}_i) \subseteq n^{n-1}\mathfrak{j}. \qquad \textbf{q.e.d.}$$

10.5 Theorem. *Suppose that $\mathfrak{g}$ is a Lie ring and that α is an automorphism of $\mathfrak{g}$ of finite order n. If $\mathfrak{j}$ is an ideal of $\mathfrak{g}$ containing all elements of $\mathfrak{g}$ which are invariant under α,*

$$n^{n-1}(n\mathfrak{g})^{(2^{n-1}-1)} \subseteq n^{n-1}\mathfrak{j}.$$

Proof. Let ω be a primitive n-th root of unity in a field of characteristic 0 and let $\mathsf{A} = \mathbb{Z}[\omega]$ be the set of polynomials

$$c_0 + c_1\omega + \cdots + c_{n-1}\omega^{n-1},$$

where the c_i are rational integers. Thus A is a ring and A is a free Abelian group. Let $\mathfrak{g}^*$ be the Lie algebra $\mathfrak{g} \otimes_{\mathbb{Z}} \mathsf{A}$. By 8.13b), $\alpha^* = \alpha \otimes 1$ is an automorphism of $\mathfrak{g}^*$; clearly the order of α^* is n. Let $\mathfrak{j}^*$ be the ideal of $\mathfrak{g}^*$ generated by all elements of $\mathfrak{g}^*$ which are invariant under α^*. By 10.3c) and 10.4c), there exists a subalgebra $\mathfrak{h}$ of $\mathfrak{g}^*$ such that $\mathfrak{h} \supseteq n\mathfrak{g}^*$ and

$$n^{n-1}\mathfrak{h}^{(2^{n-1}-1)} \subseteq n^{n-1}\mathfrak{j}^*.$$

But by 8.13d), $\mathfrak{j}^* \subseteq \mathfrak{j} \otimes \mathsf{A}$. Thus

$$n^{n-1}(n\mathfrak{g}^*)^{(2^{n-1}-1)} \subseteq n^{n-1}(\mathfrak{j} \otimes \mathsf{A}) = n^{n-1}\mathfrak{j} \otimes \mathsf{A}.$$

It follows from 8.13c) that

$$(n^{n-1}(n\mathfrak{g})^{(2^{n-1}-1)}) \otimes \mathsf{A} \subseteq n^{n-1}\mathfrak{j} \otimes \mathsf{A},$$

and hence the assertion. **q.e.d.**

In order to deal with the case $\mathfrak{j} = 0$, we need the following elementary lemma.

10.6 Lemma. *Suppose that* V *is an additive Abelian group, that* α *is an automorphism of* V *of finite order n and that the only element of* V *left fixed by* α *is* 0.

a) $\alpha^{n-1} + \cdots + \alpha + 1 = 0$ *(cf. V, 8.9d)).*

b) *If n is prime,* $a \in \mathsf{V}$ *and* $na = 0$, *then* $a = 0$.

c) *If the additive order of every element of* V *is a power of the prime p, then the only element of* V *left fixed by* α^p *is* 0.

Proof. a) If $a \in \mathsf{V}$ and $b = a(\alpha^{n-1} + \cdots + \alpha + 1)$, then

$$b\alpha - b = a(\alpha^n - 1) = 0.$$

Thus $b = 0$, so $\alpha^{n-1} + \cdots + \alpha + 1 = 0$.

b) Let B be the group generated by $a, a\alpha, \ldots, a\alpha^{n-1}$. Then B is a finite Abelian group since $na = 0$. Further α induces an automorphism $\bar{\alpha}$ on B, and $\bar{\alpha}$ leaves fixed only the zero element of B. Decomposition of B into orbits under $\langle\bar{\alpha}\rangle$ gives the congruence $|\mathsf{B}| \equiv 1\ (n)$, since n is a prime. Hence $\mathsf{B} = 0$ and $a = 0$.

c) Let $\mathsf{C} = \{a | a \in \mathsf{V}, a\alpha^p = a\}$. Then C is a subgroup of V and α induces an automorphism $\tilde{\alpha}$ on C of order a divisor of p. Also $\tilde{\alpha}$ leaves fixed no non-zero elements of C. It follows from b) and the fact that the order of every element of C is a power of p that $\mathsf{C} = 0$. **q.e.d.**

10.7 Theorem (KREKNIN [1]). *Suppose that the Lie ring* $\mathfrak{g}$ *possesses an automorphism* α *of order n, which leaves only the zero element of* $\mathfrak{g}$ *fixed. Then* $\mathfrak{g}$ *is soluble and* $\mathfrak{g}^{(2^n-2)} = 0$.

More precisely, let

$$\mathfrak{t} = \{a | a \in \mathfrak{g}, n^k a = 0 \quad \textit{for some integer } k \geq 0\}.$$

Then $\mathfrak{t}$ *is an ideal of* $\mathfrak{g}$, *and*

a) $\mathfrak{g}^{(2^{n-1}-1)} \subseteq \mathfrak{t}$;

b) $\mathfrak{t}^{(2^{n-1}-1)} = 0$.

Proof. It is clear that $\mathfrak{t}$ is an ideal of $\mathfrak{g}$.

a) We may apply 10.5 with $\mathfrak{j} = 0$. Thus

$$n^{n-1}(n\mathfrak{g})^{(2^{n-1}-1)} = 0$$

and $\mathfrak{g}^{(2^{n-1}-1)} \subseteq \mathfrak{t}$.

b) Let $p_1, \ldots, p_r$ be the distinct prime divisors of n, and write

$$\mathfrak{t}_i = \{a \mid a \in \mathfrak{g},\ p_i^k a = 0 \quad \text{for some } k \geq 0\}.$$

By I, 13.9,

$$\mathfrak{t} = \mathfrak{t}_1 \oplus \cdots \oplus \mathfrak{t}_r.$$

Also each $\mathfrak{t}_i$ is an ideal of $\mathfrak{g}$ and $\mathfrak{t}_i\alpha = \mathfrak{t}_i$. Write $n = p_i^k m_i$, where $(p_i, m_i) = 1$. If α_i is the restriction of α to $\mathfrak{t}_i$, write $\beta_i = \alpha_i^{p_i^k}$. By 10.6c), the only element of $\mathfrak{t}_i$ fixed by β_i is 0. Since the order of β_i divides m_i, a) gives

$$\mathfrak{t}_i^{(2^{m_i-1}-1)} \subseteq \{a \mid a \in \mathfrak{t}_i,\ m_i^l a = 0 \quad \text{for some integer } l \geq 0\}.$$

Since the order of any element of $\mathfrak{t}_i$ is a power of p_i and $(p_i, m_i) = 1$, $m_i^l a = 0$ implies $a = 0$, for any $a \in \mathfrak{t}_i$. Hence $\mathfrak{t}_i^{(2^{m_i-1}-1)} = 0$. Since $n \geq m_i$, $\mathfrak{t}_i^{(2^{n-1}-1)} = 0$. But since each $\mathfrak{t}_i$ is an ideal of $\mathfrak{g}$,

$$\mathfrak{t}^{(d)} = \mathfrak{t}_1^{(d)} \oplus \cdots \oplus \mathfrak{t}_r^{(d)}$$

for any integer d; hence $\mathfrak{t}^{(2^{n-1}-1)} = 0$.

Thus $\mathfrak{g}^{(2^n-2)} = (\mathfrak{g}^{(2^{n-1}-1)})^{(2^{n-1}-1)} \subseteq \mathfrak{t}^{(2^{n-1}-1)} = 0$. **q.e.d.**

10.8 Lemma. *Suppose that p is a prime.*

a) *Suppose that $1 \leq r < p$ and that $x_1, \ldots, x_r$ are not necessarily distinct non-zero elements of an additive group of order p. Then $|X| \geq r + 1$, where*

$$X = \{x_{i_1} + \cdots + x_{i_s} \mid 0 \leq s \leq r, i_1 < \cdots < i_s\}.$$

b) *Suppose that $n_1, \ldots, n_{p-1}$ are not necessarily distinct integers for which $(n_i, p) = 1$. Given any integer n, there exist $i_1, \ldots, i_s$ $(s \geq 0)$ such that $i_1 < \cdots < i_s$ and $n \equiv n_{i_1} + \cdots + n_{i_s}$ (p).*

Proof. a) This is proved by induction on r. For $r = 1$, it is clear, since $X = \{0, x_1\}$ and $x_1 \neq 0$. For $r > 1$, let

$$Y = \{x_{i_1} + \cdots + x_{i_s} | 0 \leq s \leq r-1, i_1 < \cdots < i_s \leq r-1\}.$$

By the inductive hypothesis $|Y| \geq r$. Now

$$X = Y \cup \{y + x_r | y \in Y\}.$$

Thus, if $|X| \leq r$, $y + x_r \in Y$ for all $y \in Y$. Hence $y + nx_r \in Y$ for every integer $n > 0$. Since $x_r \neq 0$, this implies that $|Y| = p$, so $|X| = p > r$. Thus $|X| \geq r + 1$.

b) This follows at once from the case $r = p - 1$ of a), applied to the additive group of integral residue classes modulo p. **q.e.d.**

10.9 Lemma. *Suppose that $\mathfrak{i}$ is an ideal of a Lie algebra $\mathfrak{h}$. Suppose that $u \in \mathfrak{i}$ and $v_1, \ldots, v_n$ are elements of $\mathfrak{h}$. If π is a permutation of $\{1, \ldots, n\}$,*

$$[u, v_{1\pi}, \ldots, v_{n\pi}] + [\mathfrak{i}, \mathfrak{h}^2]$$

is independent of π.

Proof. It is clearly sufficient to prove this in the case when π is the transposition $(i, i+1)$. Put $x = [u, v_1, \ldots, v_{i-1}]$, $y = v_i$, $z = v_{i+1}$; we must show that

$$[x, y, z, \ldots] - [x, z, y, \ldots] \in [\mathfrak{i}, \mathfrak{h}^2].$$

But by the Jacobi identity,

$$[x, y, z] - [x, z, y] = -[y, z, x] = [x, [y, z]] \in [\mathfrak{i}, \mathfrak{h}^2].$$

The assertion follows because $[\mathfrak{i}, \mathfrak{h}^2]$ is an ideal of $\mathfrak{h}$ (8.5c)). **q.e.d.**

10.10 Theorem. *Suppose that p is a prime and that A is a commutative ring which contains a primitive p-th root of unity ω. Suppose that $\mathfrak{g}$ is a Lie algebra over A and that α is an automorphism of $\mathfrak{g}$ of order p. For each integer j, write*

$$\mathsf{G}_j = \{a | a \in \mathfrak{g}, a\alpha = \omega^j a\},$$

and let $\mathfrak{h} = \mathsf{G}_0 + \cdots + \mathsf{G}_{p-1}$. Let $\mathfrak{j}$ be the ideal of $\mathfrak{h}$ generated by G_0.

a) $[(\mathfrak{h}^2)^n, \underbrace{\mathfrak{h}, \ldots, \mathfrak{h}}_{p-1}] \subseteq (\mathfrak{h}^2)^{n+1} + \mathfrak{j}$ *for all $n \geq 1$.*

b) $\mathfrak{h}^{(p-1)n+2} \subseteq (\mathfrak{h}^2)^{n+1} + \mathfrak{j}$ *for all $n \geq 0$.*

c) $\mathfrak{h}^{\xi(n)+1} \subseteq \mathfrak{h}^{(n)} + \mathfrak{j}$ *for all* $n \geq 1$, *where*

$$\xi(n) = 1 + (p-1) + \cdots + (p-1)^{n-1}.$$

Proof. a) By 10.3d), $\mathfrak{h}^2 = \sum_{k=0}^{p-1} (\mathfrak{h}^2 \cap \mathsf{G}_k)$. It follows that

$$(\mathfrak{h}^2)^n = \sum [\mathfrak{h}^2 \cap \mathsf{G}_{i_1}, \ldots, \mathfrak{h}^2 \cap \mathsf{G}_{i_n}].$$

By 10.3b), the summand displayed here is contained in $(\mathfrak{h}^2)^n \cap \mathsf{G}_{i_1+\cdots+i_n}$; hence

$$(\mathfrak{h}^2)^n = \sum_{i=0}^{p-1} ((\mathfrak{h}^2)^n \cap \mathsf{G}_i).$$

To prove a), it must therefore be shown that

$$[(\mathfrak{h}^2)^n \cap \mathsf{G}_i, \mathsf{G}_{j_1}, \ldots, \mathsf{G}_{j_{p-1}}] \subseteq (\mathfrak{h}^2)^{n+1} + \mathfrak{j},$$

where $0 \leq i < p$ and $0 \leq j_k < p$. This is clear if any j_k is 0, since $\mathsf{G}_0 \subseteq \mathfrak{j}$. Suppose, then, that $0 < j_k < p$ for $k = 1, \ldots, p-1$. We show that if $u \in (\mathfrak{h}^2)^n \cap \mathsf{G}_i$ and $v_l \in \mathsf{G}_{j_l}$ $(l = 1, \ldots, p-1)$, then

$$[u, v_1, \ldots, v_{p-1}] \in (\mathfrak{h}^2)^{n+1} + \mathfrak{j}.$$

We observe first that $(\mathfrak{h}^2)^n$ is an ideal of $\mathfrak{h}$ by 8.5c); it therefore follows from 10.9 that if π is any permutation of $\{1, \ldots, p-1\}$,

$$[u, v_{1\pi}, \ldots, v_{(p-1)\pi}] - [u, v_1, \ldots, v_{p-1}] \in [(\mathfrak{h}^2)^n, \mathfrak{h}^2] = (\mathfrak{h}^2)^{n+1}.$$

It therefore suffices to prove that

$$[u, v_{1\pi}, \ldots, v_{(p-1)\pi}] \in (\mathfrak{h}^2)^{n+1} + \mathfrak{j}$$

for *some* permutation π. But by 10.8b), there exist $k_1, \ldots, k_s$ $(s \leq p-1)$ such that $k_1 < \cdots < k_s$ and $-i \equiv j_{k_1} + \cdots + j_{k_s}$ (p). Let π be a permutation of $\{1, \ldots, p\}$ such that $l\pi = k_l$ for $l = 1, \ldots, s$. Then

$$i + j_{1\pi} + \cdots + j_{s\pi} \equiv 0 \ (p);$$

by 10.3b),

$$[u, v_{1\pi}, \ldots, v_{s\pi}] \in [\mathsf{G}_i, \mathsf{G}_{j_{1\pi}}, \ldots, \mathsf{G}_{j_{s\pi}}] \subseteq \mathsf{G}_0 \subseteq \mathfrak{j}.$$

Thus $[u, v_{1\pi}, \ldots, v_{(p-1)\pi}] \in \mathfrak{j}$, as required.

b) This follows easily from a) by induction on n: for $n > 0$,

$$\mathfrak{h}^{(p-1)n+2} = [\mathfrak{h}^{(p-1)(n-1)+2}, \underbrace{\mathfrak{h}, \ldots, \mathfrak{h}}_{p-1}] \subseteq [(\mathfrak{h}^2)^n + \mathfrak{j}, \underbrace{\mathfrak{h}, \ldots, \mathfrak{h}}_{p-1}],$$

by the inductive hypothesis. The result follows by applying a).

c) This is also proved by induction on n. The case $n = 1$ is trivial since $\xi(1) = 1$. For $n > 1$, we apply the inductive hypothesis to $\mathfrak{h}^2$. Since $\mathfrak{h}^2 = \sum_{k=0}^{p-1}(\mathfrak{h}^2 \cap \mathsf{G}_k)$, this gives

$$(\mathfrak{h}^2)^{\xi(n-1)+1} \subseteq (\mathfrak{h}^2)^{(n-1)} + \mathfrak{j} = \mathfrak{h}^{(n)} + \mathfrak{j}.$$

But

$$\mathfrak{h}^{\xi(n)+1} = \mathfrak{h}^{2+(p-1)\xi(n-1)} \subseteq (\mathfrak{h}^2)^{\xi(n-1)+1} + \mathfrak{j}$$

by b), whence the assertion. **q.e.d.**

10.11 Theorem. *Suppose that* $\mathfrak{g}$ *is a Lie ring and that* α *is an automorphism of* $\mathfrak{g}$ *of prime order* p. *Let* $\mathfrak{j}$ *be an ideal containing all elements of* $\mathfrak{g}$ *which are invariant under* α. *Then*

a) $(p\mathfrak{g})^{\xi(n)+1} \subseteq \mathfrak{g}^{(n)} + \mathfrak{j}$ *for all* $n \geq 1$;

b) $p^{p-1}(p\mathfrak{g})^{\xi(2^{p-1}-1)+1} \subseteq p^{p-1}\mathfrak{j}$,

where $\xi(n) = 1 + (p-1) + \cdots + (p-1)^{n-1}$.

Proof. As in 10.5, it may be assumed that $\mathfrak{g}$ is a Lie algebra over $\mathsf{A} = \mathbb{Z}[\omega]$, where ω is a primitive p-th root of unity. Let $\mathfrak{h}$ be the subalgebra of $\mathfrak{g}$ defined in 10.3; thus $\mathfrak{h} \supseteq p\mathfrak{g}$. By 10.10c),

$$\mathfrak{h}^{\xi(n)+1} \subseteq \mathfrak{h}^{(n)} + \mathfrak{j}$$

for all $n \geq 1$, from which a) follows at once. It also follows from this that

$$p^{p-1}(p\mathfrak{g})^{\xi(2^{p-1}-1)+1} \subseteq p^{p-1}\mathfrak{h}^{\xi(2^{p-1}-1)+1} \subseteq p^{p-1}(\mathfrak{h}^{(2^{p-1}-1)} + \mathfrak{j}).$$

But by 10.4c),

$$p^{p-1}\mathfrak{h}^{(2^{p-1}-1)} \subseteq p^{p-1}\mathfrak{j}.$$

Hence $p^{p-1}(p\mathfrak{g})^{\xi(2^{p-1}-1)+1} \subseteq p^{p-1}\mathfrak{j}$. **q.e.d.**

10.12 Theorem (G. HIGMAN [1], KREKNIN and KOSTRIKIN [1]). *Suppose that the Lie ring* $\mathfrak{g}$ *possesses an automorphism* α *of prime order* p, *which leaves only the zero element of* $\mathfrak{g}$ *fixed. Then* $\mathfrak{g}$ *is nilpotent of class at most* $\xi(2^{p-1} - 1)$. *If* $\mathfrak{g}^{(d)} = 0$, *the class of* $\mathfrak{g}$ *is at most* $\xi(d)$. *Here* $\xi(n) = 1 + (p - 1) + \cdots + (p - 1)^{n-1}$.

Proof. By 10.11b),

$$p^{p-1}(p\mathfrak{g})^{\xi(2^{p-1}-1)+1} = 0.$$

It follows from 10.6b) that

$$\mathfrak{g}^{\xi(2^{p-1}-1)+1} = 0,$$

so $\mathfrak{g}$ is nilpotent of class at most $\xi(2^{p-1} - 1)$. If $\mathfrak{g}^{(d)} = 0$, then by 10.11a),

$$(p\mathfrak{g})^{\xi(d)+1} = 0.$$

Again the result follows from 10.6b). **q.e.d.**

10.13 Remarks. a) MEIXNER [1] has shown that in the previous theorems, $\xi(n)$ can be replaced by $(p - 1)^{n-1}$. To do this, the essential point is to replace 10.10b) by $\mathfrak{h}^{(p-1)n+1} \subseteq (\mathfrak{h}^2)^{n+1} + \mathfrak{j}$.

b) As we see from the case $p = 3$, the bound on the class given in 10.12 is far from being sharp. HIGMAN [1] showed that the bound is at least $\frac{1}{4}(p^2 - 1)$ for all $p \geq 5$; this has been shown to be sharp for $p = 5$ by Higman and for $p = 7$ by Scimemi.

We shall now use 9.3 and 9.4 to apply 10.11 and 10.12 to groups. For this the following lemma is needed.

10.14 Lemma. a) *Suppose that* α *is an automorphism of a finite group* $\mathfrak{G}$ *and that* $\mathfrak{N}$ *is a normal subgroup of* $\mathfrak{G}$ *for which* $\mathfrak{N}\alpha = \mathfrak{N}$. *Let* $\bar{\alpha}$ *be the automorphism of* $\mathfrak{G}/\mathfrak{N}$ *induced by* α. *If* $\mathfrak{C} = \{x | x \in \mathfrak{G}, x\alpha = x\}$ *and* $\overline{\mathfrak{C}} = \{\bar{x} | \bar{x} \in \mathfrak{G}/\mathfrak{N}, \bar{x}\bar{\alpha} = \bar{x}\}$, *then* $|\mathfrak{C}| \geq |\overline{\mathfrak{C}}|$.

b) *(V, 8.10) Suppose that* α *is an automorphism of a group* $\mathfrak{G}$, *which leaves no non-identity element of* $\mathfrak{G}$ *fixed. If* $\mathfrak{N}$ *is a normal subgroup of* $\mathfrak{G}$ *for which* $\mathfrak{N}\alpha = \mathfrak{N}$, *the automorphism of* $\mathfrak{G}/\mathfrak{N}$ *induced by* α *leaves fixed no non-identity element of* $\mathfrak{G}/\mathfrak{N}$.

Proof. a) Suppose that $\overline{\mathfrak{C}} = \mathfrak{D}/\mathfrak{N}$. If $x \in \mathfrak{D}$, put $x\beta = x^{-1}(x\alpha)$; thus β is a mapping of $\mathfrak{D}$ into $\mathfrak{N}$. Now β carries every element of the coset

$\mathfrak{C}x$ into $x^{-1}(x\alpha)$. Conversely, suppose that $x\beta = y\beta$. Then $x^{-1}(x\alpha) = y^{-1}(y\alpha)$, so $yx^{-1} = (yx^{-1})\alpha$, $yx^{-1} \in \mathfrak{C}$ and $y \in \mathfrak{C}x$. Hence $|\operatorname{im}\beta| = |\mathfrak{D}:\mathfrak{C}|$. Since $\operatorname{im}\beta \subseteq \mathfrak{N}$, $|\mathfrak{D}:\mathfrak{C}| \leq |\mathfrak{N}|$, and $|\overline{\mathfrak{C}}| = |\mathfrak{D}:\mathfrak{N}| \leq |\mathfrak{C}|$.

b) This follows at once from a). **q.e.d.**

10.15 Theorem. *Suppose that $\mathfrak{G}$ is a finite group and that $\mathfrak{G}$ has an automorphism α of prime order p, which leaves no non-identity element of $\mathfrak{G}$ fixed. Then $\mathfrak{G}$ is nilpotent. Let c be the class of $\mathfrak{G}$.*

a) *If $p = 2$, $c \leq 1$. If $p = 3$, $c \leq 2$. In general, $c \leq \xi(2^{p-1} - 1)$, where $\xi(n) = 1 + (p-1) + \cdots + (p-1)^{n-1}$.*

b) *If $\mathfrak{G}^{(d)} = 1$, $c \leq \xi(d)$.*

Proof. By V, 8.14, $\mathfrak{G}$ is nilpotent. Let c be the class of $\mathfrak{G}$.

By 9.3, there exists a Lie ring $\mathfrak{g} = \bigoplus_{i \geq 1} \mathsf{G}_i$ and isomorphisms σ_i of $\gamma_i(\mathfrak{G})/\gamma_{i+1}(\mathfrak{G})$ onto G_i $(i \geq 1)$ such that

$$(1) \qquad [(x\gamma_{i+1}(\mathfrak{G}))\sigma_i, (y\gamma_{j+1}(\mathfrak{G}))\sigma_j] = ([x, y]\gamma_{i+j+1}(\mathfrak{G}))\sigma_{i+j}$$

for all $x \in \gamma_i(\mathfrak{G})$, $y \in \gamma_j(\mathfrak{G})$; further $\mathfrak{g}$ possesses an automorphism ρ for which

$$(x\gamma_{i+1}(\mathfrak{G}))\sigma_i\rho = ((x\alpha)\gamma_{i+1}(\mathfrak{G}))\sigma_i \quad (x \in \gamma_i(\mathfrak{G})).$$

By 10.14b), α induces an automorphism on $\mathfrak{G}/\gamma_{i+1}(\mathfrak{G})$ which leaves no non-identity element fixed. Thus ρ leaves no non-zero element of G_i fixed. Hence ρ leaves no non-zero element of $\mathfrak{g}$ fixed. Also the order of ρ is p.

By 9.4, $\mathfrak{g}$ is nilpotent of class c. The assertions in a) about c now follow from 10.1, 10.2, 10.6 and 10.12. If $\mathfrak{G}^{(d)} = 1$, then $\mathfrak{g}^{(d)} = 0$ by 9.4d), so by 10.12, $c \leq \xi(d)$. **q.e.d.**

10.16 Remarks. a) The assertion of 10.15 for $p = 2$ can, of course, be proved much more easily (see V, 8.18). But it is not true for infinite groups: if $\mathfrak{F}$ is a free group with group-basis $\{x, y\}$ and α is the automorphism of $\mathfrak{F}$ such that $x\alpha = y$, $y\alpha = x$, then α leaves only the identity element fixed.

b) A direct proof of the assertions for $p = 3$ is given in BURNSIDE ([1], p. 90). The analogue of this proof for Lie rings is given in Exercises 15, 16.

c) Theorem 10.15 remains valid if the word "finite" is replaced by "nilpotent" (G. HIGMAN [1]).

d) In analogy with 10.7, it is conjectured that if the finite group $\mathfrak{G}$

possesses an automorphism α which leaves only the identity element fixed, then $\mathfrak{G}$ is soluble. This is true if the order n of α is a prime (by V, 8.14) or a power of 2 (by the Feit-Thompson theorem). An alternative proof in the case $n = 4$ was given by GORENSTEIN and HERSTEIN [1] (see GORENSTEIN [1], 10.4; also IX, 6.8).

It is also conjectured that if the finite group $\mathfrak{G}$ possesses a group $\mathfrak{A}$ of automorphisms such that $(|\mathfrak{G}|, |\mathfrak{A}|) = 1$ and only the identity element of $\mathfrak{G}$ is left fixed by every element of $\mathfrak{A}$, then $\mathfrak{G}$ is soluble. This will be discussed in Chapter X.

An application will now be given in which the number of fixed points is greater than 1. For this we need the following results.

10.17 Lemma. *Let p be a prime. Suppose that V is a finite Abelian p-group, that α is an automorphism of V of order p, and that the group of elements of V left fixed by α is of order p^n. Then V can be generated by pn elements.*

Proof. We write V in the additive notation. If B is a subgroup of V invariant under α, then $\mathsf{B}(\alpha - 1)$ is contained in B and is isomorphic to $\mathsf{B}/(\mathsf{B} \cap \ker(\alpha - 1))$, so $|\mathsf{B} : \mathsf{B}(\alpha - 1)| \leq |\ker(\alpha - 1)| = p^n$. It follows that $|\mathsf{V} : \mathsf{V}(\alpha - 1)^p| \leq p^{np}$. But

$$(\alpha - 1)^p = \alpha^p - p\alpha^{p-1} + \cdots + (-1)^p.$$

Hence

$$\mathsf{V}(\alpha - 1)^p \subseteq \mathsf{V}(\alpha^p - 1) + p\mathsf{V} = p\mathsf{V},$$

since p is the order of α. Thus $|\mathsf{V} : p\mathsf{V}| \leq p^{np}$. By III, 3.14, $p\mathsf{V}$ is the Frattini subgroup of V. Hence by III, 3.15, V can be generated by pn elements. **q.e.d.**

The following result may be compared with III, 7.10.

10.18 Theorem (P. HALL [1]). *Suppose that $\mathfrak{N}$ is a normal subgroup of the p-group $\mathfrak{G}$ and that $\mathfrak{N} \leq \gamma_n(\mathfrak{G})$. If $i > 1$ and $\gamma_{i+1}(\mathfrak{N}) \neq 1$, then $|\gamma_i(\mathfrak{N}) : \gamma_{i+1}(\mathfrak{N})| \geq p^n$.*

Proof. Since $\gamma_{i+1}(\mathfrak{N}) \neq 1$, there exists a normal subgroup $\mathfrak{M}$ of $\mathfrak{G}$ such that $\mathfrak{M} < \gamma_{i+1}(\mathfrak{N})$ and $|\gamma_{i+1}(\mathfrak{N}) : \mathfrak{M}| = p$. Thus $[\gamma_i(\mathfrak{N}), \mathfrak{N}] \nleq \mathfrak{M}$. It follows since $\mathfrak{N} \leq \gamma_n(\mathfrak{G})$ that $[\gamma_i(\mathfrak{N}), \gamma_n(\mathfrak{G})] \not\leq \mathfrak{M}$. Let $\mathfrak{Z}/\mathfrak{M}$ be the n-th term of the upper central series of $\mathfrak{G}/\mathfrak{M}$; thus by III, 2.11c), $[\mathfrak{Z}, \gamma_n(\mathfrak{G})] \leq \mathfrak{M}$. Hence $\gamma_i(\mathfrak{N}) \not\leq \mathfrak{Z}$. By III, 7.2b), $|\gamma_i(\mathfrak{N}) : \mathfrak{M}| \geq p^{n+1}$. Thus $|\gamma_i(\mathfrak{N}) : \gamma_{i+1}(\mathfrak{N})| \geq p^n$. **q.e.d.**

10.19 Theorem (ALPERIN [1]). *For each prime p and each non-negative integer n, there exists an integer $k(p, n)$ having the following property. If $\mathfrak{G}$ is a finite p-group, α is an automorphism of $\mathfrak{G}$ of order p and the group of elements of $\mathfrak{G}$ left fixed by α is of order p^n, then $\mathfrak{G}^{(k(p,n))} = 1$.*

Proof. Let $q = \xi(2^{p-1} - 1)$, where
$\xi(n) = 1 + (p - 1) + \cdots + (p - 1)^{n-1}$. Let d be the least integer such that $2^d > (n + p + q)pn$ and put $k(p, n) = d + q + 2$.

$$\begin{array}{ccc} \gamma_i(\mathfrak{H})/\gamma_{i+1}(\mathfrak{H}) & \xrightarrow{\sigma_i} & \mathsf{H}_i \\ \downarrow \alpha & & \downarrow \rho \\ \gamma_i(\mathfrak{H})/\gamma_{i+1}(\mathfrak{H}) & \xrightarrow{\sigma_i} & \mathsf{H}_i \end{array}$$

Suppose that α is an automorphism of order p of the finite p-group $\mathfrak{G}$ and that the group of elements of $\mathfrak{G}$ left fixed by α is of order p^n. Let $\mathfrak{H} = \mathfrak{G}^{(d)}$. By 9.3, there exist a Lie ring $\mathfrak{h} = \bigoplus_{i \geq 1} \mathsf{H}_i$ and isomorphisms σ_i of $\gamma_i(\mathfrak{H})/\gamma_{i+1}(\mathfrak{H})$ onto H_i; also the restriction of α to $\mathfrak{H}$ induces an automorphism ρ of $\mathfrak{h}$ such that $\sigma_i \rho \sigma_i^{-1}$ is the automorphism of $\gamma_i(\mathfrak{H})/\gamma_{i+1}(\mathfrak{H})$ induced by α. By 10.14a), the group of elements of H_i left fixed by ρ is of order at most p^n. It follows that if $a \in \mathfrak{h}$ and $a\rho = a$, then $p^n a = 0$. Let $\mathfrak{j} = \{b \in \mathfrak{h}, p^n b = 0\}$. Then $\mathfrak{j}$ is an ideal containing all elements of $\mathfrak{h}$ which are invariant under ρ. Hence by 10.11b),

$$p^{p-1}(p\mathfrak{h})^{q+1} \subseteq p^{p-1}\mathfrak{j}.$$

Hence $p^{n+p-1+(q+1)}\mathfrak{h}^{q+1} \subseteq p^{n+p-1}\mathfrak{j} = 0$. But by 9.4,
$\mathfrak{h}^{q+1} = \mathsf{H}_{q+1} + \mathsf{H}_{q+2} + \cdots$, so $p^{n+p+q}\mathsf{H}_{q+1} = 0$. By 10.17, H_{q+1} can be generated by np elements. Hence

$$|\gamma_{q+1}(\mathfrak{H}) : \gamma_{q+2}(\mathfrak{H})| = |\mathsf{H}_{q+1}| \leq p^{(n+p+q)pn} < p^{2^d}.$$

But by III, 2.12a), $\mathfrak{H} = \mathfrak{G}^{(d)} \leq \gamma_{2^d}(\mathfrak{G})$. Hence by 10.18, $\gamma_{q+2}(\mathfrak{H}) = 1$. Hence $\mathfrak{G}^{(k(p,n))} = \mathfrak{H}^{(q+2)} \leq \gamma_{q+2}(\mathfrak{H}) = 1$. **q.e.d.**

It is clear that the estimate given in the proof of 10.19 is in general a very crude one.

We conclude this section by showing that 10.19 is valid also for soluble groups.

10.20 Theorem (ALPERIN [1]). *For each prime p and each non-negative integer n, there exists an integer $l(p, n)$ having the following property. If $\mathfrak{G}$ is a finite soluble group, α is an automorphism of $\mathfrak{G}$ of order p, and the group of elements of $\mathfrak{G}$ left fixed by α is of order p^n, then $\mathfrak{G}^{(l(p,n))} = 1$.*

Proof. Let $\mathfrak{M} = \mathbf{O}_{p'}(\mathfrak{G})$, $\mathfrak{N} = \mathbf{O}_{p',p}(\mathfrak{G})$. Thus $\mathfrak{M}\alpha = \mathfrak{M}$, $\mathfrak{N}\alpha = \mathfrak{N}$, and if $\mathfrak{S} = \mathfrak{N}/\mathfrak{M}$, α induces an automorphism $\bar{\alpha}$ on the p-group $\mathfrak{S}$. By 10.14 a), the automorphism of $\mathfrak{S}/\Phi(\mathfrak{S})$ induced by $\bar{\alpha}$ leaves at most p^n elements fixed. Thus if $\mathfrak{F}/\mathfrak{M} = \Phi(\mathfrak{S})$, $\mathfrak{N}/\mathfrak{F}$ is an elementary Abelian p-group of order at most p^{pn}, by 10.17. By VI, 6.5, $\mathfrak{N}/\mathfrak{F}$ is self-centralizing; hence there exists a monomorphism of $\mathfrak{G}/\mathfrak{N}$ into the group of automorphisms of $\mathfrak{N}/\mathfrak{F}$. Thus $\mathfrak{G}/\mathfrak{N}$ is a soluble subgroup of $GL(r, p)$, where $r \leq pn$. There is therefore a bound on the derived length of $\mathfrak{G}/\mathfrak{N}$, which depends only on p and n. But by 10.19, the same is true of the p-group $\mathfrak{N}/\mathfrak{M}$, for α induces on $\mathfrak{N}/\mathfrak{M}$ an automorphism of order a divisor of p and this automorphism leaves at most p^n elements fixed, by 10.14a). Finally, since the group of elements left fixed by α is a p-group and $\mathfrak{M}$ is a p'-group, α leaves no non-identity element of $\mathfrak{M}$ fixed. Hence by 10.15, there is a bound on the derived length of $\mathfrak{M}$ which depends only on p. Combining these statements, there is a bound on the derived length of $\mathfrak{G}$ which depends only on p and n. **q.e.d.**

Exercises

15) Suppose that α is an automorphism of a Lie ring $\mathfrak{g}$. For any elements a, b of $\mathfrak{g}$, write

$$f(a) = [a, a\alpha], \quad g(a, b) = [a\alpha, b] + [b\alpha, a].$$

Verify the identity

$$g(a, [b\alpha - b, a]) - [f(a), b\alpha - b] - 3[g(a, b\alpha), a]$$
$$= [b(\alpha^2 + \alpha + 1), a\alpha, a] - 3[b\alpha^2, a, a].$$

16) Suppose that the Lie ring $\mathfrak{g}$ possesses an automorphism α of order 3, which leaves only the zero element fixed. Prove (using 10.6a)) that $f(a) = g(a, b) = 0$ for all a, b in $\mathfrak{g}$. Deduce that $\mathfrak{g}$ is nilpotent of class at most 2.

17) (G. HIGMAN) If the (associative) ring $\mathfrak{A}$ has an automorphism α of prime order p and α leaves no non-zero element of the ideal $\mathfrak{J}$ fixed, then $\mathfrak{J}^p = 0$. (Extend the ground-ring to $\mathbb{Z}[\omega]$, where ω is a primitive p-th root of unity. Show as in the Lie algebra case that if $\mathfrak{J}^p \neq 0$, there is a product $x_1 \cdots x_p \neq 0$ with $x_i\alpha = \omega^{k_i}x_i$. There exist r, s such that $k_r + \cdots + k_s \equiv 0\,(p)$.)

18) Let δ be a linear transformation of a vector space V over a field K. Show that $\mathsf{V} \oplus \mathsf{K}$ is a Lie algebra, if the product is defined by

$$[v_1 + \lambda_1, v_2 + \lambda_2] = \lambda_2(v_1\delta) - \lambda_1(v_2\delta).$$

19) (KREKNIN and KOSTRIKIN [1]) Let n be a composite integer. There exists a Lie algebra $\mathfrak{g}$ which is not nilpotent, such that $\mathfrak{g}$ has an automorphism α of order n, which leaves no non-zero element fixed. (If $n = lm$, $l > 1$, $m > 1$ and V is a vector space with basis $\{y_1, \ldots, y_m\}$ over the algebraically closed field K of characteristic 0, construct $\mathfrak{g} = \mathsf{V} \oplus \mathsf{K}$ as in **18)** with $y_i\delta = y_{i+1}$, $y_m\delta = y_1$. Let $(y_i + \lambda)\alpha = \omega^{1+(i-1)l}y_i + \lambda\omega^l$, where ω is a primitive n-th root of unity.)

20) (KREKNIN and KOSTRIKIN [1]) Let n be a composite integer. Given any integer c, there exists a nilpotent Lie algebra $\mathfrak{g}$ of class c and an automorphism α of $\mathfrak{g}$ of order n, such that α leaves no non-zero element of $\mathfrak{g}$ fixed. (Proceed as in **19)**, with V having a basis $\{y_i^j | 1 \leq j \leq c, 1 \leq i \leq m\}$ and $y_i^j\delta = y_{i+1}^{j+1}$, $y_m^j\delta = y_1^{j+1}$, $y_i^c\delta = 0$. Then $\mathfrak{g}^r$ is spanned by all y_i^j with $j \geq r$. Let $(y_i^j + \lambda)\alpha = \omega^{(i-1)l+1}y_i^j + \lambda\omega^l$.)

§ 11. The Lower Central Series of Free Groups

The Lie ring method was originally developed for its use in a theory, due mainly to Magnus and Witt, which gives some insight into the structure of the lower central series of a free group. This theory will be described in this section. It uses the fact that a free associative algebra can be embedded in the group-ring of a free group, and so it is necessary to study the Lie structure of the free associative algebra. The basic result used to do this is the Birkhoff-Witt theorem (11.2).

In order to unify the terminology, we shall use the word "basis" in connection with free groups and algebras. Thus a *group-basis* of a free group $\mathfrak{F}$ is a subset X of $\mathfrak{F}$ such that X generates $\mathfrak{F}$ and any mapping of X into a group is the restriction of some homomorphism of $\mathfrak{F}$.

We recall that with each associative algebra $\mathfrak{A}$ there is associated a Lie algebra $\mathfrak{l}(\mathfrak{A})$ (see 8.2).

11.1 Lemma. *Let $\mathfrak{g}$ be a Lie algebra over a commutative ring* A *and let $\mathfrak{A}$ be an associative* A*-algebra. Suppose that α is a Lie homomorphism of $\mathfrak{g}$ into $\mathfrak{l}(\mathfrak{A})$ and that $\mathfrak{A}$ is the associative algebra generated by* $\operatorname{im}\alpha$. *For $g \in \mathfrak{g}$, write $\bar{g} = g\alpha$. Let* B *be a fully ordered set such that $\mathfrak{g}$ is the* A*-module spanned by* B.

a) *For each integer $n \geq 0$, let V_n be the* A*-module spanned by all products $\bar{g}_1 \cdots \bar{g}_m$ with $g_i \in \mathfrak{g}$ and $m \leq n$. (In particular, $V_0 = A1$.) Given $g_i \in \mathfrak{g}$ $(i = 1, \ldots, n; n > 0)$, then*

$$\bar{g}_{1\pi} \cdots \bar{g}_{n\pi} - \bar{g}_1 \cdots \bar{g}_n \in V_{n-1}$$

for any permutation π of $\{1, \ldots, n\}$. Also for $n > 0$, V_n is the A*-module spanned by V_{n-1} and all products $\bar{b}_1 \cdots \bar{b}_n$, where $b_i \in B$ and $b_1 \leq \cdots \leq b_n$.*

b) *$\mathfrak{A}$ is the* A*-module spanned by all products $\bar{b}_1 \cdots \bar{b}_n$, where $n \geq 0$, $b_i \in B$ and $b_1 \leq \cdots \leq b_n$.*

c) *If $\mathfrak{g}$ is the Lie algebra generated by X, $\mathfrak{A}$ is the associative algebra generated by $X\alpha$.*

Proof. a) To prove the first assertion, it is clearly sufficient to show that

$$\bar{g}_1 \cdots \bar{g}_{i+1}\bar{g}_i \cdots \bar{g}_n - \bar{g}_1 \cdots \bar{g}_i\bar{g}_{i+1} \cdots \bar{g}_n \in V_{n-1}.$$

But $\bar{g}_{i+1}\bar{g}_i - \bar{g}_i\bar{g}_{i+1} = [g_{i+1}\alpha, g_i\alpha] = [g_{i+1}, g_i]\alpha \in V_1$, so this is clear.

Since every element of $\mathfrak{g}$ is an A-linear combination of elements of B, V_n is spanned by V_{n-1} and all products $\bar{b}_1 \cdots \bar{b}_n$ with $b_i \in B$. By the previous statement, we may add the condition $b_1 \leq b_2 \leq \cdots \leq b_n$.

b) Since $\mathfrak{A}$ is the associative algebra generated by all $\bar{g}$ $(g \in \mathfrak{g})$, $\mathfrak{A}$ is the A-module spanned by all V_n $(n \geq 0)$. The assertion thus follows from a).

c) If $\mathfrak{A}^*$ is the associative algebra generated by $X\alpha$, then $\operatorname{im}\alpha \subseteq \mathfrak{A}^*$, so all $\bar{b}_1 \cdots \bar{b}_n \in \mathfrak{A}^*$. Thus $\mathfrak{A}^* = \mathfrak{A}$ by b). **q.e.d.**

11.2 Theorem (BIRKHOFF-WITT). *Let $\mathfrak{g}$ be a Lie algebra over a commutative ring* A. *Suppose that $\mathfrak{g}$ has an* A*-basis* B *and that* B *is fully ordered. Then there exists an associative algebra $\mathfrak{A}$ over* A *and a monomorphism ε of $\mathfrak{g}$ into $\mathfrak{l}(\mathfrak{A})$ such that*

$$\mathsf{B}^* = \{(b_1\varepsilon)\cdots(b_n\varepsilon)\,|\,n \geq 0, b_i \in \mathsf{B}, b_1 \leq \cdots \leq b_n\}$$

is an A*-basis of* $\mathfrak{A}$.

Proof. Let $\mathfrak{P}$ be the A-algebra of polynomials in a set B of independent indeterminates for which there exists a bijective mapping $b \to \bar{b}$ of B onto B. We assert that it is sufficient to define an A-bilinear mapping (,) on $\mathfrak{P} \times \mathfrak{g}$ into $\mathfrak{P}$ such that

a) $(\bar{b}_1 \cdots \bar{b}_n, b) = \bar{b}_1 \cdots \bar{b}_n\bar{b}$ if $b_1 \leq \cdots \leq b_n \leq b \quad (b_i \in \mathsf{B}, b \in \mathsf{B})$, and

b) $(f, [b, b']) = ((f, b), b') - ((f, b'), b) \quad (f \in \mathfrak{P}, b \in \mathsf{B}, b' \in \mathsf{B})$.

For suppose that such a mapping exists. Given $g \in \mathfrak{g}$, $f \to (f, g)$ is an element $g\varepsilon$ of $\mathrm{Hom}_{\mathsf{A}}(\mathfrak{P}, \mathfrak{P})$; thus $f(g\varepsilon) = (f, g)$. Now ε is a Lie homomorphism of $\mathfrak{g}$ into $\mathrm{I}(\mathrm{Hom}_{\mathsf{A}}(\mathfrak{P}, \mathfrak{P}))$, for

$$\begin{aligned} f([g_1, g_2]\varepsilon) &= (f, [g_1, g_2]) \\ &= ((f, g_1), g_2) - ((f, g_2), g_1) \quad (\text{by } b)) \\ &= f((g_1\varepsilon)(g_2\varepsilon) - (g_2\varepsilon)(g_1\varepsilon)) \\ &= f[g_1\varepsilon, g_2\varepsilon]. \end{aligned}$$

By 11.1, the associative algebra $\mathfrak{A}$ generated by $\mathrm{im}\,\varepsilon$ is spanned over A by all $(b_1\varepsilon)\cdots(b_n\varepsilon)$ with $b_i \in \mathsf{B}$ and $b_1 \leq \cdots \leq b_n$. In fact, these elements form an A-basis of $\mathfrak{A}$, for the elements $\bar{b}_1 \cdots \bar{b}_n$ with $b_1 \leq \cdots \leq b_n$ are linearly independent and

$$1(b_1\varepsilon)\cdots(b_n\varepsilon) = \bar{b}_1 \cdots \bar{b}_n,$$

as is easily seen from *a)* by induction on n. Finally, ε is a monomorphism, since the $1(b\varepsilon) = \bar{b}$ are certainly linearly independent.

For each $n \geq 0$, let Q_n denote the A-module spanned by all $\bar{b}_1 \cdots \bar{b}_m$ with $m \leq n$. We shall define a bilinear mapping $(\ ,\ )_n$ on $\mathsf{Q}_n \times \mathfrak{g}$ into $\mathfrak{P}$ by induction on n, and it will satisfy the following.

a_n) $(\bar{b}_1 \cdots \bar{b}_m, b)_n = \bar{b}_1 \cdots \bar{b}_m\bar{b}$ if $m \leq n$ and $b_1 \leq \cdots \leq b_m \leq b$.

b_n) If $m \leq n$ and $f \in \mathsf{Q}_m$, $(f, b)_n - f\bar{b} \in \mathsf{Q}_m$; hence $(f, b)_n \in \mathsf{Q}_{m+1}$.

c_n) If $f \in \mathsf{Q}_{n-1}$, $(f, [b, b'])_n = ((f, b)_n, b')_n - ((f, b')_n, b)_n$.

d_n) On $\mathsf{Q}_{n-1} \times \mathfrak{g}$, $(\ ,\ )_n$ coincides with $(\ ,\ )_{n-1}$.

It is clearly sufficient to do this, for then by d_n), a form may be defined on $\mathfrak{P} \times \mathfrak{g}$ which satisfies *a)* and *b)* on account of a_n) and c_n).

For $n = 0$, put $(\lambda, b)_0 = \lambda b \ (\lambda \in \mathsf{A})$. Then a_0), b_0) are satisfied trivially, c_0), d_0) vacuously.

For $n > 0$, we define $(f, g)_n$ for $f \in \mathsf{Q}_{n-1}$ and $f = \bar{b}_1 \cdots \bar{b}_n$; the definition is then completed by linearity. For $f \in \mathsf{Q}_{n-1}$, put $(f, g)_n = (f, g)_{n-1}$.

If $b_1 \le \cdots \le b_n \le b$, put $(\bar{b}_1 \cdots \bar{b}_n, b)_n = \bar{b}_1 \cdots \bar{b}_n \bar{b}$. If $b_1 \le \cdots \le b_n$ and $b < b_n$, then by b_{n-1}), $u = (\bar{b}_1 \cdots \bar{b}_{n-1}, b)_{n-1} - \bar{b}_1 \cdots \bar{b}_{n-1}\bar{b} \in \mathsf{Q}_{n-1}$, and we may put

$$(\bar{b}_1 \cdots \bar{b}_n, b)_n = \bar{b}_1 \cdots \bar{b}_n \bar{b} + (\bar{b}_1 \cdots \bar{b}_{n-1}, [b_n, b])_{n-1} + (u, b_n)_{n-1}.$$

Then $(\ ,\ )_n$ is bilinear and a_n), b_n), d_n) are satisfied trivially. It only remains to verify c_n) with $f = \bar{b}_1 \cdots \bar{b}_{n-1}$. We may suppose that $b < b'$ and $b_1 \le \cdots \le b_{n-1}$.

We prove it first when $b_{n-1} \le b'$. The definition then gives

$$\begin{aligned}((\bar{b}_1 \cdots \bar{b}_{n-1}, b')_n, b)_n &= (\bar{b}_1 \cdots \bar{b}_{n-1}\bar{b}', b)_n \\ &= \bar{b}_1 \cdots \bar{b}_{n-1}\bar{b}\bar{b}' + (\bar{b}_1 \cdots \bar{b}_{n-1}, [b', b])_{n-1} \\ &\quad + (u_1 - u_2, b')_{n-1},\end{aligned}$$

where $u_1 = (\bar{b}_1 \cdots \bar{b}_{n-1}, b)_n \in \mathsf{Q}_n$, $u_2 = \bar{b}_1 \cdots \bar{b}_{n-1}\bar{b} \in \mathsf{Q}_n$ and $u_1 - u_2 \in \mathsf{Q}_{n-1}$. Hence $(u_2, b')_n = \bar{b}_1 \cdots \bar{b}_{n-1}\bar{b}\bar{b}'$ and

$$\begin{aligned}((\bar{b}_1 \cdots \bar{b}_{n-1}, b')_n, b)_n &= \bar{b}_1 \cdots \bar{b}_{n-1}\bar{b}\bar{b}' + (\bar{b}_1 \cdots \bar{b}_{n-1}, [b', b])_{n-1} \\ &\quad + (u_1, b')_n - (u_2, b')_n \\ &= ((\bar{b}_1 \cdots \bar{b}_{n-1}, b)_n, b')_n - (\bar{b}_1 \cdots \bar{b}_{n-1}, [b, b'])_{n-1},\end{aligned}$$

as required.

Now suppose that $b < b' < b_{n-1}$. Then $n > 1$. We put $f = \bar{b}_1 \cdots \bar{b}_{n-2}$, $a = b_{n-1}$ and $c = b'$. We then have $f \in \mathsf{Q}_{n-2}$, $b < a$ and $c < a$, and we must prove that

$$(f\bar{a}, [b, c])_n = ((f\bar{a}, b)_n, c)_n - ((f\bar{a}, c)_n, b)_n.$$

By a_{n-1}), c_{n-1}) and b_{n-1}) we have

$$\begin{aligned}(f\bar{a}, b)_n &= ((f, a)_{n-1}, b)_{n-1} \\ &= ((f, b)_{n-1}, a)_{n-1} + (f, [a, b])_{n-1} \\ &= (v + f\bar{b}, a)_{n-1} + (f, [a, b])_{n-1},\end{aligned}$$

where $v = (f, b)_{n-1} - f\bar{b} \in \mathsf{Q}_{n-2}$. Hence

$$((f\bar{a}, b)_n, c)_n = ((v, a)_{n-1}, c)_{n-1} + ((f\bar{b}, a)_n, c)_n + ((f, [a, b])_{n-1}, c)_{n-1}.$$

We apply c_{n-1}) to the first and third terms, and we apply the case of c_n) already proved to the second. Thus

$$\begin{aligned}((f\bar{a}, b)_n, c)_n &= ((v, c)_n, a)_n + (v, [a, c])_n + ((f\bar{b}, c)_n, a)_n \\ &\quad + (f\bar{b}, [a, c])_n + ((f, c)_n, [a, b])_n + (f, [a, b, c])_n.\end{aligned}$$

Since $v + f\bar{b} = (f, b)_{n-1}$,

$$\begin{aligned}((f\bar{a}, b)_n, c)_n &= (((f, b)_n, c)_n, a)_n + ((f, b)_n, [a, c])_n \\ &\quad + ((f, c)_n, [a, b])_n + (f, [a, b, c])_n.\end{aligned}$$

Next we interchange b and c and subtract. This gives

$$\begin{aligned}&((f\bar{a}, b)_n, c)_n - ((f\bar{a}, c)_n, b)_n \\ &= (((f, b)_n, c)_n, a)_n - (((f, c)_n, b)_n, a)_n + (f, [a, b, c] - [a, c, b])_n.\end{aligned}$$

But by two applications of c_{n-1}),

$$\begin{aligned}(((f, b)_n, c)_n, a)_n - (((f, c)_n, b)_n, a)_n &= ((f, [b, c])_n, a)_{n-1} \\ &= ((f, a)_n, [b, c])_n + (f, [b, c, a])_n.\end{aligned}$$

Hence

$$\begin{aligned}&((f\bar{a}, b)_n, c)_n - ((f\bar{a}, c)_n, b)_n \\ &= ((f, a)_n, [b, c])_n + (f, [b, c, a] + [c, a, b] + [a, b, c])_n \\ &= ((f, a)_n, [b, c])_n = (f\bar{a}, [b, c])_n,\end{aligned}$$

by the Jacobi identity. **q.e.d.**

Theorem 11.2 is not true if it is not assumed that $\mathfrak{g}$ is a free A-module. A counterexample was given by CARTIER [1].

11.3 Definitions. a) Let $\mathfrak{A}$ be an associative algebra over A. The subset X of $\mathfrak{A}$ is called an *associative basis* of $\mathfrak{A}$ if (i) $\mathfrak{A}$ is the associative algebra generated by X, and (ii) any mapping of X into any associative algebra over A is the restriction of a homomorphism of $\mathfrak{A}$. An associative algebra which possesses an associative basis is called a *free associative algebra.*

b) Let $\mathfrak{g}$ be a Lie algebra over A. The subset X of $\mathfrak{g}$ is called a *Lie basis* of $\mathfrak{g}$ if (i) $\mathfrak{g}$ is the Lie algebra generated by X, and (ii) any mapping of X into any Lie algebra over A is the restriction of a homomorphism of $\mathfrak{g}$. A Lie algebra which possesses a Lie basis is called a *free Lie algebra.*

11.4 Lemma. a) *Let $\mathfrak{L}$, $\mathfrak{M}$ be Lie or associative algebras over A and let ρ be a homomorphism of $\mathfrak{L}$ into $\mathfrak{M}$. Suppose that X generates $\mathfrak{L}$, that ρ is injective on X and that $X\rho$ is a Lie or associative basis of $\mathfrak{M}$. Then ρ is an isomorphism of $\mathfrak{L}$ onto $\mathfrak{M}$.*

b) *Let $\mathfrak{L}$, $\mathfrak{M}$ be free Lie or associative algebras over A. Suppose that X is a basis of $\mathfrak{L}$, Y is a basis of $\mathfrak{M}$ and that α is an injective mapping of X onto Y. Then α is the restriction of an isomorphism of $\mathfrak{L}$ onto $\mathfrak{M}$.*

Proof. a) Since $X\rho$ is a basis of $\mathfrak{M}$, $X\rho$ generates $\mathfrak{M}$. Hence ρ is an epimorphism. Since $X\rho$ is a basis of $\mathfrak{M}$ and ρ is injective on X, there exists a homomorphism α of $\mathfrak{M}$ into $\mathfrak{L}$ such that $\rho\alpha$ is the identity mapping on X. Since X generates $\mathfrak{L}$, it follows that $\rho\alpha = 1$. Hence ρ is also a monomorphism.

b) Since X is a basis of $\mathfrak{L}$, α is the restriction of a homomorphism ρ of $\mathfrak{L}$ into $\mathfrak{M}$. By a), ρ is an isomorphism. **q.e.d.**

11.5 Remarks. a) Given a set X, let $\mathfrak{B}$ be a free A-module with basis $\{(x_1, \ldots, x_n) | n \geq 0, x_i \in X\}$. It is easy to see that $\mathfrak{B}$ has the structure of an associative algebra in which $(x_1, \ldots, x_m)(x_{m+1}, \ldots, x_n) = (x_1, \ldots, x_n)$ and that $\mathfrak{B}$ is a free associative algebra with basis $\{(x) | x \in X\}$. Thus there exists a free associative algebra having a basis with given cardinal number. By 11.4b), it is unique to within isomorphism.

b) The proof of the existence of free Lie algebras is not quite so simple. In fact, however, if $\mathfrak{A}$ is a free associative algebra over A with basis X, the Lie subalgebra $\mathfrak{f}$ of $\mathrm{I}(\mathfrak{A})$ generated by X is a free Lie algebra with basis X. This is proved in 11.10 for the case $\mathsf{A} = \mathbb{Z}$. In the case when A is a field K, it can be deduced from the Birkhoff-Witt theorem as follows.

Suppose that α is a mapping of X into a Lie algebra $\mathfrak{g}$. Since K is a field $\mathfrak{g}$ has a K-basis. Hence by 11.2, there is a monomorphism ε of $\mathfrak{g}$ into $\mathrm{I}(\mathfrak{B})$ for some associative algebra $\mathfrak{B}$. By 11.3a), $\alpha\varepsilon$ is the restriction to X of an associative homomorphism ρ of $\mathfrak{A}$ into $\mathfrak{B}$.

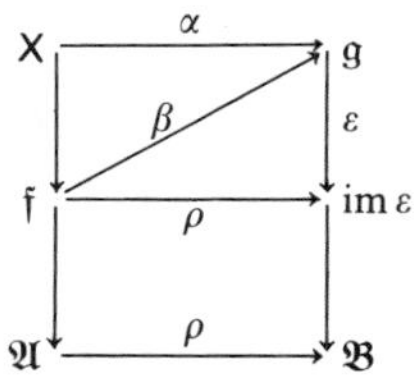

Now $\operatorname{im}\varepsilon$ is a Lie subalgebra of $\mathfrak{l}(\mathfrak{B})$ containing $X\alpha\varepsilon = X\rho$. Since $\mathfrak{f}$ is the Lie algebra generated by X, it follows that $\operatorname{im}\varepsilon \supseteq \mathfrak{f}\rho$. Since ε is a monomorphism, there exists a mapping β of $\mathfrak{f}$ into $\mathfrak{g}$ such that $\beta\varepsilon = \rho$. β is a Lie homomorphism, since if a, b are in $\mathfrak{f}$,

$$([a, b]\beta - [a\beta, b\beta])\varepsilon = [a, b]\rho - [a\rho, b\rho] = 0.$$

And the restriction of β to X is α, since if $x \in X$, $x\beta\varepsilon = x\rho = x\alpha\varepsilon$.

c) Let $\mathfrak{g}$ be a Lie algebra over A. By a), there exists a free associative algebra $\mathfrak{F}$ having a basis X for which there exists an injective mapping ξ of $\mathfrak{g}$ onto X. Let $\mathfrak{J}$ be the ideal of $\mathfrak{F}$ generated by all $[g, h]\xi - [g\xi, h\xi]$, as g, h run through $\mathfrak{g}$. The algebra $\mathfrak{E}(\mathfrak{g}) = \mathfrak{F}/\mathfrak{J}$ is called the *universal enveloping algebra* of $\mathfrak{g}$. Let α be the mapping of $\mathfrak{g}$ into $\mathfrak{E}(\mathfrak{g})$ given by $g\alpha = g\xi + \mathfrak{J}$; thus α is a Lie homomorphism of $\mathfrak{g}$ into $\mathfrak{l}(\mathfrak{E}(\mathfrak{g}))$. $\mathfrak{E}(\mathfrak{g})$ is characterized by the following properties.

(i) $\mathfrak{E}(\mathfrak{g})$ is generated by $\operatorname{im}\alpha$.

(ii) If $\mathfrak{A}$ is an associative algebra and β is a Lie homomorphism of $\mathfrak{g}$ into $\mathfrak{l}(\mathfrak{A})$, $\beta = \alpha\eta$ for some associative homomorphism η of $\mathfrak{E}(\mathfrak{g})$ into $\mathfrak{A}$.

The Birkhoff-Witt theorem asserts that if B is a fully ordered A-basis of $\mathfrak{g}$,

$$\{(b_1\alpha)\cdots(b_n\alpha) \mid n \geq 0,\ b_i \in \mathsf{B},\ b_1 \leq \cdots \leq b_n\}$$

is an A-basis of $\mathfrak{E}(\mathfrak{g})$.

11.6 Definition. Let $\mathfrak{A}$ be a free associative algebra over A with basis X. Let $\mathsf{H}_0 = \mathsf{A}1$, and for each integer $n > 0$, let H_n be the A-module generated by all products $x_1 \cdots x_n$ with $x_i \in X$. H_n is called *the homogeneous component of* $\mathfrak{A}$ *with respect to* X *of degree* n. The H_n have the following properties.

(i) H_n is a free A-module with basis $\{x_1 \cdots x_n \mid x_i \in X\}$.

(ii) $\mathfrak{A} = \mathsf{H}_0 \oplus \mathsf{H}_1 \oplus \mathsf{H}_2 \oplus \cdots$.

(iii) If $a \in \mathsf{H}_m$ and $b \in \mathsf{H}_n$, then $ab \in \mathsf{H}_{m+n}$. If further A is an integral domain and $a \neq 0$, $b \neq 0$, then $ab \neq 0$.

Of these, (iii) is an immediate consequence of (i) and (ii). To prove these, observe that they hold trivially in the algebra $\mathfrak{B}$ constructed in 11.5a) and that, by 11.4b), there is an isomorphism between $\mathfrak{B}$ and $\mathfrak{A}$ in which (x) and x correspond $(x \in X)$.

11.7 Theorem. a) *Given a set* X, *there exists a free Lie algebra over* A *with a Lie basis in* $(1, 1)$ *correspondence with* X.

b) *Let* $\mathfrak{g}$ *be a free Lie algebra with Lie basis* X. *For each* $n \geq 1$, *let* U_n *be the* A*-module spanned by all products* $[a_1, \ldots, a_n]$ *with* $a_i \in X$. *Then* $\mathfrak{g}^n = \mathsf{U}_n \oplus \mathsf{U}_{n+1} \oplus \cdots$ *and* $\bigcap_{n \geq 1} \mathfrak{g}^n = 0$.

Proof. By 11.4b), it is sufficient to prove b) for a free Lie algebra with a basis in (1, 1) correspondence with X. Both parts of the theorem will therefore be proved if we (i) establish the existence of a free Lie algebra $\mathfrak{f}$ having a Lie basis $\overline{X}$ in (1, 1) correspondence with X, and (ii) prove that $\mathfrak{f}$ and $\overline{X}$ satisfy the assertion of b).

First we construct the non-associative words in X. For $n \geq 1$, we define a set X_n by induction on n, with $X_1 = X$ and

$$X_n = \bigcup_{i=1}^{n-1} X_i \times X_{n-i}$$

for $n > 1$. Thus $X_m \cap X_n = \emptyset$ for $m \neq n$. Let M_n be the free A-module with basis X_n and let

$$\mathfrak{M} = \mathsf{M}_1 \oplus \mathsf{M}_2 \oplus \cdots.$$

$\mathfrak{M}$ is thus the free A-module with basis $\mathsf{X} = \bigcup_{n \geq 1} X_n$, and if $u \in \mathsf{X}$, $v \in \mathsf{X}$, the ordered pair (u, v) also lies in X. We define multiplication in $\mathfrak{M}$ by putting

$$\left(\sum_{u \in \mathsf{X}} \lambda_u u\right) \cdot \left(\sum_{v \in \mathsf{X}} \mu_v v\right) = \sum_{u, v} \lambda_u \mu_v (u, v).$$

Obviously the distribution laws hold.

We define a set $\mathscr{S}$ of A-submodules of $\mathfrak{M}$ as follows: the A-submodule N lies in $\mathscr{S}$ if

(iii) $\eta(a) = a \cdot a$ and $\zeta(a, b, c) = ((a \cdot b) \cdot c) + ((b \cdot c) \cdot a) + ((c \cdot a) \cdot b)$ lie in N for all a, b, c in $\mathfrak{M}$; and

(iv) $a \cdot b \in \mathsf{N}$ and $b \cdot a \in \mathsf{N}$ if $a \in \mathsf{N}$ and $b \in \mathfrak{M}$.

$\mathscr{S}$ is non-empty since $\mathfrak{M} \in \mathscr{S}$. Let $\mathfrak{J} = \bigcap_{\mathsf{N} \in \mathscr{S}} \mathsf{N}$. Then $\mathfrak{J}$ also satisfies conditions (iii) and (iv). From (iv) it follows that there exists a multiplication in the module $\mathfrak{M}/\mathfrak{J}$ in which

$$[a + \mathfrak{J}, b + \mathfrak{J}] = a \cdot b + \mathfrak{J}.$$

The distributive laws hold, and by (iii), $\mathfrak{M}/\mathfrak{J}$ is a Lie algebra $\mathfrak{f}$.

Define a mapping ξ of X into $\mathfrak{f}$ by putting $x\xi = x + \mathfrak{J}$ $(x \in X)$. We show that $X\xi$ generates $\mathfrak{f}$. Certainly the subalgebra of $\mathfrak{f}$ generated by $X\xi$ is an A-submodule of $\mathfrak{M}/\mathfrak{J}$ and is thus of the form $\mathsf{U}/\mathfrak{J}$ for some A-submodule U of $\mathfrak{M}$. If $\mathsf{U} \neq \mathfrak{M}$, there exists an integer n such that $X_n \nsubseteq \mathsf{U}$. Let n be the smallest such integer. Then $n \neq 1$, for if $x \in X$, then $x\xi \in \mathsf{U}/\mathfrak{J}$ and $x \in \mathsf{U}$. Suppose that $w \in X_n$, $w \notin \mathsf{U}$. Since $n > 1$, $w = (u, v)$ for $u \in X_i$, $v \in X_{n-i}$ $(1 \leq i \leq n - 1)$. Then $u \in \mathsf{U}$, $v \in \mathsf{U}$, so $u + \mathfrak{J} \in \mathsf{U}/\mathfrak{J}$, $v + \mathfrak{J} \in \mathsf{U}/\mathfrak{J}$, and since $\mathsf{U}/\mathfrak{J}$ is a subalgebra, $[u + \mathfrak{J}, v + \mathfrak{J}] \in \mathsf{U}/\mathfrak{J}$. Thus $u \cdot v \in \mathsf{U}$, and since $u \cdot v = (u, v) = w$, we have a contradiction. Hence $\mathsf{U} = \mathfrak{M}$ and $\mathfrak{f}$ is generated by $X\xi$.

We now show that given a mapping β of X into a Lie algebra $\mathfrak{g}$ over A, there exists a homomorphism σ of $\mathfrak{f}$ into $\mathfrak{g}$ such that $\beta = \xi\sigma$. To do this we first define a mapping β_n of X_n into $\mathfrak{g}$ by induction on n, with $\beta_1 = \beta$ and $(u, v)\beta_n = [u\beta_i, v\beta_{n-i}]$ if $u \in X_i$, $v \in X_{n-i}$. There exists an A-module homomorphism ρ of $\mathfrak{M}$ into $\mathfrak{g}$ such that $u\rho = u\beta_n$ for all $u \in X_n$. Thus $(u \cdot v)\rho = [u\rho, v\rho]$ if $u \in X_m$, $v \in X_n$, and, indeed, if u, v are any elements of $\mathfrak{M}$. It follows easily that $\ker \rho \in \mathscr{S}$. Thus $\ker \rho \supseteq \mathfrak{J}$ and there exists an A-homomorphism σ of $\mathfrak{M}/\mathfrak{J} = \mathfrak{f}$ into $\mathfrak{g}$ such that $(a + \mathfrak{J})\sigma = a\rho$ for all $a \in \mathfrak{M}$. It is easy to see that σ is a Lie algebra homomorphism and that $x\xi\sigma = x\rho = x\beta$ for all $x \in X$.

By choosing β to be injective, we see that ξ is necessarily injective. Thus if $\overline{X} = X\xi$, $\overline{X}$ is a Lie basis of $\mathfrak{f}$ in (1, 1) correspondence with X. Hence (i) is proved.

To prove (ii) let $\mathsf{J}_n = \mathfrak{J} \cap \mathsf{M}_n$ and let

$$\mathsf{J} = \mathsf{J}_1 \oplus \mathsf{J}_2 \oplus \cdots.$$

Thus J is an A-submodule of $\mathfrak{J}$, and J satisfies (iv), for if $a \in \mathsf{J}_m$ and $b \in \mathsf{M}_n$, then $a \cdot b \in \mathfrak{J} \cap \mathsf{M}_{m+n} = \mathsf{J}_{m+n}$ and $b \cdot a \in \mathsf{J}_{m+n}$. Also J satisfies (iii). Indeed $\zeta(a, b, c)$ is a linear combination of elements $\zeta(a', b', c')$ with $a' \in \mathsf{M}_l$, $b' \in \mathsf{M}_m$, $c' \in \mathsf{M}_n$, and $\zeta(a', b', c') \in \mathsf{M}_{l+m+n} \cap \mathfrak{J}$. As for $\eta(a)$, $\eta(a)$ is a linear combination of elements $\eta(a')$ and $\overline{\eta}(a', b') = [a', b'] + [b', a']$ with $a' \in \mathsf{M}_l$, $b' \in \mathsf{M}_m$. Now $\eta(a') \in \mathfrak{J} \cap \mathsf{M}_{2l}$. Also $\overline{\eta}(a', b') \in \mathsf{M}_{l+m}$, and

$$\overline{\eta}(a', b') = \eta(a' + b') - \eta(a') - \eta(b') \in \mathfrak{J},$$

so $\overline{\eta}(a', b') \in \mathfrak{J} \cap \mathsf{M}_{l+m}$. Thus J satisfies (iii). Hence $\mathsf{J} \supseteq \mathfrak{J}$. Since obviously $\mathsf{J} \subseteq \mathfrak{J}$, we have $\mathsf{J} = \mathfrak{J}$ and

$$\mathfrak{J} = \mathsf{J}_1 \oplus \mathsf{J}_2 \oplus \cdots.$$

It follows at once that

$$\mathfrak{f} = \mathfrak{M}/\mathfrak{J} = (\mathsf{M}_1 + \mathfrak{J}/\mathfrak{J}) \oplus (\mathsf{M}_2 + \mathfrak{J}/\mathfrak{J}) \oplus \cdots.$$

Now let V_n be the A-module spanned by all products $[b_1, \ldots, b_n]$ with $b_i \in \overline{X}$. By a trivial induction, $[b_1, \ldots, b_n] = u + \mathfrak{J}$ for some $u \in X_n$, so $\mathsf{V}_n \subseteq (\mathsf{M}_n + \mathfrak{J})/\mathfrak{J}$. By 8.3b), $\mathfrak{f} = \mathsf{V}_1 + \mathsf{V}_2 + \cdots$. Hence $\mathfrak{f} = \mathsf{V}_1 \oplus \mathsf{V}_2 \oplus \cdots$. By 8.7b), $\mathfrak{f}^n = \mathsf{V}_n \oplus \mathsf{V}_{n+1} \oplus \cdots$. It follows at once that $\bigcap_{n \geq 1} \mathfrak{f}^n = 0$, so (ii) is proved. **q.e.d.**

11.8 Theorem. *Let $\mathfrak{A}$ be a free associative algebra over $\mathbb{Z}$ with associative basis X and let $\mathfrak{g}$ be the Lie subalgebra of $\mathfrak{l}(\mathfrak{A})$ generated by X. For $n \geq 0$, let H_n be the homogeneous component of $\mathfrak{A}$ with respect to X of degree n (11.6), and let $\mathsf{G}_n = \mathfrak{g} \cap \mathsf{H}_n$.*

a) $\mathfrak{g} = \mathsf{G}_1 \oplus \mathsf{G}_2 \oplus \cdots$, *and G_n is the additive group generated by all products $[x_1, \ldots, x_n]$ with $x_i \in X$.*

b) *G_n and $\mathsf{H}_n/\mathsf{G}_n$ are free Abelian groups.*

c) *If X is finite, G_n is a free Abelian group of rank $d_n(|X|)$. Here the integer $d_n(q)$ is defined for each positive integer q by induction on n by Witt's formula*

$$q^n = \sum_{r|n} r d_r(q).$$

Proof. By 8.3b), $\mathfrak{g}$ is the additive group generated by all $[x_1, \ldots, x_n]$ with $n \geq 1$ and $x_i \in X$. Since $[x_1, \ldots, x_n] \in \mathsf{H}_n \cap \mathfrak{g} = \mathsf{G}_n$, $\mathfrak{g} = \mathsf{G}_1 + \mathsf{G}_2 + \cdots$. Since $\mathsf{G}_n \subseteq \mathsf{H}_n$ and $\mathfrak{A} = \mathsf{H}_0 \oplus \mathsf{H}_1 \oplus \mathsf{H}_2 \oplus \cdots$, a) follows at once.

By 11.6, H_n is a free Abelian group. Since every subgroup of a free Abelian group is free (cf. I, 13.4), G_n is a free Abelian group. Let B_n be a $\mathbb{Z}$-basis of G_n and let $\mathsf{B} = \bigcup_{n>0} \mathsf{B}_n$. By a), B is a $\mathbb{Z}$-basis of $\mathfrak{g}$. We suppose B to be fully ordered in such a way that $b_1 < b_2$ whenever $b_1 \in \mathsf{H}_m$, $b_2 \in \mathsf{H}_n$ and $m < n$. Since $\mathfrak{A}$ is the associative algebra generated by $\mathfrak{g}$, it follows from 11.1 that $\mathfrak{A}$ is the additive group generated by

$$\mathsf{B}^* = \{b_1 b_2 \cdots b_n \mid n \geq 0, b_i \in \mathsf{B}, b_1 \leq \cdots \leq b_n\}.$$

However by 11.2 there is an associative algebra $\mathfrak{C}$ and a monomorphism ε of $\mathfrak{g}$ into $\mathfrak{l}(\mathfrak{C})$ such that

$$\{(b_1\varepsilon)(b_2\varepsilon) \cdots (b_n\varepsilon) \mid n \geq 0, b_i \in \mathsf{B}, b_1 \leq \cdots \leq b_n\}$$

is linearly independent over $\mathbb{Z}$. Since X is an associative basis of $\mathfrak{A}$, there is a homomorphism σ of $\mathfrak{A}$ into $\mathfrak{C}$ such that $x\sigma = x\varepsilon$ for all $x \in X$. Hence if $x_i \in X$ $(i = 1, \ldots, n)$,

$$[x_1, \ldots, x_n]\sigma = [x_1\sigma, \ldots, x_n\sigma] = [x_1\varepsilon, \ldots, x_n\varepsilon] = [x_1, \ldots, x_n]\varepsilon.$$

It follows from 8.3b) that $a\sigma = a\varepsilon$ for all $a \in \mathfrak{g}$. Thus $(b_1 \cdots b_n)\sigma = (b_1\varepsilon) \cdots (b_n\varepsilon)$, so B^* is linearly independent over $\mathbb{Z}$. Hence B^* is a $\mathbb{Z}$-basis of $\mathfrak{A}$.

Now each element of B, and hence each element of B^*, lies in some H_n. It follows, since $\mathfrak{A} = \mathsf{H}_0 \oplus \mathsf{H}_1 \oplus \cdots$, that $\mathsf{B}_n^* = \mathsf{B}^* \cap \mathsf{H}_n$ is a $\mathbb{Z}$-basis of H_n $(n \geq 0)$. Since $\mathsf{B}_n \subseteq \mathsf{B}_n^*$, $\mathsf{H}_n = \mathsf{G}_n \oplus \mathsf{K}_n$, where K_n is the additive group generated by $\mathsf{B}_n^* - \mathsf{B}_n$. Thus $\mathsf{H}_n/\mathsf{G}_n$ is a free Abelian group, and b) is proved.

Suppose that $|X| = q$. Thus by 11.6, H_n is a free Abelian group of rank q^n. Hence $|\mathsf{B}_n^*| = q^n$. But

$$\mathsf{B}_n^* = \{b_1 b_2 \cdots b_m | b_i \in \mathsf{B}, b_1 \leq \cdots \leq b_m, b_1 b_2 \cdots b_m \in \mathsf{H}_n\},$$

so $|\mathsf{B}_n^*|$ is the coefficient of t^n in the formal power series

$$(1 + t + t^2 + \cdots)^{d_1}(1 + t^2 + t^4 + \cdots)^{d_2} \cdots$$
$$= \prod_{r \geq 1} (1 + t^r + t^{2r} + \cdots)^{d_r},$$

where $d_r = |\mathsf{B}_r|$. Hence

$$\prod_{r \geq 1} (1 + t^r + t^{2r} + \cdots)^{d_r} = \sum_n q^n t^n,$$

and

$$\prod_{r \geq 1} \frac{1}{(1 - t^r)^{d_r}} = \frac{1}{1 - qt}.$$

Taking logarithms,

$$\sum_{r=1}^{\infty} \sum_{s=1}^{\infty} \frac{d_r t^{rs}}{s} = \sum_{n=1}^{\infty} \frac{q^n t^n}{n}.$$

Hence

$$q^n = n \sum_{rs=n} \frac{d_r}{s} = \sum_{r|n} r d_r.$$

Hence $d_r = d_r(q)$. Since $d_r = |\mathsf{B}_r|$ is the rank of G_r, c) is proved. **q.e.d.**

11.9 Lemma. *Suppose that* U *is a finitely generated additive (Abelian) group but that* U *is not free. Then there exist primes* p, q *such that* $|\mathsf{U}/p\mathsf{U}| = p^m$, $|\mathsf{U}/q\mathsf{U}| = q^n$ *and* $m \neq n$.

Proof. By I, 13.12, U is the direct sum of a finite number of cyclic groups $\mathsf{U}_1, \ldots, \mathsf{U}_r$. Since U is not free, at least one of these cyclic groups is finite, say U_1. If p is a prime divisor of $|\mathsf{U}_1|$ and q is a prime for which U has no element of order q, $m - n$ is the number of U_i for which p divides $|\mathsf{U}_i|$, so $m - n > 0$. **q.e.d.**

11.10 Theorem. *Let* $\mathfrak{A}$ *be a free associative algebra over* $\mathbb{Z}$ *with associative basis* X, *and let* $\mathfrak{g}$ *be the Lie subalgebra of* $\mathfrak{l}(\mathfrak{A})$ *generated by* X. *Then* $\mathfrak{g}$ *is a free Lie algebra with Lie basis* X.

Proof. Suppose this is false. By 11.7a), there exists a free Lie algebra $\mathfrak{f}_1$ having a Lie basis Y_1 in (1, 1) correspondence with X. Hence there is a Lie epimorphism ρ of $\mathfrak{f}_1$ onto $\mathfrak{g}$ such that the restriction of ρ to Y_1 is a bijection onto X. Since the theorem is false, $\ker \rho \neq 0$. Suppose $u \in \mathfrak{f}_1$, $u \neq 0$, $u\rho = 0$. By 8.3, u lies in the Lie algebra $\mathfrak{f}$ generated by a finite subset Y of Y_1. It is clear from the definition (11.3b)) that $\mathfrak{f}$ is a free Lie algebra with basis Y. Let $T = Y\rho$. If $\mathfrak{B}$ is the associative subalgebra of $\mathfrak{A}$ generated by T, $\mathfrak{B}$ is free with basis T.

Suppose that $\mathfrak{f}$ has a $\mathbb{Z}$-basis. By 11.2, there exists an associative algebra $\mathfrak{C}$ and a monomorphism ε of $\mathfrak{f}$ into $\mathfrak{l}(\mathfrak{C})$. If $x \in T$, $x = y\rho$ for a unique $y \in Y$. Thus there is an associative homomorphism α of $\mathfrak{B}$ into $\mathfrak{C}$ such that $x\alpha = y\varepsilon$ and $\rho\alpha = \varepsilon$ on $\mathfrak{f}$. But then $u\varepsilon = u\rho\alpha = 0$. Since ε is a monomorphism, this implies $u = 0$, a contradiction. Hence $\mathfrak{f}$ has no $\mathbb{Z}$-basis.

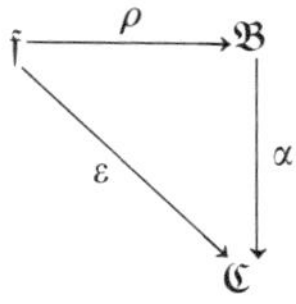

For each $n \geq 1$, let U_n be the additive group spanned by all products $[y_1, \ldots, y_n]$ with $y_i \in Y$. By 11.7b),

$$\mathfrak{f} = \mathsf{U}_1 \oplus \mathsf{U}_2 \oplus \cdots.$$

Hence U_n has no $\mathbb{Z}$-basis for some n. By 11.9, there exist primes p_1, p_2 such that $|\mathsf{U}_n/p_i\mathsf{U}_n| = p_i^{m_i}$ with $m_1 \neq m_2$. We shall obtain a contradiction by proving that for any prime p, $|\mathsf{U}_n/p\mathsf{U}_n| = p^{d_n(q)}$, where $q = |Y|$.

Let $\mathsf{K} = GF(p)$. Then $\mathfrak{f}/p\mathfrak{f}$ can be regarded as a Lie algebra over K. Now $\mathfrak{B}/p\mathfrak{B}$ is a free associative algebra over K with basis $X_p = \{x + p\mathfrak{B} | x \in T\}$. Hence by 11.5b), the Lie subalgebra $\mathfrak{f}_p$ of $\mathfrak{l}(\mathfrak{B}/p\mathfrak{B})$ generated by X_p is a free Lie algebra with basis X_p. But since $\mathfrak{f}$ is a free Lie ring, there is a homomorphism σ of $\mathfrak{f}$ into $\mathfrak{f}_p$ such that $y\sigma = y\rho + p\mathfrak{B}$ for all $y \in Y$. Clearly $p\mathfrak{f} \subseteq \ker \sigma$, so σ induces a K-homomorphism $\bar{\sigma}$ of $\mathfrak{f}/p\mathfrak{f}$ into $\mathfrak{f}_p$ such that $(y + p\mathfrak{f})\bar{\sigma} = y\rho + p\mathfrak{B}$. By 11.4a), $\bar{\sigma}$ is an isomorphism of $\mathfrak{f}/p\mathfrak{f}$ onto $\mathfrak{f}_p$. Thus $(\mathsf{U}_n + p\mathfrak{f})/p\mathfrak{f}$ is isomorphic to the additive subgroup generated by all $[x_1, \ldots, x_n] + p\mathfrak{B}$ with $x_i \in T$. Hence if G_n is the additive group generated by all $[x_1, \ldots, x_n]$, $\mathsf{U}_n/p\mathsf{U}_n$ is isomorphic to $(\mathsf{G}_n + p\mathfrak{B})/p\mathfrak{B}$. By 11.8a), G_n is a direct summand of $\mathfrak{B}$, so $\mathsf{G}_n \cap p\mathfrak{B} = p\mathsf{G}_n$. Hence $\mathsf{U}_n/p\mathsf{U}_n \cong \mathsf{G}_n/p\mathsf{G}_n$. But by 11.8, G_n is a free Abelian group of rank $d_n(q)$, so $|\mathsf{U}_n/p\mathsf{U}_n| = p^{d_n(q)}$. **q.e.d.**

Next we consider the augmentation ideal of the group-ring of a free group.

11.11 Theorem. *Let $\mathfrak{F}$ be a free group with basis X and let $\mathfrak{J}$ be the augmentation ideal (see 2.2) of $\mathsf{A}\mathfrak{F}$.*

a) *For each integer $n > 0$, $\mathfrak{J}^n$ is a free $\mathsf{A}\mathfrak{F}$-module with basis the set B_n of all products $(x_1 - 1) \cdots (x_n - 1)$ with $x_i \in X$.*

b) $\bigcap_{n>0} \mathfrak{J}^n = 0$.

Proof. a) This is proved by induction on n. The case $n = 1$ is III, 18.5. For $n > 1$, observe that $\mathfrak{J}^{n-1}$ is the $\mathsf{A}\mathfrak{F}$-module generated by all $(x_1 - 1) \cdots (x_{n-1} - 1)$, by the inductive hypothesis. Hence $\mathfrak{J}^{n-1}$ is the A-module generated by all $(x_1 - 1) \cdots (x_{n-1} - 1)a$ with $a \in \mathsf{A}\mathfrak{F}$. By definition (2.2), $\mathfrak{J}^n = \mathfrak{J}^{n-1}\mathfrak{J}$ is the additive group generated by all $(x_1 - 1) \cdots (x_{n-1} - 1)b$ with $b \in \mathfrak{J}$. From the case $n = 1$, $\mathfrak{J}$ is the $\mathsf{A}\mathfrak{F}$-module generated by all $x - 1$ with $x \in X$. Hence $\mathfrak{J}^n$ is the $\mathsf{A}\mathfrak{F}$-module generated by B_n. If

$$\sum_{x_i} (x_1 - 1) \cdots (x_n - 1) b_{x_1, \ldots, x_n} = 0$$

with $b_{x_1, \ldots, x_n} \in \mathsf{A}\mathfrak{F}$, it follows from the fact that B_1 is an $\mathsf{A}\mathfrak{F}$-basis of $\mathfrak{J}$ that for each $x_1 \in X$,

$$\sum_{x_2, \ldots, x_n} (x_2 - 1) \cdots (x_n - 1) b_{x_1, \ldots, x_n} = 0.$$

Hence by the inductive hypothesis, all $b_{x_1, \ldots, x_n}$ are zero.

b) (R. H. FOX [1]). For $f \in \mathfrak{F}$, let $l(f)$ denote the smallest integer l for which there exists a sequence $(x_1^{\varepsilon_1}, \ldots, x_l^{\varepsilon_l})$ with $x_j \in X$, $\varepsilon_j = \pm 1$ and $f = x_1^{\varepsilon_1} \cdots x_l^{\varepsilon_l}$. For $u \in \mathsf{A}\mathfrak{F}$, $u \neq 0$ and $u = \sum_{f \in \mathfrak{F}} \lambda_f f$ ($\lambda_f \in \mathsf{A}$), let $l(u)$ denote the greatest integer l for which there exists $f \in \mathfrak{F}$ such that $\lambda_f \neq 0$ and $l(f) = l$. Finally, put $l(0) = 0$.

We prove that every non-zero element u of $\mathfrak{I}$ is expressible in the form

$$u = \sum_{x \in X} (x - 1)(x^{-1}u_x + v_x), \tag{1}$$

where u_x, v_x are elements of $\mathsf{A}\mathfrak{F}$ for which $l(u_x) < l(u)$ and $l(v_x) < l(u)$. Since $\mathfrak{I}$ is the additive group generated by all $f - 1$ with $f \in \mathfrak{F}$, it suffices to establish this when $u = f - 1$, where $f \in \mathfrak{F}$ and $f \neq 1$. We use induction on $l(f)$. Since $l(f) > 0$, we may write $f = x^\varepsilon g$, where $x \in X$, $\varepsilon = \pm 1$, $g \in \mathfrak{F}$, $l(g) < l(f)$. If $\varepsilon = 1$, we have $f - 1 = (x - 1)g + (g - 1)$; the result follows from this, trivially if $g = 1$, and by application of the inductive hypothesis to $g - 1$ if $g \neq 1$. If $\varepsilon = -1$, $f - 1 = -(x - 1)x^{-1}g + (g - 1)$, and the assertion follows in the same way.

Now let $\mathfrak{J} = \bigcap_{n>0} \mathfrak{I}^n$ and suppose that $\mathfrak{J} \neq 0$. We observe that if $a = \sum_{x \in X} (x - 1)a_x \in \mathfrak{J}$ with $a_x \in \mathsf{A}\mathfrak{F}$, then $a_x \in \mathfrak{J}$. Indeed for any positive integer n, $a \in \mathfrak{I}^{n+1}$, so by a), a can be written as an A-linear combination of elements of the form $(x - 1)(x_1 - 1) \cdots (x_n - 1)b$, with x, $x_1, \ldots, x_n$ in X and b in $\mathsf{A}\mathfrak{F}$. Since the $x - 1$ form an $\mathsf{A}\mathfrak{F}$-basis of $\mathfrak{I}$, it follows that a_x is an A-linear combination of products of the form $(x_1 - 1) \cdots (x_n - 1)b$. Thus $a_x \in \mathfrak{I}^n$ for all $n > 0$ and $a_x \in \mathfrak{J}$.

Now choose a non-zero element u of $\mathfrak{J}$ for which $l(u)$ is minimal. Then $l(u) \neq 0$. By (1), we can write

$$u = \sum_{x \in X} (x - 1)(x^{-1}u_x + v_x),$$

where $l(u_x) < l(u)$, $l(v_x) < l(u)$. By the above remark, $x^{-1}u_x + v_x \in \mathfrak{J}$. Thus if λ_x is the sum of the coefficients of u_x, $u_x - \lambda_x 1 \in \mathfrak{I}$ and

$$v_x + \lambda_x 1 = (x^{-1}u_x + v_x) - x^{-1}(u_x - \lambda_x 1) - \lambda_x x^{-1}(1 - x) \in \mathfrak{I}.$$

By (1), we may write

$$u_x - \lambda_x 1 = \sum_{y \in X} (y - 1)u_{x,y}, \quad v_x + \lambda_x 1 = \sum_{y \in X} (y - 1)v_{x,y},$$

where $l(u_{x,y}) \leq l(u_x - \lambda_x 1) < l(u)$ and $l(v_{x,y}) \leq l(v_x + \lambda_x 1) < l(u)$. Now $u_x + xv_x = x(x^{-1}u_x + v_x) \in \mathfrak{J}$, and

$$u_x + xv_x = \left\{\sum_{y \neq x} (y - 1)(u_{x,y} + v_{x,y})\right\} + (x - 1)(v_x + u_{x,x} + v_{x,x}).$$

Since $u_x + xv_x \in \mathfrak{J}$, it follows from the above remark that $u_{x,y} + v_{x,y} \in \mathfrak{J}$ $(y \neq x)$ and $u_{x,x} + v_{x,x} + v_x \in \mathfrak{J}$. By minimality of $l(u)$, it follows that $u_{x,y} + v_{x,y} = u_{x,x} + v_{x,x} + v_x = 0$. Hence $u_x + xv_x = 0$ for all x, and $u = 0$, a contradiction. **q.e.d.**

11.12 Theorem. *Let $\mathfrak{F}$ be a free group.*

a) *If p is any prime and the subgroups $\kappa_n(\mathfrak{F})$ are defined as in 1.10, then*

$$\bigcap_{n \geq 1} \kappa_n(\mathfrak{F}) = 1.$$

b) (MAGNUS) $\bigcap_{n \geq 1} \gamma_n(\mathfrak{F}) = 1.$

Proof. Let $\mathsf{K} = GF(p)$, and let $\mathfrak{J}$ be the augmentation ideal of the group-ring $\mathsf{K}\mathfrak{F}$. If $f \in \kappa_n(\mathfrak{F})$, then $f - 1 \in \mathfrak{J}^n$ by 2.7. It follows that if $f \in \bigcap_{n \geq 1} \kappa_n(\mathfrak{F})$, then $f - 1 \in \bigcap_{n \geq 1} \mathfrak{J}^n$; hence $f - 1 = 0$ by 11.11b). This gives a), and b) follows at once, as $\kappa_n(\mathfrak{G}) \geq \gamma_n(\mathfrak{G})$. **q.e.d.**

We now describe the embedding of the free associative algebra in the group-ring of a free group, which was mentioned earlier.

11.13 Theorem. *Let $\mathfrak{F}$ be a free group with basis X.*

a) *The subalgebra $\mathfrak{A}$ of $\mathsf{A}\mathfrak{F}$ generated by the $x - 1$ $(x \in X)$ is a free associative algebra with basis $\mathsf{B} = \{x - 1 | x \in X\}$.*

b) *If H_n is the homogeneous component of $\mathfrak{A}$ of degree n with respect to B,*

$$\mathfrak{J}^n = \mathsf{H}_n \oplus \mathfrak{J}^{n+1},$$

where $\mathfrak{J}$ is the augmentation ideal of $\mathsf{A}\mathfrak{F}$.

Proof. For each $n \geq 0$, let B_n be the set of all products $(x_1 - 1) \cdots (x_n - 1)$ with $x_i \in X$. Let H'_n be the A-module spanned by B_n. Thus

$$\mathfrak{A} = \mathsf{H}'_0 + \mathsf{H}'_1 + \cdots.$$

By 11.11a), B_n is an $\mathsf{A}\mathfrak{F}$-basis of $\mathfrak{I}^n$. *A fortiori*, B_n is an A-basis of H_n'. Also $\mathfrak{I}^n = \mathsf{H}_n' + \mathfrak{I}^{n+1}$, for if $b \in \mathsf{B}_n$ and $u \in \mathfrak{F}$,

$$bu = b + b(u - 1) \in \mathsf{H}_n' + \mathfrak{I}^{n+1}.$$

If $v \in \mathsf{H}_n' \cap \mathfrak{I}^{n+1}$, then

$$v = \sum_{b \in \mathsf{B}_n} \lambda_b b = \sum_{b \in \mathsf{B}_n, x \in X} b(x - 1)u_{b,x},$$

with $\lambda_b \in \mathsf{A}$ and $u_{b,x} \in \mathsf{A}\mathfrak{F}$. Hence

$$\lambda_b = \sum_{x \in X} (x - 1)u_{b,x},$$

$\lambda_b = 0$ and $v = 0$. Hence $\mathfrak{I}^n = \mathsf{H}_n' \oplus \mathfrak{I}^{n+1}$ and

$$\mathfrak{A} = \mathsf{H}_0' \oplus \mathsf{H}_1' \oplus \cdots.$$

Since B_n is an A-basis of H_n', it follows at once that $\mathfrak{A}$ is a free associative algebra and that $\mathsf{H}_n = \mathsf{H}_n'$. **q.e.d.**

By 11.13b),

$$\mathfrak{I} = \mathsf{H}_1 \oplus \mathsf{H}_2 \oplus \cdots \oplus \mathsf{H}_m \oplus \mathfrak{I}^{m+1}.$$

Thus if $u \in \mathfrak{I}$, u has an expansion of the form

$$u = h_1 + h_2 + \cdots + h_m + r_m,$$

where $h_i \in \mathsf{H}_i$ and $r_m \in \mathfrak{I}^{m+1}$. We examine the first non-zero term of this expansion in the case when $\mathsf{A} = \mathbb{Z}$ and $u = f - 1$ with $f \in \mathfrak{F} - \{1\}$. We shall see that if $f \in \gamma_n(\mathfrak{F}) - \gamma_{n+1}(\mathfrak{F})$ $(n \leq m)$, the first non-zero term is h_n and it lies in G_n, (defined as in 11.8). Further, every non-zero element of G_n arises in this way.

11.14 Theorem (MAGNUS, WITT). *Let $\mathfrak{F}$ be a free group with basis X, let $\mathfrak{I}$ be the augmentation ideal of the integral group-ring $\mathbb{Z}\mathfrak{F}$ and, for each $n \geq 1$, let G_n be the additive group generated by all Lie products $[x_1 - 1, \ldots, x_n - 1]$ with $x_i \in X$. There exists an isomorphism ρ_n of $\gamma_n(\mathfrak{F})/\gamma_{n+1}(\mathfrak{F})$ onto G_n such that if $f \in \gamma_n(\mathfrak{F})$,*

$$f - 1 - (f\gamma_{n+1}(\mathfrak{F}))\rho_n \in \mathfrak{I}^{n+1}.$$

Proof. Let $\mathfrak{A}$ be the associative algebra generated by $B = \{x - 1 | x \in X\}$. By 11.13a), $\mathfrak{A}$ is a free associative algebra with basis B. For each $n \geq 0$, let H_n be the homogeneous component of $\mathfrak{A}$ with respect to B; thus $\mathfrak{I}^n = \mathsf{H}_n \oplus \mathfrak{I}^{n+1}$ by 11.13b). By 11.10, the Lie algebra $\mathfrak{g}$ generated by B is free with basis B, and by 11.8a), $\mathsf{G}_n = \mathfrak{g} \cap \mathsf{H}_n$.

For $n \geq 1$, let

$$\mathfrak{F}_n = \{f | f \in \mathfrak{F}, f - 1 \in \mathfrak{I}^n\}.$$

By 2.3a), $\mathfrak{F}_n \trianglelefteq \mathfrak{F}$, and by 2.3b),

$$[\mathfrak{F}_m, \mathfrak{F}_n] \leq \{f | f \in \mathfrak{F}, f - 1 \in \mathfrak{I}^{m+n}\} = \mathfrak{F}_{m+n}.$$

Finally, by 2.3c), if $f \in \mathfrak{F}_m$ and $g \in \mathfrak{F}_n$, then

$$(2) \qquad [f, g] - 1 \equiv [f - 1, g - 1] \mod \mathfrak{I}^{m+n+1}.$$

Hence $\mathfrak{F} = \mathfrak{F}_1 \geq \mathfrak{F}_2 \geq \mathfrak{F}_3 \geq \cdots$ is a strongly central series of $\mathfrak{F}$. Thus $\mathfrak{F}_n \geq \gamma_n(\mathfrak{F})$. Thus if $f \in \gamma_n(\mathfrak{F})$, $f - 1 \in \mathfrak{I}^n = \mathsf{H}_n \oplus \mathfrak{I}^{n+1}$. Hence there exists a unique element $u \in \mathsf{H}_n$ such that $f - 1 - u \in \mathfrak{I}^{n+1}$. If also $g \in \gamma_n(\mathfrak{F})$, $v \in \mathsf{H}_n$ and $g - 1 - v \in \mathfrak{I}^{n+1}$, then $u + v \in \mathsf{H}_n$ and

$$fg - 1 - (u + v) = f(g - 1 - v) + (f - 1 - u)(1 + v) + uv \in \mathfrak{I}^{n+1},$$

since $uv \in \mathsf{H}_{2n} \subseteq \mathfrak{I}^{2n} \subseteq \mathfrak{I}^{n+1}$. Hence the mapping $f \to u$ is a homomorphism of $\gamma_n(\mathfrak{F})$ into H_n. $\gamma_{n+1}(\mathfrak{F})$ is contained in the kernel, since $\gamma_{n+1}(\mathfrak{F}) \leq \mathfrak{F}_{n+1}$. Hence there exists a homomorphism ρ_n of $\gamma_n(\mathfrak{F})/\gamma_{n+1}(\mathfrak{F})$ into H_n such that for any $f \in \gamma_n(\mathfrak{F})$,

$$(3) \qquad f - 1 - (f\gamma_{n+1}(\mathfrak{F}))\rho_n \in \mathfrak{I}^{n+1}.$$

Note in particular that if $x \in X$, $(x\gamma_2(\mathfrak{F}))\rho_1 = x - 1$. It follows from (2) and (3) that if $f \in \gamma_m(\mathfrak{F})$ and $g \in \gamma_n(\mathfrak{F})$,

$$[f, g] - 1 \equiv [(f\gamma_{m+1}(\mathfrak{F}))\rho_m, (g\gamma_{n+1}(\mathfrak{F}))\rho_n] \mod \mathfrak{I}^{m+n+1}.$$

Since the term on the right-hand side lies in H_{m+n} and $[f, g] \in \gamma_{m+n}(\mathfrak{F})$, it follows that

$$(4) \qquad ([f, g]\gamma_{m+n+1}(\mathfrak{F}))\rho_{m+n} = [(f\gamma_{m+1}(\mathfrak{F}))\rho_m, (g\gamma_{n+1}(\mathfrak{F}))\rho_n].$$

By 9.3, there exists a Lie ring $\mathfrak{f} = \mathsf{F}_1 \oplus \mathsf{F}_2 \oplus \cdots$ and, for each $n \geq 1$, an isomorphism σ_n of $\gamma_n(\mathfrak{F})/\gamma_{n+1}(\mathfrak{F})$ onto F_n such that

$$(5) \qquad ([f, g]\gamma_{m+n+1}(\mathfrak{F}))\sigma_{m+n} = [(f\gamma_{m+1}(\mathfrak{F}))\sigma_m, (g\gamma_{n+1}(\mathfrak{F}))\sigma_n]$$

for $f \in \gamma_m(\mathfrak{F})$, $g \in \gamma_n(\mathfrak{F})$. Also, by 9.3b), there exists a Lie homomorphism θ of $\mathfrak{f}$ into $\mathrm{I}(\mathfrak{A})$ such that $\sigma_n\theta = \rho_n$ for all $n \geq 1$.

By 9.4, $\mathfrak{f}$ is generated as a Lie ring by $\{(x\gamma_2(\mathfrak{F}))\sigma_1 | x \in X\}$. Now θ carries this set bijectively onto B, since

$$(x\gamma_2(\mathfrak{F}))\sigma_1\theta = (x\gamma_2(\mathfrak{F}))\rho_1 = x - 1.$$

Hence $\operatorname{im}\theta$ is contained in $\mathfrak{g}$. Since $\mathfrak{g}$ is a free Lie ring with basis B, it follows from 11.4a) that θ is an isomorphism of $\mathfrak{f}$ onto $\mathfrak{g}$.

Since σ_n, θ are both monomorphisms, so is ρ_n. Since $x - 1 \in \operatorname{im}\theta$ for $x \in X$, it follows from the definition of G_n that $\mathsf{G}_n \subseteq \operatorname{im}\theta$. But

$$\mathsf{F}_n\theta = (\gamma_n(\mathfrak{F})/\gamma_{n+1}(\mathfrak{F}))\sigma_n\theta = (\gamma_n(\mathfrak{F})/\gamma_{n+1}(\mathfrak{F}))\rho_n \subseteq \mathsf{H}_n,$$

so $\mathsf{F}_n\theta \subseteq \mathsf{H}_n \cap \mathfrak{g} = \mathsf{G}_n$. Since $\mathfrak{g} = \mathsf{G}_1 \oplus \mathsf{G}_2 \oplus \cdots$, it follows that $\mathsf{G}_n = \mathsf{F}_n\theta$. Hence ρ_n is an epimorphism of $\gamma_n(\mathfrak{F})/\gamma_{n+1}(\mathfrak{F})$ onto G_n. Hence ρ_n is an isomorphism. **q.e.d.**

11.15 Theorem. *Let $\mathfrak{F}$ be a free group.*

a) *For each $n \geq 1$, $\gamma_n(\mathfrak{F})/\gamma_{n+1}(\mathfrak{F})$ is a free Abelian group, and if $\mathfrak{F}$ has a finite basis with q elements, the rank of $\gamma_n(\mathfrak{F})/\gamma_{n+1}(\mathfrak{F})$ is $d_n(q)$.*

b) *Let $\mathfrak{J}$ be the augmentation ideal of $\mathbb{Z}\mathfrak{F}$ and suppose that $f \in \mathfrak{F}$. Then $f - 1 \in \mathfrak{J}^n$ if and only if $f \in \gamma_n(\mathfrak{F})$ $(n \geq 1)$.*

Proof. Let $\mathfrak{A}$ be the associative algebra generated by $B = \{x - 1 | x \in X\}$, where X is a basis of $\mathfrak{F}$. By 11.13a), $\mathfrak{A}$ is a free associative algebra with basis B. If G_n is the additive group generated by all Lie products $[x_1 - 1, \ldots, x_n - 1]$ with $x_i \in X$, then by 11.8, G_n is a free Abelian group, and if $|X| = q$, the rank of G_n is $d_n(q)$. By 11.14, there exists an isomorphism ρ_n of $\gamma_n(\mathfrak{F})/\gamma_{n+1}(\mathfrak{F})$ onto G_n and

$$f - 1 - (f\gamma_{n+1}(\mathfrak{F}))\rho_n \in \mathfrak{J}^{n+1}$$

for all $f \in \gamma_n(\mathfrak{F})$. Thus a) follows at once.

If $f \in \gamma_n(\mathfrak{F}) - \gamma_{n+1}(\mathfrak{F})$, then $f\gamma_{n+1}(\mathfrak{F}) \neq 1$ and $u = (f\gamma_{n+1}(\mathfrak{F}))\rho_n \neq 0$, since ρ_n is an isomorphism. Thus $u \in \mathfrak{J}^n - \mathfrak{J}^{n+1}$. Since $f - 1 - u \in \mathfrak{J}^{n+1}$, $f - 1 \in \mathfrak{J}^n - \mathfrak{J}^{n+1}$. This implies b). **q.e.d.**

11.16 Remarks. a) For any group $\mathfrak{G}$, the dimension subgroup $\tilde{\gamma}_n(\mathfrak{G})$ is defined by

$$\tilde{\gamma}_n(\mathfrak{G}) = \{x \mid x \in \mathfrak{G}, x - 1 \in \mathfrak{J}^n\} \quad (n = 1, 2, \ldots),$$

where $\mathfrak{J}$ is the augmentation ideal of $\mathbb{Z}\mathfrak{G}$. By 2.3, $\gamma_n(\mathfrak{G}) \leq \tilde{\gamma}_n(\mathfrak{G})$. Theorem 11.15b) asserts that $\gamma_n(\mathfrak{F}) = \tilde{\gamma}_n(\mathfrak{F})$ for a free group $\mathfrak{F}$, and it was conjectured that this is the case for all groups. This is true for $n = 1, 2, 3$, but RIPS [1] gave a counterexample for $n = 4$. SJOGREN [1] proved that there is a bound on the exponent of $\tilde{\gamma}_n(\mathfrak{G})/\gamma_n(\mathfrak{G})$ independent of $\mathfrak{G}$.

b) Let $\mathfrak{R}$ be a normal subgroup of the free group $\mathfrak{F}$, let $\mathfrak{I}$ be the kernel of the natural homomorphism of $\mathbb{Z}\mathfrak{F}$ onto $\mathbb{Z}(\mathfrak{F}/\mathfrak{R})$ and let

$$\mathfrak{R}_n = \{f \mid f \in \mathfrak{F}, f - 1 \in \mathfrak{I}\mathfrak{J}^n\} \quad (n \geq 0),$$

where $\mathfrak{J}$ is the augmentation ideal of $\mathbb{Z}\mathfrak{F}$. The question of the characterization of $\mathfrak{R}_n$ has been raised by FOX [1]. In the case when $\mathfrak{R} = \mathfrak{F}$, the answer is given by 11.15, and the answer is trivial if $n = 0$. It was proved by Schumann that $\mathfrak{R}_1 = \gamma_2(\mathfrak{R})$ (see FOX [1]) and by HURLEY [1] that

$$\mathfrak{R}_2 = (\mathfrak{R} \cap \mathfrak{F}')'\gamma_3(\mathfrak{F}).$$

Exercises

21) Show that the function $d_n(q)$ defined in 11.8 is given by

$$d_n(q) = \frac{1}{n}\sum_{r|n} \mu(r)q^{n/r};$$

here μ is the Möbius function, defined by $\mu(1) = 1$, $\mu(r) = 0$ if r is divisible by the square of a prime and $\mu(p_1 \cdots p_m) = (-1)^m$ if $p_1, \ldots, p_m$ are distinct primes. (First prove this for $q = 1$.)

(It is remarkable that this formula coincides with that of Gauss for the number of irreducible polynomials of degree n over $GF(q)$.)

22) If A is a commutative ring with identity and $\mathfrak{f}$ is a free Lie algebra over $\mathbb{Z}$, $\mathfrak{f} \otimes_{\mathbb{Z}} \mathsf{A}$ is a free Lie algebra over A.

Deduce 11.8 and 11.10 for algebras over A.

23) Suppose that $\mathfrak{F}$ is a free group and that p is a prime. Let $\mathfrak{J}$ be the augmentation ideal of $\mathbb{Z}\mathfrak{F}$. Suppose that $f \in \gamma_n(\mathfrak{F})$ but that $f\gamma_{n+1}(\mathfrak{F})$ is not the p-th power of any element of $\gamma_n(\mathfrak{F})/\gamma_{n+1}(\mathfrak{F})$. Prove that $f - 1 + \mathfrak{J}^{n+1} \notin p(\mathfrak{J}^n/\mathfrak{J}^{n+1})$.

24) Let $\mathfrak{M}$ be the *Magnus algebra* on a set X over A, that is, the set of all (possibly infinite) formal sums

$$\sum \lambda_{i_1 \dots i_n} x_{i_1} \cdots x_{i_n} \quad (\lambda_{i_1 \dots i_n} \in \mathsf{A}, x_i \in X)$$

with X as a set of non-commuting indeterminates. Show that if $x \in X$, $1 + x$ is a unit of $\mathfrak{M}$. Let $\mathfrak{F}$ be the subgroup of the group of units of $\mathfrak{M}$ generated by all $1 + x$ ($x \in X$). Prove that $\mathfrak{F}$ is a free group and that the subalgebra of $\mathfrak{M}$ generated by $\mathfrak{F}$ is the group-ring $\mathsf{A}\mathfrak{F}$.

(Hint: This follows from 11.11 and 11.13.)

25) Let $\mathfrak{F}$ be a free group and let $\mathfrak{D}$ be the ideal of $\mathbb{Z}\mathfrak{F}$ consisting of all elements $\sum_{f \in \mathfrak{F}} a_f f$ for which $\sum_{f \in \mathfrak{F}} a_f$ is divisible by the prime p. Show that

$$\lambda_n(\mathfrak{F}) = \{f \mid f \in \mathfrak{F}, f - 1 \in \mathfrak{D}^n\}.$$

§ 12. Remarks on the Burnside Problem

It has been proved by Adyan and Novikov (see, for example, ADYAN [1]) that the answer to the Burnside problem is negative. In this book we shall be more concerned with the restricted Burnside problem (III, 6.7). In the case of prime-power exponent, linear methods have been used in investigations of this, and we shall now sketch some of the ways in which this has been done. We begin by interpreting 2.16 in the Lie ring of a group of exponent p. The following lemmas are needed (cf. III, 1.11).

12.1 Lemma. *Suppose that* $\mathfrak{G} = \langle X \rangle$.

a) *If* $a \in \mathfrak{G}$ *and* $b \in \mathfrak{G}$, *then* $a^b = ac_1 \cdots c_n$, *where each* c_i *is a commutator* $[a, y_1, \dots, y_r]$ *for which* $r \geq 1$ *and* $y_i \in X \cup X^{-1}$.

b) $\gamma_n(\mathfrak{G})$ *is generated by all commutators* $[y_1, \dots, y_r]$ *for which* $r \geq n$ *and* $y_i \in X \cup X^{-1}$.

Proof. a) Suppose that $b = y_1 \cdots y_m$, where $y_i \in X \cup X^{-1}$ and $m \geq 0$. We use induction on m. If $m = 0$, the assertion is trivial with $n = 0$. If $m > 0$, then

$$\begin{aligned} a^b = (a^{y_1})^{y_2 \cdots y_m} &= (a[a, y_1])^{y_2 \cdots y_m} \\ &= a^{y_2 \cdots y_m}[a, y_1]^{y_2 \cdots y_m}. \end{aligned}$$

The inductive hypothesis then gives the result at once.

b) By III, 1.11a), $\gamma_n(\mathfrak{G})$ is generated by all $[y_1, \ldots, y_n]^b$, where $y_i \in X$. The assertion follows at once by applying a). **q.e.d.**

12.2 Lemma. *Suppose that $\mathfrak{G}$ is a group of prime exponent p. If $u \in \gamma_n(\mathfrak{G})$ and $u_i \in \gamma_{n_i}(\mathfrak{G})$ $(i = 1, \ldots, p-1)$, then*

$$\prod_{\sigma \in \mathfrak{S}_{p-1}} [u, u_{1\sigma}, \ldots, u_{(p-1)\sigma}] \in \gamma_{m+1}(\mathfrak{G}),$$

where $m = n + n_1 + \cdots + n_{p-1}$.

Proof. This is obvious for $p = 2$ on account of the identity

$$[u, u_1] = u^{-2}(uu_1^{-1})^2 u_1^2.$$

Henceforth we suppose p odd. Let $\mathfrak{F}$ be a free group with basis $x, y_1, \ldots, y_{p-1}$. If $y = y_1 \cdots y_{p-1}$, by 2.16,

$$[x, y, \underset{p-1}{\ldots}, y] \prod_{i=0}^{p-1} (xy^i)^p \in \mathfrak{F}^{p^2} \gamma_2(\mathfrak{F})^p \gamma_{p+1}(\mathfrak{F}).$$

Hence

$$[x, y, \underset{p-1}{\ldots}, y] \in \mathfrak{F}^p \gamma_{p+1}(\mathfrak{F}).$$

By a trivial induction based on III, 1.2,

$$[x, y, \underset{n}{\ldots}, y] \equiv \prod [x, y_{i_1}, \ldots, y_{i_n}] \mod \gamma_{n+2}(\mathfrak{F}),$$

where the product is taken with any order of the factors over all suffixes i_j with $1 \leq i_j \leq p - 1$. Taking $n = p - 1$, we find that

$$\prod [x, y_{i_1}, \ldots, y_{i_{p-1}}] \in \mathfrak{F}^p \gamma_{p+1}(\mathfrak{F}). \tag{1}$$

Since $x, y_1, \ldots, y_{p-1}$ is a basis of $\mathfrak{F}$, there is a homomorphism of $\mathfrak{F}$ into $\mathfrak{F}$ which carries each of $x, y_1, \ldots, y_{p-1}$ into a preassigned element of $\mathfrak{F}$. Such a homomorphism always carries $\mathfrak{F}^p \gamma_{p+1}(\mathfrak{F})$ into itself, so (1) remains true if $x, y_1, \ldots, y_{p-1}$ are replaced by other elements of $\mathfrak{F}$. Thus, putting $y_1 = 1$, we see that if z is the product of those factors of the left-hand side of (1) for which no i_k is equal to 1, then $z \in \mathfrak{F}^p\gamma_{p+1}(\mathfrak{F})$. Removing all such factors from (1), we see that (1) remains true if the product is taken over those factors for which at least one i_k is equal to 1. We now repeat this argument on the formula thus obtained for the suffix 2, then for 3, and so on. The conclusion is thereby reached that (1) is valid if the product is restricted to those suffixes for which $(i_1, \ldots, i_{p-1})$ is a permutation of $(1, 2, \ldots, p-1)$. Thus

$$\prod_{\sigma \in \mathfrak{S}_{p-1}} [x, y_{1\sigma}, \ldots, y_{(p-1)\sigma}] \in \mathfrak{F}^p\gamma_{p+1}(\mathfrak{F}).$$

By 12.1b), there exist commutators $g_i = [z_1, \ldots, z_k]$ $(i = 1, \ldots, r)$ such that $k \geq p + 1$, $z_i = x^{\pm 1}$ or $y_j^{\pm 1}$ and

(2) $$\prod_{\sigma \in \mathfrak{S}_{p-1}} [x, y_{1\sigma}, \ldots, y_{(p-1)\sigma}] \equiv g_1^{\varepsilon_1} \cdots g_r^{\varepsilon_r} \mod \mathfrak{F}^p,$$

where $\varepsilon_i = \pm 1$. Putting $x = 1$, we see that if q is the product of those $g_i^{\varepsilon_i}$ for which no z_j is $x^{\pm 1}$, taken in the correct order, then $q \in \mathfrak{F}^p$. We may therefore replace the right-hand side of (2) by $q^{-1}g_1^{\varepsilon_1} \cdots g_r^{\varepsilon_r}$. But it is easy to see that $q^{-1}g_1^{\varepsilon_1} \cdots g_r^{\varepsilon_r}$ is a product of factors $(g_i^{b_i})^{\varepsilon_i}$, where only those g_i occur for which some z_j is $x^{\pm 1}$. By 12.1a), $g_i^{b_i}$ is a product of commutators of the same form as g_i, but with the additional condition that some z_j is $x^{\pm 1}$. Thus in (2) we may assume that all g_i satisfy this condition. Indeed, by repeating this argument, we find that it may be assumed that in each g_i, all of $x^{\pm 1}, y_1^{\pm 1}, \ldots, y_{p-1}^{\pm 1}$ occur among $z_1, \ldots, z_k$. And since $k \geq p + 1$, at least one of $x, y_1, \ldots, y_{p-1}$ occurs twice.

Let ρ be the homomorphism of $\mathfrak{F}$ into $\mathfrak{G}$ for which $x\rho = u$ and $y_i\rho = u_i$ $(i = 1, \ldots, p-1)$. By III, 2.11b), it follows that $g_i\rho \in \gamma_{m+1}(\mathfrak{G})$, since $u_i \in \gamma_{n_i}(\mathfrak{G})$ and $u \in \gamma_n(\mathfrak{G})$. Since $\mathfrak{G}$ is of exponent p, $\mathfrak{F}^p\rho = 1$. The assertion therefore follows by applying ρ to (2). **q.e.d.**

12.3 Theorem. *Suppose that $\mathfrak{G}$ is a group of prime exponent p, and that $\mathfrak{g}$ is a Lie ring. Suppose that for each $n \geq 1$, a group homomorphism σ_n of $\gamma_n(\mathfrak{G})/\gamma_{n+1}(\mathfrak{G})$ into $\mathfrak{g}$ is defined, and that the following conditions are satisfied:*

a) $\mathfrak{g} = \sum_{n\geq 1} \mathsf{G}_n$, *where* $\mathsf{G}_n = \operatorname{im} \sigma_n$;

b) $[(x\gamma_{m+1}(\mathfrak{G}))\sigma_m, (y\gamma_{n+1}(\mathfrak{G}))\sigma_n] = ([x, y]\gamma_{m+n+1}(\mathfrak{G}))\sigma_{m+n}$, *whenever* $x \in \gamma_m(\mathfrak{G})$, $y \in \gamma_n(\mathfrak{G})$.

Then $(\operatorname{ad} a)^{p-1} = 0$ *for any* $a \in \mathfrak{g}$.

Proof. Suppose that $a_{n_i} \in \mathsf{G}_{n_i}$ $(i = 1, \ldots, p - 1)$ and that $b \in \mathsf{G}_n$. There exist $u \in \gamma_n(\mathfrak{G})$, $u_i \in \gamma_{n_i}(\mathfrak{G})$ such that

$$b = (u\gamma_{n+1}(\mathfrak{G}))\sigma_n, \quad a_{n_i} = (u_i\gamma_{n_i+1}(\mathfrak{G}))\sigma_{n_i}.$$

By induction on k $(k \geq 0)$, it follows from b) that

$$b(\operatorname{ad} a_{n_1}) \cdots (\operatorname{ad} a_{n_k}) = ([u, u_1, \ldots, u_k]\gamma_{m_k+1}(\mathfrak{G}))\sigma_{m_k},$$

where $m_k = n + n_1 + \cdots + n_k$. Taking $k = p - 1$,

$$b(\operatorname{ad} a_{n_1}) \cdots (\operatorname{ad} a_{n_{p-1}}) = ([u, u_1, \ldots, u_{p-1}]\gamma_{m+1}(\mathfrak{G}))\sigma_m,$$

where $m = n + n_1 + \cdots + n_{p-1}$. It follows from 12.2 that

$$\sum_{\tau \in \mathfrak{S}_{p-1}} b(\operatorname{ad} a_{n_{1\tau}}) \cdots (\operatorname{ad} a_{n_{(p-1)\tau}}) = 0.$$

Since $\mathfrak{g} = \sum_{n \geq 1} \mathsf{G}_n$ and the left-hand side is linear in b and in each a_{n_i}, it follows that

$$\sum_{\tau \in \mathfrak{S}_{p-1}} b(\operatorname{ad} a_{1\tau}) \cdots (\operatorname{ad} a_{(p-1)\tau}) = 0$$

for any elements a_i in $\mathfrak{g}$. Putting $a_1 = \cdots = a_{p-1} = a$, we find that $(p - 1)!(\operatorname{ad} a)^{p-1} = 0$. Since the additive group of $\mathfrak{g}$ is an elementary Abelian p-group, it follows that $(\operatorname{ad} a)^{p-1} = 0$. **q.e.d.**

A condition on Lie algebras about some power of $\operatorname{ad} a$ being zero is called an *Engel condition*; such a condition sometimes implies that the algebra is nilpotent. The deepest known result of this nature is the following theorem.

12.4 Theorem (KOSTRIKIN [1]). a) *Suppose that $\mathfrak{g}$ is a finitely generated Lie algebra over a field of characteristic p. If $(\operatorname{ad} a)^{p-1} = 0$ for all $a \in \mathfrak{g}$, then $\mathfrak{g}$ is nilpotent.*

b) *Suppose that p is a prime and d is a positive integer. Then there exists a finite group $\mathfrak{G}$ generated by d elements and of exponent p such that every finite group generated by d elements and of exponent p is an epimorphic image of $\mathfrak{G}$.*

Kostrikin's proof of a) is an extremely difficult calculation which cannot be reproduced here. We shall however show how b) may be deduced from it.

Proof. b) Suppose that this is false. Let $\mathfrak{F}$ be a free group of rank d. Let $\mathfrak{F}_n = \gamma_n(\mathfrak{F})\mathfrak{F}^p$ $(n = 1, 2, \ldots)$. By 2.1, $\mathfrak{F}/\mathfrak{F}_n$ is finite. Since $\mathfrak{F}/\mathfrak{F}_n$ does not have the property described in the assertion of the theorem, there exists a finite group $\mathfrak{H}$ generated by d elements and of exponent p, such that $\mathfrak{H}$ is not an epimorphic image of $\mathfrak{F}/\mathfrak{F}_n$. But since $\mathfrak{F}$ is free of rank d, there is an epimorphism of $\mathfrak{F}$ onto $\mathfrak{H}$; thus if $\mathfrak{M}$ is the kernel, $\mathfrak{M} \not\geq \mathfrak{F}_n$. Since $\mathfrak{F}/\mathfrak{M} \cong \mathfrak{H}$, $\mathfrak{F}/\mathfrak{M}$ is nilpotent and $\mathfrak{M} \geq \mathfrak{F}^p$. Hence $\mathfrak{F}_n > \mathfrak{M} \cap \mathfrak{F}_n \geq \mathfrak{F}^p$ and $\mathfrak{F}/(\mathfrak{M} \cap \mathfrak{F}_n)$ is nilpotent (III, 2.5c)). By III, 2.6 applied to $\mathfrak{F}/(\mathfrak{M} \cap \mathfrak{F}_n)$, $[\mathfrak{F}_n, \mathfrak{F}]\mathfrak{F}^p < \mathfrak{F}_n$. Thus

$$\mathfrak{F}_{n+1}/\mathfrak{F}^p = [\gamma_n(\mathfrak{F}), \mathfrak{F}]\mathfrak{F}^p/\mathfrak{F}^p = [\mathfrak{F}_n/\mathfrak{F}^p, \mathfrak{F}/\mathfrak{F}^p] < \mathfrak{F}_n/\mathfrak{F}^p,$$

so $\mathfrak{F}_{n+1} < \mathfrak{F}_n$. Let $\mathfrak{N} = \bigcap_{n \geq 1} \mathfrak{F}_n$. Then $\mathfrak{N} \geq \mathfrak{F}^p$, but $\mathfrak{F}/\mathfrak{N}$ is infinite. The lower central series of $\mathfrak{F}/\mathfrak{N}$ is

$$\mathfrak{F}_1/\mathfrak{N} > \mathfrak{F}_2/\mathfrak{N} > \mathfrak{F}_3/\mathfrak{N} > \cdots.$$

By 9.3, there exists a Lie ring $\mathfrak{g} = \mathsf{G}_1 \oplus \mathsf{G}_2 \oplus \mathsf{G}_3 \oplus \cdots$ and, for each $n \geq 1$, an isomorphism σ_n of $\mathfrak{F}_n/\mathfrak{F}_{n+1}$ onto G_n such that the conditions of 12.3 are satisfied. By 12.3, $(\operatorname{ad} a)^{p-1} = 0$ for any $a \in \mathfrak{g}$. By a), $\mathfrak{g}$ is nilpotent. By 9.4, $\mathfrak{g}^n = \mathsf{G}_n \oplus \mathsf{G}_{n+1} \oplus \cdots$. Thus there exists n such that $\mathsf{G}_n = 0$. Hence $\mathfrak{F}_n = \mathfrak{F}_{n+1}$, a contradiction. **q.e.d.**

12.5 Remark. Let $\mathfrak{F}$ be a free group with n generators. The restricted Burnside problem (III, 6.7) may be stated as finding whether or not $\mathfrak{F}/\mathfrak{F}^m$ has a maximal finite homomorphic image. Theorem 12.4b) states that if m is a prime, the answer is affirmative.

Let $\mathfrak{G}$ be the maximal finite homomorphic image of $\mathfrak{F}/\mathfrak{F}^p$. Let $\mathfrak{f}$ be a free Lie algebra over $\mathsf{K} = GF(p)$ with n generators, and let $\mathfrak{j}$ be the ideal of $\mathfrak{f}$ generated by all $b(\operatorname{ad} a)^{p-1}$ $(a \in \mathfrak{f}, b \in \mathfrak{f})$. Then there is an epimorphism of $\mathfrak{f}/\mathfrak{j}$ onto the Lie ring defined by the lower central series of $\mathfrak{G}$ (in the sense of 9.3). WALL [1] has investigated the kernel of this epimorphism and, in particular, he has shown that for $p = 5$, 7 or 11 and $n \geq 3$, it is non-zero.

In the negative direction, we now show that there do exist non-nilpotent finitely generated groups in which the order of every element is a power of p. To construct these, the following lemma will be used.

12.6 Lemma. *Let* K *be a field. Let* $\mathfrak{A}$ *be a free associative algebra over* K *with associative basis* $x_1, \ldots, x_d$, *and for each* $n \geq 0$, *let* H_n *be the homogeneous component of* $\mathfrak{A}$ *of degree n with respect to this basis. For each* $n > 1$, *let* R_n *be a finite subset of* H_n, *let* $r_n = |R_n|$ *and let* $R = \bigcup_{n>1} R_n$.

Let $\mathfrak{J}$ be the ideal of $\mathfrak{A}$ generated by R, let $\mathfrak{B} = \mathfrak{A}/\mathfrak{J}$, let $\mathsf{B}_n = (\mathsf{H}_n + \mathfrak{J})/\mathfrak{J}$ and let $b_n = \dim_{\mathsf{K}} \mathsf{B}_n$.

a) $\mathfrak{J} = (\mathfrak{J} \cap \mathsf{H}_2) \oplus (\mathfrak{J} \cap \mathsf{H}_3) \oplus \cdots$ *and* $\mathfrak{B} = \mathsf{B}_0 \oplus \mathsf{B}_1 \oplus \mathsf{B}_2 \oplus \cdots$.

b) $b_n \geq db_{n-1} - \sum_{i=2}^{n} r_i b_{n-i}$ $(n \geq 2)$.

c) *If $s_n \geq r_n$ and the coefficient of t^n in the formal power-series*

$$\left(1 - dt + \sum_{n=2}^{\infty} s_n t^n\right)^{-1}$$

is non-negative for all $n \geq 2$, then $\mathfrak{B}$ is not of finite dimension over K.

d) *If there exists a real number e such that $0 < e < \frac{1}{2}d$ and $r_n \leq e^2(d - 2e)^{n-2}$ for all $n \geq 2$, then $\mathfrak{B}$ is not of finite dimension over* K.

Proof. Let $\mathsf{J}_n = \mathfrak{J} \cap \mathsf{H}_n$. Suppose that $u \in \mathsf{J}_n$. Since $\mathfrak{J}$ is the ideal generated by R, $\mathfrak{J}$ is spanned over K by all products $P = x_{i_1} \cdots x_{i_r} z x_{j_1} \cdots x_{j_s}$, where $r \geq 0$, $s \geq 0$ and $z \in R$. Thus u is a linear combination of such elements P. But $z \in R_m$ for some m and then $x_{i_1} \cdots x_{i_r} z x_{j_1} \cdots x_{j_s} \in \mathsf{H}_{r+s+m}$. Since $u \in \mathsf{H}_n$ and $\mathfrak{A}$ is the direct sum of its homogeneous components, it follows by comparing terms in H_n that u is a linear combination of those P for which $z \in R_{n-r-s}$. Thus

(1) J_n is spanned over K by all products $x_{i_1} \cdots x_{i_r} z x_{j_1} \cdots x_{j_s}$, where $r \geq 0$, $s \geq 0$ and $z \in R_{n-r-s}$.

In particular, since R_n is defined only for $n > 1$,

(2) $\mathsf{J}_0 = \mathsf{J}_1 = 0$.

Observe also that since $\mathfrak{J}$ is spanned by all products P, $\mathfrak{J}$ is the sum of the J_n. Since also $\mathsf{J}_n \subseteq \mathsf{H}_n$,

(3) $\mathfrak{J} = \mathsf{J}_2 \oplus \mathsf{J}_3 \oplus \cdots$.

a) It is clear that $\mathfrak{B}$ is the sum of the B_n, since $\mathfrak{A}$ is the sum of the H_n. Suppose that $\sum_{n\geq 0} v_n = 0$, where $v_n \in \mathsf{B}_n$. Then $v_n = u_n + \mathfrak{J}$ for some $u_n \in \mathsf{H}_n$, and $\sum_{n\geq 0} u_n \in \mathfrak{J}$. It follows at once from (3) that $u_n \in \mathsf{J}_n$, and so $v_n = 0$. Thus $\mathfrak{B}$ is the direct sum of the B_n.

b) Let L_m be a subspace of H_m such that $\mathsf{H}_m = \mathsf{L}_m \oplus \mathsf{J}_m$ $(m \geq 0)$. For $n \geq 2$, let V be the subspace spanned by all elements ux_j, where $u \in \mathsf{J}_{n-1}$ and $1 \leq j \leq d$. Clearly

(4) $\dim_{\mathsf{K}} \mathsf{V} \leq d(\dim_{\mathsf{K}} \mathsf{J}_{n-1})$.

By (1), $\mathsf{J}_n = \mathsf{V} + \mathsf{V}^*$, where V^* is the vector space spanned over K by all products $P = x_{i_1} \cdots x_{i_r} z$, where $z \in R_{n-r}$ $(0 \leq r \leq n - 2)$. Write $x_{i_1} \cdots x_{i_r} = y_1 + y_2$, where $y_1 \in \mathsf{L}_r$, $y_2 \in \mathsf{J}_r$. Thus $P = y_1 z + y_2 z$. But it is clear from the definition of V that $y_2 z \in \mathsf{V}$. Thus $\mathsf{V}^* \subseteq \mathsf{U} + \mathsf{V}$, where U is the vector space over K spanned by all products yz with $y \in \mathsf{L}_r$, $z \in R_{n-r}$ $(0 \leq r \leq n - 2)$. Since $\mathsf{U} \subseteq \mathsf{J}_n$, we have $\mathsf{J}_n = \mathsf{V} + \mathsf{U}$.

Clearly $\dim_K \mathsf{U} \leq \sum_{i=2}^{n} |R_i|(\dim_K \mathsf{L}_{n-i})$. Since
$\mathsf{L}_m \cong \mathsf{H}_m/\mathsf{J}_m \cong \mathsf{H}_m/(\mathfrak{J} \cap \mathsf{H}_m) \cong (\mathsf{H}_m + \mathfrak{J})/\mathfrak{J} = \mathsf{B}_m$, $\dim_K \mathsf{L}_m = b_m$. Hence

(5) $\dim_K \dot{\mathsf{U}} \leq \sum_{i=2}^{n} r_i b_{n-i}$.

Since $\mathsf{J}_n = \mathsf{V} + \mathsf{U}$, it follows from (4) and (5) that

(6) $\dim_K \mathsf{J}_n \leq \dim_K \mathsf{V} + \dim_K \mathsf{U} \leq d(\dim_K \mathsf{J}_{n-1}) + \sum_{i=2}^{n} r_i b_{n-i}$.

But $\dim_K \mathsf{J}_m = \dim_K \mathsf{H}_m - \dim_K \mathsf{L}_m = d^m - b_m$. Substitution in (6) now yields the desired inequality.

c) For $n \geq 1$, let

$$a_n = b_n - db_{n-1} + \sum_{i=2}^{n} r_i b_{n-i}.$$

Thus $a_1 = 0$ and, for $n > 1$, $a_n \geq 0$ by b). Let

$$B(t) = 1 + \sum_{n \geq 1} b_n t^n,$$

$$D(t) = 1 - dt + \sum_{n \geq 2} r_n t^n,$$

$$A(t) = 1 + \sum_{n \geq 1} a_n t^n.$$

Then $B(t)D(t) = A(t)$ since $b_0 = 1$, $b_1 = d$. Write

$$D(t)^{-1} = 1 + \sum_{n \geq 1} c_n t^n.$$

Since $B(t) = A(t)D(t)^{-1}$,

(7) $b_n = c_n + \sum_{i=1}^{n-1} c_{n-i} a_i + a_n \quad (n \geq 1)$.

Similarly write

$$D^*(t) = 1 - dt + \sum_{n \geq 2} s_n t^n,$$

$$U(t) = \sum_{n \geq 2} (s_n - r_n) t^n.$$

Thus $U(t) = D^*(t) - D(t)$. By hypothesis the coefficients of t^n in $D^*(t)^{-1}$ and in $U(t)$ are non-negative. But

$$D(t)^{-1} = D^*(t)^{-1}(1 - D^*(t)^{-1} U(t))^{-1} = \sum_{n=0}^{\infty} D^*(t)^{-n-1} U(t)^n;$$

thus the coefficient of t^n in $D(t)^{-1}$ is non-negative, that is, $c_n \geq 0$. Since also $a_n \geq 0$, it follows from (7) that $b_n \geq c_n$.

Suppose there exists n_0 such that $b_n = 0$ for all $n > n_0$. Then $c_n = 0$ for all $n > n_0$, and

$$D(t)^{-1} = 1 + \sum_{n=1}^{n_0} c_n t^n.$$

On account of the definition of $D(t)$,

$$\left(1 + \sum_{n\geq 2} r_n t^n\right)\left(1 + \sum_{n=1}^{n_0} c_n t^n\right) = 1 + dt + \sum_{n=1}^{n_0} dc_n t^{n+1}.$$

Since $c_n \geq 0$ and $r_n \geq 0$ for all $n \geq 1$, this is obviously impossible.

Hence $b_n \neq 0$ for infinitely many n. By a), $\mathfrak{B}$ is not of finite dimension over K.

d) Let $s_n = e^2(d - 2e)^{n-2}$ $(n \geq 2)$. Then

$$\begin{aligned}\left(1 - dt + \sum_{n=2}^{\infty} s_n t^n\right)^{-1} &= \left\{1 - dt + t^2e^2 \frac{1}{1 - (d - 2e)t}\right\}^{-1} \\ &= \frac{1 - (d - 2e)t}{(1 - (d - e)t)^2} \\ &= 1 + \sum_{n=1}^{\infty} t^n\{(n + 1)(d - e)^n \\ &\qquad - (d - 2e)n(d - e)^{n-1}\} \\ &= 1 + \sum_{n=1}^{\infty} t^n(d - e)^{n-1}(d + (n - 1)e).\end{aligned}$$

The coefficient of t^n in this is positive. Since also $r_n \leq s_n$ for $n \geq 2$, the result follows at once from c). **q.e.d.**

12.7 Theorem (GOLOD and ŠAFAREVIČ). *Let* K *be a countable field and let* d *be an integer greater than 1. There exists an associative algebra* $\mathfrak{B}$ *over* K *which has the following properties.*

a) $\mathfrak{B}$ *is generated by* d *elements.*

b) $\mathfrak{B}$ *is not of finite dimension over* K.

c) *There exist subspaces* B_n *of* $\mathfrak{B}$ $(n \geq 0)$ *such that* $\mathsf{B}_0 = \{\lambda 1_{\mathfrak{B}} \mid \lambda \in \mathsf{K}\}$,

$$\mathfrak{B} = \mathsf{B}_0 \oplus \mathsf{B}_1 \oplus \mathsf{B}_2 \oplus \cdots$$

and $\mathsf{B}_m\mathsf{B}_n \subseteq \mathsf{B}_{m+n}$ *for any m, n. In particular* $\overline{\mathfrak{I}} = \mathsf{B}_1 \oplus \mathsf{B}_2 \oplus \cdots$ *is an ideal of* $\mathfrak{B}$.

d) *Let V be a finite subset of* $\overline{\mathfrak{I}}$ *having fewer than d elements. Then there exists an integer r such that the product of any r elements of V is 0.*

Proof. Let $\mathfrak{A}$ be a free associative algebra over K with associative basis $x_1, \ldots, x_d$. For each $n \geq 0$, let H_n be the homogeneous component of $\mathfrak{A}$ of degree n with respect to this basis. Then $\mathfrak{A} = \mathsf{H}_0 \oplus \mathsf{H}_1 \oplus \cdots$. If $a \in \mathfrak{A}$ and $a = a_1 + a_2 + \cdots$, where $a_n \in \mathsf{H}_n$, we refer to a_n as the *homogeneous component* of a of degree n. Let $\mathfrak{I}$ be the ideal $\mathsf{H}_1 \oplus \mathsf{H}_2 \oplus \cdots$ of $\mathfrak{A}$. Let $\mathscr{A}$ be the set of all finite subsets U of $\mathfrak{I}$ for which $|U| = d - 1$. Then $\mathscr{A}$ is countable; we write

$$\mathscr{A} = \{U_1, U_2, \ldots\}.$$

Let q be a fixed integer such that $q \geq 4d$ and $q^2 \leq \left(1 + \frac{1}{2d}\right)^{q-2}$; clearly such an integer exists. We construct certain finite non-empty subsets S_m of $\bigcup_{r>q} \mathsf{H}_r$ for each $m \geq 0$ satisfying the following conditions.

(i) If $S_m \cap \mathsf{H}_r \neq \varnothing$ and $S_{m+1} \cap \mathsf{H}_s \neq \varnothing$, then $r < s$.

(ii) For each $m > 0$ there exists an integer $q_m > q$ such that S_m is the set of all homogeneous components of all products of q_m elements of U_m.

To do this we use induction on m. Put $S_0 = \varnothing$. For $m > 0$, let q_m be the smallest integer greater than q such that every element of S_{m-1} lies in some H_r with $r < q_m$; this is possible since S_{m-1} is finite. We then define S_m to be the set of all homogeneous components of all products of q_m elements of U_m. Since $U_m \subseteq \mathfrak{I}$, it is clear that if $\mathsf{H}_r \cap S_m \neq \varnothing$, then $r \geq q_m$. Thus the conditions (i), (ii) are satisfied.

Let $R = \bigcup_{m>0} S_m$. By (i), $R \cap \mathsf{H}_n$ is finite; let $r_n = |R \cap \mathsf{H}_n|$. Thus $r_n = 0$ if $n \leq q$. If $r_n \neq 0$, there exists a unique integer m such that $R \cap \mathsf{H}_n \subseteq S_m$ by (i). Then each element of $R \cap \mathsf{H}_n$ is the homogeneous component of degree n of a product $u_1 \cdots u_{q_m}$, where $u_i \in U_m$. Since $|U_m| = d - 1$, it follows that

$$r_n \leq (d-1)^{q_m}.$$

Since $r_n \neq 0$, $n \geq q_m$, so $r_n \leq (d-1)^n$. Thus

$$r_n < d^2(d-1)^{n-2} = d^n\left(1 - \frac{1}{d}\right)^{n-2} < d^n\left\{\left(1 - \frac{1}{2d}\right)\left(1 + \frac{1}{2d}\right)^{-1}\right\}^{n-2}.$$

Since $q \geq 4d$ and $n > q$, it follows that

$$r_n < d^n\left(1 - \frac{2}{q}\right)^{n-2}\left(1 + \frac{1}{2d}\right)^{-(q-2)}.$$

On account of the choice of q this implies that

$$r_n < d^n\left(1 - \frac{2}{q}\right)^{n-2}\frac{1}{q^2} = \left(\frac{d}{q}\right)^2\left(d - 2\frac{d}{q}\right)^{n-2}.$$

Let $\mathfrak{J}$ be the ideal of $\mathfrak{A}$ generated by R, let $\mathfrak{B} = \mathfrak{A}/\mathfrak{J}$ and let $\mathsf{B}_n = (\mathsf{H}_n + \mathfrak{J})/\mathfrak{J}$ $(n \geq 0)$. By 12.6d) with $e = d/q$, $\mathfrak{B}$ is not of finite dimension over K. By 12.6a),

$$\mathfrak{B} = \mathsf{B}_0 \oplus \mathsf{B}_1 \oplus \cdots;$$

clearly $\mathsf{B}_m\mathsf{B}_n \subseteq \mathsf{B}_{m+n}$.

Let V be a finite subset of $\overline{\mathfrak{J}} = \mathsf{B}_1 \oplus \mathsf{B}_2 \oplus \cdots$ for which $|V| < d$. Then there exists U_m such that $V \subseteq \{u + \mathfrak{J} | u \in U_m\}$. If z is a product of any q_m elements of U_m, then by (ii) every homogeneous component of z lies in S_m. Hence $z \in \mathfrak{J}$. Thus the product of any q_m elements of V is 0.

q.e.d.

12.8 Theorem. *Let p be a prime and let d be an integer greater than 1. There exists an infinite group $\mathfrak{G}$ which has the following properties.*

a) *$\mathfrak{G}$ is generated by d elements.*

b) *The order of any element of $\mathfrak{G}$ is a power of p.*

c) $\bigcap_{n\geq 1} \gamma_n(\mathfrak{G}) = 1$.

d) *$\mathfrak{G}$ is not nilpotent, but any subgroup of $\mathfrak{G}$ which can be generated by fewer than d elements is nilpotent.*

Proof. Let $\mathsf{K} = GF(p)$ and let $\mathfrak{B}$ be the associative algebra constructed in 12.7. Let

$$\mathfrak{H} = \{1 + v | v \in \overline{\mathfrak{J}}\}.$$

Now if $v \in \overline{\mathfrak{J}}$, then by 12.7d), there exists an integer r such that $v^r = 0$. If p^s is the smallest power of p greater than r, $(1 + v)^{p^s} = 1$ since K is of characteristic p. Thus every element of $\mathfrak{H}$ has an inverse in $\mathfrak{H}$. Hence $\mathfrak{H}$ is a group and the order of any element of $\mathfrak{H}$ is a power of p. By 12.7c), $\overline{\mathfrak{J}}_n = \mathsf{B}_n \oplus \mathsf{B}_{n+1} \oplus \cdots$ is an ideal of $\mathfrak{B}$. Let

$$\mathfrak{H}_n = \{1 + v | v \in \overline{\mathfrak{J}}_n\} = \{x \in \mathfrak{H} | x - 1 \in \overline{\mathfrak{J}}_n\}.$$

By 2.3, $\mathfrak{H}_n \trianglelefteq \mathfrak{H}$ and $[\mathfrak{H}_n, \mathfrak{H}] \leq \mathfrak{H}_{n+1}$. Thus

$$\mathfrak{H} = \mathfrak{H}_1 \geq \mathfrak{H}_2 \geq \cdots$$

is a central series of $\mathfrak{H}$. Hence $\mathfrak{H}_n \geq \gamma_n(\mathfrak{H})$.

By 12.7a), $\mathfrak{B}$ is generated by d elements $y_1, \ldots, y_d$; by 12.7c), we may suppose that $y_i \in \overline{\mathfrak{J}}$. Let $\mathfrak{G}$ be the subgroup of $\mathfrak{H}$ generated by $1 + y_1, \ldots, 1 + y_d$. Then $\gamma_n(\mathfrak{G}) \leq \gamma_n(\mathfrak{H}) \leq \mathfrak{H}_n$. By 12.7c), $\bigcap_{n\geq 1} \overline{\mathfrak{J}}_n = 0$, so $\bigcap_{n\geq 1} \gamma_n(\mathfrak{G}) = 1$. Thus $\mathfrak{G}$ satisfies a), b), c) of 12.8.

We prove next that $\mathfrak{G}$ is infinite. There exists an associative homomorphism ρ of $\mathsf{K}\mathfrak{G}$ into $\mathfrak{B}$ induced by the embedding of $\mathfrak{G}$ in $\mathfrak{B}$. Now $\operatorname{im}\rho$ contains $1 + y_i$ and 1; hence $\operatorname{im}\rho$ contains y_i and ρ is an epimorphism. Since $\mathfrak{B}$ is not of finite dimension over K, neither is $\mathsf{K}\mathfrak{G}$. Thus $\mathfrak{G}$ is infinite.

It follows from 2.1 that $\mathfrak{G}$ is not nilpotent. Let $\mathfrak{K}$ be a subgroup of $\mathfrak{G}$ generated by fewer than d elements. Then there is a subset V of $\overline{\mathfrak{J}}$ having fewer than d elements, such that $\mathfrak{K} = \langle 1 + v | v \in V \rangle$. By 12.7d), there exists an integer r such that the product of any r elements of V is 0. If σ is the homomorphism of $\mathsf{K}\mathfrak{K}$ into $\mathfrak{B}$ induced by the embedding of $\mathfrak{K}$ in $\mathfrak{B}$ and $\mathfrak{J}$ is the augmentation ideal of $\mathsf{K}\mathfrak{K}$, then $\mathfrak{J}\sigma$ consists of sums of products of elements of V. Hence $\mathfrak{J}^r\sigma = 0$. But it follows from 2.3 that if $x \in \gamma_s(\mathfrak{K})$, then $x - 1 \in \mathfrak{J}^s$. Since σ is injective on $\mathfrak{K}$, it follows that $\gamma_r(\mathfrak{K}) = 1$ and $\mathfrak{K}$ is nilpotent. Thus $\mathfrak{G}$ also satisfies d). **q.e.d.**

12.9 Theorem. *Let* K *be a countable field and let* d *be an integer greater than 1. There exists a Lie algebra* $\mathfrak{g}$ *over* K *which has the following properties.*

a) $\mathfrak{g}$ *is generated by* d *elements.*

b) $\mathfrak{g}$ *is not of finite dimension over* K.

c) $\bigcap_{n\geq 1} \mathfrak{g}^n = 0$.

d) $\mathfrak{g}$ *is not nilpotent, but any subalgebra of* $\mathfrak{g}$ *which can be generated by fewer than* d *elements is nilpotent.*

Proof. Let $\mathfrak{B}$ be the associative algebra constructed in 12.7. Suppose that $\mathfrak{B}$ is generated by $y_1, \ldots, y_d$ and let $\mathfrak{g}$ be the Lie subalgebra of $\mathfrak{l}(\mathfrak{B})$ generated by $y_1, \ldots, y_d$. Since $\mathfrak{g}^n \subseteq \mathsf{B}_n \oplus \mathsf{B}_{n+1} \oplus \cdots$, $\bigcap_{n\geq 1} \mathfrak{g}^n = 0$. It follows at once from 12.7d) that any subalgebra of $\mathfrak{g}$ which can be generated by fewer than d elements is nilpotent.

Suppose that $\mathfrak{g}$ is of finite dimension over K; let $g_1, \ldots, g_n$ be a basis of $\mathfrak{g}$. Clearly the associative subalgebra of $\mathfrak{B}$ generated by $\mathfrak{g}$ is $\mathfrak{B}$, since $\mathfrak{g}$ contains $y_1, \ldots, y_d$. By 11.1, it follows that $\mathfrak{B}$ is spanned by the set of elements $g_1^{i_1} \cdots g_n^{i_n}$. But by 12.7d), there is an integer r_i such that

$g_i^{r_i} = 0$ ($i = 1, 2, \ldots, n$). Hence $\mathfrak{B}$ is spanned by a finite set, contrary to 12.7b). Hence $\mathfrak{g}$ is not of finite dimension over K. By 8.7, $\mathfrak{g}^m/\mathfrak{g}^{m+1}$ is finitely generated and hence of finite dimension for all m. Thus $\mathfrak{g}$ is not nilpotent. **q.e.d.**

12.10 Remarks. a) Suppose that $d > 2$ in 12.9. If x, y are elements of $\mathfrak{g}$, x and y generate a nilpotent subalgebra, so there is an integer n such that $x(\operatorname{ad} y)^n = 0$. Similarly, if $d > 2$ in 12.8 and x, y are elements of $\mathfrak{G}$, there is an integer n such that $[x, \underbrace{y, \ldots, y}_{n}] = 1$. Thus there exist non-nilpotent groups and Lie rings in which the Engel condition $[x, y, \ldots, y] = 1$ (or 0) holds.

b) Another construction of finite p-groups from the free associative algebra was given by WALL [2] to prove that for $2 \leq d \leq p^m$, there is a finite p-group $\mathfrak{G}$ with the following properties.

(i) $|\mathfrak{G} : \Phi(\mathfrak{G})| = p^d$.

(ii) The exponent of $\mathfrak{G}$ is p^{m+1}.

(iii) The centre $\mathfrak{Z}$ of $\mathfrak{G}$ is cyclic, and if $\Omega_1(\mathfrak{Z}) = \langle z \rangle$, z is a power of every element of $\mathfrak{G} - \Phi(\mathfrak{G})$.

§ 13. Automorphisms of p-Groups

If $\mathfrak{G}$ is a finite p-group, $\mathfrak{G}/\Phi(\mathfrak{G})$ is an elementary Abelian p-group and can therefore be regarded as a vector space V over $GF(p)$. Thus the group of automorphisms of $\mathfrak{G}$ induces a subgroup $\mathfrak{H}$ of $GL(n, p)$ on V. In this section we shall prove a theorem of Bryant and Kovács which asserts that any subgroup $\mathfrak{H}$ of $GL(n, p)$ can arise in this way. We begin with an elementary lemma.

13.1 Lemma. *Let* V *be a vector space over an infinite field* K *and let S be a finite set of linear transformations of* V *containing no scalar multiplications. Then there exists $v \in$ V such that v, vs are linearly independent for all $s \in S$.*

Proof. Suppose that this is false. Then given $v \in$ V, there exists $s \in S$ such that $vs = \lambda v$ for some $\lambda \in$ K. Hence V is the union of the eigen-spaces of the elements of S. Since no element of S is a scalar multiplication, no eigen-space of an element of S is the whole of V. Since S is finite, it follows that V is the union of a finite number of proper subspaces. But this is impossible, since K is infinite. **q.e.d.**

The proof of the main theorem is by Lie methods; the theorem about Lie algebras that we need is the following.

13.2 Theorem (BRYANT and KOVÁCS [1]). *Let $\mathfrak{g}$ be a free Lie algebra over a field K with Lie basis X. For each $n \geq 1$, let G_n be the K-space spanned by all products $[x_1, \ldots, x_n]$ with $x_i \in X$. Let $\mathfrak{G}$ be a group for which G_1 is a faithful $\mathsf{K}\mathfrak{G}$-module.*

a) *$\mathfrak{g}$ has the structure of a $\mathsf{K}\mathfrak{G}$-module in which G_1 is a $\mathsf{K}\mathfrak{G}$-submodule and given $g \in \mathfrak{G}$, the mapping $a \to ag$ ($a \in \mathfrak{g}$) is an automorphism of the Lie algebra $\mathfrak{g}$. Each G_n is a $\mathsf{K}\mathfrak{G}$-submodule of $\mathfrak{g}$.*

b) *Suppose that $|X| > 1$ and $\mathfrak{G}$ is finite. Let $\mathfrak{Z}$ be the set of elements of $\mathfrak{G}$ which induce scalar multiplications on G_1, and put $q = |\mathfrak{Z}|$. Then there exists n_0 such that for all $n > n_0$, $\mathsf{G}_n \oplus \mathsf{G}_{n+1} \oplus \cdots \oplus \mathsf{G}_{n+q-1}$ has a $\mathsf{K}\mathfrak{G}$-submodule for the regular representation of $\mathfrak{G}$.*

Proof. By 11.5b), $\mathfrak{g}$ is the Lie subalgebra of $\mathrm{I}(\mathfrak{A})$ generated by X, where $\mathfrak{A}$ is the free associative algebra with basis X. Let H_n be the homogeneous component of $\mathfrak{A}$ of degree n ($n \geq 0$). Then

$$\mathfrak{A} = \mathsf{H}_0 \oplus \mathsf{H}_1 \oplus \cdots.$$

Since $\mathsf{G}_n \subseteq \mathsf{H}_n$, we have

$$\mathfrak{g} = \mathsf{G}_1 \oplus \mathsf{G}_2 \oplus \cdots$$

and $\mathsf{G}_n = \mathsf{H}_n \cap \mathfrak{g}$. Suppose that $g \in \mathfrak{G}$. Since $\mathfrak{A}$ is free, there is a unique endomorphism α_g of $\mathfrak{A}$ such that $x\alpha_g = xg$ for each $x \in X$. Then $\alpha_{g_1}\alpha_{g_2} = \alpha_{g_1 g_2}$ and $\alpha_1 = 1$, so α_g is an automorphism of $\mathfrak{A}$. Also $y\alpha_g = yg$ for all $y \in \mathsf{H}_1 = \mathsf{G}_1$. We put $yg = y\alpha_g$ for all $y \in \mathfrak{A}$. Then $\mathfrak{A}$ is a $\mathsf{K}\mathfrak{G}$-module and G_1 is a $\mathsf{K}\mathfrak{G}$-submodule of $\mathfrak{A}$. Also H_n, $\mathfrak{g}$, G_n are all $\mathsf{K}\mathfrak{G}$-submodules, and α_g induces a Lie automorphism on $\mathfrak{g}$. Thus a) is proved.

We observe next that if b) holds for free Lie algebras over infinite fields, then it holds in general. For if K is any field, there exists an infinite field L containing K. Then $\mathfrak{A} \otimes_{\mathsf{K}} \mathsf{L}$ is the free associative algebra over L with basis X and $\mathfrak{g} \otimes_{\mathsf{K}} \mathsf{L}$ is the Lie subalgebra of $\mathrm{I}(\mathfrak{A} \otimes_{\mathsf{K}} \mathsf{L})$ generated by X. Also $\mathsf{G}_n \otimes_{\mathsf{K}} \mathsf{L}$ is the L-space spanned by all products $[x_1, \ldots, x_n]$ with $x_i \in X$, and $\mathsf{G}_1 \otimes_{\mathsf{K}} \mathsf{L}$ is an $\mathsf{L}\mathfrak{G}$-module. Finally, $\mathfrak{Z}$ is the set of elements of $\mathfrak{G}$ which induce scalar multiplications on $\mathsf{G}_1 \otimes_{\mathsf{K}} \mathsf{L}$. Assuming, then, that the theorem holds for Lie algebras over L, there exists n_0 such that for all $n > n_0$, $(\mathsf{G}_n \oplus \cdots \oplus \mathsf{G}_{n+q-1}) \otimes_{\mathsf{K}} \mathsf{L}$ contains an $\mathsf{L}\mathfrak{G}$-submodule for the regular representation of $\mathfrak{G}$. By VII, 7.23, (or, if X is infinite, by its (immediate) extension to the infinite-dimensional case), $\mathsf{G}_n \oplus \cdots \oplus \mathsf{G}_{n+q-1}$ contains a $\mathsf{K}\mathfrak{G}$-submodule for the regular representation of $\mathfrak{G}$.

We shall suppose, then, that K is infinite. This will enable us to use 13.1.

Let T be a transversal of $\mathfrak{Z}$ in $\mathfrak{G}$ containing 1. We prove the following.

(*) Suppose that $k > 3(|\mathfrak{G}:\mathfrak{Z}| - 1)$. Then there exists $a \in \mathsf{G}_k$ such that $\{at | t \in \mathsf{T}\}$ is linearly independent.

If $\mathfrak{Z} = \mathfrak{G}$, there is nothing to prove, since $|X| > 1$, so we suppose that $\mathfrak{Z} < \mathfrak{G}$. By 13.1, there exists $b \in \mathsf{H}_1$ such that b, bg are linearly independent for all $g \in \mathfrak{G} - \mathfrak{Z}$. Choose $c \in \mathsf{H}_1$ such that b, c are linearly independent. Then there exist x, y in X and a non-singular linear transformation α of H_1 such that $x\alpha = b$, $y\alpha = c$. But

$$y(\operatorname{ad} x)^{k-1} = \sum_{i=0}^{k-1} \binom{k-1}{i}(-1)^i x^i y x^{k-i-1} \neq 0.$$

By a), α can be extended to an automorphism of $\mathfrak{g}$. Hence if

$$(1) \qquad a = c(\operatorname{ad} b)^{k-1} = \sum_{i=0}^{k-1} \binom{k-1}{i}(-1)^i b^i c b^{k-i-1},$$

a is a non-zero element of G_k.

We shall define two subspaces B, C of H_k such that $\mathsf{H}_k = \mathsf{B} \oplus \mathsf{C}$, $a \in \mathsf{B}$ and $at \in \mathsf{C}$ for all $t \in \mathsf{T} - \{1\}$. To do this, let $t_1, \ldots, t_m$ be the non-identity elements of T. Thus $m + 1 = |\mathfrak{G}:\mathfrak{Z}|$ and $k > 3m$. Since b, bt_j are linearly independent ($j = 1, \ldots, m$), there exist subspaces X_j, Y_j of H_1 such that $b \in \mathsf{X}_j$, $bt_j \in \mathsf{Y}_j$ and $\mathsf{H}_1 = \mathsf{X}_j \oplus \mathsf{Y}_j$. Then H_k is the direct sum of all its subspaces of the form

$$\mathsf{Z}_1\mathsf{Z}_2 \cdots \mathsf{Z}_{3m}\mathsf{H}_1 \cdots \mathsf{H}_1,$$

where Z_{3j-2}, Z_{3j-1}, Z_{3j} are each either X_j or Y_j ($j = 1, \ldots, m$) and $\mathsf{Z}_1\mathsf{Z}_2 \cdots \mathsf{Z}_{3m}\mathsf{H}_1 \cdots \mathsf{H}_1$ is the space spanned by

$$\{z_1 z_2 \cdots z_{3m} a_1 \cdots a_{k-3m} | z_i \in \mathsf{Z}_i, a_l \in \mathsf{H}_1\}.$$

Let B be the sum of all these subspaces for which at most one of the Z_i is a Y_j and let C be the sum of the remaining ones. Thus $\mathsf{H}_k = \mathsf{B} \oplus \mathsf{C}$ and, since $b \in \mathsf{X}_j$, $b^i c b^{k-i-1} \in \mathsf{B}$ ($i = 0, \ldots, k-1$). Thus $a \in \mathsf{B}$. But $bt_j \in \mathsf{Y}_j$, so at least two of the terms in the $(3j-2)$-*th*, $(3j-1)$-*th* or $(3j)$-*th* places of $(bt_j)^i(ct_j)(bt_j)^{k-i-1}$ lie in Y_j. Hence $(b^i c b^{k-i-1})t_j \in \mathsf{C}$ and, by (1), $at_j \in \mathsf{C}$.

Since $a \neq 0$, it follows that $a \notin \sum_{h \notin \mathfrak{Z}} \mathsf{K}ah$. Hence for any $t \in \mathsf{T}$, $at \notin \sum_{h \notin \mathfrak{Z}t} \mathsf{K}ah$, and $\{at | t \in \mathsf{T}\}$ is linearly independent. Thus (*) is proved.

Now G_1 is a faithful $\mathsf{K}\mathfrak{G}$-module. Thus $\mathfrak{Z}$, being isomorphic to a finite subgroup of $\mathsf{K}^\times$, is cyclic. Suppose that $\mathfrak{Z} = \langle z \rangle$ and that $az = \lambda a$ for all $a \in \mathsf{H}_1$. Thus $az = \lambda^k a$ for all $a \in \mathsf{H}_k$.

We take $n_0 = 3(|\mathfrak{G} : \mathfrak{Z}| - 1)$. If $n > n_0$, then for $k = n, \dots, n + q - 1$, (*) shows that G_k contains an element a_k such that $\{a_k t \mid t \in \mathsf{T}\}$ is linearly independent. Let $\mathsf{U}_k = \mathsf{K}a_k$ and

$$\mathsf{N} = \mathsf{U}_n \oplus \cdots \oplus \mathsf{U}_{n+q-1};$$

also let $\mathsf{V}_k = \sum_{t \in \mathsf{T}} \mathsf{K}a_k t$ and

$$\mathsf{M} = \mathsf{V}_n \oplus \cdots \oplus \mathsf{V}_{n+q-1}.$$

Then U_k is a $\mathsf{K}\mathfrak{Z}$-module for the representation $z \to \lambda^k$ of $\mathfrak{Z}$, so N is a $\mathsf{K}\mathfrak{Z}$-module for the regular representation of $\mathfrak{Z}$. Also, V_k is a $\mathsf{K}\mathfrak{G}$-module isomorphic to the induced module $\mathsf{U}_k^{\mathfrak{G}}$, so M is a $\mathsf{K}\mathfrak{G}$-module isomorphic to $\mathsf{N}^{\mathfrak{G}}$. Hence M is a $\mathsf{K}\mathfrak{G}$-submodule of $\mathsf{G}_n \oplus \cdots \oplus \mathsf{G}_{n+q-1}$ for the regular representation of $\mathfrak{G}$. **q.e.d.**

We shall give an application of 13.2 which involves the subgroups $\lambda_n(\mathfrak{G})$ defined in 1.4. For this the following lemma is needed.

13.3 Lemma. *Suppose that $\mathfrak{F}$ is a free group.*

a) *If $\mathfrak{N} \trianglelefteq \mathfrak{F}$ and α is an endomorphism of $\mathfrak{F}/\mathfrak{N}$, there exists an endomorphism β of $\mathfrak{F}$ which carries $\mathfrak{N}$ into itself and induces α on $\mathfrak{F}/\mathfrak{N}$.*

b) *Suppose that p is a prime and β is an endomorphism of $\mathfrak{F}$. If β induces an automorphism on $\mathfrak{F}/\lambda_2(\mathfrak{F})$, then β induces an automorphism on $\mathfrak{F}/\lambda_{n+1}(\mathfrak{F})$ for all $n \geq 1$.*

Proof. a) Let X be a basis of $\mathfrak{F}$. For each $x \in X$, choose $x' \in \mathfrak{F}$ such that $(x\mathfrak{N})\alpha = x'\mathfrak{N}$. There exists an endomorphism β of $\mathfrak{F}$ such that $x' = x\beta$. Thus $(x\mathfrak{N})\alpha = (x\beta)\mathfrak{N}$ for all $x \in X$. Hence for any $f \in \mathfrak{F}$, $(f\mathfrak{N})\alpha = (f\beta)\mathfrak{N}$, so β has the required properties.

b) Let α be the automorphism of $\mathfrak{F}/\lambda_2(\mathfrak{F})$ induced by β. By a), there is an endomorphism β' of $\mathfrak{F}$ which induces α^{-1} on $\mathfrak{F}/\lambda_2(\mathfrak{F})$. Thus $\beta\beta'$ and $\beta'\beta$ induce the identity automorphism on $\mathfrak{F}/\lambda_2(\mathfrak{F})$. By 1.7a), $\beta\beta'$ and $\beta'\beta$ induce the identity automorphism on $\lambda_i(\mathfrak{F})/\lambda_{i+1}(\mathfrak{F})$ for all $i \geq 1$. Hence β induces an automorphism on $\lambda_i(\mathfrak{F})/\lambda_{i+1}(\mathfrak{F})$. It follows by a simple induction on n that β induces an automorphism on $\mathfrak{F}/\lambda_{n+1}(\mathfrak{F})$. **q.e.d.**

We shall now use the main theorem of § 11 to deduce the following from 13.2.

13.4 Theorem (BRYANT and KOVÁCS [1]). *Let $\mathfrak{F}$ be a free group, let p be a prime and let $\mathsf{K} = GF(p)$. Let $\mathfrak{G}$ be a group for which $\mathfrak{F}/\lambda_2(\mathfrak{F})$ is a $\mathsf{K}\mathfrak{G}$-module.*

a) *For each $n \geq 1$, $\lambda_n(\mathfrak{F})/\lambda_{n+1}(\mathfrak{F})$ has the structure of a $\mathsf{K}\mathfrak{G}$-module such that whenever β is an endomorphism of $\mathfrak{F}$, $g \in \mathfrak{G}$ and $(f\lambda_2(\mathfrak{F}))g = (f\beta)\lambda_2(\mathfrak{F})$ for all $f \in \mathfrak{F}$, then $(f\lambda_{n+1}(\mathfrak{F}))g = (f\beta)\lambda_{n+1}(\mathfrak{F})$ for all $f \in \lambda_n(\mathfrak{F})$.*

b) *Suppose that $\mathfrak{F}$ is non-cyclic, $\mathfrak{G}$ is finite and that $\mathfrak{F}/\lambda_2(\mathfrak{F})$ is a faithful $\mathsf{K}\mathfrak{G}$-module. Then given a positive integer r, there exists n_0 such that for all $n > n_0$, $\lambda_n(\mathfrak{F})/\lambda_{n+1}(\mathfrak{F})$ contains a free $\mathsf{K}\mathfrak{G}$-submodule of rank r.*

Proof. a) Suppose that $g \in \mathfrak{G}$. By 13.3a), there exists an endomorphism β of $\mathfrak{F}$ such that $(f\beta)\lambda_2(\mathfrak{F}) = (f\lambda_2(\mathfrak{F}))g$ for all $f \in \mathfrak{F}$. Then $\lambda_i(\mathfrak{F})\beta \subseteq \lambda_i(\mathfrak{F})$, so we may put

$$(f\lambda_{n+1}(\mathfrak{F}))g = (f\beta)\lambda_{n+1}(\mathfrak{F})$$

for all $f \in \lambda_n(\mathfrak{F})$. It follows from 1.7a) that if β' is *any* endomorphism of $\mathfrak{F}$ such that $(f\beta')\lambda_2(\mathfrak{F}) = (f\lambda_2(\mathfrak{F}))g$, then

$$(f\lambda_{n+1}(\mathfrak{F}))g = (f\beta')\lambda_{n+1}(\mathfrak{F}).$$

It follows at once that $\lambda_n(\mathfrak{F})/\lambda_{n+1}(\mathfrak{F})$ is a $\mathsf{K}\mathfrak{G}$-module.

b) Let X be a basis of $\mathfrak{F}$. By 11.14, there exists an isomorphism ρ_i of $\gamma_i(\mathfrak{F})/\gamma_{i+1}(\mathfrak{F})$ onto the additive subgroup generated by all Lie products $[x_1 - 1, \ldots, x_i - 1]$ $(x_j \in X)$ in the group-ring $\mathbb{Z}\mathfrak{F}$ of $\mathfrak{F}$, and ρ_i is determined by the fact that if $f \in \gamma_i(\mathfrak{F})$,

$$f - 1 - (f\gamma_{i+1}(\mathfrak{F}))\rho_i \in \mathfrak{J}^{i+1},$$

where $\mathfrak{J}$ is the augmentation ideal of $\mathbb{Z}\mathfrak{F}$. Let

$$\mathfrak{C}_i = \gamma_i(\mathfrak{F})/\gamma_i(\mathfrak{F})^p\gamma_{i+1}(\mathfrak{F}) \quad (i = 1, 2, \ldots).$$

Since $\mathsf{K}\mathfrak{F} \cong \mathbb{Z}\mathfrak{F}/p\mathbb{Z}\mathfrak{F}$, it follows that there exists an isomorphism σ_i of $\mathfrak{C}_i$ onto the additive subgroup G_i of $\mathsf{K}\mathfrak{F}$ generated by all elements $[x_1 - 1, \ldots, x_i - 1]$ $(x_j \in X)$, determined by the fact that if $f \in \gamma_i(\mathfrak{F})$,

$$f - 1 - (f\gamma_i(\mathfrak{F})^p\gamma_{i+1}(\mathfrak{F}))\sigma_i \in \mathfrak{J}^{i+1},$$

where $\mathfrak{J}$ is the augmentation ideal of $\mathsf{K}\mathfrak{F}$. In particular, σ_1 is an isomorphism of $\mathfrak{C}_1 = \mathfrak{F}/\lambda_2(\mathfrak{F})$ onto G_1 and if $x \in X$, $(x\lambda_2(\mathfrak{F}))\sigma_1 = x - 1$. If $f_1 \in \gamma_i(\mathfrak{F})$, $f_2 \in \gamma_j(\mathfrak{F})$,

$$[f_1 - 1, f_2 - 1]$$
$$\equiv [(f_1\gamma_i(\mathfrak{F})^p\gamma_{i+1}(\mathfrak{F}))\sigma_i, (f_2\gamma_j(\mathfrak{F})^p\gamma_{j+1}(\mathfrak{F}))\sigma_j] \mod \mathfrak{J}^{i+j+1},$$

and by 2.3c),

$$[f_1 - 1, f_2 - 1] \equiv [f_1, f_2] - 1 \mod \mathfrak{J}^{i+j+1}.$$

Hence since $\mathfrak{J}^{i+j+1} \cap \mathsf{G}_{i+j} = 0$,

$$\text{(1)} \qquad ([f_1, f_2]\gamma_{i+j}(\mathfrak{F})^p\gamma_{i+j+1}(\mathfrak{F}))\sigma_{i+j} = [(f_1\gamma_i(\mathfrak{F})^p\gamma_{i+1}(\mathfrak{F}))\sigma_i, (f_2\gamma_j(\mathfrak{F})^p\gamma_{j+1}(\mathfrak{F}))\sigma_j].$$

By 11.13, the associative subalgebra $\mathfrak{A}$ of $\mathsf{K}\mathfrak{F}$ generated by $B = \{x - 1 | x \in X\}$ is free with basis B, and, by 11.10, the Lie subalgebra $\mathfrak{g}$ of $\mathrm{I}(\mathfrak{A})$ generated by B is a free Lie algebra with basis B. We make G_1 into a $\mathsf{K}\mathfrak{G}$-module in such a way that σ_1 is a $\mathsf{K}\mathfrak{G}$-isomorphism. Thus G_1 is a faithful $\mathsf{K}\mathfrak{G}$-module. By 13.2, $\mathfrak{g}$ has the structure of a $\mathsf{K}\mathfrak{G}$-module such that G_1 is a $\mathsf{K}\mathfrak{G}$-submodule of $\mathfrak{g}$ and, given $g \in \mathfrak{G}$, the mapping $a \to ag$ ($a \in \mathfrak{g}$) is an automorphism of the Lie algebra $\mathfrak{g}$. Each G_i is a $\mathsf{K}\mathfrak{G}$-submodule of $\mathfrak{g}$, so $\mathfrak{C}_i$ can be made into a $\mathsf{K}\mathfrak{G}$-module in such a way that σ_i is a $\mathsf{K}\mathfrak{G}$-isomorphism. We prove by induction on i that if β is an endomorphism of $\mathfrak{F}$, $g \in \mathfrak{G}$ and $(x\lambda_2(\mathfrak{F}))g = (x\beta)\lambda_2(\mathfrak{F})$ for all $x \in \mathfrak{F}$, then

$$\text{(2)} \qquad (y\gamma_i(\mathfrak{F})^p\gamma_{i+1}(\mathfrak{F}))g = (y\beta)\gamma_i(\mathfrak{F})^p\gamma_{i+1}(\mathfrak{F})$$

for all $y \in \gamma_i(\mathfrak{F})$. This is trivial for $i = 1$, and for $i > 1$, it is sufficient to prove it for $y = [f_1, f_2]$ with $f_1 \in \gamma_{i-1}(\mathfrak{F})$, $f_2 \in \mathfrak{F}$. Then by (1),

$$(y\gamma_i(\mathfrak{F})^p\gamma_{i+1}(\mathfrak{F}))\sigma_i = [(f_1\gamma_{i-1}(\mathfrak{F})^p\gamma_i(\mathfrak{F}))\sigma_{i-1}, (f_2\mathfrak{F}^p\gamma_2(\mathfrak{F}))\sigma_1].$$

Since $a \to ag$ is a Lie automorphism,

$$(y\gamma_i(\mathfrak{F})^p\gamma_{i+1}(\mathfrak{F}))\sigma_i g = [(f_1\gamma_{i-1}(\mathfrak{F})^p\gamma_i(\mathfrak{F}))\sigma_{i-1}g, (f_2\mathfrak{F}^p\gamma_2(\mathfrak{F}))\sigma_1 g].$$

Since the σ_j are $\mathsf{K}\mathfrak{G}$-isomorphisms,

$$((y\gamma_i(\mathfrak{F})^p\gamma_{i+1}(\mathfrak{F}))g)\sigma_i = [((f_1\gamma_{i-1}(\mathfrak{F})^p\gamma_i(\mathfrak{F}))g)\sigma_{i-1}, ((f_2\mathfrak{F}^p\gamma_2(\mathfrak{F}))g)\sigma_1].$$

Using the inductive hypothesis and (1), this gives

$$\begin{aligned}((y\gamma_i(\mathfrak{F})^p\gamma_{i+1}(\mathfrak{F}))g)\sigma_i &= [((f_1\beta)\gamma_{i-1}(\mathfrak{F})^p\gamma_i(\mathfrak{F}))\sigma_{i-1}, ((f_2\beta)\mathfrak{F}^p\gamma_2(\mathfrak{F}))\sigma_1] \\ &= ([f_1\beta, f_2\beta]\gamma_i(\mathfrak{F})^p\gamma_{i+1}(\mathfrak{F}))\sigma_i \\ &= (([f_1, f_2]\beta)\gamma_i(\mathfrak{F})^p\gamma_{i+1}(\mathfrak{F}))\sigma_i \\ &= ((y\beta)\gamma_i(\mathfrak{F})^p\gamma_{i+1}(\mathfrak{F}))\sigma_i.\end{aligned}$$

This implies (2).

By 13.2, there exist integers n_0, q such that for all $n > n_0$, $\mathsf{G}_{n-rq+1} \oplus \cdots \oplus \mathsf{G}_{n-(r-1)q}$ has a $\mathsf{K}\mathfrak{G}$-submodule for the regular representation of $\mathfrak{G}$. Hence $\mathsf{G}_{n-rq+1} \oplus \cdots \oplus \mathsf{G}_n$ has a free $\mathsf{K}\mathfrak{G}$-submodule of rank r. Thus the direct product $\mathfrak{C}_{n-rq+1} \times \cdots \times \mathfrak{C}_n$ is a $\mathsf{K}\mathfrak{G}$-module and has a free $\mathsf{K}\mathfrak{G}$-submodule of rank r. But since $\gamma_i(\mathfrak{F})/\gamma_{i+1}(\mathfrak{F})$ is torsion-free, it follows from 1.9 that, provided $n - rq + 1 > 1$, there is a monomorphism α of $\mathfrak{C}_{n-rq+1} \times \cdots \times \mathfrak{C}_n$ into $\lambda_n(\mathfrak{F})/\lambda_{n+1}(\mathfrak{F})$ given by

$$(\bar{a}_{n-rq+1}, \ldots, \bar{a}_n)\alpha = a_{n-rq+1}^{p^{rq-1}} \cdots a_n\lambda_{n+1}(\mathfrak{F}),$$

where $a_i \in \gamma_i(\mathfrak{F})$ and $\bar{a}_i = a_i\gamma_i(\mathfrak{F})^p\gamma_{i+1}(\mathfrak{F})$. To prove the theorem, it is thus sufficient to show that α is a $\mathsf{K}\mathfrak{G}$-homomorphism.

Suppose that $g \in \mathfrak{G}$. By 13.3a), there is an endomorphism β of $\mathfrak{F}$ such that

$$(f\beta)\lambda_2(\mathfrak{F}) = (f\lambda_2(\mathfrak{F}))g$$

for all $f \in \mathfrak{F}$. It follows from (2) that

$$\bar{a}_i g = (a_i\beta)\gamma_i(\mathfrak{F})^p\gamma_{i+1}(\mathfrak{F}).$$

Hence

$$\begin{aligned}((\bar{a}_{n-rq+1}, \ldots, \bar{a}_n)g)\alpha &= (a_{n-rq+1}\beta)^{p^{rq-1}} \cdots (a_n\beta)\lambda_{n+1}(\mathfrak{F}) \\ &= ((a_{n-rq+1}^{p^{rq-1}} \cdots a_n)\beta)\lambda_{n+1}(\mathfrak{F}).\end{aligned}$$

By a), the right-hand side is

$$(a_{n-rq+1}^{p^{rq-1}} \cdots a_n\lambda_{n+1}(\mathfrak{F}))g = ((\bar{a}_{n-rq+1}, \ldots, \bar{a}_n)\alpha)g.$$

Thus b) is proved. **q.e.d.**

The main result of this section will be deduced from 13.4.

13.5 Theorem (BRYANT and KOVÁCS [1]). *Let* V *be a vector space of dimension greater than 1 over* $GF(p)$*, and let* $\mathfrak{H}$ *be a subgroup of the group of non-singular linear transformations of* V*. Then there exists a finite* p*-group* $\mathfrak{P}$ *such that* $\mathfrak{P}/\Phi(\mathfrak{P})$ *is isomorphic to* V *and the group of automorphisms of* $\mathfrak{P}/\Phi(\mathfrak{P})$ *induced by all the automorphisms of* $\mathfrak{P}$ *corresponds to* $\mathfrak{H}$.

Proof. We may suppose that $\mathsf{V} = \mathfrak{F}/\lambda_2(\mathfrak{F})$ for some non-cyclic finitely generated free group $\mathfrak{F}$. Let $\mathfrak{G}$ be the group of all automorphisms of $\mathfrak{F}/\lambda_2(\mathfrak{F})$. Thus $\mathfrak{H} \leq \mathfrak{G}$ and $\mathfrak{F}/\lambda_2(\mathfrak{F})$ can be regarded as a faithful $\mathsf{K}\mathfrak{G}$-module, where $\mathsf{K} = GF(p)$. By 13.4, each $\lambda_n(\mathfrak{F})/\lambda_{n+1}(\mathfrak{F})$ has the structure of a $\mathsf{K}\mathfrak{G}$-module such that whenever $g \in \mathfrak{G}$ and β is an endomorphism of $\mathfrak{F}$, the equation

$$(f\lambda_{n+1}(\mathfrak{F}))g = (f\beta)\lambda_{n+1}(\mathfrak{F}) \quad (f \in \lambda_n(\mathfrak{F}))$$

holds for all n provided that it holds for $n = 1$. Further, there exists $n > 1$ such that $\lambda_n(\mathfrak{F})/\lambda_{n+1}(\mathfrak{F})$ has a $\mathsf{K}\mathfrak{G}$-submodule for the regular representation of $\mathfrak{G}$. Thus there is a $\mathsf{K}\mathfrak{G}$-isomorphism θ of $\mathsf{K}\mathfrak{G}$ into $\lambda_n(\mathfrak{F})/\lambda_{n+1}(\mathfrak{F})$. Let $\mathfrak{N}/\lambda_{n+1}(\mathfrak{F}) = (\mathsf{K}\mathfrak{H})\theta$. Thus $\mathfrak{N}/\lambda_{n+1}(\mathfrak{F})$ is a $\mathsf{K}\mathfrak{H}$-submodule of $\lambda_n(\mathfrak{F})/\lambda_{n+1}(\mathfrak{F})$ and an element g of $\mathfrak{G}$ lies in $\mathfrak{H}$ if and only if $(\mathfrak{N}/\lambda_{n+1}(\mathfrak{F}))g \leq \mathfrak{N}/\lambda_{n+1}(\mathfrak{F})$. Since $\lambda_{n+1}(\mathfrak{F}) \leq \mathfrak{N} \leq \lambda_n(\mathfrak{F})$, $\mathfrak{N} \trianglelefteq \mathfrak{F}$ and $\mathfrak{P} = \mathfrak{F}/\mathfrak{N}$ is a finite p-group, by 2.1.

Let α be an automorphism of $\mathfrak{P}$ and let g denote the automorphism of $\mathfrak{F}/\lambda_2(\mathfrak{F})$ induced by α. By 13.3a), there exists an endomorphism β of $\mathfrak{F}$ which carries $\mathfrak{N}$ into itself and induces α on $\mathfrak{F}/\mathfrak{N}$. Also $(f\lambda_2(\mathfrak{F}))g = (f\beta)\lambda_2(\mathfrak{F})$ for all $f \in \mathfrak{F}$, so

$$(f\lambda_{n+1}(\mathfrak{F}))g = (f\beta)\lambda_{n+1}(\mathfrak{F})$$

for all $f \in \lambda_n(\mathfrak{F})$. Since $\mathfrak{N}\beta \leq \mathfrak{N}$,

$$(\mathfrak{N}/\lambda_{n+1}(\mathfrak{F}))g \leq \mathfrak{N}/\lambda_{n+1}(\mathfrak{F})$$

and $g \in \mathfrak{H}$.

Conversely, suppose that $h \in \mathfrak{H}$. By 13.3a), there is an endomorphism β of $\mathfrak{F}$ which induces h on $\mathfrak{F}/\lambda_2(\mathfrak{F})$. Then if $f \in \mathfrak{N}$,

$$(f\beta)\lambda_{n+1}(\mathfrak{F}) = (f\lambda_{n+1}(\mathfrak{F}))h \in \mathfrak{N}/\lambda_{n+1}(\mathfrak{F}),$$

since $\mathfrak{N}/\lambda_{n+1}(\mathfrak{F})$ is a $\mathsf{K}\mathfrak{H}$-module. Thus $f\beta \in \mathfrak{N}$ and β carries $\mathfrak{N}$ into itself. By 13.3b), β induces an automorphism on $\mathfrak{F}/\lambda_{n+1}(\mathfrak{F})$ and hence on $\mathfrak{F}/\mathfrak{N} = \mathfrak{P}$.

Hence the group of automorphisms of $\mathfrak{F}/\lambda_2(\mathfrak{F})$ induced by all the automorphisms of $\mathfrak{P}$ is precisely $\mathfrak{H}$. **q.e.d.**

13.6 Remarks. a) Theorem 13.5 shows that any subgroup of $GL(n, p)$ is the linear group induced on $\mathfrak{P}/\Phi(\mathfrak{P})$ by **Aut** $\mathfrak{P}$ for some p-group $\mathfrak{P}$. Since any finite group is isomorphic to a subgroup of $GL(n, p)$ for some n, every finite group is isomorphic to the group induced on $\mathfrak{P}/\Phi(\mathfrak{P})$ by **Aut** $\mathfrak{P}$ for some p-group $\mathfrak{P}$. It has been proved by HEINEKEN and LIEBECK [1] that for p odd, any finite group is isomorphic to the group induced on $\mathfrak{G}/\mathbf{Z}(\mathfrak{G})$ by **Aut** $\mathfrak{G}$ for some p-group $\mathfrak{G}$ of class 2 and exponent p^2.

b) HARTLEY and ROBINSON [1] have used Theorem 13.4 to prove that given a non-identity p'-group $\mathfrak{H}$, there exists a group $\mathfrak{G}$ such that $\mathbf{F}(\mathfrak{G})$ is a p-group, $\mathfrak{G}/\mathbf{F}(\mathfrak{G}) \cong \mathfrak{H}$, $\mathbf{Z}(\mathfrak{G}) = 1$ and every automorphism of $\mathfrak{G}$ is inner.

c) Taking $\mathfrak{H} = 1$ in Theorem 13.5, we obtain a p-group in which every maximal subgroup is characteristic. However Heineken has constructed a p-group in which every normal subgroup is characteristic.

Notes on Chapter VIII

§ **1:** The two central series discussed here are among many which were given by LAZARD [1]. The κ-series had been introduced earlier by JENNINGS [1] and ZASSENHAUS [4]. The treatment of the λ-series follows BLACKBURN and EVENS [1].
§ **3:** The proof of Theorem 3.5 is that given in PASSMAN [3].
§ **4:** The method for solving the congruences in 4.1–4.5 is taken from SHAW [1].
§ **5:** Theorems 5.1–5.5 follow BLACKBURN [1]. Theorems 5.13 and 5.15 originated in G. HIGMAN [2].
§ **7:** The results in this section are based on G. HIGMAN [2].
§ **9:** Theorem 9.12 was first conjectured by J. WIEGOLD [1]. It was proved in various special cases prior to Vaughan-Lee's proof.
§ **10:** The fact that $\mathfrak{g}$ is nilpotent in 10.12 was proved by G. HIGMAN [1]. The bound on the class given here was obtained by Kreknin and Kostrikin for Lie algebras over a field.
§ **11:** An alternative treatment of free Lie algebras is given by BOURBAKI [3]. The approach to the lower central series of free groups was given by MAGNUS [1, 2]. The treatment given here is derived from WITT [1].
§ **12:** The Golod-Šafarevič construction follows FISCHER and STRUIK [1].
§ **13:** We are indebted to R. M. Bryant for some helpful comments on this section.

Chapter IX

Linear Methods and Soluble Groups

Linear methods have been used extensively for quantitative investigations of soluble groups; for soluble linear groups, bounds in terms of the degree are known for such invariants as the order, derived length, etc. In 1956, P. Hall and G. Higman developed these ideas to obtain upper bounds for the p-length of a p-soluble group $\mathfrak{G}$ in terms of various invariants of the Sylow p-subgroup of $\mathfrak{G}$. There emerged from this a body of techniques which have come to be known as Hall-Higman methods. The present chapter is an introduction to these methods. They are given in an elementary form in § 1, which is already sufficient to solve the restricted Burnside problem of exponent 6.

Let $\mathfrak{G}$ be a p-soluble group and let $\mathfrak{A}$, $\mathfrak{B}$ be normal subgroups of $\mathfrak{G}$ such that $\mathfrak{A} \geq \mathfrak{B}$ and $\mathfrak{A}/\mathfrak{B}$ is an elementary Abelian p-group. Then $\mathfrak{A}/\mathfrak{B}$ can be regarded as a vector space V over $GF(p)$ and there is a natural representation of $\mathfrak{G}$ on V. Representations over larger fields of characteristic p can be obtained by extension of the ground-field. Such representations were studied in general terms in Chapter VII, but here we are concerned with specific problems involving above all the minimum polynomial of the p-elements. These problems can be reduced to questions about faithful irreducible representations of groups of the form $\mathfrak{P}\mathfrak{Q}$, where $\mathfrak{P}$ is nilpotent and $\mathfrak{Q}$ is normal, in algebraically closed fields. Further conditions are usually imposed; in almost all such situations that have been studied, $(|\mathfrak{P}|, |\mathfrak{Q}|) = 1$ and the characteristic of the ground-field does not divide $|\mathfrak{Q}|$. Also, an extraspecial group often occurs in $\mathfrak{Q}$. The question asked is whether or not the representation module has a direct summand isomorphic to the group-ring of $\mathfrak{P}$. If, for instance, this is the case and g is an element of $\mathfrak{P}$ of order k, the minimum polynomial of g is $t^k - 1$.

Hall and Higman's Theorem B (Theorem 2.9) is the classical case of such a theorem; it states that in a p-soluble group $\mathfrak{G}$ of linear transformations of a vector space over a field of characteristic p satisfying the condition $\mathbf{O}_p(\mathfrak{G}) = 1$, the minimum polynomial of an element of order p^n is *in general* $(t - 1)^{p^n}$. But there are exceptional cases and these are studied in § 3, again with specific applications in mind. One of these is a bound on the p-length of a p-soluble group in terms of the exponent of a Sylow p-subgroup, and this can be used to reduce the restricted Burnside problem for soluble groups to the case of prime-power exponent. The effectiveness of this reduction is evident in view of Kostrikin's theorem on groups of prime exponent (VIII, 12.4).

Further applications of Theorem B are given in § 5. Among these are the theorem that for p odd, the p-length of a p-soluble group is at most the derived length of a Sylow p-subgroup, and similar bounds for the p-length in terms of the class and order of the Sylow p-subgroup are given. It is a consequence that the derived length of a soluble group is bounded in terms of the derived lengths of the Sylow subgroups. Another application is the proof that a simple group in which every proper subgroup is p-soluble for some odd prime divisor p of the order can always be generated by 3 elements.

The global invariant of a soluble group $\mathfrak{G}$ corresponding to the p-length is the Fitting height $h(\mathfrak{G})$, and the Hall-Higman methods have been used to give upper bounds for this. For instance, Dade found such a bound for $h(\mathfrak{G})$ in terms of the composition length of a Carter subgroup. And if $\mathfrak{A}$ is a soluble group of automorphisms of $\mathfrak{G}$ for which $(|\mathfrak{A}|, |\mathfrak{G}|) = 1$ and $|\mathfrak{A}| = p_1 p_2 \cdots p_n$, where $p_1, \ldots, p_n$ are not necessarily distinct primes, Thompson and Kurzweil found bounds for $h(\mathfrak{G})$ in terms of $h(\mathbf{C}_{\mathfrak{G}}(\mathfrak{A}))$ and n. If, further, $\mathbf{C}_{\mathfrak{G}}(\mathfrak{A}) = 1$, this reduces to a bound on $h(\mathfrak{G})$ in terms of n alone, and it is conjectured that $h(\mathfrak{G}) \leq n$. Substantial progress towards the proof of this has been made by Berger. Unfortunately, simple proofs are known for none of these results. But the analogue of Theorem B, in which the characteristic of the field does not divide the order of the p-soluble group, can be used to prove some special cases of this conjecture (§ 6).

Further applications of the Hall-Higman theorems will be found in the next chapter, where they are used to obtained local information about possibly insoluble groups. Many applications of this nature require conditions under which $[x, y, y] = 1$ implies $[x, y] = 1$ for elements x, y of a group. In order to formalize this, the technical notion of p-stability is introduced. For $p > 3$, p-soluble groups are p-stable. This follows from Theorem B, but a more general result is that for p odd, all sections of a group $\mathfrak{G}$ are p-stable if and only if $\mathfrak{G}$ has no section isomorphic to the special affine group $SA(2, p)$. The proof of this makes use of the Dickson list of subgroups of $SL(2, p)$ rather than the representation theory techniques of the earlier part of the chapter.

Finally, a further criterion for p-length 1 is given in § 8.

§ 1. Introduction

The main aim in this chapter is the derivation of bounds for the p-length of a p-soluble group in terms of invariants of the Sylow p-subgroup. Some preliminary remarks about this theory were made in VI, § 6. We begin by enlarging on these.

1.1 Notation. If π is a set of primes, the product of all the normal π-subgroups of a finite group $\mathfrak{G}$ is a characteristic π-subgroup of $\mathfrak{G}$ denoted by $\mathbf{O}_\pi(\mathfrak{G})$. Thus $\mathbf{O}_\pi(\mathfrak{G})$ is the *maximal normal π-subgroup* of $\mathfrak{G}$, and $\mathbf{O}_\pi(\mathfrak{G}/\mathbf{O}_\pi(\mathfrak{G})) = 1$.

More generally, if $\pi_1, \pi_2, \ldots$ are sets of primes, we define a normal subgroup $\mathbf{O}_{\pi_1,\ldots,\pi_i}(\mathfrak{G})$ of $\mathfrak{G}$ by induction on i: for $i > 1$,

$$\mathbf{O}_{\pi_1,\ldots,\pi_i}(\mathfrak{G})/\mathbf{O}_{\pi_1,\ldots,\pi_{i-1}}(\mathfrak{G}) = \mathbf{O}_{\pi_i}(\mathfrak{G}/\mathbf{O}_{\pi_1,\ldots,\pi_{i-1}}(\mathfrak{G})).$$

Thus $\mathbf{O}_{\pi_1,\ldots,\pi_i}(\mathfrak{G})$ is a characteristic subgroup of $\mathfrak{G}$.

For example, if p is a prime, the *upper p-series* of $\mathfrak{G}$ is

$$1 \leq \mathbf{O}_{p'}(\mathfrak{G}) \leq \mathbf{O}_{p',p}(\mathfrak{G}) \leq \mathbf{O}_{p',p,p'}(\mathfrak{G}) \leq \cdots$$

(cf. VI, 6.1).

Observe also that any normal p-nilpotent subgroup $\mathfrak{N}$ of $\mathfrak{G}$ is contained in $\mathbf{O}_{p',p}(\mathfrak{G})$, for the normal p-complement of $\mathfrak{N}$ is contained in $\mathbf{O}_{p'}(\mathfrak{G})$. Thus $\mathbf{O}_{p',p}(\mathfrak{G})$ is the maximal normal p-nilpotent subgroup of $\mathfrak{G}$. The Fitting subgroup of $\mathfrak{G}$ is $\prod_p \mathbf{O}_p(\mathfrak{G})$.

Dually, $\mathbf{O}^\pi(\mathfrak{G})$ is defined to be the intersection of all normal subgroups $\mathfrak{N}$ of $\mathfrak{G}$ for which $\mathfrak{G}/\mathfrak{N}$ is a π-group. Thus $\mathfrak{G}/\mathbf{O}^\pi(\mathfrak{G})$ is the maximal π-factor group of $\mathfrak{G}$, and $\mathbf{O}^\pi(\mathfrak{G})$ is a characteristic subgroup of $\mathfrak{G}$.

For example, $\mathbf{O}^p(\mathfrak{G})$ is the normal p-complement of a p-nilpotent group $\mathfrak{G}$.

The basic lemma in VI, § 6 is Lemma 6.5 of Hall and Higman. We shall need slightly different versions of this.

1.2 Lemma. *Suppose that π is a set of primes and $\mathfrak{P} = \mathbf{O}_\pi(\mathfrak{G})$. If $\mathfrak{K} \trianglelefteq \mathfrak{G}$, $\mathfrak{K} \leq \mathbf{C}_\mathfrak{G}(\mathfrak{P})$ and $\mathfrak{P}\mathfrak{K}/\mathfrak{P}$ is a π'-group, then there exists a π'-subgroup $\mathfrak{C}$ of $\mathbf{C}_\mathfrak{G}(\mathfrak{P})$ such that $\mathfrak{C} \trianglelefteq \mathfrak{G}$ and $\mathfrak{K} = \mathfrak{C} \times (\mathfrak{K} \cap \mathfrak{P})$.*

Proof. $\mathfrak{P}\mathfrak{K}/\mathfrak{P} \cong \mathfrak{K}/(\mathfrak{K} \cap \mathfrak{P})$ is a π'-group and $\mathfrak{K} \cap \mathfrak{P}$ is a π-group. It follows from the Schur-Zassenhaus theorem (I, 18.1) that there exists a subgroup $\mathfrak{C}$ of $\mathfrak{K}$ such that $\mathfrak{K} = \mathfrak{C}(\mathfrak{K} \cap \mathfrak{P})$ and $\mathfrak{C} \cap \mathfrak{P} = 1$. Evidently,

$\mathfrak{C}$ is a π'-subgroup of $\mathbf{C}_{\mathfrak{G}}(\mathfrak{P})$. Thus $[\mathfrak{C}, \mathfrak{P}] = 1$ and $\mathfrak{K} = \mathfrak{C} \times (\mathfrak{K} \cap \mathfrak{P})$. Hence $\mathfrak{C}$ is a normal Hall π'-subgroup of $\mathfrak{K}$, whence $\mathfrak{C}$ is a characteristic subgroup of $\mathfrak{K}$ and $\mathfrak{C} \trianglelefteq \mathfrak{G}$. **q.e.d.**

1.3 Lemma. *Suppose that $\mathfrak{G}$ is a finite group and that π is a set of primes. Suppose that every chief factor of $\mathfrak{G}$ is either a π-group or a π'-group. Then*

$$\mathbf{C}_{\mathfrak{G}}(\mathbf{O}_{\pi,\pi'}(\mathfrak{G})/\mathbf{O}_{\pi}(\mathfrak{G})) \leq \mathbf{O}_{\pi,\pi'}(\mathfrak{G}).$$

Proof. Let $\mathfrak{H} = \mathfrak{G}/\mathbf{O}_{\pi}(\mathfrak{G})$, so that $\mathbf{O}_{\pi}(\mathfrak{H}) = 1$. The assertion is that $\mathfrak{C} \leq \mathfrak{D}$, where $\mathfrak{C} = \mathbf{C}_{\mathfrak{H}}(\mathbf{O}_{\pi'}(\mathfrak{H}))$ and $\mathfrak{D} = \mathbf{O}_{\pi'}(\mathfrak{H})$. Suppose that $\mathfrak{C} \not\leq \mathfrak{D}$. Then $\mathfrak{C} > \mathfrak{C} \cap \mathfrak{D}$. Let $\mathfrak{B}/(\mathfrak{C} \cap \mathfrak{D})$ be a minimal normal subgroup of $\mathfrak{H}/(\mathfrak{C} \cap \mathfrak{D})$ for which $\mathfrak{B} \leq \mathfrak{C}$. Then $\mathfrak{B}/(\mathfrak{C} \cap \mathfrak{D})$ is a chief factor of $\mathfrak{G}$ and is thus either a π-group or a π'-group. If $\mathfrak{B}/(\mathfrak{C} \cap \mathfrak{D})$ is a π'-group, then $\mathfrak{B}\mathfrak{D}/\mathfrak{D}$ and $\mathfrak{B}\mathfrak{D}$ are normal π'-groups, whence $\mathfrak{B} \leq \mathbf{O}_{\pi'}(\mathfrak{H}) = \mathfrak{D}$ and $\mathfrak{B} \leq \mathfrak{C} \cap \mathfrak{D}$, a contradiction. Hence $\mathfrak{B}/(\mathfrak{C} \cap \mathfrak{D})$ is a π-group. Thus $\mathfrak{B}\mathfrak{D}/\mathfrak{D} \cong \mathfrak{B}/(\mathfrak{B} \cap \mathfrak{D}) = \mathfrak{B}/(\mathfrak{C} \cap \mathfrak{D})$ is a π-group. By 1.2, there exists a π-subgroup $\mathfrak{A}$ of $\mathfrak{C}$ such that $\mathfrak{A} \trianglelefteq \mathfrak{H}$ and $\mathfrak{B} = \mathfrak{A} \times (\mathfrak{B} \cap \mathfrak{D})$. Since $\mathbf{O}_{\pi}(\mathfrak{H}) = 1$, $\mathfrak{A} = 1$ and $\mathfrak{B} \leq \mathfrak{D}$, a contradiction. **q.e.d.**

We recall (VII, 13.3) that a group $\mathfrak{G}$ is called *p-constrained* if $\mathbf{C}_{\mathfrak{G}}(\mathbf{O}_{p',p}(\mathfrak{G})/\mathbf{O}_{p'}(\mathfrak{G})) \leq \mathbf{O}_{p',p}(\mathfrak{G})$.

1.4 Corollary. *A p-soluble group is p-constrained.*

Proof. This follows at once from 1.3. **q.e.d.**

1.5 Lemma. *Suppose that $\mathfrak{G}$ is p-constrained.*

a) *If $\mathbf{O}_{p'}(\mathfrak{G}) = 1$ and $\mathbf{O}_p(\mathfrak{G}) \leq \mathfrak{H} \leq \mathfrak{G}$, then $\mathbf{O}_{p'}(\mathfrak{H}) = 1$.*

b) *If $\mathfrak{S} \in S_p(\mathfrak{G})$, $\mathfrak{K}$ is a p'-subgroup of $\mathfrak{G}$ and $\mathfrak{S} \leq \mathbf{N}_{\mathfrak{G}}(\mathfrak{K})$, then $\mathfrak{K} \leq \mathbf{O}_{p'}(\mathfrak{G})$.*

Proof. a) Since $\mathbf{O}_p(\mathfrak{G})$ and $\mathbf{O}_{p'}(\mathfrak{H})$ are normal subgroups of $\mathfrak{H}$ of coprime orders, $\mathbf{O}_{p'}(\mathfrak{H}) \leq \mathbf{C}_{\mathfrak{G}}(\mathbf{O}_p(\mathfrak{G}))$. Now $\mathbf{C}_{\mathfrak{G}}(\mathbf{O}_p(\mathfrak{G})) \leq \mathbf{O}_p(\mathfrak{G})$, since $\mathbf{O}_{p'}(\mathfrak{G}) = 1$. Thus $\mathbf{O}_{p'}(\mathfrak{H})$ is a p'-subgroup of $\mathbf{O}_p(\mathfrak{G})$ and so $\mathbf{O}_{p'}(\mathfrak{H}) = 1$.

b) Let $\overline{\mathfrak{G}} = \mathfrak{G}/\mathbf{O}_{p'}(\mathfrak{G})$, $\overline{\mathfrak{S}} = \mathfrak{S}\mathbf{O}_{p'}(\mathfrak{G})/\mathbf{O}_{p'}(\mathfrak{G})$, $\overline{\mathfrak{K}} = \mathfrak{K}\mathbf{O}_{p'}(\mathfrak{G})/\mathbf{O}_{p'}(\mathfrak{G})$. Thus $\mathbf{O}_{p'}(\overline{\mathfrak{G}}) = 1$ and $\overline{\mathfrak{G}}$ is p-constrained. Since $\overline{\mathfrak{S}} \in S_p(\overline{\mathfrak{G}})$, $\mathbf{O}_p(\overline{\mathfrak{G}}) \leq \overline{\mathfrak{S}}$. Also $\overline{\mathfrak{S}} \leq \mathbf{N}_{\overline{\mathfrak{G}}}(\overline{\mathfrak{K}})$, so $\overline{\mathfrak{S}}\overline{\mathfrak{K}}$ is a subgroup of $\overline{\mathfrak{G}}$ containing $\mathbf{O}_p(\overline{\mathfrak{G}})$. By a), $\mathbf{O}_{p'}(\overline{\mathfrak{S}}\overline{\mathfrak{K}}) = 1$. Since $\mathfrak{K}$ is a p'-group, $\mathbf{O}_{p'}(\overline{\mathfrak{S}}\overline{\mathfrak{K}}) = \overline{\mathfrak{K}}$. Hence $\overline{\mathfrak{K}} = 1$ and $\mathfrak{K} \leq \mathbf{O}_{p'}(\mathfrak{G})$. **q.e.d.**

Lemma VI, 6.5 follows from 1.4 and the following.

1.6 Lemma. *Suppose that $\mathfrak{G}$ is a p-constrained group and that the subgroup $\mathfrak{U}$ of $\mathfrak{G}$ is defined by*

$$\mathfrak{U}/\mathbf{O}_{p'}(\mathfrak{G}) = \Phi(\mathbf{O}_{p',p}(\mathfrak{G})/\mathbf{O}_{p'}(\mathfrak{G})).$$

Then $\mathbf{C}_{\mathfrak{G}}(\mathbf{O}_{p',p}(\mathfrak{G})/\mathfrak{U}) = \mathbf{O}_{p',p}(\mathfrak{G})$.

Proof. We may assume that $\mathbf{O}_{p'}(\mathfrak{G}) = 1$. Then $\mathfrak{U} = \Phi(\mathbf{O}_p(\mathfrak{G}))$, so $\mathbf{O}_p(\mathfrak{G})/\mathfrak{U}$ is elementary Abelian. Thus $\mathbf{C}_{\mathfrak{G}}(\mathbf{O}_p(\mathfrak{G})/\mathfrak{U}) \geq \mathbf{O}_p(\mathfrak{G})$. Suppose that $\mathfrak{C} = \mathbf{C}_{\mathfrak{G}}(\mathbf{O}_p(\mathfrak{G})/\mathfrak{U}) > \mathbf{O}_p(\mathfrak{G})$. Since $\mathfrak{C} \trianglelefteq \mathfrak{G}$, $\mathfrak{C}/\mathbf{O}_p(\mathfrak{G})$ is not a p-group. Let a be an element of $\mathfrak{C} - \mathbf{O}_p(\mathfrak{G})$ of order prime to p, and let α be the automorphism of $\mathbf{O}_p(\mathfrak{G})$ given by $x\alpha = a^{-1}xa$ $(x \in \mathbf{O}_p(\mathfrak{G}))$. Since $a \in \mathbf{C}_{\mathfrak{G}}(\mathbf{O}_p(\mathfrak{G})/\Phi(\mathbf{O}_p(\mathfrak{G})))$, α induces the identity automorphism on $\mathbf{O}_p(\mathfrak{G})/\Phi(\mathbf{O}_p(\mathfrak{G}))$. By III, 3.18, $\alpha = 1$, since the order of α is prime to p. Thus $a \in \mathbf{C}_{\mathfrak{G}}(\mathbf{O}_p(\mathfrak{G}))$. Since $\mathfrak{G}$ is p-constrained and $\mathbf{O}_{p'}(\mathfrak{G}) = 1$, $\mathbf{C}_{\mathfrak{G}}(\mathbf{O}_p(\mathfrak{G})) \leq \mathbf{O}_p(\mathfrak{G})$. Thus $a \in \mathbf{O}_p(\mathfrak{G})$, a contradiction. Hence $\mathfrak{C} = \mathbf{O}_p(\mathfrak{G})$. **q.e.d.**

1.7 Notation. Suppose that $\mathfrak{A}$ is a group of operators on a group $\mathfrak{G}$, that $\mathfrak{B} \trianglelefteq \mathfrak{A}$ and that $gb = g$ for all $b \in \mathfrak{B}$. Given $a \in \mathfrak{A}$, the mapping $g \rightarrow ga$ is an automorphism of $\mathfrak{G}$ which depends only on $a\mathfrak{B}$: we denote this automorphism by $\rho(a\mathfrak{B})$. Then ρ is a homomorphism of $\mathfrak{A}/\mathfrak{B}$ into the group of automorphisms of $\mathfrak{G}$. We write $(\mathfrak{A}/\mathfrak{B}$ *on* $\mathfrak{G})$ for ρ.

This notation is particularly useful in the following context. Suppose that $\mathfrak{H} \leq \mathfrak{G}$, $\mathfrak{K} \trianglelefteq \mathfrak{H}$, $\mathfrak{A} \leq \mathbf{N}_{\mathfrak{G}}(\mathfrak{H})$, $\mathfrak{A} \leq \mathbf{N}_{\mathfrak{G}}(\mathfrak{K})$, $\mathfrak{B} \trianglelefteq \mathfrak{A}$ and $[\mathfrak{H}, \mathfrak{B}] \leq \mathfrak{K}$. Then $\mathfrak{A}$ acts as a group of operators on $\mathfrak{H}/\mathfrak{K}$, with $(h\mathfrak{K})a = h^a\mathfrak{K}$ $(h \in \mathfrak{H}, a \in \mathfrak{A})$, and $(h\mathfrak{K})b = h\mathfrak{K}$ for all $b \in \mathfrak{B}$. We may thus speak of $(\mathfrak{A}/\mathfrak{B}$ on $\mathfrak{H}/\mathfrak{K})$. Note that

$$\ker(\mathfrak{A}/\mathfrak{B} \text{ on } \mathfrak{H}/\mathfrak{K}) = \mathbf{C}_{\mathfrak{A}/\mathfrak{B}}(\mathfrak{H}/\mathfrak{K}).$$

The conclusion of 1.6 is that $(\mathfrak{G}/\mathbf{O}_{p',p}(\mathfrak{G})$ on $\mathbf{O}_{p',p}(\mathfrak{G})/\mathfrak{U})$ is faithful. Now $\mathbf{O}_{p',p}(\mathfrak{G})/\mathfrak{U}$ is an elementary Abelian p-group and can therefore be regarded as a vector space over $GF(p)$. The following lemma is an interpretation of commutators and powers in linear terms.

1.8 Lemma. *Suppose that $\mathfrak{U} \trianglelefteq \mathfrak{G}$, $\mathfrak{V} \trianglelefteq \mathfrak{G}$, $\mathfrak{V} \leq \mathfrak{U}$ and $\mathfrak{U}/\mathfrak{V}$ is an elementary Abelian p-group. We regard $\mathsf{U} = \mathfrak{U}/\mathfrak{V}$ as a vector space over $GF(p)$ and write $\rho = (\mathfrak{G}$ on $\mathsf{U})$.*

a) *If $u \in \mathfrak{U}$ and $x_1, \ldots, x_n$ are elements of $\mathfrak{G}$,*

$$(u\mathfrak{V})(\rho(x_1) - 1) \cdots (\rho(x_n) - 1) = [u, x_1, \ldots, x_n]\mathfrak{V}.$$

b) *If* $u \in \mathfrak{U}$, $x \in \mathfrak{G}$ *and* n *is a positive integer,*

$$x^{-n}(xu)^n\mathfrak{V} = (u\mathfrak{V})(1 + \rho(x) + \cdots + \rho(x)^{n-1}).$$

Proof. a) We have

$$(u\mathfrak{V})(\rho(x_1) - 1) = (u^{x_1}\mathfrak{V})(u^{-1}\mathfrak{V}) = u^{-1}u^{x_1}\mathfrak{V} = [u, x_1]\mathfrak{V},$$

and the assertion follows by induction on n.

$$\text{b)}\ (u\mathfrak{V})(1 + \rho(x) + \cdots + \rho(x)^{n-1}) = (u\mathfrak{V})(u^x\mathfrak{V}) \cdots (u^{x^{n-1}}\mathfrak{V})$$
$$= u^{x^{n-1}} \cdots u^x u\mathfrak{V}$$
$$= x^{-n}(xu)^n\mathfrak{V}. \qquad \textbf{q.e.d.}$$

From 1.8, we see that $u\mathfrak{V}$ is carried by $(\rho(x) - 1)^n$ into $[u, x, \ldots, x]\mathfrak{V}$ and by $1 + \rho(x) + \cdots + \rho(x)^{n-1}$ into $x^{-n}(xu)^n\mathfrak{V}$. Thus whether or not certain powers and commutators are equal to 1 depends on the vanishing of certain polynomials in $\rho(x)$, and it is therefore necessary to study the minimum polynomial of $\rho(x)$. To do this we prove the following lemma.

1.9 Lemma. *Let* V *be a vector space of finite dimension over a field* K, *and let* $\mathfrak{G}$ *be a group of non-singular linear transformations of* V. *Suppose that* $\mathfrak{N}$ *is a normal Abelian subgroup of* $\mathfrak{G}$, *that* char K *does not divide* $|\mathfrak{N}|$ *and that* K *is a splitting field for* $\mathfrak{N}$. *Suppose further that* $\mathfrak{P} \leq \mathfrak{G}$ *and* $\mathbf{C}_{\mathfrak{P}}(\mathfrak{N}) = 1$. *Then there exist* $\mathsf{K}\mathfrak{N}$*-submodules* $\mathsf{V}_1, \ldots, \mathsf{V}_m$ *of* V *such that*

$$\mathsf{V} = \mathsf{V}_1 \oplus \cdots \oplus \mathsf{V}_m,$$

and there exists a faithful permutation representation σ *of* $\mathfrak{P}$ *on* $\{1, \ldots, m\}$ *such that* $\mathsf{V}_i g = \mathsf{V}_{i\sigma(g)}$ *for all* $g \in \mathfrak{P}$ $(i = 1, \ldots, m)$.

Proof. Since K is a splitting field for $\mathfrak{N}$, any irreducible representation of $\mathfrak{N}$ in K is of degree 1 (e.g. V, 11.4). Let $\rho_1, \ldots, \rho_m$ be the distinct irreducible representations of $\mathfrak{N}$ in K, and let

$$\mathsf{V}_i = \{v \mid v \in \mathsf{V},\ vx = \rho_i(x)v \quad \text{for all } x \in \mathfrak{N}\}.$$

By the Maschke-Schur theorem, V is the direct sum of irreducible $\mathsf{K}\mathfrak{N}$-submodules. Since each such submodule is contained in some V_i,

$$\mathsf{V} = \mathsf{V}_1 + \cdots + \mathsf{V}_m.$$

By VIII, 5.16,

$$\mathsf{V} = \mathsf{V}_1 \oplus \cdots \oplus \mathsf{V}_m.$$

Now if $g \in \mathfrak{P}$, $x \to \rho_i(gxg^{-1})$ is an irreducible representation of $\mathfrak{N}$ in K, so there exists j such that $\rho_i(gxg^{-1}) = \rho_j(x)$ for all $x \in \mathfrak{N}$. We put $j = i\sigma(g)$, so

$$\rho_i(gxg^{-1}) = \rho_{i\sigma(g)}(x) \quad (x \in \mathfrak{N}).$$

If also $h \in \mathfrak{P}$,

$$\rho_{i\sigma(gh)}(x) = \rho_i(ghxh^{-1}g^{-1}) = \rho_{i\sigma(g)}(hxh^{-1}) = \rho_{i\sigma(g)\sigma(h)}(x),$$

whence $\sigma(gh) = \sigma(g)\sigma(h)$. Also $\sigma(1) = 1$, so σ is a permutation representation of $\mathfrak{P}$ on $\{1, \ldots, m\}$. If $v \in \mathsf{V}_i$ and $g \in \mathfrak{P}$,

$$(vg)x = v(gxg^{-1})g = \rho_i(gxg^{-1})vg = \rho_{i\sigma(g)}(x)vg$$

for all $x \in \mathfrak{N}$, so $vg \in \mathsf{V}_{i\sigma(g)}$. Thus $\mathsf{V}_i g = \mathsf{V}_{i\sigma(g)}$ for all $g \in \mathfrak{P}$. It remains to show that σ is faithful.

Suppose that $g \in \mathfrak{P}$ and $g \neq 1$. By hypothesis, $g \notin \mathbf{C}_{\mathfrak{P}}(\mathfrak{N})$, so there exists $y \in \mathfrak{N}$ such that $z = [y, g^{-1}] \neq 1$. Now $z \in \mathfrak{N}$ and there is some j such that z does not induce the identity mapping on V_j. Thus $\rho_j(z) \neq 1$, and since $yz = gyg^{-1}$,

$$\rho_{j\sigma(g)}(y) = \rho_j(gyg^{-1}) = \rho_j(yz) = \rho_j(y)\rho_j(z) \neq \rho_j(y).$$

Hence $j\sigma(g) \neq j$ and $\sigma(g) \neq 1$. Hence σ is faithful. **q.e.d.**

1.10 Theorem (G. HIGMAN). *Let* V *be a vector space of finite dimension over a field* K, *and let* $\mathfrak{G}$ *be a group of non-singular linear transformations of* V. *Let* $\mathfrak{N}$ *be a normal, Abelian subgroup of* $\mathfrak{G}$ *such that* char K *does not divide* $|\mathfrak{N}|$. *Suppose that* a *is an element of* $\mathfrak{G}$ *of order* p^n *and that* $a^{p^{n-1}} \notin \mathbf{C}_{\mathfrak{G}}(\mathfrak{N})$. *Then the minimum polynomial of* a *is* $t^{p^n} - 1$.

Proof. Since the minimum polynomial of a linear transformation is unaltered by an extension of the ground-field, we may suppose that K is a splitting field for $\mathfrak{N}$. We apply 1.9 with $\mathfrak{P} = \langle a \rangle$. Thus there exist $\mathsf{K}\mathfrak{N}$-submodules $\mathsf{V}_1, \ldots, \mathsf{V}_m$ of V such that

$$\mathsf{V} = \mathsf{V}_1 \oplus \cdots \oplus \mathsf{V}_m,$$

and there exists a permutation α of $\{1, \ldots, m\}$ such that α is of order p^n and $\mathsf{V}_i a = \mathsf{V}_{i\alpha}$ $(i = 1, \ldots, m)$. Now α has a cycle $(i, i\alpha, \ldots, i\alpha^{p^n-1})$ of length p^n. If v is a non-zero element of V_i, $va^j \in \mathsf{V}_{i\alpha^j}$, so $v, va, \ldots, va^{p^n-1}$ are linearly independent. Hence $1, a, \ldots, a^{p^n-1}$ are linearly independent, and the degree of the minimum polynomial of a is at least p^n. Since, however, $a^{p^n} = 1$, the minimum polynomial of a is a divisor of $t^{p^n} - 1$. The minimum polynomial of a is thus precisely $t^{p^n} - 1$. **q.e.d.**

Theorem 1.10 can be applied, for example, to replace step r) in the proof of IV, 6.2. In 1.13, it will be seen how results of this kind may be used to obtain bounds on the p-length of the p-soluble group $\mathfrak{G}$. We need the following, which will be much used in the sequel.

1.11 Theorem. *Suppose that π is a set of primes, $\mathfrak{G}$ is a π-group and that $\mathfrak{A}$ is a π'-group of operators on $\mathfrak{G}$. Suppose that $\sigma \subseteq \pi$. Suppose further that either* (1) *$\mathfrak{G}$ is soluble, or* (2) *σ consists of just one prime p and either $\mathfrak{G}$ or $\mathfrak{A}$ is soluble.*

a) *There exists an $\mathfrak{A}$-invariant Hall σ-subgroup of $\mathfrak{G}$.*

b) *If $\mathfrak{H}_1$, $\mathfrak{H}_2$ are $\mathfrak{A}$-invariant Hall σ-subgroups of $\mathfrak{G}$, there exists $x \in \mathbf{C}_{\mathfrak{G}}(\mathfrak{A})$ such that $\mathfrak{H}_1^x = \mathfrak{H}_2$.*

c) *Any $\mathfrak{A}$-invariant σ-subgroup of $\mathfrak{G}$ is contained in an $\mathfrak{A}$-invariant Hall σ-subgroup of $\mathfrak{G}$, provided that, in case* (1), *$\mathfrak{A}$ is soluble.*

Proof. Let $\overline{\mathfrak{G}}$ be the semidirect product of $\mathfrak{G}$ and $\mathfrak{A}$; thus $\overline{\mathfrak{G}} = \mathfrak{G}\mathfrak{A}$, $\mathfrak{G} \trianglelefteq \overline{\mathfrak{G}}$ and $\mathfrak{G} \cap \mathfrak{A} = 1$.

a) By VI, 1.8 in case (1), or Sylow's theorem in case (2), $\mathfrak{G}$ possesses a Hall σ-subgroup $\mathfrak{Q}$ and all Hall σ-subgroups of $\mathfrak{G}$ are conjugate. By the Frattini argument, $\overline{\mathfrak{G}} = \mathfrak{K}\mathfrak{G}$, where $\mathfrak{K} = \mathbf{N}_{\overline{\mathfrak{G}}}(\mathfrak{Q})$. Thus

$$\mathfrak{A} \cong \overline{\mathfrak{G}}/\mathfrak{G} \cong \mathfrak{K}\mathfrak{G}/\mathfrak{G} \cong \mathfrak{K}/(\mathfrak{K} \cap \mathfrak{G})$$

and $\mathfrak{K}/(\mathfrak{K} \cap \mathfrak{G})$ is a π'-group. Since $\mathfrak{K} \cap \mathfrak{G}$ is a π-group, it follows from the Schur-Zassenhaus theorem (I, 18.1) that there exists $\mathfrak{L} \leq \mathfrak{K}$ such that $\mathfrak{K} = \mathfrak{L}(\mathfrak{K} \cap \mathfrak{G})$ and $\mathfrak{G} \cap \mathfrak{L} = 1$. Thus $\overline{\mathfrak{G}} = \mathfrak{K}\mathfrak{G} = \mathfrak{L}(\mathfrak{K} \cap \mathfrak{G})\mathfrak{G} = \mathfrak{L}\mathfrak{G}$, so $\mathfrak{L}$, $\mathfrak{A}$ are both complements of $\mathfrak{G}$ in $\overline{\mathfrak{G}}$. Since either $\mathfrak{G}$ or $\mathfrak{A}$ is soluble, it follows from I, 18.2 that $\mathfrak{A} = \mathfrak{L}^x$ for some $x \in \overline{\mathfrak{G}}$. Since $\mathfrak{L} \leq \mathfrak{K} = \mathbf{N}_{\overline{\mathfrak{G}}}(\mathfrak{Q})$, $\mathfrak{A} = \mathfrak{L}^x \leq \mathbf{N}_{\overline{\mathfrak{G}}}(\mathfrak{Q}^x)$. Also $\mathfrak{Q}^x \leq \mathfrak{G}$ since $\mathfrak{G} \trianglelefteq \overline{\mathfrak{G}}$, so $\mathfrak{Q}^x$ is an $\mathfrak{A}$-invariant Hall σ-subgroup of $\mathfrak{G}$.

b) Let $\overline{\mathfrak{N}} = \mathbf{N}_{\overline{\mathfrak{G}}}(\mathfrak{H}_1)$, $\mathfrak{N} = \mathbf{N}_{\mathfrak{G}}(\mathfrak{H}_1)$. Since $\mathfrak{A} \leq \overline{\mathfrak{N}}$, $\overline{\mathfrak{N}} = (\mathfrak{G} \cap \overline{\mathfrak{N}})\mathfrak{A} = \mathfrak{N}\mathfrak{A}$. Now by VI, 1.8 or Sylow's theorem, $\mathfrak{H}_1 = \mathfrak{H}_2^g$ for some $g \in \mathfrak{G}$. Hence $\mathfrak{A}^g \leq \mathbf{N}_{\overline{\mathfrak{G}}}(\mathfrak{H}_2^g) = \overline{\mathfrak{N}}$ and $|\overline{\mathfrak{N}}| = |\mathfrak{N}||\mathfrak{A}| = |\mathfrak{N}\mathfrak{A}^g|$, so $\overline{\mathfrak{N}} = \mathfrak{N}\mathfrak{A}^g$ also. Thus $\mathfrak{A}$, $\mathfrak{A}^g$ are complements of the normal Hall π-subgroup $\mathfrak{N}$ of

$\overline{\mathfrak{N}}$, and by I, 18.2, $\mathfrak{A}^g = \mathfrak{A}^u$ for some $u \in \overline{\mathfrak{N}}$. Write $u = av$ for $a \in \mathfrak{A}$, $v \in \mathfrak{N}$; then $\mathfrak{A}^v = \mathfrak{A}^{av} = \mathfrak{A}^u = \mathfrak{A}^g$. Thus $v = xg$ for some $x \in \mathbf{N}_{\mathfrak{G}}(\mathfrak{A})$. Hence $[\mathfrak{A}, x] \leq \mathfrak{A} \cap \mathfrak{G} = 1$, so $x \in \mathbf{C}_{\mathfrak{G}}(\mathfrak{A})$. Also $\mathfrak{H}_1^x = \mathfrak{H}_1^{vg^{-1}} = \mathfrak{H}_1^{g^{-1}} = \mathfrak{H}_2$.

c) We give different proofs in the two cases. Let $\mathfrak{X}$ be an $\mathfrak{A}$-invariant σ-subgroup of $\mathfrak{G}$.

In case (1), $\overline{\mathfrak{G}}$ is soluble. Thus $\mathfrak{X}\mathfrak{A}$ is contained in some Hall $(\sigma \cup \pi')$-subgroup $\mathfrak{K}$ of $\overline{\mathfrak{G}}$, by VI, 1.8. Thus $\mathfrak{X} \leq \mathfrak{K} \cap \mathfrak{G}$ and $\mathfrak{K} \cap \mathfrak{G}$ is an $\mathfrak{A}$-invariant Hall σ-subgroup of $\mathfrak{G}$.

In case (2), we proceed by induction on $|\mathfrak{G} : \mathfrak{X}|$. If $|\mathfrak{G} : \mathfrak{X}|$ is prime to p, $\mathfrak{X}$ is an $\mathfrak{A}$-invariant Sylow p-subgroup of $\mathfrak{G}$ and there is nothing more to prove. Otherwise there exists $\mathfrak{S} \in S_p(\mathfrak{G})$ such that $\mathfrak{X} < \mathfrak{S}$. By I, 8.8, $\mathfrak{X} < \mathbf{N}_{\mathfrak{S}}(\mathfrak{X})$, so the order of a Sylow p-subgroup of $\mathbf{N}_{\mathfrak{G}}(\mathfrak{X})$ is greater than $|\mathfrak{X}|$. But $\mathbf{N}_{\mathfrak{G}}(\mathfrak{X})$ is $\mathfrak{A}$-invariant, so by a), $\mathbf{N}_{\mathfrak{G}}(\mathfrak{X})$ has an $\mathfrak{A}$-invariant Sylow p-subgroup $\mathfrak{T}$. Thus $\mathfrak{T} \geq \mathfrak{X}$ and $|\mathfrak{T}| > |\mathfrak{X}|$, so $\mathfrak{T} > \mathfrak{X}$. By the inductive hypothesis, $\mathfrak{T}$ is contained in an $\mathfrak{A}$-invariant Sylow p-subgroup of $\mathfrak{G}$. Hence, so is $\mathfrak{X}$. **q.e.d.**

1.12 Corollary. *Let π be a set of primes, let $\mathfrak{H}$ be a π'-subgroup of a group $\mathfrak{G}$ and let $\mathfrak{K}$ be a subgroup of $\mathfrak{H}$. Suppose that*

(i) *$\mathfrak{N}$ is a π-subgroup of $\mathfrak{G}$;*
(ii) *$\mathfrak{H} \leq \mathbf{N}_{\mathfrak{G}}(\mathfrak{N})$ but $\mathfrak{K} \nleq \mathbf{C}_{\mathfrak{G}}(\mathfrak{N})$;*
(iii) *either $\mathfrak{H}$ or $\mathfrak{N}$ is soluble.*

Then there exists a Sylow subgroup $\mathfrak{P}$ of $\mathfrak{N}$ such that $\mathfrak{H} \leq \mathbf{N}_{\mathfrak{G}}(\mathfrak{P})$ but $\mathfrak{K} \nleq \mathbf{C}_{\mathfrak{G}}(\mathfrak{P})$.

Proof. For any $p \in \pi$, there exists $\mathfrak{P} \in S_p(\mathfrak{N})$ such that $\mathfrak{H} \leq \mathbf{N}_{\mathfrak{G}}(\mathfrak{P})$, by 1.11. But since $|\mathfrak{N}| = \prod_{p \in \pi} |\mathfrak{P}|$, $\mathfrak{N} = \langle \mathfrak{P} | p \in \pi \rangle$. Since $\mathfrak{K} \nleq \mathbf{C}_{\mathfrak{G}}(\mathfrak{N})$, $\mathfrak{K} \nleq \mathbf{C}_{\mathfrak{G}}(\mathfrak{P})$ for at least one $\mathfrak{P}$. **q.e.d.**

1.13 Theorem. *Suppose that $\mathfrak{G}$ is a p-soluble group and that for each prime $q \neq p$, the Sylow q-subgroups of $\mathfrak{G}$ are Abelian. If the exponent of the Sylow p-subgroups is p^n, the p-length of $\mathfrak{G}$ is at most n.*

Proof. This is proved by induction on the p-length l of $\mathfrak{G}$. If $l = 1$, $\mathfrak{G}$ is not a p'-group and there is nothing further to prove. Suppose that $l > 1$; let $\mathfrak{N} = \mathbf{O}_{p',p}(\mathfrak{G})$ and $\mathfrak{H} = \mathfrak{G}/\mathfrak{N}$. Then $l_p(\mathfrak{H}) = l - 1$. By the inductive hypothesis, $\mathfrak{H}$ possesses an element a of order p^{l-1}. Let $b = a^{p^{l-2}}$.

Since $\mathfrak{N} = \mathbf{O}_{p',p}(\mathfrak{G})$, $\mathbf{O}_p(\mathfrak{H}) = 1$, so by 1.3, $\mathbf{C}_{\mathfrak{H}}(\mathbf{O}_{p'}(\mathfrak{H})) \leq \mathbf{O}_{p'}(\mathfrak{H})$. Thus b does not centralize $\mathbf{O}_{p'}(\mathfrak{H})$. Let $\mathfrak{Q}$ be a minimal subgroup of $\mathbf{O}_{p'}(\mathfrak{H})$ which is normalised by a but not centralized by b. By 1.12, $\mathfrak{Q}$ is a q-group for some prime $q \neq p$. By hypothesis, $\mathfrak{Q}$ is Abelian. Let $\mathfrak{R} = \langle a, \mathfrak{Q} \rangle$.

By 1.4, $\mathfrak{G}$ is p-constrained, so by 1.6, $\mathbf{C}_{\mathfrak{G}}(\mathfrak{N}/\mathfrak{U}) = \mathfrak{N}$, where $\mathfrak{U}/\mathbf{O}_{p'}(\mathfrak{G}) = \Phi(\mathbf{O}_{p',p}(\mathfrak{G})/\mathbf{O}_{p'}(\mathfrak{G}))$. Hence, if $\rho = (\mathfrak{H}$ on $\mathfrak{N}/\mathfrak{U})$, ρ is faithful. We may apply 1.10 to $\rho(\mathfrak{R})$, since $p \nmid |\mathfrak{Q}|$. Thus the minimum polynomial of $\rho(a)$ is $t^{p^{l-1}} - 1$. Hence $1 + \rho(a) + \cdots + \rho(a)^{p^{l-1}-1} \neq 0$, and there exists $\bar{x} \in \mathfrak{N}/\mathfrak{U}$ such that $\bar{x}(1 + \rho(a) + \cdots + \rho(a)^{p^{l-1}-1}) \neq 1$. We may choose a p-element $g \in \mathfrak{G}$ such that $a = g\mathfrak{N}$. If $\mathfrak{S} \in S_p(\mathfrak{G})$ and $g \in \mathfrak{S}$, then $(\mathfrak{S} \cap \mathfrak{N})\mathbf{O}_{p'}(\mathfrak{G}) = \mathfrak{N}$ and we may choose $x \in \mathfrak{S} \cap \mathfrak{N}$ such that $x\mathfrak{U} = \bar{x}$. By 1.8b), $g^{-p^{l-1}}(gx)^{p^{l-1}} \notin \mathfrak{U}$. Since g and gx are both p-elements, it follows that either g or gx is of order at least p^l. Thus the exponent of the Sylow p-subgroups of $\mathfrak{G}$ is at least p^l. **q.e.d.**

We can use this to obtain the best possible bound for the order of a finite group of exponent 6 with d generators. For this, we also need I, 19.11, which will now be proved.

1.14 Theorem (SCHREIER). *Let $\mathfrak{F}$ be a free group with group-basis X and suppose $\mathfrak{H} \leq \mathfrak{F}$. Then $\mathfrak{H}$ is free. More precisely, there exists a transversal S of $\mathfrak{H}$ in $\mathfrak{F}$ with the following properties.*

a) $1 \in \mathsf{S}$.

b) *If $s \in \mathsf{S}$ and $x \in X$, $sxs'^{-1} \in \mathfrak{H}$ for a unique $s' \in \mathsf{S}$; put $\beta(s, x) = sxs'^{-1}$. There exists an injective mapping π of $\mathsf{S} - \{1\}$ into the Cartesian product $\mathsf{S} \times X$ such that if $s\pi = (s_1, x_1)$ with $s_1 \in \mathsf{S}$, $x_1 \in X$, then $\beta(s_1, x_1) = 1$ and either $s_1 = s$ or $s_1x_1 = s$.*

c) *Let $Y = \{\beta(s, x) | s \in \mathsf{S}, x \in X, (s, x) \neq s'\pi$ for all $s' \in \mathsf{S}\}$. Then Y is a group-basis of $\mathfrak{H}$.*

In particular, if $\mathfrak{F}$ is of rank d and $|\mathfrak{F} : \mathfrak{H}| = j$, the rank of $\mathfrak{H}$ is $(d - 1)j + 1$.

Proof. Since $\mathfrak{F} = \langle X \rangle$, every element f of $\mathfrak{F}$ is expressible in the form

$$f = x_1^{\varepsilon_1} \cdots x_n^{\varepsilon_n},$$

where $n \geq 0$, $x_i \in X$ and $\varepsilon_i = \pm 1$. We choose such an expression with n minimal and write $n = l(f)$. Thus $l(f) = 0$ if and only if $f = 1$. For any non-empty subset U of $\mathfrak{F}$, let

$$l(U) = \min\{l(f) | f \in U\}.$$

We construct a transversal S of $\mathfrak{H}$ in $\mathfrak{F}$ with the following properties.

(i) $1 \in \mathsf{S}$; more generally, $l(s) = l(\mathfrak{H}s)$ for all $s \in \mathsf{S}$.

(ii) There exists a mapping λ of $\mathsf{S} - \{1\}$ into $X \cup X^{-1}$ such that if $s \in \mathsf{S}$ and $s \neq 1$, then $s\lambda(s)^{-1} \in \mathsf{S}$ and $l(s\lambda(s)^{-1}) < l(s)$.

We define the representative of the coset $\mathfrak{H}f\,(f \in \mathfrak{F})$ by induction on $l(\mathfrak{H}f)$. If $l(\mathfrak{H}f) = 0$, let 1 be the representative. If $l(\mathfrak{H}f) = n > 0$, $\mathfrak{H}f$ contains an element f' for which $l(f') = n$. Hence $f' = f_1 x^{\varepsilon}$, where $l(f_1) = n - 1$, $x \in X$ and $\varepsilon = \pm 1$. By the inductive hypothesis, a representative s_1 of $\mathfrak{H}f_1$ has already been chosen and $l(s_1) \leq n - 1$. Since $\mathfrak{H}f = \mathfrak{H}f' = \mathfrak{H}f_1 x^{\varepsilon} = \mathfrak{H}s_1 x^{\varepsilon}$, we may choose $s = s_1 x^{\varepsilon}$ for our representative of $\mathfrak{H}f$ and put $\lambda(s) = x^{\varepsilon}$. Since $l(s_1) \leq n - 1$, $l(s) \leq n$, but since $s \in \mathfrak{H}f$, $l(s) \geq n$; thus $l(s) = n$. Also $s\lambda(s)^{-1} = s_1 \in \mathsf{S}$ and $l(s\lambda(s)^{-1}) = l(s_1) < n$. Thus if S is a transversal constructed in this way, S has the properties (i) and (ii).

Suppose that $s \in \mathsf{S}$ and $x \in X$. Then $sx \in \mathfrak{H}s'$ for a unique $s' \in \mathsf{S}$. Put $\beta(s, x) = sxs'^{-1}$; thus $\beta(s, x) \in \mathfrak{H}$. We define the mapping π of $\mathsf{S} - \{1\}$ into $\mathsf{S} \times X$ as follows. If $s \in \mathsf{S} - \{1\}$ and $\lambda(s) = x^{-1}$ for some $x \in X$, put $s\pi = (s, x)$; since $sx = s\lambda(s)^{-1} \in \mathsf{S}$, $\mathfrak{H}sx \cap \mathsf{S} = \{sx\}$ and $\beta(s, x) = 1$. If $s \in \mathsf{S} - \{1\}$ and $\lambda(s) = y \in X$, put $s\pi = (s\lambda(s)^{-1}, y)$; since $s\lambda(s)^{-1}y = s \in \mathsf{S}$, $\beta(s\lambda(s)^{-1}, y) = 1$. We prove that π is injective. Suppose, then, that $s_1\pi = s_2\pi$, where s_1, s_2 are in $\mathsf{S} - \{1\}$. If $\lambda(s_i) = x_i^{-1}$ with $x_i \in X$ $(i = 1, 2)$, then $(s_1, x_1) = s_1\pi = s_2\pi = (s_2, x_2)$ and $s_1 = s_2$. If $\lambda(s_i) = y_i \in X$ $(i = 1, 2)$, $(s_1 y_1^{-1}, y_1) = s_1\pi = s_2\pi = (s_2 y_2^{-1}, y_2)$ and again $s_1 = s_2$. We must therefore show that it is not possible to have $\lambda(s_1) = x^{-1}$, $\lambda(s_2) = y$ with x, y in X. If this is the case, $(s_1, x) = s_1\pi = s_2\pi = (s_2 y^{-1}, y)$, so $y = x$ and $s_2 = s_1 x$. But then by (ii),

$$l(s_1) > l(s_1\lambda(s_1)^{-1}) = l(s_1 x) = l(s_2)$$

and

$$l(s_2) > l(s_2\lambda(s_2)^{-1}) = l(s_2 y^{-1}) = l(s_1),$$

a contradiction. Thus a) and b) are proved.

By I, 19.2, there exists a free group $\mathfrak{G}$ having a group-basis consisting of elements $g(s, x)$, where $s \in \mathsf{S}$, $x \in X$ and $(s, x) \neq s'\pi$ for all $s' \in \mathsf{S}$. Thus there is a homomorphism η of $\mathfrak{G}$ into $\mathfrak{H}$ such that $g(s, x)\eta = \beta(s, x)$. If $(s, x) = s'\pi$ for some $s' \in \mathsf{S}$, put $g(s, x) = 1_{\mathfrak{G}}$; thus $g(s, x)\eta = \beta(s, x)$ for all $s \in \mathsf{S}$, $x \in X$.

Our aim is to prove that η is an isomorphism. The fact that it is an epimorphism is equivalent to the assertion

$$\mathfrak{H} = \langle \beta(s, x) \,|\, s \in \mathsf{S}, x \in X \rangle.$$

To prove this, take $h \in \mathfrak{H}$ and write

$$h = x_1^{\varepsilon_1} \cdots x_n^{\varepsilon_n} \quad (x_i \in X, \varepsilon_i = \pm 1).$$

For $i = 0, \ldots, n$, there exists $s_i \in \mathsf{S}$ such that

$$\mathfrak{H} x_1^{\varepsilon_1} \cdots x_i^{\varepsilon_i} = \mathfrak{H} s_i.$$

In particular, $s_0 = s_n = 1$, since $h \in \mathfrak{H}$. Then if $1 \leq i \leq n$, $\mathfrak{H} s_i = \mathfrak{H} s_{i-1} x_i^{\varepsilon_i}$, so $s_{i-1} x_i^{\varepsilon_i} s_i^{-1} \in \mathfrak{H}$. Thus $s_{i-1} x_i^{\varepsilon_i} s_i^{-1} = \beta(s_{i-1}, x_i)$ if $\varepsilon_i = +1$ and $s_{i-1} x_i^{\varepsilon_i} s_i^{-1} = \beta(s_i, x_i)^{-1}$ if $\varepsilon_i = -1$. Since

$$h = s_0 h = (s_0 x_1^{\varepsilon_1} s_1^{-1})(s_1 x_2^{\varepsilon_1} s_2^{-1}) \cdots (s_{n-1} x_n^{\varepsilon_n} s_n^{-1}),$$

$h \in \langle \beta(s, x) | s \in \mathsf{S}, x \in X \rangle$. Hence η is an epimorphism.

Now let Ω be the Cartesian product $\mathfrak{G} \times \mathsf{S}$. For each $x \in X$, we define mappings $x\xi$, $x\xi'$ of Ω into Ω as follows. If $u \in \mathfrak{G}$, $s \in \mathsf{S}$ and $sx \in \mathfrak{H} s'$ ($s' \in \mathsf{S}$), put

$$(u, s)(x\xi) = (ug(s, x), s').$$

If $u \in \mathfrak{G}$, $s \in \mathsf{S}$ and $sx^{-1} \in \mathfrak{H} s''$ ($s'' \in \mathsf{S}$), put

$$(u, s)(x\xi') = (ug(s'', x)^{-1}, s'').$$

Thus

$$(u, s)(x\xi)(x\xi') = (ug(s, x)g(t, x)^{-1}, t),$$

where $t \in \mathsf{S}$ and $s'x^{-1} \in \mathfrak{H} t$; but then $\mathfrak{H} t = \mathfrak{H} s' x^{-1} = \mathfrak{H} s$, so $t = s$ and $(x\xi)(x\xi') = 1$. Similarly $(x\xi')(x\xi) = 1$, so $x\xi$ is a permutation of Ω. Since X is a group-basis of $\mathfrak{F}$, ξ can be extended to a homomorphism of $\mathfrak{F}$ into the group of all permutations of Ω.

We prove by induction on $l(s)$ that if $u \in \mathfrak{G}$ and $s \in \mathsf{S}$,

$$(1) \qquad (u, 1)(s\xi) = (u, s).$$

This is trivial if $l(s) = 0$. If $l(s) > 0$, we have $s = s_1 x^{\varepsilon}$, where $x \in X$, $x^{\varepsilon} = \lambda(s)$, $s_1 \in \mathsf{S}$ and $l(s_1) < l(s)$ by (ii). By the inductive hypothesis,

$$(u, 1)(s_1 \xi) = (u, s_1).$$

Thus

$$(u, 1)(s\xi) = (u, 1)(s_1\xi)(x\xi)^\varepsilon = (u, s_1)(x\xi)^\varepsilon.$$

If $\varepsilon = 1$, then $s\pi = (s_1, x)$ and $g(s_1, x) = 1$; thus

$$(u, 1)(s\xi) = (ug(s_1, x), s) = (u, s).$$

If $\varepsilon = -1$, then $s\pi = (s, x)$, so $g(s, x) = 1$; thus

$$(u, 1)(s\xi) = (u, s_1)(x\xi)^{-1} = (u, s_1)(x\xi') = (ug(s, x)^{-1}, s) = (u, s).$$

Thus (1) is proved.

Next, we show that for any $u \in \mathfrak{G}$, $g \in \mathfrak{G}$,

$$(2) \qquad (u, 1)(g\eta\xi) = (ug, 1).$$

First note that the set of $g \in \mathfrak{G}$ for which (2) is true is a subgroup of $\mathfrak{G}$. Thus it is sufficient to prove (2) for $g = g(s, x)$. Suppose that $sx \in \mathfrak{H}s'$ ($s' \in \mathsf{S}$). Then

$$g(s, x)\eta = \beta(s, x) = sxs'^{-1},$$

and

$$\begin{aligned} (u, 1)(g(s, x)\eta\xi) &= (u, 1)(s\xi)(x\xi)(s'\xi)^{-1} \\ &= (u, s)(x\xi)(s'\xi)^{-1} && \text{by (1)} \\ &= (ug(s, x), s')(s'\xi)^{-1} \\ &= (ug(s, x), 1) && \text{by (1).} \end{aligned}$$

Thus (2) is proved.

It follows at once from (2) that η is a monomorphism. Hence η is an isomorphism of $\mathfrak{G}$ onto $\mathfrak{H}$. This implies c).

If $\mathfrak{F}$ is of rank d and $|\mathfrak{F} : \mathfrak{H}| = j$, $|X| = d$ and $|\mathsf{S}| = j$. Hence the total number of pairs (s, x) is jd. Since π is injective, precisely $j - 1$ of these are of the form $s'\pi$ for some $s' \in \mathsf{S}$. There remain $jd - j + 1$, so $|Y| = jd - j + 1 = (d - 1)j + 1$. **q.e.d.**

1.15 Theorem (HALL and HIGMAN [1]). *Suppose that $\mathfrak{G}$ is a finite group of exponent 6 with d generators. Then $|\mathfrak{G}|$ is a divisor of*

$$2^m \cdot 3^{n+\binom{n}{2}+\binom{n}{3}},$$

where

$$m = 1 + (d - 1)\cdot 3^{d+\binom{d}{2}+\binom{d}{3}}, \quad n = 1 + (d - 1)\cdot 2^d.$$

Proof. This is proved along the same lines as VI, 6.7. Since $\mathfrak{G}$ is of exponent 6, $|\mathfrak{G}| = 2^a 3^b$ for suitable a, b. By V, 7.3, $\mathfrak{G}$ is soluble. The Sylow 2-subgroups of $\mathfrak{G}$ are of exponent 2 and are therefore Abelian. Hence by VI, 6.6a), $l_2(\mathfrak{G}) \leq 1$. Thus $\mathfrak{G}/\mathbf{O}_{3,2}(\mathfrak{G})$ is a group of exponent 3 with d generators. Hence by III, 6.6, $|\mathfrak{G}/\mathbf{O}_{3,2}(\mathfrak{G})| \leq 3^{d+\binom{d}{2}+\binom{d}{3}}$. By 1.14, $\mathbf{O}_{3,2}(\mathfrak{G})$ can be generated by m elements. Now if $\mathfrak{S} \in S_2(\mathfrak{G})$, $\mathbf{O}_{3,2}(\mathfrak{G}) = \mathbf{O}_3(\mathfrak{G})\mathfrak{S}$, since $l_2(\mathfrak{G}) \leq 1$. Thus $\mathfrak{S}$ is generated by m elements. Since $\mathfrak{S}$ is Abelian and of exponent 2, $|\mathfrak{S}| = 2^a$ with $a \leq m$.

Since $\mathfrak{S}$ is Abelian, $l_3(\mathfrak{G}) \leq 1$ by 1.13. Thus $\mathfrak{G}/\mathbf{O}_{2,3}(\mathfrak{G})$ is an elementary Abelian 2-group with d generators. Hence $|\mathfrak{G}/\mathbf{O}_{2,3}(\mathfrak{G})| \leq 2^d$. By 1.14, $\mathbf{O}_{2,3}(\mathfrak{G})$ can be generated by n elements. If $\mathfrak{T} \in S_3(\mathfrak{G})$, $\mathbf{O}_{2,3}(\mathfrak{G}) = \mathbf{O}_2(\mathfrak{G})\mathfrak{T}$, so $\mathfrak{T}$ is generated by n elements. Since $\mathfrak{T}$ is of exponent 3, $|\mathfrak{T}| = 3^b$ with $b \leq n + \binom{n}{2} + \binom{n}{3}$. Thus $|\mathfrak{G}| = 2^a 3^b$ divides $2^m \cdot 3^{n+\binom{n}{2}+\binom{n}{3}}$

q.e.d.

1.16 Remark. The bound in 1.15 is best possible. To see this, we need the fact that the bound for the order of a group with d generators and exponent 3 given in III, 6.6 is best possible. We shall not prove this, but we interpret it in the following way. Let $\mathfrak{F}$ be a free group of rank d and let

$$\mathfrak{F}^3 = \langle x^3 | x \in \mathfrak{F} \rangle.$$

Then any group of exponent 3 with d generators is a homomorphic image of $\mathfrak{F}/\mathfrak{F}^3$, so $|\mathfrak{F}/\mathfrak{F}^3|$ is the best bound for the order of a group with d generators and exponent 3. Thus

$$|\mathfrak{F} : \mathfrak{F}^3| = 3^{d+\binom{d}{2}+\binom{d}{3}}.$$

Hence by 1.14, $\mathfrak{F}^3$ is a free group of rank m, where

$$m = 1 + (d - 1)\cdot 3^{d+\binom{d}{2}+\binom{d}{3}}.$$

Now if $\mathfrak{M} = \langle y^2 | y \in \mathfrak{F}^3 \rangle$, $\mathfrak{F}^3/\mathfrak{M}$ is the largest group of exponent 2 with m generators. Hence $|\mathfrak{F}^3/\mathfrak{M}| = 2^m$. Note that $\mathfrak{M}$ is a characteristic subgroup of $\mathfrak{F}^3$ and hence of $\mathfrak{F}$. Also $\mathfrak{F}/\mathfrak{M}$ is of exponent 6.

Similarly, if

$$\mathfrak{F}^2 = \langle x^2 | x \in \mathfrak{F} \rangle,$$

$\mathfrak{F}/\mathfrak{F}^2$ is of order 2^d and $\mathfrak{F}^2$ is a free group of rank n, where

$$n = 1 + (d - 1) \cdot 2^d.$$

Finally, if $\mathfrak{N} = \langle y^3 | y \in \mathfrak{F}^2 \rangle$, $|\mathfrak{F}^2/\mathfrak{N}| = 3^{n+\binom{n}{2}+\binom{n}{3}}$, $\mathfrak{N}$ is a characteristic subgroup of $\mathfrak{F}$ and $\mathfrak{F}/\mathfrak{N}$ is of exponent 6. Let $\mathfrak{K} = \mathfrak{M} \cap \mathfrak{N}$. Then $\mathfrak{F}/\mathfrak{K}$ is a finite group of exponent 6, and $|\mathfrak{F}/\mathfrak{K}|$ is divisible by $|\mathfrak{F}^3/\mathfrak{M}| = 2^m$ and by $|\mathfrak{F}^2/\mathfrak{N}| = 3^{n+\binom{n}{2}+\binom{n}{3}}$. Using 1.15, it follows that

$$|\mathfrak{F}/\mathfrak{K}| = 2^m \cdot 3^{n+\binom{n}{2}+\binom{n}{3}}.$$

In fact, $\mathfrak{K} = \langle z^6 | z \in \mathfrak{F} \rangle$. For there is no infinite group of exponent 6 with d generators (M. HALL [1]), and no finite one is larger than $\mathfrak{F}/\mathfrak{K}$.

§ 2. Hall and Higman's Theorem B

In order to apply results of the form of Theorem 1.10 more generally, it is necessary to remove the hypothesis that $\mathfrak{N}$ is Abelian. Theorem 1.10 does not remain true if this is done, but it is clear from 1.12 and III, 13.5 that in a minimal counterexample, $\mathfrak{N}$ is a non-Abelian special q-group for some prime q (that is, $\Phi(\mathfrak{N}) = \mathfrak{N}' = \mathbf{Z}(\mathfrak{N})$ (III, 13.1)). By further reduction, additional restrictions are obtained, and we are led to consider the groups described in the following definition.

2.1 Definition. A group $\mathfrak{G}$ is called *critical* if it satisfies the following conditions.

a) $\mathfrak{G}$ has an extraspecial normal q-subgroup $\mathfrak{Q}$ for some prime q.

b) There exists a cyclic non-identity q'-subgroup $\mathfrak{S}$ of $\mathfrak{G}$ such that $\mathfrak{G} = \mathfrak{S}\mathfrak{Q}$ and $\mathfrak{S} \cap \mathfrak{Q} = 1$.

c) $[\mathfrak{Q}', \mathfrak{S}] = 1$ and each non-identity element $y \in \mathfrak{S}$ induces a fixed-point free automorphism on $\mathfrak{Q}/\mathfrak{Q}'$. Thus if $1 \neq y \in \mathfrak{S}$ and $x \in \mathfrak{Q} - \mathfrak{Q}'$, $x^y\mathfrak{Q}' \neq x\mathfrak{Q}'$.

(For the structure of extraspecial groups, see III, 13.7 and 13.8.)

2.2 Remarks. Suppose $\mathfrak{G} = \mathfrak{Q}\mathfrak{S}$ is a critical group and $|\mathfrak{Q}| = q^{2m+1}$.

a) By 2.1c), $\mathfrak{G}/\mathfrak{Q}'$ is a Frobenius group (V, 8.1) with Frobenius kernel $\mathfrak{Q}/\mathfrak{Q}'$ and Frobenius partition

$$(\mathfrak{Q}/\mathfrak{Q}') \cup \left(\bigcup_{t \in \mathsf{T}} \mathfrak{S}^t\mathfrak{Q}'/\mathfrak{Q}'\right).$$

where T is a transversal of $\mathfrak{Q}'$ in $\mathfrak{Q}$.

b) Since $\mathfrak{Q}$ is extraspecial, $|\mathfrak{Q}'| = q$; thus there is an isomorphism α of $\mathfrak{Q}'$ onto the additive group of $GF(q)$. We may write

$$(x\mathfrak{Q}', y\mathfrak{Q}') = [x, y]\alpha \quad (x \in \mathfrak{Q}, y \in \mathfrak{Q});$$

then $\mathfrak{Q}/\mathfrak{Q}'$ becomes a non-singular symplectic vector space over $GF(q)$, by III, 13.7b). Since $[\mathfrak{Q}', \mathfrak{S}] = 1$,

$$\begin{aligned}(x^s\mathfrak{Q}', y^s\mathfrak{Q}') &= [x^s, y^s]\alpha = [x, y]^s\alpha \\ &= [x, y]\alpha = (x\mathfrak{Q}', y\mathfrak{Q}')\end{aligned}$$

for all $x \in \mathfrak{Q}$, $y \in \mathfrak{Q}$, $s \in \mathfrak{S}$. Thus $\mathfrak{S}$ induces a subgroup of the symplectic group $Sp(2m, q)$ on $\mathfrak{Q}/\mathfrak{Q}'$.

In V, 17.13, the complex characters of $\mathfrak{G}$ were determined, and it was a side-result that $q^m \equiv \pm 1$ ($|\mathfrak{S}|$). We give here a more direct proof of this, using the following lemma.

2.3 Lemma. *Suppose that* V *is a non-singular symplectic vector space over* K, $\mathfrak{G}$ *is a subgroup of the symplectic group on* V *and* char K *does not divide* $|\mathfrak{G}|$. *If every irreducible* $\mathsf{K}\mathfrak{G}$*-submodule of* V *is isotropic, then* V *is the direct sum of an even number of irreducible* $\mathsf{K}\mathfrak{G}$*-submodules.*

Proof. This is proved by induction on $\dim_{\mathsf{K}}\mathsf{V}$. Since char K does not divide $|\mathfrak{G}|$, we can write

$$\mathsf{V} = \mathsf{V}_1 \oplus \cdots \oplus \mathsf{V}_r,$$

for irreducible $\mathsf{K}\mathfrak{G}$-modules V_i. By hypothesis, each V_i is isotropic, so $r > 1$. If $r = 2$, there is nothing to prove, so we suppose that $r > 2$. Suppose that $0 \neq u \in \mathsf{V}_1$. Then $(u, v) \neq 0$ for some $v \in \mathsf{V}$. Write $v = v_1 + \cdots + v_r$, with $v_i \in \mathsf{V}_i$. Since $(u, v_1) = 0$, there exists $i > 1$ such that $(u, v_i) \neq 0$. Let $\mathsf{U} = \mathsf{V}_1 \oplus \mathsf{V}_i$. Then U is not isotropic and, since $r > 2$, $\mathsf{U} \neq \mathsf{V}$. Let T be the radical of U. Since U is a $\mathsf{K}\mathfrak{G}$-submodule, so is T.

Hence $\mathsf{U} = \mathsf{T} \oplus \mathsf{W}$ for some $\mathsf{K}\mathfrak{G}$-submodule W. If $0 \neq w \in \mathsf{W}$, then $w \notin \mathsf{T}$, so $(w, u') \neq 0$ for some $u' \in \mathsf{U}$. Write $u' = t' + w'$ with $t' \in \mathsf{T}$. $w' \in \mathsf{W}$. Then $(w, w') \neq 0$. Hence W is a non-singular proper $\mathsf{K}\mathfrak{G}$-submodule of V. Since U is not isotropic, $\mathsf{W} \neq 0$. By II, 9.4d), $\mathsf{V} = \mathsf{W} \oplus \mathsf{W}^{\perp}$. Since $\mathsf{W}^{\perp}$ is also non-singular, the inductive hypothesis may be applied to W and $\mathsf{W}^{\perp}$; the assertion then follows at once. **q.e.d.**

2.4 Lemma. *Suppose that $\mathfrak{G} = \mathfrak{Q}\mathfrak{S}$ is a critical group, where $\mathfrak{Q}$ is a normal extraspecial q-subgroup of $\mathfrak{G}$ and $\mathfrak{S}$ is a cyclic q'-subgroup. If $|\mathfrak{Q}| = q^{2m+1}$, $q^m \equiv \pm 1\ (|\mathfrak{S}|)$.*

Proof. Let V be an additive group isomorphic to $\mathfrak{Q}/\mathfrak{Q}'$. By 2.2b), V can be regarded as a non-singular symplectic space over $\mathsf{K} = GF(q)$. Also, V is a $\mathsf{K}\mathfrak{S}$-module and $\mathfrak{S}$ induces a subgroup of the symplectic group on V. And by 2.1c), $vy \neq v$ if $v \in \mathsf{V} - \{0\}$ and $y \in \mathfrak{S} - \{1\}$. Thus $\mathfrak{S}$ is represented faithfully on any non-zero $\mathsf{K}\mathfrak{S}$-submodule of V.

Let $\mathfrak{S} = \langle s \rangle$ and let l be the smallest positive integer such that $q^l \equiv 1\ (|\mathfrak{S}|)$. Then by II, 3.10, any irreducible $\mathsf{K}\mathfrak{S}$-submodule of V is of dimension l, since $\mathfrak{S}$ is represented faithfully on it. By the Maschke-Schur theorem,

$$\mathsf{V} = \mathsf{U}_1 \oplus \cdots \oplus \mathsf{U}_r,$$

where the U_i are irreducible $\mathsf{K}\mathfrak{S}$-submodules. Thus $\dim_{\mathsf{K}} \mathsf{U}_i = l$ and $lr = \dim_{\mathsf{K}} \mathsf{V} = 2m$.

If r is even, say $r = 2r'$, then $lr' = m$ and $q^m = (q^l)^{r'} \equiv 1\ (|\mathfrak{S}|)$, as required. If r is odd, then by 2.3, there is an irreducible $\mathsf{K}\mathfrak{S}$-submodule U of V which is not isotropic. Thus $\mathsf{U} \nsubseteq \mathsf{U}^{\perp}$ and $\mathsf{U} \cap \mathsf{U}^{\perp} \subset \mathsf{U}$. But $\mathsf{U}^{\perp}$ is a $\mathsf{K}\mathfrak{S}$-submodule and U is irreducible. Thus $\mathsf{U} \cap \mathsf{U}^{\perp} = 0$ and U is non-singular. Since $\dim_{\mathsf{K}} \mathsf{U} = l$, l is even, by II, 9.6b). If $l = 2l'$, $m = l'r$. By II, 9.23, $|\mathfrak{S}|$ divides $q^{l'} + 1$. Since r is odd, $q^{l'} + 1$ divides $q^{l'r} + 1 = q^m + 1$. Thus $q^m \equiv -1\ (|\mathfrak{S}|)$. **q.e.d.**

2.5 Lemma. *Suppose that $\mathfrak{G} = \mathfrak{Q}\mathfrak{S}$, where $\mathfrak{Q}$ is a normal extraspecial q-subgroup of order q^{2m+1}, $\mathfrak{S} = \langle s \rangle$ is cyclic and $[\mathfrak{S}, \mathbf{Z}(\mathfrak{Q})] = 1$. Let K be an algebraically closed field with* char $\mathsf{K} \neq q$, *and let V be an irreducible $\mathsf{K}\mathfrak{G}$-module on which $\mathfrak{Q}$ is represented faithfully. Then V is an irreducible $\mathsf{K}\mathfrak{Q}$-module and* $\dim_{\mathsf{K}} \mathsf{V} = q^m$.

Proof. Since $\mathbf{Z}(\mathfrak{Q}) \leq \mathbf{Z}(\mathfrak{G})$, V is irreducible and K is algebraically closed, $\mathbf{Z}(\mathfrak{Q})$ is represented on V by scalar multiples of the unit matrix (V, 4.3).

Thus there exists a non-trivial homomorphism α of $\mathbf{Z}(\mathfrak{Q})$ into $\mathsf{K}^\times$ such that

$$vx = \alpha(x)v \quad (v \in \mathsf{V}, x \in \mathbf{Z}(\mathfrak{Q})).$$

Hence if U is any irreducible $\mathsf{K}\mathfrak{Q}$-submodule of V,

$$ux = \alpha(x)u \quad (u \in \mathsf{U}, x \in \mathbf{Z}(\mathfrak{Q})).$$

It follows from V, 16.14 that all irreducible $\mathsf{K}\mathfrak{Q}$-submodules of V are isomorphic. By V, 17.3, V is the direct sum of k isomorphic irreducible $\mathsf{K}\mathfrak{Q}$-submodules, and by V, 17.5, $\mathfrak{S} \cong \mathfrak{G}/\mathfrak{Q}$ has an irreducible projective representation ρ of degree k. If n is the order of s and $\rho(s) = A$, then $A^n = \lambda I$, where $0 \neq \lambda \in \mathsf{K}$. But since K is algebraically closed, $\lambda = \mu^n$ for some $\mu \in \mathsf{K}$. Hence $\mathfrak{S}$ has an ordinary irreducible representation $s \to \mu^{-1}A$ of degree k. By V, 6.1, $k = 1$. Thus V is a faithful irreducible $\mathsf{K}\mathfrak{Q}$-module. By V, 16.14, $\dim_{\mathsf{K}} \mathsf{V} = q^m$. **q.e.d.**

Next we obtain the analogue of V, 17.13 in the case when $\mathfrak{S}$ is a p-group and char $\mathsf{K} = p$.

2.6 Theorem (HALL and HIGMAN). *Suppose that $\mathfrak{G} = \mathfrak{Q}\mathfrak{S}$ is a critical group, $\mathfrak{Q}$ is a normal extraspecial q-subgroup of order q^{2m+1} and $\mathfrak{S} = \langle s \rangle$ is a cyclic group of order p^a. Let K be an algebraically closed field of characteristic p and let V be an irreducible $\mathsf{K}\mathfrak{G}$-module on which $\mathfrak{Q}$ is represented faithfully. Then V is an irreducible $\mathsf{K}\mathfrak{Q}$-module and $\dim_{\mathsf{K}} \mathsf{V} = q^m$. As a $\mathsf{K}\mathfrak{S}$-module, $\mathsf{V} = \mathsf{F} \oplus \mathsf{W}$, where F is a free $\mathsf{K}\mathfrak{S}$-module, W is indecomposable and $\dim_{\mathsf{K}} \mathsf{W}$ is either 1 or $p^a - 1$. (It is possible that $\mathsf{F} = 0$.)*

In particular, the minimum polynomial of the linear transformation $v \to vs$ of V is $(t - 1)^r$, where either $r = p^a$ or $r = p^a - 1 = q^m$.

Proof. By 2.5, V is an irreducible $\mathsf{K}\mathfrak{Q}$-module and $\dim_{\mathsf{K}} \mathsf{V} = q^m$.

Let $\mathfrak{Z} = \mathbf{Z}(\mathfrak{Q})$, $\mathfrak{H} = \mathfrak{S}\mathfrak{Z}$. Thus $|\mathfrak{G} : \mathfrak{H}| = q^{2m}$. Denote V, regarded as a $\mathsf{K}\mathfrak{H}$-module, by U. By VII, 4.3, V is an indecomposable direct summand of $\mathsf{U}^{\mathfrak{G}}$. But if

$$\mathsf{U} = \mathsf{U}_1 \oplus \cdots \oplus \mathsf{U}_k,$$

where the U_i are indecomposable $\mathsf{K}\mathfrak{H}$-submodules of U,

$$\mathsf{U}^{\mathfrak{G}} = \mathsf{U}_1^{\mathfrak{G}} \oplus \cdots \oplus \mathsf{U}_k^{\mathfrak{G}}.$$

It follows from the Krull-Schmidt theorem (I, 12.3) that V is a direct summand of $\mathsf{U}_i^{\mathfrak{G}}$ for some i. Put $\mathsf{W} = \mathsf{U}_i$. Thus W is an indecomposable $\mathsf{K}\mathfrak{H}$-submodule of U, and V is a direct summand of $\mathsf{W}^{\mathfrak{G}}$.

If $x \in \mathfrak{G} - \mathfrak{H}$, the cosets $\mathfrak{H}x, \mathfrak{H}xs, \ldots, \mathfrak{H}xs^{p^a-1}$ are distinct. For $x = ys^r$ with $y \in \mathfrak{Q}$. Thus if $\mathfrak{H}xs^i = \mathfrak{H}xs^j$, $ys^{i-j}y^{-1} = xs^{i-j}x^{-1} \in \mathfrak{H}$, whence $s^{-(i-j)}ys^{i-j}y^{-1} \in \mathfrak{H} \cap \mathfrak{Q} = \mathfrak{Z}$ and $\mathfrak{Z}y = \mathfrak{Z}y^{s^{i-j}}$. By 2.1c), $s^i = s^j$, since otherwise $y \in \mathfrak{Z}$ and $x \in \mathfrak{H}$.

Hence there exists a set T such that

$$\{1\} \cup \{ts^i | 0 \leq i < p^a, t \in \mathsf{T}\}$$

is a transversal of $\mathfrak{H}$ in $\mathfrak{G}$. Thus

$$\mathsf{W}^{\mathfrak{G}} = \mathsf{W}_1 \oplus \left(\bigoplus_{t \in \mathsf{T}} \mathsf{W}_t\right),$$

where $\mathsf{W}_1 = \mathsf{W} \otimes 1$ and, for each $t \in \mathsf{T}$,

$$\mathsf{W}_t = (\mathsf{W} \otimes t) \oplus (\mathsf{W} \otimes ts) \oplus \cdots \oplus (\mathsf{W} \otimes ts^{p^a-1}).$$

Now it is clear that W_t is a free $\mathsf{K}\mathfrak{S}$-module, for a K-basis of $\mathsf{W} \otimes t$ is a $\mathsf{K}\mathfrak{S}$-basis of W_t. Hence

$$\mathsf{W}^{\mathfrak{G}} = \mathsf{W}_1 \oplus \mathsf{F}_1 \oplus \cdots \oplus \mathsf{F}_l,$$

where each F_i is a $\mathsf{K}\mathfrak{S}$-submodule of $\mathsf{W}^{\mathfrak{G}}$ isomorphic to $\mathsf{K}\mathfrak{S}$. By VII, 5.3, F_i is an indecomposable $\mathsf{K}\mathfrak{S}$-module. Also, W_1 is an indecomposable $\mathsf{K}\mathfrak{S}$-module. For as a $\mathsf{K}\mathfrak{H}$-module, W_1 is indecomposable, being isomorphic to W. But since W is a submodule of U, $\mathfrak{Z}$ is represented on W by scalar multiplications, so any $\mathsf{K}\mathfrak{S}$-submodule of W is also a $\mathsf{K}\mathfrak{H}$-module.

Since V is a direct summand of $\mathsf{W}^{\mathfrak{G}}$, it follows from the Krull-Schmidt theorem that if

$$\mathsf{V} = \mathsf{V}_1 \oplus \cdots \oplus \mathsf{V}_h$$

is a direct decomposition of V into indecomposable $\mathsf{K}\mathfrak{S}$-modules, at most one summand is isomorphic to W and the rest are isomorphic to $\mathsf{K}\mathfrak{S}$. If all are isomorphic to $\mathsf{K}\mathfrak{S}$,

$$q^m = \dim_{\mathsf{K}} \mathsf{V} = h|\mathfrak{S}| = hp^a,$$

which is impossible since $q \neq p$. Thus we may suppose that $\mathsf{V}_1 \cong \mathsf{W}$, so

$$q^m = \dim_{\mathsf{K}} \mathsf{V} = \dim_{\mathsf{K}} \mathsf{V}_1 + (h - 1)p^a.$$

By 2.4, $q^m \equiv \pm 1\ (p^a)$, so $\dim_{\mathsf{K}} \mathsf{V}_1 \equiv \pm 1\ (p^a)$. By VII, 5.3, either $\dim_{\mathsf{K}} \mathsf{V}_1 = 1$ or V_1 is the unique indecomposable $\mathsf{K}\mathfrak{S}$-module of dimension $p^a - 1$. In the latter case, the minimum polynomial of the mapping $u \to us$ ($u \in \mathsf{V}_1$) on V_1 is $(t - 1)^{p^a - 1}$. And if $\dim_{\mathsf{K}} \mathsf{V}_1 = 1$, $h > 1$. Hence the minimum polynomial of the mapping $v \to vs$ on V is $(t - 1)^{p^a}$, except when $h = 1$, in which case it is $(t - 1)^{p^a - 1}$ and $\dim_{\mathsf{K}} \mathsf{V} = q^m = p^a - 1$. **q.e.d.**

The equation $q^m = p^a - 1$, which arises in the exceptional case of 2.6, will now be analysed.

2.7 Lemma. *Suppose that $p^a = q^b + 1$, where p, q are primes and a, b are positive integers. Then either*

a) $p = 2$, $b = 1$, *a is a prime and $q = 2^a - 1$ is a Mersenne prime, or*

b) $q = 2$, $a = 1$, $b = 2^m$ *and* $p = 2^{2^m} + 1$ *is a Fermat prime, or*

c) $p^a = 9$, $q^b = 8$.

Proof. Obviously $p = 2$ or $q = 2$.

a) Suppose that $p = 2$. Then $a > 1$. If $b = 2c$,

$$q^{2c} - 1 = p^a - 2 \not\equiv 0\ (4).$$

This is a contradiction, since q is odd. Thus b is odd and

$$2^a = (q + 1)(q^{b-1} - q^{b-2} + \cdots + 1).$$

Hence $q + 1 = 2^d$ for some d, and

$$2^{a-d} = q^{b-1} - q^{b-2} + \cdots + 1 \equiv b \equiv 1\ (2).$$

Thus $a = d$ and $b = 1$. If $a = ef$,

$$q = 2^a - 1 = (2^e - 1)(2^{e(f-1)} + \cdots + 1),$$

so either $2^e - 1 = 1$ and $e = 1$, or $2^e - 1 = q$ and $e = a$. Hence a is prime.

b) Suppose that $q = 2$. Thus $p^a = 2^b + 1$ and

$$2^b = p^a - 1 = (p-1)(p^{a-1} + \cdots + 1).$$

Thus $p - 1 = 2^c$ for some c, and $p^{a-1} + \cdots + 1 = 2^{b-c}$.

If $a > 1$, $2^{b-c} = p^{a-1} + \cdots + 1 > 1$, so $a \equiv 2^{b-c} \equiv 0\,(2)$. If $a = 2d$,

$$2^b = (p^d - 1)(p^d + 1),$$

so $p^d - 1 = 2^e$ and $p^d + 1 = 2^f$ for suitable e, f. Thus $2^f - 2^e = 2$, and $e = 1, f = 2$. Thus $p^d = 3$, $p^a = 9$ and $q^b = 8$.

If $a = 1$, $p = 2^b + 1$. If $b = 2^c d$, where d is odd,

$$p = (2^{2^c} + 1)(2^{(d-1)2^c} - \cdots + 1).$$

Thus $2^{2^c} + 1 = p$ and $b = 2^c$. **q.e.d.**

In order to apply 2.6 to more general situations we need the following lemma.

2.8 Lemma. *Let* V *be a finite-dimensional vector space over a field* K, *and let* $\mathfrak{G}$ *be a p-soluble group of non-singular linear transformations of* V *for which* $\mathbf{O}_p(\mathfrak{G}) = 1$ *and* char K *does not divide* $|\mathbf{O}_{p'}(\mathfrak{G})|$. *Let* $\mathfrak{P}$ *be a p-subgroup of* $\mathfrak{G}$ *and suppose that* $1 \neq y \in \mathfrak{P}$. *Then there exists a section* $\mathfrak{H} = \mathfrak{G}_1/\mathfrak{G}_2$ *of* $\mathfrak{G}$ *with the following properties.*

a) $\mathfrak{G}_1 \geq \mathfrak{P}$, *and* $y \notin \mathfrak{G}_2$. *Let* $\bar{y} = y\mathfrak{G}_2$.

b) $\mathfrak{H} = \mathfrak{Q}\mathfrak{S}$, *where* $\mathfrak{Q}$ *is a normal, special q-subgroup of* $\mathfrak{H}$ *for some prime* $q \neq p$, *and* $\mathfrak{S} = \mathfrak{P}\mathfrak{G}_2/\mathfrak{G}_2 \in S_p(\mathfrak{H})$.

c) $\mathfrak{Q}/\Phi(\mathfrak{Q})$ *is a minimal normal subgroup of* $\mathfrak{H}/\Phi(\mathfrak{Q})$, *and* $\bar{y}$ *centralizes* $\Phi(\mathfrak{Q})$ *but not* $\mathfrak{Q}$.

d) $\mathfrak{H}$ *has a faithful, irreducible representation on a section* $\mathsf{V}_1/\mathsf{V}_2$ *of* V. *Further, if the representation of* $\mathfrak{Q}$ *on* $\mathsf{V}_1/\mathsf{V}_2$ *is absolutely irreducible,* $\mathfrak{Q}$ *is extraspecial.*

Proof. By 1.3, $\mathbf{C}_{\mathfrak{G}}(\mathbf{O}_{p'}(\mathfrak{G})) \leq \mathbf{O}_{p'}(\mathfrak{G})$. Hence y does not centralize $\mathbf{O}_{p'}(\mathfrak{G})$. Thus, if $\mathscr{S}$ is the set of subgroups $\mathfrak{X}$ of $\mathbf{O}_{p'}(\mathfrak{G})$ for which $\mathfrak{P} \leq \mathbf{N}_{\mathfrak{G}}(\mathfrak{X})$ and $y \notin \mathbf{C}_{\mathfrak{G}}(\mathfrak{X})$, $\mathbf{O}_{p'}(\mathfrak{G}) \in \mathscr{S}$. Let $\mathfrak{X}$ be an element of $\mathscr{S}$ of minimal order, and choose $x \in \mathfrak{X}$ such that $z = [x, y] \neq 1$. Let $\mathfrak{G}_1 = \mathfrak{P}\mathfrak{X}$ and let

$$\mathsf{V} = \mathsf{V}_0 > \mathsf{V}_1 > \cdots > \mathsf{V}_n = 0$$

be a $\mathsf{K}\mathfrak{G}_1$-composition series of V. Since char K does not divide the order of z, there exists a $\mathsf{K}\langle z\rangle$-submodule U_i of V_{i-1} such that $\mathsf{V}_{i-1} = \mathsf{U}_i \oplus \mathsf{V}_i$ $(i = 1, \ldots, n)$, by I, 17.7. Thus

$$\mathsf{V} = \mathsf{U}_1 \oplus \mathsf{U}_2 \oplus \cdots \oplus \mathsf{U}_n.$$

Since $z \neq 1$, there exists some U_i on which z induces a non-identity linear transformation. Let $\mathfrak{G}_2$ be the kernel of the representation of $\mathfrak{G}_1$ induced on $\mathsf{V}_{i-1}/\mathsf{V}_i$. Thus $\mathfrak{H} = \mathfrak{G}_1/\mathfrak{G}_2$ has a faithful irreducible representation on $\mathsf{V}_{i-1}/\mathsf{V}_i$. Since U_i is $\mathsf{K}\langle z\rangle$-isomorphic to $\mathsf{V}_{i-1}/\mathsf{V}_i$, $z \notin \mathfrak{G}_2$. Since $\mathfrak{G}_2 \trianglelefteq \mathfrak{G}_1$, it follows that $y \notin \mathfrak{G}_2$. Thus a) is satisfied.

Let $\mathfrak{Q} = \mathfrak{X}\mathfrak{G}_2/\mathfrak{G}_2$, $\mathfrak{S} = \mathfrak{P}\mathfrak{G}_2/\mathfrak{G}_2$. Then $\mathfrak{Q} \trianglelefteq \mathfrak{H}$, $\mathfrak{H} = \mathfrak{Q}\mathfrak{S}$ and, since $\mathfrak{P} \in S_p(\mathfrak{G}_1)$, $\mathfrak{S} \in S_p(\mathfrak{H})$. If $\mathfrak{Y} < \mathfrak{X}$ and $\mathfrak{P} \leq \mathbf{N}_{\mathfrak{G}}(\mathfrak{Y})$, then $\mathfrak{Y} \notin \mathscr{S}$, since $\mathfrak{X}$ is an element of $\mathscr{S}$ of minimal order, so $y \in \mathbf{C}_{\mathfrak{G}}(\mathfrak{Y})$. Hence by 1.12, $\mathfrak{X}$ is a q-group for some prime $q \neq p$. Thus $\mathfrak{Q}$ is a q-group. Since $z \notin \mathfrak{G}_2$ and $x\mathfrak{G}_2 \in \mathfrak{Q}$, $\bar{y}$ does not centralize $\mathfrak{Q}$.

Let $\bar{\mathfrak{S}}$ be the normal closure of $\bar{y}$ in $\mathfrak{S}$. We show that $\bar{\mathfrak{S}}$ centralizes any proper $\mathfrak{S}$-invariant subgroup $\mathfrak{Q}_0$ of $\mathfrak{Q}$. We have $\mathfrak{Q}_0 = \mathfrak{X}_0\mathfrak{G}_2/\mathfrak{G}_2$, where $\mathfrak{X}_0 < \mathfrak{X}$ and $\mathfrak{P} \leq \mathbf{N}_{\mathfrak{G}}(\mathfrak{X}_0)$. Thus $y \in \mathbf{C}_{\mathfrak{P}}(\mathfrak{X}_0)$ and $\bar{y} \in \mathbf{C}_{\mathfrak{S}}(\mathfrak{Q}_0)$. But since $\mathfrak{S} \leq \mathbf{N}_{\mathfrak{H}}(\mathfrak{Q}_0)$, $\mathbf{C}_{\mathfrak{S}}(\mathfrak{Q}_0) \trianglelefteq \mathfrak{S}$. Thus $\bar{\mathfrak{S}} \leq \mathbf{C}_{\mathfrak{S}}(\mathfrak{Q}_0)$.

By III, 13.5, $\mathfrak{Q}/\Phi(\mathfrak{Q})$ is a minimal normal subgroup of $\mathfrak{H}/\Phi(\mathfrak{Q})$ and $[\Phi(\mathfrak{Q}), \bar{\mathfrak{S}}] = 1$. Thus c) is proved. Also $\mathfrak{Q}$ is a special q-group. Now if the representation of $\mathfrak{Q}$ on $\mathsf{V}_1/\mathsf{V}_2$ is absolutely irreducible, then by Schur's lemma, $\mathbf{Z}(\mathfrak{Q})$ is represented on $\mathsf{V}_1/\mathsf{V}_2$ by scalar multiples of the identity linear transformation. Since the representation is faithful and $\bar{y} \notin \mathbf{C}_{\mathfrak{H}}(\mathfrak{Q})$, it follows that $\mathfrak{Q}$ is non-Abelian and that $\mathbf{Z}(\mathfrak{Q})$ is cyclic. Thus $\Phi(\mathfrak{Q}) = \mathfrak{Q}' = \mathbf{Z}(\mathfrak{Q})$ is of order q (III, 13.1), and $\mathfrak{Q}$ is extraspecial. **q.e.d.**

The celebrated Theorem B of Hall and Higman will now be proved.

2.9 Theorem (HALL and HIGMAN). *Let* K *be a field of characteristic* p *and let* V *be a finite-dimensional vector space over* K. *Let* $\mathfrak{G}$ *be a finite* p*-soluble group of non-singular linear transformations of* V *such that* $\mathbf{O}_p(\mathfrak{G}) = 1$. *If* g *is an element of* $\mathfrak{G}$ *of order* p^n, *the minimum polynomial of* g *is* $(t-1)^r$, *where* $r \leq p^n$. *If* $r < p^n$, *there exist integers* $n_0 \leq n$ *for which* $p^{n_0} - 1$ *is a power of a prime* q *and the Sylow* q*-subgroups of* $\mathfrak{G}$ *are non-Abelian; further, if* n_0 *is the least such integer,* $r \geq p^{n-n_0}(p^{n_0} - 1)$.

Proof. Since extension of the ground field does not affect the minimum polynomial of a linear transformation, we may suppose that K is algebraically closed. The theorem will be proved by induction on n. Since $0 = g^{p^n} - 1 = (g-1)^{p^n}$, the minimum polynomial of g is $(t-1)^r$, where $r \leq p^n$. Suppose that $r < p^n$.

We apply 2.8 with $\mathfrak{P} = \langle g \rangle$, $y = g^{p^{n-1}}$. Thus there exists a section $\mathfrak{H} = \mathfrak{G}_1/\mathfrak{G}_2$ of $\mathfrak{G}$ such that $g \in \mathfrak{G}_1$ and, if $s = g\mathfrak{G}_2$, $s^{p^{n-1}} \neq 1$. Thus the order of s is p^n. Also $\mathfrak{H} = \mathfrak{Q}\mathfrak{S}$, where $\mathfrak{Q}$ is a normal q-subgroup of $\mathfrak{H}$ for some prime $q \neq p$, $\mathfrak{S} = \langle s \rangle$, $\mathfrak{Q}/\Phi(\mathfrak{Q})$ is a minimal normal subgroup of $\mathfrak{H}/\Phi(\mathfrak{Q})$ and $s^{p^{n-1}}$ centralizes $\Phi(\mathfrak{Q})$. Finally, $\mathfrak{H}$ has a faithful irreducible representation on a section U of V. Thus $\mathbf{O}_p(\mathfrak{H}) = 1$ by V, 5.17, and by 1.3, $\mathfrak{Q} \geq \mathbf{C}_{\mathfrak{H}}(\mathfrak{Q})$. Since $r < p^n$, $u(s-1)^{p^n-1} = 0$ for all $u \in \mathsf{U}$. It suffices to prove that

(i) $\mathfrak{Q}$ is non-Abelian,

(ii) $q^{m_1} = p^{n_1} - 1$ for some m_1, n_1 with $n_1 \leq n$, and

(iii) $u(s-1)^{p^{n-n_1}(p^{n_1}-1)-1} \neq 0$ for some $u \in \mathsf{U}$.

For then, the Sylow q-subgroup of $\mathfrak{G}$ is non-Abelian, so there certainly exist integers $n_0 \leq n$ for which $p^{n_0} - 1$ is a power of a prime q_0 and the Sylow q_0-subgroups of $\mathfrak{G}$ are non-Abelian. Further, if n_0 is the least such integer, $n_0 \leq n_1$, so $p^{n-n_0}(p^{n_0} - 1) - 1 \leq p^{n-n_1}(p^{n_1} - 1) - 1$ and $(g-1)^{p^{n-n_0}(p^{n_0}-1)-1} \neq 0$.

First suppose that U is an irreducible $\mathsf{K}\mathfrak{Q}$-module. We prove that $\mathfrak{H}$ is a critical group in the sense of 2.1. By 2.8, $\mathfrak{Q}$ is extraspecial. Now $\mathfrak{Q}/\Phi(\mathfrak{Q})$ may be regarded as a vector space over the field $GF(q)$, and if $\rho = (\mathfrak{S} \text{ on } \mathfrak{Q}/\Phi(\mathfrak{Q}))$, ρ is irreducible, since $\mathfrak{Q}/\Phi(\mathfrak{Q})$ is a minimal normal subgroup and $\mathfrak{H} = \mathfrak{Q}\mathfrak{S}$. By II, 3.10, $\mathfrak{Q}/\Phi(\mathfrak{Q})$ is isomorphic to the additive group of a certain field, and $\rho(s)$ corresponds to multiplication by an element λ of this field. By I, 4.4, $\rho(s^{p^{n-1}}) \neq 1$, so $\lambda^{p^{n-1}} \neq 1$. It follows that $\rho(s^i)$ is fixed point free ($i = 1, \ldots, p^n - 1$). Thus $\mathfrak{H}$ is a critical group (2.1).

Since $u(s-1)^{p^n-1} = 0$ for all $u \in \mathsf{U}$, it follows from 2.6 that $p^n - 1$ is a power of q and $u(s-1)^{p^n-2} \neq 0$ for some $u \in \mathsf{U}$. Hence (i), (ii) and (iii) are proved.

Now suppose that U is a reducible $\mathsf{K}\mathfrak{Q}$-module. By V, 17.3b),

$$\mathsf{U} = \mathsf{U}_1 \oplus \cdots \oplus \mathsf{U}_l,$$

where U_i is the sum of all $\mathsf{K}\mathfrak{Q}$-submodules of U isomorphic to some irreducible $\mathsf{K}\mathfrak{Q}$-module. By V, 17.3d), there is a transitive permutation representation σ of $\mathfrak{H}$ on $\{\mathsf{U}_1, \ldots, \mathsf{U}_l\}$ such that $\mathsf{U}_i\sigma(h) = \mathsf{U}_i h$ ($h \in \mathfrak{H}$). Let $\mathfrak{H}_0 = \ker \sigma$. Then $\mathfrak{Q} \leq \mathfrak{H}_0$, so $\mathfrak{H}_0 = \mathfrak{S}_0\mathfrak{Q}$, where $\mathfrak{S}_0 = \mathfrak{S} \cap \mathfrak{H}_0$. The restriction σ' of σ to $\mathfrak{S}$ is transitive, and since $\mathfrak{S}$ is Abelian, $\mathfrak{S}_0 = \ker \sigma'$ is the stabiliser of any U_i in σ'. Thus $l = |\mathfrak{S} : \mathfrak{S}_0| = |\mathfrak{H} : \mathfrak{H}_0|$. Since U is reducible, U possesses non-isomorphic irreducible $\mathsf{K}\mathfrak{Q}$-submodules, by VII, 9.19. Thus $l > 1$ and $\mathfrak{S}_0 < \mathfrak{S}$. We observe also that $\mathfrak{S}_0 \neq 1$. For if $\mathfrak{S}_0 = 1$, $\sigma(s)^{p^{n-1}} \neq 1$, so $\sigma(s)$ has a cycle $(\mathsf{U}_i, \mathsf{U}_i s, \ldots)$ of length p^n. But then, if u is a non-zero element of U_i, $u, us, \ldots, us^{p^n-1}$ lie in distinct summands of the direct sum $\mathsf{U} = \mathsf{U}_1 \oplus \cdots \oplus \mathsf{U}_l$. Thus these elements are linearly independent, contrary to the fact that $u(s-1)^{p^n-1} = 0$. Hence $1 < \mathfrak{S}_0 < \mathfrak{S}$. Thus $s^{p^{n-1}} \in \mathfrak{H}_0$. Suppose that $\mathfrak{S}_0 = \langle s^{p^m} \rangle$, so that $1 \le m < n$.

Since $\mathfrak{Q} \ge \mathbf{C}_{\mathfrak{H}}(\mathfrak{Q})$, there exists $x \in \mathfrak{Q}$ such that $z = [x, s^{p^{n-1}}] \ne 1$. Then $z \in \mathfrak{Q} \le \mathfrak{H}_0$, but there exists U_j on which z induces a non-identity linear transformation. Now U_j is a $\mathsf{K}\mathfrak{H}_0$-module; let W be an irreducible $\mathsf{K}\mathfrak{H}_0$-submodule of U_j. If $\tilde{\mathsf{W}}$ is an irreducible $\mathsf{K}\mathfrak{Q}$-submodule of W, U_j is the direct sum of $\mathsf{K}\mathfrak{Q}$-submodules isomorphic to $\tilde{\mathsf{W}}$, so z induces a non-identity linear transformation on $\tilde{\mathsf{W}}$ and on W. Thus, if $\mathfrak{N}$ is the kernel of the representation of $\mathfrak{H}_0$ on W, $z \notin \mathfrak{N}$ and $s^{p^{n-1}} \notin \mathfrak{N}$.

Since $\mathfrak{S}_0 = \langle s^{p^m} \rangle$ is the stabiliser of U_j, $\mathsf{W}, \mathsf{W}s, \dots, \mathsf{W}s^{p^m-1}$ are contained in distinct summands of the direct sum $\mathsf{U} = \mathsf{U}_1 \oplus \cdots \oplus \mathsf{U}_l$. It follows that if $w \in \mathsf{W}$ and $w \ne 0$, then $w(s-1)^{p^m-1} \ne 0$.

Let $\overline{\mathfrak{H}} = \mathfrak{H}_0/\mathfrak{N}$, $h = s^{p^m}\mathfrak{N}$. Then W is a faithful, irreducible $\mathsf{K}\overline{\mathfrak{H}}$-module, so $\mathbf{O}_p(\overline{\mathfrak{H}}) = 1$. We wish to apply the inductive hypothesis to $\overline{\mathfrak{H}}$ and the element h, which is of order p^{n-m} since $s^{p^{n-1}} \notin \mathfrak{N}$. If $w \in \mathsf{W}$,

$$\begin{aligned}(w(h-1)^{p^{n-m}-1})(s-1)^{p^m-1} &= w(s-1)^{p^n-p^m}(s-1)^{p^m-1}\\ &= w(s-1)^{p^n-1} = 0.\end{aligned}$$

Thus by the above remark, $w(h-1)^{p^{n-m}-1} = 0$ for all $w \in \mathsf{W}$. By the inductive hypothesis, $p^{n_1} - 1 = q^{m_1}$ for some m_1, n_1 with $n_1 \le n - m$; also $\mathfrak{Q}/\mathfrak{N}$ is non-Abelian, and there exists $w \in \mathsf{W}$ such that

$$w_1 = w(h-1)^{p^{n-m}-p^{n-m-n_1}-1} \ne 0.$$

Again, we have $w_1(s-1)^{p^m-1} \ne 0$. But

$$\begin{aligned}w_1(s-1)^{p^m-1} &= w(s^{p^m}-1)^{p^{n-m}-p^{n-m-n_1}-1}(s-1)^{p^m-1}\\ &= w(s-1)^{p^n-p^{n-n_1}-1},\end{aligned}$$

so (i), (ii), (iii) are again proved. **q.e.d.**

2.10 Remarks. a) The result corresponding to the crucial theorem 2.6 in the case when char K does not divide $|\mathfrak{G}|$ is V, 17.13. This will be generalized along the lines of 2.9 in 6.2.

b) Several other proofs of 2.6 are known. The original proof of HALL and HIGMAN [1, Theorem 2.5.1] is not quite so general. Dade gave a proof of it along lines similar to his proof of V, 17.13. Another proof for q odd was given by BERGER [1, IV]. THOMPSON [4] gave a proof using vertices and sources similar to the one presented here.

c) A number of situations similar to that of 2.6 have been studied. Examples may be found in DADE [5] or BERGER [1].

§ 3. The Exceptional Case

Let $\langle g \rangle$ be a cyclic group of non-singular linear transformations of a vector space V and suppose that g is of order n. The minimum polynomial $f(t)$ of g is then a divisor of $t^n - 1$. We have seen, for example in 1.13, that the case when $f(t) = t^n - 1$ leads to the desired theorems. However, this does not always follow from the appropriate hypotheses. To study the exceptional case, it is necessary to make some rather detailed calculations.

3.1 Definitions. Let α be an automorphism of an extraspecial q-group $\mathfrak{Q}$. We say that α *bisects* $\mathfrak{Q}$ if there exist α-invariant subgroups $\mathfrak{Q}_1$, $\mathfrak{Q}_2$ of $\mathfrak{Q}$ with the following properties.

a) $\mathfrak{Q} = \mathfrak{Q}_1\mathfrak{Q}_2$, $[\mathfrak{Q}_1, \mathfrak{Q}_2] = 1$ and $\mathfrak{Q}_1 \cap \mathfrak{Q}_2 = \mathbf{Z}(\mathfrak{Q})$.

b) $(\langle\alpha\rangle$ on $\mathfrak{Q}_1/\mathbf{Z}(\mathfrak{Q}))$ is irreducible, and $x\alpha = x$ for all $x \in \mathfrak{Q}_2$.

In this case, we also have the following.

c) $\mathfrak{Q}_1$ is extraspecial and either $\mathfrak{Q}_2$ is extraspecial or $\mathfrak{Q}_2 = \mathbf{Z}(\mathfrak{Q})$.

To see this, observe that since $[\mathfrak{Q}_1, \mathfrak{Q}_2] = 1$, $\mathbf{Z}(\mathfrak{Q}_1) \leq \mathbf{Z}(\mathfrak{Q})$ and $\mathbf{Z}(\mathfrak{Q}_2) \leq \mathbf{Z}(\mathfrak{Q})$. Hence $\mathbf{Z}(\mathfrak{Q}_1) = \mathbf{Z}(\mathfrak{Q}_2) = \mathbf{Z}(\mathfrak{Q})$. Since $(\langle\alpha\rangle$ on $\mathfrak{Q}_1/\mathbf{Z}(\mathfrak{Q}))$ is irreducible, $\mathfrak{Q}_1 \neq \mathbf{Z}(\mathfrak{Q})$. Thus $\mathfrak{Q}_1$ is extraspecial. Similarly, if $\mathfrak{Q}_2 \neq \mathbf{Z}(\mathfrak{Q})$, $\mathfrak{Q}_2$ is extraspecial.

We also observe that $\mathfrak{Q}_1$, $\mathfrak{Q}_2$ are uniquely determined by the conditions a), b). For $\alpha \neq 1$, so $\mathfrak{Q}_2 = \mathbf{C}_{\mathfrak{Q}}(\alpha)$ and $\mathfrak{Q}_1/\mathbf{Z}(\mathfrak{Q}) = [\mathfrak{Q}/\mathbf{Z}(\mathfrak{Q}), \langle\alpha\rangle]$.

3.2 Theorem. *Suppose that $\mathfrak{G} = \mathfrak{Q}\mathfrak{S}$, where $\mathfrak{Q}$ is a normal, extraspecial q-group, $\mathfrak{S} = \langle s \rangle$ is cyclic and $[\mathbf{Z}(\mathfrak{Q}), \mathfrak{S}] = 1$. Suppose that s is of order p^n for some prime $p \neq q$, and that $s^{p^{n-1}} \notin \mathbf{C}_{\mathfrak{G}}(\mathfrak{Q})$. Let V be a faithful, irreducible $\mathsf{K}\mathfrak{G}$-module, where K is an algebraically closed field and* char $\mathsf{K} \neq q$, *and suppose that the degree m of the minimum polynomial of the linear transformation $v \to vs$ ($v \in \mathsf{V}$) is less than p^n. Then the following assertions hold.*

a) *$p^n - 1 = q^d$ for some d.*

b) *The automorphism of $\mathfrak{Q}$ induced by s bisects $\mathfrak{Q}$, and* $|\mathfrak{Q} : \mathbf{C}_{\mathfrak{Q}}(s)| = q^{2d}$.

c) $m = p^n - 1$.

Proof. Let $\mathsf{F} = GF(q)$. The elementary Abelian q-group $\mathfrak{Q}/\mathbf{Z}(\mathfrak{Q})$ may be regarded as an $\mathsf{F}\mathfrak{S}$-module, so by the Maschke-Schur theorem, $\mathfrak{Q}/\mathbf{Z}(\mathfrak{Q})$ is the direct product of irreducible $\mathsf{F}\mathfrak{S}$-submodules. If $s^{p^{n-1}}$ centralizes all these submodules, $s^{p^{n-1}}$ centralizes $\mathfrak{Q}/\mathbf{Z}(\mathfrak{Q}) = \mathfrak{Q}/\Phi(\mathfrak{Q})$, and by III, 3.18, $s^{p^{n-1}} \in \mathbf{C}_{\mathfrak{G}}(\mathfrak{Q})$. This is contrary to the hypothesis, so there exists an irreducible $\mathsf{F}\mathfrak{S}$-submodule $\mathfrak{Q}_1/\mathbf{Z}(\mathfrak{Q})$ which is not centralized by $s^{p^{n-1}}$. Thus $\mathfrak{Q}_1$ is invariant under s and $(\langle s\rangle$ on $\mathfrak{Q}_1/\mathbf{Z}(\mathfrak{Q}))$ is irreducible. Hence the centre of $\mathfrak{Q}_1$ is either $\mathfrak{Q}_1$ or $\mathbf{Z}(\mathfrak{Q})$. If it is $\mathfrak{Q}_1$, $\mathfrak{Q}_1$ is Abelian and $s^{p^{n-1}} \notin \mathbf{C}_{\mathfrak{G}}(\mathfrak{Q}_1)$. But then, by 1.10 applied to $\operatorname{im}(\mathfrak{S}\mathfrak{Q}_1$ on $\mathsf{V})$, the minimum polynomial of the linear transformation $v \to vs$ $(v \in \mathsf{V})$ is $t^{p^n} - 1$. This is not the case, so the centre of $\mathfrak{Q}_1$ is $\mathbf{Z}(\mathfrak{Q})$ and $\mathfrak{Q}_1$ is extraspecial. Let $\mathfrak{Q}_2 = \mathbf{C}_{\mathfrak{Q}}(\mathfrak{Q}_1)$. Then $\mathfrak{Q}_2$ is $\mathfrak{S}$-invariant, $[\mathfrak{Q}_1, \mathfrak{Q}_2] = 1$ and $\mathfrak{Q}_1 \cap \mathfrak{Q}_2 = \mathbf{Z}(\mathfrak{Q})$. If $\mathfrak{Q}/\mathbf{Z}(\mathfrak{Q})$ is regarded as a symplectic space, $\mathfrak{Q}_1/\mathbf{Z}(\mathfrak{Q})$ is a non-singular subspace, and $\mathfrak{Q}_2/\mathbf{Z}(\mathfrak{Q})$ is the orthogonal complement $(\mathfrak{Q}_1/\mathbf{Z}(\mathfrak{Q}))^{\perp}$ of $\mathfrak{Q}_1/\mathbf{Z}(\mathfrak{Q})$. Thus by II, 9.4d), $\mathfrak{Q} = \mathfrak{Q}_1\mathfrak{Q}_2$.

Since $\mathbf{Z}(\mathfrak{Q}) \leq \mathbf{Z}(\mathfrak{G})$ and $\mathfrak{G}$ is irreducible, there exists a non-identity linear character χ of $\mathbf{Z}(\mathfrak{Q})$ such that

$$vx = \chi(x)v \quad (v \in \mathsf{V},\, x \in \mathbf{Z}(\mathfrak{Q})).$$

By V, 17.3, V is the direct sum of irreducible $\mathsf{K}\mathfrak{Q}_1$-submodules, since $\mathfrak{Q}_1 \trianglelefteq \mathfrak{G}$. By V, 16.14, all of these irreducible $\mathsf{K}\mathfrak{Q}_1$-submodules are isomorphic. Hence by the proof of V, 17.5, we may suppose that $\mathsf{V} = \mathsf{V}_1 \otimes_{\mathsf{K}} \mathsf{V}_2$, where V_1 is an irreducible $\mathsf{K}\mathfrak{Q}_1$-module, and that

$$(v_1 \otimes v_2)x = v_1\rho_1(x) \otimes v_2\rho_2(x) \quad (v_i \in \mathsf{V}_i),$$

where ρ_1, ρ_2 are irreducible projective representations of $\mathfrak{G}$ on V_1, V_2 respectively; further, $v_1\rho_1(y) = v_1 y$ and $\rho_2(y) = 1_{\mathsf{V}_2}$ for all $y \in \mathfrak{Q}_1$.

If $z \in \mathfrak{Q}_2$, $zy = yz$ for all $y \in \mathfrak{Q}_1$ and

$$\begin{aligned} v_1\rho_1(y)\rho_1(z) \otimes v_2\rho_2(z) &= (v_1 \otimes v_2)yz = (v_1 \otimes v_2)zy \\ &= v_1\rho_1(z)\rho_1(y) \otimes v_2\rho_2(z). \end{aligned}$$

Thus $\rho_1(y)\rho_1(z) = \rho_1(z)\rho_1(y)$, and $\rho_1(z)$ commutes with every element of the group $\mathfrak{R} = \langle \rho_1(y) | y \in \mathfrak{Q}_1\rangle$. Since $v_1\rho_1(y) = v_1 y$ for $y \in \mathfrak{Q}_1$, $\mathfrak{R}$ is an irreducible group of linear transformations of V_1. Thus $\rho_1(z)$ is a scalar multiplication.

Since

$$v_1 \otimes v_2 = (v_1 \otimes v_2)s^{p^n} = v_1\rho_1(s)^{p^n} \otimes v_2\rho_2(s)^{p^n},$$

$\rho_1(s)^{p^n} = \xi 1_{V_1}$ and $\rho_2(s)^{p^n} = \xi^{-1} 1_{V_2}$, where $0 \neq \xi \in K$. Since K is algebraically closed, $\xi = \eta^{p^n}$ for some $\eta \in K$. Thus, if $\sigma_1 = \eta^{-1}\rho_1(s)$, $\sigma_2 = \eta\rho_2(s)$,

$$\sigma_1^{p^n} = 1_{V_1}, \quad \sigma_2^{p^n} = 1_{V_2}, \quad (v_1 \otimes v_2)s = v_1\sigma_1 \otimes v_2\sigma_2.$$

Since $v_1\rho_1(y) = v_1 y$ and $\rho_2(y) = 1_{V_2}$ for all $y \in \mathfrak{Q}_1$, $\mathfrak{R}$ is a group isomorphic to $\mathfrak{Q}_1$. Also

$$v_1\sigma_1^{-1}\rho_1(y)\sigma_1 \otimes v_2 = (v_1 \otimes v_2)s^{-1}ys = v_1\rho_1(s^{-1}ys) \otimes v_2,$$

so $\sigma_1^{-1}\rho_1(y)\sigma_1 = \rho_1(s^{-1}ys)$ for all $y \in \mathfrak{Q}_1$. Let $\tilde{\mathfrak{G}} = \langle \sigma_1, \mathfrak{R} \rangle$. Thus $\tilde{\mathfrak{G}}$ is a group of non-singular linear transformations of V_1, $\mathfrak{R}$ is a normal, extra-special q-subgroup of $\tilde{\mathfrak{G}}$, $\tilde{\mathfrak{G}} = \langle \sigma_1 \rangle \mathfrak{R}$ and $[\mathbf{Z}(\mathfrak{R}), \langle \sigma_1 \rangle] = 1$. Since $s^{p^{n-1}}$ does not centralize $\mathfrak{Q}_1/\mathbf{Z}(\mathfrak{Q})$ and ($\langle s \rangle$ on $\mathfrak{Q}_1/\mathbf{Z}(\mathfrak{Q})$) is irreducible, $\mathbf{C}_{\mathfrak{Q}_1/\mathbf{Z}(\mathfrak{Q})}(s^{p^{n-1}}) = 1$. Hence $\mathbf{C}_{\mathfrak{R}/\mathbf{Z}(\mathfrak{R})}(\sigma_1^{p^{n-1}}) = 1$ and $\sigma_1^{p^{n-1}}$ induces a fixed point free automorphism on $\mathfrak{R}/\mathbf{Z}(\mathfrak{R})$. Thus $\tilde{\mathfrak{G}}$ is a critical group. Also V_1 is an irreducible $K\mathfrak{R}$-module, and the degree of the minimum polynomial of σ_1 is at most m. For if $1, \sigma_1, \ldots, \sigma_1^m$ are linearly independent elements of $\mathrm{Hom}_K(V_1, V_1)$, then by V, 9.11, $1, \sigma_1 \otimes \sigma_2, \ldots, \sigma_1^m \otimes \sigma_2^m$ are linearly independent elements of $\mathrm{Hom}_K(V_1, V_1) \otimes_K \mathrm{Hom}_K(V_2, V_2)$, and by V, 9.14, $1, \sigma, \ldots, \sigma^m$ are linearly independent elements of $\mathrm{Hom}_K(V, V)$, where $v\sigma = vs$ ($v \in V$).

We deduce that if $|\mathfrak{R}| = q^{2d+1}$, then $q^d = p^n - 1$ and V_1 is $K\langle \sigma_1 \rangle$-isomorphic to $K\langle \sigma_1 \rangle/R$, where R is a $K\langle \sigma_1 \rangle$-submodule of $K\langle \sigma_1 \rangle$ and $\dim_K R = 1$. This follows from V, 17.13 if char $K \neq p$ and from 2.6 and VII, 5.3 in the case when char $K = p$. Now if α is a linear transformation of a vector space X and X_0 is an α-invariant subspace of X, the minimum polynomial of α divides the product of the minimum polynomials of the linear transformations of X/X_0 and X_0 induced by α. Thus the degree of the minimum polynomial of σ_1 is $p^n - 1$; hence $p^n - 1 \leq m < p^n$ and $m = p^n - 1$. It only remains to show that $\mathfrak{Q}_2 = \mathbf{C}_{\mathfrak{Q}}(s)$. If $\mathfrak{Q}_2 \not\leq \mathbf{C}_{\mathfrak{Q}}(s)$, there exists $z \in \mathfrak{Q}_2$ and $w_i \in V_i$ ($i = 1, 2$) such that $(w_1 \otimes w_2)sz \neq (w_1 \otimes w_2)zs$, or

$$w_1\sigma_1\rho_1(z) \otimes w_2\sigma_2\rho_2(z) \neq w_1\rho_1(z)\sigma_1 \otimes w_2\rho_2(z)\sigma_2.$$

Since $\rho_1(z)$ is a scalar multiplication, it follows that $\sigma_2\rho_2(z) \neq \rho_2(z)\sigma_2$. We now use different methods according to whether char $K \neq p$ or char $K = p$.

If char $K \neq p$, K contains a primitive p^n-*th* root of unity ω, and σ_1, σ_2 can be diagonalized. Since V_1 is an epimorphic image of $K\langle \sigma_1 \rangle$, the eigen-values of σ_1 are distinct. Thus the eigen-values of σ_1 are 1, $\omega, \ldots, \omega^{h-1}, \omega^{h+1}, \ldots, \omega^{p^n-1}$ for some h. Now since $\sigma_2\rho_2(z) \neq \rho_2(z)\sigma_2$, σ_2 has at least two distinct eigen-values ω^i, ω^j. Thus all ω^{i+k} and ω^{j+k} with $k \neq h$ are eigen-values of σ. Hence every power of ω is an eigen-value of σ. This shows that the minimum polynomial of σ is divisible by $t^{p^n} - 1$, a contradiction.

If char $\mathsf{K} = p$, we observe that $\sigma_2 \neq 1$, since $\sigma_2\rho_2(z) \neq \rho_2(z)\sigma_2$. Thus there is an indecomposable $\mathsf{K}\langle\sigma_2\rangle$-submodule of V_2 of dimension greater than 1, and by VII, 5.3, there is an element u_2 of this submodule such that $u_2\sigma_2 = u_2 + u_2'$, $u_2' \neq 0$, $u_2'\sigma_2 = u_2'$. Since the degree of the minimum polynomial of σ_1 is $p^n - 1$, there exists $u_1 \in \mathsf{V}_1$ such that $u_1(\sigma_1 - 1)^{p^n-2} \neq 0$. But for $j \geq 1$, by induction,

$$\begin{aligned}&(u_1 \otimes u_2)(s - 1)^j \\ &= u_1(\sigma_1 - 1)^j \otimes u_2 + ju_1(\sigma_1 - 1)^j \otimes u_2' + ju_1(\sigma_1 - 1)^{j-1} \otimes u_2',\end{aligned}$$

so

$$(u_1 \otimes u_2)(s - 1)^{p^n-1} = -u_1(\sigma_1 - 1)^{p^n-2} \otimes u_2' \neq 0,$$

a contradiction. Hence $\mathfrak{Q}_2 \leq \mathbf{C}_{\mathfrak{Q}}(s)$.

Thus $\mathbf{C}_{\mathfrak{Q}}(s) = \mathbf{C}_{\mathfrak{Q}_1}(s)\mathfrak{Q}_2$. Since $\mathbf{C}_{\mathfrak{Q}_1/\mathbf{Z}(\mathfrak{Q})}(s^{p^{n-1}}) = 1$, $\mathbf{C}_{\mathfrak{Q}_1}(s) = \mathbf{Z}(\mathfrak{Q})$, so $\mathbf{C}_{\mathfrak{Q}}(s) = \mathfrak{Q}_2$. **q.e.d.**

Suppose now that $\mathfrak{G} = \mathfrak{S}\mathfrak{Q}$, where $\mathfrak{Q}$ is a normal, extraspecial q-group, $\mathfrak{S}$ is a p-group ($p \neq q$) and $[\mathfrak{Q}', \mathfrak{S}] = 1$. As in 2.2, $\mathfrak{Q}/\mathbf{Z}(\mathfrak{Q})$ is a non-singular symplectic space over $GF(q)$ and $\mathfrak{S}$ induces a p-subgroup of a symplectic group on $\mathfrak{Q}/\mathbf{Z}(\mathfrak{Q})$. Accordingly, we need to describe certain Sylow p-subgroups of $Sp(2m, q)$.

3.3 Lemma. *Suppose that p, q are distinct primes and that m is the smallest positive integer such that $q^{2m} \equiv 1(p)$. Let $\mathsf{F} = GF(q^{2m})$. Let $(\ ,\ )$ be a non-singular, symplectic bilinear form on F over $GF(q)$, and let $\mathfrak{L}$ be a Sylow p-subgroup of the corresponding symplectic group $Sp(2m, q)$ on F.*

Let X be a vector space over F with F-basis $\{y_1, \ldots, y_r\}$, and let $\mathfrak{P}$ be the set of all mappings α of X onto X of the form

$$\left(\sum_{i=1}^{r} \lambda_i y_i\right)\alpha = \sum_{i=1}^{r} (\lambda_i\sigma_i)y_{i\pi},$$

where $\sigma_1, \ldots, \sigma_r$ run through $\mathfrak{L}$ and π runs through a fixed Sylow p-subgroup of the symmetric group of degree r. Define a bilinear form $(\ ,\ )'$ on X over $GF(q)$ by putting

$$\left(\sum_{i=1}^{r} \lambda_i y_i, \sum_{i=1}^{r} \mu_i y_i\right)' = \sum_{i=1}^{r} (\lambda_i, \mu_i).$$

Then $(\ ,\ )'$ is non-singular and symplectic, and $\mathfrak{P}$ is a Sylow p-subgroup of the corresponding symplectic group $Sp(2mr, q)$ on X.

Proof. It is clear that $(\ ,\ )'$ is a non-singular, symplectic bilinear form on X over $GF(q)$. $\mathfrak{P}$ is a subgroup of the corresponding symplectic group on X, for if $\alpha \in \mathfrak{P}$ and α has the form displayed in the statement of the lemma,

$$\begin{aligned}\left(\left(\sum_{i=1}^{r} \lambda_i y_i\right)\alpha, \left(\sum_{i=1}^{r} \mu_i y_i\right)\alpha\right)' &= \left(\sum_{i=1}^{r} (\lambda_i \sigma_i) y_{i\pi}, \sum_{i=1}^{r} (\mu_i \sigma_i) y_{i\pi}\right)' \\ &= \sum_{i=1}^{r} (\lambda_i \sigma_i, \mu_i \sigma_i) \\ &= \sum_{i=1}^{r} (\lambda_i, \mu_i) \\ &= \left(\sum_{i=1}^{r} \lambda_i y_i, \sum_{i=1}^{r} \mu_i y_i\right)'.\end{aligned}$$

Thus it need only be shown that $|\mathfrak{P}|$ is the order of a Sylow p-subgroup of $Sp(2mr, q)$. By II, 9.13b),

$$|Sp(2mr, q)| = (q^{2mr} - 1)(q^{2(mr-1)} - 1) \cdots (q^2 - 1) q^{m^2 r^2}.$$

For any positive integer x, let $p^{\nu(x)}$ be the highest power of p which divides x. Then by definition of m, $\nu(q^{2i} - 1) > 0$ if and only if i is divisible by m. Thus

$$\nu(|Sp(2mr, q)|) = \sum_{i=1}^{r} \nu(q^{2mi} - 1).$$

In particular, $|\mathfrak{L}| = p^f$, where $f = \nu(q^{2m} - 1)$. Of course, $f > 0$, but note that if $p = 2$, $f \geq 2$. Now $|\mathfrak{P}| = p^{rf + \nu(r!)}$, so it is sufficient to show that

$$v(q^{2mi} - 1) = f + v(i) \quad (i = 1, \dots, r).$$

Suppose that $q^{2m} - 1 = p^f j$; thus j is not divisible by p. Let $i = sp^k$, where s is not divisible by p and $k = v(i)$. Then

$$q^{2mi} - 1 = (1 + p^f j)^i - 1 = p^{f+k} sj + \sum_{h=2}^{i} \binom{sp^k}{h} p^{fh} j^h.$$

Using I, 13.18 and the fact that for $p = 2$, $f \geq 2$, it is easy to check that all the summands for $h \geq 2$ are divisible by p^{f+k+1}. Thus

$$v(q^{2mi} - 1) = f + k = f + v(i). \qquad \textbf{q.e.d.}$$

3.4 Definition. Let V be a vector space over a field K. We say that a non-singular linear transformation α of V *bisects* V (over K) if $\mathsf{V} = \mathbf{C}_{\mathsf{V}}(\alpha) \oplus \mathsf{U}$ for some irreducible $\mathsf{K}\langle\alpha\rangle$-module U. If this is the case, U is unique, and U is a faithful $\mathsf{K}\langle\alpha\rangle$-module.

3.5 Lemma. *Let $\mathsf{K} = GF(q)$, $\mathsf{F} = GF(q^m)$ $(m > 1)$, and let X be a vector space over F with F-basis $\{y_1, \dots, y_r\}$. Let $\mathfrak{L}$ be a group of K-linear transformations of F, let $\mathfrak{R}$ be a subgroup of the symmetric group of degree r and let $\mathfrak{P}$ be the set of all K-linear transformations ξ of X for which there exist $\sigma_i \in \mathfrak{L}$ $(i = 1, \dots, r)$ and $\pi = \pi(\xi) \in \mathfrak{R}$ such that*

$$\left(\sum_{i=1}^{r} \lambda_i y_i\right)\xi = \sum_{i=1}^{r} (\lambda_i \sigma_i) y_{i\pi} \quad (\lambda_i \in \mathsf{F}).$$

a) *If ξ is an element of $\mathfrak{P}$ of prime-power order $p^k > 1$ and ξ bisects X (over K), $\pi^{p^{k-1}} = 1$, where $\pi = \pi(\xi)$.*

b) *Suppose that $\mathfrak{L}$ is Abelian. If ζ is an element of $\mathfrak{P}'$ of prime order p, ζ does not bisect X.*

c) *Suppose that p is a prime and that any two elements of $\mathfrak{L}$ of order p commute. Suppose that ξ, η are elements of $\mathfrak{P}$ which bisect X and that the orders of ξ, η are p^m, p^n respectively. Then $\xi^{p^{m-1}}$ and $\eta^{p^{n-1}}$ commute.*

d) *Suppose that $m = 2$, $q = 2^n - 1$ is a Mersenne prime and $\mathfrak{L}$ is a Sylow 2-subgroup of $SL(2, q)$ (acting on F). Suppose that ξ is an element of $\mathfrak{P}$ of order 2^n and that ξ bisects X. Then $\pi(\xi) = 1$ and*

$$[\xi^{2^{n-1}}, \eta] = [\xi^{2^{n-2}}, \eta]^2$$

for all $\eta \in \mathfrak{P}$.

Proof. a) If $\pi^{p^{k-1}} \neq 1$, π has a cycle $(i_1, \ldots, i_{p^k})$ of length p^k. Let

$$\mathsf{Y} = \mathsf{F}y_{i_1} \oplus \cdots \oplus \mathsf{F}y_{i_{p^k}}.$$

Thus Y is a $\mathsf{K}(\langle\xi\rangle)$-submodule. Now $\mathsf{X} = \mathbf{C}_{\mathsf{X}}(\xi) \oplus \mathsf{X}_1$ for some irreducible $\mathsf{K}(\langle\xi\rangle)$-module X_1. Thus if $\mathsf{Y} \cap \mathsf{X}_1 = 0$, Y is $\mathsf{K}\langle\xi\rangle$-isomorphic to a submodule of X/X_1 and of $\mathbf{C}_{\mathsf{X}}(\xi)$, which is impossible since $p^k > 1$. Hence $\mathsf{Y} \cap \mathsf{X}_1 = \mathsf{X}_1$, $\mathsf{X}_1 \leq \mathsf{Y}$ and $\mathsf{Y} = \mathbf{C}_{\mathsf{Y}}(\xi) \oplus \mathsf{X}_1$. Now it is easy to see that

$$\mathbf{C}_{\mathsf{Y}}(\xi) = \{u(1 + \xi + \cdots + \xi^{p^k-1}) | u \in \mathsf{F}y_{i_1}\}.$$

Thus $\dim_{\mathsf{K}} \mathbf{C}_{\mathsf{Y}}(\xi) = \dim_{\mathsf{K}} \mathsf{F} = m$. Since $\dim_{\mathsf{K}} \mathsf{Y} = p^k m$, it follows that

$$\dim_{\mathsf{K}} \mathsf{X}_1 = m(p^k - 1).$$

Since X_1 is a faithful, irreducible $\mathsf{K}\langle\xi\rangle$-module, it follows from II, 3.10 that $m(p^k - 1)$ is the smallest of the set of integers l for which $q^l \equiv 1(p^k)$. Hence $p^{k-1}(p-1)$ is divisible by $m(p^k - 1)$ and $p^{k-1}(p-1) \geq m(p^k - 1)$. This is impossible since $m > 1$. Hence $\pi^{p^{k-1}} = 1$.

b) If $\xi \in \mathfrak{P}$ is as displayed in the statement of the lemma, write

$$\delta(\xi) = \sigma_1 \sigma_2 \cdots \sigma_r;$$

since $\mathfrak{L}$ is Abelian, the order of the factors in this product is immaterial. It is easy to see that $\delta(\xi)\delta(\eta) = \delta(\xi\eta)$, so δ is a homomorphism of $\mathfrak{P}$ into the Abelian group $\mathfrak{L}$. Hence $\mathfrak{P}' \leq \ker \delta$ and $\delta(\zeta) = 1$. Now if ζ bisects X, it follows from a) that

$$\left(\sum_{i=1}^{r} \lambda_i y_i\right)\zeta = \sum_{i=1}^{r} (\lambda_i \tau_i) y_i$$

for certain $\tau_i \in \mathfrak{L}$, since ζ is of order p. Thus

$$\tau_1 \tau_2 \cdots \tau_r = \delta(\zeta) = 1,$$

but not all τ_i are 1. Suppose $\mathsf{X} = \mathbf{C}_{\mathsf{X}}(\zeta) \oplus \mathsf{X}_1$, where X_1 is an irreducible $\mathsf{K}(\langle\zeta\rangle)$-module. If $\tau_j \neq 1$, $\mathsf{F}y_j \cap \mathsf{X}_1 \neq 0$, so $\mathsf{F}y_j \geq \mathsf{X}_1$. Thus if $i \neq j$, $\mathsf{F}y_i \cap \mathsf{X}_1 = 0$, $\mathsf{F}y_i \leq \mathbf{C}_{\mathsf{X}}(\zeta)$ and $\tau_i = 1$. Hence $1 = \tau_1 \cdots \tau_r = \tau_j \neq 1$, a contradiction.

c) By a),

$$(\sum \lambda_i y_i)\xi^{p^{m-1}} = \sum (\lambda_i \sigma_i) y_i,$$

$$(\sum \lambda_i y_i)\eta^{p^{n-1}} = \sum (\lambda_i \tau_i) y_i$$

for elements σ_i, τ_i of $\mathfrak{L}$ such that $\sigma_i^p = \tau_i^p = 1$. It follows from the hypothesis that σ_i and τ_i commute. Hence $\xi^{p^{m-1}}$ and $\eta^{p^{n-1}}$ commute.

d) Note that $n > 1$; thus $q \not\equiv 1\ (2^n)$ but $q^2 \equiv 1\ (2^n)$. By II, 3.10, the dimension of any faithful, irreducible $\mathsf{K}\langle\xi\rangle$-module is 2. Thus $\mathsf{X} = \mathbf{C}_\mathsf{X}(\xi) \oplus \mathsf{X}_1$, where X_1 is an irreducible $\mathsf{K}\langle\xi\rangle$-module and $\dim_\mathsf{K} \mathsf{X}_1 = 2$. Let $\pi = \pi(\xi)$.

First suppose that $\pi^2 \neq 1$. Then π has a cycle of length greater than 2, so π^2 has two distinct cycles $(i_1, i_2, \dots)$, $(j_1, j_2, \dots)$ of length greater than 1. Hence if $\mathsf{Y}_1 = \mathsf{F}y_{i_1} \oplus \mathsf{F}y_{i_2} \oplus \cdots$ and $\mathsf{Y}_2 = \mathsf{F}y_{j_1} \oplus \mathsf{F}y_{j_2} \oplus \cdots$, Y_1 and Y_2 are disjoint $\mathsf{K}\langle\xi^2\rangle$-modules on which ξ^2 operates non-trivially. Thus $\mathsf{Y}_i \cap \mathsf{X}_1 \neq 0$ $(i = 1, 2)$. Since $\mathsf{Y}_1 \cap \mathsf{Y}_2 = 0$ and $\dim_\mathsf{K} \mathsf{X}_1 = 2$, $\mathsf{X}_1 \cap \mathsf{Y}_1$, $\mathsf{X}_1 \cap \mathsf{Y}_2$ are both of dimension 1, and

$$\mathsf{X}_1 = (\mathsf{X}_1 \cap \mathsf{Y}_1) \oplus (\mathsf{X}_1 \cap \mathsf{Y}_2).$$

It follows that ξ^2 operates faithfully on either $\mathsf{X}_1 \cap \mathsf{Y}_1$ or $\mathsf{X}_1 \cap \mathsf{Y}_2$, whence $q \equiv 1\ (2^{n-1})$. Hence $n = 2$ and by a), $\pi^2 = 1$, a contradiction.

Thus $\pi^2 = 1$. If $\pi \neq 1$, π has a cycle (j, l) of length 2. Then $\mathsf{X}_1 \leq \mathsf{F}y_j \oplus \mathsf{F}y_l$ and $\mathsf{F}y_i \leq \mathbf{C}_\mathsf{X}(\xi)$ if $i \neq l$ and $i \neq j$. Suppose that

$$(\lambda y_j + \mu y_l)\xi = (\lambda\sigma)y_l + (\mu\sigma')y_j.$$

Then

$$(\lambda y_j + \mu y_l)\xi^2 = (\lambda\sigma\sigma')y_j + (\mu\sigma'\sigma)y_l.$$

Since $\xi^2 \neq 1$, either $\sigma\sigma' \neq 1$ or $\sigma'\sigma \neq 1$. But in either case, $\sigma' \neq \sigma^{-1}$, so $\sigma\sigma' \neq 1$ and $\sigma'\sigma \neq 1$. Hence $\mathsf{F}y_j \cap \mathsf{X}_1 \neq 0$ and $\mathsf{F}y_l \cap \mathsf{X}_1 \neq 0$. Thus $\mathsf{F}y_j \cap \mathsf{X}_1$ is a 1-dimensional $\mathsf{K}\langle\xi^2\rangle$-submodule of $\mathsf{F}y_j$, and ξ^2 operates trivially on $\mathsf{F}y_j/(\mathsf{F}y_j \cap \mathsf{X}_1)$. Since $\mathfrak{L} \leq SL(2, q)$, $\det\sigma\sigma' = 1$; thus ξ^2 operates trivially on $\mathsf{F}y_j \cap \mathsf{X}_1$. Similarly, ξ^2 operates trivially on $\mathsf{F}y_l \cap \mathsf{X}_1$. Hence $\xi^2 = 1$, contrary to $n > 1$. Hence $\pi = 1$.

It follows that each $\mathsf{F}y_i$ is a $\mathsf{K}(\langle\xi\rangle)$-submodule of X, so either ξ operates trivially on $\mathsf{F}y_i$ or $\mathsf{F}y_i \supseteq \mathsf{X}_1$. Therefore

$$(\sum \lambda_i y_i)\xi = (\lambda_j\sigma)y_j + \sum_{i \neq j} \lambda_i y_i$$

for some j and some element σ of $\mathfrak{L}$ of order 2^n. Put $\rho = \sigma^{2^{n-2}}$. Then ρ^2 is an element of $SL(2, q)$ of order 2, so $\rho^2 = -1$. Write

$$(\sum \lambda_i y_i)\eta = \sum (\lambda_i \tau_i) y_{i\omega}.$$

We prove that

$$[\xi^{2^{n-1}}, \eta] = [\xi^{2^{n-2}}, \eta]^2$$

by direct calculation, distinguishing between the cases when $j\omega \neq j$ and $j\omega = j$. If $j\omega \neq j$, one verifies that $[\xi^{2^{n-1}}, \eta]$ and $[\xi^{2^{n-2}}, \eta]^2$ both carry $\sum \lambda_i y_i$ into

$$-\lambda_j y_j - \lambda_{j\omega} y_{j\omega} + \sum_{i \neq j, i \neq j\omega} \lambda_i y_i.$$

If $j\omega = j$, $\xi^{2^{n-1}}$ and η commute, whereas

$$(\sum \lambda_i y_i)[\xi^{2^{n-2}}, \eta] = (\lambda_j[\sigma^{2^{n-2}}, \tau_j]) y_j + \sum_{i \neq j,\, i \neq j\omega} \lambda_i y_i.$$

But by II, 8.10, $\mathfrak{L}$ is a generalized quaternion group of order 2^{n+1}, so

$$[\sigma^{2^{n-2}}, \tau_j] = \sigma^{2^{n-1}} \quad \text{or} \quad 1.$$

Thus $[\xi^{2^{n-2}}, \eta]^2 = 1$. **q.e.d.**

3.6 Definition. Let V be a $\mathsf{K}\langle g\rangle$-module, where $\langle g\rangle$ is a cyclic group of order n. We say that g is *exceptional on* V if the degree of the minimum polynomial of the linear transformation $v \to vg$ ($v \in \mathsf{V}$) is less than n.

3.7 Lemma. *Suppose that $\mathfrak{G} = \mathfrak{S}\mathfrak{Q}$, where $\mathfrak{Q}$ is a normal, extraspecial q-group and $\mathfrak{S} \in S_p(\mathfrak{G})$ ($p \neq q$). Let V be a faithful $\mathsf{K}\mathfrak{G}$-module, where K is algebraically closed and* char $\mathsf{K} \neq q$. *Suppose that all irreducible $\mathsf{K}\mathfrak{Q}$-submodules of V are isomorphic.*

a) *If p is odd, s is an element of $\mathfrak{S}$ of order p, $s \notin \mathbf{O}_p(\mathfrak{G})$ and s is exceptional on V, then $s \notin \mathfrak{S}'\mathbf{O}_p(\mathfrak{G})$.*

b) *Suppose that s, s' are elements of $\mathfrak{S}$, s is of order p^n, s' is of order $p^{n'}$ and s, s' are both exceptional on V. Then $s^{p^{n-1}}\mathbf{O}_p(\mathfrak{G})$ and $s'^{p^{n'-1}}\mathbf{O}_p(\mathfrak{G})$ commute.*

c) *Suppose that $p = 2$ and $q = 2^n - 1$ is a Mersenne prime. If s is an element of $\mathfrak{S}$ of order 2^n, $s^{2^{n-1}} \notin \mathbf{O}_2(\mathfrak{G})$ and s is exceptional on V, then*

$$[s^{2^{n-1}}, x] \equiv [s^{2^{n-2}}, x]^2 \quad \operatorname{mod} \mathbf{O}_2(\mathfrak{G})$$

for all $x \in \mathfrak{S}$.

Proof. Since all irreducible $\mathsf{K}\mathfrak{Q}$-submodules of V are isomorphic, $\mathbf{Z}(\mathfrak{Q})$ is faithfully represented on V by scalar multiples of the identity mapping. Thus $\mathbf{Z}(\mathfrak{Q}) \leq \mathbf{Z}(\mathfrak{G})$. Hence, as in 2.2, $\mathfrak{Q}/\mathbf{Z}(\mathfrak{Q})$ is a symplectic space, and if $\rho' = (\mathfrak{S}$ on $\mathfrak{Q}/\mathbf{Z}(\mathfrak{Q}))$, $\operatorname{im} \rho'$ is a subgroup of the symplectic group on $\mathfrak{Q}/\mathbf{Z}(\mathfrak{Q})$. By III, 3.18, $\ker \rho' = \mathbf{C}_{\mathfrak{S}}(\mathfrak{Q})$. Since $\mathbf{C}_{\mathfrak{S}}(\mathfrak{Q}) \trianglelefteq \mathfrak{Q}\mathfrak{S} = \mathfrak{G}$, $\mathbf{C}_{\mathfrak{S}}(\mathfrak{Q}) = \mathbf{O}_p(\mathfrak{G})$.

In a), we have $s \notin \mathbf{O}_p(\mathfrak{G})$ and in c), $s^{2^{n-1}} \notin \mathbf{O}_2(\mathfrak{G})$. In b), we may suppose that $s^{p^{n-1}} \notin \mathbf{O}_p(\mathfrak{G})$ and $s'^{p^{n'-1}} \notin \mathbf{O}_p(\mathfrak{G})$. In all cases, let U be an irreducible $\mathsf{K}\langle s \rangle \mathfrak{Q}$-submodule of V. Then U and V are both direct sums of $\mathsf{K}\mathfrak{Q}$-submodules isomorphic to some irreducible $\mathsf{K}\mathfrak{Q}$-module, so $\ker(\mathfrak{Q}$ on $\mathsf{U}) = \ker(\mathfrak{Q}$ on $\mathsf{V}) = 1$. Now if $\mathfrak{X}$ is the subgroup of $\langle s \rangle$ of order p, $\mathfrak{X} \cap \mathbf{O}_p(\mathfrak{G}) = 1$, so $\mathfrak{X} \not\leq \mathbf{C}_{\mathfrak{G}}(\mathfrak{Q})$ and $[\mathfrak{Q}, \mathfrak{X}] \neq 1$. Thus $([\mathfrak{Q}, \mathfrak{X}]$ on $\mathsf{U})$ is non-trivial and $(\mathfrak{X}$ on $\mathsf{U})$ is non-trivial. Therefore $\langle s \rangle$ is faithfully represented on U. Since $\ker(\mathfrak{Q}$ on $\mathsf{U}) = 1$, it follows that U is a faithful, irreducible $\mathsf{K}(\langle s \rangle \mathfrak{Q})$-module. Also s is exceptional on U, so by 3.2, $p^i - 1 = q^j$ for certain positive integers i, j; also the automorphism of $\mathfrak{Q}$ induced by s bisects $\mathfrak{Q}$. Similarly, in b), the automorphism of $\mathfrak{Q}$ induced by s' bisects $\mathfrak{Q}$. Thus $\rho'(s)$ and, in b), $\rho'(s')$ bisect $\mathfrak{Q}/\mathbf{Z}(\mathfrak{Q})$.

Let m be the smallest positive integer such that $q^{2m} \equiv 1\ (p)$ and let $\mathsf{F} = GF(q^{2m})$. Let $(\ ,\ )$ be a non-singular, symplectic bilinear form on F over $GF(q)$, and let $\mathfrak{L}$ be a Sylow p-subgroup of the corresponding symplectic group $Sp(2m, q)$. Let X be a vector space over F with F-basis $\{y_1, \ldots, y_r\}$, where r is an integer for which $q^{2mr} \geq |\mathfrak{Q} : \mathbf{Z}(\mathfrak{Q})|$. Let $\mathfrak{P}$ be the set of all mappings α of X into X of the form

$$\left(\sum_{i=1}^{r} \lambda_i y_i\right)\alpha = \sum_{i=1}^{r} (\lambda_i \sigma_i) y_{i\pi},$$

where $\sigma_1, \ldots, \sigma_r$ run through $\mathfrak{L}$ and π runs through a fixed Sylow p-subgroup of the symmetric group of degree r. By 3.3, there exists a non-singular, symplectic bilinear form on X over $\mathsf{L} = GF(q)$ such that $\mathfrak{P}$ is a Sylow p-subgroup of the corresponding symplectic group $Sp(2mr, q)$ on X. By II, 9.6, $\mathsf{X} = \mathsf{Y} \oplus \mathsf{Z}$, where Y, Z are orthogonal, non-singular L-subspaces of X and there exists a symplectic isomorphism θ of Y onto $\mathfrak{Q}/\mathbf{Z}(\mathfrak{Q})$.

Given $x \in \mathfrak{S}$, there exists a L-linear transformation $\rho''(x)$ of X such that $z\rho''(x) = z$ for all $z \in \mathsf{Z}$ and

$$y\rho''(x) = y\theta\rho'(x)\theta^{-1}$$

for all $y \in \mathsf{Y}$. Then ρ'' is a monomorphism of $\mathfrak{S}$ into $Sp(2mr, q)$ and $\operatorname{im}\rho''$ is a p-subgroup of $Sp(2mr, q)$. By Sylow's theorem, there exists $\beta \in Sp(2mr, q)$ such that $\operatorname{im}\rho'' \leq \beta^{-1}\mathfrak{P}\beta$. For $x \in \mathfrak{S}$, write

$$\rho(x) = \beta\rho''(x)\beta^{-1};$$

thus ρ is a homomorphism of $\mathfrak{S}$ into $\mathfrak{P}$. Note that if $x \in \mathfrak{S}$ and $\rho'(x)$ bisects $\mathfrak{Q}/\mathbf{Z}(\mathfrak{Q})$, then $\rho(x)$ bisects X, for if Q is a $\rho'(x)$-invariant subspace of $\mathfrak{Q}/\mathbf{Z}(\mathfrak{Q})$ such that $(\langle\rho'(x)\rangle$ on $\mathsf{Q})$ is irreducible, $(\langle\rho(x)\rangle$ on $\mathsf{Q}\theta^{-1}\beta^{-1})$ is irreducible.

If p is odd, then by 2.7, $p = 2^e + 1$ is a Fermat prime and $q = 2$. In this case $m = e$ and $\mathfrak{L}$ is a Sylow p-subgroup of $Sp(2e, 2)$. Since p^2 does not divide $2^{2e} - 1$, $|\mathfrak{L}| = p$ and $\mathfrak{L}$ is Abelian. On the other hand, if $p = 2$, then $m = 1$ and $\mathfrak{L}$, being a Sylow 2-subgroup of $Sp(2, q) = SL(2, q)$ (II, 9.12)), is a generalized quaternion group.

a) In this case $\rho(s)$ bisects X and $\mathfrak{L}$ is Abelian. By 3.5b), $\rho(s) \notin \mathfrak{P}'$, so $s \notin \mathfrak{S}'\mathbf{O}_p(\mathfrak{G})$, since $\ker\rho = \ker\rho' = \mathbf{O}_p(\mathfrak{G})$.

b) Since $\mathfrak{L}$ has only one subgroup of order p, any two elements of order p commute. By 3.5c), $\rho(s)^{p^{n-1}}$ and $\rho(s')^{p^{n'-1}}$ commute. Thus $s^{p^{n-1}}\mathbf{O}_p(\mathfrak{G})$ and $s'^{p^{n'-1}}\mathbf{O}_p(\mathfrak{G})$ commute.

c) This follows from 3.5d). **q.e.d.**

We now build these assertions into more general theorems.

3.8 Theorem (HALL and HIGMAN). *Let K be a field of characteristic p and let V be a finite-dimensional vector space over K. Let $\mathfrak{G}$ be a finite, p-soluble group of non-singular linear transformations of V such that $\mathbf{O}_p(\mathfrak{G}) = 1$. Suppose that g, h are elements of a Sylow p-subgroup of $\mathfrak{G}$, and let $c = [g, h]$. If $c \neq 1$ and $(c - 1)^{p-1} = 0$, then either $(g - 1)^{p-1}(c - 1)^{p-2} \neq 0$ or $(h - 1)^{p-1}(c - 1)^{p-2} \neq 0$.*

Proof. We may suppose that K is algebraically closed, and we apply 2.8 with $\mathfrak{P} = \langle g, h\rangle$, $y = c$. Thus there exists a section $\mathfrak{H} = \mathfrak{G}_1/\mathfrak{G}_2$ of $\mathfrak{G}$ with the following properties. First, $g \in \mathfrak{G}_1$, $h \in \mathfrak{G}_1$ and $c \notin \mathfrak{G}_2$. Put $s_1 = h\mathfrak{G}_2$, $s_2 = g\mathfrak{G}_2$. $s = c\mathfrak{G}_2$. Then $s = [s_2, s_1]$ and $s \neq 1$. Also $\mathfrak{H} = \mathfrak{S}\mathfrak{Q}$, where $\mathfrak{Q}$ is a special normal q-subgroup of $\mathfrak{H}$ for some prime $q \neq p$ and $\mathfrak{S} = \langle s_1, s_2\rangle$. Further, $\mathfrak{Q}/\mathbf{Z}(\mathfrak{Q})$ is a minimal normal subgroup of $\mathfrak{H}/\mathbf{Z}(\mathfrak{Q})$ and s centralizes $\mathbf{Z}(\mathfrak{Q})$ but not $\mathfrak{Q}$. Finally, $\mathfrak{H}$ has a faithful irreducible representation on a section U of V, and $\mathfrak{Q}$ is extra-special if U is an irreducible $\mathsf{K}\mathfrak{Q}$-module.

Thus $\mathbf{O}_p(\mathfrak{H}) = 1$ (V, 5.17), $\mathfrak{Q} \geq \mathbf{C}_{\mathfrak{H}}(\mathfrak{Q})$ (1.4) and $u(s-1)^{p-1} = 0$ for all $u \in \mathsf{U}$. Since $s \neq 1$, p is odd. Also $u(s^p - 1) = u(s-1)^p = 0$ and $s^p = 1$. If U is an irreducible $\mathsf{K}\mathfrak{Q}$-module, $\mathfrak{Q}$ is extraspecial; but then 3.7a) leads to a contradiction. Hence U is a reducible $\mathsf{K}\mathfrak{Q}$-module. By V, 17.3b),

$$\mathsf{U} = \mathsf{U}_1 \oplus \cdots \oplus \mathsf{U}_l,$$

where U_i is the sum of all $\mathsf{K}\mathfrak{Q}$-submodules of U isomorphic to some irreducible $\mathsf{K}\mathfrak{Q}$-module. By VII, 9.19, $l > 1$. By V, 17.3d), there is a transitive permutation representation σ of $\mathfrak{H}$ on $\{\mathsf{U}_1, \ldots, \mathsf{U}_l\}$ such that $\mathsf{U}_i\sigma(u) = \mathsf{U}_i u$ $(u \in \mathfrak{H})$. Since $\mathfrak{Q} \geq \mathbf{C}_{\mathfrak{H}}(\mathfrak{Q})$, $s \notin \mathbf{C}_{\mathfrak{H}}(\mathfrak{Q})$. Thus there exists $x \in \mathfrak{Q}$ such that $z = [s, x] \neq 1$. Thus the representation of z on one of the U_j is non-trivial. If $\mathfrak{H}_0$ is the stabiliser of U_j, $\mathfrak{Q} \leq \mathfrak{H}_0 < \mathfrak{H}$, since $l > 1$. Since $\mathfrak{H} = \mathfrak{Q}\mathfrak{S}$ and $\mathfrak{S} = \langle s_1, s_2 \rangle$, it follows that $s_i \notin \mathfrak{H}_0$ for $i = 1$ or $i = 2$. Hence the length of the cycle of $\sigma(s_i)$ containing U_j is at least p, and $\mathsf{U}_j, \mathsf{U}_j s_i, \ldots, \mathsf{U}_j s_i^{p-1}$ are distinct.

Since $u(s-1)^{p-1} = 0$ for all $u \in \mathsf{U}$, the length of any cycle of $\sigma(s)$ is less than p and is therefore 1. Hence $s \in \mathfrak{H}_0$ and $\mathsf{U}_j s_i^k s = \mathsf{U}_j s_i^k$ $(k = 0, \ldots, p-1)$. Let W be an irreducible $\mathsf{K}\mathfrak{H}_0$-submodule of U_j, let $\mathfrak{N} = \ker(\mathfrak{H}_0 \text{ on } \mathsf{W})$ and let $\overline{\mathfrak{H}} = \mathfrak{H}_0/\mathfrak{N}$. Since $\mathfrak{G}$ is p-soluble, $\overline{\mathfrak{H}}$ is p-constrained, by 1.4. Since $\overline{\mathfrak{H}}$ has a faithful, irreducible representation on W, $\mathbf{O}_p(\overline{\mathfrak{H}}) = 1$. Also z is represented non-trivially on W, since U_j is the direct sum of isomorphic $\mathsf{K}\mathfrak{Q}$-submodules. Thus if $\bar{s} = s\mathfrak{N}$, $\bar{s}$ is of order p. By 2.9, the minimum polynomial of the mapping $w \to w\bar{s}$ $(w \in \mathsf{W})$ is $(t-1)^r$, where $p - 1 \leq r \leq p$. Thus there exists $w \in \mathsf{W}$ such that $w(s-1)^{p-2} \neq 0$. Let $u = w + ws_i + \cdots + ws_i^{p-1}$; thus

$$u \in \mathsf{U}_j \oplus \mathsf{U}_j s_i \oplus \cdots \oplus \mathsf{U}_j s_i^{p-1}.$$

Since $\mathsf{U}_j s_i^k (s-1)^{p-2} \subseteq \mathsf{U}_j s_i^k$, it follows that $u(s-1)^{p-2} \neq 0$. But $u = w(s_i - 1)^{p-1}$, so $w(s_i - 1)^{p-1}(s-1)^{p-2} \neq 0$. Hence either $(g-1)^{p-1}(c-1)^{p-2} \neq 0$ or $(h-1)^{p-1}(c-1)^{p-2} \neq 0$. **q.e.d.**

3.9 Theorem (HALL and HIGMAN). *Let* K *be a field of characteristic p and let* V *be a finite-dimensional vector space over* K. *Let* $\mathfrak{G}$ *be a finite, p-soluble group of non-singular linear transformations of* V *such that* $\mathbf{O}_p(\mathfrak{G}) = 1$. *Suppose that g, h are elements of a Sylow p-subgroup of* $\mathfrak{G}$ *and that* $g^{p^{m-1}}$, $h^{p^{n-1}}$ *do not commute. Then either* $(g-1)^{p^m-1} \neq 0$ *or* $(h-1)^{p^n-1} \neq 0$.

Proof. We may suppose that K is algebraically closed. The theorem will be proved by induction on $m + n$. Suppose that $(g - 1)^{p^m - 1} = (h - 1)^{p^n - 1} = 0$. We apply 2.8 with $\mathfrak{P} = \langle g, h \rangle$, $y = [g^{p^{m-1}}, h^{p^{n-1}}]$. Thus there exists a section $\mathfrak{H} = \mathfrak{G}_1/\mathfrak{G}_2$ of $\mathfrak{G}$ with the following properties. First, $g \in \mathfrak{G}_1$, $h \in \mathfrak{G}_1$ and $y \notin \mathfrak{G}_2$. Put $s_1 = h\mathfrak{G}_2$, $s_2 = g\mathfrak{G}_2$, $s = y\mathfrak{G}_2$. Thus $s = [s_2^{p^{m-1}}, s_1^{p^{n-1}}] \neq 1$. Also $\mathfrak{H} = \mathfrak{Q}\mathfrak{S}$, where $\mathfrak{Q}$ is a normal q-subgroup of $\mathfrak{H}$ for some prime $q \neq p$ and $\mathfrak{S} = \langle s_1, s_2 \rangle$. Further $\mathfrak{Q}/\mathbf{Z}(\mathfrak{Q})$ is a minimal normal subgroup of $\mathfrak{H}/\mathbf{Z}(\mathfrak{Q})$ and s centralizes $\mathbf{Z}(\mathfrak{Q})$ but not $\mathfrak{Q}$. And $\mathfrak{H}$ has a faithful, irreducible representation on a section U of V. Thus $\mathbf{O}_p(\mathfrak{H}) = 1$ (V, 5.17), $\mathfrak{Q} \geq \mathbf{C}_{\mathfrak{H}}(\mathfrak{Q})$ (1.4) and $u(s_2 - 1)^{p^m - 1} = u(s_1 - 1)^{p^n - 1} = 0$ for all $u \in$ U. Then $s_2^{p^m} = s_1^{p^n} = 1$. Since $s_2^{p^{m-1}}$ and $s_1^{p^{n-1}}$ do not commute, it follows that the orders of s_2, s_1 are p^m, p^n respectively. Finally, if U is an irreducible K$\mathfrak{Q}$-module, $\mathfrak{Q}$ is extraspecial. But then 3.7b) leads to a contradiction. Hence U is a reducible K$\mathfrak{Q}$-module. By V, 17.3b),

$$\mathsf{U} = \mathsf{U}_1 \oplus \cdots \oplus \mathsf{U}_l,$$

where U_i is the sum of all K$\mathfrak{Q}$-submodules of U isomorphic to some irreducible K$\mathfrak{Q}$-module. By VII, 9.19, $l > 1$. By V, 17.3d), there is a transitive permutation representation σ of $\mathfrak{H}$ on $\{\mathsf{U}_1, \ldots, \mathsf{U}_l\}$ such that $\mathsf{U}_i\sigma(u) = \mathsf{U}_i u$ $(u \in \mathfrak{H})$. Since $\mathfrak{Q} \geq \mathbf{C}_{\mathfrak{H}}(\mathfrak{Q})$, $s \notin \mathbf{C}_{\mathfrak{H}}(\mathfrak{Q})$. Thus there exists $x \in \mathfrak{Q}$ such that $z = [s, x] \neq 1$. Thus the representation of z on one of the U_j is non-trivial. Let $\mathfrak{H}_0$ be the stabiliser of U_j. Since $l > 1$, $\mathfrak{Q} \leq \mathfrak{H}_0 < \mathfrak{H}$.

Let p^a, p^b be the lengths of the cycles of $\sigma(s_1)$, $\sigma(s_2)$ containing U_j respectively. Since $u(s_2 - 1)^{p^m - 1} = u(s_1 - 1)^{p^n - 1} = 0$ for all $u \in$ U, $a < n$ and $b < m$. Hence $s_1^{p^a} \in \mathfrak{H}_0$ and $s_2^{p^b} \in \mathfrak{H}_0$. Since $\mathfrak{Q} \leq \mathfrak{H}_0 < \mathfrak{H}$ and $\mathfrak{S} = \langle s_1, s_2 \rangle$, either $a > 0$ or $b > 0$. Let W be an irreducible K$\mathfrak{H}_0$-submodule of U_j, let $\mathfrak{N} = \ker(\mathfrak{H}_0$ on W) and let $\overline{\mathfrak{H}} = \mathfrak{H}_0/\mathfrak{N}$. Thus W is a faithful, irreducible K$\overline{\mathfrak{H}}$-module. Hence $\mathbf{O}_p(\overline{\mathfrak{H}}) = 1$. Let $\bar{g} = s_2^{p^b}\mathfrak{N}$, $\bar{h} = s_1^{p^a}\mathfrak{N}$. Thus, $\bar{g}$, $\bar{h}$ are elements of a Sylow p-subgroup of $\overline{\mathfrak{H}}$. Also, since $a < n$ and $b < m$, $s \in \mathfrak{H}_0$ and

$$s\mathfrak{N} = [\bar{g}^{p^{m-b-1}}, \bar{h}^{p^{n-a-1}}].$$

Since z is represented non-trivially on U_j, $z \in \mathfrak{Q}$ and all irreducible K$\mathfrak{Q}$-submodules of U_j are isomorphic, z is represented non-trivially on W. Thus $z \notin \mathfrak{N}$ and $s \notin \mathfrak{N}$.

It follows from the inductive hypothesis, applied to im($\overline{\mathfrak{H}}$ on W), that either $w(\bar{g} - 1)^{p^{m-b} - 1} \neq 0$ or $w(\bar{h} - 1)^{p^{n-a} - 1} \neq 0$, for some $w \in$ W.

Suppose that $w' = w(\bar{g} - 1)^{p^{m-b}-1} \neq 0$. By definition of b, the elements $w', w's_2, \ldots, w's_2^{p^b-1}$ lie in distinct U_i, so

$$w'(1 + s_2 + \cdots + s_2^{p^b-1}) \neq 0.$$

But

$$\begin{aligned} w'(1 + s_2 + \cdots + s_2^{p^b-1}) &= w(s_2^{p^b} - 1)^{p^{m-b}-1}(s_2 - 1)^{p^b-1} \\ &= w(s_2 - 1)^{p^m-1}, \end{aligned}$$

so $(g - 1)^{p^m-1} \neq 0$. Similarly, if $w(\bar{h} - 1)^{p^{n-a}-1} \neq 0$ for some $w \in \mathsf{W}$, $(h - 1)^{p^n-1} \neq 0$. **q.e.d.**

In order to generalize 3.7c), two lemmas are needed.

3.10 Lemma. *Let* $\mathfrak{G} = \mathfrak{S}\mathfrak{Q}$, *where* $\mathfrak{Q}$ *is a normal* q*-subgroup of* $\mathfrak{G}$ *for some odd prime* q, $\mathfrak{S} = \langle g \rangle$ *is cyclic of order* 2^n $(n > 1)$ *and* $g^{2^{n-1}} \notin \mathbf{C}_{\mathfrak{G}}(\mathfrak{Q})$. *Suppose that* V *is a faithful* $\mathsf{K}\mathfrak{G}$*-module, where* char $\mathsf{K} \neq q$, *and that the degree of the minimum polynomial of the linear transformation* $v \to vg$ *of* V *is at most 3. Then* $q = 3$ *and* $n = 2$.

Proof. Without loss of generality, we may suppose that K is algebraically closed. By 2.8, there exists a section $\mathfrak{H} = \mathfrak{G}_1/\mathfrak{G}_2$ of $\mathfrak{G}$ with the following properties. First, $\mathfrak{G}_1 \geq \mathfrak{S}$ and $g^{2^{n-1}} \notin \mathfrak{G}_2$. Also $\mathfrak{H} = \mathfrak{Q}_1\mathfrak{S}_1$, where $\mathfrak{Q}_1$ is a special q-group and $\mathfrak{S}_1 = \langle \bar{g} \rangle$, where $\bar{g} = g\mathfrak{G}_2$. Further, $\bar{g}^{2^{n-1}}$ does not centralize $\mathfrak{Q}_1$, and $\mathfrak{H}$ has a faithful, irreducible representation on a section U of V. By V, 17.3,

$$\mathsf{U} = \mathsf{U}_1 \oplus \cdots \oplus \mathsf{U}_l,$$

where U_i is the sum of all the submodules of U which are $\mathsf{K}\mathfrak{Q}_1$-isomorphic to an irreducible $\mathsf{K}\mathfrak{Q}_1$-module, and there exists a transitive permutation representation σ of $\mathfrak{S}_1$ on $\{1, \ldots, l\}$ such that $\mathsf{U}_i x = \mathsf{U}_{i\sigma(x)}$ for all $x \in \mathfrak{S}_1$. Now for any $u \in \mathsf{U}$, $u, u\bar{g}, u\bar{g}^2, u\bar{g}^3$ are linearly dependent. Thus $i, i\sigma(\bar{g}), i\sigma(\bar{g})^2, i\sigma(\bar{g})^3$ cannot be distinct, from which it follows that $\sigma(\bar{g})^2 = 1$. If $\sigma(\bar{g}) \neq 1$, $l = 2$ since σ is transitive. Now $\mathsf{U}_i\bar{g}^2 = \mathsf{U}_i$, but since $\bar{g}^2 \notin \mathbf{C}_{\mathfrak{H}}(\mathfrak{Q}_1)$, $\bar{g}^2$ cannot be represented by scalar multiples of the identity mapping on both U_1 and U_2. Hence there exists u in either U_1 or U_2 such that $u, u\bar{g}^2$ are linearly independent. But then $u, u\bar{g}, u\bar{g}^2, u\bar{g}^3$ are linearly independent, a contradiction. Therefore $\sigma(\bar{g}) = 1$.

Hence $l = 1$ and U is the direct sum of isomorphic, irreducible $\mathsf{K}\mathfrak{Q}_1$-modules. $\mathfrak{Q}_1$ is faithfully represented on any one of these, so

$[\mathbf{Z}(\mathfrak{Q}_1), \mathfrak{S}_1] = 1$, $\mathbf{Z}(\mathfrak{Q}_1)$ is cyclic and $\mathfrak{Q}_1$ is extraspecial. By 3.2, $2^n - 1 = q^d$ for some d, and $2^n - 1 \leq 3$. Thus $q = 3$ and $n = 2$. **q.e.d.**

3.11 Lemma. *Suppose that $\mathfrak{G}$ is a soluble group and that the Fitting subgroup $\mathbf{F}(\mathfrak{G})$ of $\mathfrak{G}$ is of odd order. If $g \in \mathfrak{G}$ and $[g, x] = 1$ for every element x of $\mathbf{F}(\mathfrak{G})$ of prime order, $g \in \mathbf{F}(\mathfrak{G})$.*

Proof. Let $\mathfrak{N}$ be the set of elements $g \in \mathfrak{G}$ such that $[g, x] = 1$ for every element x of $\mathbf{F}(\mathfrak{G})$ of prime order. Clearly $\mathfrak{N}$ is a normal subgroup of $\mathfrak{G}$. If $\mathfrak{N} \not\leq \mathbf{F}(\mathfrak{G})$, choose a minimal normal subgroup $\mathfrak{M}/(\mathfrak{N} \cap \mathbf{F}(\mathfrak{G}))$ of $\mathfrak{G}/(\mathfrak{N} \cap \mathbf{F}(\mathfrak{G}))$ contained in $\mathfrak{N}/(\mathfrak{N} \cap \mathbf{F}(\mathfrak{G}))$. Then $\mathfrak{M}/(\mathfrak{N} \cap \mathbf{F}(\mathfrak{G}))$ is an elementary Abelian p-group for some prime p. For every prime $q \neq p$, the Sylow q-subgroup $\mathfrak{Q}$ of $\mathfrak{M}$ is also a Sylow subgroup of $\mathfrak{N} \cap \mathbf{F}(\mathfrak{G})$. Thus $\mathfrak{Q} \trianglelefteq \mathfrak{M}$, since $\mathbf{F}(\mathfrak{G})$ is nilpotent. Also, if $\mathfrak{P}$ is a Sylow p-subgroup of $\mathfrak{M}$, it follows from IV, 5.12 that $\mathfrak{P} \leq \mathbf{C}_{\mathfrak{M}}(\mathfrak{Q})$, since $\mathfrak{P} \leq \mathfrak{N}$ and $|\mathbf{F}(\mathfrak{G})|$ is odd. Thus $\mathfrak{Q} \leq \mathbf{N}_{\mathfrak{M}}(\mathfrak{P})$ for all $q \neq p$. Since also $\mathfrak{P} \leq \mathbf{N}_{\mathfrak{M}}(\mathfrak{P})$, $\mathfrak{P} \trianglelefteq \mathfrak{M}$. Thus all Sylow subgroups of $\mathfrak{M}$ are normal and $\mathfrak{M}$ is nilpotent. But $\mathfrak{M} \trianglelefteq \mathfrak{G}$, so $\mathfrak{M} \leq \mathbf{F}(\mathfrak{G})$ and $\mathfrak{M} \leq \mathfrak{N} \cap \mathbf{F}(\mathfrak{G})$, a contradiction. Hence $\mathfrak{N} \leq \mathbf{F}(\mathfrak{G})$. **q.e.d.**

3.12 Theorem (GROSS). *Let V be a finite-dimensional vector space over a field K. Let $\mathfrak{G}$ be a finite soluble group of non-singular linear transformations of V such that $|\mathbf{F}(\mathfrak{G})|$ is odd and is not divisible by* char K. *Suppose that g is an element of $\mathfrak{G}$ of order 2^n and that g is exceptional on V. Then*

$$g^{2^{n-1}}\mathbf{F}(\mathfrak{G}) \in \mathbf{F}(\mathfrak{G}/\mathbf{F}(\mathfrak{G})).$$

Proof. We may suppose without loss of generality that K is algebraically closed. If $n = 1$, the fact that g is exceptional on V means that g is a scalar multiple of the identity mapping; thus $g \in \mathbf{Z}(\mathfrak{G}) \leq \mathbf{F}(\mathfrak{G})$, which is impossible since $|\mathbf{F}(\mathfrak{G})|$ is odd. Hence $n > 1$.

Let $\mathfrak{L}$ be any normal, nilpotent subgroup of $\mathfrak{G}$. Thus $\mathfrak{L} \leq \mathbf{F}(\mathfrak{G})$, so by hypothesis, char K does not divide $|\mathfrak{L}|$. Hence V is a completely reducible $\mathsf{K}\mathfrak{L}$-module. Write

$$\text{(1)} \qquad \mathsf{V} = \mathsf{V}_1 \oplus \cdots \oplus \mathsf{V}_l,$$

where V_i is the direct sum of isomorphic irreducible $\mathsf{K}\mathfrak{L}$-submodules and, for $i \neq j$, the summands of V_i are not isomorphic to those of V_j. Then it is easily deduced from the Jordan-Hölder theorem that every irreducible $\mathsf{K}\mathfrak{L}$-submodule of V is contained in some V_i, and V_i is the sum of all $\mathsf{K}\mathfrak{L}$-submodules of V isomorphic to a given irreducible $\mathsf{K}\mathfrak{L}$-module. We call (1) the *Wedderburn decomposition* of V for $\mathfrak{L}$.

If $x \in \mathfrak{G}$ and U, U' are $\mathsf{K}\mathfrak{L}$-isomorphic submodules of V, $\mathsf{U}x$ and $\mathsf{U}'x$ are $\mathsf{K}\mathfrak{L}$-isomorphic. Thus $\mathsf{V}_i x = \mathsf{V}_j$ for some j. Put $j = i\sigma(x)$; then $\sigma(x)$ is a mapping of $\{1, \ldots, l\}$ into itself. Since $\sigma(1) = 1$ and $\sigma(xy) = \sigma(x)\sigma(y)$ $(x, y \in \mathfrak{G})$, σ is a permutation representation of $\mathfrak{G}$. Let $k(\mathfrak{L}) = \ker \sigma$.

We put

$$\mathfrak{H} = \bigcap_{\mathfrak{L}} k(\mathfrak{L}),$$

where the intersection is taken over all normal, nilpotent subgroups $\mathfrak{L}$ of $\mathfrak{G}$. Thus $\mathfrak{H} \trianglelefteq \mathfrak{G}$.

a) $g^{2^{n-1}} \in \mathfrak{H}$.

For suppose that $\mathfrak{L}$, σ are as above. Then since g is exceptional on V, $\sigma(g)$ cannot have a cycle of length 2^n. Hence

$$\sigma(g^{2^{n-1}}) = \sigma(g)^{2^{n-1}} = 1$$

and $g^{2^{n-1}} \in k(\mathfrak{L})$. Thus $g^{2^{n-1}} \in \mathfrak{H}$.

b) If $\mathfrak{A}$ is a normal Abelian subgroup of $\mathfrak{G}$, $[\mathfrak{A}, \mathfrak{H}] = 1$.

Suppose that (1) is the Wedderburn decomposition of V for $\mathfrak{A}$. Since $\mathfrak{A}$ is Abelian and K is algebraically closed, any irreducible representation of $\mathfrak{A}$ in K is of degree 1; thus each element x of $\mathfrak{A}$ is represented on each V_i by a scalar multiple of the identity mapping. Now if $y \in \mathfrak{H}$, $\mathsf{V}_i y = \mathsf{V}_i$, so $[x, y]$ is represented on each V_i by the identity mapping. Hence $[x, y] = 1$. Thus $[\mathfrak{A}, \mathfrak{H}] = 1$.

c) The class of $\mathfrak{F} = \mathbf{F}(\mathfrak{H})$ is 2.

Let $\mathfrak{A} = \mathfrak{F}' \cap \mathbf{Z}_2(\mathfrak{F})$. By III, 2.11, $\mathfrak{A}$ is Abelian. By b), $[\mathfrak{A}, \mathfrak{H}] = 1$. Thus $\mathfrak{A} \leq \mathbf{Z}(\mathfrak{F})$. By III, 2.6, $\mathfrak{F}' \leq \mathbf{Z}(\mathfrak{F})$, so the class of $\mathfrak{F}$ is at most 2. If $\mathfrak{F}$ is Abelian, then $\mathfrak{F} \leq \mathbf{Z}(\mathfrak{H})$ by b). Thus $\mathfrak{H} \leq \mathbf{C}_{\mathfrak{H}}(\mathfrak{F}) \leq \mathfrak{F}$, since $\mathfrak{H}$ is soluble (III, 4.2). Hence $\mathfrak{F} = \mathfrak{H}$. But by a), $|\mathfrak{H}|$ is even, whereas since $\mathfrak{F} \leq \mathbf{F}(\mathfrak{G})$, $|\mathfrak{F}|$ is odd. Thus $\mathfrak{F}$ is non-Abelian, and the class of $\mathfrak{F}$ is precisely 2.

d) If p is odd, $g^{2^{n-1}}$ centralizes $\mathbf{O}_p(\mathfrak{H}/\mathfrak{F})$.

Let $\mathbf{O}_p(\mathfrak{H}/\mathfrak{F}) = \mathfrak{X}/\mathfrak{F}$. Thus $|\mathfrak{X}|$ is odd, so by 1.11, there exists $\mathfrak{S} \in S_p(\mathfrak{X})$ such that $\mathfrak{S}^g = \mathfrak{S}$. Let $\mathfrak{S}_0 = [\mathfrak{S}, \langle g^{2^{n-1}} \rangle]$. We show that if q is any prime divisor of $|\mathfrak{F}|$ other than p, $[\mathbf{O}_q(\mathfrak{H}), \mathfrak{S}_0] = 1$.

For suppose that this is false. Then by IV, 5.12, there exists an element of order q in $\mathbf{O}_q(\mathfrak{H})$ which is not centralized by $\mathfrak{S}_0$. By c) and III, 10.2, $\mathbf{O}_q(\mathfrak{H})$ is a regular q-group; hence by III, 10.5, the elements of order at most q in $\mathbf{O}_q(\mathfrak{H})$ form a characteristic subgroup $\mathfrak{Q}$. We have $[\mathfrak{Q}, \mathfrak{S}_0] \neq 1$. Hence there is a summand U in the Wedderburn decomposition of V for $\mathfrak{Q}$ such that ($[\mathfrak{Q}, \mathfrak{S}_0]$ on U) is non-trivial. Since $g^{2^{n-1}} \in \mathfrak{H}$ and $\mathfrak{S} \leq \mathfrak{H}$, $g^{2^{n-1}}$ and $\mathfrak{S}$ leave U fixed. It follows that $g^{2^{n-1}}$ cannot be represented on U by a scalar multiple of the identity mapping.

Let 2^m be the length of the orbit containing U in the permutation representation of $\langle g \rangle$ on $\{\mathsf{U}a \mid a \in \mathfrak{G}\}$. Thus if $h = g^{2^m}$, h leaves U fixed, but U, Ug, ..., Ug^{2^m-1} form a direct sum. Thus $m < n$. Since

$$\sum_{i=0}^{2^n-1} \lambda_i g^i = \sum_{i=0}^{2^m-1} \left(\sum_{j=0}^{2^{n-m}-1} \lambda_{i+2^m j} h^j \right) g^i \quad (\lambda_i \in \mathsf{K}),$$

the linear dependence of $1, g, \ldots, g^{2^n-1}$ implies that of $1, h, \ldots, h^{2^{n-m}-1}$ on U. Thus h is exceptional on U. Also, if $m = n - 1$, then the restrictions to U of $1, h$ are linearly dependent, and h is represented on U by a scalar multiple of the identity mapping, a contradiction. Thus $m < n - 1$.

Let $\mathfrak{G}_0 = \langle h, \mathfrak{Q} \rangle$ and let $\mathfrak{K} = \ker(\mathfrak{G}_0 \text{ on } \mathsf{U})$. Thus $[\mathfrak{Q}, \mathfrak{S}_0] \nleq \mathfrak{K}$ and $g^{2^{n-1}} \notin \mathfrak{K}$. Since $\mathfrak{K} \trianglelefteq \mathfrak{G}_0$ and all the involutions in $\mathfrak{G}_0$ are conjugate, it follows that $\mathfrak{K}$ contains no involutions. Hence $|\mathfrak{K}|$ is odd and $\mathfrak{K} \leq \mathfrak{Q}$. Thus $\mathfrak{K} = \ker(\mathfrak{Q} \text{ on } \mathsf{U})$. Since $\mathfrak{S}$ leaves U fixed and $\mathfrak{S} \leq \mathbf{N}_{\mathfrak{G}}(\mathfrak{Q})$, we have $\mathfrak{S} \leq \mathbf{N}_{\mathfrak{G}}(\mathfrak{K})$. We deduce that

$$[\langle g^{2^{n-1}} \rangle, \mathfrak{Q}] \nleq \mathfrak{K}.$$

For otherwise, we would have

$$[\langle g^{2^{n-1}} \rangle, \mathfrak{Q}, \mathfrak{S}] \leq [\mathfrak{K}, \mathfrak{S}] \leq \mathfrak{K}$$

and

$$[\mathfrak{Q}, \mathfrak{S}, \langle g^{2^{n-1}} \rangle] \leq [\mathfrak{Q}, \langle g^{2^{n-1}} \rangle] \leq \mathfrak{K}.$$

But then it follows from III, 1.10 that

$$[\mathfrak{S}_0, \mathfrak{Q}] = [\mathfrak{S}, \langle g^{2^{n-1}} \rangle, \mathfrak{Q}] \leq \mathfrak{K},$$

since $\mathfrak{K} \trianglelefteq \langle h, \mathfrak{S}, \mathfrak{Q} \rangle$. This is a contradiction, so

(2) $$[\langle h^{2^{n-m-1}} \rangle, \mathfrak{Q}] \nleq \mathfrak{K}.$$

Now U is the direct sum of isomorphic irreducible K$\mathfrak{Q}$-modules, so $\mathfrak{Q}/\mathfrak{K}$ has a faithful, irreducible representation. Thus the centre $\mathfrak{Z}/\mathfrak{K}$ of $\mathfrak{Q}/\mathfrak{K}$ is cyclic, and $\mathfrak{Z}$ is represented on U by scalar multiples of the identity mapping. Thus $[\langle h\mathfrak{K}\rangle, \mathfrak{Z}/\mathfrak{K}] = 1$. By (2), $\mathfrak{Z} \neq \mathfrak{Q}$, so $\mathfrak{Q}/\mathfrak{K}$ is non-Abelian. By c), the class of $\mathfrak{Q}/\mathfrak{K}$ is 2. Also $\mathfrak{Q}$ is of exponent q, and the centre of $\mathfrak{Q}/\mathfrak{K}$ is cyclic so $\mathfrak{Q}/\mathfrak{K}$ is extraspecial.

Let U_1 be an irreducible $\mathsf{K}\mathfrak{G}_0$-submodule of U. Then U_1 is the direct sum of faithful irreducible $\mathsf{K}(\mathfrak{Q}/\mathfrak{K})$-modules, so by (2), $([\langle h^{2^{n-m-1}}\rangle, \mathfrak{Q}]$ on $\mathsf{U}_1)$ is non-trivial. Hence if $\mathfrak{K}_1 = \ker(\mathfrak{G}_0$ on $\mathsf{U}_1)$, $h^{2^{n-m-1}} \notin \mathfrak{K}_1$ and $\mathfrak{K}_1 \leq \mathfrak{Q}$. Thus $\mathfrak{K}_1 = \ker(\mathfrak{Q}$ on $\mathsf{U}_1) = \mathfrak{K}$. Hence U_1 is a faithful $\mathsf{K}(\mathfrak{G}_0/\mathfrak{K})$-module. By 3.2, $2^{n-m} - 1 = q^d$ and

$$|\mathfrak{Q}/\mathfrak{K} : \mathbf{C}_{\mathfrak{Q}/\mathfrak{K}}(h)| = q^{2d}$$

for some d. By 2.7, $d = 1$. Hence if W is the elementary Abelian q-group $\mathfrak{Q}/\mathfrak{Z}$ regarded as a vector space over $GF(q)$,

$$|\mathsf{W} : \mathbf{C}_{\mathsf{W}}(h)| \leq q^2.$$

Hence the degree of the minimum polynomial of the linear transformation $w \to wh$ of W is at most 3. Also since $\mathfrak{S} \leq \mathbf{N}_{\mathfrak{G}}(\mathfrak{K})$, $\mathfrak{S}$ is represented on W. Now ($\mathfrak{S}_0$ on W) is non-trivial. For otherwise, since $\mathfrak{Z}$ is represented on U by scalar multiples of the identity mapping, $\mathfrak{S}_0$ operates trivially on $\mathfrak{Q}/\mathfrak{Z}$ and on $\mathfrak{Z}/\mathfrak{K}$. Thus by I, 4.4, ($\mathfrak{S}_0$ on $\mathfrak{Q}/\mathfrak{K}$) is trivial and $[\mathfrak{Q}, \mathfrak{S}_0] \leq \mathfrak{K}$, a contradiction. It follows from 3.10 that $p = 3$ and $n - m = 2$. But then

$$q = q^d = 2^{n-m} - 1 = 3 = p,$$

a contradiction.

Hence $[\mathbf{O}_q(\mathfrak{H}), \mathfrak{S}_0] = 1$ for every prime divisor q of $|\mathfrak{F}|$ other than p. Hence $\mathfrak{S}_0 \leq \mathbf{C}_{\mathfrak{S}}(\mathbf{O}_{p'}(\mathfrak{F}))$, since $\mathfrak{F}$ is nilpotent. Now if $\mathfrak{C} = \mathbf{C}_{\mathfrak{X}}(\mathbf{O}_{p'}(\mathfrak{F}))$, $\mathfrak{C} \trianglelefteq \mathfrak{X}$, so $\mathfrak{S} \cap \mathfrak{C} \in S_p(\mathfrak{C})$ and $\mathfrak{S} \cap \mathfrak{C} \trianglelefteq \mathfrak{S}$. Thus $\mathfrak{S} \cap \mathfrak{C} \trianglelefteq \mathfrak{S}\mathbf{O}_{p'}(\mathfrak{F}) = \mathfrak{X}$. Hence the Sylow p-subgroup of $\mathfrak{C}$ is normal in $\mathfrak{C}$. But if $q \neq p$, the Sylow q-subgroup of $\mathfrak{X}$ is normal; hence so is that of $\mathfrak{C}$. Thus $\mathfrak{C}$ is a nilpotent normal subgroup of $\mathfrak{H}$ and $\mathfrak{C} \leq \mathbf{F}(\mathfrak{H}) = \mathfrak{F}$. Thus $[\mathfrak{S}, \langle g^{2^{n-1}}\rangle] = \mathfrak{S}_0 \leq \mathfrak{C} \leq \mathfrak{F}$. Since $\mathfrak{X} = \mathfrak{S}\mathfrak{F}$, $g^{2^{n-1}}$ centralizes $\mathfrak{X}/\mathfrak{F} = \mathbf{O}_p(\mathfrak{H}/\mathfrak{F})$. Thus d) is proved.

Since we cannot prove d) for $p = 2$, we introduce another normal subgroup $\mathfrak{H}_1$. For each prime divisor p of $|\mathfrak{F}|$, $\mathbf{O}_p(\mathfrak{H})$ is regular by c) and III, 10.2; thus the elements of $\mathbf{O}_p(\mathfrak{H})$ of order at most p form a characteristic subgroup $\Omega_1(\mathbf{O}_p(\mathfrak{H}))$, by III, 10.5. Let

$$\mathsf{V} = \mathsf{V}_1^{(p)} \oplus \mathsf{V}_2^{(p)} \oplus \cdots$$

be the Wedderburn decomposition of V for $\Omega_1(\mathbf{O}_p(\mathfrak{H}))$. Let $\mathfrak{K}_j^{(p)} = \ker(\Omega_1(\mathbf{O}_p(\mathfrak{H}))$ on $\mathsf{V}_j^{(p)})$. Since $\mathfrak{H}$ leaves $\mathsf{V}_j^{(p)}$ invariant, $\mathfrak{K}_j^{(p)} \trianglelefteq \mathfrak{H}$. Let

$$\mathfrak{C}_j^{(p)} = \mathbf{C}_{\mathfrak{H}}(\Omega_1(\mathbf{O}_p(\mathfrak{H}))/\mathfrak{K}_j^{(p)}),$$

so $\mathfrak{C}_j^{(p)} \trianglelefteq \mathfrak{H}$. Let $\mathfrak{H}_1$ be the intersection of those $\mathfrak{C}_j^{(p)}$ which contain $g^{2^{n-1}}$. (If $g^{2^{n-1}}$ lies in no $\mathfrak{C}_j^{(p)}$, put $\mathfrak{H}_1 = \mathfrak{H}$). In any event, $\mathfrak{H}_1 \trianglelefteq \mathfrak{H}$ and, by a), $g^{2^{n-1}} \in \mathfrak{H}_1$. Also $\mathsf{V}_j^{(p)}g = \mathsf{V}_i^{(p)}$ for some i, and

$$g^{-1}\mathfrak{K}_j^{(p)}g = \mathfrak{K}_i^{(p)}, \quad g^{-1}\mathfrak{C}_j^{(p)}g = \mathfrak{C}_i^{(p)}.$$

Thus $g \in \mathbf{N}_{\mathfrak{G}}(\mathfrak{H}_1)$, and $\mathfrak{H}_1 \trianglelefteq \langle g, \mathfrak{H}\rangle$. Note also that $\mathbf{F}(\mathfrak{H}_1) = \mathfrak{F} \cap \mathfrak{H}_1$, for $\mathbf{F}(\mathfrak{H}_1)$ is a normal, nilpotent subgroup of $\mathfrak{H}$.

e) If p is odd, $g^{2^{n-1}}$ centralizes $\mathbf{O}_p(\mathfrak{H}_1/\mathbf{F}(\mathfrak{H}_1))$.

If $\mathfrak{X}_1/\mathbf{F}(\mathfrak{H}_1) = \mathbf{O}_p(\mathfrak{H}_1/\mathbf{F}(\mathfrak{H}_1))$, $\mathfrak{X}_1$ is a characteristic subgroup of $\mathfrak{H}_1$, and $\mathfrak{X}_1 \cap \mathfrak{F} = \mathfrak{F} \cap \mathfrak{H}_1$. Thus $\mathfrak{X}_1 \trianglelefteq \mathfrak{H}$ and $\mathfrak{X}_1\mathfrak{F}/\mathfrak{F}$ is a normal p-subgroup of $\mathfrak{H}/\mathfrak{F}$. By d), $g^{2^{n-1}}$ centralizes $\mathfrak{X}_1\mathfrak{F}/\mathfrak{F}$. Hence $g^{2^{n-1}}$ centralizes $\mathfrak{X}_1/(\mathfrak{X}_1 \cap \mathfrak{F}) = \mathbf{O}_p(\mathfrak{H}_1/\mathbf{F}(\mathfrak{H}_1))$.

Now let $\mathfrak{Y}/\mathbf{F}(\mathfrak{H}_1) = \mathbf{O}_2(\mathfrak{H}_1/\mathbf{F}(\mathfrak{H}_1))$. Let $\mathfrak{P}^*$ be a Sylow 2-subgroup of $\langle g, \mathfrak{Y}\rangle$ containing g, and put $\mathfrak{P} = \mathfrak{P}^* \cap \mathfrak{Y}$. Thus $\mathfrak{P} \in S_2(\mathfrak{Y})$ and $g \in \mathbf{N}_{\mathfrak{G}}(\mathfrak{P})$.

f) If $x \in \mathfrak{P}$, $[g^{2^{n-1}}, x] = [g^{2^{n-2}}, x]^2$.

Let $y = [g^{2^{n-1}}, x]\ [g^{2^{n-2}}, x]^{-2}$, and suppose that $y \neq 1$. Since $g \in \mathbf{N}_{\mathfrak{G}}(\mathfrak{P})$, $y \in \mathfrak{P} \leq \mathfrak{H}_1$. Thus y is a 2-element. Since $|\mathfrak{F}|$ is odd, $y \notin \mathfrak{F}$. By 3.11, there exists a prime q such that $y \notin \mathbf{C}_{\mathfrak{G}}(\mathfrak{Q})$, where $\mathfrak{Q} = \Omega_1(\mathbf{O}_q(\mathfrak{H}))$. Thus ($[\mathfrak{Q}, \langle y\rangle]$ on $\mathsf{V}_j^{(q)}$) is non-trivial for some j, and

$$[\mathfrak{Q}, \langle y\rangle] \nleq \mathfrak{K}_j^{(q)}.$$

Hence $y \notin \mathfrak{C}_j^{(q)}$. But $y \in \mathfrak{H}_1$, so it follows from the definition of $\mathfrak{H}_1$ that $g^{2^{n-1}} \notin \mathfrak{C}_j^{(q)}$. Hence ($[\mathfrak{Q}, \langle g^{2^{n-1}}\rangle]$ on $\mathsf{V}_j^{(q)}$) is non-trivial.

Write $\mathsf{U} = \mathsf{V}_j^{(q)}$. Thus $[\mathfrak{Q}, \langle y\rangle]$ and $[\mathfrak{Q}, \langle g^{2^{n-1}}\rangle]$ are both represented non-trivially on U. Let 2^m be the length of the orbit containing U in the permutation representation of $\langle g\rangle$ on $\{\mathsf{U}a \mid a \in \mathfrak{G}\}$, and let $h = g^{2^m}$. As in the proof of d), (p. 445), $m < n$ and h is exceptional on U. Since ($[\mathfrak{Q}, g^{2^{n-1}}]$ on U) is non-trivial, $g^{2^{n-1}}$ is not exceptional on U, and $m < n - 1$. Let $\mathfrak{G}_0 = \langle h, \mathfrak{P}, \mathfrak{Q}\rangle$ and let $\mathfrak{K} = \ker(\mathfrak{G}_0$ on $\mathsf{U})$. Then $h^{2^{n-m-1}} = g^{2^{n-1}} \notin \mathfrak{K}$.

Now U is the direct sum of isomorphic, irreducible $\mathsf{K}\mathfrak{Q}$-modules, so $\mathfrak{Q}\mathfrak{K}/\mathfrak{K}$ has a faithful irreducible representation. Thus the centre $\mathfrak{Z}/\mathfrak{K}$ of $\mathfrak{Q}\mathfrak{K}/\mathfrak{K}$ is cyclic, and $\mathfrak{Z}$ is represented on U by scalar multiples of the identity mapping. Thus $[\mathfrak{Z}, \mathfrak{G}_0] \leq \mathfrak{K}$. Since $[\mathfrak{Q}, \mathfrak{G}_0] \nleq \mathfrak{K}$, $\mathfrak{Z} \neq \mathfrak{Q}\mathfrak{K}$.

By c), the class of $\mathfrak{Q}\mathfrak{R}/\mathfrak{R}$ is 2, and $\mathfrak{Q}\mathfrak{R}/\mathfrak{R}$ is extraspecial, since $\mathfrak{Q}$ is of exponent q and $\mathfrak{Z}/\mathfrak{R}$ is cyclic.

If U_1 is an irreducible $\mathsf{K}(\langle h\rangle\mathfrak{Q})$-submodule of U, then U_1 and U are direct sums of isomorphic irreducible $\mathsf{K}\mathfrak{Q}$-submodules, so ker($\mathfrak{Q}$ on U_1) = ker($\mathfrak{Q}$ on U). Hence ($[\mathfrak{Q}, h^{2^{n-m-1}}]$ on U_1) is non-trivial. By 3.2, $2^{n-m} - 1 = q^d$ for some d. By 2.7, $d = 1$. By 3.7c), $y\mathfrak{R} \in \mathbf{O}_2(\mathfrak{G}_0/\mathfrak{R})$. Thus $y\mathfrak{R}$ centralizes $\mathfrak{Q}\mathfrak{R}/\mathfrak{R}$ and $[\mathfrak{Q}, \langle y\rangle] \leq \mathfrak{R}$. This is a contradiction, so f) is proved.

g) $g^{2^{n-1}}\mathbf{F}(\mathfrak{H}_1) \in \mathbf{F}(\mathfrak{H}_1/\mathbf{F}(\mathfrak{H}_1))$.

Let $\mathfrak{E}/\mathbf{F}(\mathfrak{H}_1) = \mathbf{F}(\mathfrak{H}_1/\mathbf{F}(\mathfrak{H}_1))$. Now $\mathfrak{Y}/\mathbf{F}(\mathfrak{H}_1) \in S_2(\mathfrak{E}/\mathbf{F}(\mathfrak{H}_1))$, so $\mathfrak{P} \in S_2(\mathfrak{E})$. Let $\mathfrak{D} = \Phi(\mathfrak{P})\mathbf{F}(\mathfrak{H}_1)$. If $x \in \mathfrak{P}$, $[g^{2^{n-2}}, x] \in \mathfrak{P}$, so by f), $[g^{2^{n-1}}, x] \in \mathfrak{D}$. Thus $g^{2^{n-1}}$ centralizes $\mathfrak{P}\mathbf{F}(\mathfrak{H}_1)/\mathfrak{D}$. But by e), $g^{2^{n-1}}$ centralizes all the other Sylow subgroups of $\mathfrak{E}/\mathfrak{D}$. Thus $g^{2^{n-1}}$ centralizes $\mathfrak{E}/\mathfrak{D}$.

Now $\mathfrak{E}/\mathfrak{D}$ is nilpotent, so if $\mathfrak{E}^*/\mathfrak{D} = \mathbf{F}(\mathfrak{H}_1/\mathfrak{D})$, $\mathfrak{E} \leq \mathfrak{E}^*$.
But $\mathfrak{D}/\mathbf{F}(\mathfrak{H}_1) \leq \Phi(\mathfrak{E}^*/\mathbf{F}(\mathfrak{H}_1))$, since $\mathfrak{P}\mathbf{F}(\mathfrak{H}_1)/\mathbf{F}(\mathfrak{H}_1) \trianglelefteq \mathfrak{E}^*/\mathbf{F}(\mathfrak{H}_1)$. Thus by III, 3.7, $\mathfrak{E}^*/\mathbf{F}(\mathfrak{H}_1)$ is nilpotent. Thus $\mathfrak{E}^* = \mathfrak{E}$ and

$$\mathfrak{E}/\mathfrak{D} = \mathbf{F}(\mathfrak{H}_1/\mathfrak{D}).$$

Hence $g^{2^{n-1}}\mathfrak{D} \in \mathbf{C}_{\mathfrak{H}_1/\mathfrak{D}}(\mathbf{F}(\mathfrak{H}_1/\mathfrak{D}))$. By III, 4.2b), $g^{2^{n-1}} \in \mathfrak{E}$.

h) $g^{2^{n-1}}\mathbf{F}(\mathfrak{G}) \in \mathbf{F}(\mathfrak{G}/\mathbf{F}(\mathfrak{G}))$.

By g), $g^{2^{n-1}} \in \mathfrak{E}$, where $\mathfrak{E}/\mathbf{F}(\mathfrak{H}_1) = \mathbf{F}(\mathfrak{H}_1/\mathbf{F}(\mathfrak{H}_1))$. Since $\mathfrak{E}$ is a characteristic subgroup of $\mathfrak{H}_1$, $\mathfrak{E} \trianglelefteq \mathfrak{H}$. Since $\mathfrak{E} \cap \mathfrak{F} = \mathbf{F}(\mathfrak{H}_1)$, $\mathfrak{E}/(\mathfrak{E} \cap \mathfrak{F})$ is nilpotent. Thus $\mathfrak{E}\mathfrak{F}/\mathfrak{F}$ is a normal nilpotent subgroup of $\mathfrak{H}/\mathfrak{F}$. Hence if $\mathfrak{E}^*/\mathbf{F}(\mathfrak{H}) = \mathbf{F}(\mathfrak{H}/\mathbf{F}(\mathfrak{H}))$, $\mathfrak{E} \leq \mathfrak{E}^*$ and $g^{2^{n-1}} \in \mathfrak{E}^*$. But $\mathbf{F}(\mathfrak{H}) \leq \mathbf{F}(\mathfrak{G})$, so $\mathfrak{E}^*\mathbf{F}(\mathfrak{G})/\mathbf{F}(\mathfrak{G})$ is a normal nilpotent subgroup of $\mathfrak{G}/\mathbf{F}(\mathfrak{G})$. Thus

$$g^{2^{n-1}}\mathbf{F}(\mathfrak{G}) \in \mathbf{F}(\mathfrak{G}/\mathbf{F}(\mathfrak{G})).$$ **q.e.d.**

3.13 Remark. HOARE [1] proved the following improvement of 3.9.

Let $\mathfrak{G}$ be a p-soluble group of non-singular linear transformations of a vector space V over a field K of characteristic p. Suppose that $\mathbf{O}_p(\mathfrak{G}) = 1$. If g, h are elements of a Sylow p-subgroup of $\mathfrak{G}$ and $g^{p^{m-1}}$, $h^{p^{n-1}}$ do not commute, either $(g - 1)^{p^m - 1} \neq 0$ or $(h - 1)^{p^n - 1}(g - 1)^{p^{m-1}} \neq 0$.

Exercises

1) Let $p = 2^e + 1$ be a Fermat prime and let $\mathsf{F} = GF(2^{2e})$. For $a \in \mathsf{F}$, put

$$\operatorname{tr} a = \operatorname{tr}_{\mathsf{F}:GF(2)}(a) = a + a^2 + a^4 + \cdots + a^{2^{2e-1}},$$

and for a, b in F, put $(a, b) = \mathrm{tr}(ab^{2^e})$. Verify that $(\ , \ )$ is a bilinear form on F over $GF(2)$, and that $(\ , \ )$ is non-singular and symplectic. F has a primitive p-th root of unity ω; let $\overline{\omega}$ be the mapping $a \to a\omega$ on F. Then $(a\overline{\omega}, b\overline{\omega}) = (a, b)$, so $\langle \overline{\omega} \rangle$ is a Sylow p-subgroup of the symplectic group $Sp(2e, 2)$ on F.

2) Let $q = 2^f - 1$ be a Mersenne prime and let $\mathsf{F} = GF(q^2)$. Let ω be a primitive 2^{f+1}-th root of unity in F and define the mappings σ, τ of F into F by

$$x\sigma = \omega^2 x, \quad x\tau = \omega x^q \quad (x \in \mathsf{F}).$$

Then σ, τ are linear over $GF(q)$ and $\mathfrak{Q} = \langle \sigma, \tau \rangle$ is a generalized quaternion group of order 2^{f+1}. If $(a, b) = ab^q - a^q b$ $(a \in \mathsf{F}, b \in \mathsf{F})$, $(\ , \ )$ is a bilinear form on F over $GF(q)$ and $(\ , \ )$ is non-singular and symplectic. Also $(a\alpha, b\alpha) = (a, b)$ for all $\alpha \in \mathfrak{Q}$, so $\mathfrak{Q}$ is a Sylow 2-subgroup of the symplectic group $Sp(2, q)$ on F.

§ 4. Reduction Theorems for Burnside's Problem

The results of § 2 and § 3 will now be used to improve Theorem 1.13.

4.1 Definition. Suppose that $\mathfrak{P} \in S_p(\mathfrak{G})$.

a) The exponent of $\mathfrak{P}$ will be denoted by $p^{e_p(\mathfrak{G})}$.

b) For $n \geq 0$, let $\tau_{2n}(\mathfrak{P}) = \langle x^{p^n} | x \in \mathfrak{P} \rangle$, and for $n \geq 1$, let

$$\tau_{2n-1}(\mathfrak{P}) = \langle [x^{p^{n-1}}, y^{p^{n-1}}], z^{p^n} | x \in \mathfrak{P}, y \in \mathfrak{P}, z \in \mathfrak{P} \rangle.$$

Then

$$\mathfrak{P} = \tau_0(\mathfrak{P}) \geq \tau_1(\mathfrak{P}) \geq \cdots \geq \tau_{2n}(\mathfrak{P}) \geq \tau_{2n+1}(\mathfrak{P}) \geq \cdots.$$

Let $e_p^*(\mathfrak{G})$ be the smallest positive integer m for which $\tau_m(\mathfrak{P}) = 1$. Hence

$$e_p(\mathfrak{G}) = [\tfrac{1}{2}(1 + e_p^*(\mathfrak{G}))].$$

The invariant $e_p^*(\mathfrak{G})$ is introduced since it is not possible to prove that the p-length of any p-soluble group $\mathfrak{G}$ is at most $\epsilon_p(\mathfrak{G})$. We now investigate what happens to e_p, e_p^* on passage from $\mathfrak{G}$ to $\mathfrak{G}/\mathbf{O}_{p',p}(\mathfrak{G})$.

4.2 Lemma. *Suppose that $\mathfrak{G}$ is a p-soluble group and p divides $|\mathfrak{G}|$.*

a) $e_p(\mathfrak{G}/\mathbf{O}_{p',p}(\mathfrak{G})) \leq e_p(\mathfrak{G}) - 1$, *provided that one of the following three conditions is satisfied:*

(i) *p is odd and p is not a Fermat prime.*

(ii) *p is a Fermat prime and the Sylow 2-subgroups of $\mathfrak{G}$ are Abelian.*

(iii) $p = 2$ *and the Sylow q-subgroups of $\mathfrak{G}$ are Abelian for every Mersenne prime q.*

b) *If* $p \geq 3$, $e_p^*(\mathfrak{G}/\mathbf{O}_{p',p}(\mathfrak{G})) \leq e_p^*(\mathfrak{G}) - 1$.

Proof. Suppose that $\mathfrak{P} \in S_p(\mathfrak{G})$, and write

$$\mathfrak{U}/\mathbf{O}_{p'}(\mathfrak{G}) = \Phi(\mathbf{O}_{p',p}(\mathfrak{G})/\mathbf{O}_{p'}(\mathfrak{G})).$$

Thus $\mathsf{V} = \mathbf{O}_{p',p}(\mathfrak{G})/\mathfrak{U}$ may be regarded as a vector space over $GF(p)$. Let $\rho = (\mathfrak{G} \text{ on } \mathsf{V})$. By 1.4 and 1.6, $\ker \rho = \mathbf{C}_{\mathfrak{G}}(\mathbf{O}_{p',p}(\mathfrak{G})/\mathfrak{U}) = \mathbf{O}_{p',p}(\mathfrak{G})$. Thus if $\overline{\mathfrak{G}} = \operatorname{im} \rho$, $\overline{\mathfrak{G}}$ is a group of non-singular linear transformations of V and $\mathbf{O}_p(\overline{\mathfrak{G}}) = 1$. We shall apply 2.9 and 3.9 to $\overline{\mathfrak{G}}$.

First note the following consequences of 1.8.

α) If $y \in \mathfrak{P} \cap \mathbf{O}_{p',p}(\mathfrak{G})$, $x \in \mathfrak{P}$ and

$$(y\mathfrak{U})(\rho(x) - 1)^{p^n-1} \neq 1,$$

then either $x^{p^n} \neq 1$ or $(xy)^{p^n} \neq 1$.

For by 1.8b),

$$\begin{aligned} x^{-p^n}(xy)^{p^n}\mathfrak{U} &= (y\mathfrak{U})(1 + \rho(x) + \cdots + \rho(x)^{p^n-1}) \\ &= (y\mathfrak{U})(\rho(x) - 1)^{p^n-1} \neq 1. \end{aligned}$$

β) If $y \in \mathfrak{P} \cap \mathbf{O}_{p',p}(\mathfrak{G})$, $x \in \mathfrak{P}$ and

$$(y\mathfrak{U})(\rho(x) - 1)^{2p^n-1} \neq 1,$$

then $[(xy)^{p^n}, x^{p^n}] \neq 1$.

For if $v = (y\mathfrak{U})(\rho(x) - 1)^{p^n-1}$, then by 1.8b),

$$\begin{aligned} v &= (y\mathfrak{U})(1 + \rho(x) + \cdots + \rho(x)^{p^n-1}) \\ &= x^{-p^n}(xy)^{p^n}\mathfrak{U}. \end{aligned}$$

Hence by 1.8a),

$$\begin{aligned} 1 \neq v(\rho(x) - 1)^{p^n} &= v(\rho(x^{p^n}) - 1) \\ &= [x^{-p^n}(xy)^{p^n}, x^{p^n}]\mathfrak{U} \\ &= [(xy)^{p^n}, x^{p^n}]\mathfrak{U}. \end{aligned}$$

a) Let $e_p(\mathfrak{G}/\mathbf{O}_{p',p}(\mathfrak{G})) = n$. Thus there exists $x \in \mathfrak{P}$ such that the order of $x\mathbf{O}_{p',p}(\mathfrak{G})$ and $\rho(x)$ is p^n. By 2.7, there is no integer n_0 such that $p^{n_0} - 1$ is a power of a prime q and the Sylow q-subgroups of $\mathfrak{G}$ are non-Abelian, since one of (i), (ii) or (iii) holds. Hence, by 2.9, the minimum polynomial of $\rho(x)$ is $(t - 1)^{p^n}$. Since $\mathbf{O}_{p',p}(\mathfrak{G}) = \mathfrak{U}(\mathfrak{P} \cap \mathbf{O}_{p',p}(\mathfrak{G}))$, it follows that there exists $y \in \mathfrak{P} \cap \mathbf{O}_{p',p}(\mathfrak{G})$ such that

$$(y\mathfrak{U})(\rho(x) - 1)^{p^n - 1} \neq 1.$$

By α), either $x^{p^n} \neq 1$ or $(xy)^{p^n} \neq 1$. Since x and xy are in $\mathfrak{P}$, it follows that $e_p(\mathfrak{G}) > n$.

b) Suppose first that $e_p^*(\mathfrak{G}/\mathbf{O}_{p',p}(\mathfrak{G})) = 2n$ is even. If $2n = 0$, there is nothing to prove, since $\mathfrak{P} \neq 1$. Thus we suppose that $2n > 0$. Hence $z^{p^n} \in \mathbf{O}_{p',p}(\mathfrak{G})$ for all $z \in \mathfrak{P}$, but there exist $x_1 \in \mathfrak{P}$, $x_2 \in \mathfrak{P}$ such that $[x_1^{p^{n-1}}, x_2^{p^{n-1}}] \notin \mathbf{O}_{p',p}(\mathfrak{G})$. Thus $\rho(x_1)^{p^{n-1}}$ and $\rho(x_2)^{p^{n-1}}$ do not commute. By 3.9, $(\rho(x) - 1)^{p^n - 1} \neq 0$, where $x = x_1$ or $x = x_2$. Thus there exists $y \in \mathfrak{P} \cap \mathbf{O}_{p',p}(\mathfrak{G})$ such that

$$(y\mathfrak{U})(\rho(x) - 1)^{p^n - 1} \neq 1.$$

By α), either $x^{p^n} \neq 1$ or $(xy)^{p^n} \neq 1$. Thus $e_p^*(\mathfrak{G}) > 2n$.

Next suppose that $e_p^*(\mathfrak{G}/\mathbf{O}_{p',p}(\mathfrak{G})) = 2n + 1$ is odd. Then there is an element x in $\mathfrak{P}$ such that the order of $x\mathbf{O}_{p',p}(\mathfrak{G})$ and $\rho(x)$ is p^{n+1}. By 2.9, the minimum polynomial of $\rho(x)$ is $(t - 1)^r$, where $r \geq p^{n+1} - p^n$. Since $p \geq 3$, $r > 2p^n - 1$, so there exists $y \in \mathfrak{P} \cap \mathbf{O}_{p',p}(\mathfrak{G})$ such that

$$(y\mathfrak{U})(\rho(x) - 1)^{2p^n - 1} \neq 1.$$

By β), $[(xy)^{p^n}, x^{p^n}] \neq 1$, so $e_p^*(\mathfrak{G}) > 2n + 1$. **q.e.d.**

This brings us to one of the main theorems on bounds for the p-length.

4.3 Theorem. *Suppose that $\mathfrak{G}$ is a p-soluble group of p-length l and that p^e is the exponent of a Sylow p-subgroup of $\mathfrak{G}$.*

a) *$l \leq e$, provided that one of the following three conditions holds:*

(i) *p is odd and p is not a Fermat prime.*

(ii) *p is a Fermat prime and the Sylow 2-subgroups of $\mathfrak{G}$ are Abelian.*

(iii) *$p = 2$ and the Sylow q-subgroups of $\mathfrak{G}$ are Abelian for every Mersenne prime q.*

b) *If p is a Fermat prime, $l \leq 2e$.*

Proof. a) This is proved by induction on l. If $l = 0$ or $l = 1$, it is trivial. If $l > 1$, the p-length of $\overline{\mathfrak{G}} = \mathfrak{G}/\mathbf{O}_{p',p}(\mathfrak{G})$ is $l - 1$. By the inductive hypothesis, $l - 1 \leq e_p(\overline{\mathfrak{G}})$. By 4.2a), $e_p(\overline{\mathfrak{G}}) \leq e_p(\mathfrak{G}) - 1$, so $l \leq e_p(\mathfrak{G}) = e$.

b) We prove by induction on l that $l \leq e_p^*(\mathfrak{G})$. This is clear for $l \leq 1$. If $l > 1$, $l - 1 \leq e_p^*(\mathfrak{G}/\mathbf{O}_{p',p}(\mathfrak{G})) \leq e_p^*(\mathfrak{G}) - 1$ by the inductive hypothesis and 4.2b). Thus $l \leq e_p^*(\mathfrak{G})$. Hence $l \leq 2e$, since by 4.1, $e_p^*(\mathfrak{G}) < 2e + 1$. **q.e.d.**

4.4 Remark. HOARE [1] has introduced the invariant $e_p'(\mathfrak{G})$ defined as follows. For $\mathfrak{P} \in S_p(\mathfrak{G})$, let $\sigma_0(\mathfrak{P}) = \mathfrak{P}$, and for $n \geq 1$, let

$$\sigma_{3n-2}(\mathfrak{P}) = \langle x^{p^n} | x \in \mathfrak{P} \rangle,$$

$$\sigma_{3n-1}(\mathfrak{P}) = \langle [x^{p^n}, y^{p^{n-1}}], z^{p^{n+1}} | x \in \mathfrak{P}, y \in \mathfrak{P}, z \in \mathfrak{P} \rangle,$$

$$\sigma_{3n}(\mathfrak{P}) = \langle [x^{p^n}, y^{p^n}], z^{p^{n+1}} | x \in \mathfrak{P}, y \in \mathfrak{P}, z \in \mathfrak{P} \rangle.$$

Then $e_p'(\mathfrak{G})$ is the smallest positive integer m for which $\sigma_m(\mathfrak{P}) = 1$. It follows from 3.13 that if $\mathfrak{G}$ is a 2-soluble group of even order, $e_2'(\mathfrak{G}/\mathbf{O}_{2',2}(\mathfrak{G})) \leq e_2'(\mathfrak{G}) - 1$. Hence the 2-length of $\mathfrak{G}$ is at most $e_2'(\mathfrak{G})$.

Since $e_2(\mathfrak{G}) = [\frac{1}{3}(4 + e_2'(\mathfrak{G}))]$, this implies that the 2-length of $\mathfrak{G}$ is at most $3e_2(\mathfrak{G}) - 2$. However, for a soluble group $\mathfrak{G}$ of order divisible by 4, it has been proved by GROSS [1, 2] that the 2-length is at most $2e_2(\mathfrak{G}) - 2$. A slightly weaker result will now be deduced from 3.12.

4.5 Theorem (GROSS). *Let $\mathfrak{G}$ be a soluble group of 2-length l, and let 2^e be the exponent of a Sylow 2-subgroup of $\mathfrak{G}$.*

a) *If $l > 0$, $e_2(\mathfrak{G}/\mathbf{O}_{2',2,2',2}(\mathfrak{G})) \leq e - 1$.*

b) *If $l > 0$, $l \leq 2e - 1$.*

Proof. a) This is trivial if $l = 1$ or $l = 2$, so we suppose that $l > 2$. Let

$$\mathfrak{U}/\mathbf{O}_{2'}(\mathfrak{G}) = \Phi(\mathbf{O}_{2',2}(\mathfrak{G})/\mathbf{O}_{2'}(\mathfrak{G}))$$

and let $\mathsf{V} = \mathbf{O}_{2',2}(\mathfrak{G})/\mathfrak{U}$, regarded as a vector space over $GF(2)$. Let $\rho = (\mathfrak{G}$ on $\mathsf{V})$ and let $\overline{\mathfrak{G}} = \operatorname{im} \rho$. By 1.4 and 1.6, $\ker \rho = \mathbf{O}_{2',2}(\mathfrak{G})$, so $\mathbf{O}_2(\overline{\mathfrak{G}}) = 1$ and $|\mathbf{F}(\overline{\mathfrak{G}})|$ is odd. Suppose that $\mathfrak{P} \in S_2(\mathfrak{G})$ and that $x \in \mathfrak{P}$. For any $y \in \mathfrak{P} \cap \mathbf{O}_{2',2}(\mathfrak{G})$, $(y\mathfrak{U})(\rho(x) - 1)^{2^e - 1} = x^{-2^e}(xy)^{2^e}\mathfrak{U} = 1$, by 1.8b). Thus either $\rho(x)^{2^{e-1}} = 1$ or $\rho(x)$ is exceptional. It follows from 3.12 that in either case, $\rho(x)^{2^{e-1}} \in \mathfrak{L}$, where

$$\mathfrak{L}/\mathbf{F}(\overline{\mathfrak{G}}) = \mathbf{F}(\overline{\mathfrak{G}}/\mathbf{F}(\overline{\mathfrak{G}})).$$

But $|\mathbf{F}(\overline{\mathfrak{G}})|$ is odd, so $\mathbf{F}(\overline{\mathfrak{G}}) \leq \mathbf{O}_{2'}(\overline{\mathfrak{G}})$. Thus $\mathfrak{L}\mathbf{O}_{2'}(\overline{\mathfrak{G}})/\mathbf{O}_{2'}(\overline{\mathfrak{G}})$ is nilpotent. Since $\rho(x)^{2^{e-1}}$ is a 2-element, it follows that $\rho(x)^{2^{e-1}} \in \mathbf{O}_{2',2}(\overline{\mathfrak{G}})$. Hence $x^{2^{e-1}} \in \mathbf{O}_{2',2,2',2}(\mathfrak{G})$ for all $x \in \mathfrak{P}$. Thus

$$e_2(\mathfrak{G}/\mathbf{O}_{2',2,2',2}(\mathfrak{G})) \leq e - 1.$$

b) This is proved by induction on l. If $l = 1$, it is trivial. If $l = 2$, it must be proved that $e > 1$. But if $e = 1$, the Sylow 2-subgroup of $\mathfrak{G}$ is Abelian and $l = 1$ by VI, 6.6a). If $l > 2$, the inductive hypothesis and a) give

$$l - 2 \leq 2e_2(\mathfrak{G}/\mathbf{O}_{2',2,2',2}(\mathfrak{G})) - 1 \leq 2e - 3,$$

so $l \leq 2e - 1$. **q.e.d.**

We now discuss the question of the extent to which these results are best possible.

4.6 Example (HALL and HIGMAN). A good deal of the proof of the theorems leading to 4.3a) was devoted to the investigation of a minimal counterexample to the proposition that $l \leq e$. To show, then, that Fermat primes really are exceptional in this regard, we simply have to verify that this minimal counterexample exists.

a) The first step is to construct a group of the kind discussed in 2.6 when p is a Fermat prime and $q = 2$. This could be done, for example, by observing that the order of an appropriate orthogonal group is divisible by p. A more concrete approach is as follows.

Let $\mathsf{K} = GF(2)$, $\mathsf{L} = GF(2^{2m})$ $(m \geq 1)$. Let ε be a primitive $(2^m + 1)$-th root of unity in L. Then $\varepsilon^{2^i - 1} \neq 1$ for $i = 1, \ldots, 2m - 1$, so $\mathsf{L} = \mathsf{K}(\varepsilon)$. Let α be the automorphism $x \to x^{2^m}$ of L; thus α is of order 2. Also, for $x \in \mathsf{L}$, write

$$\operatorname{tr} x = \operatorname{tr}_{\mathsf{L}:\mathsf{K}} x = x + x^2 + x^4 + \cdots + x^{2^{2m-1}}.$$

Thus $\operatorname{tr} x \in \mathsf{K}$, $\operatorname{tr}(x + y) = \operatorname{tr} x + \operatorname{tr} y$ and $\operatorname{tr} x^2 = \operatorname{tr} x$. As is well-known, there exists $c \in \mathsf{L}$ such that $\operatorname{tr} c = 1$; (this can be seen from the properties of the Vandermonde determinant, for example). Put $x\tau = \operatorname{tr}(cx(x\alpha))$. Thus

$$(x + y)\tau + x\tau + y\tau = \operatorname{tr}(c(x(y\alpha) + y(x\alpha))).$$

Hence by VIII, 6.10, there exists a 2-group $\mathfrak{Q}$, a subgroup $\mathfrak{N}$ of the centre of $\mathfrak{Q}$ and isomorphisms ρ, σ of $\mathfrak{Q}/\mathfrak{N}$, $\mathfrak{N}$ onto L, K respectively such that if $(u\mathfrak{N})\rho = x$, then $(u^2)\sigma = x\tau$. Also $(\varepsilon x)\tau = \operatorname{tr}(c\varepsilon^{1+2^m}x(x\alpha)) = x\tau$, so by VIII, 6.6, there is an automorphism β of $\mathfrak{Q}$ such that $((u\beta)\mathfrak{N})\rho = \varepsilon((u\mathfrak{N})\rho)$ for all $u \in \mathfrak{Q}$ and $u\beta = u$ for all $u \in \mathfrak{N}$. Since $\mathsf{L} = \mathsf{K}(\varepsilon)$, β induces an irreducible automorphism on $\mathfrak{Q}/\mathfrak{N}$. Hence either $\mathbf{Z}(\mathfrak{Q}) = \mathfrak{Q}$ or $\mathbf{Z}(\mathfrak{Q}) = \mathfrak{N}$. But in fact $\mathbf{Z}(\mathfrak{Q}) = \mathfrak{N}$. For otherwise $\mathfrak{Q}$ is Abelian, and since $\mathfrak{N} < \Omega_1(\mathfrak{Q}) \leq \mathfrak{Q}$, $\mathfrak{Q}$ is elementary Abelian. This, however, is not the case, since $1\tau = \operatorname{tr} c = 1$. Thus $\mathbf{Z}(\mathfrak{Q}) = \mathfrak{N}$ and $\mathfrak{Q}$ is extraspecial. Also β^{2^m+1} induces the identity automorphism on $\mathfrak{Q}/\mathfrak{N}$ and on $\mathfrak{N}$. By I, 4.4, β^{2^m+1} is of order 2^a for some a. If $\varepsilon' = \varepsilon^{2^a}$, ε' is also a primitive $(2^m + 1)$-th root of unity and $((u\beta^{2^a})\mathfrak{N})\rho = \varepsilon'((u\mathfrak{N})\rho)$ for all $u \in \mathfrak{Q}$. Thus we may assume that $\beta^{2^m+1} = 1$. Let $\mathfrak{H}$ be the semidirect product $\mathfrak{Q}\langle s\rangle$, where $\mathfrak{Q} \trianglelefteq \mathfrak{H}$ and $u^s = u\beta$ for all $u \in \mathfrak{Q}$. Then $|\mathfrak{H}| = 2^{2m+1}(2^m + 1)$, $\mathbf{C}_{\mathfrak{H}}(\mathfrak{Q}/\mathfrak{N}) = \mathfrak{Q}$ and $\mathfrak{N}$ is the only minimal normal subgroup of $\mathfrak{H}$.

b) Next we construct a suitable representation of $\mathfrak{H}$. If F is a field and $\operatorname{char} \mathsf{F} \neq 2$, consider the group-ring $\mathsf{F}\mathfrak{H}$ as an $\mathsf{F}\mathfrak{H}$-module, and let

$$\mathsf{F}\mathfrak{H} = \mathsf{V}_0 > \mathsf{V}_1 > \cdots > \mathsf{V}_{l-1} > \mathsf{V}_l = 0$$

be an $\mathsf{F}\mathfrak{H}$-composition series of $\mathsf{F}\mathfrak{H}$. Since $\operatorname{char} \mathsf{F}$ does not divide $|\mathfrak{N}|$, $\mathsf{V}_{i-1} = \mathsf{V}_i \oplus \mathsf{W}_i$ for some $\mathsf{F}\mathfrak{N}$-module W_i, and $\mathsf{V}_0 = \mathsf{W}_1 \oplus \cdots \oplus \mathsf{W}_l$. Now $\mathfrak{N}$ is not represented trivially on $\mathsf{F}\mathfrak{H}$, so there exists some W_i on which the representation of $\mathfrak{N}$ is non-trivial. Since $\mathfrak{N}$ is the only minimal normal subgroup of $\mathfrak{H}$, $\mathfrak{H}$ is represented faithfully on $\mathsf{V}_{i-1}/\mathsf{V}_i$. Thus $\mathsf{V} = \mathsf{V}_{i-1}/\mathsf{V}_i$ is a faithful irreducible $\mathsf{K}\mathfrak{H}$-module. If F is algebraically closed, it follows from 2.5 that V is an irreducible $\mathsf{K}\mathfrak{Q}$-module and $\dim_{\mathsf{F}} \mathsf{V} = 2^m$.

If F is the algebraic closure of $GF(p)$, $\mathfrak{H}$ is represented on V by a finite number of matrices, the coefficients of all of which lie in some finite field $\mathsf{F}_0 = GF(p^n)$. Thus there exists a faithful, irreducible $\mathsf{F}_0\mathfrak{H}$-module U of degree 2^m. Let $\mathfrak{G}$ be the corresponding semidirect product $\mathfrak{H}\mathsf{U}$. Thus U is an elementary Abelian minimal normal p-subgroup of $\mathfrak{G}$ and $\mathbf{C}_{\mathfrak{G}}(\mathsf{U}) = \mathsf{U}$. Thus $\mathbf{O}_{p'}(\mathfrak{G}) = 1$ and $\mathbf{O}_p(\mathfrak{G}) = \mathsf{U}$.

If p is a Fermat prime, we may take $2^m = p - 1$. Thus $|\mathfrak{H}/\mathfrak{Q}| = p$ and $\mathfrak{G}$ is of p-length 2. But if g is a p-element of $\mathfrak{G}$, $g = hu$, where $h \in \mathfrak{H}$, $u \in \mathsf{U}$ and $h^p = 1$. Thus the minimum polynomial of the linear transformation ξ of U induced by h is $(t - 1)^r$ for some r. But $r \leq \dim_{\mathsf{F}_0} \mathsf{U} = 2^m = p - 1$, so $(\xi - 1)^{p-1} = 0$. Hence

$$\begin{aligned} g^p &= (hu)^p \\ &= h^p u^{h^{p-1}} u^{h^{p-2}} \cdots u \\ &= h^p(u(1 + \xi + \cdots + \xi^{p-1})) \\ &= h^p(u(\xi - 1)^{p-1}) \\ &= h^p = 1. \end{aligned}$$

Thus the exponent of the Sylow p-subgroup of $\mathfrak{G}$ is p.

c) We show that F_0 may be chosen to be $GF(p)$.

First suppose $p = 3$. Then $\mathfrak{Q}$ is the quaternion group of order 8, for the dihedral group of order 8 has no automorphism of order 3. Thus $\mathfrak{H} \cong SL(2, 3)$, since $SL(2, 3)$ is an extension of a quaternion group (consisting of ± 1 and the matrices of trace 0 and determinant 1) by a cyclic group of order 3. Thus $\mathfrak{H}$ certainly has a faithful, irreducible representation of degree 2 in $GF(3)$.

If $p > 3$, we show first that the matrices representing elements of $\mathfrak{Q}$ may be chosen with coefficients in $GF(p)$. Since $p - 1 = 2^m > 2$, $GF(p)$ possesses a primitive 4-th root of unity i. By V, 16.14, the representation of $\mathfrak{Q}$ on V is induced from the representation of degree 1 of a maximal normal Abelian subgroup $\mathfrak{A}$ of $\mathfrak{Q}$. But since the exponent of $\mathfrak{A}$ is at most 4, the matrices of such a representation are all of the form (i^n). Thus we may suppose that $\mathfrak{Q}$ is represented by matrices with coefficients in $GF(p)$. Denote this matrix representation of $\mathfrak{H}$ by ρ.

We know that the coefficients of all $\rho(x)$ ($x \in \mathfrak{H}$) lie in some finite field $GF(p^k)$. Let α be an automorphism of $GF(p^k)$. If $x \in \mathfrak{Q}$,

$$\rho(x)^{\rho(s)} = \rho(x^s) = \rho(x^s)\alpha = (\rho(x)\alpha)^{\rho(s)\alpha} = \rho(x)^{\rho(s)\alpha}.$$

Thus $(\rho(s)\alpha)\rho(s)^{-1}$ commutes with all $\rho(x)$. Since the restriction of ρ to $\mathfrak{Q}$ is absolutely irreducible, it follows from Schur's lemma that $\rho(s)\alpha = \lambda\rho(s)$ for some $\lambda \in GF(p^k)$. Taking p-th powers, $\lambda^p = 1$. Since the characteristic is p, $\lambda = 1$. Thus $\rho(s)\alpha = \rho(s)$ for every automorphism α of $GF(p^k)$. Thus all the coefficients of $\rho(s)$ lie in $GF(p)$.

With this choice of F_0, $|\mathfrak{G}| = p^p \cdot 2^{2m+1}$.

4.7 Remarks. a) Let p, q be primes such that q divides $p - 1$ and let $\mathfrak{H}$ be the permutation group of order pq and degree p. Let

$$\mathfrak{G}_n = \mathfrak{H} \wr \mathfrak{H} \wr \cdots \wr \mathfrak{H}$$

be the wreath product of n copies of $\mathfrak{H}$ (I, § 15). Thus $\mathfrak{G}_n$ is a permutation group of degree p^n. By I, 15.5, if $\mathfrak{P} \in S_p(\mathfrak{H})$ and

$$\mathfrak{P}_n = \mathfrak{P} \wr \mathfrak{P} \wr \cdots \wr \mathfrak{P},$$

$\mathfrak{P}_n \in S_p(\mathfrak{G}_n)$. By III, 15.3b), the exponent of $\mathfrak{P}_n$ is p^n.

But also, the p-length of $\mathfrak{G}_n$ is n. To prove this, one observes that if

(i) $\mathfrak{H}_1$, $\mathfrak{H}_2$ are p-soluble groups of p-lengths l_1, l_2 respectively,
(ii) $\mathfrak{H}_2$ is a transitive permutation group, and
(iii) either $\mathbf{O}^p(\mathfrak{H}_1) = \mathfrak{H}_1$ or $\mathbf{O}_p(\mathfrak{H}_2) = 1$,

then $\mathfrak{H}_1 \wr \mathfrak{H}_2$ is p-soluble of p-length $l_1 + l_2$.

Thus the inequality $l \leq e$ in 4.3a) cannot be improved.

b) Let p be a Fermat prime, and let $\mathfrak{G}_1$ be a group of p-length 2 in which the exponent of the Sylow p-subgroup is p (cf. 4.6). We define $\mathfrak{G}_n$ by induction on n: for $n > 1$,

$$\mathfrak{G}_n = \mathfrak{G}_{n-1} \wr (\mathfrak{Z}_2 \wr \mathfrak{G}_1).$$

(Here, the regular wreath product is meant, and $\mathfrak{Z}_2$ denotes the cyclic group of order 2.) Since $\mathbf{O}_p(\mathfrak{Z}_2 \wr \mathfrak{G}_1) = 1$, it follows from the observation in a) that the p-length of $\mathfrak{G}_n$ is $2n$. But the exponent of a Sylow p-subgroup is at most p^n, so the inequality $l \leq 2e$ in 4.3b) cannot be improved.

c) For $p = 2$, it appears that no example is known for which $l > e$. As was remarked earlier (4.4), the inequality $l \leq e$ has been proved by Gross if $e = 2$.

We use 4.3 to prove the following.

4.8 Theorem. *Suppose that $\mathfrak{G}$ is p-soluble of p-length l and that $\mathfrak{P} \in S_p(\mathfrak{G})$.*

a) *Suppose that $p > 2$ and that p is not a Fermat prime. If the wreath product of two cyclic groups of order p does not occur among the sections of $\mathfrak{P}$, $l \leq 1$.*

b) *Suppose that $\mathfrak{P}$ is regular (in the sense of III, §10). If p is not a Fermat prime, $l \leq 1$; in any case, $l \leq 2$.*

Proof. Both assertions are proved by supposing that they are false and considering a counterexample of minimal order. Since the properties of $\mathfrak{P}$ involved in the hypotheses are inherited by homomorphic images, the p-length of any proper homomorphic image of $\mathfrak{G}$ is less than l. It follows from VI, 6.9 that $\mathfrak{N} = \mathbf{O}_{p',p}(\mathfrak{G})$ is an elementary Abelian p-group, $\mathfrak{N}$ is the unique minimal normal subgroup of $\mathfrak{G}$, $\mathbf{C}_{\mathfrak{G}}(\mathfrak{N}) = \mathfrak{N}$ and there exists $\mathfrak{H} \leq \mathfrak{G}$ such that $\mathfrak{G} = \mathfrak{H}\mathfrak{N}$, $\mathfrak{H} \cap \mathfrak{N} = 1$. Since $\mathfrak{N} = \mathbf{O}_p(\mathfrak{G})$, $\mathfrak{N} \leq \mathfrak{P}$; thus $\mathfrak{P} = \mathfrak{B}\mathfrak{N}$, where $\mathfrak{B} = \mathfrak{P} \cap \mathfrak{H}$.

a) Since $l > 1$, $\mathbf{O}_p(\mathfrak{G}) \neq \mathfrak{G}$; thus there exists a maximal normal subgroup $\mathfrak{M}$ of $\mathfrak{G}$ such that $\mathbf{O}_p(\mathfrak{G}) \leq \mathfrak{M} < \mathfrak{G}$. Since the property of $\mathfrak{P}$ involved in the hypothesis is inherited by subgroups and $\mathfrak{G}$ is a counterexample of minimal order, the p-length of $\mathfrak{M}$ is 1. Since $\mathfrak{M} \trianglelefteq \mathfrak{G}$, $\mathbf{O}_{p',p}(\mathfrak{M}) \leq \mathbf{O}_{p',p}(\mathfrak{G}) = \mathfrak{N}$, so $\mathfrak{M}/\mathfrak{N}$ is a p'-group. Thus $\mathfrak{M} \cap \mathfrak{P} = \mathfrak{N}$, and since $l > 1$, $\mathfrak{G}/\mathfrak{M}$ is of order p. Hence $|\mathfrak{B}| = p$.

By 4.3a), $\mathfrak{P}$ possesses an element x of order p^2. Write $x = yz$ with $y \in \mathfrak{B}$, $z \in \mathfrak{N}$. Then $y^p = 1$. Let $\mathfrak{N}_0 = \langle z, z^y, \ldots, z^{y^{p-1}} \rangle$. Then $y \in \mathbf{N}_{\mathfrak{G}}(\mathfrak{N}_0)$, so y induces an automorphism η on $\mathfrak{N}_0$. By 1.8,

$$z(\eta - 1)^{p-1} = z(\eta^{p-1} + \cdots + \eta + 1) = y^{-p}(yz)^p = x^p \neq 1,$$

so $(\eta - 1)^{p-1} \neq 0$. But since $y^p = 1$, $\eta^p = 1$ and $(\eta - 1)^p = 0$. Hence the degree of the minimum polynomial of η is p, so $p \leq \dim_{GF(p)} \mathfrak{N}_0$. Thus $|\mathfrak{N}_0| = p^p$ and $\langle y, \mathfrak{N}_0 \rangle$ is the wreath product of two cyclic groups of order p. By hypothesis, this is impossible.

b) If $p = 2$, $l \leq 1$ by VI, 6.6a), for a regular 2-group is Abelian (III, 10.3a)). If p is not a Fermat prime, $l \leq 1$ by a), for regularity is inherited by subgroups and factor groups and the wreath product of two cyclic groups of order p is irregular (III, 10.3d)). Suppose, then, that p is a Fermat prime. We have $\mathbf{C}_{\mathfrak{G}}(\mathfrak{N}) = \mathfrak{N}$. But by III, 10.8f),

$$[\mho_1(\mathfrak{P}), \mathfrak{N}] \leq [\mho_1(\mathfrak{P}), \Omega_1(\mathfrak{P})] = 1.$$

Hence $\mho_1(\mathfrak{P}) \leq \mathbf{C}_{\mathfrak{G}}(\mathfrak{N}) = \mathfrak{N}$, and

$$\mho_1(\mathfrak{B}) \leq \mathfrak{B} \cap \mathfrak{N} = 1.$$

Thus $\mathfrak{B}$ is of exponent p. But $\mathfrak{P} = \mathfrak{B}\mathfrak{N}$, so $\mathfrak{P}$ is of exponent p, by III, 10.5. By 4.3b), $l \leq 2$. **q.e.d.**

4.9 Remarks. a) The inequality $l \leq 2$ for Fermat primes in 4.8b) is best possible, since, by 4.6, there exist groups of p-length 2 in which the Sylow p-subgroup is of exponent p and is therefore regular.

b) DADE [1] has given examples, for p odd, of p-soluble groups of p-length 3 in which no section is isomorphic to the Sylow p-subgroup of the symmetric group of degree p^3.

We now prove the reduction theorem for Burnside's problem.

4.10 Theorem (HALL and HIGMAN [1]). *For any positive integers m, d, let $\mathfrak{S}(m, d)$ denote the class of all finite soluble groups of exponent a divisor of m and with d generators. Also, let $s(m, d)$ be the maximum of the orders of the groups in $\mathfrak{S}(m, d)$.*

Suppose that p is a prime and p does not divide m. If $s(p^n, d)$ and $s(m, d)$ are finite for all d, $s(p^n m, d)$ is finite for all d.

Proof. Suppose that $\mathfrak{G} \in \mathfrak{S}(p^n m, d)$ and that $\mathfrak{P} \in S_p(\mathfrak{G})$. Then the exponent of $\mathfrak{P}$ is at most p^n. By 4.3 and 4.5, the p-length of $\mathfrak{G}$ is at most $2n$. Thus, it is sufficient to prove that there is a bound on the orders of the groups of p-length l in $\mathfrak{S}(p^n m, d)$ for all $l \geq 0$, $d \geq 1$. This we do by induction on l. For $l = 0$, any p'-group in $\mathfrak{S}(p^n m, d)$ lies in $\mathfrak{S}(m, d)$; the assertion thus follows since $s(m, d)$ is finite. If $l > 0$, there exist terms $\mathfrak{H}$, $\mathfrak{N}$ of the upper p-series of $\mathfrak{G}$ (VI, 6.1) such that $\mathfrak{H} \leq \mathfrak{N} \leq \mathfrak{G}$, the p-length of $\mathfrak{H}$ is $l - 1$, $\mathfrak{N}/\mathfrak{H}$ is a p-group and $\mathfrak{G}/\mathfrak{N}$ is a p'-group. Thus $\mathfrak{G}/\mathfrak{N} \in \mathfrak{S}(m, d)$, so $|\mathfrak{G} : \mathfrak{N}| \leq s$, where $s = s(m, d)$. By 1.14, $\mathfrak{N}$ is generated by

$$d' = 1 + (d - 1)s$$

elements. Thus $\mathfrak{N}/\mathfrak{H} \in \mathfrak{S}(p^n, d')$ and $|\mathfrak{N} : \mathfrak{H}| \leq s' = s(p^n, d')$. By 1.14, $\mathfrak{H} \in \mathfrak{S}(p^n m, 1 + (d' - 1)s')$. But $\mathfrak{H}$ is of p-length $l - 1$. By the inductive hypothesis there is a bound r on the orders of the groups of p-length $l - 1$ in $\mathfrak{S}(p^n m, 1 + (d' - 1)s')$. Thus $|\mathfrak{H}| \leq r$ and $|\mathfrak{G}| \leq ss'r$. **q.e.d.**

To apply this with $m = 4$, we need the solution of the Burnside problem for exponent 4.

4.11 Lemma. *Suppose that $\mathfrak{G} = \langle \mathfrak{H}, g \rangle$, where $\mathfrak{H}$ is finite and $g^2 \in \mathfrak{H}$. Suppose that given $h \in \mathfrak{H}$, there exist h', h'' in $\mathfrak{H}$ such that*

$$(hg)^2 = gh'gh''.$$

Then $\mathfrak{G}$ is finite.

Proof. Since $\mathfrak{G} = \langle \mathfrak{H}, g \rangle$ and $g^2 \in \mathfrak{H}$, any element x of $\mathfrak{G}$ is expressible in the form

$$x = h_1 g h_2 g \cdots h_{n-1} g h_n \quad (h_i \in \mathfrak{H}).$$

We choose such an expression with n minimal. Thus $h_i \neq 1$ $(2 \leq i \leq n-1)$.

If $2 \leq j \leq i < n$, define $h(j, i) \in \mathfrak{H}$ by induction on $i - j$; we put $h(i, i) = h_i$, and if $j < i$,

$$(1) \qquad h(j, i) = h_j h(j+1, i)^{-1}.$$

We now prove by induction on $i - j$ that

$$(2) \qquad x = h_1 g \cdots h_{j-1} g h(j, i) g h^*_{j+1} \cdots h^*_{n-1} g h^*_n \quad (2 \leq j \leq i < n)$$

for certain $h^*_k \in \mathfrak{H}$. This is obvious if $j = i$. If $i - j > 0$, we have

$$x = h_1 g \cdots h_j g h g h'_{j+2} \cdots h'_{n-1} g h'_n$$

with $h'_k \in \mathfrak{H}$ and $h = h(j+1, i)$, by the inductive hypothesis. Now there exist h', h'' in $\mathfrak{H}$ such that $(hg)^2 = gh'gh''$. Thus $ghg = h^{-1}gh'gh''$, so

$$x = h_1 g \cdots g h_j h^{-1} g h' g h'' h'_{j+2} g \cdots g h'_n.$$

By (1), $h_j h^{-1} = h(j, i)$, so (2) holds.

Consider the elements $h(2, 2), h(2, 3), \ldots, h(2, n-1)$.
If $n - 2 > |\mathfrak{H}|$, there exist k, l such that $2 \leq k < l < n$ and $h(2, k) = h(2, l)$. By (1), $h(j, k) = h(j, l)$ for $j = 2, \ldots, k$. Thus

$$h_k = h(k, k) = h(k, l) = h_k h(k+1, l)^{-1},$$

so $h(k+1, l) = 1$. Hence by (2),

$$x = h_1 g \cdots h_k g^2 h^*_{k+2} \cdots g h^*_n,$$

contrary to the definition of n. Thus $n - 2 \leq |\mathfrak{H}|$. Hence if $|\mathfrak{H}| = m$, $|\mathfrak{G}| \leq m^{m+2}$. **q.e.d.**

4.12 Theorem (Sanov [1]). *A finitely generated group of exponent 4 is finite.*

Proof. Suppose that $\mathfrak{G} = \langle x_1, \ldots, x_n \rangle$ is of exponent 4. We use induction on n. If $n = 1$, $|\mathfrak{G}| \leq 4$. If $n > 1$, $\mathfrak{H}_1 = \langle x_1, \ldots, x_{n-1} \rangle$ is finite by the inductive hypothesis. We apply 4.11 twice, first with $(\mathfrak{H}, g) = (\mathfrak{H}_1, x_n^2)$, then with $(\mathfrak{H}, g) = (\langle \mathfrak{H}_1, x_n^2 \rangle, x_n)$. In both cases, since $(hg)^4 = 1$,

$$(hg)^2 = (g^{-1}h^{-1})^2 = g(g^{-2}h^{-1}g^{-2})gh^{-1},$$

and $g^{-2}h^{-1}g^{-2} \in \mathfrak{H}$ since $g^2 \in \mathfrak{H}$. Thus $\langle \mathfrak{H}_1, x_n^2 \rangle$ and $\langle \mathfrak{H}_1, x_n^2, x_n \rangle = \mathfrak{G}$ are finite. **q.e.d.**

We shall now apply 4.10 to the *restricted Burnside problem* (VIII, 12.5), which may be formulated as follows. Given positive integers m, d, let $b(m, d)$ be the maximum of the orders of finite groups with d generators and exponent a divisor of m. We ask, what is $b(m, d)$? In particular, is $b(m, d)$ finite? In the proof of the next theorem, we shall use the theorem of Kostrikin (VIII, 12.4b)), which asserts that $b(p, d)$ is finite for any prime p.

4.13 Theorem (HALL and HIGMAN [1]). *If p is an odd prime, $b(2p, d)$ and $b(4p, d)$ are finite for all $d \geq 1$.*

Proof. By a theorem of Burnside (V, 7.3), any finite group of exponent a divisor of $4p$ is soluble. Thus $b(4p, d) = s(4p, d)$, in the notation of 4.10. By 4.12, $s(4, d)$ is finite, and by Kostrikin's theorem, $s(p, d)$ is finite. The assertion thus follows from 4.10. **q.e.d.**

4.14 Remark. Taking $p = 3$, the value of $b(6, d)$ was given in 1.16. It is not possible, however, to give a good estimate for $b(12, d)$, since no good estimate for $b(4, d)$ is known. The proof of Sanov's theorem (Lemma 4.11) gives a very poor estimate for $b(4, 2)$ already; in fact, $b(4, 2) = 2^{12}$ (III, 6.7).

The reduction theorem (4.10) effectively reduces the restricted Burnside problem to questions on p-groups and simple groups.

4.15 Theorem. *Suppose that $m = p_1^{n_1} \cdots p_r^{n_r}$, where $p_1, \ldots, p_r$ are distinct primes. Suppose that*

(1) *for every positive integer d, $b(p_i^{n_i}, d)$ is finite $(i = 1, \ldots, r)$; and*

(2) *the number of isomorphism types of finite simple groups of exponent a divisor of m is finite.*

Then $b(m, d)$ is finite for every $d \geq 1$.

Proof. Let $\mathfrak{B}(m, d)$ be the class of finite groups with d generators and of exponent a divisor of m. We prove by induction on m that $\mathfrak{B}(m, d)$ contains only a finite number of non-isomorphic groups. If m is a prime-power, this follows at once from (1). Suppose that m is not a prime-power.

a) $\mathfrak{B}(m, d)$ contains only a finite number of non-isomorphic groups $\mathfrak{G}$ having an insoluble minimal normal subgroup $\mathfrak{N}$ for which $\mathbf{C}_{\mathfrak{G}}(\mathfrak{N}) = 1$.

For such a group $\mathfrak{G}$, $\mathfrak{N} = \mathfrak{K}_1 \times \cdots \times \mathfrak{K}_r$, where $\mathfrak{K}_1, \ldots, \mathfrak{K}_r$ are isomorphic simple groups and $\mathfrak{K}_i^g \in \{\mathfrak{K}_1, \ldots, \mathfrak{K}_r\}$ for all $g \in \mathfrak{G}$, $1 \leq i \leq r$ (I, 9.12). Thus a permutation representation ρ of $\mathfrak{G}$ on $\{\mathfrak{K}_1, \ldots, \mathfrak{K}_r\}$ is defined by $\mathfrak{K}_i \rho(g) = \mathfrak{K}_i^g$. The stabiliser of $\mathfrak{K}_1$ is $\mathbf{N}_{\mathfrak{G}}(\mathfrak{K}_1)$. Also ρ is transitive, since $\mathfrak{N}$ is a minimal normal subgroup. Thus $|\mathfrak{G} : \mathbf{N}_{\mathfrak{G}}(\mathfrak{K}_1)| = r$.

Let p be a prime divisor of $|\mathfrak{K}_1|$ and write $m = p^n m'$, where $(m', p) = 1$. Suppose that g is an element of $\mathfrak{G}$ of order p^n. Then $g \in \mathfrak{S}$ for some $\mathfrak{S} \in S_p(\mathfrak{G})$ and $\mathfrak{S} \cap \mathfrak{K}_1 \in S_p(\mathfrak{K}_1)$. Hence $\mathfrak{S} \cap \mathfrak{K}_1 \neq 1$; choose $x \in \mathfrak{S} \cap \mathfrak{K}_1$ with $x \neq 1$. Then $(gx)^{p^n} = 1$, since the exponent of $\mathfrak{G}$ divides m. But

$$(gx)^{p^n} = g^{p^n} x^{g^{p^n-1}} x^{g^{p^n-2}} \cdots x^g x.$$

Hence

$$x^{g^{p^n-1}} x^{g^{p^n-2}} \cdots x^g x = 1.$$

But $x^{g^i} \in \mathfrak{K}_1 \rho(g)^i$ $(i \geq 0)$. It follows that $\mathfrak{K}_1, \mathfrak{K}_1 \rho(g), \ldots, \mathfrak{K}_1 \rho(g)^{p^n-1}$ cannot be distinct factors of $\mathfrak{N}$, since $x \neq 1$. Thus the length of the orbit of $\rho(g)$ containing $\mathfrak{K}_1$ is less than p^n. Since g is a p-element, $\mathfrak{K}_1 \rho(g)^{p^{n-1}} = \mathfrak{K}_1$. Thus $g^{p^{n-1}} \in \mathbf{N}_{\mathfrak{G}}(\mathfrak{K}_1)$ for every p-element g in $\mathfrak{G}$.

Let $\mathfrak{M}$ be the group generated by all $g^{p^{n-1}}$, as g runs through the p-elements of $\mathfrak{G}$. Thus $\mathfrak{M} \leq \mathbf{N}_{\mathfrak{G}}(\mathfrak{K}_1)$ and $\mathfrak{M} \trianglelefteq \mathfrak{G}$. But $\mathfrak{G}/\mathfrak{M} \in \mathfrak{B}(p^{n-1}m', d)$, and by the inductive hypothesis, $b(p^{n-1}m', d)$ is finite. Thus

$$b(p^{n-1}m', d) \geq |\mathfrak{G}/\mathfrak{M}| \geq |\mathfrak{G} : \mathbf{N}_{\mathfrak{G}}(\mathfrak{K}_1)| = r.$$

By (2), there is an integer n_0 such that the order of every finite simple group of exponent a divisor of m is at most n_0. Then $|\mathfrak{K}_1| \leq n_0$, and

$$|\mathfrak{N}| = |\mathfrak{K}_1|^r \leq n_0^{b(p^{n-1}m', d)}.$$

Since $\mathbf{C}_{\mathfrak{G}}(\mathfrak{N}) = 1$, $\mathfrak{G}$ is isomorphic to a subgroup of the group of automorphisms of $\mathfrak{N}$. The number of possibilities for $\mathfrak{G}$ is therefore finite, and a) is proved.

b) $\mathfrak{B}(m, d)$ contains only a finite number of non-isomorphic groups which have no non-identity soluble normal subgroup.

By a), there exists a finite number of groups $\mathfrak{H}_1, \ldots, \mathfrak{H}_k$ such that every group $\mathfrak{G} \in \mathfrak{B}(m, d)$ having an insoluble minimal normal subgroup $\mathfrak{N}$ for which $\mathbf{C}_{\mathfrak{G}}(\mathfrak{N}) = 1$ is isomorphic to one of the $\mathfrak{H}_i$. Let h be the maximum of the $|\mathfrak{H}_i|$.

Suppose that $\mathfrak{H}$ is any group with d generators. If $\mathfrak{H}/\mathfrak{M} \cong \mathfrak{H}_j$, there is an epimorphism θ of $\mathfrak{H}$ onto $\mathfrak{H}_j$ and $\mathfrak{M} = \ker\theta$. Now if $\mathfrak{H} = \langle y_1, \ldots, y_d \rangle$, there are at most h possibilities for $y_i\theta$, and since θ is determined by $y_1\theta, \ldots, y_d\theta$, there are at most h^d possibilities for θ. Hence there are at most h^d possibilities for $\mathfrak{M}$. Thus there are at most kh^d normal subgroups $\mathfrak{L}$ of $\mathfrak{H}$ for which $\mathfrak{H}/\mathfrak{L}$ is isomorphic to one of the $\mathfrak{H}_i$.

Now suppose that $\mathfrak{G} \in \mathfrak{B}(m, d)$, $\mathfrak{G} \neq 1$ and $\mathfrak{G}$ has no non-identity soluble normal subgroup. Let $\mathfrak{N}_1, \ldots, \mathfrak{N}_r$ be the distinct minimal normal subgroups of $\mathfrak{G}$, and let $\mathfrak{C}_i = \mathbf{C}_{\mathfrak{G}}(\mathfrak{N}_i)$. Since $\mathfrak{N}_i$ is insoluble, $\mathfrak{C}_i \cap \mathfrak{N}_i = 1$. Thus $\mathfrak{N}_i\mathfrak{C}_i/\mathfrak{C}_i$ is an insoluble minimal normal subgroup of $\mathfrak{G}/\mathfrak{C}_i$. If $g\mathfrak{C}_i$ commutes with $x\mathfrak{C}_i$ for all $x \in \mathfrak{N}_i$, then $[g, x] \in \mathfrak{N}_i \cap \mathfrak{C}_i = 1$ for all $x \in \mathfrak{N}_i$, whence $g \in \mathbf{C}_{\mathfrak{G}}(\mathfrak{N}_i) = \mathfrak{C}_i$ and $g\mathfrak{C}_i = 1$. Thus $\mathbf{C}_{\mathfrak{G}/\mathfrak{C}_i}(\mathfrak{N}_i\mathfrak{C}_i/\mathfrak{C}_i) = 1$. Hence $\mathfrak{G}/\mathfrak{C}_i$ is isomorphic to one of the $\mathfrak{H}_j$.

It follows that all the $\mathfrak{C}_i$ $(i = 1, \ldots, r)$ occur among at most kh^d normal subgroups of $\mathfrak{G}$ of index at most h. But $\bigcap_i \mathfrak{C}_i = 1$, for otherwise, there would exist $\mathfrak{N}_j$ such that $\mathfrak{N}_j \leq \bigcap_i \mathfrak{C}_i \leq \mathfrak{C}_j$. Therefore $\mathfrak{G}$ is isomorphic to a subgroup of the direct product of kh^d subgroups all of order at most h. The number of possibilities for $\mathfrak{G}$ is therefore finite.

c) $\mathfrak{B}(m, d)$ contains only a finite number of non-isomorphic groups.

By b), there exists an integer n_1 such that if $\mathfrak{G} \in \mathfrak{B}(m, d)$ and $\mathfrak{G}$ has no non-identity soluble normal subgroup, $|\mathfrak{G}| \leq n_1$.

Now suppose $\mathfrak{G} \in \mathfrak{B}(m, d)$. Let $\mathfrak{K}$ be the maximal soluble normal subgroup of $\mathfrak{G}$. Then $|\mathfrak{G}/\mathfrak{K}| \leq n_1$. By 1.14, $\mathfrak{K}$ can be generated by $d' = 1 + (d - 1)n_1$ elements, so $\mathfrak{K} \in \mathfrak{S}(m, d')$, in the notation of 4.10. But by 4.10, $s(m, d')$ is finite and $|\mathfrak{G}| \leq n_1 s(m, d')$. Thus there is only a finite number of possibilities for $\mathfrak{G}$. **q.e.d.**

The hypotheses (1) and (2) of 4.15 are satisfied for square-free exponents. To prove this, we need (i) the theorem of Kostrikin on the finiteness of $b(p, d)$, (ii) the theorem of Feit and Thompson on the solubility of groups of odd order, and (iii) the following theorem of WALTER [2]; (see also BENDER [4]).

4.16 Theorem (WALTER). *Suppose that the Sylow 2-subgroups of the finite group $\mathfrak{G}$ are Abelian. Then $\mathbf{O}^{2'}(\mathfrak{G}/\mathbf{O}_{2'}(\mathfrak{G}))$ is the direct product of a 2-group and some of the following simple groups.*

(1) *$PSL(2, 2^n)$, where $n > 1$.*

(2) *$PSL(2, p^n)$, where $p^n \equiv 3$ or 5 (8) and $p^n > 3$.*

(3) *The first simple group* $\mathfrak{J}_1$ *of Janko.*

(4) *A simple group* $\mathfrak{R}$ *of order* $(r^3+1)r^3(r-1)$, *where* $r = 3^{2n+1}$ $(n > 0)$, *in which the centralizer of any involution t is* $\langle t \rangle \times \mathfrak{K}$, *where* $\mathfrak{K}$ *is isomorphic to* $PSL(2, 3^{2n+1})$.

The groups arising in (3) and (4) will be discussed in XI, § 13.

4.17 Theorem. *If m is square-free,* $b(m, d)$ *is finite.*

Proof. $b(p, d)$ is finite for every prime p, by Kostrikin's theorem; thus, by 4.15, it is sufficient to show that there is only a finite number of simple groups of exponent a divisor of m, to within isomorphism.

Let $\mathfrak{G}$ be a simple group of exponent m. Since m is square-free, 4 does not divide m and the Sylow 2-subgroups of $\mathfrak{G}$ are of exponent at most 2. Thus the Sylow 2-subgroups of $\mathfrak{G}$ are Abelian. It follows from 4.16 that $\mathfrak{G}$ is isomorphic to one of the simple groups mentioned in 4.16.

However, only a finite number of these groups are of exponent a divisor of m, for by II, 8.3, $PSL(2, 2^n)$ has an element of order $2^n - 1$ and $PSL(2, p^n)$ (p odd) has an element of order $\frac{1}{2}(p^n - 1)$. **q.e.d.**

We conclude by proving that the Burnside problem has an affirmative answer for groups of linear transformations of a finite-dimensional vector space.

4.18 Theorem (BURNSIDE). a) *Let* K *be an algebraically closed field of arbitrary characteristic and let* $\mathfrak{G}$ *be an irreducible subgroup of* $GL(n, \mathsf{K})$. *If* $x^m = 1$ *for all* $x \in \mathfrak{G}$, $|\mathfrak{G}| \leq m^{n^3}$.

b) *Let* K *be any field and let* $\mathfrak{G}$ *be a finitely generated subgroup of* $GL(n, \mathsf{K})$. *If* $x^m = 1$ *for all* $x \in \mathfrak{G}$, $\mathfrak{G}$ *is finite.*

Proof. a) By V, 5.14, $\mathfrak{G}$ contains n^2 linearly independent elements $g^{(1)}, \ldots, g^{(n^2)}$. Write $g^{(k)} = (g_{ij}^{(k)})$. Then if $x \in GL(n, \mathsf{K})$ and $x = (x_{ij})$,

$$\operatorname{tr} g^{(k)}x = \sum_{i,j=1}^{n} g_{ij}^{(k)} x_{ji} \quad (k = 1, \ldots, n^2).$$

Now the coefficients $(g_{ij}^{(k)})$ in these n^2 equations form a non-singular matrix, since $g^{(1)}, \ldots, g^{(n^2)}$ are linearly independent. Hence there exists at most one $x \in GL(n, \mathsf{K})$ for which $\operatorname{tr} g^{(k)}x$ takes a preassigned value for all $k = 1, \ldots, n^2$. But if $x \in \mathfrak{G}$, $(g^{(k)}x)^m = 1$, the eigen-values of $g^{(k)}x$ are m-th roots of unity and there are at most m^n possible values for $\operatorname{tr} g^{(k)}x$ $(k = 1, \ldots, n^2)$. Hence there are at most m^{n^3} possibilities for x.

b) We may suppose that K is algebraically closed. We proceed by induction on n. If $\mathfrak{G}$ is irreducible, the assertion follows from a). If $\mathfrak{G}$ is reducible, we may suppose that every element of $\mathfrak{G}$ is of the form

$$g = \begin{pmatrix} g_1 & 0 \\ \bar{g} & g_2 \end{pmatrix},$$

where g_1 runs through a finitely generated subgroup $\mathfrak{G}_1$ of $GL(r, \mathsf{K})$ $(1 \leq r < n)$ and g_2 runs through a finitely generated subgroup $\mathfrak{G}_2$ of $GL(n-r, \mathsf{K})$. By the inductive hypothesis, $\mathfrak{G}_1$ and $\mathfrak{G}_2$ are finite. Now

$$g \rightarrow \begin{pmatrix} g_1 & 0 \\ 0 & g_2 \end{pmatrix}$$

is a homomorphism θ of $\mathfrak{G}$ into the direct product $\mathfrak{G}_1 \times \mathfrak{G}_2$. If $\mathfrak{N} = \ker \theta$, $\mathfrak{G}/\mathfrak{N}$ is finite. By 1.14, $\mathfrak{N}$ is finitely generated. But also $\mathfrak{N}$ is Abelian. Since $x^m = 1$ for all $x \in \mathfrak{N}$, $\mathfrak{N}$ is finite. Thus $\mathfrak{G}$ is finite. **q.e.d.**

§ 5. Other Consequences of Theorem B

5.1 Lemma. *Suppose that $\mathfrak{G}$ is p-soluble and $\mathfrak{P} \in S_p(\mathfrak{G})$.*

a) *Suppose that $\mathfrak{H} \leq \mathfrak{P}$ and $\mathfrak{H} \not\leq \mathbf{O}_{p',p}(\mathfrak{G})$. If $p > 3$, there exist $x \in \mathfrak{H}$ and $y \in \mathfrak{P}$ such that $[y, x, x, x] \neq 1$. If $p = 3$ and the Sylow 2-subgroups of $\mathfrak{G}$ are Abelian, there exist $x \in \mathfrak{H}$ and $y \in \mathfrak{P}$ such that $[y, x, x] \neq 1$.*

b) *Suppose that $\mathfrak{N} \trianglelefteq \mathfrak{P}$. If $p > 3$ and the class of $\mathfrak{N}$ is at most 2, $\mathfrak{N} \leq \mathbf{O}_{p',p}(\mathfrak{G})$. If $p = 3$, $\mathfrak{N}$ is Abelian and the Sylow 2-subgroups of $\mathfrak{G}$ are Abelian, then $\mathfrak{N} \leq \mathbf{O}_{p',p}(\mathfrak{G})$.*

Proof. a) Write

$$\mathfrak{U}/\mathbf{O}_{p'}(\mathfrak{G}) = \Phi(\mathbf{O}_{p',p}(\mathfrak{G})/\mathbf{O}_{p'}(\mathfrak{G})),$$

and let $\rho = (\mathfrak{G}$ on $\mathbf{O}_{p',p}(\mathfrak{G})/\mathfrak{U})$. By 1.4 and 1.6, $\ker \rho = \mathbf{O}_{p',p}(\mathfrak{G})$, and by 1.8,

$$(v\mathfrak{U})(\rho(x) - 1)^m = [v, \underbrace{x, \ldots, x}_{m}]\mathfrak{U} \quad (v \in \mathbf{O}_{p',p}(\mathfrak{G}),\ x \in \mathfrak{P}).$$

Choose $x \in \mathfrak{H}$, $x \notin \mathbf{O}_{p',p}(\mathfrak{G})$ and let $(t-1)^r$ be the minimum polynomial of $\rho(x)$. By 2.9, $r \geq p-1$ and, if $p = 3$ and the Sylow 2-subgroups of $\mathfrak{G}$ are Abelian, $r \geq 3$. Thus if $p > 3$, $r \geq 4$, $(\rho(x)-1)^3 \neq 0$ and there exists $v \in \mathbf{O}_{p',p}(\mathfrak{G})$ such that $(v\mathfrak{U})(\rho(x)-1)^3 \neq 0$. If $p = 3$ and the Sylow 2-subgroups are Abelian, there exists $v \in \mathbf{O}_{p',p}(\mathfrak{G})$ such that $(v\mathfrak{U})(\rho(x)-1)^2 \neq 0$. Since $\mathbf{O}_{p',p}(\mathfrak{G}) = \mathfrak{U}(\mathfrak{P} \cap \mathbf{O}_{p',p}(\mathfrak{G}))$, $v\mathfrak{U} = y\mathfrak{U}$ for some $y \in \mathfrak{P}$; thus $[y, x, x, x] \neq 1$ in the first case and $[y, x, x] \neq 1$ in the second.

b) Suppose that $\mathfrak{N} \not\leq \mathbf{O}_{p',p}(\mathfrak{G})$. Since $[\mathfrak{P}, \mathfrak{N}] \leq \mathfrak{N}$, it follows from a) that $[\mathfrak{N}, \mathfrak{N}, \mathfrak{N}] \neq 1$ if $p > 3$ and $[\mathfrak{N}, \mathfrak{N}] \neq 1$ if $p = 3$ and the Sylow 2-subgroups are Abelian. This is contrary to the hypothesis. **q.e.d.**

In the semi-direct product $\mathfrak{G}$ of $SL(2, 3)$ with an elementary Abelian group $\mathfrak{E}$ of order 9, there are Abelian normal subgroups of a Sylow 3-subgroup which do not lie in $\mathbf{O}_{3',3}(\mathfrak{G}) = \mathfrak{E}$. Thus the condition on the Sylow 2-subgroups in 5.1b) cannot be dropped. However, we have the following.

5.2 Theorem. *Suppose that $\mathfrak{G}$ is a p-soluble group, that $\mathfrak{P} \in S_p(\mathfrak{G})$ and that $\mathfrak{A}$ is a cyclic normal subgroup of $\mathfrak{P}$. Suppose further that either*

(1) *p is odd, or*

(2) *no section of $\mathfrak{G}$ is isomorphic to the symmetric group $\mathfrak{S}_4$.*

Then $\mathfrak{A} \leq \mathbf{O}_{p',p}(\mathfrak{G})$.

Proof. Suppose that $\mathfrak{G}$ is a counterexample of minimal order.

Let $\mathfrak{R}$ be the normal closure of $\mathfrak{A}$ in $\mathfrak{G}$. Then $\mathbf{O}_{p',p}(\mathfrak{R}) \trianglelefteq \mathfrak{G}$, so $\mathbf{O}_{p',p}(\mathfrak{R}) \leq \mathbf{O}_{p',p}(\mathfrak{G})$. Hence $\mathfrak{A} \not\leq \mathbf{O}_{p',p}(\mathfrak{R})$. But $\mathfrak{A} \trianglelefteq \mathfrak{P} \cap \mathfrak{R} \in S_p(\mathfrak{R})$, so by minimality of $|\mathfrak{G}|$, $\mathfrak{R} = \mathfrak{G}$.

Since $\mathfrak{A} \not\leq \mathbf{O}_{p',p}(\mathfrak{G})$, $\mathbf{O}_{p',p}(\mathfrak{G}) \neq \mathfrak{G}$ and $\mathfrak{G}$ is not p-nilpotent. It follows that $\mathfrak{G}$ is not of p-length 1. For otherwise, $\mathfrak{G}$ would have a proper normal subgroup $\mathfrak{L}$ of index prime to p; but then $\mathfrak{A} \leq \mathfrak{L} \trianglelefteq \mathfrak{G}$ and so $\mathfrak{L} \geq \mathfrak{R} = \mathfrak{G}$. But if $1 < \mathfrak{N} \trianglelefteq \mathfrak{G}$, then $\mathfrak{A}\mathfrak{N}/\mathfrak{N} \leq \mathbf{O}_{p',p}(\mathfrak{G}/\mathfrak{N})$ by minimality of $|\mathfrak{G}|$. Hence $\mathfrak{G}/\mathfrak{N} = \mathfrak{R}\mathfrak{N}/\mathfrak{N} = \mathbf{O}_{p',p}(\mathfrak{G}/\mathfrak{N})$, so the p-length of $\mathfrak{G}/\mathfrak{N}$ is 1. By VI, 6.9, $\mathbf{O}_{p'}(\mathfrak{G}) = 1$ and $\mathfrak{G}$ has just one minimal normal subgroup $\mathfrak{M}$, which is elementary Abelian; further $\mathbf{O}_p(\mathfrak{G}) = \mathfrak{M}$, $\mathbf{C}_{\mathfrak{G}}(\mathfrak{M}) = \mathfrak{M}$ and there exists a subgroup $\mathfrak{H}$ of $\mathfrak{G}$ such that $\mathfrak{G} = \mathfrak{M}\mathfrak{H}$ and $\mathfrak{M} \cap \mathfrak{H} = 1$. Also $\mathfrak{G}/\mathfrak{M}$ is p-nilpotent. Let $\mathfrak{K}/\mathfrak{M} = \mathbf{O}_{p'}(\mathfrak{G}/\mathfrak{M})$; thus $\mathfrak{A}\mathfrak{K} \trianglelefteq \mathfrak{P}\mathfrak{K} = \mathfrak{G}$ and, since $\mathfrak{R} = \mathfrak{G}$, $\mathfrak{A}\mathfrak{K} = \mathfrak{G}$. Hence
$\mathfrak{P} = \mathfrak{A}(\mathfrak{K} \cap \mathfrak{P}) = \mathfrak{A}\mathfrak{M}$. Since $\mathfrak{A} \trianglelefteq \mathfrak{P}$ and $\mathfrak{A}$, $\mathfrak{M}$ are both Abelian, it follows that $\mathfrak{P}' \leq [\mathfrak{A}, \mathfrak{M}] \leq \mathfrak{A} \cap \mathfrak{M}$. Since $\mathfrak{G}$ is not of p-length 1, $\mathfrak{P}' \neq 1$. But $|\mathfrak{A} \cap \mathfrak{M}| \leq p$, since $\mathfrak{A}$ is cyclic; thus $\mathfrak{P}' = \mathfrak{A} \cap \mathfrak{M}$ is of order p. In particular, the class of $\mathfrak{P}$ is 2.

If p is odd, $\mathfrak{P}$ is a regular p-group and the elements of $\mathfrak{P}$ of order at most p form a characteristic subgroup $\mathfrak{P}_1$ of $\mathfrak{P}$. Since $\mathfrak{M} \leq \mathfrak{P}_1$, $\mathfrak{P}_1 = (\mathfrak{A} \cap \mathfrak{P}_1)\mathfrak{M}$. But $\mathfrak{A} \cap \mathfrak{P}_1 = \mathfrak{A} \cap \mathfrak{M}$, since $|\mathfrak{A} \cap \mathfrak{M}| = p$. Thus $\mathfrak{P}_1 = \mathfrak{M}$. Hence $\mathfrak{P}_1 \cap \mathfrak{H} = \mathfrak{M} \cap \mathfrak{H} = 1$. Since the elements of $\mathfrak{P} \cap \mathfrak{H}$ of order p lie in $\mathfrak{P}_1 \cap \mathfrak{H}$, it follows that $\mathfrak{P} \cap \mathfrak{H} = 1$. Since $\mathfrak{G} = \mathfrak{M}\mathfrak{H}$, $\mathfrak{P} = \mathfrak{M}(\mathfrak{P} \cap \mathfrak{H}) = \mathfrak{M}$ and $\mathfrak{A} \leq \mathfrak{M}$, a contradiction.

Suppose then that $p = 2$. Let $\mathfrak{K}_0/\mathfrak{K}$ be the subgroup of $\mathfrak{G}/\mathfrak{K}$ of order 2. Then $\mathbf{O}_{2',2}(\mathfrak{K}_0) \trianglelefteq \mathfrak{G}$ and $\mathbf{O}_{2',2}(\mathfrak{K}_0) \leq \mathbf{O}_{2',2}(\mathfrak{G}) = \mathfrak{M}$. But $\mathfrak{K}_0 = (\mathfrak{A} \cap \mathfrak{K}_0)\mathfrak{K}$, so $\mathfrak{A} \cap \mathfrak{K}_0 \nleq \mathbf{O}_{2',2}(\mathfrak{K}_0)$. Since $\mathfrak{A} \cap \mathfrak{K}_0 \trianglelefteq \mathfrak{P} \cap \mathfrak{K}_0 \in S_2(\mathfrak{K}_0)$, it follows from the minimality of $|\mathfrak{G}|$ that $\mathfrak{K}_0 = \mathfrak{G}$. Thus $|\mathfrak{G} : \mathfrak{K}| = |\mathfrak{P} : \mathfrak{M}| = 2$.

Of course, $\mathfrak{P}$ is not normal in $\mathfrak{G}$; let $\mathfrak{P}_1$ be another Sylow p-subgroup of $\mathfrak{G}$ and let $\mathfrak{U} = \langle \mathfrak{P}, \mathfrak{P}_1 \rangle$. Since $\mathfrak{U} \geq \mathfrak{M} = \mathbf{C}_{\mathfrak{G}}(\mathfrak{M})$, $\mathbf{O}_{2'}(\mathfrak{U}) = 1$. Also $\mathfrak{M} \leq \mathbf{O}_2(\mathfrak{U}) \leq \mathfrak{P} \cap \mathfrak{P}_1$, and since $|\mathfrak{P} : \mathfrak{M}| = 2$, $\mathfrak{M} = \mathbf{O}_2(\mathfrak{U})$. Thus $\mathfrak{A} \nleq \mathbf{O}_{2',2}(\mathfrak{U})$, so by minimality of $|\mathfrak{G}|$, $\mathfrak{G} = \mathfrak{U} = \langle \mathfrak{P}, \mathfrak{P}_1 \rangle$. Thus $\mathbf{Z}(\mathfrak{P}) \cap \mathbf{Z}(\mathfrak{P}_1) \leq \mathbf{Z}(\mathfrak{G})$. Since $\mathfrak{M}$ is the only minimal normal subgroup of $\mathfrak{G}$ and $\mathfrak{M} = \mathbf{C}_{\mathfrak{G}}(\mathfrak{M})$, $\mathbf{Z}(\mathfrak{G}) = 1$. Thus $\mathbf{Z}(\mathfrak{P}) \cap \mathbf{Z}(\mathfrak{P}_1) = 1$. But since $\mathfrak{G} = \mathfrak{M}\mathfrak{H}$, $\mathfrak{P} = \mathfrak{M}\mathfrak{D}$, where $\mathfrak{D} = \mathfrak{P} \cap \mathfrak{H}$ and $\mathbf{Z}(\mathfrak{P}) = \mathbf{C}_{\mathfrak{M}}(\mathfrak{D})$. Since $\mathfrak{D} \cap \mathfrak{M} = \mathfrak{P} \cap \mathfrak{H} \cap \mathfrak{M} = 1$, $|\mathfrak{D}| = 2$. Since also $|\mathfrak{P}'| = 2$, $|\mathfrak{M} : \mathbf{C}_{\mathfrak{M}}(\mathfrak{D})| = 2$. Thus $|\mathfrak{M} : \mathbf{Z}(\mathfrak{P})| = 2$. Similarly $|\mathfrak{M} : \mathbf{Z}(\mathfrak{P}_1)| = 2$, so $|\mathfrak{M}| = |\mathfrak{M} : \mathbf{Z}(\mathfrak{P}) \cap \mathbf{Z}(\mathfrak{P}_1)| \leq 4$. Since $\mathfrak{H}$ is isomorphic to a subgroup of $\mathbf{Aut}\,\mathfrak{M}$, $\mathfrak{H} \cong \mathfrak{S}_3$ and $\mathfrak{G} \cong \mathfrak{S}_4$. This is contrary to the hypothesis. **q.e.d.**

Our next result (Theorem 5.4) will be deduced from 5.1 for $p > 3$, but for $p = 3$, we need the following replacement of it.

5.3 Lemma. *Suppose that $\mathfrak{G}$ is 3-soluble and $\mathfrak{P} \in S_3(\mathfrak{G})$. If $\mathfrak{N} \trianglelefteq \mathfrak{P}$ and $\mathfrak{N}' \nleq \mathbf{O}_{3',3}(\mathfrak{G})$, then $[\mathfrak{P}, \mathfrak{N}, \mathfrak{N}, \mathfrak{N}'] \neq 1$.*

Proof. Let $\mathfrak{U}/\mathbf{O}_{3'}(\mathfrak{G}) = \Phi(\mathbf{O}_{3',3}(\mathfrak{G})/\mathbf{O}_{3'}(\mathfrak{G}))$ and let $\rho = (\mathfrak{G}$ on $\mathbf{O}_{3',3}(\mathfrak{G})/\mathfrak{U})$. By 1.4 and 1.6, $\ker \rho = \mathbf{O}_{3',3}(\mathfrak{G})$, and by 1.8,

$$(v\mathfrak{U})(\rho(x_1) - 1) \cdots (\rho(x_n) - 1)$$
$$= [v, x_1, \ldots, x_n]\mathfrak{U} \quad (v \in \mathbf{O}_{3',3}(\mathfrak{G}),\ x_i \in \mathfrak{P}).$$

Since $\mathfrak{N}' \nleq \mathbf{O}_{3',3}(\mathfrak{G})$, there exist y_1, y_2 in $\mathfrak{N}$ such that $z = [y_1, y_2] \notin \mathbf{O}_{3',3}(\mathfrak{G})$. Thus $[\rho(y_1), \rho(y_2)] = \rho(z) \neq 1$. By 3.8, at least one of $(\rho(z) - 1)^2$, $(\rho(y_1) - 1)^2(\rho(z) - 1)$ or $(\rho(y_2) - 1)^2(\rho(z) - 1)$ is non-zero.

If $(\rho(z) - 1)^2 \neq 0$, there exists $u \in \mathfrak{P} \cap \mathbf{O}_{3',3}(\mathfrak{G})$ such that

$$[u, z, z]\mathfrak{U} = (u\mathfrak{U})(\rho(z) - 1)^2 \neq 1,$$

so $[\mathfrak{P}, \mathfrak{N}', \mathfrak{N}'] \neq 1$. But

$$[\mathfrak{P}, \mathfrak{N}'] = [\mathfrak{N}', \mathfrak{P}] = [\mathfrak{N}, \mathfrak{N}, \mathfrak{P}]$$
$$\leq [\mathfrak{N}, \mathfrak{P}, \mathfrak{N}][\mathfrak{P}, \mathfrak{N}, \mathfrak{N}] = [\mathfrak{P}, \mathfrak{N}, \mathfrak{N}],$$

by III, 1.10b). Thus $[\mathfrak{P}, \mathfrak{N}', \mathfrak{N}'] \leq [\mathfrak{P}, \mathfrak{N}, \mathfrak{N}, \mathfrak{N}']$ and $[\mathfrak{P}, \mathfrak{N}, \mathfrak{N}, \mathfrak{N}'] \neq 1$.

Otherwise, there exists $v \in \mathfrak{P} \cap \mathbf{O}_{3',3}(\mathfrak{G})$ such that

$$[v, y_i, y_i, z]\mathfrak{U} = (v\mathfrak{U})(\rho(y_i) - 1)^2(\rho(z) - 1) \neq 1$$

for $i = 1$ or $i = 2$. Thus $[v, y_i, y_i, z] \neq 1$ and $[\mathfrak{P}, \mathfrak{N}, \mathfrak{N}, \mathfrak{N}'] \neq 1$.

q.e.d.

5.4 Theorem. *Suppose that $\mathfrak{G}$ is a p-soluble group, where $p \geq 3$, and that $\mathfrak{P} \in S_p(\mathfrak{G})$.*

a) *If $\mathfrak{P} \neq 1$, the derived length of $\mathfrak{P}$ exceeds that of $\mathfrak{P}\mathbf{O}_{p',p}(\mathfrak{G})/\mathbf{O}_{p',p}(\mathfrak{G})$.*

b) *The p-length of $\mathfrak{G}$ is not greater than the derived length of $\mathfrak{P}$.*

Proof. a) Let k be the derived length of $\mathfrak{P}$. Since $\mathfrak{P} \neq 1$, $k > 0$.

Suppose that $p > 3$. By 5.1b), $\mathfrak{P}^{(k-1)} \leq \mathbf{O}_{p',p}(\mathfrak{G})$, since $\mathfrak{P}^{(k-1)}$ is a normal Abelian subgroup of $\mathfrak{P}$, so $(\mathfrak{P}\mathbf{O}_{p',p}(\mathfrak{G})/\mathbf{O}_{p',p}(\mathfrak{G}))^{(k-1)} = 1$. Thus the derived length of $\mathfrak{P}\mathbf{O}_{p',p}(\mathfrak{G})/\mathbf{O}_{p',p}(\mathfrak{G})$ is less than k.

If $p = 3$, we observe that for $k = 1$, the assertion follows from VI, 6.6a). If $k > 1$, put $\mathfrak{N} = \mathfrak{P}^{(k-2)}$. Then $[\mathfrak{P}, \mathfrak{N}] \leq \mathfrak{N}$, $[\mathfrak{P}, \mathfrak{N}, \mathfrak{N}] \leq \mathfrak{N}'$ and

$$[\mathfrak{P}, \mathfrak{N}, \mathfrak{N}, \mathfrak{N}'] \leq \mathfrak{N}'' = \mathfrak{P}^{(k)} = 1.$$

By 5.3, $\mathfrak{N}' \leq \mathbf{O}_{3',3}(\mathfrak{G})$; that is, $\mathfrak{P}^{(k-1)} \leq \mathbf{O}_{3',3}(\mathfrak{G})$. Hence the derived length of $\mathfrak{P}\mathbf{O}_{3',3}(\mathfrak{G})/\mathbf{O}_{3',3}(\mathfrak{G})$ is less than k.

b) This is proved by induction on the p-length l of $\mathfrak{G}$. If $l \leq 1$, it is clear. If $l > 1$, the p-length of $\mathfrak{G}/\mathbf{O}_{p',p}(\mathfrak{G})$ is $l - 1$. Thus, by the inductive hypothesis, the derived length of $\mathfrak{P}\mathbf{O}_{p',p}(\mathfrak{G})/\mathbf{O}_{p',p}(\mathfrak{G})$ is at least $l - 1$. Hence by a), the derived length of $\mathfrak{P}$ is at least l. **q.e.d.**

The example given in 4.7a) shows that the inequality in 5.4b) is best possible, for the derived length of the wreath product of n cyclic groups of order p is n (III, 15.3d)). BERGER and GROSS [1] have shown that if the derived length of the Sylow 2-subgroup of a soluble group $\mathfrak{G}$ is $d \geq 2$, the 2-length of $\mathfrak{G}$ is at most $2d - 2$. It is conjectured that the 2-length is at most d, as in the case when p is odd.

Arguments along the same lines give inequalities involving other invariants of $\mathfrak{P}$.

5.5 Theorem. *Let $\mathfrak{G}$ be a p-soluble group of p-length l, and suppose that $1 \neq \mathfrak{P} \in S_p(\mathfrak{G})$. Let c be the class of $\mathfrak{P}$ and let $|\mathfrak{P}| = p^b$.*

a) *If $p \neq 2$ and p is not a Fermat prime, $c \geq p^{l-1}$ and $b \geq (p^l - 1)/(p - 1)$.*

b) *If p is a Fermat prime and $p > 3$, $c \geq ((p - 2)^l - 1)/(p - 3)$ and $b \geq ((p - 2)^{l+1} - l(p - 3) - p + 2)/(p - 3)^2$.*

c) *If $p = 3$, $c \geq 2^{l-1}$ and $b \geq 2^{l-1} + l - 1$.*

Proof. These assertions are proved by induction on l and are all trivial if $l = 1$. For $l > 1$, let $\mathfrak{U}/\mathbf{O}_{p'}(\mathfrak{G}) = \Phi(\mathbf{O}_{p',p}(\mathfrak{G})/\mathbf{O}_{p'}(\mathfrak{G}))$ and let $\rho = (\mathfrak{G}$ on $\mathbf{O}_{p',p}(\mathfrak{G})/\mathfrak{U})$. By 1.4 and 1.6, $\ker \rho = \mathbf{O}_{p',p}(\mathfrak{G})$.

a) Suppose that the exponent of $\mathfrak{P}\mathbf{O}_{p',p}(\mathfrak{G})/\mathbf{O}_{p',p}(\mathfrak{G})$ is p^e. Thus there exists $x \in \mathfrak{P}$ such that the order of $x\mathbf{O}_{p',p}(\mathfrak{G})$ and $\rho(x)$ is p^e. By 2.9, the minimum polynomial of $\rho(x)$ is of degree p^e, so $(\rho(x) - 1)^{p^e - 1} \neq 0$. By 1.8, there is a commutator of weight p^e which is not 1; hence $c \geq p^e$. By 4.3a), $e \geq l - 1$, since $l - 1$ is the p-length of $\mathfrak{G}/\mathbf{O}_{p',p}(\mathfrak{G})$. Hence $c \geq p^{l-1}$.

Also, since p^e is the degree of the minimum polynomial of a linear transformation of $\mathbf{O}_{p',p}(\mathfrak{G})/\mathfrak{U}$,

$$|\mathbf{O}_{p',p}(\mathfrak{G})/\mathfrak{U}| \geq p^{p^e} \geq p^{p^{l-1}}.$$

Using the inductive hypothesis, it follows that if $|\mathfrak{P}\mathfrak{U}/\mathfrak{U}| = p^n$,

$$n \geq (p^{l-1} - 1)/(p - 1) + p^{l-1} = (p^l - 1)/(p - 1).$$

Hence $b \geq (p^l - 1)/(p - 1)$.

b, c) Let c' be the class of $\mathfrak{P}\mathbf{O}_{p',p}(\mathfrak{G})/\mathbf{O}_{p',p}(\mathfrak{G})$. Since $l > 1$, $c' \geq 1$, and if

$$\mathfrak{P} = \mathfrak{P}_0 \geq \mathfrak{P}_1 \geq \cdots$$

is the lower central series of $\mathfrak{P}$, $\mathfrak{P}_{c'} \not\leq \mathbf{O}_{p',p}(\mathfrak{G})$. Suppose $x \in \mathfrak{P}_{c'}$ and $x \notin \mathbf{O}_{p',p}(\mathfrak{G})$. Then the order of $\rho(x)$ is at least p, so by 2.9, $(\rho(x) - 1)^{p-2} \neq 0$. By 1.8, $z = [y, \underbrace{x, \ldots, x}_{p-2}] \neq 1$ for some $y \in \mathfrak{P}$. But by III, 2.11, $z \in \mathfrak{P}_{1+(p-2)c'}$, so $c \geq 1 + (p - 2)c'$. If $p > 3$, $c' \geq ((p - 2)^{l-1} - 1)/(p - 3)$ by the inductive hypothesis, whence

$$c \geq ((p - 2)^l - 1)/(p - 3).$$

If $p = 3$, observe that by 5.4, $\mathfrak{P}^{(l-1)} \neq 1$, so by III, 2.12, $c \geq 2^{l-1}$.

For any p, we define a series

$$\mathfrak{U} = \mathfrak{U}_0 \leq \mathfrak{U}_1 \leq \cdots \leq \mathfrak{U}_n \leq \cdots$$

of $\mathfrak{P}$-invariant subgroups of $\mathbf{O}_{p',p}(\mathfrak{G})$ inductively; if $n > 0$, $\mathfrak{P}$ operates on $\mathbf{O}_{p',p}(\mathfrak{G})/\mathfrak{U}_{n-1}$ and $\mathfrak{U}_n/\mathfrak{U}_{n-1}$ is defined to be the set of elements of $\mathbf{O}_{p',p}(\mathfrak{G})/\mathfrak{U}_{n-1}$ left fixed by $\mathfrak{P}$. By III, 2.8, $[\mathfrak{U}_n, \mathfrak{P}_i] \leq \mathfrak{U}_{n-i}$ if $i \leq n$; thus $\mathfrak{P}_{c'}$ centralizes $\mathfrak{U}_{c'}/\mathfrak{U}$, $\mathfrak{U}_{2c'}/\mathfrak{U}_{c'}$, ... Hence $\rho(x) - 1$ induces the zero linear transformation on these spaces, and $(\rho(x) - 1)^{p-2}$ induces the zero linear transformation on $\mathfrak{U}_{(p-2)c'}/\mathfrak{U}$. But $(\rho(x) - 1)^{p-2} \neq 0$, so $\mathfrak{U}_{(p-2)c'} < \mathbf{O}_{p',p}(\mathfrak{G})$. Hence $\mathfrak{U}_{i-1} < \mathfrak{U}_i$ for $i = 1, \ldots, (p-2)c'$, and if $|\mathbf{O}_{p',p}(\mathfrak{G})/\mathfrak{U}| = p^m$, $m \geq (p-2)c' + 1$.

If $p > 3$, it follows by using the inductive hypothesis that

$$\begin{aligned} b &\geq m + ((p-2)^l - (l-1)(p-3) - p + 2)/(p-3)^2 \\ &\geq ((p-3)^2(p-2)c' + (p-2)^l - l(p-3) + p^2 - 6p + 8)/(p-3)^2. \end{aligned}$$

But by the first part, $(p-3)c' \geq (p-2)^{l-1} - 1$, so

$$b \geq ((p-2)^{l+1} - l(p-3) - p + 2)/(p-3)^2.$$

For $p = 3$, it follows by using the inductive hypothesis that

$$\begin{aligned} b &\geq m + 2^{l-2} + l - 2 \\ &\geq c' + 1 + 2^{l-2} + l - 2 \\ &\geq 2^{l-1} + l - 1. \end{aligned}$$

q.e.d.

The group constructed in 4.7 shows that the inequalities in 5.5a) are best possible.

5.6 Lemma. *Let $\mathfrak{G}$ be a soluble group of p-length l. Suppose that $p_1, \ldots, p_r$ are the prime divisors of $|\mathfrak{G}|$ other than p and that l_i is the p_i-length of $\mathfrak{G}$. Then*

$$l \leq \prod_{i=1}^{r} (1 + l_i).$$

Proof. This is proved by induction on r and is trivial for $r = 0$. If $r > 0$, let

$$1 = \mathfrak{P}_0 \leq \mathfrak{N}_0 < \mathfrak{P}_1 < \mathfrak{N}_1 < \cdots < \mathfrak{N}_{l_r - 1} < \mathfrak{P}_{l_r} \leq \mathfrak{N}_{l_r} = \mathfrak{G}$$

be the upper p_r-series of $\mathfrak{G}$. If k_j is the p-length of $\mathfrak{N}_j/\mathfrak{P}_j$ ($j = 0, 1, \ldots, l_r$),

$$l \leq k_0 + k_1 + \cdots + k_{l_r}.$$

But the p_i-length of $\mathfrak{N}_j/\mathfrak{P}_j$ is at most l_i ($i = 1, \ldots, r - 1$) and $\mathfrak{N}_j/\mathfrak{P}_j$ is a p_r'-group. Hence by the inductive hypothesis,

$$k_j \leq \prod_{i=1}^{r-1} (1 + l_i),$$

so

$$l \leq (1 + l_r) \prod_{i=1}^{r-1} (1 + l_i) = \prod_{i=1}^{r} (1 + l_i).$$ **q.e.d.**

5.7 Theorem (HALL and HIGMAN). *Let $\mathfrak{G}$ be a soluble group of derived length d. For each prime divisor p of $|\mathfrak{G}|$, let d_p be the derived length of a Sylow p-subgroup of $\mathfrak{G}$ and let l_p be the p-length of $\mathfrak{G}$.*

a) $d \leq \sum_p d_p l_p$.

b) $d \leq \sum_{p>2} d_p^2 + d_2 \prod_{p>2} (1 + d_p)$.

c) *If c is the class of the Sylow 2-subgroups of $\mathfrak{G}$,*

$$d \leq c + \sum_{p>2} \tfrac{1}{2} d_p(d_p + 1).$$

Proof. a) This is proved by induction on $|\mathfrak{G}|$. Suppose that $\mathfrak{G} \neq 1$. If $\mathfrak{N}_1, \mathfrak{N}_2$ are distinct minimal normal subgroups of $\mathfrak{G}$ and d_i' is the derived length of $\mathfrak{G}/\mathfrak{N}_i$, $d \leq \max(d_1', d_2')$, since $\mathfrak{G}$ is isomorphic to a subgroup of the direct product $(\mathfrak{G}/\mathfrak{N}_1) \times (\mathfrak{G}/\mathfrak{N}_2)$. But by the inductive hypothesis, $d_i' \leq \sum_p d_p l_p$, so the assertion follows at once. We may suppose, then, that $\mathfrak{G}$ has a unique minimal normal subgroup $\mathfrak{N}$. Since $\mathfrak{G}$ is soluble, $|\mathfrak{N}|$ is a power of a prime p; also $\mathbf{O}_{p'}(\mathfrak{G}) = 1$. If $\mathfrak{P} = \mathbf{O}_p(\mathfrak{G})$ and d' is the derived length of $\mathfrak{G}/\mathfrak{P}$,

$$d \leq d' + d_p,$$

since the derived length of $\mathfrak{P}$ is at most d_p. Since $\mathfrak{P} = \mathbf{O}_{p',p}(\mathfrak{G})$, the p-length of $\mathfrak{G}/\mathfrak{P}$ is $l_p - 1$, so by the inductive hypothesis,

$$d' \leq d_p(l_p - 1) + \sum_{q \neq p} d_q l_q.$$

Thus

$$d \leq \sum_q d_q l_q.$$

b) By Theorem 5.4, $l_p \leq d_p$ for p odd, so by a),

$$d \leq \sum_{p>2} d_p^2 + d_2 l_2.$$

But by 5.6,

$$l_2 \leq \prod_{p>2} (1 + l_p) \leq \prod_{p>2} (1 + d_p),$$

so

$$d \leq \sum_{p>2} d_p^2 + d_2 \prod_{p>2} (1 + d_p).$$

c) This is proved by induction on $|\mathfrak{G}|$. As in a), we may suppose that $\mathfrak{G}$ has a unique minimal normal subgroup $\mathfrak{N}$. If $|\mathfrak{N}|$ is a power of p, $\mathbf{O}_{p'}(\mathfrak{G}) = 1$; let $\mathfrak{P} = \mathbf{O}_p(\mathfrak{G})$ and let $\mathfrak{S} \in S_p(\mathfrak{G})$.

Suppose that $p = 2$. Since $\mathfrak{P} \leq \mathfrak{S}$, $\mathbf{Z}(\mathfrak{S}) \leq \mathbf{C}_{\mathfrak{G}}(\mathfrak{P})$. By 1.4, $\mathfrak{P} \geq \mathbf{C}_{\mathfrak{G}}(\mathfrak{P})$, so $\mathbf{Z}(\mathfrak{S}) \leq \mathfrak{P}$ and $\mathbf{Z}(\mathfrak{S}) \leq \mathbf{Z}(\mathfrak{P})$. But $\mathbf{Z}(\mathfrak{P}) \trianglelefteq \mathfrak{G}$ and $\mathfrak{S}/\mathbf{Z}(\mathfrak{P}) \in S_p(\mathfrak{G}/\mathbf{Z}(\mathfrak{P}))$, so the class of the Sylow 2-subgroup of $\mathfrak{G}/\mathbf{Z}(\mathfrak{P})$ is less than c. By the inductive hypothesis, the derived length of $\mathfrak{G}/\mathbf{Z}(\mathfrak{P})$ is at most

$$c - 1 + \sum_{q>2} \tfrac{1}{2} d_q(d_q + 1).$$

Since $\mathbf{Z}(\mathfrak{P})$ is Abelian, the assertion follows at once.

If $p > 2$, then by 5.4a), the derived length of $\mathfrak{S}/\mathfrak{P}$ is less than d_p. Hence by the inductive hypothesis, the derived length of $\mathfrak{G}/\mathfrak{P}$ is at most

$$c + \sum_{\substack{q>2 \\ q \neq p}} \tfrac{1}{2} d_q(d_q + 1) + \tfrac{1}{2} d_p(d_p - 1).$$

Since the derived length of $\mathfrak{P}$ is at most d_p,

$$d \leq c + \sum_{\substack{q>2 \\ q \neq p}} \tfrac{1}{2} d_q(d_q + 1) + \tfrac{1}{2} d_p(d_p - 1) + d_p = c + \sum_{q>2} \tfrac{1}{2} d_q(d_q + 1).$$

q.e.d.

In many ways, the most elementary way of using Theorem B is by means of the following lemma.

5.8 Lemma. *Suppose that a minimal normal subgroup $\mathfrak{N}$ of the group $\mathfrak{G}$ is an elementary Abelian p-group. Let $\mathfrak{C} = \mathbf{C}_{\mathfrak{G}}(\mathfrak{N})$. Suppose that $\mathfrak{G}/\mathfrak{C}$ is p-soluble, and that either $p > 3$ or $p = 3$ and the Sylow 2-subgroups of $\mathfrak{G}/\mathfrak{C}$ are Abelian. If $g \in \mathfrak{G}$ and $[\mathfrak{N}, g, g] = 1$, then $g \in \mathfrak{C}$.*

Proof. Let $\rho = (\mathfrak{G}$ on $\mathfrak{N})$; thus $\ker \rho = \mathfrak{C}$. Since $\mathfrak{N}$ is a minimal normal subgroup of $\mathfrak{G}$, ρ is irreducible. Thus by V, 5.17, $\mathbf{O}_p(\mathfrak{G}/\mathfrak{C}) = 1$ and $\mathbf{O}_p(\operatorname{im} \rho) = 1$. By 1.8a), $(\rho(g) - 1)^2 = 0$, since $[\mathfrak{N}, g, g] = 1$. Thus

$$\rho(g)^p - 1 = (\rho(g) - 1)^p = 0,$$

and $\rho(g)^p = 1$. If $\rho(g) \neq 1$, the order of $\rho(g)$ is p and the minimum polynomial of $\rho(g)$ is $(t - 1)^2$. By 2.9, $p > 3$ and $2 \geq p - 1$, a contradiction. Hence $\rho(g) = 1$ and $g \in \mathfrak{C}$. **q.e.d.**

5.9 Theorem. *Let $\mathfrak{G}$ be a finite group.*

a) *If $\mathfrak{S} \in S_p(\mathfrak{G})$ and $\mathfrak{A}$ is a maximal normal Abelian subgroup of $\mathfrak{S}$, $\mathbf{C}_{\mathfrak{G}}(\mathfrak{A}) = \mathfrak{A} \times \mathfrak{D}$ for some p'-subgroup $\mathfrak{D}$.*

b) (THOMPSON) *Suppose that $\mathfrak{G}$ is soluble, π is a set of primes greater than 3 and $\mathfrak{H}$ is a Hall π-subgroup of $\mathfrak{G}$. If $\mathfrak{A}$ is a maximal normal Abelian subgroup of the Fitting subgroup $\mathbf{F}(\mathfrak{H})$ of $\mathfrak{H}$, $\mathbf{C}_{\mathfrak{G}}(\mathfrak{A}) = \mathfrak{A} \times \mathfrak{D}$ for some subgroup $\mathfrak{D}$ of $\mathbf{O}_{\pi'}(\mathfrak{G})$.*

Proof. a) Since $\mathfrak{A} \trianglelefteq \mathfrak{S}$, $\mathfrak{S} \in S_p(\mathbf{N}_{\mathfrak{G}}(\mathfrak{A}))$. Since $\mathbf{C}_{\mathfrak{G}}(\mathfrak{A}) \trianglelefteq \mathbf{N}_{\mathfrak{G}}(\mathfrak{A})$, $\mathfrak{S} \cap \mathbf{C}_{\mathfrak{G}}(\mathfrak{A}) \in S_p(\mathbf{C}_{\mathfrak{G}}(\mathfrak{A}))$. But $\mathfrak{S} \cap \mathbf{C}_{\mathfrak{G}}(\mathfrak{A}) = \mathbf{C}_{\mathfrak{S}}(\mathfrak{A}) = \mathfrak{A}$ by III, 7.3, so $\mathfrak{A} \in S_p(\mathbf{C}_{\mathfrak{G}}(\mathfrak{A}))$. By IV, 2.6, there exists $\mathfrak{D} \leq \mathbf{C}_{\mathfrak{G}}(\mathfrak{A})$ such that $\mathbf{C}_{\mathfrak{G}}(\mathfrak{A}) = \mathfrak{A}\mathfrak{D}$ and $\mathfrak{A} \cap \mathfrak{D} = 1$. Thus $\mathbf{C}_{\mathfrak{G}}(\mathfrak{A}) = \mathfrak{A} \times \mathfrak{D}$.

b) We use induction on $|\mathfrak{G}|$. Let $\mathfrak{N} = \mathbf{O}_{\pi'}(\mathfrak{G})$. Then $\mathfrak{H}\mathfrak{N}/\mathfrak{N}$ is a Hall π-subgroup of $\mathfrak{G}/\mathfrak{N}$ and $\mathfrak{A}\mathfrak{N}/\mathfrak{N}$ is a maximal normal Abelian subgroup of $\mathbf{F}(\mathfrak{H}\mathfrak{N}/\mathfrak{N}) = \mathbf{F}(\mathfrak{H})\mathfrak{N}/\mathfrak{N}$. If $\mathfrak{N} \neq 1$, it follows from the inductive hypothesis that

$$\mathbf{C}_{\mathfrak{G}/\mathfrak{N}}(\mathfrak{A}\mathfrak{N}/\mathfrak{N}) = \mathfrak{A}\mathfrak{N}/\mathfrak{N},$$

since $\mathbf{O}_{\pi'}(\mathfrak{G}/\mathfrak{N}) = 1$. Hence $\mathbf{C}_{\mathfrak{G}}(\mathfrak{A}) \leq \mathfrak{A}\mathfrak{N}$ and $\mathbf{C}_{\mathfrak{G}}(\mathfrak{A}) = \mathfrak{A} \times \mathbf{C}_{\mathfrak{N}}(\mathfrak{A})$.

If $\mathfrak{N} = 1$, then $\mathbf{O}_p(\mathfrak{G}) = 1$ for all $p \in \pi'$. Hence the nilpotent normal subgroup $\mathbf{F}(\mathfrak{G})$ is a π-group. Thus $\mathbf{F}(\mathfrak{G}) \leq \mathfrak{H}$ (IV, 7.2), and $\mathbf{F}(\mathfrak{G}) \leq \mathbf{F}(\mathfrak{H})$. Hence $\mathbf{F}(\mathfrak{G}) \leq \mathbf{N}_{\mathfrak{G}}(\mathfrak{A})$. Since $\mathbf{C}_{\mathfrak{G}}(\mathfrak{A}) \trianglelefteq \mathbf{N}_{\mathfrak{G}}(\mathfrak{A})$, $\mathbf{F}(\mathfrak{G}) \leq \mathbf{N}_{\mathfrak{G}}(\mathbf{C}_{\mathfrak{G}}(\mathfrak{A}))$ and

$$[\mathbf{F}(\mathfrak{G}), \mathbf{C}_{\mathfrak{G}}(\mathfrak{A})] \leq \mathbf{C}_{\mathfrak{G}}(\mathfrak{A}) \cap \mathbf{F}(\mathfrak{G}).$$

Since $\mathbf{F}(\mathfrak{G}) \leq \mathbf{F}(\mathfrak{H})$,

$$[\mathbf{F}(\mathfrak{G}), \mathbf{C}_{\mathfrak{G}}(\mathfrak{A})] \leq \mathbf{C}_{\mathbf{F}(\mathfrak{H})}(\mathfrak{A}).$$

Now for each prime p, the Sylow p-subgroup $\mathfrak{A}_p$ of $\mathfrak{A}$ is a maximal normal Abelian subgroup of the Sylow p-subgroup $\mathfrak{S}_p$ of $\mathbf{F}(\mathfrak{H})$, so $\mathbf{C}_{\mathfrak{S}_p}(\mathfrak{A}_p) = \mathfrak{A}_p$, by III, 7.3. Thus $\mathbf{C}_{\mathbf{F}(\mathfrak{H})}(\mathfrak{A}) = \mathfrak{A}$. Therefore

$$[\mathbf{F}(\mathfrak{G}), \mathbf{C}_{\mathfrak{G}}(\mathfrak{A})] \leq \mathfrak{A}$$

and

$$[\mathbf{F}(\mathfrak{G}), \mathbf{C}_{\mathfrak{G}}(\mathfrak{A}), \mathbf{C}_{\mathfrak{G}}(\mathfrak{A})] = 1.$$

Now let

$$\mathbf{F}(\mathfrak{G}) = \mathfrak{F}_0 > \mathfrak{F}_1 > \cdots > \mathfrak{F}_m = 1$$

be the part below $\mathbf{F}(\mathfrak{G})$ of a chief series of $\mathfrak{G}$, and let $\mathfrak{C}_i = \mathbf{C}_{\mathfrak{G}}(\mathfrak{F}_{i-1}/\mathfrak{F}_i)$ $(i = 1, \ldots, m)$. Hence $\mathbf{F}(\mathfrak{G}) \leq \mathfrak{C}_i$, and for each $i = 1, \ldots, m$, $\mathfrak{F}_{i-1}/\mathfrak{F}_i$ is a p-group for some $p \in \pi$, since $\mathbf{F}(\mathfrak{G}) \leq \mathfrak{H}$. Since $2 \notin \pi$, $3 \notin \pi$ and

$$[\mathfrak{F}_{i-1}, \mathbf{C}_{\mathfrak{G}}(\mathfrak{A}), \mathbf{C}_{\mathfrak{G}}(\mathfrak{A})] \leq \mathfrak{F}_i,$$

it follows from 5.8 that $\mathbf{C}_{\mathfrak{G}}(\mathfrak{A}) \leq \mathfrak{C}_i$ $(i = 1, \ldots, m)$. Hence $\mathbf{C}_{\mathfrak{G}}(\mathfrak{A}) \leq \mathfrak{C}$, where $\mathfrak{C} = \bigcap_{i=1}^m \mathfrak{C}_i$.

Also $\mathbf{F}(\mathfrak{G}) \leq \mathfrak{C}$. Now if $\mathfrak{C} \neq \mathbf{F}(\mathfrak{G})$, there is a non-identity normal Abelian subgroup $\mathfrak{M}/\mathbf{F}(\mathfrak{G})$ contained in $\mathfrak{C}/\mathbf{F}(\mathfrak{G})$, since $\mathfrak{C} \trianglelefteq \mathfrak{G}$ and $\mathfrak{G}$ is soluble. But $\mathfrak{C}$ centralizes each $\mathfrak{F}_{i-1}/\mathfrak{F}_i$, so

$$\mathfrak{M} > \mathfrak{F}_0 > \mathfrak{F}_1 > \cdots > \mathfrak{F}_m = 1$$

is a central series of $\mathfrak{M}$, and $\mathfrak{M}$ is nilpotent. Since $\mathfrak{M} \trianglelefteq \mathfrak{G}$, it follows that $\mathfrak{M} \leq \mathbf{F}(\mathfrak{G})$, which is a contradiction. Hence $\mathfrak{C} = \mathbf{F}(\mathfrak{G})$ and $\mathbf{C}_{\mathfrak{G}}(\mathfrak{A}) \leq \mathbf{F}(\mathfrak{G}) \leq \mathbf{F}(\mathfrak{H})$. Thus $\mathbf{C}_{\mathfrak{G}}(\mathfrak{A}) = \mathbf{C}_{\mathbf{F}(\mathfrak{H})}(\mathfrak{A}) = \mathfrak{A}$. **q.e.d.**

We shall prove next an application of 4.2 to the minimal number of generators of minimal simple groups. For this several lemmas are needed.

5.10 Lemma. *Suppose that* $\mathfrak{G} = \mathfrak{P}\mathfrak{V}$, *where* $\mathfrak{V}$ *is a minimal normal elementary Abelian q-subgroup of* $\mathfrak{G}$ *and* $\mathfrak{P}$ *is a p-subgroup* ($p \neq q$). *If* $\mathfrak{P} = \langle a_1, \ldots, a_n \rangle$, *where* $n > 1$, *then* $\mathfrak{G} = \langle a_1, \ldots, a_i z, \ldots, a_n \rangle$ *for some* $z \in \mathfrak{V} - \{1\}$ *and some i.*

Proof. Suppose that $x \in \mathfrak{V} - \{1\}$ and $\mathfrak{H} = \langle a_1 x, a_2, \ldots, a_n \rangle$. If $\mathfrak{H} = \mathfrak{G}$, there is nothing to prove, so we suppose that $\mathfrak{H} \neq \mathfrak{G}$. Now $\mathfrak{H}\mathfrak{V}$ contains $a_1 = (a_1 x)x^{-1}, a_2, \ldots, a_n$ and $\mathfrak{V}$, so $\mathfrak{H}\mathfrak{V} = \mathfrak{G}$. Since $\mathfrak{H} \neq \mathfrak{G}$, $\mathfrak{H} \ngeq \mathfrak{V}$. Thus $\mathfrak{H} \cap \mathfrak{V} < \mathfrak{V}$. But $\mathfrak{H} \cap \mathfrak{V} \trianglelefteq \mathfrak{H}\mathfrak{V} = \mathfrak{G}$, since $\mathfrak{V}$ is Abelian. Since $\mathfrak{V}$ is a minimal normal subgroup, $\mathfrak{H} \cap \mathfrak{V} = 1$. Hence $|\mathfrak{H}| = |\mathfrak{G} : \mathfrak{V}| = |\mathfrak{P}|$, and $\mathfrak{H}$, $\mathfrak{P}$ are Sylow p-subgroups of $\mathfrak{G}$. Thus $\mathfrak{H} = \mathfrak{P}^y$ for some $y \in \mathfrak{V}$, since $\mathfrak{G} = \mathfrak{P}\mathfrak{V}$. Hence $a_2^y \in \mathfrak{H}$ and

$$[a_2, y] = a_2^{-1} a_2^y \in \mathfrak{V} \cap \mathfrak{H} = 1.$$

Thus y and a_2 commute. Since their orders are coprime, $\langle a_2 y \rangle = \langle a_2, y \rangle$. Hence if $\mathfrak{K} = \langle a_1, a_2 y, \ldots, a_n \rangle$, $\mathfrak{P} \leq \mathfrak{K}$ and $y \in \mathfrak{K}$. Hence $\mathfrak{K} \cap \mathfrak{V} \trianglelefteq \mathfrak{P}\mathfrak{V} = \mathfrak{G}$ and either $\mathfrak{K} \cap \mathfrak{V} = 1$ or $\mathfrak{V} \leq \mathfrak{K}$. If $\mathfrak{K} \cap \mathfrak{V} = 1$, then $y = 1$, $\mathfrak{H} = \mathfrak{P}$ and $1 \neq x = a_1^{-1}(a_1 x) \in \mathfrak{P} \cap \mathfrak{V} = 1$, a contradiction. So $\mathfrak{K} \geq \mathfrak{V}$ and $\mathfrak{K} \geq \mathfrak{P}\mathfrak{V} = \mathfrak{G}$. Thus $\mathfrak{G} = \langle a_1, a_2 y, \ldots, a_n \rangle$. **q.e.d.**

5.11 Lemma. *Suppose that* $\mathfrak{G}$ *is a group,* π *is a set of primes and* $\mathbf{O}_\pi(\mathfrak{G}) = 1$. *Let* $\mathfrak{V}$ *be a subgroup of* $\mathfrak{G}$ *such that* $\bigcap_{\mathfrak{V} \leq \mathfrak{H} < \mathfrak{G}} \mathbf{O}_\pi(\mathfrak{H}) \neq 1$. *Then* $\mathfrak{G} = \langle \mathfrak{V}, g \rangle$ *for some* $g \in \mathfrak{G}$.

Proof. Suppose that $u \in \bigcap_{\mathfrak{V} \leq \mathfrak{H} < \mathfrak{G}} \mathbf{O}_\pi(\mathfrak{H})$ and $u \neq 1$. Since $\mathbf{O}_\pi(\mathfrak{G}) = 1$, $u \notin \mathbf{O}_\pi(\mathfrak{G})$. Hence the group generated by the conjugates of u in $\mathfrak{G}$ is not a π-group. Let k be the smallest integer for which there exist elements $a_1, \ldots, a_k$ of $\mathfrak{G}$ such that $y = u^{a_1} \cdots u^{a_k}$ is not a π-element. Since $u \in \mathbf{O}_\pi(\mathfrak{V})$, u is a π-element. Thus $k > 1$. Let $g = (u^{a_2} \cdots u^{a_k})^{a_1^{-1}}$. By definition of k, g is a π-element. Hence if $\mathfrak{H} = \langle \mathfrak{V}, g \rangle$, $\langle g, \mathbf{O}_\pi(\mathfrak{H}) \rangle$ is a π-group. But $ug = y^{a_1^{-1}}$ is not a π-element, so $ug \notin \langle g, \mathbf{O}_\pi(\mathfrak{H}) \rangle$. Hence $u \notin \mathbf{O}_\pi(\mathfrak{H})$. It follows from the definition of u that $\mathfrak{H} = \mathfrak{G}$. Hence $\mathfrak{G} = \langle \mathfrak{V}, g \rangle$. **q.e.d.**

5.12 Lemma. *Suppose that* $\mathfrak{G}$ *is a group of order divisible by the odd prime p, and suppose that every proper subgroup of* $\mathfrak{G}$ *is p-soluble. Then there is a p-subgroup* $\mathfrak{U}$ *of* $\mathfrak{G}$ *such that* $\mathfrak{U}$ *has 2 generators and*

$$\bigcap_{\mathfrak{U} \leq \mathfrak{H} < \mathfrak{G}} \mathbf{O}_{p',p}(\mathfrak{H}) \neq 1.$$

Proof. Let x be a p-element of maximal order p^e, and put $\mathfrak{U} = \langle x \rangle$. We may suppose that $\mathfrak{U}$ does not have the stated property. Thus there is a subgroup $\mathfrak{H}$ of $\mathfrak{G}$ such that $\mathfrak{U} \leq \mathfrak{H} < \mathfrak{G}$ and $x^{p^{e-1}} \notin \mathbf{O}_{p',p}(\mathfrak{H})$. Now $\mathfrak{H}$ is p-soluble and, in the notation of 4.1, $e_p^*(\mathfrak{H}/\mathbf{O}_{p',p}(\mathfrak{H})) > 2e - 2$. By 4.2b), $e_p^*(\mathfrak{H}) > 2e - 1$. Hence there exist elements x_1, x_2 in a Sylow p-subgroup of $\mathfrak{H}$ such that $v = [x_1^{p^{e-1}}, x_2^{p^{e-1}}] \neq 1$. Let $\mathfrak{V} = \langle x_1, x_2 \rangle$. If $\mathfrak{V} \leq \mathfrak{K} < \mathfrak{G}$, then $\mathfrak{K}$ is p-soluble and $e_p^*(\mathfrak{K}) \leq 2e$. By 4.2b), $e_p^*(\mathfrak{K}/\mathbf{O}_{p',p}(\mathfrak{K})) \leq 2e - 1$. Thus $v \in \mathbf{O}_{p',p}(\mathfrak{K})$. Hence

$$v \in \bigcap_{\mathfrak{V} \leq \mathfrak{K} < \mathfrak{G}} \mathbf{O}_{p',p}(\mathfrak{K}) \neq 1,$$

and $\mathfrak{V}$ has the required properties. **q.e.d.**

5.13 Lemma. *Suppose that $\mathfrak{G}$ is a group of order divisible by the odd prime p, and suppose that every proper subgroup of $\mathfrak{G}$ is p-soluble. Then there exists a subgroup $\mathfrak{V}$ generated by 2 elements, such that either*

$$\bigcap_{\mathfrak{V} \leq \mathfrak{H} < \mathfrak{G}} \mathbf{O}_p(\mathfrak{H}) \neq 1 \text{ or } \bigcap_{\mathfrak{V} \leq \mathfrak{H} < \mathfrak{G}} \mathbf{O}_{p'}(\mathfrak{H}) \neq 1.$$

Proof. By 5.12, there is a p-subgroup $\mathfrak{U}$ of $\mathfrak{G}$ such that $\mathfrak{U}$ has 2 generators and $\mathfrak{U}^* = \bigcap_{\mathfrak{U} \leq \mathfrak{H} < \mathfrak{G}} \mathbf{O}_{p',p}(\mathfrak{H})$ is a normal, non-identity subgroup of $\mathfrak{U}$. By III, 7.2a), $\mathbf{Z}(\mathfrak{U}) \cap \mathfrak{U}^*$ contains an element $u \neq 1$. If $\bigcap_{\mathfrak{U} \leq \mathfrak{H} < \mathfrak{G}} \mathbf{O}_p(\mathfrak{H}) \neq 1$, there is nothing further to prove, so we suppose that this is not the case. Thus

$$\mathscr{S} = \{\mathfrak{X} \mid \mathfrak{U} \leq \mathfrak{X} < \mathfrak{G}, u \notin \mathbf{O}_p(\mathfrak{X})\}$$

is non-empty. Let $\mathfrak{V}$ be an element of $\mathscr{S}$ of minimal order, and let $\mathfrak{Q} = \mathbf{O}_{p'}(\mathfrak{V})$. By 1.2 (with $(\pi, \mathfrak{G}, \mathfrak{P}, \mathfrak{K})$ replaced by $(p', \mathfrak{V}, \mathfrak{Q}, \mathbf{O}_{p',p}(\mathfrak{V}) \cap \mathbf{C}_{\mathfrak{V}}(\mathfrak{Q}))$,

$$\mathbf{O}_{p',p}(\mathfrak{V}) \cap \mathbf{C}_{\mathfrak{V}}(\mathfrak{Q}) = \mathfrak{C} \times \mathfrak{D}$$

for some normal p-subgroup $\mathfrak{C}$ of $\mathfrak{V}$ and some p'-subgroup $\mathfrak{D}$. Since $u \notin \mathbf{O}_p(\mathfrak{V})$, $u \notin \mathfrak{C}$; since u is a p-element, $u \notin \mathfrak{C} \times \mathfrak{D}$. And since $u \in \mathbf{O}_{p',p}(\mathfrak{V})$, $u \notin \mathbf{C}_{\mathfrak{V}}(\mathfrak{Q})$. In particular, $u \notin \mathbf{O}_p(\mathfrak{Q}\mathfrak{U})$, so $\mathfrak{Q}\mathfrak{U} \in \mathscr{S}$. By minimality of $|\mathfrak{V}|$, $\mathfrak{V} = \mathfrak{Q}\mathfrak{U}$. But if $\mathfrak{Q}_0$ is a $\mathfrak{U}$-invariant proper subgroup of $\mathfrak{Q}$, $\mathfrak{Q}_0\mathfrak{U} \notin \mathscr{S}$, so $u \in \mathbf{O}_p(\mathfrak{Q}_0\mathfrak{U})$ and $u \in \mathbf{C}_{\mathfrak{V}}(\mathfrak{Q}_0)$. By 1.12, $\mathfrak{Q}$ is a q-group for some prime q. Since $\langle u \rangle \trianglelefteq \mathfrak{U}$, it follows from III, 13.5 that $\mathfrak{Q}$ is a special q-group and that $(\mathfrak{U}$ on $\mathfrak{Q}/\mathfrak{Q}')$ is irreducible. Since $\mathfrak{U}$ has 2 generators, it follows from 5.10 that $\mathfrak{V}/\mathfrak{Q}' = \mathfrak{U}\mathfrak{Q}/\mathfrak{Q}'$ has 2 generators. Since $\mathfrak{Q}' \leq \Phi(\mathfrak{Q}) \leq \Phi(\mathfrak{V})$, it follows that $\mathfrak{V}$ has 2 generators.

Since $u \notin \mathbf{C}_{\mathfrak{V}}(\mathfrak{Q})$, there exists $y \in \mathfrak{Q}$ such that $z = [u, y] \neq 1$. Now if $\mathfrak{V} \leq \mathfrak{H} < \mathfrak{G}$, then $\mathfrak{U} \leq \mathfrak{H} < \mathfrak{G}$ and $u \in \mathfrak{U}^* \leq \mathbf{O}_{p',p}(\mathfrak{H})$.
Thus $z \in \mathbf{O}_{p',p}(\mathfrak{H}) \cap \mathfrak{Q}$, since $\mathfrak{Q} \trianglelefteq \mathfrak{V}$. But since $\mathfrak{Q}$ is a p'-group, $\mathbf{O}_{p',p}(\mathfrak{H}) \cap \mathfrak{Q} \leq \mathbf{O}_{p'}(\mathfrak{H})$. Hence $z \in \mathbf{O}_{p'}(\mathfrak{H})$ for every proper subgroup $\mathfrak{H}$ of $\mathfrak{G}$ containing $\mathfrak{V}$. Thus

$$\bigcap_{\mathfrak{V} \leq \mathfrak{H} < \mathfrak{G}} \mathbf{O}_{p'}(\mathfrak{H}) \neq 1. \qquad \textbf{q.e.d.}$$

5.14 Theorem (POWELL). *Suppose that $\mathfrak{G}$ is a simple non-Abelian group, p is an odd prime divisor of $|\mathfrak{G}|$ and that every proper subgroup of $\mathfrak{G}$ is p-soluble. Then $\mathfrak{G}$ can be generated by 3 elements.*

Proof. By 5.13, there exists a subgroup $\mathfrak{V}$ generated by 2 elements such that $\bigcap_{\mathfrak{V} \leq \mathfrak{H} < \mathfrak{G}} \mathbf{O}_{\pi}(\mathfrak{H}) \neq 1$, where π is either $\{p\}$ or p'. But $\mathbf{O}_{\pi}(\mathfrak{G}) = 1$, since $\mathfrak{G}$ is simple and p divides $|\mathfrak{G}|$. By 5.11, $\mathfrak{G} = \langle \mathfrak{V}, g \rangle$ for some $g \in \mathfrak{G}$, so $\mathfrak{G}$ has 3 generators. **q.e.d.**

Exercises

3) Show that the following are counterexamples to Theorem 5.9b) with the omission of the words 'greater than 3'.

(i) $\mathfrak{G}$ is the semidirect product $\mathfrak{X}\mathfrak{Y}$, where $\mathfrak{X}$ is the quaternion group of order 8 and $\mathfrak{Y}$ is a group isomorphic to $GL(2, 2)$ operating faithfully on $\mathfrak{X}$.

(ii) $\mathfrak{G}$ is the semidirect product $\mathfrak{X}\mathfrak{Y}$, where $\mathfrak{X}$ is the non-Abelian group of order 27 and exponent 3 and $\mathfrak{Y}$ is a group isomorphic to $SL(2, 3)$ operating faithfully on $\mathfrak{X}$.

§ 6. Fixed Point Free Automorphism Groups

The technique of Hall and Higman has been used in a number of other investigations. In this section, we use it to prove some theorems about fixed point free automorphism groups. First we need the following lemma.

6.1 Lemma. *Suppose that $p = 2^s + 1$ $(s > 0)$ is a Fermat prime. If $k > 0$, $2^n \equiv 1$ (p^k) if and only if $n \equiv 0$ $(2sp^{k-1})$.*

Proof. Since $(p-1)^2 \equiv 1\,(p)$, $(p-1)^{2p^{k-1}} \equiv 1\,(p^k)$. Thus $2^{2sp^{k-1}} \equiv 1\,(p^k)$, and if $n \equiv 0\,(2sp^{k-1})$, $2^n \equiv 1\,(p^k)$.

Conversely, suppose that $2^n \equiv 1\ (p^k)$. Write $n = sq + r$, where $0 \leq r < s$. Then

$$1 \equiv 2^n \equiv 2^{sq+r} \equiv (-1)^q 2^r (p).$$

Thus p divides $2^r - (-1)^q$, so if $r > 0$,

$$2^s - (-1)^q > 2^r - (-1)^q \geq p = 2^s + 1.$$

This is impossible, so $r = 0$, $(-1)^q \equiv 1\,(p)$ and, since p is odd, q is even. Hence $n = 2ms$ for some integer m. Thus

$$1 \equiv 2^n \equiv 2^{2sm} \equiv (p-1)^{2m} \quad (p^k).$$

Now if $m = p^l m'$ and $(m', p) = 1$, it follows from I, 13.18a) that

$$(p-1)^{2m} - 1 + 2mp \equiv 0\,(p^{l+2}).$$

Putting $(p-1)^{2m} - 1 = p^k u$, we obtain

$$p^k u + 2m'p^{l+1} \equiv 0\,(p^{l+2}).$$

Since $(m', p) = 1$, it follows that $k \leq l + 1$. Thus $m \equiv 0\ (p^{k-1})$ and $n \equiv 0\,(2sp^{k-1})$. **q.e.d.**

We prove next an analogue of Theorem 2.9 in which the characteristic is prime to the order of the group.

6.2 Theorem. *Let* V *be a non-zero, finite-dimensional vector space over a field* K, *and let* $\mathfrak{G}$ *be a p-soluble group of linear transformations for which* $\mathbf{O}_p(\mathfrak{G}) = 1$ *and* char $\mathsf{K} \nmid |\mathfrak{G}|$. *Let g be an element of* $\mathfrak{G}$ *of order* p^n. *Then one of the following occurs.*

(1) *The minimum polynomial of g is* $t^{p^n} - 1$, *and* $\mathbf{C}_{\mathsf{V}}(g) \neq 0$.

(2) *The Sylow 2-subgroups of* $\mathfrak{G}$ *are non-Abelian, p is a Fermat prime and* $\mathbf{C}_{\mathfrak{G}}(g) \cap \mathbf{O}_{p'}(\mathfrak{G}) \neq 1$.

(3) $p = 2$ *and there exists a Mersenne prime* $q < 2^n$ *such that the Sylow q-subgroups of* $\mathfrak{G}$ *are non-Abelian.*

Proof. Since the conditions in (1) remain unchanged under an extension of the ground-field, we may suppose that K is algebraically closed. We suppose that the theorem is false and choose a counterexample for which $\dim_K V + |\mathfrak{G}|$ is minimal. We have $\mathbf{O}_{p'}(\mathfrak{G}) \neq 1$.

a) The minimum polynomial of g is not $t^{p^n} - 1$.

For otherwise, $v = w(1 + g + \cdots + g^{p^{n-1}}) \neq 0$ for some $w \in V$. But $v \in \mathbf{C}_V(g)$, so $\mathbf{C}_V(g) \neq 0$ and (1) holds.

b) $\mathfrak{G} = \mathfrak{P}\mathfrak{H}$, where $\mathfrak{P} = \langle g \rangle$, $\mathfrak{H} = \mathbf{O}_{p'}(\mathfrak{G})$.

For suppose that $\mathfrak{G}_0 = \langle g \rangle \mathbf{O}_{p'}(\mathfrak{G})$ and $\mathfrak{G}_0 < \mathfrak{G}$. Then $\mathbf{O}_p(\mathfrak{G}_0) \leq \mathbf{C}_{\mathfrak{G}}(\mathbf{O}_{p'}(\mathfrak{G})) \leq \mathbf{O}_{p'}(\mathfrak{G})$ by 1.3, since $\mathbf{O}_p(\mathfrak{G}) = 1$. Thus $\mathbf{O}_p(\mathfrak{G}_0) = 1$. By minimality of the counterexample, the theorem is valid for $(\mathfrak{G}_0$ on $V)$. But this implies that the theorem is true. Hence $\mathfrak{G}_0 = \mathfrak{G}$.

c) V is an irreducible $K\mathfrak{G}$-module.

If not, then by the Maschke-Schur theorem, $V = V_1 \oplus V_2$ for certain non-zero $K\mathfrak{G}$-submodules V_1, V_2 of V. If $\mathfrak{R}_i = \ker(\mathfrak{G}$ on $V_i)$ $(i = 1, 2)$, $\mathfrak{R}_1 \cap \mathfrak{R}_2 = 1$. Let $\mathfrak{S}_i/\mathfrak{R}_i = \mathbf{O}_p(\mathfrak{G}/\mathfrak{R}_i)$ $(i = 1, 2)$. Then $\mathfrak{S}_1 \cap \mathfrak{S}_2 \leq \mathbf{O}_p(\mathfrak{G}) = 1$. Thus $g^{p^{n-1}} \notin \mathfrak{S}_i$ for either $i = 1$ or $i = 2$, and $\mathfrak{P} \cap \mathfrak{S}_i = 1$. Hence $\mathfrak{R}_i$ is a p'-group, $\mathfrak{S}_i = \mathfrak{R}_i$ and $\mathbf{O}_p(\mathfrak{G}/\mathfrak{R}_i) = 1$. By minimality of the counterexample, the theorem is valid for $(\mathfrak{G}/\mathfrak{R}_i$ on $V_i)$. By a), (1) cannot hold in $\mathfrak{G}/\mathfrak{R}_i$; also (3) cannot hold in $\mathfrak{G}/\mathfrak{R}_i$, since otherwise it holds in $\mathfrak{G}$. Thus (2) holds in $\mathfrak{G}/\mathfrak{R}_i$. Hence the Sylow 2-subgroups of $\mathfrak{G}$ are non-Abelian, p is a Fermat prime and there exists $x \in \mathfrak{H}$ for which $x^g \mathfrak{R}_i = x \mathfrak{R}_i \neq 1$. By V, 8.10, $\mathbf{C}_{\mathfrak{H}}(g) \neq 1$, and (2) holds in $\mathfrak{G}$, a contradiction.

d) Let $b = g^{p^{n-1}}$. $\mathfrak{H}$ is a special q-group for some prime $q \neq p$, b centralizes $\mathfrak{H}'$, $\mathbf{C}_{\mathfrak{H}/\mathfrak{H}'}(b) = 1$ and $(\mathfrak{P}$ on $\mathfrak{H}/\mathfrak{H}')$ is irreducible.

Let $\mathfrak{H}_0$ be a proper $\mathfrak{P}$-invariant subgroup of $\mathfrak{H}$. By minimality of the counterexample, the theorem holds in $\mathfrak{P}\mathfrak{H}_0$. But the validity of (1), (2) or (3) in $\mathfrak{P}\mathfrak{H}_0$ would imply the validity of the same in $\mathfrak{G}$, so $\mathbf{O}_p(\mathfrak{P}\mathfrak{H}_0) \neq 1$. Since $\mathbf{O}_p(\mathfrak{P}\mathfrak{H}_0) \leq \mathfrak{P}$, $b \in \mathbf{O}_p(\mathfrak{P}\mathfrak{H}_0)$ and $b \in \mathbf{C}_{\mathfrak{G}}(\mathfrak{H}_0)$. By 1.3, $b \notin \mathbf{C}_{\mathfrak{G}}(\mathfrak{H})$. Thus by 1.12, $\mathfrak{H}$ is a q-group for some prime $q \neq p$. And by III, 13.5, $\mathfrak{H}$ is a special q-group, b centralizes $\mathfrak{H}'$ and $(\mathfrak{P}$ on $\mathfrak{H}/\mathfrak{H}')$ is irreducible. Since $\mathbf{C}_{\mathfrak{H}/\mathfrak{H}'}(b)$ is $\mathfrak{P}$-invariant and $\mathbf{C}_{\mathfrak{H}/\mathfrak{H}'}(b) \neq \mathfrak{H}/\mathfrak{H}'$ by I, 4.4, $\mathbf{C}_{\mathfrak{H}/\mathfrak{H}'}(b) = 1$.

By V, 17.3,

$$V = V_1 \oplus \cdots \oplus V_l,$$

where V_i is the direct sum of isomorphic irreducible $K\mathfrak{H}$-submodules of V and V_i, V_j have no isomorphic irreducible $K\mathfrak{H}$-submodules for $i \neq j$; also there is a permutation representation σ of $\mathfrak{P}$ on $\{1, \ldots, l\}$ such that $V_i g = V_{i\sigma(g)}$. Since V is an irreducible $K\mathfrak{G}$-module, σ is transitive. Thus if $\ker \sigma = \langle g^{p^m} \rangle$, $\langle g^{p^m} \rangle$ is the stabiliser of each i and $l = p^m$. Let $c = g^{p^m}$ and $\mathfrak{H}_0 = \langle c, \mathfrak{H} \rangle$. Thus $\mathfrak{H}_0 \trianglelefteq \mathfrak{G}$.

e) $\mathfrak{H}_0 = \{x | x \in \mathfrak{G}, V_i x = V_i\}$ and V_i is an irreducible $K\mathfrak{H}_0$-module.

Since $i\sigma(c) = i$, $\mathsf{V}_i c = \mathsf{V}_i$. Conversely, if $\mathsf{V}_i y g^j = \mathsf{V}_i$ $(y \in \mathfrak{H})$, then $\mathsf{V}_i g^j = \mathsf{V}_i$, $g^j \in \langle c \rangle$ and $y g^j \in \mathfrak{H}_0$. V_i is an irreducible $\mathsf{K}\mathfrak{H}_0$-module by V, 17.3e).

f) $c \neq 1$ and the minimum polynomial of the linear transformation $u \to uc$ of V_i is not $t^{p^{n-m}} - 1$. In particular, $b \in \mathfrak{H}_0$.

Suppose that the minimum polynomial of the linear transformation $u \to uc$ of V_i is $t^{p^{n-m}} - 1$. Then, as is well-known, there exists $u \in \mathsf{V}_i$ such that $u, uc, \ldots, uc^{p^{n-m}-1}$ are linearly independent. But by a), $u, ug, \ldots, ug^{p^n-1}$ are linearly dependent, so there exist $\lambda_{jk} \in \mathsf{K}$ $(j = 0, \ldots, p^{n-m} - 1;$ $k = 0, \ldots, p^m - 1)$ such that

$$\sum_{j,k} \lambda_{jk} u g^{jp^m + k} = 0$$

and $\lambda_{jk} \neq 0$ for some j, k. Since $ug^{jp^m+k} = uc^j g^k \in \mathsf{V}_i\sigma(g^k)$, it follows that

$$\sum_{j=0}^{p^{n-m}-1} \lambda_{jk} u g^{jp^m+k} = 0 \quad (k = 0, \ldots, p^m - 1)$$

and hence

$$\sum_{j=0}^{p^{n-m}-1} \lambda_{jk} u c^j = 0,$$

contrary to the linear independence of $u, uc, \ldots, uc^{p^{n-m}-1}$.

In particular, $c \neq 1$ and $b = c^{p^{n-m-1}} \in \mathfrak{H}_0$.

Let $\mathfrak{K}_i = \ker(\mathfrak{H}_0 \text{ on } \mathsf{V}_i)$, and let $\mathfrak{G}_i = \mathfrak{H}_0/\mathfrak{K}_i$, $c_i = c\mathfrak{K}_i$ $(i = 1, \ldots, p^m)$. Thus by e), V_i is a faithful irreducible $\mathsf{K}\mathfrak{G}_i$-module.

g) $\bigcap_i \mathfrak{K}_i = 1$, $\mathfrak{K}_i^g = \mathfrak{K}_{i\sigma(g)}$ and $\mathfrak{K}_i < \mathfrak{H}$ $(i = 1, \ldots, p^m)$.

For $\bigcap_i \mathfrak{K}_i \leq \ker(\mathfrak{H}_0 \text{ on } \mathsf{V}) = 1$. Since $\mathfrak{K}_i = \ker(\mathfrak{H}_0 \text{ on } \mathsf{V}_i)$, $\mathfrak{K}_i^g = \ker(\mathfrak{H}_0^g \text{ on } \mathsf{V}_i g) = \ker(\mathfrak{H}_0 \text{ on } \mathsf{V}_{i\sigma(g)}) = \mathfrak{K}_{i\sigma(g)}$. If p divides $|\mathfrak{K}_1|$, $\langle c \rangle \cap \mathfrak{K}_1 \neq 1$, since $\langle c \rangle \in S_p(\mathfrak{H}_0)$ and $\mathfrak{K}_1 \trianglelefteq \mathfrak{H}_0$. Thus $b \in \mathfrak{K}_1$. Since σ is transitive and $\mathfrak{K}_i^g = \mathfrak{K}_{i\sigma(g)}$, it follows that $b \in \mathfrak{K}_i$ for all i and that $b = 1$, a contradiction. Thus $\mathfrak{K}_1$ is a p'-group and $\mathfrak{K}_1 \leq \mathfrak{H}$. If $\mathfrak{K}_1 = \mathfrak{H}$, then $\mathfrak{K}_i = \mathfrak{H}$ for all $i = 1, \ldots, p^m$ and $\mathfrak{H} = \bigcap_i \mathfrak{K}_i = 1$, a contradiction. Hence $\mathfrak{K}_1 < \mathfrak{H}$ and $\mathfrak{K}_i < \mathfrak{H}$ for all $i = 1, \ldots, p^m$.

Let $\mathfrak{Q}_i = \mathfrak{H}/\mathfrak{K}_i$ $(i = 1, \ldots, p^m)$.

h) $\mathfrak{Q}_i$ is extraspecial, $\mathbf{Z}(\mathfrak{Q}_i) \leq \mathbf{Z}(\mathfrak{G}_i)$, $\mathbf{C}_{\mathfrak{Q}_i/\mathfrak{Q}_i'}(b) = 1$ and V_i is a faithful irreducible $\mathsf{K}\mathfrak{Q}_i$-module.

Suppose that V_i is the direct sum of s_i isomorphic irreducible $\mathsf{K}\mathfrak{H}$-modules. By V, 17.5, there exists an irreducible projective representation ρ of $\mathfrak{H}_0/\mathfrak{H}$ of degree s_i. Then $\rho(c\mathfrak{H})^{p^{n-m}} = \lambda 1$ for some $\lambda \in \mathsf{K}$, and since K is algebraically closed, $\lambda = \mu^{p^{n-m}}$ for some $\mu \in \mathsf{K}$. Hence $c \to \mu^{-1}\rho(c\mathfrak{H})$ is an ordinary irreducible representation of $\mathfrak{H}_0/\mathfrak{H}$ of degree s_i. By V, 6.1, $s_i = 1$. Thus V_i is an irreducible $\mathsf{K}\mathfrak{H}$-module. Since $\mathfrak{K}_i = \ker(\mathfrak{H}$ on $\mathsf{V}_i)$ and $\mathfrak{Q}_i = \mathfrak{H}/\mathfrak{K}_i$, V_i is a faithful irreducible $\mathsf{K}\mathfrak{Q}_i$-module.

By Schur's lemma, $\mathbf{Z}(\mathfrak{Q}_i)$ is represented on V_i by scalar multiples of the unit mapping. Thus, since $(\mathfrak{G}_i$ on $\mathsf{V}_i)$ is faithful, $\mathbf{Z}(\mathfrak{Q}_i) \leq \mathbf{Z}(\mathfrak{G}_i)$ and since $b \in \mathfrak{H}_0$, $b\mathfrak{K}_i$ centralizes $\mathbf{Z}(\mathfrak{Q}_i)$. But by d), $\mathbf{C}_{\mathfrak{H}/\mathfrak{H}'}(b) = 1$, so by V, 8.10, $\mathbf{C}_{\mathfrak{H}/\mathfrak{H}'\mathfrak{K}_i}(b) = 1$. Hence $\mathbf{Z}(\mathfrak{Q}_i) \leq \mathfrak{H}'\mathfrak{K}_i/\mathfrak{K}_i$. Since $\mathfrak{H}$ is special, $\mathfrak{H}'\mathfrak{K}_i/\mathfrak{K}_i \leq \mathbf{Z}(\mathfrak{H}/\mathfrak{K}_i) = \mathbf{Z}(\mathfrak{Q}_i)$, and $\Phi(\mathfrak{Q}_i) = \Phi(\mathfrak{H})\mathfrak{K}_i/\mathfrak{K}_i = \mathfrak{H}'\mathfrak{K}_i/\mathfrak{K}_i = \mathfrak{Q}_i'$. Hence $\mathbf{Z}(\mathfrak{Q}_i) = \mathfrak{Q}_i' = \Phi(\mathfrak{Q}_i)$. By g), $\mathfrak{Q}_i \neq 1$, so $\mathfrak{Q}_i$ is extraspecial. Since $\mathfrak{Q}_i/\mathfrak{Q}_i' \cong \mathfrak{H}/\mathfrak{H}'\mathfrak{K}_i$, $\mathbf{C}_{\mathfrak{Q}_i/\mathfrak{Q}_i'}(b) = 1$.

Let $|\mathfrak{Q}_i| = q^{2d+1}$. By g), d is independent of i.

i) p is a Fermat prime and $q = 2$. Also $|\mathfrak{H}/\mathfrak{H}'| = 2^{2dp^m}$ and $(\langle c_i \rangle$ on $\mathfrak{Q}_i/\mathfrak{Q}_i')$ is irreducible.

By h), $\mathfrak{Q}_i'$ is centralized by c_i and $\mathbf{C}_{\mathfrak{Q}_i/\mathfrak{Q}_i'}(b) = 1$. We may therefore apply V, 17.13 to $(\mathfrak{G}_i$ on $\mathsf{V}_i)$, which is faithful and irreducible. Now by f), $(\langle c_i \rangle$ on $\mathsf{V}_i)$ has no component isomorphic to the regular representation $(\langle c_i \rangle$ on $\mathsf{K}\langle c_i \rangle)$. Hence V, 17.13b) shows that $q^d + 1 = |\langle c_i \rangle| = p^{n-m}$. In particular $q < p^n$. Hence if $p = 2$ and q is a Mersenne prime, the conclusion (3) of the theorem holds. This is not the case, so by 2.7, $q = 2$ and p is a Fermat prime; also either $n - m = 1$ or $n - m + 1 = p = d = 3$. It follows from 6.1 that $2^x \equiv 1\ (p^{n-m})$ if and only if $x \equiv 0\ (2d)$ and $2^y \equiv 1\ (p^n)$ if and only if $y \equiv 0\ (2dp^m)$.

It follows from II, 3.10 that $2d$ is the degree of any faithful irreducible representation of $\langle c_i \rangle$ in $GF(2)$. Since $\mathbf{C}_{\mathfrak{Q}_i/\mathfrak{Q}_i'}(b) = 1$, any irreducible component of $(\langle c_i \rangle$ on $\mathfrak{Q}_i/\mathfrak{Q}_i')$ is faithful and hence of degree $2d$. Thus $(\langle c_i \rangle$ on $\mathfrak{Q}_i/\mathfrak{Q}_i')$ is irreducible.

Also by II, 3.10, $2dp^m$ is the degree of any faithful irreducible representation of $\mathfrak{P} = \langle g \rangle$ in $GF(2)$. By d), $|\mathfrak{H}/\mathfrak{H}'| = 2^{2dp^m}$.

Note that since (2) does not hold, $\mathbf{C}_{\mathfrak{H}}(g) = 1$. It follows that $p^m > 1$, for otherwise $\mathfrak{K}_1 = 1$ by g), $\mathfrak{Q}_1 = \mathfrak{H}$, $\mathfrak{G}_1 = \mathfrak{G}$ and, by h), $\mathbf{Z}(\mathfrak{H}) \leq \mathbf{Z}(\mathfrak{G})$. Let

$$\mathfrak{L}_i = \bigcap_{j \neq i} \mathfrak{K}_j\mathfrak{H}' \quad (i = 1, \ldots, p^m).$$

j) If $i \neq j$, $vx = v$ for all $v \in \mathsf{V}_j$, $x \in [\mathfrak{L}_i, \mathfrak{H}]$.

For $[\mathfrak{L}_i, \mathfrak{H}] \leq [\mathfrak{K}_j\mathfrak{H}', \mathfrak{H}]$. By d), $\mathfrak{H}' \leq \mathbf{Z}(\mathfrak{H})$, so $[\mathfrak{L}_i, \mathfrak{H}] \leq [\mathfrak{K}_j, \mathfrak{H}] \leq \mathfrak{K}_j$. Thus j) follows at once.

k) $\mathfrak{H}' = \mathfrak{L}'_1 \times \cdots \times \mathfrak{L}'_{p^m}$, and $|\mathfrak{L}'_i| = 2$.

Since $|\mathfrak{H} : \mathfrak{K}_j \mathfrak{H}'| = |\mathfrak{Q}_j : \mathfrak{Q}'_j| = 2^{2d}$, it follows from I, 2.13 that $|\mathfrak{H} : \mathfrak{L}_i| \leq 2^{2d(p^m - 1)}$. From i), $\mathfrak{L}_i \not\leq \mathfrak{H}'$, so $\mathfrak{L}_i > \mathfrak{H}'$. Thus

$$\mathfrak{H} \geq \mathfrak{L}_1 \cdots \mathfrak{L}_{p^m} > \mathfrak{H}'.$$

But $\mathfrak{L}_i^g = \bigcap_{j \neq i} \mathfrak{K}_j^g \mathfrak{H}' = \bigcap_{j \neq i} \mathfrak{K}_{j\sigma(g)} \mathfrak{H}' = \mathfrak{L}_{i\sigma(g)}$, by g). Thus

$$(\mathfrak{L}_1 \cdots \mathfrak{L}_{p^m})^g = \mathfrak{L}_1 \cdots \mathfrak{L}_{p^m}$$

and $\mathfrak{L}_1 \cdots \mathfrak{L}_{p^m}$ is $\mathfrak{P}$-invariant. It follows from d) that

$$\mathfrak{L}_1 \cdots \mathfrak{L}_{p^m} = \mathfrak{H}.$$

Hence

$$\mathfrak{H}' = \mathfrak{L}'_1 \cdots \mathfrak{L}'_{p^m} \prod_{i \neq j} [\mathfrak{L}_i, \mathfrak{L}_j].$$

But by j), ($[\mathfrak{L}_i, \mathfrak{L}_j]$ on V_k) is trivial for all k ($i \neq j$), so $[\mathfrak{L}_i, \mathfrak{L}_j] = 1$. Thus

$$\mathfrak{H}' = \mathfrak{L}'_1 \cdots \mathfrak{L}'_{p^m}.$$

Now by j), ($\mathfrak{L}'_i$ on V_j) is trivial for all $j \neq i$. Hence ($\prod_{j \neq i} \mathfrak{L}'_j$ on V_i) is trivial and (($\mathfrak{L}'_i \cap \prod_{j \neq i} \mathfrak{L}'_j$) on V_k) is trivial for all k. Thus $\mathfrak{L}'_i \cap \prod_{j \neq i} \mathfrak{L}'_j = 1$, and

$$\mathfrak{H}' = \mathfrak{L}'_1 \times \cdots \times \mathfrak{L}'_{p^m}.$$

If $x \in \mathfrak{L}'_i$, then $x \in \mathfrak{H}'$, $x\mathfrak{K}_i \in \mathfrak{Q}'_i$ and $x\mathfrak{K}_i$ is represented on V_i by ± 1. Since x is represented on V_j by 1 for all $j \neq i$, it follows that $|\mathfrak{L}'_i| \leq 2$. Since $\mathfrak{L}'^g_i = \mathfrak{L}'_{i\sigma(g)}$, all the $\mathfrak{L}'_j$ are conjugate. Thus if $\mathfrak{L}'_i = 1$ for some i, $\mathfrak{L}'_i = 1$ for all i and $\mathfrak{H}' = 1$, contrary to h). Therefore $|\mathfrak{L}'_i| = 2$ for all i.

By k),

$$|\mathfrak{H}'| = 2^{p^m} \equiv 2\ (p).$$

Thus the length of some orbit of ($\langle g \rangle$ on $\mathfrak{H}'$) other than $\{1\}$ is not divisible by p and $\mathbf{C}_{\mathfrak{H}'}(g) \neq 1$. Thus (2) holds in $\mathfrak{G}$. **q.e.d.**

In our application of this, we shall need the following elementary lemma.

6.3 Lemma. *Suppose that* $\mathfrak{G}$ *is a group,* $\mathfrak{N} \trianglelefteq \mathfrak{G}$ *and* α *is an automorphism of* $\mathfrak{G}$ *which leaves fixed every element of* $\mathfrak{N}$*. Then* α *leaves* $\mathbf{C}_{\mathfrak{G}}(\mathfrak{N})$ *fixed and* α *induces the identity automorphism on* $\mathfrak{G}/\mathbf{C}_{\mathfrak{G}}(\mathfrak{N})$.

Proof. If $x \in \mathfrak{N}$ and $g \in \mathfrak{G}$, then α leaves fixed the elements x, x^g of $\mathfrak{N}$. Hence $x^g = (x^g)\alpha = (x\alpha)^{g\alpha} = x^{g\alpha}$ and $x^{(g\alpha)g^{-1}} = x$. Thus, for any $g \in \mathfrak{G}$, $(g\alpha)g^{-1} \in \mathbf{C}_{\mathfrak{G}}(\mathfrak{N})$. Hence α leaves $\mathbf{C}_{\mathfrak{G}}(\mathfrak{N})$ fixed and induces the identity automorphism on $\mathfrak{G}/\mathbf{C}_{\mathfrak{G}}(\mathfrak{N})$. **q.e.d.**

6.4 Theorem (SHULT [1], GROSS [3]). *Let p be an odd prime. If* $\mathfrak{G}$ *is a soluble group,* α *is an automorphism of* $\mathfrak{G}$ *of order* p^n *and* $\mathbf{C}_{\mathfrak{G}}(\alpha) = 1$*, the Fitting height of* $\mathfrak{G}$ *is at most n.*

(By the Fitting height of $\mathfrak{X}$ is meant the *nilpotente Länge* in the sense of III, 4.7; it will be denoted by $h(\mathfrak{X})$.)

Proof. This is proved by induction on $|\mathfrak{G}|$. It is trivial for $\mathfrak{G} = 1$. Suppose, then, that $\mathfrak{G} \neq 1$. Since $\mathbf{C}_{\mathfrak{G}}(\alpha) = 1$, the length of each orbit of ($\langle\alpha\rangle$ on $\mathfrak{G}$) other than $\{1\}$ is divisible by p. Thus $(p, |\mathfrak{G}|) = 1$.

Let $\mathfrak{N}$ be a minimal normal subgroup of $\mathfrak{G}$. By V, 8.10, $\mathbf{C}_{\mathfrak{G}/\mathfrak{N}}(\alpha) = 1$, so by the inductive hypothesis, $h(\mathfrak{G}/\mathfrak{N})$ is at most n. As in III, 4.6, we define a series

$$\mathfrak{G} = \mathfrak{K}_0 \geq \mathfrak{K}_1 \geq \cdots$$

of characteristic subgroups of $\mathfrak{G}$ inductively by the rules that $\mathfrak{K}_0 = \mathfrak{G}$ and, for $i > 0$, $\mathfrak{K}_{i-1}/\mathfrak{K}_i$ is the maximal nilpotent factor group of $\mathfrak{K}_{i-1}$. Since $h(\mathfrak{G}/\mathfrak{N}) \leq n$, $\mathfrak{K}_n \leq \mathfrak{N}$. Hence either $\mathfrak{K}_n = 1$ or $\mathfrak{K}_n = \mathfrak{N}$. If $\mathfrak{K}_n = 1$, there is nothing to prove. We therefore assume that $\mathfrak{K}_n = \mathfrak{N}$. Thus $\mathfrak{N}$ is the only minimal normal subgroup of $\mathfrak{G}$. Since $\mathfrak{G}$ is soluble, $\mathfrak{N}$ is an elementary Abelian r-group for some prime r. It follows that $\mathbf{O}_{r'}(\mathfrak{G}) = 1$ and $\mathbf{O}_r(\mathfrak{G}) = \mathbf{F}(\mathfrak{G})$. Let

$$\mathfrak{U} = \Phi(\mathbf{O}_r(\mathfrak{G})).$$

By 1.4 and 1.6, $\mathbf{C}_{\mathfrak{G}}(\mathbf{O}_r(\mathfrak{G})/\mathfrak{U}) = \mathbf{O}_r(\mathfrak{G})$. Hence $\mathbf{O}_{r'}(\mathfrak{G}/\mathfrak{U}) = 1$ and $\mathbf{O}_r(\mathfrak{G})/\mathfrak{U} = \mathbf{F}(\mathfrak{G}/\mathfrak{U})$. If $\mathfrak{U} \neq 1$, $\mathfrak{U} \geq \mathfrak{N}$ and $h(\mathfrak{G}/\mathfrak{U})$ is at most n. By III, 4.6, the Fitting height of $(\mathfrak{G}/\mathfrak{U})/\mathbf{F}(\mathfrak{G}/\mathfrak{U}) \cong \mathfrak{G}/\mathbf{O}_r(\mathfrak{G})$ is at most $n - 1$. Hence $h(\mathfrak{G}) \leq n$, and there is nothing further to prove. We therefore suppose that $\mathfrak{U} = 1$ and write $\mathfrak{B} = \mathbf{O}_r(\mathfrak{G})$. Thus $\mathfrak{B}$ is an elementary Abelian r-group.

Let $\mathfrak{M} = \mathbf{O}_{r,r'}(\mathfrak{G})$. We show that $\beta = \alpha^{p^{n-1}}$ induces the identity mapping on $\mathfrak{M}/\mathfrak{B}$. Suppose that this is not the case, and let $\mathfrak{H}$ be the semi-direct product $\mathfrak{M}\langle\alpha\rangle$. Since $p \neq r$, $\mathfrak{B} = \mathbf{O}_r(\mathfrak{H})$. If $\mathfrak{P}/\mathfrak{B} = \mathbf{O}_p(\mathfrak{H}/\mathfrak{B})$, $[\mathfrak{M}, \mathfrak{P}] \leq \mathfrak{B}$, since p does not divide $|\mathfrak{M}|$. Hence $\beta \notin \mathfrak{P}$ and $\langle\alpha\mathfrak{B}\rangle \cap (\mathfrak{P}/\mathfrak{B}) = 1$. Since $\langle\alpha\mathfrak{B}\rangle$ is a Sylow p-subgroup of $\mathfrak{H}/\mathfrak{B}$ and $\mathfrak{P}/\mathfrak{B}$ is a normal p-subgroup of $\mathfrak{H}/\mathfrak{B}$, it follows that $\mathfrak{P}/\mathfrak{B} = 1$. Thus $\mathbf{O}_p(\mathfrak{H}/\mathfrak{B}) = 1$. Also, since $\mathbf{O}_{r'}(\mathfrak{G}) = 1$, $\mathbf{O}_{r'}(\mathfrak{H}) \cap \mathfrak{M} \leq \mathbf{O}_{r'}(\mathfrak{M}) = 1$ and $\mathbf{O}_{r'}(\mathfrak{H})$ is a p-group. Thus $\mathbf{O}_{r'}(\mathfrak{H})\mathfrak{B}/\mathfrak{B} \leq \mathbf{O}_p(\mathfrak{H}/\mathfrak{B}) = 1$, $\mathbf{O}_{r'}(\mathfrak{H}) \leq \mathfrak{B}$ and, since $\mathfrak{B}$ is an r-group, $\mathbf{O}_{r'}(\mathfrak{H}) = 1$. By 1.3, $\mathbf{C}_{\mathfrak{H}}(\mathfrak{B}) = \mathfrak{B}$. Hence ($\mathfrak{H}/\mathfrak{B}$ on $\mathfrak{B}$) is faithful. By 6.2, either (1) $\mathbf{C}_{\mathfrak{B}}(\alpha) \neq 1$, or (2) $\mathbf{C}_{\mathfrak{H}/\mathfrak{B}}(\alpha) \neq 1$, since p is odd. But $\mathbf{C}_{\mathfrak{G}}(\alpha) = 1$, so (1) is impossible. Also (2) is impossible, by V, 8.10. Thus β induces the identity mapping on $\mathfrak{M}/\mathfrak{B}$.

If $\mathfrak{C}/\mathfrak{B} = \mathbf{C}_{\mathfrak{G}/\mathfrak{B}}(\mathfrak{M}/\mathfrak{B})$, it follows from 6.3 that β induces the identity mapping on $\mathfrak{G}/\mathfrak{C}$. But by 1.3, $\mathfrak{C} \leq \mathfrak{M}$, so β induces the identity mapping on $\mathfrak{G}/\mathfrak{M}$ and on $\mathfrak{M}/\mathfrak{B}$. By I, 4.4, β induces the identity mapping on $\mathfrak{G}/\mathfrak{B}$. Thus α induces an automorphism $\bar{\alpha}$ on $\mathfrak{G}/\mathfrak{B}$ of order at most p^{n-1}, and by V, 8.10, $\mathbf{C}_{\mathfrak{G}/\mathfrak{B}}(\bar{\alpha}) = 1$. By the inductive hypothesis, $h(\mathfrak{G}/\mathfrak{B}) \leq n - 1$. Since $\mathfrak{B}$ is nilpotent, $h(\mathfrak{G}) \leq n$. **q.e.d.**

6.5 Remark. SHULT [1] has shown that the bound in 6.4 is best possible. More generally, GROSS [4] has shown that if $\mathfrak{A}$ is a soluble group of order $p_1 p_2 \cdots p_n$, where $p_1, p_2, \ldots, p_n$ are (not necessarily distinct) primes, there exists a finite soluble group $\mathfrak{G}$ of Fitting height n for which $\mathfrak{A}$ is a group of operators and $\mathbf{C}_{\mathfrak{G}}(\mathfrak{A}) = 1$.

To obtain results corresponding to 6.4 for $p = 2$, we need the following lemmas.

6.6 Lemma. *Suppose that $\mathfrak{G}$ is of odd order and τ is an automorphism of $\mathfrak{G}$ such that $\tau^2 = 1$.*

a) *If $x \in \mathfrak{G}$, there exist unique elements y, z of $\mathfrak{G}$ such that $x = yz$, $y\tau = y$ and $z\tau = z^{-1}$.*

b) *Suppose that x, y are elements of $\mathfrak{G}$ such that $x\tau = x^{\varepsilon}$, $(x^y)\tau = (x^y)^{\varepsilon}$ and $(y\tau)y \in \mathbf{C}_{\mathfrak{G}}(x)$, where $\varepsilon = \pm 1$. Then x and y commute.*

Proof. a) Since $|\mathfrak{G}|$ is odd, there exists $z \in \mathfrak{G}$ such that $z^2 = (x\tau)^{-1}x$. Then

$$(z\tau)^2 = z^2\tau = x^{-1}(x\tau) = z^{-2}.$$

Since $|\mathfrak{G}|$ is odd, $z\tau = z^{-1}$. If $y = xz^{-1}$,

$$y\tau = (x\tau)(z\tau)^{-1} = xz^{-2}z = xz^{-1} = y.$$

Clearly, y and z are unique.

b) We have

$$x^{y^2} = (x^y)^y = (((x^y)\tau)^y)^\varepsilon = ((x\tau)^{(y\tau)y})^\varepsilon = x^{(y\tau)y} = x,$$

so x commutes with y^2. Since $|\mathfrak{G}|$ is odd, x commutes with y. **q.e.d.**

Note that by 6.6a), if $\mathbf{C}_{\mathfrak{G}}(\tau) = 1$, then $x\tau = x^{-1}$ for all $x \in \mathfrak{G}$, whence $\mathfrak{G}$ is Abelian (V, 8.18a)).

6.7 Lemma. *Suppose that $\mathfrak{G}$ is soluble, $\mathfrak{L} \trianglelefteq \mathfrak{G}$, $\mathfrak{L}$ is not nilpotent but $\mathfrak{L}\mathfrak{K}/\mathfrak{K}$ is nilpotent for every non-identity characteristic subgroup $\mathfrak{K}$ of $\mathfrak{G}$. Then there exists a prime p such that $\mathbf{O}_{p'}(\mathfrak{G}) = 1$ and $\mathbf{O}_p(\mathfrak{G})$ is elementary Abelian.*

Proof. Let p be the prime divisor of an elementary Abelian normal subgroup of $\mathfrak{G}$. Then $\mathfrak{P} = \mathbf{O}_p(\mathfrak{G}) \neq 1$. Suppose that $\mathfrak{N} = \mathbf{O}_{p'}(\mathfrak{G}) \neq 1$. Then $\mathfrak{L}\mathfrak{P}/\mathfrak{P}$ and $\mathfrak{L}\mathfrak{N}/\mathfrak{N}$ are nilpotent, so $\mathfrak{L}/(\mathfrak{L} \cap \mathfrak{P})$ and $\mathfrak{L}/(\mathfrak{L} \cap \mathfrak{N})$ are nilpotent. By III, 2.5, $\mathfrak{L}/(\mathfrak{L} \cap \mathfrak{P} \cap \mathfrak{N})$ is nilpotent. But $\mathfrak{P} \cap \mathfrak{N} = 1$, so $\mathfrak{L}$ is nilpotent, contrary to hypothesis. Thus $\mathbf{O}_{p'}(\mathfrak{G}) = 1$.

If $\mathfrak{P} = \mathbf{O}_p(\mathfrak{G})$ is not elementary Abelian, $\Phi(\mathfrak{P}) \neq 1$. By III, 3.3b), $\Phi(\mathfrak{P}) \leq \Phi(\mathfrak{G})$, so $\Phi(\mathfrak{G}) \neq 1$. Hence $\mathfrak{L}\Phi(\mathfrak{G})/\Phi(\mathfrak{G})$ is nilpotent. By III, 3.5, $\mathfrak{L}\Phi(\mathfrak{G})$ is nilpotent. Thus $\mathfrak{L}$ is nilpotent, a contradiction. Hence $\mathbf{O}_p(\mathfrak{G})$ is elementary Abelian. **q.e.d.**

6.8 Theorem (GORENSTEIN and HERSTEIN [1]). *Let σ be an automorphism of the soluble group $\mathfrak{G}$. If $\sigma^4 = 1$ and $\mathbf{C}_{\mathfrak{G}}(\sigma) = 1$, $\mathfrak{G}'$ is nilpotent.*

Proof. Suppose that this is false and let $\mathfrak{G}$ be a counterexample of minimal order.

a) $|\mathfrak{G}|$ is odd.

If l is the length of any orbit of σ on $\mathfrak{G} - \{1\}$, l divides 4 and $l > 1$, so l is even. Hence $|\mathfrak{G}| - 1$ is even and $|\mathfrak{G}|$ is odd.

b) There exists a prime p such that $\mathbf{O}_{p'}(\mathfrak{G}) = 1$ and $\mathfrak{P} = \mathbf{O}_p(\mathfrak{G})$ is elementary Abelian.

Let $\mathfrak{X}$ be any non-identity characteristic subgroup of $\mathfrak{G}$. By V, 8.10, σ induces a fixed point free automorphism on $\mathfrak{G}/\mathfrak{X}$. Since $\mathfrak{G}$ is a counterexample of minimal order, $\mathfrak{G}'\mathfrak{X}/\mathfrak{X}$ is nilpotent. The assertion b) thus follows from 6.7.

Let $\tau = \sigma^2$, $\mathfrak{C} = \mathbf{C}_{\mathfrak{G}}(\tau)$.

c) $\mathfrak{C}$ is Abelian.

For $\mathfrak{C}$ is σ-invariant and the restriction σ' to $\mathfrak{C}$ of σ is an automorphism of $\mathfrak{C}$ of order 2 such that $\mathbf{C}_{\mathfrak{C}}(\sigma') = 1$. Hence $\mathfrak{C}$ is Abelian.

By 1.11 applied to the semi-direct product $\mathfrak{G}\langle\sigma\rangle$, $\mathfrak{G}$ has a σ-invariant Hall p'-subgroup $\mathfrak{H}$.

d) $\mathfrak{H} \not\leq \mathfrak{C}$.

Let $\mathfrak{R} = \mathbf{O}_{p,p'}(\mathfrak{G})$. By 1.3, $\mathbf{C}_{\mathfrak{G}}(\mathfrak{R}/\mathfrak{P}) \leq \mathfrak{R}$. Since $\mathfrak{R} \trianglelefteq \mathfrak{G}$, $\mathfrak{H} \cap \mathfrak{R}$ is a Hall p'-subgroup of $\mathfrak{R}$ and $\mathfrak{R} = \mathfrak{P}(\mathfrak{H} \cap \mathfrak{R})$. By c), $\mathfrak{C}$ is Abelian, so if $\mathfrak{H} \leq \mathfrak{C}$, $\mathfrak{C}$ centralizes $\mathfrak{H} \cap \mathfrak{R}$ and $\mathfrak{R}/\mathfrak{P}$. Thus $\mathfrak{C} \leq \mathfrak{R}$. Hence $\mathfrak{P}\mathfrak{C} \leq \mathfrak{R} = \mathfrak{P}(\mathfrak{H} \cap \mathfrak{R}) \leq \mathfrak{P}\mathfrak{C}$ and $\mathfrak{R} = \mathfrak{P}\mathfrak{C}$. Thus $\mathfrak{R}' \leq \mathfrak{P}$ and $\mathfrak{R}'$ is nilpotent. Hence $\mathfrak{R} < \mathfrak{G}$. If $y \in \mathfrak{G}$ and $y \notin \mathfrak{R}$, $y = uz$, where $u \in \mathfrak{C}$ and $z\tau = z^{-1}$, by 6.6a). Thus $u \in \mathfrak{R}$ and $z \notin \mathfrak{R}$. Hence there exists $\bar{x} \in \mathfrak{R}/\mathfrak{P}$ such that $\bar{x}$ and $\bar{z} = z\mathfrak{P}$ do not commute. But since $\mathfrak{R} = \mathfrak{P}\mathfrak{C}$, $\bar{x}\bar{\tau} = \bar{x}$ and $(\bar{x}^{\bar{z}})\bar{\tau} = \bar{x}^{\bar{z}}$, where $\bar{\tau}$ is the automorphism of $\mathfrak{G}/\mathfrak{P}$ induced by τ. Since $\bar{z}\bar{\tau} = \bar{z}^{-1}$, 6.6b) shows that $\bar{x}$ and $\bar{z}$ commute, a contradiction. Thus $\mathfrak{H} \not\leq \mathfrak{C}$.

By d), we may choose a minimal σ-invariant subgroup $\mathfrak{J}$ of $\mathfrak{H}$ such that $\mathfrak{J} \not\leq \mathfrak{C}$. Let $\mathfrak{K}$ be a maximal normal σ-invariant subgroup of $\mathfrak{J}$. Thus $\mathfrak{J}/\mathfrak{K}$ is an elementary Abelian group and ($\langle\sigma\rangle$ on $\mathfrak{J}/\mathfrak{K}$) is irreducible. Also $\mathfrak{K} \leq \mathfrak{C}$, by definition of $\mathfrak{J}$. It follows from c) that $\mathfrak{K}$ is Abelian.

By b) and the Maschke-Schur theorem, applied to $\langle\sigma\rangle\mathfrak{J}$,

$$\mathfrak{P} = \mathfrak{P}_1 \times \cdots \times \mathfrak{P}_n,$$

where $\langle\sigma\rangle\mathfrak{J} \leq \mathbf{N}_{\mathfrak{G}}(\mathfrak{P}_i)$ and ($\langle\sigma\rangle\mathfrak{J}$ on $\mathfrak{P}_i$) is irreducible. Also, since $\mathfrak{J} \not\leq \mathfrak{C}$, it follows by applying 6.6a) to $\mathfrak{J}$ that there exists $z \in \mathfrak{J} - \{1\}$ such that $z\tau = z^{-1}$. By b) and 1.3, $\mathfrak{P} = \mathbf{C}_{\mathfrak{G}}(\mathfrak{P})$, so z does not centralize some $\mathfrak{P}_i$. Denote such a $\mathfrak{P}_i$ by $\mathfrak{B}$; thus ($\langle\sigma\rangle\mathfrak{J}$ on $\mathfrak{B}$) is irreducible and $z \notin \mathbf{C}_{\mathfrak{G}}(\mathfrak{B})$.

e) $[\mathfrak{K}, \mathfrak{B}] = 1$.

If not, $\mathbf{C}_{\mathfrak{B}}(\mathfrak{K})$ is a proper $\langle\sigma\rangle\mathfrak{J}$-invariant subgroup of $\mathfrak{B}$, so $\mathbf{C}_{\mathfrak{B}}(\mathfrak{K}) = 1$. But since $\mathfrak{K} \leq \mathfrak{C}$ and $\mathfrak{C}$ is Abelian, $\mathfrak{C} \cap \mathfrak{B} \leq \mathbf{C}_{\mathfrak{B}}(\mathfrak{K})$, so $\mathfrak{C} \cap \mathfrak{B} = 1$. Hence $\mathbf{C}_{\mathfrak{B}}(\tau) = 1$ and $v\tau = v^{-1}$ for all $v \in \mathfrak{B}$. Hence $(v^z)\tau = (v^z)^{-1}$ and by 6.6b), $z \in \mathbf{C}_{\mathfrak{G}}(\mathfrak{B})$, a contradiction. Hence $[\mathfrak{K}, \mathfrak{B}] = 1$.

Let $\rho = (\langle\sigma\rangle\mathfrak{J}$ on $\mathfrak{B})$. By e), $\ker\rho \geq \mathfrak{K}$. Since $z \notin \mathbf{C}_{\mathfrak{G}}(\mathfrak{B})$, $\mathfrak{J} \not\leq \ker\rho$. Thus $(\ker\rho) \cap \mathfrak{J}$ is a proper, normal σ-invariant subgroup of $\mathfrak{J}$ containing $\mathfrak{K}$. By definition of $\mathfrak{K}$, $\mathfrak{K} = (\ker\rho) \cap \mathfrak{J}$. But $\sigma^2 = \tau$ does not centralize $\mathfrak{J}/\mathfrak{K}$. Also $z \notin \mathbf{C}_{\mathfrak{G}}(\mathfrak{B})$, so by 6.6b), $\tau \notin \ker\rho$. Hence by 1.10 applied to $\operatorname{im}\rho$, the minimum polynomial of the linear transformation of $\mathfrak{B}$ induced by σ is of degree 4. Thus there exists $v \in \mathfrak{B}$ such that $u = v(v\sigma)(v\sigma^2)(v\sigma^3) \neq 1$. But $u\sigma = u$, so $\mathbf{C}_{\mathfrak{G}}(\sigma) \neq 1$, a contradiction.

q.e.d.

6.9 Theorem (GROSS). *Suppose that $\mathfrak{G}$ is a soluble group and σ is an automorphism of $\mathfrak{G}$ such that $\mathbf{C}_{\mathfrak{G}}(\sigma) = 1$. If the order of σ is 2^n ($n > 1$), the Fitting height of $\mathfrak{G}$ is at most $2n - 2$.*

Proof. Suppose that this is false and let $\mathfrak{G}$ be a counterexample of minimal order. By 6.8, $n \geq 3$.

a) $|\mathfrak{G}|$ is odd.

As in 6.4, the length of any orbit of σ on $\mathfrak{G} - \{1\}$ is even, so $|\mathfrak{G}| - 1$ is even and $|\mathfrak{G}|$ is odd.

b) There exists a prime p such that $\mathbf{O}_{p'}(\mathfrak{G}) = 1$ and $\mathfrak{P} = \mathbf{O}_p(\mathfrak{G})$ is elementary Abelian.

Define the series

$$\mathfrak{G} = \mathfrak{G}_0 \geq \mathfrak{G}_1 \geq \cdots$$

of characteristic subgroups $\mathfrak{G}_i$ of $\mathfrak{G}$ by the fact that for $i > 0$, $\mathfrak{G}_{i-1}/\mathfrak{G}_i$ is the maximal nilpotent factor group of $\mathfrak{G}_{i-1}$. Then $\mathfrak{G}_{2n-2} \neq 1$, by III, 4.6, and $\mathfrak{G}_{2n-3}$ is not nilpotent. But if $\mathfrak{X}$ is any non-identity characteristic subgroup of $\mathfrak{G}$, $\mathbf{C}_{\mathfrak{G}/\mathfrak{X}}(\sigma) = 1$ by V, 8.10. Since $\mathfrak{G}$ is a counterexample of minimal order, the Fitting height of $\mathfrak{G}/\mathfrak{X}$ is at most $2n - 2$, so $\mathfrak{G}_{2n-3}\mathfrak{X}/\mathfrak{X}$ is nilpotent. The assertion b) thus follows from 6.7.

Let $\tau = \sigma^{2^{n-1}}$.

c) The restriction of τ to $\mathfrak{P}$ is not the identity mapping.

Since $\tau \neq 1$, it follows from 6.6a) that there exists $z \in \mathfrak{G}$ such that $z\tau = z^{-1} \neq 1$. But if c) is false, $x\tau = x$ and $(x^z)\tau = x^z$ for all $x \in \mathfrak{P}$, so by 6.6b), $z \in \mathbf{C}_{\mathfrak{G}}(\mathfrak{P})$. But by b) and 1.3, $\mathbf{C}_{\mathfrak{G}}(\mathfrak{P}) = \mathfrak{P}$. Thus $z \in \mathfrak{P}$. Since $z\tau = z^{-1}$, the restriction of τ to $\mathfrak{P}$ is not the identity mapping.

Let $\overline{\mathfrak{G}}$ be the semi-direct product $\langle\sigma\rangle\mathfrak{G}$, let $\mathfrak{F}_1/\mathfrak{P} = \mathbf{F}(\overline{\mathfrak{G}}/\mathfrak{P})$, $\mathfrak{F}_2/\mathfrak{F}_1 = \mathbf{F}(\overline{\mathfrak{G}}/\mathfrak{F}_1)$.

d) $\tau \in \mathfrak{F}_2$.

If $\mathfrak{N} \trianglelefteq \overline{\mathfrak{G}}$ and $|\mathfrak{N}|$ is even, then $\mathfrak{N}$ intersects every Sylow 2-subgroup of $\overline{\mathfrak{G}}$ non-trivially, so $\tau \in \mathfrak{N}$. By c), $\tau \notin \mathbf{C}_{\overline{\mathfrak{G}}}(\mathfrak{P})$, so $|\mathbf{C}_{\overline{\mathfrak{G}}}(\mathfrak{P})|$ is odd. Hence $\mathbf{C}_{\overline{\mathfrak{G}}}(\mathfrak{P}) = \mathbf{C}_{\mathfrak{G}}(\mathfrak{P}) = \mathfrak{P}$.

Thus if $\rho = (\overline{\mathfrak{G}}/\mathfrak{P}$ on $\mathfrak{P})$, ρ is faithful. If $|\mathfrak{F}_1/\mathfrak{P}|$ is even, $\tau \in \mathfrak{F}_1$ and the assertion is clear; thus we may suppose that $|\mathfrak{F}_1/\mathfrak{P}|$ is odd. Clearly $|\mathfrak{F}_1/\mathfrak{P}|$ is not divisible by p, since $\mathfrak{F}_1/\mathfrak{P}$ is nilpotent and $\mathfrak{P} = \mathbf{O}_p(\mathfrak{G})$. We also observe that $\rho(\sigma\mathfrak{P})$ is exceptional on $\mathfrak{P}$, in the sense of 3.6. For otherwise the degree of the minimum polynomial of $\rho(\sigma\mathfrak{P})$ is 2^n. Thus there exists $v \in \mathfrak{P}$ such that

$$u = v(v\sigma)\cdots(v\sigma^{2^n-1}) \neq 1.$$

But then $u\sigma = u$, so $\mathbf{C}_{\mathfrak{G}}(\sigma) \neq 1$, a contradiction.

It follows from 3.12 that $\tau = \sigma^{2^{n-1}} \in \mathfrak{F}_2$.

e) τ induces the identity automorphism on $\mathfrak{G}/(\mathfrak{F}_1 \cap \mathfrak{G})$.

By d), $\tau \in \mathfrak{F}_2$; also, $\mathfrak{F}_2/\mathfrak{F}_1$ is nilpotent. Since $|(\mathfrak{F}_2 \cap \mathfrak{G})\mathfrak{F}_1/\mathfrak{F}_1|$ is odd, it follows that τ centralizes $(\mathfrak{F}_2 \cap \mathfrak{G})\mathfrak{F}_1/\mathfrak{F}_1$. Thus τ induces the identity automorphism on $(\mathfrak{F}_2 \cap \mathfrak{G})/(\mathfrak{F}_1 \cap \mathfrak{G})$. But since $\mathfrak{G}$ and $\mathfrak{F}_2$ are normal in $\overline{\mathfrak{G}}$, $[\mathfrak{G}, \tau] \leq \mathfrak{F}_2 \cap \mathfrak{G}$, so τ induces the identity automorphism on $\mathfrak{G}/(\mathfrak{F}_2 \cap \mathfrak{G})$. By I, 4.4, τ induces the identity automorphism on $\mathfrak{G}/(\mathfrak{F}_1 \cap \mathfrak{G})$.

It follows from e) that if σ' is the automorphism of $\mathfrak{G}/(\mathfrak{F}_1 \cap \mathfrak{G})$ induced by σ, $\sigma'^{2^{n-1}} = 1$. By V, 8.10, σ' is fixed point free. Since $n \geq 3$ and $\mathfrak{G}$ is a counterexample of minimal order, it follows that the Fitting height of $\mathfrak{G}/(\mathfrak{F}_1 \cap \mathfrak{G})$ is at most $2n - 4$. Since $(\mathfrak{F}_1 \cap \mathfrak{G})/\mathfrak{P}$ and $\mathfrak{P}$ are nilpotent, the Fitting height of $\mathfrak{G}$ is at most $2n - 2$. **q.e.d.**

6.10 Remark. The hypothesis of solubility in 6.8 and 6.9 can be omitted. For the proof that $|\mathfrak{G}|$ is odd in both theorems does not utilize the solubility of $\mathfrak{G}$, and this then follows from the theorem of Feit and Thompson. A proof of the solubility of $\mathfrak{G}$ in 6.8, independent of the theorem of Feit and Thompson, was given by Gorenstein and Herstein (see GORENSTEIN [1], p. 342).

It has been conjectured that if the group $\mathfrak{G}$ has an automorphism σ such that $\mathbf{C}_{\mathfrak{G}}(\sigma) = 1$, then $\mathfrak{G}$ is soluble.

We conclude this section with a theorem of J. N. WARD [1] on soluble groups having an elementary Abelian fixed point free automorphism group of order p^2. For this we need the following lemmas.

6.11 Lemma. *If $\mathfrak{N}$ is a normal p'-subgroup and $\mathfrak{P}$ is a p-subgroup of $\mathfrak{G}$, $\mathbf{N}_{\mathfrak{G}/\mathfrak{N}}(\mathfrak{P}\mathfrak{N}/\mathfrak{N}) = \mathbf{N}_{\mathfrak{G}}(\mathfrak{P})\mathfrak{N}/\mathfrak{N}$ and $\mathbf{C}_{\mathfrak{G}/\mathfrak{N}}(\mathfrak{P}\mathfrak{N}/\mathfrak{N}) = \mathbf{C}_{\mathfrak{G}}(\mathfrak{P})\mathfrak{N}/\mathfrak{N}$.*

Proof. Let $\mathfrak{K}/\mathfrak{N} = \mathbf{N}_{\mathfrak{G}/\mathfrak{N}}(\mathfrak{P}\mathfrak{N}/\mathfrak{N})$. Then $\mathfrak{P}\mathfrak{N} \trianglelefteq \mathfrak{K}$ and $\mathfrak{P} \in S_p(\mathfrak{P}\mathfrak{N})$, so by the Frattini argument, $\mathfrak{K} = \mathbf{N}_{\mathfrak{K}}(\mathfrak{P})(\mathfrak{P}\mathfrak{N}) \leq \mathbf{N}_{\mathfrak{G}}(\mathfrak{P})\mathfrak{N}$. Clearly $\mathbf{N}_{\mathfrak{G}}(\mathfrak{P}) \leq \mathfrak{K}$, so $\mathfrak{K} = \mathbf{N}_{\mathfrak{G}}(\mathfrak{P})\mathfrak{N}$.

If $\mathfrak{L}/\mathfrak{N} = \mathbf{C}_{\mathfrak{G}/\mathfrak{N}}(\mathfrak{P}\mathfrak{N}/\mathfrak{N})$, then $\mathfrak{N} \leq \mathfrak{L} \leq \mathfrak{K} = \mathbf{N}_{\mathfrak{G}}(\mathfrak{P})\mathfrak{N}$. Hence $\mathfrak{L} = \mathfrak{M}\mathfrak{N}$, where $\mathfrak{M} = \mathfrak{L} \cap \mathbf{N}_{\mathfrak{G}}(\mathfrak{P})$. Thus $[\mathfrak{P}, \mathfrak{M}] \leq \mathfrak{P}$. But also

$$[\mathfrak{P}, \mathfrak{M}] \leq [\mathfrak{P}, \mathfrak{L}] \leq \mathfrak{N},$$

so $[\mathfrak{P}, \mathfrak{M}] \leq \mathfrak{P} \cap \mathfrak{N} = 1$ and $\mathfrak{M} \leq \mathbf{C}_{\mathfrak{G}}(\mathfrak{P})$. Thus $\mathfrak{L} = \mathbf{C}_{\mathfrak{G}}(\mathfrak{P})\mathfrak{N}$. **q.e.d.**

6.12 Theorem. *Suppose that $\mathfrak{G} = \mathfrak{A}\mathfrak{H}$, where $\mathfrak{A}$ is elementary Abelian of order p^2 and $\mathfrak{H}$ is a soluble, normal p'-subgroup of $\mathfrak{G}$. Let V be a faithful $\mathsf{K}\mathfrak{G}$-module, where K is a field of finite characteristic q such that $q \neq p$ and $\mathbf{O}_q(\mathfrak{H}) = 1$. Suppose that for any non-identity element a of $\mathfrak{A}$, $\mathbf{C}_{\mathfrak{H}}(a)$ is nilpotent and any q'-element of $\mathbf{C}_{\mathfrak{H}}(a)$ leaves fixed every element of $\mathbf{C}_{\mathsf{V}}(a)$. Then $\mathfrak{H}$ is nilpotent.*

Proof. If L is an extension of K and $\mathsf{V}_\mathsf{L} = \mathsf{V} \otimes_\mathsf{K} \mathsf{L}$, $\mathbf{C}_{\mathsf{V}_\mathsf{L}}(g) = \mathbf{C}_\mathsf{V}(g) \otimes_\mathsf{K} \mathsf{L}$ for all $g \in \mathfrak{G}$. If $a \in \mathfrak{A} - \{1\}$, $\mathbf{C}_\mathsf{V}(a) \leq \mathbf{C}_\mathsf{V}(g)$ for every q'-element g of $\mathbf{C}_\mathfrak{H}(a)$, so $\mathbf{C}_{\mathsf{V}_\mathsf{L}}(a) \leq \mathbf{C}_{\mathsf{V}_\mathsf{L}}(g)$. We may therefore suppose without loss of generality that K is algebraically closed.

Suppose now that the theorem is false. Let $\mathfrak{G}$ be a counterexample of minimal order, and let V be a faithful $\mathsf{K}\mathfrak{G}$-module of minimal dimension, such that $\mathbf{C}_\mathsf{V}(a) \leq \mathbf{C}_\mathsf{V}(g)$ whenever $a \in \mathfrak{A} - \{1\}$ and g is a q'-element of $\mathbf{C}_\mathfrak{H}(a)$.

a) V is an irreducible $\mathsf{K}\mathfrak{G}$-module.

To prove this, let

$$\mathsf{V} = \mathsf{V}_0 > \mathsf{V}_1 > \cdots > \mathsf{V}_r = 0$$

be a $\mathsf{K}\mathfrak{G}$-composition series of V. If $\mathfrak{K}_i = \ker(\mathfrak{G} \text{ on } \mathsf{V}_{i-1}/\mathsf{V}_i)$ $(i = 1, \ldots, r)$ and $\mathfrak{K} = \bigcap_i \mathfrak{K}_i$, $\mathfrak{K}$ is a q-group (cf. I, 4.4). Since $\mathbf{O}_q(\mathfrak{G}) = 1$, $\mathfrak{K} = 1$. Thus $\mathfrak{H}/(\bigcap_i \mathfrak{K}_i)$ is not nilpotent. By III, 2.5, $\mathfrak{H}\mathfrak{K}_i/\mathfrak{K}_i$ is not nilpotent for some i. We prove that the conditions of the theorem are satisfied by the $\mathsf{K}(\mathfrak{G}/\mathfrak{K}_i)$-module $\mathsf{V}_{i-1}/\mathsf{V}_i$. First, $\mathbf{O}_q(\mathfrak{G}/\mathfrak{K}_i) = 1$, since $\mathsf{V}_{i-1}/\mathsf{V}_i$ is a faithful irreducible $\mathsf{K}(\mathfrak{G}/\mathfrak{K}_i)$-module. Next, suppose that $a \in \mathfrak{A} - \{1\}$. By 6.11,

$$\mathbf{C}_{\mathfrak{H}\mathfrak{K}_i/\mathfrak{K}_i}(a) \cong \mathbf{C}_{\mathfrak{H}/(\mathfrak{H} \cap \mathfrak{K}_i)}(a) = \mathbf{C}_\mathfrak{H}(a)(\mathfrak{H} \cap \mathfrak{K}_i)/(\mathfrak{H} \cap \mathfrak{K}_i)$$

is nilpotent, and every q'-element of this group leaves fixed every element of $\mathbf{C}_{\mathsf{V}_{i-1}/\mathsf{V}_i}(a)$. By the Maschke-Schur theorem,

$$\mathbf{C}_{\mathsf{V}_{i-1}/\mathsf{V}_i}(a) = (\mathbf{C}_{\mathsf{V}_{i-1}}(a) + \mathsf{V}_i)/\mathsf{V}_i.$$

Finally $|\mathfrak{A}\mathfrak{K}_i/\mathfrak{K}_i| = p^2$; otherwise, there exists $b \in (\mathfrak{A} \cap \mathfrak{K}_i) - \{1\}$, so that $\mathfrak{H}\mathfrak{K}_i/\mathfrak{K}_i = \mathbf{C}_{\mathfrak{H}\mathfrak{K}_i/\mathfrak{K}_i}(b)$ is nilpotent, contrary to the choice of i. It thus follows from the minimality of $|\mathfrak{G}|$ and $\dim_\mathsf{K} \mathsf{V}$ that $\mathfrak{K}_i = 1$, $\mathsf{V}_{i-1} = \mathsf{V}$ and $\mathsf{V}_i = 0$. Thus V is an irreducible $\mathsf{K}\mathfrak{G}$-module.

b) V is the direct sum of isomorphic irreducible $\mathsf{K}\mathfrak{H}$-modules.

By V, 17.3,

$$\mathsf{V} = \mathsf{W}_1 \oplus \cdots \oplus \mathsf{W}_s,$$

where each W_i is the direct sum of isomorphic irreducible $\mathsf{K}\mathfrak{H}$-modules, and there is a transitive permutation representation σ of $\mathfrak{A}$ such that $\mathsf{W}_{i\sigma(a)} = \mathsf{W}_i a$ $(a \in \mathfrak{A})$. Let $\mathfrak{B} = \ker \sigma$. Since $\mathfrak{A}$ is Abelian, $\mathfrak{B}$ is the stabiliser of i for all $i = 1, \ldots, s$, so $|\mathfrak{A} : \mathfrak{B}| = s$. Thus it is to be proved that $\mathfrak{B} = \mathfrak{A}$.

Let $\mathfrak{F} = \mathbf{F}(\mathfrak{H})$. By V, 8.15, there exists $a \in \mathfrak{A} - \{1\}$ such that $\mathbf{C}_\mathfrak{F}(a) \neq 1$. On the other hand, $\mathbf{C}_\mathfrak{F}(a) \neq \mathfrak{F}$. For otherwise, a centralizes $\mathfrak{H}/\mathbf{C}_\mathfrak{H}(\mathfrak{F})$, by 6.3. But by III, 4.2, $\mathbf{C}_{\mathfrak{H}}(\mathfrak{F}) \leq \mathfrak{F}$, so a centralizes $\mathfrak{F}$ and $\mathfrak{H}/\mathfrak{F}$. By I, 4.4, a centralizes $\mathfrak{H}$. Hence $\mathbf{C}_{\mathfrak{H}}(a) = \mathfrak{H}$, which is impossible since $\mathbf{C}_{\mathfrak{H}}(a)$ is nilpotent and $\mathfrak{H}$ is not. Thus $\mathbf{C}_{\mathfrak{F}}(a) \neq \mathfrak{F}$, and $\mathbf{C}_{\mathfrak{F}}(a)$ is a proper $\mathfrak{A}$-invariant subgroup of $\mathfrak{F}$. Let $\mathfrak{F}_0$ be a minimal $\mathfrak{A}$-invariant subgroup such that $\mathbf{C}_{\mathfrak{F}}(a) < \mathfrak{F}_0 \leq \mathfrak{F}$. Since $\mathfrak{F}$ is nilpotent, $\mathbf{C}_{\mathfrak{F}}(a) < \mathbf{N}_{\mathfrak{F}_0}(\mathbf{C}_{\mathfrak{F}}(a))$, so by minimality of $\mathfrak{F}_0$, $\mathbf{C}_{\mathfrak{F}}(a) \lhd \mathfrak{F}_0$. Let $\mathfrak{T} = \mathfrak{F}_0/\mathbf{C}_{\mathfrak{F}}(a)$. If $\mathbf{C}_{\mathfrak{T}}(a) = \mathfrak{S}/\mathbf{C}_{\mathfrak{F}}(a)$, then a centralizes $\mathfrak{S}$ by I, 4.4 and $\mathfrak{S} \leq \mathbf{C}_{\mathfrak{F}}(a)$. Hence $\mathbf{C}_{\mathfrak{T}}(a) = 1$. But again by V, 18.5, there exists $b \in \mathfrak{A} - \{1\}$ such that $\mathbf{C}_{\mathfrak{T}}(b) \neq 1$. Evidently, $b \notin \langle a \rangle$, so $\mathfrak{A} = \langle a, b \rangle$. And since $\mathbf{C}_{\mathfrak{T}}(b) \neq 1$, $\mathbf{C}_{\mathfrak{F}}(b) \neq 1$, by 6.11. Thus to prove that $\mathfrak{B} = \mathfrak{A}$, it suffices to show that if $c \in \mathfrak{A}$ and $\mathbf{C}_{\mathfrak{F}}(c) \neq 1$, then $c \in \mathfrak{B}$.

To do this, suppose that $x \in \mathbf{C}_{\mathfrak{F}}(c)$ and $x \neq 1$. Then there exists an element w of some W_i such that $wx \neq w$. Let

$$u = w + wc + \cdots + wc^{p-1};$$

then $u \in \mathbf{C}_{\mathsf{V}}(c)$. Since $\mathbf{O}_q(\mathfrak{H}) = 1$, $\mathfrak{F}$ is a q'-group, so by hypothesis, u is left fixed by all elements of $\mathbf{C}_{\mathfrak{F}}(c)$. Hence $ux = u$. But since $x \in \mathbf{C}_{\mathfrak{F}}(c)$,

$$ux = wx + wxc + \cdots + wxc^{p-1},$$

and

$$0 = u(x-1) = w(x-1) + w(x-1)c + \cdots + w(x-1)c^{p-1}.$$

Now $w(x-1)c^j \in \mathsf{W}_{i\sigma(c^j)}$ and $w(x-1) \neq 0$; hence the sum

$$\mathsf{W}_i + \mathsf{W}_{i\sigma(c)} + \cdots + \mathsf{W}_{i\sigma(c^{p-1})}$$

is not direct. Thus c lies in the stabiliser $\mathfrak{B}$ of i.

c) If $\mathfrak{L}$ is any non-identity Abelian normal subgroup of $\mathfrak{G}$ and $\mathfrak{L} \leq \mathfrak{H}$, $\mathbf{C}_{\mathfrak{L}}(\mathfrak{A}) \neq 1$.

By V, 17.3,

$$\mathsf{V} = \mathsf{V}_1 \oplus \cdots \oplus \mathsf{V}_n,$$

where V_i is the direct sum of all $\mathsf{K}\mathfrak{L}$-submodules of V isomorphic to some irreducible $\mathsf{K}\mathfrak{L}$-module X_i, and there exists a permutation representation ρ of $\mathfrak{G}$ on $\{1, \ldots, n\}$ such that $\mathsf{V}_{j\rho(g)} = \mathsf{V}_j g$ for all $g \in \mathfrak{G}$. Since $\mathfrak{L}$ is Abelian and K is algebraically closed, $\dim_{\mathsf{K}} \mathsf{X}_i = 1$. Thus there exist distinct homomorphisms $\chi_1, \ldots, \chi_n$ of $\mathfrak{L}$ into $\mathsf{K}^\times$ such that if $x \in \mathfrak{L}$, $wx = \chi_j(x)w$ for all $w \in \mathsf{X}_j$ and hence for all $w \in \mathsf{V}_j$.

Let $\{j_1, \ldots, j_k\}$ be an orbit of the restriction of ρ to $\mathfrak{H}$. Then $\mathsf{V} = \mathsf{W}_1 \oplus \mathsf{W}_2$, where W_1, W_2 are $\mathsf{K}\mathfrak{H}$-submodules of V and $\mathsf{W}_1 = \bigoplus_{i=1}^k \mathsf{V}_{j_i}$. If $\mathsf{W}_2 \neq 0$, then the irreducible $\mathsf{K}\mathfrak{H}$-submodules of W_1, W_2 are isomorphic, by b), so W_1, W_2 certainly have isomorphic irreducible $\mathsf{K}\mathfrak{L}$-submodules. This, however, is impossible on account of the definition of the V_i. Hence $\mathsf{W}_2 = 0$ and the restriction of ρ to $\mathfrak{H}$ is transitive.

It follows that $|\mathfrak{H}|$ is divisible by n. Since $\mathfrak{H}$ is a p'-group, we deduce that p does not divide n. Hence the restriction of ρ to $\mathfrak{A}$ has a fixed point j and $\mathsf{V}_j a = \mathsf{V}_j$ for all $a \in \mathfrak{A}$. Thus if $a \in \mathfrak{A}$ and $x \in \mathfrak{L}$, then $va^{-1} \in \mathsf{V}_j$ and

$$va^{-1} \cdot x^{-1} = \chi_j(x)^{-1} va^{-1} = vx^{-1}a^{-1}$$

for all $v \in \mathsf{V}_j$. Hence $v[a, x] = v$. If $[\mathfrak{A}, \mathfrak{L}] = \mathfrak{L}$, it follows that $vy = v$ for all $v \in \mathsf{V}_j$, $y \in \mathfrak{L}$. But then, if $1 \leq i \leq n$, $i = j\rho(g)$ for some $g \in \mathfrak{H}$, since the restriction of ρ to $\mathfrak{H}$ is transitive. Thus if $u \in \mathsf{V}_i$, then $ug^{-1} \in \mathsf{V}_j$, $ug^{-1}y = ug^{-1}$ for all $y \in \mathfrak{L}$, so $ug^{-1}yg = u$ and $uy = u$ for all $y \in \mathfrak{L}$. Hence $\mathfrak{L}$ is represented trivially on all V_i, which is impossible since $\mathfrak{L} \neq 1$. Hence $[\mathfrak{A}, \mathfrak{L}] < \mathfrak{L}$.

Put $\mathfrak{L}_0 = [\mathfrak{A}, \mathfrak{L}]$. Then $\mathbf{C}_{\mathfrak{A}\mathfrak{L}/\mathfrak{L}_0}(\mathfrak{A}) = \mathfrak{A}\mathfrak{L}/\mathfrak{L}_0$, and by 6.11,

$$\mathbf{C}_{\mathfrak{A}\mathfrak{L}}(\mathfrak{A})\mathfrak{L}_0/\mathfrak{L}_0 = \mathbf{C}_{\mathfrak{A}\mathfrak{L}/\mathfrak{L}_0}(\mathfrak{A}) = \mathfrak{A}\mathfrak{L}/\mathfrak{L}_0.$$

Hence $\mathfrak{L} = \mathbf{C}_{\mathfrak{L}}(\mathfrak{A})\mathfrak{L}_0$. Since $\mathfrak{L}_0 < \mathfrak{L}$, $\mathbf{C}_{\mathfrak{L}}(\mathfrak{A}) \neq 1$.

d) $\mathbf{Z}(\mathfrak{H}) = 1$.

Let $\mathfrak{N} = \mathbf{C}_{\mathbf{Z}(\mathfrak{H})}(\mathfrak{A})$. Then $\mathfrak{N}$ is contained in the centre of $\mathfrak{A}\mathfrak{H} = \mathfrak{G}$. It follows from a) that each element of $\mathfrak{N}$ is represented on V by a scalar multiple of the identity mapping.

Now by V, 8.12, there exists $a \in \mathfrak{A} - \{1\}$ such that $\mathbf{C}_{\mathsf{V}}(a) \neq 0$. Since $\mathbf{O}_q(\mathfrak{H}) = 1$, $\mathbf{Z}(\mathfrak{H})$ is a q'-group, so by hypothesis $\mathbf{C}_{\mathbf{Z}(\mathfrak{H})}(a)$ leaves fixed every element of $\mathbf{C}_{\mathsf{V}}(a)$. But $\mathfrak{N} \leq \mathbf{C}_{\mathbf{Z}(\mathfrak{H})}(a)$. Thus if $x \in \mathfrak{N}$ and $vx = \lambda v$ for all $v \in \mathsf{V}$, then $\lambda v = v$ for all $v \in \mathbf{C}_{\mathsf{V}}(a)$, so $\lambda = 1$, $vx = v$ for all $v \in \mathsf{V}$ and $x = 1$. Hence $\mathfrak{N} = 1$. By c), $\mathbf{Z}(\mathfrak{H}) = 1$.

e) To obtain a contradiction, let $\mathfrak{F} = \mathbf{F}(\mathfrak{H})$. Since $\mathfrak{G}$ is a counterexample of minimal order, ($\mathfrak{A}$ on $\mathfrak{H}/\mathfrak{F}$) is irreducible. Hence $\mathfrak{H}/\mathfrak{F}$ is an elementary Abelian r-group for some prime r. By 1.11, $\mathfrak{G}$ possesses an $\mathfrak{A}$-invariant Sylow r-subgroup $\mathfrak{R}$. Thus $\mathfrak{H} = \mathfrak{R}\mathfrak{F} = \mathfrak{R}\mathbf{O}_{r'}(\mathfrak{F})$, since $\mathfrak{F}$ is nilpotent. Hence $\mathbf{Z}(\mathfrak{R}) \cap \mathfrak{F}$ centralizes $\mathfrak{H}$ and $\mathbf{Z}(\mathfrak{R}) \cap \mathfrak{F} \leq \mathbf{Z}(\mathfrak{H})$. By d), $\mathbf{Z}(\mathfrak{R}) \cap \mathfrak{F} = 1$. By III, 7.2, $\mathfrak{R} \cap \mathfrak{F} = 1$. Thus $\mathfrak{R}$ is $\mathfrak{A}$-isomorphic to $\mathfrak{H}/\mathfrak{F}$ and ($\mathfrak{A}$ on $\mathfrak{R}$) is irreducible. But by V, 8.15, there exists $a \in \mathfrak{A} - \{1\}$ such that $\mathbf{C}_{\mathfrak{R}}(a) \neq 1$. Since $\mathbf{C}_{\mathfrak{R}}(a)$ is $\mathfrak{A}$-invariant, $\mathbf{C}_{\mathfrak{R}}(a) = \mathfrak{R}$. But $\mathbf{C}_{\mathfrak{H}}(a)$

is nilpotent and $\mathbf{C}_{\mathfrak{H}}(a) = \mathfrak{N}\mathbf{C}_{\mathfrak{F}}(a)$, so $\mathfrak{N}$ centralizes $\mathbf{C}_{\mathfrak{F}}(a)$. Thus $\mathbf{C}_{\mathbf{Z}(\mathfrak{F})}(a) \leq \mathbf{Z}(\mathfrak{H}) = 1$. This contradicts c).

Hence $\mathfrak{H}$ is nilpotent. **q.e.d.**

6.13 Theorem (J. N. WARD). *Let $\mathfrak{G}$ be a soluble p'-group and let $\mathfrak{A}$ be an elementary Abelian group of automorphisms of $\mathfrak{G}$ of order p^2. If $\mathbf{C}_{\mathfrak{G}}(a)$ is nilpotent for every non-identity element a of $\mathfrak{A}$, the Fitting height of $\mathfrak{G}$ is at most 2.*

Proof. This is proved by induction on $|\mathfrak{G}|$. Let $\mathfrak{N}$ be the minimal normal subgroup of $\mathfrak{G}$ for which $\mathfrak{G}/\mathfrak{N}$ is nilpotent. If $\mathfrak{K}$ is any non-identity characteristic subgroup of $\mathfrak{G}$, $\mathbf{C}_{\mathfrak{G}/\mathfrak{K}}(a)$ is nilpotent for every non-identity element a of $\mathfrak{A}$, since by 6.11, $\mathbf{C}_{\mathfrak{G}/\mathfrak{K}}(a) = \mathbf{C}_{\mathfrak{G}}(a)\mathfrak{K}/\mathfrak{K}$. Thus by the inductive hypothesis, $\mathfrak{N}\mathfrak{K}/\mathfrak{K}$ is nilpotent. By 6.7, we may suppose that $\mathbf{O}_{q'}(\mathfrak{G}) = 1$ for some prime q and that $\mathbf{O}_q(\mathfrak{G})$ is elementary Abelian. Since $\mathfrak{G}$ is a p'-group, $q \neq p$.

Let $\mathfrak{H}$ denote the semi-direct product $\mathfrak{A}\mathfrak{G}$. Then $\mathbf{O}_{q'}(\mathfrak{H}) \cap \mathfrak{G} \leq \mathbf{O}_{q'}(\mathfrak{G}) = 1$, so $\mathbf{O}_{q'}(\mathfrak{H})$ is a p-group. Thus $\mathbf{O}_{q'}(\mathfrak{H}) \leq \mathfrak{A}$. Hence, if $\mathbf{O}_{q'}(\mathfrak{H}) \neq 1$, $\mathbf{C}_{\mathfrak{G}}(\mathbf{O}_{q'}(\mathfrak{H}))$ is nilpotent. But $\mathbf{C}_{\mathfrak{G}}(\mathbf{O}_{q'}(\mathfrak{H})) = \mathfrak{G}$, so the proof is complete in this case. We may therefore suppose that $\mathbf{O}_{q'}(\mathfrak{H}) = 1$. Thus if $\overline{\mathfrak{H}} = \mathfrak{H}/\mathbf{O}_q(\mathfrak{G})$, ($\overline{\mathfrak{H}}$ on $\mathbf{O}_q(\mathfrak{G})$) is faithful, by 1.3. If a is a non-identity element of $\mathfrak{A}$, $\mathbf{C}_{\mathfrak{G}}(a)$ is nilpotent, so $\mathbf{C}_{\mathfrak{G}/\mathbf{O}_q(\mathfrak{G})}(a)$ is nilpotent and any q'-element of $\mathbf{C}_{\mathfrak{G}/\mathbf{O}_q(\mathfrak{G})}(a)$ operates trivially on $\mathbf{C}_{\mathbf{O}_q(\mathfrak{G})}(a)$. Hence $\mathfrak{G}/\mathbf{O}_q(\mathfrak{G})$ is nilpotent, by 6.12. Thus the Fitting height of $\mathfrak{G}$ is at most 2. **q.e.d.**

6.14 Theorem (J. N. WARD). *Suppose that $\mathfrak{G}$ is a soluble group, $\mathfrak{A}$ is an elementary Abelian group of automorphisms of $\mathfrak{G}$ of order p^2 and $\mathbf{C}_{\mathfrak{G}}(\mathfrak{A}) = 1$. Then the Fitting height of $\mathfrak{G}$ is at most 2.*

Proof. Since $\mathbf{C}_{\mathfrak{G}}(\mathfrak{A}) = 1$, the length of any orbit of $\mathfrak{A}$ on $\mathfrak{G}$ other than $\{1\}$ is divisible by p. Hence $|\mathfrak{G}| \equiv 1\ (p)$ and $\mathfrak{G}$ is a p'-group.

If $a \in \mathfrak{A} - \{1\}$, there exists $b \in \mathfrak{A}$ such that $\mathfrak{A} = \langle a, b\rangle$. If $\mathfrak{C} = \mathbf{C}_{\mathfrak{G}}(a)$, $\mathfrak{C}$ is $\langle b\rangle$-invariant and $\mathbf{C}_{\mathfrak{C}}(b) = \mathbf{C}_{\mathfrak{G}}(a, b) = 1$. By V, 8.14, $\mathfrak{C}$ is nilpotent.

Thus by 6.13, the Fitting height of $\mathfrak{G}$ is at most 2. **q.e.d.**

6.15 Remarks. a) For $p = 2$, it was proved by S. Bauman that under the conditions of 6.14, $\mathfrak{G}'$ is nilpotent (see GORENSTEIN [1]). Shult has also proved that if $\mathfrak{A}$ is a group of automorphisms of $\mathfrak{G}$, $\mathfrak{A} \cong \mathfrak{S}_3$, $(|\mathfrak{G}|, 6) = 1$ and $\mathbf{C}_{\mathfrak{G}}(\mathfrak{A}) = 1$, then $\mathfrak{G}'$ is nilpotent.

b) The hypothesis of solubility in 6.14 is not necessary. This will be proved in X, 11.18, and 6.14 will be used in the proof.

c) Theorems 6.4, 6.9 and 6.14 are all special cases of the following theorem of BERGER [2].

Suppose that $\mathfrak{G}$ is a soluble group, $\mathfrak{A}$ is a group of automorphisms of $\mathfrak{G}$, $(|\mathfrak{A}|, |\mathfrak{G}|) = 1$ and $\mathbf{C}_{\mathfrak{G}}(\mathfrak{A}) = 1$. Suppose that $\mathfrak{A}$ is nilpotent and that for every prime p, $\mathfrak{A}$ has no section isomorphic to the wreath product of two cyclic groups of order p. Then, if $|\mathfrak{A}| = p_1 p_2 \cdots p_n$, where $p_1, \ldots, p_n$ are (not necessarily distinct) primes, the Fitting height of $\mathfrak{G}$ is at most n.

d) Let $\mathfrak{A}$ be a soluble group of automorphisms of the soluble group $\mathfrak{G}$ for which $(|\mathfrak{A}|, |\mathfrak{G}|) = 1$. Let $|\mathfrak{A}| = p_1 p_2 \cdots p_n$ for primes $p_1, \ldots, p_n$. We denote the Fitting height of the group $\mathfrak{X}$ by $h(\mathfrak{X})$.

THOMPSON [1] proved that

$$h(\mathfrak{G}) \leq 5^n h(\mathbf{C}_{\mathfrak{G}}(\mathfrak{A})),$$

and later KURZWEIL [1] improved this to

$$h(\mathfrak{G}) \leq h(\mathbf{C}_{\mathfrak{G}}(\mathfrak{A})) + 4n.$$

In particular, if $\mathbf{C}_{\mathfrak{G}}(\mathfrak{A}) = 1$, $h(\mathfrak{G}) \leq 4n + 1$.

e) DADE [4] proved that for any soluble group $\mathfrak{G}$,

$$h(\mathfrak{G}) \leq 10(2^l - 1) - 4l,$$

where $\mathfrak{C}$ is a Carter subgroup of $\mathfrak{G}$ and $|\mathfrak{C}| = p_1 \cdots p_l$ for primes $p_1, \ldots, p_l$.

The proof of this is very lengthy. The same method was used by HARTLEY [1] and RAE [1] to prove that if p is odd and $\mathfrak{G}$ is a p-soluble group of p-length at least $2^{k+2} - 1$, then any subgroup of $\mathfrak{G}$ of order p^k is contained in more than one Sylow p-subgroup of $\mathfrak{G}$.

§ 7. p-Stability

In order to formalize the consequences of Lemma 5.8, we make the following definition.

7.1 Definition. The finite group $\mathfrak{G}$ is called *p-stable* if, whenever $\mathfrak{P}$ is a p-subgroup of $\mathfrak{G}$, $g \in \mathbf{N}_{\mathfrak{G}}(\mathfrak{P})$ and $[\mathfrak{P}, g, g] = 1$, then
$g\mathbf{C}_{\mathfrak{G}}(\mathfrak{P}) \in \mathbf{O}_p(\mathbf{N}_{\mathfrak{G}}(\mathfrak{P})/\mathbf{C}_{\mathfrak{G}}(\mathfrak{P}))$.

7.2 Example. Any p-nilpotent group $\mathfrak{G}$ is p-stable.

For if $\mathfrak{P}$ is a p-subgroup of $\mathfrak{G}$, $\mathbf{N}_{\mathfrak{G}}(\mathfrak{P})/\mathbf{C}_{\mathfrak{G}}(\mathfrak{P})$ is a p-group.

Lemma 5.8 implies the p-stability of a large class of p-soluble groups. To prove this, we need the following.

7.3 Lemma. *Suppose that $\mathfrak{G}$ is a finite p-group and*

$$\mathfrak{G} = \mathfrak{N}_0 \geq \mathfrak{N}_1 \geq \cdots \geq \mathfrak{N}_k = 1,$$

where $\mathfrak{N}_i \trianglelefteq \mathfrak{G}$. Let $\mathfrak{A}$ be a group of automorphisms α of $\mathfrak{G}$ for which $\mathfrak{N}_i\alpha = \mathfrak{N}_i$ $(i = 0, 1, \ldots, k)$, and let

$$\mathfrak{B} = \{\beta \mid \beta \in \mathfrak{A}, (x\mathfrak{N}_i)\beta = x\mathfrak{N}_i \quad \textit{for all } x \in \mathfrak{N}_{i-1} \quad (i = 1, \ldots, k)\}.$$

Then $\mathfrak{B}$ is a normal p-subgroup of $\mathfrak{A}$.

In particular, any non-identity p'-element of $\mathfrak{A}$ induces a non-identity automorphism on $\mathfrak{N}_{i-1}/\mathfrak{N}_i$ for at least one i.

Proof. If $\alpha \in \mathfrak{A}$, then for each $i = 1, \ldots, k$, α induces an automorphism α_i on $\mathfrak{N}_{i-1}/\mathfrak{N}_i$. The mapping $\alpha \to (\alpha_1, \ldots, \alpha_k)$ is a homomorphism of $\mathfrak{A}$ into the direct product of the groups of automorphisms of the $\mathfrak{N}_{i-1}/\mathfrak{N}_i$, and $\mathfrak{B}$ is the kernel. Thus $\mathfrak{B} \trianglelefteq \mathfrak{A}$. By repeated application of I, 4.4, $\mathfrak{B}$ is a p-group. **q.e.d.**

7.4 Theorem. *Let $\mathfrak{G}$ be a p-soluble group. If $p > 3$, $\mathfrak{G}$ is p-stable. If $p = 3$ and the Sylow 2-subgroups of $\mathfrak{G}$ are Abelian, $\mathfrak{G}$ is 3-stable.*

Proof. Suppose that $\mathfrak{P}$ is a p-subgroup of $\mathfrak{G}$, $g \in \mathfrak{H} = \mathbf{N}_{\mathfrak{G}}(\mathfrak{P})$ and $[\mathfrak{P}, g, g] = 1$. Let

$$\mathfrak{P} = \mathfrak{P}_0 > \mathfrak{P}_1 > \cdots > \mathfrak{P}_m = 1$$

be the part below $\mathfrak{P}$ of a chief series of $\mathfrak{H}$. For $1 \leq i \leq m$, let $\mathfrak{C}_i = \mathbf{C}_{\mathfrak{H}}(\mathfrak{P}_{i-1}/\mathfrak{P}_i)$. By 5.8, $g \in \mathfrak{C}_i$. Thus if $\overline{\mathfrak{C}} = \bigcap_{i=1}^m \mathfrak{C}_i$, $g \in \overline{\mathfrak{C}}$. Let $\mathfrak{C} = \mathbf{C}_{\mathfrak{G}}(\mathfrak{P})$, so $\mathfrak{C} \leq \overline{\mathfrak{C}}$. Now $\mathfrak{H}/\mathfrak{C}$ is isomorphic to a group $\mathfrak{A}$ of automorphisms α of $\mathfrak{P}$ for which $\mathfrak{P}_i\alpha = \mathfrak{P}_i$ $(i = 0, \ldots, m)$. In this isomorphism, $\overline{\mathfrak{C}}/\mathfrak{C}$ corresponds to $\mathfrak{B}$, where

$$\mathfrak{B} = \{\beta \mid \beta \in \mathfrak{A}, (x\mathfrak{P}_i)\beta = x\mathfrak{P}_i \quad \text{for all } x \in \mathfrak{P}_{i-1} \quad (i = 1, \ldots, m)\}.$$

By 7.3, $\mathfrak{B}$ is a normal p-subgroup of $\mathfrak{A}$, so $\overline{\mathfrak{C}}/\mathfrak{C} \leq \mathbf{O}_p(\mathfrak{H}/\mathfrak{C})$. Thus $g\mathfrak{C} \in \mathbf{O}_p(\mathfrak{H}/\mathfrak{C})$. Hence $\mathfrak{G}$ is p-stable. **q.e.d.**

Our principal aim in this section is to obtain a generalization of 7.4 which gives a characterization of groups in which every section is p-stable, for p odd. First we give an example of a group which is not p-stable.

7.5 Example. For any prime p, let $\mathfrak{G} = SA(2, p)$ be the set of all matrices of the form

$$\begin{pmatrix} a & b & t \\ c & d & u \\ 0 & 0 & 1 \end{pmatrix}$$

with coefficients in $GF(p)$ and $ad - bc = 1$. Let $\mathfrak{H}$ be the set of all such matrices for which $t = u = 0$ and let $\mathfrak{N}$ be the set of all such matrices for which $a = d = 1, b = c = 0$. Then $\mathfrak{N} \lhd \mathfrak{G}$, $\mathfrak{G} = \mathfrak{H}\mathfrak{N}$ and $\mathfrak{H} \cap \mathfrak{N} = 1$. Also $\mathbf{C}_{\mathfrak{G}}(\mathfrak{N}) = \mathfrak{N}$ and $\mathfrak{H} \cong SL(2, p)$. Let g be a non-identity p-element of $\mathfrak{H}$. Since $|\mathfrak{N}| = p^2$, $[\mathfrak{N}, g, g] = 1$. But $\mathbf{O}_p(\mathfrak{G}/\mathbf{C}_{\mathfrak{G}}(\mathfrak{N})) = \mathbf{O}_p(\mathfrak{H}\mathfrak{N}/\mathfrak{N}) \cong \mathbf{O}_p(\mathfrak{H}) = 1$ and $g\mathfrak{N} \neq 1$, so $\mathfrak{G}$ is not p-stable.

In particular, the soluble group $SA(2, 3)$ is not 3-stable; its Sylow 2-subgroup is the quaternion group of order 8.

Note that $SA(2, 2) \cong \mathfrak{S}_4$.

7.6 Theorem. *For p odd, let $\mathfrak{M}$ be the subgroup of $SL(2, p^f)$ generated by*

$$\begin{pmatrix} 1 & 0 \\ b & 1 \end{pmatrix} \quad \text{and} \quad \begin{pmatrix} 1 & 1 \\ 0 & 1 \end{pmatrix},$$

where $b \neq 0$. Let $\mathfrak{Z} = \langle -1 \rangle \leq SL(2, p^f)$.

a) *If $p = 3$ and $b^2 = -1$, then f is even, $\mathfrak{M} \geq \mathfrak{Z}$, $\mathfrak{M} \cong SL(2, 5)$ and $\mathfrak{M} \geq SL(2, 3)^x$ for some $x \in SL(2, 9)$.*

b) *Otherwise, $\mathfrak{M} = SL(2, p^m)$, where $GF(p^m)$ is the field generated by b over $GF(p)$.*

Proof. We prove this in a number of steps.

(1) If $\mathfrak{P}_0$ is a non-identity subgroup of $SL(2, p^f)$ consisting of elements of the form $\begin{pmatrix} 1 & a \\ 0 & 1 \end{pmatrix}$, the normaliser $\mathfrak{N}_0$ of $\mathfrak{P}_0$ in $SL(2, p^f)$ consists of matrices of the form $\begin{pmatrix} x & y \\ 0 & x^{-1} \end{pmatrix}$, and the only Sylow p-subgroup of $\mathfrak{N}_0$ is $\left\{ \begin{pmatrix} 1 & y \\ 0 & 1 \end{pmatrix} \middle| y \in GF(p^f) \right\}$.

If $xv - yu = 1$,

$$\begin{pmatrix} v & -y \\ -u & x \end{pmatrix}\begin{pmatrix} 1 & a \\ 0 & 1 \end{pmatrix}\begin{pmatrix} x & y \\ u & v \end{pmatrix} = \begin{pmatrix} 1 + auv & av^2 \\ -au^2 & 1 - auv \end{pmatrix}.$$

Thus $\mathfrak{N}_0$ consists of matrices of the stated form and the set of all matrices of the form $\begin{pmatrix} 1 & y \\ 0 & 1 \end{pmatrix}$ with $y \in GF(p^f)$ is a normal Sylow p-subgroup of $\mathfrak{N}_0$.

(2) Let $\mathfrak{P}$ be the set of all matrices $\begin{pmatrix} 1 & a \\ 0 & 1 \end{pmatrix}$ which lie in $\mathfrak{M}$. Then $\mathfrak{P} \in S_p(\mathfrak{M})$, and $\mathfrak{P}$ is the only Sylow p-subgroup of $\mathfrak{M}$ which contains $\begin{pmatrix} 1 & 1 \\ 0 & 1 \end{pmatrix}$.

Suppose that $\mathfrak{P}_1 \in S_p(\mathfrak{M})$ and $\begin{pmatrix} 1 & 1 \\ 0 & 1 \end{pmatrix} \in \mathfrak{P}_1$. Since the Sylow p-subgroups of $SL(2, p^f)$ are Abelian, it follows from (1) that $\mathfrak{P}_1$ consists of elements of the form $\begin{pmatrix} x & y \\ 0 & x^{-1} \end{pmatrix}$ and, further, that $\mathfrak{P}_1 \leq \mathfrak{P}$. Then $\mathfrak{P}_1 = \mathfrak{P}$.

(3) $\mathfrak{M} \geq \mathfrak{Z}$, and either $\mathfrak{M}/\mathfrak{Z} \cong PSL(2, p^n)$ for some n, or $p = 3$ and $\mathfrak{M}/\mathfrak{Z} \cong \mathfrak{A}_5$.

It follows from (2) and the definition of $\mathfrak{M}$ that $\mathfrak{M}$ has more than one Sylow p-subgroup, for $\begin{pmatrix} 1 & 0 \\ b & 1 \end{pmatrix} \notin \mathfrak{P}$. Also, if $\mathfrak{K} \trianglelefteq \mathfrak{M}$ and $\mathfrak{M}/\mathfrak{K}$ is a p'-group, $\mathfrak{K} = \mathfrak{M}$, for $\mathfrak{M}/\mathfrak{K}$ is generated by p-elements. It follows from II, 8.27 that $\mathfrak{M}\mathfrak{Z}/\mathfrak{Z}$ is isomorphic to $\mathfrak{A}_4$, $\mathfrak{A}_5$ or $PSL(2, p^n)$ for some n. Thus the order of $\mathfrak{M}$ is even. Since -1 is the only element of order 2 in $SL(2, p^f)$, $\mathfrak{Z} \leq \mathfrak{M}$.

Suppose that $\mathfrak{M}/\mathfrak{Z}$ is not isomorphic to $PSL(2, p^r)$ for any r. Then $\mathfrak{M}/\mathfrak{Z}$ is isomorphic to $\mathfrak{A}_4$ or $\mathfrak{A}_5$. Since p divides $|\mathfrak{M}|$, $p = 3$ or $p = 5$. Since $\mathfrak{A}_4 \cong PSL(2, 3)$ and $\mathfrak{A}_5 \cong PSL(2, 5)$, the only possibility is $p = 3$ and $\mathfrak{M}/\mathfrak{Z} \cong \mathfrak{A}_5$.

(4) If $p = 3$ and $\mathfrak{M}/\mathfrak{Z} \cong \mathfrak{A}_5$, $b^2 = -1$.

Let α be an epimorphism of $\mathfrak{M}$ onto $\mathfrak{A}_5$ such that $\mathfrak{Z} = \ker \alpha$. If $\begin{pmatrix} 1 & 1 \\ 0 & 1 \end{pmatrix}\alpha = (i_1, i_2, i_3)$ and $\begin{pmatrix} 1 & 0 \\ b & 1 \end{pmatrix}\alpha = (j_1, j_2, j_3)$, then $\mathfrak{A}_5 = \langle (i_1, i_2, i_3), (j_1, j_2, j_3) \rangle$, so $\{i_1, i_2, i_3, j_1, j_2, j_3\} = \{1, 2, 3, 4, 5\}$. Hence we may suppose that $j_1 = i_3$ and that i_1, i_2, i_3, j_2, j_3 are distinct. Thus

$$(i_3, j_3, j_2)(i_1, i_2, i_3)(i_3, j_2, j_3)(i_1, i_2, i_3) = (i_1, i_3)(i_2, j_2).$$

Hence

$$\begin{pmatrix} * & (1-b)(1+b^2) \\ b^2(1+b^2) & * \end{pmatrix}$$
$$= \left\{\begin{pmatrix} 1 & 0 \\ -b & 1 \end{pmatrix}\begin{pmatrix} 1 & 1 \\ 0 & 1 \end{pmatrix}\begin{pmatrix} 1 & 0 \\ b & 1 \end{pmatrix}\begin{pmatrix} 1 & 1 \\ 0 & 1 \end{pmatrix}\right\}^2 \in \ker \alpha = \mathfrak{Z}.$$

This gives $b^2 = -1$.

(5) We now prove a). Suppose that $p = 3$ and $b^2 = -1$. Since $GF(3^f)$ contains a square root of -1, f is even. Hence by II, 8.13, $SL(2, 3^f)/\mathfrak{Z}$ has a subgroup $\mathfrak{L}/\mathfrak{Z}$ isomorphic to $\mathfrak{A}_5$. Since $\mathfrak{A}_5$ is generated by two elements of order 3, $\mathfrak{L} = \mathfrak{Z}\langle u, v\rangle$ for elements u, v of $SL(2, 3^f)$ of order 3. Now it follows from the double transitivity of $PSL(2, 3^f)$ on the projective line that the equivalent representation of $PSL(2, 3^f)$ by transformation of its Sylow 3-subgroups is doubly transitive. Hence there exists $A \in SL(2, 3^f)$ such that

$$u^A = \begin{pmatrix} 1 & x \\ 0 & 1 \end{pmatrix}, \quad v^A = \begin{pmatrix} 1 & 0 \\ y & 1 \end{pmatrix}$$

for suitable x, y in $GF(3^f)^\times$. Let $B = \begin{pmatrix} x & 0 \\ 0 & 1 \end{pmatrix}$, and replace $\mathfrak{L}$ by $\mathfrak{L}^{AB}$. Thus $\mathfrak{L} = \mathfrak{Z}\mathfrak{L}_0$, where

$$\mathfrak{L}_0 = \left\langle \begin{pmatrix} 1 & 1 \\ 0 & 1 \end{pmatrix}, \begin{pmatrix} 1 & 0 \\ xy & 1 \end{pmatrix} \right\rangle.$$

By (3), $\mathfrak{L}_0 \geq \mathfrak{Z}$, so $\mathfrak{L} = \mathfrak{L}_0$. By (4), $(xy)^2 = -1$. Thus $\mathfrak{M} \cong \mathfrak{L}_0 = \mathfrak{L}$ and $\mathfrak{M}/\mathfrak{Z}$ is isomorphic to $\mathfrak{A}_5$. Since $\mathfrak{A}_5$ is simple, it follows that $\mathfrak{Z}\mathfrak{M}' = \mathfrak{M}$. But $\mathfrak{Z} \leq \mathfrak{M}'$, since $\mathfrak{M}$ has only one element of order 2. Thus $\mathfrak{M} = \mathfrak{M}'$. By V, 25.7, $SL(2, 5)$ is a representation group of $PSL(2, 5) \cong \mathfrak{A}_5$, so the Schur multiplier of $\mathfrak{A}_5$ is of order 2. By V, 23.4, $\mathfrak{M}$ is a representation group of $\mathfrak{A}_5$. Using V, 23.6, it follows that $\mathfrak{M} \cong SL(2, 5)$.

Finally, $\mathfrak{M}$ contains

$$\begin{pmatrix} 1 & 0 \\ b & 1 \end{pmatrix}\begin{pmatrix} 1 & 1 \\ 0 & 1 \end{pmatrix}\begin{pmatrix} 1 & 0 \\ -b & 1 \end{pmatrix} = \begin{pmatrix} 1-b & 1 \\ 1 & 1+b \end{pmatrix}.$$

Thus if $B = \begin{pmatrix} 1 & -b \\ 0 & 1 \end{pmatrix}$, $\mathfrak{M}^B$ contains $\begin{pmatrix} 1 & 1 \\ 0 & 1 \end{pmatrix}^B = \begin{pmatrix} 1 & 1 \\ 0 & 1 \end{pmatrix}$ and

$\begin{pmatrix} 1-b & 1 \\ 1 & 1+b \end{pmatrix}^B = \begin{pmatrix} 1 & 1+b^2 \\ 1 & 1 \end{pmatrix} = \begin{pmatrix} 1 & 0 \\ 1 & 1 \end{pmatrix}$. Since $\begin{pmatrix} 1 & 1 \\ 0 & 1 \end{pmatrix}$ and $\begin{pmatrix} 1 & 0 \\ 1 & 1 \end{pmatrix}$ generate $SL(2, 3)$, $\mathfrak{M}^B$ contains $SL(2, 3)$.

(6) To prove b), suppose that either $p > 3$ or $b^2 \neq -1$. By (4), either $p > 3$ or $\mathfrak{M}/\mathfrak{Z} \not\cong \mathfrak{A}_5$, so by (3), $\mathfrak{M}/\mathfrak{Z} \cong PSL(2, p^n)$ for some n. Thus p^n is the order of a Sylow p-subgroup of $\mathfrak{M}$. Hence by (2), $\mathfrak{P} \in S_p(\mathfrak{M})$, where

$$\mathfrak{P} = \left\{ \begin{pmatrix} 1 & a \\ 0 & 1 \end{pmatrix} \middle| a \in \Lambda \right\}$$

for some additive subgroup Λ of $GF(p^f)$ of order p^n. If $\mathfrak{N} = \mathbf{N}_{\mathfrak{M}}(\mathfrak{P})$, $\mathfrak{N}/\mathfrak{Z}$ is the normaliser of a Sylow p-subgroup in $PSL(2, p^n)$, so $|\mathfrak{N}| = p^n(p^n - 1)$. By (1), $\mathfrak{N}$ consists of matrices of the form $\begin{pmatrix} x & y \\ 0 & x^{-1} \end{pmatrix}$, where x runs through some multiplicative subgroup Σ of $GF(p^f)^\times$. Thus $|\Sigma| = p^n - 1$ and $\Sigma = GF(p^n)^\times$. And if $x \in \Sigma$, $x^2 \in \Lambda$, since

$$\begin{pmatrix} x^{-1} & y \\ 0 & x \end{pmatrix}^{-1} \begin{pmatrix} 1 & 1 \\ 0 & 1 \end{pmatrix} \begin{pmatrix} x^{-1} & y \\ 0 & x \end{pmatrix} = \begin{pmatrix} x & -y \\ 0 & x^{-1} \end{pmatrix} \begin{pmatrix} x^{-1} & y+x \\ 0 & x \end{pmatrix} = \begin{pmatrix} 1 & x^2 \\ 0 & 1 \end{pmatrix}.$$

Thus Λ contains $x^2 + y^2$ for all x, y in $GF(p^n)$. Since $|\Lambda| = p^n$, it follows from II, 10.6 that $\Lambda = GF(p^n)$.

Now $\mathfrak{M}/\mathfrak{Z}$ has a doubly transitive permutation representation on the cosets of $\mathfrak{N}/\mathfrak{Z}$, since $\mathfrak{M}/\mathfrak{Z} \cong PSL(2, p^n)$. Hence

$$\mathfrak{M} = \mathfrak{N} \cup \mathfrak{N} \begin{pmatrix} 1 & 0 \\ b & 1 \end{pmatrix} \mathfrak{N}.$$

But

$$\begin{pmatrix} 1 & 0 \\ b & 1 \end{pmatrix} \begin{pmatrix} 1 & 1 \\ 0 & 1 \end{pmatrix} \begin{pmatrix} 1 & 0 \\ -b & 1 \end{pmatrix} = \begin{pmatrix} 1-b & 1 \\ -b^2 & 1+b \end{pmatrix}$$

does not lie in $\mathfrak{N}$, so there exist x, y in $\Sigma = GF(p^n)^\times$ such that

$$\begin{pmatrix} 1-b & 1 \\ -b^2 & 1+b \end{pmatrix} = \begin{pmatrix} x & x' \\ 0 & x^{-1} \end{pmatrix} \begin{pmatrix} 1 & 0 \\ b & 1 \end{pmatrix} \begin{pmatrix} y & y' \\ 0 & y^{-1} \end{pmatrix}$$
$$= \begin{pmatrix} * & * \\ bx^{-1}y & * \end{pmatrix}.$$

Thus $b = -x^{-1}y \in GF(p^n)$ and $m \leq n$. But $\mathfrak{M} \leq SL(2, p^m)$ and

$$|SL(2, p^m)| \leq |SL(2, p^n)| = |\mathfrak{M}|,$$

so $\mathfrak{M} = SL(2, p^m)$. **q.e.d.**

We use this to prove the following.

7.7 Lemma. *Suppose that $\mathfrak{G} \neq 1$ and that V is a faithful irreducible $\mathsf{K}\mathfrak{G}$-module, where $\mathsf{K} = GF(p)$ for an odd prime p. Suppose that there exist elements a, b of $\mathfrak{G}$ such that $\mathfrak{G} = \langle a, a^b \rangle$ and $\mathsf{V}(a-1)^2 = 0$. Then there exists a K-subspace U of V such that $\dim_{\mathsf{K}} \mathsf{U} = 2$ and every linear transformation of U of determinant 1 is the restriction to U of some element of $\mathfrak{G}$.*

Proof. This is proved in several steps.

(1) Put $\mathsf{V}_1 = \mathsf{V}(a-1)$, $\mathsf{V}_2 = \mathsf{V}(a^b-1)$. Then $\mathsf{V}_2 = \mathsf{V}_1 b$, $\mathsf{V}_1 = \mathbf{C}_{\mathsf{V}}(a)$, $\mathsf{V}_2 = \mathbf{C}_{\mathsf{V}}(a^b)$ and $\mathsf{V} = \mathsf{V}_1 \oplus \mathsf{V}_2$.

Clearly $\mathsf{V}_2 = \mathsf{V}_1 b$. The mapping $x \to x(a-1)$ is a K-linear mapping of V onto $\mathsf{V}_1 = \mathsf{V}(a-1)$ with kernel $\mathbf{C}_{\mathsf{V}}(a)$. Thus if $\dim_{\mathsf{K}} \mathsf{V}_1 = f$ and $\dim_{\mathsf{K}} \mathbf{C}_{\mathsf{V}}(a) = f'$, $\dim_{\mathsf{K}} \mathsf{V} = f + f'$. Since $\mathsf{V}(a-1)^2 = 0$, $\mathsf{V}_1 \leq \mathbf{C}_{\mathsf{V}}(a)$, so $f \leq f'$. But $\mathbf{C}_{\mathsf{V}}(a) \cap \mathbf{C}_{\mathsf{V}}(a^b) \leq \mathbf{C}_{\mathsf{V}}(\mathfrak{G}) = 0$, since $\mathfrak{G} = \langle a, a^b \rangle \neq 1$ and V is a faithful irreducible $\mathsf{K}\mathfrak{G}$-module. Thus $\dim_{\mathsf{K}} \mathsf{V} \geq \dim_{\mathsf{K}} \mathbf{C}_{\mathsf{V}}(a) + \dim_{\mathsf{K}} \mathbf{C}_{\mathsf{V}}(a^b)$. Since $\mathbf{C}_{\mathsf{V}}(a^b) = \mathbf{C}_{\mathsf{V}}(a)b$, $\dim_{\mathsf{K}} \mathbf{C}_{\mathsf{V}}(a^b) = f'$. It follows that $f + f' \geq 2f'$, or $f \geq f'$. Hence $f' = f$, $\mathsf{V}_1 = \mathbf{C}_{\mathsf{V}}(a)$ and $\mathsf{V}_2 = \mathbf{C}_{\mathsf{V}}(a^b)$. Thus $\mathsf{V}_1 \cap \mathsf{V}_2 = 0$; also $\dim_{\mathsf{K}} \mathsf{V}_1 + \dim_{\mathsf{K}} \mathsf{V}_2 = \dim_{\mathsf{K}} \mathsf{V}$, so $\mathsf{V} = \mathsf{V}_1 \oplus \mathsf{V}_2$.

(2) If $x \in \mathsf{V}_1$, put $x\alpha = x(a^b - 1)$, and if $y \in \mathsf{V}_2$, put $y\beta = y(a-1)$. Then α is a K-linear isomorphism of V_1 onto V_2 and β is a K-linear isomorphism of V_2 onto V_1.

It is clear that α, β are K-linear mappings of V_1 into V_2 and V_2 into V_1 respectively. If $x \in \mathsf{V}_1$ and $x\alpha = 0$, then $x(a-1) = x(a^b-1) = 0$, since $\mathsf{V}_1 = \mathbf{C}_{\mathsf{V}}(a)$. Thus $x \in \mathbf{C}_{\mathsf{V}}(\mathfrak{G}) = 0$. Hence α is a monomorphism. Since $\dim_{\mathsf{K}} \mathsf{V}_1 = \dim_{\mathsf{K}} \mathsf{V}_2$, α is an isomorphism. Similarly, β is an isomorphism.

(3) Let $\gamma = \alpha\beta$; thus γ is a non-singular K-linear transformation of V_1. $\langle \gamma \rangle$ is irreducible on V_1, and the set F of polynomials in γ with coefficients in K is a field.

We show first that $\langle \gamma \rangle$ is irreducible on V_1. Suppose that $0 < \mathsf{V}_0 \leq \mathsf{V}_1$ and $\mathsf{V}_0\gamma = \mathsf{V}_0$. Then $(\mathsf{V}_0 \oplus \mathsf{V}_0\alpha)(a-1) = 0 + \mathsf{V}_0\alpha\beta = \mathsf{V}_0\gamma = \mathsf{V}_0$, so $(\mathsf{V}_0 \oplus \mathsf{V}_0\alpha)a = \mathsf{V}_0 \oplus \mathsf{V}_0\alpha$. Again $(\mathsf{V}_0 \oplus \mathsf{V}_0\alpha)(a^b-1) = \mathsf{V}_0\alpha + 0$, so $(\mathsf{V}_0 \oplus \mathsf{V}_0\alpha)a^b = \mathsf{V}_0 \oplus \mathsf{V}_0\alpha$. Thus $\mathsf{V}_0 \oplus \mathsf{V}_0\alpha$ is a K-subspace of V invariant under $\mathfrak{G}$. Since V is irreducible, $\mathsf{V} = \mathsf{V}_0 \oplus \mathsf{V}_0\alpha$ and $\mathsf{V}_1 = \mathsf{V}_0$.

By Schur's lemma (I, 10.5), any polynomial in γ is either zero or non-singular. Hence the minimum polynomial of γ is irreducible. Therefore any non-zero element of the set F of polynomials in γ has an inverse in F. Hence F is a field.

(4) Take a fixed, non-zero element z in V_1. Given $x \in \mathsf{V}_1$, there exists a unique $\lambda \in \mathsf{F}$ such that $x = z\lambda$. In particular $|\mathsf{F}| = |\mathsf{V}_1| = p^f$.

For $\{z\lambda | \lambda \in \mathsf{F}\}$ is a non-zero γ-invariant K-subspace of V_1, so $\{z\lambda | \lambda \in \mathsf{F}\} = \mathsf{V}_1$, since $\langle\gamma\rangle$ is irreducible on V_1. And if $z\lambda_1 = z\lambda_2$ with $\lambda_i \in \mathsf{F}$ $(i = 1, 2)$, then $\lambda_1 - \lambda_2$ is a singular linear transformation. By (3), $\lambda_1 - \lambda_2 = 0$.

(5) V has the structure of a vector space over F, in which $\lambda x = x\lambda$ for all $x \in \mathsf{V}_1$, $\lambda \in \mathsf{F}$ and $\lambda y = y\beta\lambda\beta^{-1}$ for all $y \in \mathsf{V}_2$, $\lambda \in \mathsf{F}$. $\{z, z\alpha\}$ is an F-basis for V.

Given $v \in \mathsf{V}$, there exist a unique $x \in \mathsf{V}_1$ and $y \in \mathsf{V}_2$ such that $v = x + y$, since $\mathsf{V} = \mathsf{V}_1 \oplus \mathsf{V}_2$; if $\lambda \in \mathsf{F}$, define $\lambda v = x\lambda + y\beta\lambda\beta^{-1}$. It is easy to check that the vector space axioms are satisfied. z and $z\alpha$ are F-linearly independent, since $z \in \mathsf{V}_1$, $z\alpha \in \mathsf{V}_2$ and V_1, V_2 are F-subspaces. Since $|\mathsf{V}| = |\mathsf{F}|^2$, $\dim_{\mathsf{F}}\mathsf{V} = 2$. Hence $\{z, z\alpha\}$ is an F-basis of V.

(Note that if $k \in \mathsf{K}$ and ι is the identity mapping on V_1, then $(k\iota)v = kv$ for all $v \in \mathsf{V}$).

(6) If $g \in \mathfrak{G}$, the mapping $v \to vg$ $(v \in \mathsf{V})$ is F-linear. Thus V is an $\mathsf{F}\mathfrak{G}$-module.

If $x \in \mathsf{V}_1$, then $xa = x$ and $x(a^b - 1)(a - 1)a = x(a^b - 1)(a - 1)$. Thus $(x\gamma)a = x\gamma$ and $(\gamma x)a = (x\gamma)a = x\gamma = \gamma x = \gamma(xa)$.

Also, if $y \in \mathsf{V}_2$, then $ya = y(a - 1) + y$ and $y(a - 1) \in \mathsf{V}_1$, $y \in \mathsf{V}_2$; thus

$$\begin{aligned}\gamma(ya) &= y(a - 1)\gamma + y\beta\gamma\beta^{-1} = y\beta\gamma + y\beta\alpha \\ &= (y\beta\alpha)a = (y\beta\gamma\beta^{-1})a = (\gamma y)a.\end{aligned}$$

Thus $\gamma(va) = (\gamma v)a$ for all $v \in \mathsf{V}$. It follows that $(\lambda v)a = \lambda(va)$ for all $\lambda \in \mathsf{F}$.

If $x \in \mathsf{V}_1$, then $xa^b = x + x(a^b - 1)$ and $x \in \mathsf{V}_1$, $x(a^b - 1) \in \mathsf{V}_2$; hence

$$\gamma(xa^b) = x\gamma + x(a^b - 1)\beta\gamma\beta^{-1} = x\alpha\beta + x\alpha\beta\alpha = (x\gamma)a^b = (\gamma x)a^b.$$

If $y \in \mathsf{V}_2$, then $ya^b = y$ and $\gamma y \in \mathsf{V}_2$, so

$$\gamma(ya^b) = \gamma y = (\gamma y)a^b.$$

Hence $\lambda(va^b) = (\lambda v)a^b$ for all $\lambda \in \mathsf{F}$, $v \in \mathsf{V}$.

Since $\mathfrak{G} = \langle a, a^b \rangle$, it follows that $\lambda(vg) = (\lambda v)g$ for all $g \in \mathfrak{G}$. Thus the mapping $v \to vg$ is F-linear. Hence V is an $\mathsf{F}\mathfrak{G}$-module.

(7) With respect to the F-basis $\{z, z\alpha\}$ of V, the matrices of the F-linear transformations $v \to va^b$, $v \to va$ are respectively

$$\begin{pmatrix} 1 & 1 \\ 0 & 1 \end{pmatrix}, \begin{pmatrix} 1 & 0 \\ \gamma & 1 \end{pmatrix},$$

since, for example,

$$(z\alpha)a = z\alpha + (z\alpha)(a - 1) = z\alpha + z\alpha\beta = z\gamma + z\alpha = \gamma z + z\alpha.$$

By 7.6, the group generated by these matrices contains a conjugate of $SL(2, p)$ in $SL(2, p^f)$. Hence there exists an F-basis $\{u_1, u_2\}$ of V such that given any matrix A of determinant 1 with coefficients in K, there exists $g \in \mathfrak{G}$ such that A is the matrix of the F-linear transformation $v \to vg$ with respect to the F-basis $\{u_1, u_2\}$ of V. If U is the K-space spanned by u_1 and u_2, $\dim_{\mathsf{K}} \mathsf{U} = 2$ and every K-linear transformation of U of determinant 1 is induced by some element of $\mathfrak{G}$. **q.e.d.**

In our application of 7.7, the stringent condition $\mathfrak{G} = \langle a, a^b \rangle$ will be obtained by applying an important theorem of Baer (III, 6.15). We state the relevant consequence of that theorem.

7.8 Theorem. *Suppose that a is an element of the finite group $\mathfrak{G}$ and that $\langle a, a^b \rangle$ is a p-group for every $b \in \mathfrak{G}$. Then $a \in \mathbf{O}_p(\mathfrak{G})$.*

Proof. If $b \in \mathfrak{G}$, $[b, a] = (a^b)^{-1}a \in \langle a, a^b \rangle$. By hypothesis, $\langle a, a^b \rangle$ is nilpotent. Hence $[b, \underbrace{a, \ldots, a}_{n}] = 1$ for sufficiently large n. Thus a is a right Engel element (III, 6.12). By III, 6.15, $a \in \mathbf{F}(\mathfrak{G})$. But a is a p-element, so $a \in \mathbf{O}_p(\mathfrak{G})$. **q.e.d.**

7.9 Lemma. *Let p be an odd prime, and let $\mathfrak{U}$, $\mathfrak{V}$ be subgroups of a group $\mathfrak{G}$. Suppose that $\mathfrak{U} \trianglelefteq \mathfrak{V}$ and $\mathfrak{V}/\mathfrak{U}$ is elementary Abelian of order p^2. If the group of automorphisms of $\mathfrak{V}/\mathfrak{U}$ induced by $\mathbf{N}_{\mathfrak{G}}(\mathfrak{U}) \cap \mathbf{N}_{\mathfrak{G}}(\mathfrak{V})$ contains $SL(2, p)$, then $\mathfrak{G}$ has a section isomorphic to $SA(2, p)$.*

Proof. Let $\mathsf{V} = \mathfrak{V}/\mathfrak{U}$, and let $\mathfrak{H}$ be a minimal subgroup of $(\mathbf{N}_{\mathfrak{G}}(\mathfrak{U}) \cap \mathbf{N}_{\mathfrak{G}}(\mathfrak{V}))/\mathfrak{U}$ such that $\mathfrak{H} \geq \mathsf{V}$ and the group of automorphisms of V induced by $\mathfrak{H}$ contains $SL(2, p)$. Then, if $\mathfrak{C} = \mathbf{C}_{\mathfrak{H}}(\mathsf{V})$,

$\mathfrak{H}/\mathfrak{C} \cong SL(2, p)$. Thus $\mathfrak{H}/\mathfrak{C}$ possesses a unique subgroup $\mathfrak{T}/\mathfrak{C}$ of order 2; we have $v^t = v^{-1}$ for all $v \in \mathsf{V}$ and $t \in \mathfrak{T} - \mathfrak{C}$, and $\mathfrak{T} \trianglelefteq \mathfrak{H}$. Let $\mathfrak{S}$ be a Sylow 2-subgroup of $\mathfrak{T}$. Then $\mathfrak{T} = \mathfrak{C}\mathfrak{S}$. By the Frattini argument.

$$\mathfrak{H} = \mathfrak{T}\mathbf{N}_{\mathfrak{H}}(\mathfrak{S}) = \mathfrak{C}\mathfrak{S}\mathbf{N}_{\mathfrak{H}}(\mathfrak{S}) = \mathfrak{C}\mathbf{N}_{\mathfrak{H}}(\mathfrak{S}).$$

Thus the group of automorphisms of V induced by $\mathbf{N}_{\mathfrak{H}}(\mathfrak{S})$ is $SL(2, p)$. By minimality of $\mathfrak{H}$, $\mathfrak{H} = \mathsf{V}\mathbf{N}_{\mathfrak{H}}(\mathfrak{S})$. But if $\mathfrak{N} = \mathbf{N}_{\mathfrak{C}}(\mathfrak{S})$, $\mathfrak{N} = \mathfrak{C} \cap \mathbf{N}_{\mathfrak{H}}(\mathfrak{S}) \trianglelefteq \mathbf{N}_{\mathfrak{H}}(\mathfrak{S})$, so $\mathfrak{N} \trianglelefteq \mathsf{V}\mathbf{N}_{\mathfrak{H}}(\mathfrak{S}) = \mathfrak{H}$. Since $v^t = v^{-1}$ for all $v \in \mathsf{V}$ and some $t \in \mathfrak{S}$, $\mathsf{V} \cap \mathbf{N}_{\mathfrak{H}}(\mathfrak{S}) = 1$. Thus $\mathsf{V} \cap \mathfrak{N} = 1$. Put $\overline{\mathsf{V}} = \mathsf{V}\mathfrak{N}/\mathfrak{N}$, $\overline{\mathfrak{H}} = \mathfrak{H}/\mathfrak{N}$. Then every linear transformation of $\overline{\mathsf{V}}$ of determinant 1 is induced by some element of $\overline{\mathfrak{H}}$ and hence of $\mathbf{N}_{\mathfrak{H}}(\mathfrak{S})/\mathfrak{N}$. Since

$$\mathbf{N}_{\mathfrak{H}}(\mathfrak{S})/\mathfrak{N} \cong \mathfrak{C}\mathbf{N}_{\mathfrak{H}}(\mathfrak{S})/\mathfrak{C} = \mathfrak{H}/\mathfrak{C} \cong SL(2, p),$$

and $\overline{\mathfrak{H}}$ is the split extension of $\overline{\mathsf{V}}$ by $\mathbf{N}_{\mathfrak{H}}(\mathfrak{S})/\mathfrak{N}$, $\overline{\mathfrak{H}} \cong SA(2, p)$. **q.e.d.**

7.10 Theorem. *Let $\mathfrak{G}$ be a finite group and let p be an odd prime. Then every section of $\mathfrak{G}$ is p-stable if and only if no section of $\mathfrak{G}$ is isomorphic to $SA(2, p)$.*

Proof. If every section of $\mathfrak{G}$ is p-stable, then no section of $\mathfrak{G}$ is isomorphic to $SA(2, p)$, since $SA(2, p)$ is not p-stable (7.5).

Suppose that the converse is false and that $\mathfrak{G}$ is a counterexample of minimal order. Thus no section of $\mathfrak{G}$ is isomorphic to $SA(2, p)$, but $\mathfrak{G}$ has a section which is not p-stable. By minimality of $|\mathfrak{G}|$,

(1) every proper section of $\mathfrak{G}$ is p-stable.

Thus $\mathfrak{G}$ itself is not p-stable. Let $\mathscr{P}$ be the set of p-subgroups $\mathfrak{R}$ of $\mathfrak{G}$ for which there exists $g \in \mathbf{N}_{\mathfrak{G}}(\mathfrak{R})$ such that $[\mathfrak{R}, g, g] = 1$ and $g\mathbf{C}_{\mathfrak{G}}(\mathfrak{R}) \notin \mathbf{O}_p(\mathbf{N}_{\mathfrak{G}}(\mathfrak{R})/\mathbf{C}_{\mathfrak{G}}(\mathfrak{R}))$. By 7.1, $\mathscr{P}$ is non-empty. Let $\mathfrak{P}$ be an element of $\mathscr{P}$ of minimal order. Thus there exists $a \in \mathbf{N}_{\mathfrak{G}}(\mathfrak{P})$ such that

(2) $[\mathfrak{P}, a, a] = 1$

and $a\mathfrak{C} \notin \mathbf{O}_p(\mathbf{N}_{\mathfrak{G}}(\mathfrak{P})/\mathfrak{C})$, where $\mathfrak{C} = \mathbf{C}_{\mathfrak{G}}(\mathfrak{P})$. Hence $\mathbf{N}_{\mathfrak{G}}(\mathfrak{P})$ is not p-stable. By (1), $\mathbf{N}_{\mathfrak{G}}(\mathfrak{P}) = \mathfrak{G}$. Thus

(3) $\mathfrak{P} \trianglelefteq \mathfrak{G}$, $\mathfrak{C} \trianglelefteq \mathfrak{G}$ and $a\mathfrak{C} \notin \mathbf{O}_p(\mathfrak{G}/\mathfrak{C})$.

We now make a number of routine reductions.

(4) $\mathbf{O}_{p'}(\mathfrak{G}) = 1$.

By 6.11, $\mathfrak{C}/\mathbf{O}_{p'}(\mathfrak{G}) = \mathbf{C}_{\mathfrak{G}/\mathbf{O}_{p'}(\mathfrak{G})}(\mathfrak{P}\mathbf{O}_{p'}(\mathfrak{G})/\mathbf{O}_{p'}(\mathfrak{G}))$, and by (2),

$$[\mathfrak{P}\mathbf{O}_{p'}(\mathfrak{G})/\mathbf{O}_{p'}(\mathfrak{G}), a\mathbf{O}_{p'}(\mathfrak{G}), a\mathbf{O}_{p'}(\mathfrak{G})] = 1.$$

Hence $\mathfrak{G}/\mathbf{O}_{p'}(\mathfrak{G})$ is not p-stable. Thus $\mathbf{O}_{p'}(\mathfrak{G}) = 1$ by (1).

(5) If $\mathfrak{H} \leq \mathfrak{G}$ and $\mathfrak{G} = \mathfrak{H}\mathfrak{C}$, then $\mathfrak{G} = \mathfrak{H}\mathfrak{P}$.

Since $\mathfrak{G} = \mathfrak{H}\mathfrak{C}$, there exist $h \in \mathfrak{H}$ and $c \in \mathfrak{C}$ such that $a = hc$. If $g \in \mathfrak{P}$, $[g, h] = [g, ac^{-1}] = [g, a]$ since $c \in \mathbf{C}_{\mathfrak{G}}(\mathfrak{P})$. Since $[g, a] \in \mathfrak{P}$, $[g, h, h] = [g, a, h] = [g, a, a] = 1$. Thus h commutes with every element of $[\mathfrak{P}, h]$ and $[\mathfrak{P}, h, h] = 1$.

Suppose $\mathfrak{D}/\mathfrak{C} = \mathbf{O}_p(\mathfrak{G}/\mathfrak{C})$. Since $\mathfrak{G}/\mathfrak{C} = \mathfrak{H}\mathfrak{C}/\mathfrak{C} = (\mathfrak{H}\mathfrak{P})\mathfrak{C}/\mathfrak{C} \cong \mathfrak{H}\mathfrak{P}/(\mathfrak{H}\mathfrak{P} \cap \mathfrak{C})$, $\mathbf{O}_p(\mathfrak{H}\mathfrak{P}/\mathfrak{H}\mathfrak{P} \cap \mathfrak{C}) = (\mathfrak{D} \cap \mathfrak{H}\mathfrak{P})/(\mathfrak{H}\mathfrak{P} \cap \mathfrak{C})$. Hence if $h\mathbf{C}_{\mathfrak{H}\mathfrak{P}}(\mathfrak{P}) \in \mathbf{O}_p(\mathfrak{H}\mathfrak{P}/\mathbf{C}_{\mathfrak{H}\mathfrak{P}}(\mathfrak{P}))$, then $h \in \mathfrak{D}$ and $a = hc \in \mathfrak{D}\mathfrak{C} = \mathfrak{D}$, contrary to (3). Thus $h\mathbf{C}_{\mathfrak{H}\mathfrak{P}}(\mathfrak{P}) \notin \mathbf{O}_p(\mathfrak{H}\mathfrak{P}/\mathbf{C}_{\mathfrak{H}\mathfrak{P}}(\mathfrak{P}))$. Hence $\mathfrak{H}\mathfrak{P}$ is not p-stable. By (1), $\mathfrak{H}\mathfrak{P} = \mathfrak{G}$.

(6) $\mathfrak{C}$ is a p-group.

Suppose that this is false and that q is a prime divisor of $|\mathfrak{C}|$ for which $q \neq p$. If $\mathfrak{Q} \in S_q(\mathfrak{C})$, $\mathfrak{Q} \neq 1$. By the Frattini argument, $\mathfrak{G} = \mathfrak{C}\mathbf{N}_{\mathfrak{G}}(\mathfrak{Q})$. By (5), $\mathfrak{G} = \mathfrak{P}\mathbf{N}_{\mathfrak{G}}(\mathfrak{Q})$. But $[\mathfrak{P}, \mathfrak{Q}] \leq [\mathfrak{P}, \mathfrak{C}] = 1$, so $\mathfrak{P} \leq \mathbf{N}_{\mathfrak{G}}(\mathfrak{Q})$. Thus $\mathfrak{G} = \mathbf{N}_{\mathfrak{G}}(\mathfrak{Q})$ and $\mathfrak{Q} \trianglelefteq \mathfrak{G}$. Since $\mathfrak{Q} \neq 1$, it follows that $\mathbf{O}_{p'}(\mathfrak{G}) \neq 1$, contrary to (4). Hence $\mathfrak{C}$ is a p-group.

(7) $\mathfrak{P}$ is a minimal normal subgroup of $\mathfrak{G}$, and $\mathfrak{P} \leq \mathfrak{C}$.

Suppose that $\mathfrak{P}_1$ is a minimal normal subgroup of $\mathfrak{G}$ and that $1 < \mathfrak{P}_1 \leq \mathfrak{P}$. Let $\mathfrak{C}_1 = \mathbf{C}_{\mathfrak{G}}(\mathfrak{P}_1)$, $\mathfrak{C}_2/\mathfrak{P}_1 = \mathbf{C}_{\mathfrak{G}/\mathfrak{P}_1}(\mathfrak{P}/\mathfrak{P}_1)$, and let $\mathfrak{D}_i/\mathfrak{C}_i = \mathbf{O}_p(\mathfrak{G}/\mathfrak{C}_i)$ $(i = 1, 2)$. Thus $(\mathfrak{D}_1 \cap \mathfrak{D}_2)/(\mathfrak{C}_1 \cap \mathfrak{C}_2)$ is a normal p-subgroup of $\mathfrak{G}/(\mathfrak{C}_1 \cap \mathfrak{C}_2)$. Now there is an isomorphism between $\mathfrak{G}/\mathfrak{C}$ and a group of automorphisms $\mathfrak{A}$ of $\mathfrak{P}$ which leave $\mathfrak{P}_1$ fixed. In this isomorphism, $(\mathfrak{C}_1 \cap \mathfrak{C}_2)/\mathfrak{C}$ corresponds precisely to the set of those elements of $\mathfrak{A}$ which leave fixed each element of $\mathfrak{P}_1$ and of $\mathfrak{P}/\mathfrak{P}_1$. By I, 4.4, $(\mathfrak{C}_1 \cap \mathfrak{C}_2)/\mathfrak{C}$ is a normal p-subgroup of $\mathfrak{G}/\mathfrak{C}$. Since $(\mathfrak{D}_1 \cap \mathfrak{D}_2)/(\mathfrak{C}_1 \cap \mathfrak{C}_2)$ is a p-group, $(\mathfrak{D}_1 \cap \mathfrak{D}_2)/\mathfrak{C}$ is a normal p-subgroup of $\mathfrak{G}/\mathfrak{C}$. Thus $(\mathfrak{D}_1 \cap \mathfrak{D}_2)/\mathfrak{C} \leq \mathbf{O}_p(\mathfrak{G}/\mathfrak{C})$. By (3), $a \notin \mathfrak{D}_1 \cap \mathfrak{D}_2$. But $\mathfrak{G}/\mathfrak{P}_1$ is a proper section of $\mathfrak{G}$ and is therefore p-stable. Since $[\mathfrak{P}/\mathfrak{P}_1, a\mathfrak{P}_1, a\mathfrak{P}_1] = 1$, it follows that $a \in \mathfrak{D}_2$. Thus $a \notin \mathfrak{D}_1$. Since $[\mathfrak{P}_1, a, a] \leq [\mathfrak{P}, a, a] = 1$, $\mathfrak{P}_1 \in \mathscr{P}$. Since $\mathfrak{P}$ is an element of $\mathscr{P}$ of minimal order and $\mathfrak{P}_1 \leq \mathfrak{P}$, we have $\mathfrak{P}_1 = \mathfrak{P}$, and $\mathfrak{P}$ is a minimal normal subgroup of $\mathfrak{G}$. Hence $\mathfrak{P} \leq \mathfrak{C}$.

(8) There exists $b \in \mathfrak{G}$ such that $\mathfrak{G} = \langle \mathfrak{C}, a, a^b \rangle$.

By (3), $a\mathfrak{C} \notin \mathbf{O}_p(\mathfrak{G}/\mathfrak{C})$. Hence by 7.8, there exists $b \in \mathfrak{G}$ such that $\langle a\mathfrak{C}, a^b\mathfrak{C} \rangle$ is not a p-group. Hence either $a\mathfrak{C} \notin \mathbf{O}_p(\langle a, a^b, \mathfrak{C} \rangle/\mathfrak{C})$ or $a^b\mathfrak{C} \notin \mathbf{O}_p(\langle a, a^b, \mathfrak{C} \rangle/\mathfrak{C})$. But $[\mathfrak{P}, a, a] = [\mathfrak{P}, a^b, a^b] = 1$; thus $\langle a, a^b, \mathfrak{C} \rangle$ is not p-stable and $\langle a, a^b, \mathfrak{C} \rangle = \mathfrak{G}$, by (1).

(9) $\mathfrak{P}$ is an elementary Abelian group by (7), so $\mathfrak{P}$ may be regarded as a vector space V over $\mathsf{K} = GF(p)$. $\mathfrak{G}/\mathfrak{C}$ is isomorphic to a group of automorphisms of $\mathfrak{P}$, so V may be regarded as a faithful $\mathsf{K}(\mathfrak{G}/\mathfrak{C})$-module. V is irreducible by (7), and $\mathfrak{G}/\mathfrak{C} \neq 1$ by (3). By 1.8, $\mathsf{V}(a - 1)^2$ corresponds to $[\mathfrak{P}, a, a]$; hence $\mathsf{V}(a - 1)^2 = 0$. By 7.7, there exists a K-subspace U of V such that $\dim_{\mathsf{K}}\mathsf{U} = 2$ and every linear transformation of U of

determinant 1 is the restriction to U of some element of $\mathfrak{G}$. By 7.9, $\mathfrak{G}$ has a section isomorphic to $SA(2, p)$.

This is a contradiction. Thus if no section of $\mathfrak{G}$ is isomorphic to $SA(2, p)$, every section of $\mathfrak{G}$ is p-stable. **q.e.d.**

7.11 Remarks. a) GLAUBERMAN [6, Theorem 4.1] has proved the following.

Suppose that $\mathfrak{V}$ is an elementary Abelian p-group, where p is odd, and that $\mathfrak{G}$ is a subgroup of $\mathbf{Aut}\,\mathfrak{V}$ for which $\mathbf{C}_{\mathfrak{V}}(\mathfrak{G}) = 1$ and $[\mathfrak{V}, \mathfrak{G}] = \mathfrak{V}$. Suppose that $\mathbf{O}_p(\mathfrak{G}) = 1$. Suppose that $\mathfrak{M}$ is a maximal subgroup of $\mathfrak{G}$, $\mathfrak{S} \in S_p(\mathfrak{M})$, $\mathfrak{A}$ is a non-identity Abelian subgroup of $\mathfrak{S}$ and $[\mathfrak{V}, \mathfrak{A}, \mathfrak{A}] = 1$. Suppose also that $\langle \mathfrak{A}, g \rangle = \mathfrak{G}$ for all $g \in \mathfrak{G} - \mathfrak{M}$. Then $\mathfrak{G} \cong SL(2, p^n)$ for some n, and $\mathfrak{V}$ is the direct product of $\mathfrak{G}$-invariant $GF(p^n)$-spaces on each of which $\mathfrak{G}$ acts as $SL(2, p^n)$. Further, $n = 1$ if $|\mathfrak{A}| = p$.

b) Theorem 7.4 is a consequence of Theorem 7.10. For if $p > 3$, $SA(2, p)$ is not p-soluble, and the Sylow 2-subgroup of $SA(2, 3)$, that is, the quaternion group of order 8, is non-Abelian.

Exercises

4) Let $\pi = \{2, 3\}$. Show that if a is a transposition in the symmetric group $\mathfrak{S}_n$, then $\langle a, a^b \rangle$ is a π-group for every $b \in \mathfrak{S}_n$, but that for $n \geq 5$, $a \notin \mathbf{O}_\pi(\mathfrak{G})$.

§ 8. Soluble Groups with One Class of Involutions

In this section, we shall prove that the 2-length of a soluble group $\mathfrak{G}$ is 1 if all involutions in $\mathfrak{G}$ are conjugate and the Sylow 2-subgroup contains more than one involution. To do this we need two preliminary results. The first one is number-theoretical; it arises also elsewhere in group-theory (e.g. Chapter XII, § 7).

8.1 Lemma. *Let ϕ_n be the n-th cyclotomic polynomial over the field of rational numbers. Let q be a prime, let a be an integer prime to q and let f be the order of a modulo q. For each non-zero integer x, let* $w_q(x) = \max\{l \mid q^l \text{ divides } x\}$.

(1) *For $q > 2$, the following hold.*

a) $w_q(\phi_f(a)) > 0$.

b) $w_q(\phi_{fq^i}(a)) = 1$ *for all* $i \geq 1$.

c) $w_q(\phi_m(a)) = 0$ *for all other* $m \geq 1$.
d) $w_q(a^n - 1) = 0$ *if* f *does not divide* n.
e) $w_q(a^n - 1) = w_q(a^f - 1) + w_q(n)$ *if* f *divides* n.

(2) *Also*

$$w_2(\phi_{2^i}(a)) = 1 \quad \textit{for} \quad i \geq 2.$$

$$w_2(\phi_n(a)) = 0 \quad \textit{for} \quad n \neq 2^i \quad (i = 0, 1, \ldots).$$

Proof (ARTIN [1, 2]). (1) a) As f is the order of a modulo q, $w_q(a^i - 1) = 0$ if $f \nmid i$. Thus $w_q(\phi_i(a)) = 0$ if $f \nmid i$. From

$$\phi_f(a) \prod_{i \mid f,\, i \neq f} \phi_i(a) = a^f - 1,$$

it follows that $w_q(\phi_f(a)) = w_q(a^f - 1) > 0$.

b) Put $n = fq^i$, $r = fq^{i-1}$. Then

$$\frac{a^n - 1}{a^r - 1} = \frac{((a^r - 1) + 1)^q - 1}{a^r - 1}$$
$$= (a^r - 1)^{q-1} + q(a^r - 1)^{q-2} + \cdots + \binom{q}{2}(a^r - 1) + q.$$

Since $a^r - 1 \equiv 0\,(q)$ and $\binom{q}{2} \equiv 0\,(q)$,

$$\frac{a^n - 1}{a^r - 1} \equiv q\,(q^2);$$

thus

$$w_q\left(\frac{a^n - 1}{a^r - 1}\right) = 1.$$

As is well-known,

$$\frac{a^n - 1}{a^r - 1} = \prod_{d \mid n,\, d \nmid r} \phi_d(a).$$

Thus

$$\sum_{d \mid n,\, d \nmid r} w_q(\phi_d(a)) = 1.$$

Hence there exists a divisor d of n such that $d \nmid r$ and $w_q(\phi_d(a)) = 1$. Thus $w_q(a^d - 1) > 0$ and $f \mid d$. The only integer d satisfying these conditions is $d = n$; thus $w_q(\phi_n(a)) = 1$.

c) If $w_q(\phi_m(a)) \neq 0$, $w_q(a^m - 1) \neq 0$ and f divides m. Thus it is to be shown that $w_q(\phi_m(a)) = 0$ if $m = fq^i l$ with $i \geq 0$, $q \nmid l$ and $l > 1$. If $r = fq^i$, $\phi_m(a)$ is a divisor of $(a^m - 1)/(a^r - 1)$. Since $a^r \equiv 1\ (q)$,

$$\frac{a^m - 1}{a^r - 1} = \frac{((a^r - 1) + 1)^l - 1}{a^r - 1}$$
$$= (a^r - 1)^{l-1} + l(a^r - 1)^{l-2} + \cdots + l \equiv l \not\equiv 0\ (q).$$

Thus $w_q(\phi_m(a)) = 0$.

d) is obvious.

e) We have

$$\frac{a^n - 1}{a^f - 1} = \prod_{d|n,\, d \nmid f} \phi_d(a).$$

It follows from this and c) that

$$w_q\left(\frac{a^n - 1}{a^f - 1}\right) = \sum_{fq^i|n,\, i>0} w_q(\phi_{fq^i}(a)).$$

Since f divides $q - 1$, q does not divide f. Hence the range of values of i in this sum is $\{i \mid 1 \leq i \leq w_q(n)\}$. Thus by b),

$$w_q(a^n - 1) - w_q(a^f - 1) = w_q(n).$$

(2) Since $q = 2$, $f = 1$. Write $n = 2^i l$ with $i \geq 0$, l odd and $l \geq 1$. If $l > 1$, $w_2(\phi_n(a)) = 0$ exactly as in c): $w_2(\phi_n(a)) \leq w_2((a^n - 1)/(a^{2^i} - 1))$ and

$$\frac{a^n - 1}{a^{2^i} - 1} = (a^{2^i} - 1)^{l-1} + l(a^{2^i} - 1)^{l-2} + \cdots + l \not\equiv 0\ (2).$$

Also

$$\phi_{2^i}(a) = \frac{a^{2^i} - 1}{a^{2^{i-1}} - 1} = a^{2^{i-1}} + 1,$$

so for $i \geq 2$,

$$\phi_{2^i}(a) \equiv 2\,(4)$$

and $w_2(\phi_{2^i}(a)) = 1$. **q.e.d.**

8.2 Lemma. *Let $\Phi_n(x, y) = y^{\phi(n)}\,\phi_n(x/y)$, where ϕ is the Euler function and ϕ_n is the n-th cyclotomic polynomial. Let*

$$L(n) = \inf|\Phi_n(a, b)|,$$

where a, b run through all complex numbers for which $|a| \geq |b| + 1$ and $|b| \geq 1$. Then $L(n) > \prod_{p|n} p$, where p runs through all prime divisors of n, except when $n = 1, 2, 3$ or 6.

Proof (ARTIN [1, 2]). The assertion will be proved by induction on n.

a) If p divides n, $L(np) \geq L(n)$ and $L(np) \geq (1 + p)^{\phi(n)}$.

Since p divides n, $\Phi_{np}(a, b) = \Phi_n(a^p, b^p)$. Since $|a| \geq |b| + 1$ and $|b| \geq 1$, $|a|^p \geq |b|^p + 1$ and $|b|^p \geq 1$, so $L(np) \geq L(n)$. Also

$$\Phi_n(a, b) = \prod_{\varepsilon}(a - \varepsilon b),$$

where ε runs through the $\phi(n)$ primitive n-th roots of unity. This gives

$$|\Phi_n(a, b)| \geq (|a| - |b|)^{\phi(n)}.$$

But for $|a| \geq |b| + 1$ and $|b| \geq 1$,

$$\begin{aligned}|a|^p - |b|^p &= (|a| - |b| + |b|)^p - |b|^p\\ &\geq (|a| - |b|)^p + p(|a| - |b|)^{p-1}|b| \geq 1 + p.\end{aligned}$$

Hence $|\Phi_n(a^p, b^p)| \geq (1 + p)^{\phi(n)}$ and $L(np) \geq (1 + p)^{\phi(n)}$.

b) For $p \geq 5$, $L(p) > 2p$.

We have

$$|\Phi_p(a, b)| = \left|\frac{a^p - b^p}{a - b}\right| \geq \frac{|a^p| - |b^p|}{|a| + |b|} = \frac{x^p - y^p}{x + y},$$

where $x = |a|$, $y = |b|$. Thus $x \geq y + 1$ and

$$\begin{aligned}(x^p - y^p)(1 + 2y) &= x(1 + y)(x^{p-1} - (1 + y)^{p-1}) + y(x^p - (1 + y)^p)\\ &\quad + y^p(x - (1 + y)) + (x + y)((1 + y)^p - y^p)\\ &\geq (x + y)((1 + y)^p - y^p).\end{aligned}$$

Hence

$$|\Phi_p(a, b)| \geq \frac{x^p - y^p}{x + y} \geq \frac{(1 + y)^p - y^p}{1 + 2y}.$$

But $y = |b| \geq 1$, so

$$(1 + y)^p - 2^p = \sum_{i=0}^{p} \binom{p}{i}(y^i - 1) \geq y^p - 1$$

and

$$(1 + y)^p - 2^p y = \sum_{i=0}^{p} \binom{p}{i}(y^i - y) \geq y^p - y.$$

Therefore

$$\begin{aligned} 3(1 + y)^p - 2^p(1 + 2y) &= ((1 + y)^p - 2^p) + 2((1 + y)^p - 2^p y) \\ &\geq y^p - 1 + 2(y^p - y) = 3y^p - (1 + 2y). \end{aligned}$$

Hence

$$3(1 + y)^p - 3y^p \geq (2^p - 1)(1 + 2y)$$

and

$$|\Phi_p(a, b)| \geq \frac{(1 + y)^p - y^p}{1 + 2y} \geq \frac{2^p - 1}{3}.$$

Hence $L(p) \geq \dfrac{2^p - 1}{3}$ and for $p \geq 5$, $L(p) > 2p$.

c) For $p \geq 5$, $L(2p) > 2p$.
This follows from $\Phi_{2p}(a, b) = \Phi_p(a, -b)$ and b).
d) If p does not divide n, $L(np) \geq L(n)^{p-1}$.
Since p does not divide n,

$$\Phi_{np}(a, b) = \prod_{\varepsilon^p = 1 \neq \varepsilon} \Phi_n(a, b\varepsilon).$$

e) If $x \geq 3$ and $y \geq 3$, then $x^{y-1} \geq xy$.
By elementary calculus, $x^{y-1} - xy$ is an increasing function of y for $y \geq 3$. Thus

$$x^{y-1} - xy \geq x^2 - 3x \geq 0$$

for $x \geq 3$.

f) If n is divisible by the square of a prime p, then $L(n) \geq L\left(\frac{n}{p}\right)$ by a). The assertion then follows at once from the inductive hypothesis unless $\frac{n}{p}$ is 1, 2, 3 or 6. Since p divides $\frac{n}{p}$, the only possibilities are $n = 4, 9, 12$ or 18. But by the second assertion of a), $L(4) \geq 3$, $L(9) \geq 16$, $L(12) \geq 9$ and $L(18) \geq 16$.

Suppose then that n is square-free. Let p be the greatest prime divisor of n and write $n = pm$. By hypothesis, $p \geq 5$. If $m = 1$, the assertion follows from b); if $m = 2$, it follows from c). If $m = 3$, $n = 3p$ and $L(n) \geq L(p)^2 > 4p^2 > 3p$ by d) and b). Similarly if $m = 6$, $L(n) \geq L(2p)^2 > 4p^2 > 6p$ by d) and c). For other values of m, $L(m) > \prod_{q|m} q$ by the inductive hypothesis. Thus $L(m) \geq 3$, for m cannot be a power of 2 since m is square-free. Hence by e), $L(m)^{p-1} \geq pL(m)$. By d), it follows that $L(n) \geq L(m)^{p-1} \geq pL(m) > \prod_{q|n} q$. **q.e.d.**

8.3 Theorem (ZSIGMONDY [1]). *Let a, n be integers greater than 1. Then except in the cases $n = 2$, $a = 2^b - 1$ and $n = 6$, $a = 2$, there is a prime q with the following properties.*

(1) *q divides $a^n - 1$.*

(2) *q does not divide $a^i - 1$ whenever $0 < i < n$.*

(3) *q does not divide n.*

In particular, n is the order of a modulo q.

Proof (ARTIN). Suppose that for each prime divisor q of $a^n - 1$, there exists i such that $0 < i < n$ and q divides $a^i - 1$. Since $a^i - 1 = \prod_{d|i} \phi_d(a)$, it follows in particular that for each prime divisor q of $\phi_n(a)$, there exists $d < n$ such that q divides $\phi_d(a)$. Let f be the order of a modulo q. Since $w_q(\phi_n(a)) > 0$ and $w_q(\phi_d(a)) > 0$, it follows from 8.1 that if $q > 2$, $n = fq^k$ and $d = fq^j$, where $0 \leq j < k$. Thus q divides n. Also $w_q(\phi_n(a)) = 1$ by 8.1. If $w_2(\phi_n(a)) > 0$, we see from 8.1(2) that $n = 2^i$ and, if $n > 2$, $w_2(\phi_n(a)) = 1$. It follows that if $n > 2$,

$$|\phi_n(a)| = \prod_q q^{w_q(\phi_n(a))} \leq \prod_{q|n} q.$$

By 8.2, however,

$$|\phi_n(a)| = |\Phi_n(a, 1)| > \prod_{q|n} q,$$

except when n is 1, 2, 3 or 6. Further,

$$\phi_3(a) = a^2 + a + 1 > 3$$

for all $a > 1$, and

$$\phi_6(a) = a^2 - a + 1 > 6$$

for $a \geq 3$. Thus only the cases $n = 2$ and $a = 2$, $n = 6$ remain. If the assertion is false for $n = 2$, each prime divisor of $a^2 - 1$ divides $a - 1$. It follows at once that 2 is the only prime divisor of $a + 1$, since $(a + 1, a - 1) \leq 2$; thus $a = 2^b - 1$.

Apart from the stated exceptions, then, there is always a prime q such that q divides $a^n - 1$ and q does not divide $a^i - 1$ for $0 < i < n$. Thus n is the order of a modulo q. Hence n divides $q - 1$ and q does not divide n. **q.e.d.**

8.4 Corollary. *Suppose that p is a prime and $n > 1$. There exists a prime $q > n$ such that q divides $p^n - 1$, except in the case $p = 3$, $n = 2$.*

Proof. First suppose that there exists a prime q such that $q|(p^n - 1)$ but $q \nmid (p^i - 1)$ for $0 < i < n$. Since $q|p^{q-1} - 1$ by Fermat's theorem, we have $n \leq (q - 1)$; hence $q > n$.

Secondly, suppose that every prime divisor of $p^n - 1$ divides $p^i - 1$ for some i with $0 < i < n$. By 8.3, either $p = 2$ and $n = 6$, or $n = 2$. In the case $p = 2$, $n = 6$, the required prime is 7. If $n = 2$, suppose that $p^2 - 1$ is not divisible by any prime greater than 2. Then $p^2 - 1$ is a power of 2. Hence so are $p - 1$ and $p + 1$. Thus $p = 3$. **q.e.d.**

Our second preliminary result is the following.

8.5 Theorem. (ITO [1]). *Suppose that $\mathfrak{G}$ is a p-soluble group and that $\mathfrak{G}$ has a faithful representation of degree n in a field* K, *where* char K *does not divide $|\mathfrak{G}|$ and $n < p$. Then the Sylow p-subgroups of $\mathfrak{G}$ are Abelian. Further, either the Sylow p-subgroup of $\mathfrak{G}$ is normal, or $|\mathfrak{G}|$ is even and $n = p - 1 = 2^m$ for some positive integer m.*

Proof. It is clear that we may suppose that K is algebraically closed. Let V be a faithful $\mathsf{K}\mathfrak{G}$-module of dimension n over K.

a) The Sylow p-subgroups of $\mathfrak{G}$ are Abelian.

Suppose $\mathfrak{S} \in S_p(\mathfrak{G})$. By V, 12.11b), the K-dimension of every irreducible $\mathsf{K}\mathfrak{S}$-submodule of V is a power of p. Since $n < p$, it follows that the K-dimension of every irreducible $\mathsf{K}\mathfrak{S}$-submodule of V is 1. Since char K does not divide $|\mathfrak{G}|$, V is the direct sum of irreducible K$\mathfrak{S}$-submodules, so $\mathfrak{S}$ is represented on V by diagonal linear transformations. Since V is faithful, $\mathfrak{S}$ is Abelian.

Suppose that the Sylow p-subgroup of $\mathfrak{G}$ is not normal, and let $\mathfrak{G}_0$ be a subgroup of $\mathfrak{G}$ of the smallest order for which the Sylow p-subgroup of $\mathfrak{G}_0$ is not normal.

b) For some prime q, $\mathfrak{G}_0$ has a normal q-subgroup $\mathfrak{N}$ such that $\mathfrak{G}_0/\mathfrak{N}$ is a p-group.

Since $\mathfrak{G}_0$ is p-soluble and the Sylow p-subgroups of $\mathfrak{G}_0$ are Abelian, $\mathfrak{G}_0$ is of p-length 1, by VI, 6.6a). Also, by minimality of $|\mathfrak{G}_0|$, $\mathfrak{G}_0$ has no proper normal subgroup of index prime to p. Thus $\mathfrak{G}_0 = \mathbf{O}_{p',p}(\mathfrak{G}_0)$. Let $\mathfrak{S}_0$ be a Sylow p-subgroup of $\mathfrak{G}_0$. Since $\mathfrak{S}_0$ is not normal in $\mathfrak{G}_0$, there exists a prime q such that $\mathbf{N}_{\mathfrak{G}_0}(\mathfrak{S}_0)$ contains no Sylow q-subgroup of $\mathfrak{G}_0$. If $\mathfrak{N}_1 \in S_q(\mathbf{O}_{p'}(\mathfrak{G}_0))$, $\mathfrak{G}_0 = \mathbf{O}_{p'}(\mathfrak{G}_0)\mathbf{N}_{\mathfrak{G}_0}(\mathfrak{N}_1)$, by the Frattini argument. Hence $|\mathfrak{G}_0 : \mathbf{N}_{\mathfrak{G}_0}(\mathfrak{N}_1)|$ is prime to p and $\mathbf{N}_{\mathfrak{G}_0}(\mathfrak{N}_1)$ contains a Sylow p-subgroup of $\mathfrak{G}_0$. Hence $\mathbf{N}_{\mathfrak{G}_0}(\mathfrak{N}_1) \geq \mathfrak{S}_0^x$ for some x and $\mathbf{N}_{\mathfrak{G}_0}(\mathfrak{N}) \geq \mathfrak{S}_0$, where $\mathfrak{N} = \mathfrak{N}_1^{x^{-1}}$. Thus $\mathfrak{S}_0\mathfrak{N}$ is a subgroup of $\mathfrak{G}_0$, but $\mathfrak{S}_0$ is not normal in $\mathfrak{S}_0\mathfrak{N}$, since $\mathfrak{N} \in S_q(\mathfrak{G}_0)$. By minimality of $|\mathfrak{G}_0|$, $\mathfrak{G}_0 = \mathfrak{S}_0\mathfrak{N}$.

c) There exists $\mathfrak{K} \lhd \mathfrak{G}_0$ such that if $\mathfrak{H} = \mathfrak{G}_0/\mathfrak{K}$, $\mathfrak{H}$ has a faithful irreducible representation on a subspace U of V, and the Sylow p-subgroup of $\mathfrak{H}$ is not normal.

Since char K does not divide $|\mathfrak{G}_0|$, V is the direct sum of irreducible K$\mathfrak{G}_0$-submodules $\mathsf{U}_1, \mathsf{U}_2, \ldots$. Let $\mathfrak{K}_i$ be the kernel of the representation of $\mathfrak{G}_0$ on U_i. Since $\mathfrak{G}_0$ is represented faithfully on V, $\bigcap_i \mathfrak{K}_i = 1$. Thus there is a monomorphism of $\mathfrak{G}_0$ into the direct product of the $\mathfrak{G}_0/\mathfrak{K}_i$. Hence the Sylow p-subgroup of this direct product is not normal, and it follows that the Sylow p-subgroup of some $\mathfrak{G}_0/\mathfrak{K}_i$ is not normal. Thus $\mathfrak{K} = \mathfrak{K}_i$ and $\mathsf{U} = \mathsf{U}_i$ have the stated properties.

By b), $\mathfrak{H}$ has a unique Sylow q-subgroup $\mathfrak{Q}$. Suppose $\mathfrak{P} \in S_p(\mathfrak{H})$.

d) U is the direct sum of isomorphic irreducible K$\mathfrak{Q}$-submodules.

By V, 17.3, U is the direct sum of K$\mathfrak{Q}$-submodules $\mathsf{W}_1, \mathsf{W}_2, \ldots, \mathsf{W}_k$, such that each W_i is the direct sum of isomorphic irreducible K$\mathfrak{Q}$-modules and there is a transitive permutation representation of $\mathfrak{P}$ of degree k. Since $k \leq n < p$, it follows that $k = 1$.

e) $|\mathfrak{G}|$ is even and $n = p - 1 = 2^m$ for some m.

Since $\mathfrak{P} \ntrianglelefteq \mathfrak{H}$, $\mathfrak{P} \nleq \mathbf{C}_{\mathfrak{H}}(\mathfrak{Q})$. But if $\mathfrak{Q}_1 < \mathfrak{Q}$ and $\mathfrak{P} \leq \mathbf{N}_{\mathfrak{H}}(\mathfrak{Q}_1)$, then $\mathfrak{P}\mathfrak{Q}_1 = \mathfrak{G}_1/\mathfrak{K}$ for some proper subgroup $\mathfrak{G}_1$ of $\mathfrak{G}_0$. By minimality of $|\mathfrak{G}_0|$, the Sylow p-subgroup of $\mathfrak{G}_1$ is normal, so $\mathfrak{P} \leq \mathbf{C}_{\mathfrak{H}}(\mathfrak{Q}_1)$. It follows from III, 13.5 that $\mathfrak{Q}$ is a special q-group, $\Phi(\mathfrak{Q})$ is centralized by $\mathfrak{P}$ and there is an irreducible representation of $\mathfrak{P}$ on $\mathfrak{Q}/\Phi(\mathfrak{Q})$.

By d), U is the direct sum of isomorphic irreducible K$\mathfrak{Q}$-modules, so $\mathbf{Z}(\mathfrak{Q})$ is represented on U by scalar multiples of the identity mapping.

Hence $\mathbf{Z}(\mathfrak{Q}) \leq \mathbf{Z}(\mathfrak{H})$ and $\mathbf{Z}(\mathfrak{Q})$ is cyclic. Since $\mathfrak{P} \not\trianglelefteq \mathfrak{H}$, $\mathfrak{Q} \not\leq \mathbf{Z}(\mathfrak{H})$; hence $\mathfrak{Q}$ is non-Abelian. Thus $\mathfrak{Q}$ is extraspecial and $|\mathfrak{Q}| = q^{2m+1}$ for some m.

Since $\mathfrak{P}$ is represented fixed point freely on $\mathfrak{Q}/\Phi(\mathfrak{Q})$ and $|\mathfrak{Q}/\Phi(\mathfrak{Q})| = q^{2m}$, $q^{2m} - 1$ is divisible by p. By V, 16.14, the degree of any faithful irreducible representation of $\mathfrak{Q}$ in K is q^m. Hence by d), $\dim_{\mathsf{K}} \mathsf{U} = jq^m$ for some integer j. Then $jq^m \leq n < p$, so $(q^m - 1, p) = 1$. Thus p divides $q^m + 1$; indeed $p = q^m + 1$ since $q^m < p$. Thus $q = 2$ and $|\mathfrak{G}|$ is even. From $jq^m \leq n < p = q^m + 1$, it follows that $j = 1$ and $n = q^m = p - 1$.

q.e.d.

8.6 Theorem (THOMPSON). *Suppose that $\mathfrak{G}$ is a soluble group of even order and that the Sylow 2-subgroup of $\mathfrak{G}$ contains more than one involution. Suppose that all the involutions in $\mathfrak{G}$ are conjugate. Then the 2-length of $\mathfrak{G}$ is 1, and the Sylow 2-subgroups of $\mathfrak{G}$ are either homocyclic or Suzuki 2-groups.*

Proof. Since $\mathfrak{G}/\mathbf{O}_{2'}(\mathfrak{G})$ satisfies the hypotheses, we may suppose that $\mathbf{O}_{2'}(\mathfrak{G}) = 1$. Let $\mathfrak{T} = \mathbf{O}_2(\mathfrak{G})$; thus $\mathfrak{T} \neq 1$. Let $\mathfrak{J} = \Omega_1(\mathbf{Z}(\mathfrak{T}))$. Since all involutions in $\mathfrak{G}$ are conjugate and $\mathfrak{J} \neq 1$, $\mathfrak{J}$ contains all involutions in $\mathfrak{G}$. Let $|\mathfrak{J}| = q = 2^n$. Thus $n > 1$.

a) There exists a prime divisor p of $2^n - 1$ such that $p > n$ and $\mathfrak{S} \not\leq \mathbf{C}_{\mathfrak{G}}(\mathfrak{J})$, where $\mathfrak{S} \in S_p(\mathfrak{G})$.

$\mathfrak{G}$ has $|\mathfrak{J}| - 1 = 2^n - 1$ involutions and they are all conjugate; hence $|\mathfrak{G} : \mathbf{C}_{\mathfrak{G}}(t)| = 2^n - 1$ for any $t \in \mathfrak{J} - \{1\}$. By 8.4, there exists a prime divisor p of $2^n - 1$ such that $p > n$. Suppose $\mathfrak{S} \in S_p(\mathfrak{G})$. Since $|\mathfrak{S}|$ does not divide $|\mathbf{C}_{\mathfrak{G}}(t)|$, $\mathfrak{S} \not\leq \mathbf{C}_{\mathfrak{G}}(t)$. Thus $\mathfrak{S} \not\leq \mathbf{C}_{\mathfrak{G}}(\mathfrak{J})$.

Let $\mathfrak{Q}$ be a Hall 2'-subgroup of $\mathfrak{G}$ such that $\mathfrak{Q} \geq \mathfrak{S}$.

b) If t_1, t_2 are involutions, there exists $x \in \mathfrak{Q}$ such that $t_1^x = t_2$.

$\mathfrak{Q}$ and $\mathbf{C}_{\mathfrak{G}}(t_1)$ are of coprime indices in $\mathfrak{G}$, so by I, 2.13, $\mathfrak{G} = \mathbf{C}_{\mathfrak{G}}(t_1)\mathfrak{Q}$. But $t_2 = t_1^g$ for some $g \in \mathfrak{G}$, and if $g = yx$ with $y \in \mathbf{C}_{\mathfrak{G}}(t_1)$, $x \in \mathfrak{Q}$, then $t_2 = t_1^x$.

c) Any subgroup of even order normalised by $\mathfrak{Q}$ contains $\mathfrak{J}$.

If $\mathfrak{N}$ is such a subgroup, $\mathfrak{N}$ contains an involution t_1. If $t_2 \in \mathfrak{J} - \{1\}$, then by b), $t_2 = t_1^x$ for some $x \in \mathfrak{Q}$, so $t_2 \in \mathfrak{N}^x = \mathfrak{N}$.

d) Any non-identity 2-subgroup $\mathfrak{N}$ normalised by $\mathfrak{Q}$ is either homocyclic or a Suzuki 2-group.

By b), the set of involutions in $\mathfrak{N}$ is permuted transitively by the group of automorphisms $x \to x^y$ ($x \in \mathfrak{N}$, $y \in \mathfrak{Q}$) of $\mathfrak{N}$ induced by $\mathfrak{Q}$. If $\mathfrak{N}$ is non-Abelian, $\mathfrak{N}$ is a Suzuki 2-group by VIII, 7.1. If $\mathfrak{N}$ is Abelian, $\mathfrak{N}$ is homocyclic, by VIII, 5.8.

e) $\mathfrak{S}$ is a normal Abelian subgroup of $\mathfrak{Q}$.

By d), $\mathfrak{T} = \mathbf{O}_2(\mathfrak{G})$ is either homocyclic or a Suzuki 2-group. If $\mathfrak{T}$ is homocyclic, $|\mathfrak{T} : \Phi(\mathfrak{T})| = q$. If $\mathfrak{T}$ is a Suzuki 2-group, then by VIII, 7.9,

$|\mathfrak{T} : \Phi(\mathfrak{T})|$ is q or q^2. By 1.4 and 1.6, $\mathbf{C}_{\mathfrak{G}}(\mathfrak{T}/\Phi(\mathfrak{T})) = \mathfrak{T}$. Thus $\mathfrak{Q}$ has a faithful representation on $\mathsf{U} = \mathfrak{T}/\Phi(\mathfrak{T})$; the degree of this is n or $2n$. Since $n < p$, e) follows at once from 8.5 when the degree is n. Suppose then that it is $2n$.

Let $\mathsf{F} = GF(2)$. If U is a reducible $\mathsf{F}\mathfrak{Q}$-module, $\mathsf{U} = \mathsf{U}_1 \oplus \mathsf{U}_2$ for $\mathsf{F}\mathfrak{Q}$-submodules U_i $(i = 1, 2)$. If $\mathsf{U}_i = \mathfrak{T}_i/\Phi(\mathfrak{T})$, each $\mathfrak{T}_i$ is either homocyclic or a Suzuki 2-group, by d). Thus $|\mathfrak{T}_i|$ is a power of q and indeed $|\mathfrak{T}_i| = q^2$. Thus if $\mathfrak{K}_i$ is the kernel of the representation of $\mathfrak{Q}$ on U_i, the Sylow p-subgroup of $\mathfrak{Q}/\mathfrak{K}_i$ is normal and Abelian, by 8.5. Since $\mathfrak{K}_1 \cap \mathfrak{K}_2 = 1$, $\mathfrak{Q}$ is isomorphic to a subgroup of $(\mathfrak{Q}/\mathfrak{K}_1) \times (\mathfrak{Q}/\mathfrak{K}_2)$, so $\mathfrak{S}$ is a normal Abelian subgroup of $\mathfrak{Q}$.

If U is an irreducible $\mathsf{F}\mathfrak{Q}$-module, let K be a finite extension of F which is a splitting field for $\mathfrak{Q}$, and let $\mathsf{V} = \mathsf{U} \otimes_{\mathsf{F}} \mathsf{K}$. Thus V is a $\mathsf{K}\mathfrak{Q}$-module. By V, 13.3, the irreducible components of V are all of the same degree m, and by V, 12.11b), m is odd. Since m divides $2n$, we have $m \leq n$. Hence 8.5 may be applied to each component, and in the same way we find that $\mathfrak{S}$ is a normal Abelian subgroup of $\mathfrak{Q}$.

f) $\mathfrak{T} \in S_2(\mathfrak{G})$.

Let $\mathfrak{R} = \mathfrak{Q} \cap \mathbf{O}_{2,2'}(\mathfrak{G})$. Thus $\mathfrak{R}$ is a Hall $2'$-subgroup of $\mathbf{O}_{2,2'}(\mathfrak{G})$ and $\mathbf{O}_{2,2'}(\mathfrak{G}) = \mathfrak{T}\mathfrak{R}$. By 1.3, $\mathbf{O}_{2,2'}(\mathfrak{G}) \geq \mathbf{C}_{\mathfrak{G}}(\mathbf{O}_{2,2'}(\mathfrak{G})/\mathfrak{T})$; thus $\mathbf{C}_{\mathfrak{Q}}(\mathfrak{R}) \leq \mathfrak{Q} \cap \mathbf{O}_{2,2'}(\mathfrak{G}) = \mathfrak{R}$. By e), $\mathfrak{S}$ is a normal Abelian subgroup of $\mathfrak{Q}$. Hence by III, 13.4,

$$\mathfrak{S} = (\mathfrak{S} \cap \mathbf{Z}(\mathfrak{S}\mathfrak{R})) \times [\mathfrak{S}, \mathfrak{R}].$$

But $\mathfrak{S} \cap \mathbf{Z}(\mathfrak{S}\mathfrak{R}) \leq \mathbf{C}_{\mathfrak{Q}}(\mathfrak{R}) \leq \mathfrak{R}$ and $[\mathfrak{S}, \mathfrak{R}] \leq \mathfrak{R}$ since $\mathfrak{R} \trianglelefteq \mathfrak{Q}$. Hence $\mathfrak{S} \leq \mathfrak{R}$. But by a), $\mathfrak{S} \not\leq \mathbf{C}_{\mathfrak{G}}(\mathfrak{J})$. Thus $[\mathfrak{R}, \mathfrak{J}] \neq 1$. Since $[\mathfrak{R}, \mathfrak{J}] \leq \mathfrak{J}$ and $\mathfrak{R} \cap \mathfrak{J} = 1$, it follows that $[\mathfrak{R}, \mathfrak{J}] \not\leq \mathfrak{R}$, or $\mathfrak{J} \not\leq \mathbf{N}_{\mathfrak{G}}(\mathfrak{R})$. But $\mathfrak{Q}$ certainly normalises $\mathbf{N}_{\mathfrak{G}}(\mathfrak{R})$, since $\mathfrak{Q} \leq \mathbf{N}_{\mathfrak{G}}(\mathfrak{R})$. It follows from c) that $|\mathbf{N}_{\mathfrak{G}}(\mathfrak{R})|$ is odd. But by the Frattini argument applied to the Hall $2'$-subgroup $\mathfrak{R}$ of $\mathbf{O}_{2,2'}(\mathfrak{G})$,

$$\mathfrak{G} = \mathbf{N}_{\mathfrak{G}}(\mathfrak{R})\mathbf{O}_{2,2'}(\mathfrak{G}) = \mathbf{N}_{\mathfrak{G}}(\mathfrak{R})\mathfrak{T}.$$

Hence $\mathfrak{T} \in S_2(\mathfrak{G})$.

It follows at once that the 2-length of $\mathfrak{G}$ is 1, and by d), $\mathfrak{T}$ is either homocyclic or a Suzuki 2-group. **q.e.d.**

The following is a similar theorem for p odd.

8.7 Theorem (GASCHÜTZ and YEN [1]). *Suppose that $\mathfrak{G}$ is a p-soluble group, where p is an odd prime divisor of $|\mathfrak{G}|$. If the subgroups of $\mathfrak{G}$ of order p are permuted transitively by* **Aut** $\mathfrak{G}$, *the p-length of $\mathfrak{G}$ is 1.*

Proof. Let $\mathfrak{G}$ be a counterexample of minimal order. Then $\mathbf{O}_{p'}(\mathfrak{G}) = 1$, for $\mathfrak{G}/\mathbf{O}_{p'}(\mathfrak{G})$ satisfies the hypotheses of the theorem. Let $\mathfrak{M}$ be a maximal characteristic subgroup of $\mathfrak{G}$. Then $\mathfrak{G}/\mathfrak{M}$ is either a p'-group or an elementary Abelian p-group. Since $\mathfrak{G}$ is not of p-length 1, $\mathfrak{M}$ is not a p'-group. Hence the subgroups of $\mathfrak{M}$ of order p are permuted transitively by **Aut** $\mathfrak{M}$. By minimality of $|\mathfrak{G}|$, the p-length of $\mathfrak{M}$ is 1. Hence $\mathfrak{G}/\mathfrak{M}$ is a p-group. Also, since $\mathbf{O}_{p'}(\mathfrak{G}) = 1$, $\mathbf{O}_{p'}(\mathfrak{M}) = 1$. The Sylow p-subgroup $\mathfrak{P}$ of $\mathfrak{M}$ is therefore normal. Let $\mathfrak{Z} = \Omega_1(\mathbf{Z}(\mathfrak{P}))$. Then $\mathfrak{Z} \neq 1$, so it follows from the hypothesis that $\mathfrak{Z}$ is a minimal characteristic subgroup of $\mathfrak{G}$ and that $\mathfrak{Z}$ is the set of all elements of $\mathfrak{G}$ of order a divisor of p.

By the Schur-Zassenhaus theorem, $\mathfrak{P}$ has a complement $\mathfrak{Q}$ in $\mathfrak{M}$ and all such complements are conjugate in $\mathfrak{M}$. Thus by the Frattini argument, $\mathfrak{G} = \mathbf{N}_{\mathfrak{G}}(\mathfrak{Q})\mathfrak{M} = \mathbf{N}_{\mathfrak{G}}(\mathfrak{Q})\mathfrak{P}$. Hence $\mathbf{N}_{\mathfrak{G}}(\mathfrak{Q}) \neq \mathfrak{Q}$ and $\mathbf{N}_{\mathfrak{G}}(\mathfrak{Q})$ contains an element x of order p. Let $\mathfrak{S}$ be a Sylow p-subgroup of $\mathfrak{G}$ and let y be an element of $\mathbf{Z}(\mathfrak{S})$ of order p. Then $\langle x \rangle = \langle y \rangle \xi$ for some automorphism ξ of $\mathfrak{G}$, so $x \in \mathbf{Z}(\mathfrak{S}\xi)$. Also $x \in \mathfrak{Z} \cap \mathbf{N}_{\mathfrak{G}}(\mathfrak{Q}) = \mathbf{N}_{\mathfrak{Z}}(\mathfrak{Q})$ and

$$[\mathbf{N}_{\mathfrak{Z}}(\mathfrak{Q}), \mathfrak{Q}] \leq \mathfrak{Z} \cap \mathfrak{Q} = 1.$$

Thus $\mathbf{C}_{\mathfrak{G}}(x) \geq \langle \mathfrak{S}\xi, \mathfrak{Q} \rangle = \mathfrak{G}$, so $x \in \mathbf{Z}(\mathfrak{G})$. Hence $\mathbf{Z}(\mathfrak{G}) \cap \mathfrak{Z} \neq 1$, and since $\mathfrak{Z}$ is a minimal characteristic subgroup of $\mathfrak{G}$, $\mathfrak{Z} \leq \mathbf{Z}(\mathfrak{G})$.

Of course, $|\mathfrak{G}/\mathfrak{Z}|$ is divisible by p. We show that the subgroups of $\mathfrak{G}/\mathfrak{Z}$ of order p are permuted transitively by **Aut** $(\mathfrak{G}/\mathfrak{Z})$. To do this, let $u\mathfrak{Z}$ be an element of order p in $\mathbf{Z}(\mathfrak{S}/\mathfrak{Z})$; thus

$$[\langle u \rangle, \mathfrak{S}] \leq \mathfrak{Z}.$$

Now let $v\mathfrak{Z}$ be any element of $\mathfrak{G}/\mathfrak{Z}$ of order p. Then u^p, v^p are elements of $\mathfrak{G}$ of order p and $u^p = v^{pi}\zeta = (v^i\zeta)^p$ for some automorphism ζ of $\mathfrak{G}$ and some integer i with $(i, p) = 1$. Now $w = (v^i\zeta)^a \in \mathfrak{S}$ for some $a \in \mathfrak{G}$ and $w^p = (v^i\zeta)^p = u^p$ since $\mathfrak{Z} \leq \mathbf{Z}(\mathfrak{G})$. Then $[u, w] \in \mathfrak{Z}$ and

$$(uw^{-1})^p = u^p w^{-p}[u, w]^{\binom{p}{2}} = 1,$$

since p is odd. Then $uw^{-1} \in \mathfrak{Z}$ and

$$u\mathfrak{Z} = w\mathfrak{Z} = (v^i\zeta)^a\mathfrak{Z}.$$

Hence $(v^i\mathfrak{Z})\zeta' = u\mathfrak{Z}$ for some automorphism ζ' of $\mathfrak{G}/\mathfrak{Z}$. It follows from the minimality of $|\mathfrak{G}|$ that $\mathfrak{G}/\mathfrak{Z}$ is of p-length 1. If $\mathbf{O}_{p'}(\mathfrak{G}/\mathfrak{Z}) = \mathfrak{T}/\mathfrak{Z}$, $\mathfrak{T} = \mathfrak{Z} \times \mathfrak{U}$ for some $\mathfrak{U}$, since $\mathfrak{Z} \leq \mathbf{Z}(\mathfrak{G})$. Thus $\mathfrak{U} \trianglelefteq \mathfrak{G}$ and $\mathfrak{U} \leq \mathbf{O}_{p'}(\mathfrak{G}) = 1$. Hence $\mathbf{O}_{p'}(\mathfrak{G}/\mathfrak{Z}) = 1$ and $\mathfrak{S}/\mathfrak{Z} \trianglelefteq \mathfrak{G}/\mathfrak{Z}$. Thus $\mathfrak{S} \trianglelefteq \mathfrak{G}$. **q.e.d.**

It follows easily from 8.7 that the Sylow p-subgroup $\mathfrak{S}$ of $\mathfrak{G}$ has the property that its subgroups of order p are permuted transitively by **Aut** $\mathfrak{S}$. As was remarked in VIII, 7.11, this implies that $\mathfrak{S}$ is Abelian.

Notes on Chapter IX

§ **1:** This formulation of 1.10 was given in a private communication from G. Higman.
§ **2:** We thank K. Johnsen for the proof of 2.3.
§ **3:** The proof of Lemma 3.11 follows a suggestion of K. Johnsen.
§ **5:** Theorem 5.2 for p odd was proved by J. G. THOMPSON [6, Lemma 5.22]. For the version given here we are indebted to K.-U. Schaller. Sections 5.10–5.14 are due to M. B. Powell, whom we thank for a written communication of these results.
§ **6:** We are grateful to H. Kurzweil for suggesting several improvements to this section.
§ **8:** The proof of 8.2 follows Artin, but includes an improvement due to R. M. Bryant.

Bibliography

Books and lecture notes are distinguished by *

ADYAN, S. I.:* [1] The Burnside problem and identities in groups. Springer-Verlag 1978.

ALPERIN, J. L.: [1] Automorphisms of solvable groups. Proc. Amer. Math. Soc. *13*, 175–180 (1962).

— [2] Sylow intersections and fusion. J. Algebra *6*, 222–241 (1967).

ALPERIN, J. L., and D. GORENSTEIN: [1] The multiplicators of certain simple groups. Proc. Amer. Math. Soc. *17*, 515–519 (1966).

— [2] Transfer and fusion in finite groups. J. Algebra *6*, 242–255 (1967).

AMAYO, R. K., and I. STEWART:* [1] Infinite-dimensional Lie algebras. Leyden: Noordhoff International Publishing 1974.

ARAD, Z., and G. GLAUBERMAN: [1] A characteristic subgroup of a group of odd order. Pacific J. Math. *56*, 305–319 (1975).

ARTIN, E.: [1] The order of the linear groups. Comm. Pure Appl. Math. *8*, 355–366 (1955).

— [2] The order of the classical simple groups. Comm. Pure Appl. Math. *8*, 455–472 (1955).

ASCHBACHER, M.: [1] Thin finite simple groups. Bull. Amer. Math. Soc. *82*, 484 (1976).

BAŠEV, V. A.: [1] Representations of $Z_2 \times Z_2$ in the field of characteristic 2. Dokl. Akad. Nauk *141*, 1015–1018 (1961).

BEAUMONT, R. A. and R. P. PETERSON: [1] Set transitive permutation groups. Canad. J. Math. *7*, 35–42 (1955).

BENDER, H.: [1] Über den grössten p'-Normalteiler in p-auflösbaren Gruppen. Arch. Math. (Basel) *18*, 15–16 (1967).

— [2] Endliche zweifach transitive Permutationsgruppen, deren Involutionen keine Fixpunkte haben. Math. Z. *104*, 175–204 (1968).

— [3] On the uniqueness theorem. Illinois J. Math. *14*, 376–384 (1970).

— [4] On groups with Abelian Sylow 2-subgroups. Math. Z. *117*, 164–176 (1970).

— [5] Transitive Gruppen gerader Ordnung, in denen jede Involution genau einen Punkt festlässt. J. Algebra *17*, 527–554 (1971).

— [6] A group-theoretic proof of Burnside's $p^a q^b$-theorem. Math. Z. *126*, 327–338 (1972).

— [7] Finite groups with large subgroups. Illinois J. Math. *18*, 223–228 (1974).

— [8] The Brauer-Suzuki-Wall theorem. Illinois J. Math. *18*, 229–235 (1974).

— [9] On the normal p-structure of a finite group and related topics I. Hokkaido Math. J. *7*, 271–288 (1978).

BERGER, T. R.: [1] Hall-Higman type theorems, I. Canad. J. Math. *29*, 513–531 (1974); II. Trans. Amer. Math. Soc. **205**, 47–69 (1975); III. Trans. Amer. Math. Soc. **228**, 47–83 (1977); IV. Proc. Amer. Math. Soc. *37*, 317–325 (1973); V. Pacific J. Math. *73*, 1–62 (1977); VI. J. Algebra *51*, 416–424 (1978); VII. Proc. London Math. Soc. (3) *31*, 21–54 (1975).

— [2] Nilpotent fixed point free automorphism groups of solvable groups. Math. Z. *131*, 305–312 (1973).

BERGER, T. R. and F. GROSS: [1] 2-length and the derived length of a Sylow 2-subgroup. Proc. London Math. Soc. (3) *34*, 520–534 (1977).

BERMAN, S. D.: [1] The number of irreducible representations of a finite group over an arbitrary field. Dokl. Akad. Nauk *106*, 767–769 (1956).

BLACKBURN, N.: [1] Über Involutionen in 2-Gruppen. Arch. Math. (Basel) 35, 75–78 (1980).

BLACKBURN, N. and L. EVENS: [1] Schur multipliers of p-groups. J. Reine Angew. Math. *309*, 100–113 (1979).

BOMBIERI, E.: [1] Thompson's problem ($\sigma^2 = 3$). Invent. Math. *58*, 77–100 (1980).

BOURBAKI, N.:* [1] Algèbre, Chap. 4, 5. Paris: Hermann 1959.

— *[2] Algèbre Commutative, Chap. 2. Paris: Hermann 1961.

— *[3] Groupes et algèbres de Lie, Chap. 2: Algèbres de Lie libres. Paris: 1972.

BRAUER, R.: [1] Über Systeme hyperkomplexer Zahlen. Math. Z. *30*, 79–107 (1929).

— [2] Über die Darstellungen von Gruppen in Galoisschen Feldern. Act. Sci. No. 135, Paris, Hermann 1935.

— [3] On groups whose order contains a prime number to the first power I, II. Amer. J. Math. *64*, 401–420 and 421–440 (1942).

— [4] On the representation of a group of order g in the field of g-th roots of unity. Amer. J. Math. *67*, 461–471 (1945).

— [5] On the structure of groups of finite order. Proc. International Congress, Amsterdam 1954, vol. 1, 209–217.

— [6] Zur Darstellungstheorie der Gruppen endlicher Ordnung. Math. Z. *63*, 406–444 (1956).

— [7] Some applications of the theory of blocks of characters of finite groups. J. Algebra *1*, 152–167 (1964).

— [8] Some applications of the theory of blocks of characters of finite groups IV. J. Algebra *17*, 489–521 (1971).

BRAUER, R. and P. FONG: [1] A characterization of the Mathieu group M_{12}. Trans. Amer. Math. Soc. *122*, 18–47 (1966).

BRAUER, R. and K. A. FOWLER: [1] Groups of even order. Ann. of Math. *62*, 565–583 (1955).

BRAUER, R. and C. NESBITT: [1] On the modular characters of groups. Ann. of Math. *42*, 556–590 (1941).

BRAUER, R., M. SUZUKI and G. E. WALL: [1] A characterization of the one-dimensional unimodular projective groups over finite fields. Illinois J. Math. *2*, 718–745 (1958).

BRENNER, S.: [1] Modular representations of p-groups. J. Algebra *15*, 89–102 (1970).

BRYANT, R. M. and L. C. KOVÁCS: [1] Lie representations and groups of prime-power order. J. London Math. Soc. (2) *17*, 415–421 (1978).

BRYCE, N.: [1] On the Mathieu group M_{23}. J. Austral. Math. Soc. *12*, 385–392 (1971).

BUCHT. G.: [1] Die umfassendsten primitiven metazyklischen Kongruenzgruppen in drei oder vier Variablen. Ark. Mat. Astronom. Fys. *11*, (92 pages, 1917).

BURGOYNE, N. and P. FONG: [1] Multipliers of the Mathieu groups. Nagoya Math. J. *27*, 733–745 (1966). Corrections, Nagoya Math. J. *31*, 297–304 (1968).

BURKHARDT, R.: [1] Die Zerlegungsmatrizen der Gruppen PSL(2, p^f). J. Algebra *40*, 75–96 (1976).

BURNSIDE, W.:* [1] Theory of groups of finite order, 2nd edn. Cambridge, 1911; Dover Publications, 1955.

CARTIER, P.: [1] Remarques sur le théorème de Birkhoff-Witt. Ann. Scuola Norm. Sup. Pisa (3) *12*, 1–4 (1958).

COLLINS. M. J.: [1] The characterization of the Suzuki groups by their Sylow 2-subgroups. Math. Z. *123*, 432–48 (1971).

CONLON, S.: [1] The modular representation algebra of groups with Sylow 2-subgroups $Z_2 \times Z_2$. J. Austral. Math. Soc. *6*, 76–88 (1966).

COSSEY, J. and W. GASCHÜTZ: [1] A note on blocks. Proc. Second Intern. Conf. Theory of Groups, 238–240. Canberra 1973.

COXETER, H. and W. MOSER:* [1] Generators and relations for discrete groups. 2nd edition, Ergebnisse der Mathematik 14, Springer 1964.

CURTIS, C. and I. REINER:* [1] Representation theory of finite groups and associative algebras. New York: Interscience Publishers Inc. 1962.

DADE, E. C.: [1] Some p-solvable groups. J. Algebra *2*, 395–401 (1965).

— [2] Blocks with cyclic defect groups. Ann. of Math. (2) *84*, 20–48 (1966).
— [3] Degress of modular irreducible representations of p-solvable groups. Math. Z. *104*, 141–143 (1968).
— [4] Carter subgroups and Fitting heights of finite solvable groups. Illinois J. Math. *13*, 347–369 (1972).
— [5] Une extension de la théorie de Hall et Higman. J. Algebra *20*, 570–609 (1972).
DEURING, M.: [1] Galoissche Theorie und Darstellungstheorie. Math. Ann. *107*, 140–144 (1932).
— *[2] Algebren. Ergebn. der Math. 41, 2. Aufl., Springer 1968.
DIEUDONNÉ, J.:*[1] La géométrie des groupes classiques, 2nd edition. Springer-Verlag 1963.
DOLAN, S. W.: [1] Some problems in the theory of finite groups, D. Phil. dissertation, Oxford, 1975.
DORNHOFF, L.: [1] The rank of primitive solvable permutation groups. Math. Z. *109*, 205–210 (1969).
FEIT, W.: [1] On a class of doubly transitive permutation groups. Illinois J. Math. *4*, 170–186 (1960).
— [2] Group characters, exceptional characters. Summer Institute of Finite Groups, Pasadena (1960).
— *[3] Representation theory of finite groups. Lecture Notes, Yale University 1969.
FEIT, W. and J. G. THOMPSON: [1] Solvability of groups of odd order. Pacific J. Math. *13*, 773–1029 (1963).
FELSCH, W., J. NEUBÜSER and W. PLESKEN: [1] Space groups and groups of prime-power order IV. J. London Math. Soc. (2) *24*, 113–122 (1981).
FISCHER, I. and R. R. STRUIK: [1] Nil algebras and periodic groups. Amer. Math. Monthly *75*, 611–623 (1968).
FONG, P.: [1] On the characters of p-solvable groups. Trans. Amer. Math. Soc. *98*, 263–284 (1961).
— [2] On decomposition numbers of J_1 and $R(q)$. Sympos. Math. Vol. XIII, 414–422. London, New York. Acad. Press 1974.
FONG, P. and W. GASCHÜTZ: [1] A note on the modular representations of solvable groups. J. Reine Angew. Math. *208*, 73–78 (1961).
FOULSER, D.: [1] Solvable primitive permutation groups of low rank. Trans. Amer. Math. Soc. *143*, 1–54 (1969).
FOX, R. H.: [1] Free differential calculus I. Ann. of Math. *57*, 547–560 (1953).
FRASCH, H.: [1] Die Erzeugenden der Hauptkongruenzgruppen für Primzahlstufen. Math. Ann. *108*, 229–252 (1933).
FROBENIUS, G.: [1] Über Gruppencharaktere. Sitz. preuss. Akad. 1896, 985–1021.
— [2] Über Gruppen des Grades p oder $p + 1$. Sitz. preuss. Akad. 1902, 351–369.
— [3] Über die Charaktere der mehrfach transitiven Gruppen. Sitz. preuss. Akad. Berlin 1904, 558–571.
FROBENIUS, G. and I. SCHUR: [1] Über die reellen Darstellungen endlicher Gruppen. Sitz. preuss. Akad. 1906, 186–208.
GAGEN, T. M.:* [1] Topics in Finite Groups. London Mathematical Society Lecture Note Series 16, CUP 1976.
GASCHÜTZ, W.: [1] Über den Fundamentalsatz von Maschke zur Darstellungstheorie endlicher Gruppen. Math. Z. *56*, 376–387 (1952).
GASCHÜTZ, W. and T. YEN: [1] Groups with an automorphism group which is transitive on the elements of prime order. Math. Z. *86*, 123–127 (1964).
GLAUBERMAN, G.: [1] On the automorphism group of a finite group having no non-identity normal subgroups of odd order. Math. Z. *93*, 154–160 (1966).
— [2] Prime-power factor groups of finite groups. Math. Z. *107*, 159–172 (1968).
— [3] A characteristic subgroup of a p-stable group. Canad. J. Math. *20*, 1101–1135 (1968).
— [4] On a class of doubly transitive permutation groups. Illinois J. Math. *13*, 394–399 (1969).
— [5] Prime-power factor groups of finite groups II. Math. Z. *117*, 46–56 (1970).

— [6] A sufficient condition for p-stability. Proc. London Math. Soc. (3) *25*, 253–287 (1972).

— [7] Failure of factorization in p-solvable groups. Quart. J. Math. Oxford (2) *24*, 71–77 (1973); II, Quart. J. Math. Oxford (2) *26*, 257–261 (1975).

— [8] On Burnside's other $p^a q^b$ theorem. Pacific J. Math. *56*, 469–476 (1975).

— [9] On solvable signalizer functors in finite groups. Proc. London Math. Soc. (3) *33*, 1–27 (1976).

— [10] Factorizations in local subgroups of finite groups. Providence, R. I. (1977).

GLEASON, A. M.: [1] Finite Fano planes. Amer. J. Math. *78*, 797–807 (1956).

GLOVER, D. J.: [1] A study of certain modular representations. J. Algebra *51*, 425–475 (1978).

GOLDSCHMIDT, D. M.: [1] A conjugation family for finite groups. J. Algebra *16*, 138–142 (1970).

— [2] A group theoretic proof of the $p^a q^b$ theorem for odd primes. Math. Z. *113*, 373–375 (1970).

— [3] Solvable signalizer functors on finite groups. J. Algebra *21*, 137–148 (1972).

— [4] 2-fusion in finite groups. Ann. of Math. (2) *99*, 70–117 (1974).

— [5] Elements of order two in finite groups. Delta (Waukesha) *4*, 45–58 (1974/5).

GORENSTEIN, D.:* [1] Finite groups. New York, Harper and Row 1968.

GORENSTEIN, D. and I. N. HERSTEIN: [1] Finite groups admitting a fixed-point-free automorphism of order 4. Amer. J. Math. *83*, 71–78 (1961).

GORENSTEIN, D. and D. R. HUGHES: [1] Triply transitive groups in which only the identity fixes four letters. Illinois J. Math. *5*, 486–491 (1961).

GORENSTEIN, D. and J. WALTER: [1] The π-layer of a finite group. Illinois J. Math. *15*, 555–564 (1971).

— [2] Balance and generation in finite groups, J. Algebra *33*, 224–287 (1975).

GOW, R.: [1] Extensions of modular representations for relatively prime operator groups. J. Algebra *36*, 492–494 (1975).

GREEN, J. A.: [1] On the indecomposable representations of a finite group. Math. Z. *70*, 430–445 (1959).

— [2] On a theorem of P. M. Neumann. Workshop on permutation groups and indecomposable modules. Giessen, 1975.

GREEN, J. A. and R. HILL: [1] On a theorem of Fong and Gaschütz. J. London Math. Soc. (2) *1*, 573–576 (1969).

GREEN, J. A. and S. E. STONEHEWER: [1] The radicals of some group algebras. J. Algebra *13*, 137–142 (1969).

GROSS, F.: [1] The 2-length of a finite solvable group. Pacific J. Math. *15*, 1221–1237 (1965).

— [2] The 2-length of groups whose Sylow 2-groups are of exponent 4. J. Algebra *2*, 312–314 (1965).

— [3] Solvable groups admitting a fixed-point-free automorphism of prime power order. Proc. Amer. Math. Soc. *17*, 1440–1446 (1966).

— [4] A note on fixed-point-free solvable operator groups. Proc. Amer. Math. Soc. *19*, 1363–1365 (1968).

— [5] 2-automorphic 2-groups. J. Algebra *40*, 348–353 (1976).

HALL, M.: [1] Solution of the Burnside problem for exponent 6. Proc. Nat. Acad. Sci. U.S.A. *43*, 751–753 (1957); Illinois J. Math. *2*, 764–786 (1958).

— *[2] The theory of groups. New York: Macmillan Company 1959.

HALL, P.: [1] A contribution to the theory of groups of prime-power order. Proc. London Math. Soc. (2) *36*, 29–95 (1933).

HALL, P. and G. HIGMAN: [1] The p-length of a p-soluble group, and reduction theorems for Burnside's problem. Proc. London Math. Soc. (3) *6*, 1–42 (1956).

HARTLEY, B.: [1] Sylow p-subgroups and local p-solubility. J. Algebra *23*, 347–369 (1972).

HARTLEY, B. and D. J. S. ROBINSON: [1] On finite complete groups. Arch. Math. (Basel) *35*, 67–74 (1980).

HAWKES, T. O.: [1] On the automorphism group of a 2-group. Proc. London Math. Soc.

(3) *26*, 207–225 (1973).
HEINEKEN, H. and H. LIEBECK: [1] The occurrence of finite groups in the automorphism group of nilpotent groups of class 2. Arch. Math. (Basel) *25*, 8–16 (1974).
HELD, D.: [1] Eine Kennzeichnung der Mathieu Gruppe M_{22} und der alternierenden Gruppe A_{10}. J. Algebra *8*, 436–449 (1968).
— [2] A characterization of some multiply transitive permutation groups I. Illinois J. Math. *13*, 224–240 (1969).
— [3] The simple groups related to M_{24}. J. Algebra *13*, 253–296 (1969).
HELD, D. and U. SCHOENWAELDER: [1] A characterization of the simple group $\mathfrak{M}_{24}$. Math. Z. *117*, 289–308 (1970).
HELLER, A. and I. REINER: [1] Indecomposable representations. Illinois J. Math. *5*, 314–323 (1961).
HERING, C.: [1] Zweifach transitive Permutationsgruppen, in denen 2 die maximale Anzahl von Fixpunkten von Involutionen ist. Math. Z. *104*, 150–174 (1968).
HERING, C., W. M. KANTOR and G. M. SEITZ: [1] Finite groups with a split BN-pair of rank 1, I. J. Algebra *20*, 435–475 (1972).
HIGMAN, D. G.: [1] Focal series in finite groups. Canad. J. Math. *5*, 477–497 (1953).
— [2] Modules with a group of operators. Duke Math. J. *21*, 369–376 (1954).
— [3] Indecomposable representations at characteristic p. Duke Math. J. *21*, 377–381 (1954).
— [4] Induced and produced modules, Canad. J. Math. *7*, 490–508 (1955).
HIGMAN, D. G. and C. SIMS: [1] A simple group of order 44,352,000. Math. Z. *105*, 110–113 (1968).
HIGMAN, G.: [1] Groups and rings which have automorphisms without nontrivial fixed elements. J. London Math. Soc. *32*, 321–334 (1957).
— [2] Suzuki 2-groups. Illinois J. Math. *7*, 79–96 (1963).
— [3] On the simple group of D. G. Higman and C. C. Sims. Illinois J. Math. *13*, 74–80 (1969).
HILL, E. T.: [1] The annihilator of radical powers in the modular group ring of a p-group. Proc. Amer. Math. Soc. *25*, 811–815 (1970).
HOARE, A. H. M.: [1] A note on 2-soluble groups. J. London Math. Soc. *35*, 193–199 (1960).
HOLT, D. F.: [1] On the local control of Schur multipliers. Quart. J. Math. Oxford (2) *28*, 495–508 (1977).
HOPKINS, C.: [1] Metabelian groups of order p^m, $p > 2$. Trans. Amer. Math. Soc. *37*, 161–195 (1935).
HUGHES, D. R.: [1] Extensions of designs and groups: Projective, symplectic and certain affine groups. Math. Z. *89*, 199–205 (1965).
HUPPERT, B.: [1] Zweifach transitive, auflösbare Permutationsgruppen. Math. Z. *68*, 126–150 (1957).
— [2] Scharf dreifach transitive Permutationsgruppen. Arch. Math. (Basel) *13*, 61–72 (1962).
— [3] Singer-Zyklen in klassischen Gruppen. Math. Z. *117*, 141–150 (1970).
— [4] Bemerkungen zur modularen Darstellungstheorie 1. Absolut unzerlegbare Moduln. Arch. Math. (Basel) *26*, 242–249 (1975).
— [5] Zur Konstruktion der reellen Spiegelungsgruppe $\mathfrak{H}_4$. Acta Math. Szeged *26*, 331–336 (1975).
HUPPERT, B. and H. WIELANDT: [1] Normalteiler mehrfach transitiver Permutationsgruppen. Arch. Math. (Basel) *9*, 18–26 (1958).
HUPPERT, B. and W. WILLEMS: [1] Bemerkungen zur modularen Darstellungstheorie 2. Darstellungen von Normalteilern. Arch. Math. (Basel) *26*, 486–496 (1975).
HURLEY, T. C.: [1] On a problem of Fox. Invent. Math. *21*, 139–141 (1973).
ISAACS, I. M.: [1] Extensions of group representations over arbitrary fields. J. Algebra *68*, 54–74 (1981).
ITÔ, N.: [1] On a theorem of H. F. Blichfeldt. Nagoya Math. J. *5*, 75–77 (1954).
— [2] Normalteiler mehrfach transitiver Permutationsgruppen. Math. Z. *70*, 165–173 (1958).

— [3] On a class of doubly transitive permutation groups. Illinois J. Math. *6*, 341–352 (1962).
— [4] Normal subgroups of quadruply transitive permutation groups. Hokkaido Math. J. *1*, 1–6 (1972).
JACOBSON, N.:* [1] Lie algebras. New York: Interscience Publishers Inc. 1962.
JAMES, G. D.: [1] The modular characters of the Mathieu groups. J. Algebra *27*, 57–111 (1973).
JANKO, Z.: [1] A new finite simple group with Abelian Sylow 2-subgroups and its characterization. J. Algebra *3*, 147–186 (1966).
— [2] A characterization of the smallest group of Ree associated with the simple Lie algebra of type (G_2). J. Algebra *4*, 293–299 (1966).
— [3] A characterization of the Mathieu simple groups I. J. Algebra *9*, 1–19 (1968): II, J. Algebra *9*, 20–41 (1968).
JANKO, Z. and J. G. THOMPSON: [1] On a class of finite simple groups of Ree. J. Algebra *4*, 274–292 (1966).
JENNINGS, S.: [1] The structure of the group ring of a p-group over a modular field. Trans. Amer. Math. Soc. *50*, 175–185 (1941).
KANTOR, W. M.: [1] 4-homogeneous groups. Math. Z. *103*, 67–68 (1968). Corrections, Math. Z. *109*, 86 (1969).
— [2] Automorphism groups of designs. Math. Z. *109*, 246–252 (1969).
— [3] k-homogeneous groups. Math. Z. *124*, 261–265 (1972).
KAPLANSKY, I.:* [1] Fields and Rings. Chicago: University of Chicago Press 1969.
KASCH, F., KNESER, M. and H. KUPISCH: [1] Unzerlegbare modulare Darstellungen endlicher Gruppen mit zyklischer p-Sylowgruppe. Arch. Math. (Basel) *8*, 320–321 (1957).
KLEMM, M.: [1] Primitive Permutationsgruppen von Primzahlpotenzgrad. Comm. Algebra *5*, 193–205 (1977).
KOSTRIKIN, A. I.: [1] The Burnside problem. Izv. Akad. Nauk SSSR Ser. Mat. *23*, 3–34 (1959), translated in Amer. Math. Soc. Trans. (2nd ser.) *36*, 63–100 (1964).
KREKNIN, V. A.: [1] Solvability of Lie algebras with a regular automorphism of finite period. Dokl. Akad. Nauk SSSR *150*, 467–469 (1963), Soviet Mat. Dokl. *4*, 683–685 (1963).
KREKNIN, V. A. and A. I. KOSTRIKIN: [1] Lie algebras with a regular automorphism. Dokl. Akad. Nauk SSSR *149*, 249–251 (1963), Soviet Math. Dokl. *4*, 355–358 (1963).
KURZWEIL, H.: [1] p-Automorphismen von auflösbaren p'-Gruppen. Math. Z. *120*, 326–354 (1971).
LANDROCK, P.: [1] Finite groups with a quasisimple component of type $PSU(3, 2^n)$ on elementary Abelian form. Illinois J. Math. *19*, 198–230 (1975).
LANDROCK, P. and G. MICHLER: [1] Block structure of the smallest Janko group. Math. Ann. *232*, 205–238 (1978).
LAZARD, M.: [1] Sur les groupes nilpotents et les anneaux de Lie. Ann. Sci. École Norm. Sup. *71*, 101–190 (1954).
LEEDHAM-GREEN, C. R., P. M. NEUMANN and J. WIEGOLD: [1] The breadth and the class of a finite p-group. J. London Math. Soc. (2) *1*, 409–420 (1969).
LIVINGSTONE, D.: [1] On a permutation representation of the Janko group. J. Algebra *6*, 43–55 (1967).
LIVINGSTONE, D. and A. WAGNER: [1] Transitivity on unordered sets. Math. Z. *90*, 393–403 (1965).
LORENZ, F.: [1] A remark on real characters of compact groups. Proc. Amer. Math. Soc. *21*, 391–393 (1969).
LÜNEBURG, H.:* [1] Die Suzuki-Gruppen und ihre Geometrien. Springer Lecture Notes 10 (1965).
— [2] Über die Gruppen von Mathieu. J. Algebra *10*, 194–210 (1968).
— *[3] Transitive Erweiterungen endlicher Permutationsgruppen. Springer Lecture Notes 84 (1969).
MAC LANE, S.:* [1] Homology. Springer 1963.
MCDERMOTT, J. P. J.: [1] On $(t + \frac{1}{2})$-transitive permutation groups. Math. Z. *148*, 61–62 (1976).

MACDONALD, I. D.: [1] Groups of breadth four have class five. Glasgow Math. J. *19*, 141–148 (1978).
MCLAUGHLIN, J. F.: [1] Some groups generated by transvections. Arch. Math. (Basel) *18*, 364–368 (1967).
MAGNUS, W.: [1] Beziehungen zwischen Gruppen und Idealen in einem speziellen Ring. Math. Ann. *111*, 259–280 (1935).
— [2] Über Beziehungen zwischen höheren Kommutatoren. J. Reine Angew. Math. *177*, 105–115 (1937).
MARTINEAU, R. P.: [1] Solubility of groups admitting a fixed-point-free automorphism group of type (p, p). Math. Z. *124*, 67–72 (1972).
— [2] Elementary Abelian fixed point free automorphism groups. Quart. J. Math. Oxford (2) *23*, 205–212 (1972).
— [3] Solubility of groups admitting certain fixed-point-free automorphism groups. Math. Z. *130*, 143–147 (1973).
MATSUYAMA, H.: [1] Solvability of groups of order $2^a p^b$. Osaka J. Math. *10*, 375–8 (1973).
MAZET, P.: [1] Sur le multiplicateur de Schur du groupe de Mathieu M_{22}. C. R. Acad. Sci. Paris *289*, Série A, 659–661 (1979).
— [2] Some subgroups of $SL_n(F_2)$. Illinois J. Math. *13*, 108–115 (1969).
MEIXNER, T.: [1] Über endliche Gruppen mit Automorphismen, deren Fixpunktgruppen beschränkt sind. Doctoral thesis, University of Erlangen-Nürnberg.
MICHLER, G.: [1] The kernel of a block of a group algebra. Proc. Amer. Math. Soc. *37*, 47–49 (1973).
— [2] The blocks of p-nilpotent groups over arbitrary fields. J. Algebra *27*, 303–315 (1973).
NAGAO, H.: [1] On multiply transitive permutation groups I. Nagoya Math. J. *27*, 15–19 (1966).
NAKAYAMA, T.: [1] Some studies on regular representations, induced representations and modular representations. Ann. of Math. *39*, 361–369 (1938).
— [2] On Frobeniusean algebras I. Ann of Math. *40*, 611–633 (1939).
— [3] Finite groups with faithful irreducible and directly indecomposable modular representations. Proc. Japan Acad. *23*, 22–25 (1947).
NESBITT, C.: [1] On the regular representation of algebras. Ann of Math. *39*, 634–658 (1938).
NEUMANN, P. M.: [1] Transitive permutation groups of prime degree. J. London Math. Soc. (2) *5*, 202–207 (1972).
NEUMANN, P. M. and M. R. VAUGHAN-LEE: [1] An essay on BFC groups. Proc. London Math. Soc. (3) *35*, 213–237 (1977).
ONO, T.: [1] An identification of Suzuki groups with groups of generalized Lie type. Ann. of Math. *75*, 251–259 (1962); Corrigendum, Ann. of Math. *77*, 413 (1963).
OSIMA, M.: [1] Note on blocks of group characters. Math. J. Okayama Univ. *4*, 175–188 (1955).
PAHLINGS, H.: [1] Über die Kerne von Blöcken einer Gruppenalgebra. Arch. Math. (Basel) *25*, 121–124 (1974).
— [2] Groups with faithful blocks. Proc. Amer. Math. Soc. *51*, 37–40 (1975).
— [3] Minimale Kerne von Darstellungen. Arch. Math. (Basel) *32*, 431–435 (1979).
PARROTT, D.: [1] On the Mathieu groups M_{22} and M_{11}. J. Austral. Math. Soc. *11*, 69–81 (1970).
PASSMAN, D.: [1] Solvable 3/2-transitive permutation groups. J. Algebra *7*, 192–207 (1967).
— [2] p-solvable doubly transitive permutation groups. Pacific J. Math. *26*, 555–577 (1968).
— *[3] Permutation Groups. New York, Harper and Row (1968).
— [4] Exceptional 3/2-transitive permutation groups. Pacific J. Math. *29*, 669–713 (1969).
— [5] Central idempotents in group rings. Proc. Amer. Math. Soc. *22*, 555–556 (1969).
PETTET, M. R.: [1] A sufficient condition for solvability in groups admitting elementary Abelian operator groups. Canad. J. Math. *29*, 848–855 (1977).
POLLATSEK, H.: [1] First cohomology groups of some linear groups over fields of characteristic 2. Illinois J. Math. *15*, 393–417 (1971).

POWELL, M. B. and G. HIGMAN:* [1] Finite Simple Groups. London and New York: Academic Press 1971.

RAE, A.: [1] Sylow p-subgroups of finite p-soluble groups. J. London Math. Soc. (2) *7*, 117–123 (1973); Corrigendum, J. London Math. Soc. (2) *11*, 11 (1975).

RALSTON, E. W.: [1] Solvability of finite groups admitting fixed point free automorphisms of order rs. J. Algebra *23*, 164–180 (1972).

RAZMYSLOV, YU. P.: [1] On Engel Lie algebras. Algebra i Logika *10*, 33–44 (1971) translated in Algebra and Logic *10* (1971).

REE, R.: [1] A family of simple groups associated with the simple Lie algebra of type (G_2). Amer. J. Math. *83*, 432–462 (1961).

— [2] Sur une famille des groups de permutations doublement transitifs. Canad. J. Math. *16*, 797–820 (1964).

REINER, I.: [1] On the number of irreducible modular representations of a finite group. Proc. Amer. Math. Soc. *15*, 810–812 (1964).

RICKMAN, B.: [1] Groups which admit a fixed-point-free automorphism of order p^2. J. Algebra *59*, 77–171 (1959).

RINGEL, C. M.: [1] The indecomposable representations of dihedral 2-groups. Math. Ann. *214*, 19–34 (1975).

RIPS, I. A.: [1] On the fourth integer dimension subgroup. Israel J. Math. *12*, 342–346 (1972).

ROITER, A. V.: [1] Unboundedness of the dimensions of the indecomposable representations of an algebra which has infinitely many indecomposable representations. Izv. Akad. Nauk. SSSR Ser. Mat. *32*, 1275–1282 (1968).

ROTH, R. L.: [1] A dual view of Clifford theory of characters of finite groups. Canad. J. Math. *23*, 857–865 (1971).

ROWLEY, P. J.: [1] Solubility of finite groups admitting a fixed-point-free Abelian automorphism group of square-free exponent rs. Proc. London Math. Soc. (3) *37*, 385–421 (1978).

SANDLING, R.: [1] The dimension subgroup problem. J. Algebra *21*, 216–231 (1972).

SANOV, I. N.: [1] Solution of Burnside's problem for exponent 4. Leningrad State University Ann. *10*, 166–170 (1940).

SCHWARZ, W.: [1] Die Struktur modularer Gruppenringe endlicher Gruppen der p-Länge 1. J. Algebra *60*, 51–75 (1979).

SERRE, J. P.:* [1] Corps locaux. Paris: Hermann 1962.

SHAW, D.: [1] The Sylow 2-subgroups of finite soluble groups with a single class of involutions. J. Algebra *16*, 14–26 (1970).

SHULT, E. E.: [1] On groups admitting fixed point free Abelian operator groups. Illinois J. Math. *9*, 701–720 (1965).

— [2] On finite automorphic algebras. Illinois J. Math. *13*, 625–653 (1969).

— [3] On the triviality of finite automorphic algebras. Illinois J. Math. *13*, 654–659 (1969).

SIBLEY, D. A.: [1] Coherence in finite groups containing a Frobenius section. Illinois J. Math. *20*, 434–442 (1978).

SJOGREN, J. A.: [1] Dimension and lower central subgroups. J. Pure App. Algebra *14*, 175–194 (1979).

SMITH, S. D. and A. P. TYRER: [1] On finite groups with a certain Sylow normalizer. I. J. Algebra *26*, 343–365 (1973); II. J. Algebra *26*, 366–367 (1973); III, J. Algebra *29*, 489–503 (1974).

SRINIVASAN, B.: [1] On the indecomposable representations of a certain class of groups. Proc. London Math. Soc. (3) *10*, 497–513 (1960).

STANTON, R. G.: [1] The Mathieu groups. Canad. J. Math. *3*, 164–174 (1951).

SUZUKI, M.: [1] Finite groups with nilpotent centralizers. Trans. Amer. Math. Soc. *99*, 425–470 (1961).

— [2] On a finite group with a partition. Arch. Math. (Basel) *12*, 241–254 (1961).

— [3] On a class of doubly transitive groups. Ann. of Math. *75*, 104–145 (1962).

— [4] On the characterization of linear groups III. Nagoya Math. J. *21*, 159–183 (1962).

— [5] On a class of doubly transitive groups II. Ann. of Math. *79*, 514–589 (1964).
— [6] Finite groups of even order in which Sylow 2-subgroups are independent. Ann. of Math. *80*, 58–77 (1964).
— [7] A characterization of the 3-dimensional projective unitary group over a finite field of odd characteristic. J. Algebra *2*, 1–14 (1965).
— [8] Finite groups in which the centralizer of any element of order 2 is 2-closed. Ann. of Math. *82*, 191–212 (1965).
— [9] Transitive extensions of a class of doubly transitive groups. Nagoya Math. J. *27*, 159–169 (1966).
SWAN, R. G.: [1] The Grothendieck ring of a finite group. Topology *2*, 85–110 (1963).
THOMPSON, J. G.: [1] Automorphisms of solvable groups. J. Algebra *1*, 259–267 (1964).
— [2] Fixed points of *p*-groups acting on *p*-groups. Math. Z. *86*, 12–13 (1964).
— [3] Factorizations of *p*-soluble groups. Pacific J. Math. *16*, 371–372 (1966).
— [4] Vertices and sources. J. Algebra *6*, 1–6 (1967).
— [5] Towards a characterization of $E_2^*(q)$. J. Algebra *7*, 406–414 (1967); II. J. Algebra *20*, 610–621 (1972).
— [6] Nonsolvable finite groups all whose local subgroups are solvable, I. Bull. Amer. Math. Soc. *74*, 383–437 (1968); II. Pacific J. Math. *33*, 451–536 (1970); III. Pacific J. Math. *39*, 483–534 (1971); IV. Pacific J. Math. *48*, 511–592 (1973); V. Pacific J. Math. *50*, 215–297 (1974); VI. Pacific J. Math. *51*, 573–630 (1974).
— [7] A replacement theorem for *p*-groups and a conjecture. J. Algebra *13*, 149–151 (1969).
TITS, J.: [1] Généralisation des groupes projectifs, Acad. Roy. Belgique Bull. Cl. Sci. *35*, 197–208, 224–233, 568–589, 756–773 (1949).
— [2] Les groupes simples de Suzuki et de Ree. Sem. Bourbaki, décembre, 1960.
— [3] Ovoides et groupes de Suzuki. Arch. Math. (Basel) *13*, 187–198 (1962).
— [4] Une propriété caractéristique des ovoides associés aux groupes de Suzuki. Arch. Math. (Basel) *17*, 136–153 (1966).
TSUSHIMA, Y.: [1] On the annihilator ideals of the radical of a group algebra. Osaka J. Math. *8*, 91–97 (1971).
VAUGHAN-LEE, M. R.: [1] Metabelian BFC-groups. J. London Math. Soc. (2) *5*, 673–680 (1972).
— [2] Breadth and commutator subgroups of *p*-groups. J. Algebra *32*, 278–285 (1976).
VILLAMAYOR, O. E.: [1] On the semisimplicity of group algebras II. Proc. Amer. Math. Soc. *10*, 27–31 (1959).
van der WAERDEN, B. L.:* [1] Gruppen von linearen Transformationen. Ergebn. der Math. 4, Springer 1935.
— *[2] Algebra 1, 7. Aufl., Springer 1966.
— *[3] Algebra 2, 5. Aufl., Springer 1967.
WAGNER, A.: [1] Normal subgroups of triply transitive permutation groups of odd degree. Math. Z. *94*, 219–222 (1966).
WALL, G. E.: [1] On the Lie ring of a group of prime exponent. Proceedings of the second International Conference on the Theory of Groups, Canberra 1973, 667–690.
— [2] Secretive prime-power groups of large rank, Bull. Austral. Math. Soc. *12*, 363–369 (1975).
WALTER, J.: [1] Finite groups with Abelian Sylow 2-subgroups of order 8. Invent. Math. *2*, 332–376 (1967).
— [2] The characterization of finite groups with Abelian Sylow 2-subgroups. Ann. of Math. *89*, 405–514 (1969).
WARD, H. N.: [1] On Ree's series of simple groups. Trans. Amer. Math. Soc. *121*, 62–89 (1966).
— [2] The analysis of representations induced from normal subgroups. Michigan Math. J. *15*, 417–428 (1968).
WARD, J. N.: [1] Automorphisms of finite groups and their fixed-point groups. J. Austral. Math. Soc. *9*, 467–477 (1969).
WIEGOLD, J.: [1] Groups with boundedly finite classes of conjugate elements. Proc. Roy.

Soc. Ser. A. *238*, 389–401 (1957).

WIELANDT, H.: [1] Primitive Permutationsgruppen vom Grad $2p$. Math. Z. *63*, 478–485 (1956).

— [2] Beziehungen zwischen den Fixpunktzahlen von Automorphismengruppen einer endlichen Gruppe. Math. Z. *73*, 146–158 (1960).

— [3] Über den Transitivitätsgrad von Permutationsgruppen. Math. Z. *74*, 297–298 (1960).

— *[4] Finite permutation groups. New York and London, Academic Press, 1964.

— [5] On automorphisms of doubly transitive permutation groups. Proc. Intern. Conf. Theory of Groups, Canberra 1965, 389–393.

— [6] Endliche k-homogene Permutationsgruppen. Math. Z. *101*, 142 (1967).

— [7] Permutation groups through invariant relations and invariant functions. Ohio State University, Columbus 1969.

WILLEMS, W.: [1] Bemerkungen zur modularen Darstellungstheorie 3. Induzierte und eingeschränkte Moduln. Arch. Math. (Basel) *26*, 497–503 (1976).

— [2] Metrische G-Moduln über Körpern der Charakteristik 2. Math. Z. *157*, 131–139 (1977).

— [3] On the projectives of a group algebra. Math. Z. *171*, 163–174 (1980).

WITT, E.: [1] Treue Darstellung Liescher Ringe. J. Reine Angew. Math. *177*, 152–160 (1973).

— [2] Die 5-fach transitiven Gruppen von Mathieu. Abh. Math. Sem. Univ. Hamburg *12*, 256–264 (1938).

— [3] Über Steinersche Systeme. Abh. Math. Sem. Univ. Hamburg *12*, 265–275 (1938).

WONG, W.: [1] On finite groups whose 2-Sylow subgroups have cyclic subgroups of index 2. J. Austral. Math. Soc. *4*, 90–112 (1964).

— [2] A characterization of the Mathieu group $\mathfrak{M}_{12}$. Math. Z. *84*, 378–388 (1964).

ZASSENHAUS, H.: [1] Über transitive Erweiterungen gewisser Gruppen aus Automorphismen endlicher mehrdimensionaler Geometrien. Math. Ann. *111*, 748–756 (1935).

— [2] Kennzeichnung endlicher linearer Gruppen als Permutationsgruppen. Abh. Math. Sem. Univ. Hamburg *11*, 17–44 (1936).

— [3] Über endliche Fastkörper. Abh. Math. Sem. Univ. Hamburg *11*, 187–220 (1936).

— [4] Ein Verfahren, jeder endlichen p-Gruppe einen Lie-Ring mit der Charakteristik p zuzuordnen. Abh. Math. Sem. Univ. Hamburg *13*, 200–207 (1939).

ZSIGMONDY, K.: [1] Zur Theorie der Potenzreste. Monatsh. Math. Phys. *3*, 265–284 (1892).

Index of Names

Adyan, S. I. 385
Alperin, J. L. 299, 364, 365
Amayo, R. K. 239, 349
Artin, E. 504, 506, 508, 514

Baer, R. 169, 500
Baker, H. F. 239, 348
Bašev, V. A. 71
Bauman, S. F. 491
Bender, H. 462
Berger, T. R. 406, 429, 467, 492
Berman, S. D. 41
Birkhoff, G. 366, 367
Blackburn, N. 275, 404
Bourbaki, N. 15, 20, 96, 404
Brauer, R. 2, 3, 21, 31, 37, 38, 69, 180, 196, 198, 237, 315
Brenner, S. 71
Bryant, R. M. 396, 397, 400, 403, 404, 514
Burkhardt, R. 193
Burnside, W. 362, 385, 405, 449, 460, 463

Cartan, E. 2, 162, 164, 166
Cartier, P. 370
Clifford, A. H. 2, 118, 123
Conlon, S. 137
Cossey, J. 185, 186, 188, 237
Curtis, C. 13, 63, 168

Dade, E. C. 69, 144, 237, 406, 428, 429, 458, 492
Deuring, M. 13, 26, 237
Dickson, L. E. 91, 406
Dress, A. 24, 237

Eckmann, B. 161
Engel, F. 388
Evens, L. 404

Feit, W. 138
Felsch, W. 345
Fermat, P. 424
Fischer, I. 404
Fitting, H. 72
Fong, P. 108, 186, 187, 188, 230, 237
Fox, R. H. 379, 384
Frobenius, G. 165, 166, 169, 172, 173, 237, 266

Gaschütz, W. 82, 86, 185, 186, 187, 188, 204, 205, 207, 237, 512
Gauss, C. F. 384
Glauberman, G. 503
Glover, D. J. 237
Goldschmidt, D. M. 299
Golod, E. S. 392, 404
Gorenstein, D. 299, 363, 484, 487, 491
Gow, R. 106, 128, 237
Green, J. A. 74, 93, 142, 209, 223
Gross, F. 279, 299, 315, 443, 452, 456, 467, 482, 483, 485

Hall, M. 419
Hall, P. 363, 405, 407, 417, 422, 426, 428, 439, 440, 453, 458, 460, 470
Hartley, B. 404, 492
Hausdorff, F. 239, 348
Heineken, H. 404
Heller, A. 71
Herstein, I. N. 363, 484, 487
Higman, D. G. 46, 63, 68, 84, 85, 86, 237
Higman, G. 299, 313, 361, 362, 366, 404, 405, 407, 411, 417, 422, 426, 428, 439, 440, 453, 458, 460, 470, 514
Hilbert, D. 27
Hill, E. T. 261
Hill, R. 209
Hoare, A. H. M. 448, 452
Hopkins, C. 73, 349
Huppert, B. 121, 133, 237
Hurley, T. C. 384

Isaacs, I. M. 128
Ito, N. 509

Jacobi, C. G. J. 238, 316
Jacobson, N. 1, 348
James, G. D. 178
Janko, Z. 145, 463
Jennings, S. 252, 254, 258, 259, 404
Johnsen, K. 514

Kaplansky, I. 73, 90
Kasch, F. 68, 237
Killing, W. 166
Kneser, M. 68, 237
Kostrikin, A. I. 361, 366, 388, 404
Kovács, L. G. 396, 397, 400, 403
Kreknin, V. A. 356, 361, 366, 404
Kupisch, H. 68, 237
Kurzweil, H. 406, 492, 514

Landrock, P. 145, 315
Lazard, M. 404
Leedham-Green, C. R. 346
Lie, S. 316, 326, 370
Liebeck, H. 404
Loewy, W. 157, 261

Macdonald, I. D. 345
MacLane, S. 3, 57, 161
McLaughlin, J. F. 32
Magnus, W. 239, 366, 380, 381, 385, 404
Mathieu, E. 178
Meixner, T. 361
Mersenne, M. 424
Michler, G. 145, 196, 233

Nakayama, T. 44, 50, 56, 124, 152, 237
Nesbitt, C. 38, 237
Neubüser, J. 345
Neumann, P. M. 345, 346
Noether, E. 26, 237
Norton, S. 32
Novikov, P. S. 385

Osima, M. 182

Passman, D. S. 182, 404
Plesken, W. 345
Plücker, J. 118
Pollatsek, H. 116
Powell, M. B. 476, 514

Rae, A. 492
Reiner, I. 13, 63, 71, 168
Ringel, C. M. 71
Rips, I. A. 384
Robinson, D. J. S. 404
Roiter, A. V. 71
Roth, R. L. 140, 237

Šafarevič, I. R. 392, 404
Sanov, I. N. 459
Schaller, K.-U. 514
Schopf, A. 161
Schreier, O. 414
Schumann, H. G. 384
Schur, I. 21
Schwarz, W. 140, 231, 236, 237
Scimemi, B. 361
Serre, J.-P. 27
Shaw, D. 299, 313, 404
Shult, E. E. 299, 315, 482, 483, 491
Sjogren, J. A. 384
Srinivasan, B. 69, 128, 233
Stewart, I. 239, 349
Stonehewer, S. 93
Struik, R. R. 404
Suzuki, M. 116, 238, 299
Swan, R. G. 32, 143, 144

Thompson, J. G. 406, 429, 472, 492, 511, 514
Tsushima, Y. 173

Vaughan-Lee, M. R. 332, 341, 345, 404
Villamayor, O. E. 93
Voigt, D. 237

Waerden, B. L. van der 76, 77, 116
Wall, G. E. 389, 396
Walter, J. H. 462
Ward, H. N. 126, 133
Ward, J. N. 487, 491
Wedderburn, J. H. M. 10, 13, 18, 28, 443
Wiegold, J. 346, 404
Wielandt, H. 266
Willems, W. 62, 93, 107, 114, 120, 126, 128, 133, 146, 189, 192, 193, 195, 196, 203, 229, 236, 237
Witt, E. 238, 239, 366, 367, 381, 404

Yen, T. 512

Zassenhaus, H. J. 264, 404
Zsigmondy, K. 508

Index

Abelian normal subgroup of a Sylow or Hall subgroup 464, 465, 472
Abelian Lie algebra 323
Abelian Sylow subgroup 462
absolutely indecomposable module 73, 223, 229
— and extension of ground-field 74ff.
—, decomposition as direct sum of 78
absolutely irreducible module 27
adjoint mapping 338
algebra, enveloping 372
—, Frobenius 165ff.
—, Lie 316ff.
—, Magnus 385
—, quasi-Frobenius 86, 169
—, symmetric 166
algebraic conjugate of a module 15
analogue of Theorem B 477
annihilator 155
associative basis 370
augmentation ideal 65, 252
—, basis of factors of powers of 254
— of group-ring of a free group 378ff.
automorphism
— of order 2 483
— of order 3 350
—, fixed point free of order 4 484, 487
automorphism group induced on the Frattini factor group of a p-group 396, 403
automorphism of Lie algebra 324
— and derived series 355–357
— and lower central series 360
—, ideal generated by fixed points 349ff.

Baer's lemma on injective modules 169
Baer's theorem on $\mathbf{O}_p(\mathfrak{G})$ 500
Baker-Hausdorff formula 239, 348
basis of factors of powers of augmentation ideal 254
bilinear form, $\mathfrak{G}$-invariant 106
bilinear mappings on fields 288, 289
Birkhoff-Witt theorem 366, 367
bisecting automorphisms 429, 434
block 1, 180
—, kernels of 194–198
—, number of 178
—, principal 180, 184–189, 198
block character 180
block ideal 180
block idempotent 180
Burnside problem 385, 388, 394, 405, 458ff.
— for exponent 4 459
— for exponent 6 417, 418
— for linear groups 463

Cartan matrix 2, 162, 179
—, symmetry of 164, 170
— of $\mathfrak{S}_4$ 214, 218
— of $\mathfrak{A}_5$ 221, 222, 223
central idempotent 174, 182
central product 276
central series, factors of 244–247, 363
— with elementary Abelian factors 242, 248
centralizer of factors of upper π-series 408, 409
characters in prime characteristic 14
chief factors, as $\mathfrak{G}$-modules 209
chief factors and blocks 188
class of nilpotent Lie algebra 323
Clifford type theorem 2, 118, 123ff.
cohomology group 61
coinduced module 51ff.
—, universal property of 55
commutation as bilinear mapping 238, 286
commuting idempotents 283
completely reducible factor module 12
composition factors, multiplicity of 162
conjugacy classes, maximal size in p-groups 341, 345, 346
conjugate module of a normal subgroup 124
contragredient representation 98

counting argument 275
critical group 419
—, representation of 422
cyclic normal subgroup of a Sylow subgroup 465
cyclic Sylow subgroup 2, 68–71
cyclotomic polynomial 503, 506

decomposition of algebra, as right module 151
— into two-sided ideals 174
degree less than p 509
degree of irreducible module 144
derived length, and fixed points of automorphisms 364–365
— and p-length 467
— and that of Sylow subgroups 470
— of Lie algebra 323
derived series, and fixed points of automorphisms 355–357
— of Lie algebras 323
dimension subgroup problem 239, 384
direct product, representation of 136ff.
direct summand, independent of ground-field 25
dual module 97ff.
— and tensor products 100
— of projective modules 116

eigen function 285
Engel commutator 264
Engel condition 388
enveloping algebra 372
exceptional linear transformation 437
exponent and p-length 451, 452
extending a $K\mathfrak{N}$-module to a $K\mathfrak{G}$-module 128
extension of ground-field 4
— and absolute indecomposability 74ff.
—, behaviour of complete reducibility 12
—, behaviour of radical 9
— for Lie algebras 324
—, Galois 15ff., 287, 289
extraspecial group 419, 421, 422, 425

factor module, completely reducible 12
factors of central series 244–247, 363
faithful irreducible representations 267, 268
Fermat prime 424, 476
—, exceptional role in Hall-Higman theory 453
Fitting height 482
—, bounds on 492
Fitting subgroup of odd order 443

fixed point free automorphism 349, 356, 361, 362, 482, 485
— of order 4 484
fixed point free group of automorphisms of type (p, p) 491
fixed point subgroup and derived length 364–365
form, invariant 109
Frattini factor group 280–282
— and automorphisms of p-groups 396, 403
free associative algebra 370
free group, and regular representation of a group 400
—, group-ring of 378, 380
—, lower central series of 366, 380ff.
—, subgroups of 414
free Lie algebra 371, 372
— and regular representation of a group 397
Frobenius algebra 165ff.
— as injective module 169
Frobenius group 266
Frobenius reciprocity 50ff.

generators, of minimal simple group 476
— in Lie algebras 317–319, 330
group-basis of a free group 366
group, in which every automorphism is inner 404
— of class 2 regarded as Lie ring 347
— of degree less than p 509
— of exponent 6 417, 418
— of prime exponent 386–389
group-ring, ideals of 166, 167
— is symmetric 166
— of p-group 65, 252ff.
group, with cyclic Sylow subgroups 2, 66–71
— with one class of involutions 511

half-transitivity 266
Hall and Higman's Theorem B 405, 419ff.
—, elementary form of 411
Hall-Higman methods 405
Hall subgroup 412
head of module 12
— is isomorphic to socle for projective modules 170
Hom, as $K\mathfrak{G}$-module 104, 119
homocyclic p-groups 279
—, p'-automorphisms of 280
homogeneous component of a free associative algebra 372

homogeneous component of a module 18
homomorphism of Lie algebra 317

$\mathfrak{J}$-adic ring 148
$\mathfrak{J}$-adic topology 148
ideal of Lie algebra 317
idempotents, central 182, 187
—, commuting 283
— in local rings 72–73
indecomposability of tensor products 133
indecomposable Abelian operator group 280–282
indecomposable group-ring 186
indecomposable module, induced module from 126
—, number of 63ff.
— of groups with cyclic Sylow subgroups 66–71
— of large dimension 63, 68
—, principal projective 203, 208
indecomposable two-sided ideal 174ff., 234
induced module 2, 44ff.
— and projective module 91
— from indecomposable module 126
—, universal property of 49
inertia subgroup 124
inflation 91, 208
injective envelope 161
injective module 1, 4
—, same as projective for group-rings 86
invariant form 106
involution, group with unique class of 511
—, number in a 2-group 275–277
—, permuted transitively by automorphisms 266, 269, 279, 296, 299ff.
irreducibility of tensor products 133
irreducible Frattini factor group of Abelian p-group 280–282
irreducible module, degree of 144
— for $GL(2, p)$ 43
— for $SL(2, p)$ 38
—, number of 32, 37, 41–43

Jacobi identity 238, 316, 326
Jacobson radical 1, 71
— of $K\mathfrak{N}$ for $\mathfrak{N} \trianglelefteq \mathfrak{G}$ 93–95
Janko's simple group $\mathfrak{J}_1$ 145, 463
Jennings' formula 259

kernels of blocks 194–198

Lie algebra 316ff.
— associated with associative algebra 318
—, automorphism of 324
—, power of 320–323, 330, 360, 373
Lie basis 370
Lie homomorphism 317
Lie ring 316
— associated with strongly central series 327
Lie ring method, for nilpotent groups 331
lifting idempotents 1, 148, 149
linear group, of finite exponent 463
— of small degree 509
local ring 71
—, idempotents in 72–73
Loewy length 157
Loewy series 157, 172, 207, 261, 262
— of $\mathfrak{S}_4$ 214
lower central series, of free groups 380–384
— of free Lie algebras 373, 375, 377
— of groups 239
— of Lie algebras 321

McLaughlin's simple group 32
Magnus algebra 385
Mathieu group $\mathfrak{M}_{22}$ 178
maximal class, p-group of 251, 258
maximal normal π-subgroup 407
maximal π-factor group 407
maximal size of conjugacy class 341, 345, 346
Mersenne prime 424
minimal polynomial of p-element 411, 422, 426, 429
minimal simple group 476
minimal subring $\mathfrak{h}$ with $\mathfrak{h}^2 = \mathfrak{g}^2$ 338
modular representation 1
— of $SL(2, 3)$ 210ff.
— of $\mathfrak{S}_4$ 214ff.
module, absolutely indecomposable 73, 223, 229
—, coinduced 51
—, dual 97ff.
— for direct products 136ff.
—, $\mathfrak{G}$-conjugate 124
—, induced 2, 44ff., 126
—, relatively injective 82
—, relatively projective 81
—, self-dual 106ff.
multiplicity of composition factors 162

Nakayama's lemma 152
net 253

nilpotent groups and Lie ring method 331
nilpotent Lie algebra 323
normal Abelian subgroup of Sylow or Hall subgroup 464, 465, 472
normal subgroup, representation of 123ff.
number of blocks 178
number of involutions in a 2-group 275–277
number of irreducible modules 32, 37, 41–43

orbits of equal length 266
order of an integer modulo a prime 508
orthogonal group 116

p-chief factors as modules 2
p-constrained group 186, 408
p-group, group-ring of 65, 252ff.
— of maximal class 251, 258
— with all normal subgroups characteristic 404
p-length and derived length 467
p-length and regularity of Sylow p-subgroups 456
p-length bounded by p-exponent 413, 451–452
p-nilpotent group, modular representation of 233
power of a product 239, 264
prime characteristic, projective modules in 90
—, characters in 14
—, splitting field in 31
prime exponent 386–396
primitive permutation group 266
principal block 180, 184–189, 198
principal indecomposable projective module 203–208
projective envelope 230
—, dimension of 231
projective module 1, 3, 4
— and induced modules 91
— and tensor products 92
—, dual of 117
—, general properties 153ff.
—, head of 157
— in prime characteristic 90
— over an algebra 156
—, principal indecomposable 203–208
—, same as injective for group-rings 86
—, socle of 159
p-stable group 492

quadratic form, invariant 109
quasi-Frobenius algebra 86, 169

radical, Jacobson 71
reciprocity theorems 2, 50, 56, 59
regular automorphism 349, 361, 362
regularity and p-length 456
relatively injective module 82
relatively projective module 81
representation, absolutely indecomposable 73
—, absolutely irreducible 27
—, faithful irreducible 267, 268
—, modular 1
— of normal subgroup 123ff.
— of small degree 509
restricted Burnside problem 389, 405, 406, 458ff.
ring, Lie 316
—, local 71

Schur index 2, 21
second Loewy factor 207
self-dual module 106ff.
semisimplicity 1
separable algebra 13
socle of module 12, 155
—, isomorphic to head for projective modules 170
soluble Lie algebra 323
special affine group 494, 500
splitting field 27
— in prime characteristic 31
square mapping, in 2-groups 294, 297, 313–315
strongly central series 326
—, associated Lie ring 327
subgroup of free group 414ff.
subgroups of order p permuted transitively by automorphisms 266, 269, 279, 280, 296, 299ff.
subnormal subgroup 223
sums of powers of 2 271–274
Suzuki's simple groups 116
Suzuki 2-group 238, 299
Sylow subgroup of symplectic group 432–433
symmetric algebra 1, 166
symmetric form, $\mathfrak{G}$-invariant 106ff.
symplectic group 116

tensor product, and blocks 188
—, indecomposability 133
—, irreducibility 133
— of dual modules 100

— of extensions of a field 7
— of projective modules 92
Theorem B of Hall and Higman 405, 426
transitive linear groups 266ff.
transitively permuted involutions 266, 269, 279, 296, 299ff.

universal enveloping algebra of a Lie algebra 372
upper Loewy series 172
upper p-series 407

Witt identity 238
Witt's formula 375, 384